east of the town? _____

ompany that is considering building a factory
volcanoes and on observations that you make
ompany with your opinion regarding whether
explain the reasons for your recommendation.

_____
_____
_____
_____
_____

fies the location of some of the excavated ruins
I by volcanic ash and pyroclastic material from
C.E.

xt eruption of Mt. Vesuvius and it travels at
take to reach the site of ancient Pompeii?

_____

t. Vesuvius. The present conelike peak lies
Only a small portion of the rim of the larger
southern portion of the older, larger crater?

_____

highlight surface textures of basaltic lava flows.

ble at Placemark 5a?   _____

Placemark 5b?   _____

boundary between two lava flows, one older
t—the one to the east of the marker, or the one

_____

Explain the basis for your answer: _____
_____
_____
_____
_____
_____

7. What is the likely cause for the steep escarpment highlighted by the placemark labeled
Problem 7? _____

8. Both Mauna Kea and Mauna Loa are shield volcanoes. Notably, several smaller,
"parasitic" volcanoes have erupted on the flanks of Mauna Kea. Click on the
placemark labeled Problem 8 to see some of these.

(a) What kind of volcanoes are these small cones?   _____

(b) Why are these cones asymmetric?   _____

(c) With the placemark for Problem 8 in the middle of the view, tilt the image all the
way so that you are looking at Mauna Kea from the side. Now, move away from the
volcano so that you can see the whole shield. What is the steepest slope on the flank of
the volcano? (You will need to use a protractor to measure the angle, in degrees.)

_____

Mt. Shishildan, Alaska

9. Fly to Mt. Shishildan, Alaska, by clicking on Image G9.10. This is a composite
volcano. Tilt the image so that you are looking at the side of the volcano, and zoom out
enough to see the whole volcano. What is the steepest slope on side of this volcano?

_____

Which slope is steeper: Mauna Kea or Mt. Shishildan?   _____

Worksheets available on StudySpace or in the free
workbook accompany each Geotour. Answering the
questions on the worksheets ensures that students
take the time to visit the recommended sites on
their own, and learn by doing. The instructor can also
assign the worksheets as homework or quizzes.

Outstanding Art

Essentials of Geology, Third Edition, provides the most comprehensive set of illustrations and photographs of any introductory geology text. The figures are realistic enough to provide visual context, but are uncluttered so that the teaching point of the figure stands out. The book also contains hundreds of photos, produced at a pleasingly large scale. In addition, almost every chapter in *Essentials of Geology*, Third Edition, offers a two-page painting, entitled **Geology at a Glance**, by the most respected geology artist in the world—Gary Hincks. Each painting provides a synopsis of key points in a chapter. The paintings emphasize interconnections among the key concepts in an uncluttered and visually dramatic way—they make a lasting impression.

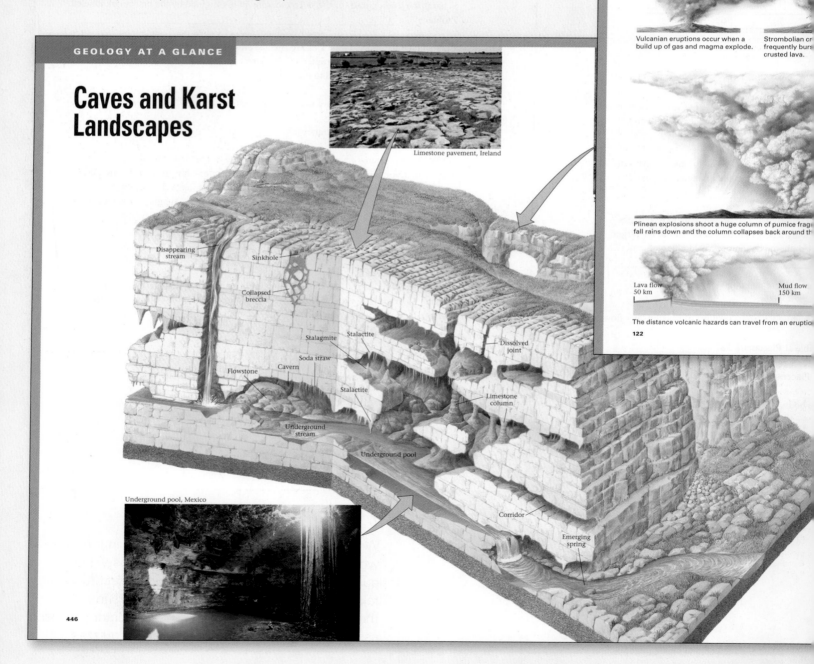

Volcanoes

Volcanic eruptions are a sight to behold, and in some hazard to fear. Beneath a volcano, magma formed in th mantle or the lower crust rises to fill a magma cham the Earth's surface. When the pressure in this magma c becomes great enough, magma forces upward throug duit, or crack, to the ground surface, and erupts.

Once molten rock has erupted at the sur- face, it is lava. Some lava spills down the side of the volcano to make a lava flow. In some cases, lava spatters or fountains out of the volcanic vent in little blobs or drops that cool quickly in the air to create fragmental igneous rock called tephra, or cinders. Larger blobs ejected by a volcano become volcanic bombs, which attain a stream- lined shape as they fall.

Vulcanian eruptions occur when a build up of gas and magma explode.

Strombolian cr frequently burs crusted lava.

Plinean explosions shoot a huge column of pumice frag fall rains down and the column collapses back around th

Lava flow 50 km

Mud flow 150 km

The distance volcanic hazards can travel from an eruptio

122

GEOLOGY AT A GLANCE

Caves and Karst Landscapes

Limestone pavement, Ireland

Disappearing stream

Sinkhole

Collapsed breccia

Stalagmite

Stalactite

Soda straw

Flowstone

Cavern

Stalactite

Dissolved joint

Limestone column

Underground stream

Underground pool

Underground pool, Mexico

Corridor

Emerging spring

446

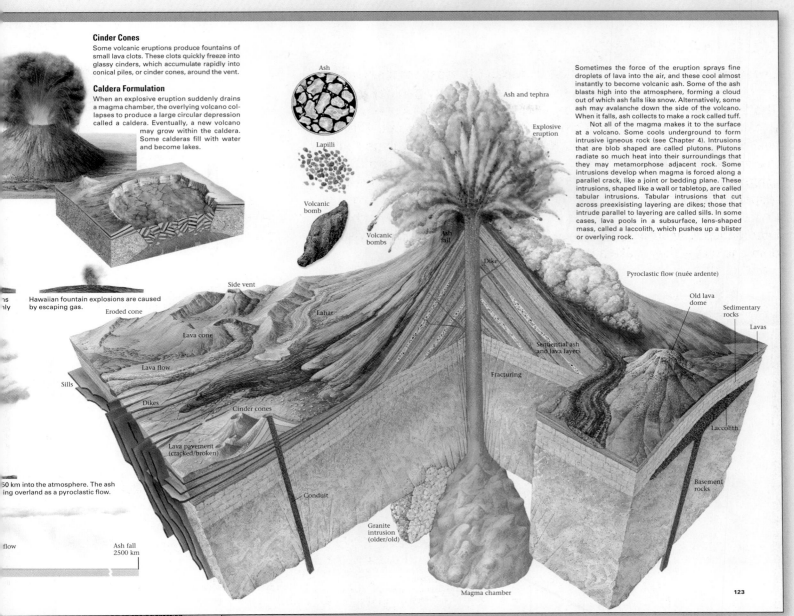

Cinder Cones

Some volcanic eruptions produce fountains of small lava clots. These clots quickly freeze into glassy cinders, which accumulate rapidly into conical piles, or cinder cones, around the vent.

Caldera Formulation

When an explosive eruption suddenly drains a magma chamber, the overlying volcano collapses to produce a large circular depression called a caldera. Eventually, a new volcano may grow within the caldera. Some calderas fill with water and become lakes.

Hawaiian fountain explosions are caused by escaping gas.

Sometimes the force of the eruption sprays fine droplets of lava into the air, and these cool almost instantly to become volcanic ash. Some of the ash blasts high into the atmosphere, forming a cloud out of which ash falls like snow. Alternatively, some ash may avalanche down the side of the volcano. When it falls, ash collects to make a rock called tuff.

Not all of the magma makes it to the surface at a volcano. Some cools underground to form intrusive igneous rock (see Chapter 4). Intrusions that are blob shaped are called plutons. Plutons radiate so much heat into their surroundings that they may metamorphose adjacent rock. Some intrusions develop when magma is forced along a parallel crack, like a joint or bedding plane. These intrusions, shaped like a wall or tabletop, are called tabular intrusions. Tabular intrusions that cut across preexisiting layering are dikes; those that intrude parallel to layering are called sills. In some cases, lava pools in a subsurface, lens-shaped mass, called a laccolith, which pushes up a blister or overlying rock.

Ash

Lapilli

Volcanic bomb

Volcanic bombs

Ash and tephra

Explosive eruption

Ash fall

Dike

Pyroclastic flow (nuée ardente)

Old lava dome

Sedimentary rocks

Lavas

Sequential ash and lava layers

Fracturing

Laccolith

Basement rocks

Side vent

Eroded cone

Lava cone

Lava flow

Lahar

Sills

Dikes

Cinder cones

Lava pavement (cracked/broken)

Conduit

Granite intrusion (older/old)

Magma chamber

50 km into the atmosphere. The ash ...ing overland as a pyroclastic flow.

flow

Ash fall
2500 km

floor. A cave's location depends on the orientation of bedding and joints, for these features localize the flow of groundwater.

Caves originally form at or near the water table (the subsurface boundary between rock or sediment in which pores contain air, and rock or sediment in which pores contain water). As the water table drops, caves empty of most water and become filled with air. In many locations, groundwater drips from the ceiling of a cave or flows along its walls. As the water evaporates and thus loses its acidity (because of the evaporation of dissolved carbon dioxide), new calcite precipitates. Over time, this calcite builds into cave formations, or speleothems, such as stalactites, stalagmites, columns, and flowstone.

Distinctive landscapes, called karst landscapes, develop at the Earth's surface over limestone bedrock. In such regions, the ground may be rough where rock has dissolved along joints; where the roofs of caves collapse, sinkholes develop. If a surface stream flows through an open joint into a cave network, we say that the stream is "disappearing." The water from such streams may flow underground for a distance and then reemerge elsewhere as a spring. In some places, the collapse of subsurface openings leaves behind natural bridges.

CHAPTER 12

Riches in Rock: Energy and Mineral Resources

Workers struggle to extinguish the inferno erupting from an oil well fire in Kuwait. Hundreds of wells were set ablaze at the end of the Gulf War in 1991. The flames are a dramatic display of the energy held in oil.

GEOPUZZLE

The average citizen of a developed nation uses about 15,000 kg of energy and mineral resources (oil, stone, salt, iron, copper, etc.) per year. Where does all this material come from and will supplies of it always exist?

321

many others. A large proportion of materials in your home have a geologic ancestry.

- Mineral resources are nonrenewable. Many are now or may soon be in short supply.

GEOPUZZLE REVISITED

The natural resources that sustain modern civilization come from the Earth and reflect the consequences of geologic processes. For example, hydrocarbons (oil and gas) come from the buried remains of algae and plankton, coal comes from the remnants of ancient plants, and metals come from special types of rocks called ores. Since natural resources can require long intervals of geologic time to form, many are being consumed faster than they can be replenished, and thus may eventually run out.

Key Terms

biofuel (p. 335)
biomass (p. 322)
cement (p. 346)
coal (p. 330)
coalbed methane (p. 332)
coal gasification (p. 333)
coal rank (p. 331)

energy (p. 322)
energy resource (p. 322)
fossil fuel (p. 322)
gas hydrate (p. 330)
geothermal energy (p. 336)
greenhouse effect (p. 340)
hydrocarbon reserve (p. 324)

349

hydrocarbon (p. 323)
kerogen (p. 324)
meltdown (of nuclear reactor) (p. 334)
metal (p. 340)
mineral resources (p. 322)
nuclear reactor (p. 334)
nuclear waste (p. 334)
Oil Age (p. 338)
oil shale (p. 324)
oil window (p. 324)
ore (p. 341)

ore deposit (p. 341)
peat (p. 330)
permeability (p. 325)
photosynthesis (p. 322)
porosity (p. 325)
reservoir rock (p. 325)
resource (p. 322)
salt dome (p. 327)
seal rock (p. 326)
source rock (p. 324)
tar sand (p. 329)
trap (p. 325)

Review Questions

1. What are the fundamental sources of energy?
2. What is the source of the organic material in oil?
3. What is the oil window, and why does oil form only there?
4. How is organic matter trapped and transformed to create an oil reserve?
5. What are tar sand and oil shale, and how can oil be extracted from them?
6. What are gas hydrates, and where do they occur?
7. Where is most of the world's oil found? At present rates of consumption, how long will it last?
8. How is coal formed?
9. Explain how coal is transformed in rank from peat to anthracite.
10. Describe the operation of a nuclear reactor.
11. Where does uranium form in the Earth's crust? Where does it usually accumulate in minable quantities?
12. What are some of the drawbacks of nuclear energy?
13. What is geothermal energy? Why is it not more widely used?
14. What is the difference between a renewable and a nonrenewable resource?
15. What is the likely future of hydrocarbon production and use in the twenty-first century?
16. Why don't we use an average granite as a source for useful metals?
17. Describe various kinds of economic mineral deposits.
18. What procedures are used to locate and mine mineral resources today?
19. How is stone cut from a quarry?
20. What are the ingredients of cement? How is Portland cement made?

Geopuzzle

Questioning is the first step in learning. Each chapter begins with a question that students can keep in mind as they read. The question prompts students to pursue answers and helps them develop a personal context for learning by thinking about what they might already know about a subject. The answer to the question, provided at the end of the chapter, serves as a brief synthesis of the chapter's theme.

Take-Home Messages

Take-Home Messages appear at the end of each section—they encourage students to pause and take stock of what they have learned, before going further. The Take-Home Messages emphasize the most important lesson of the previous section—the point that students should remember beyond the end of the course.

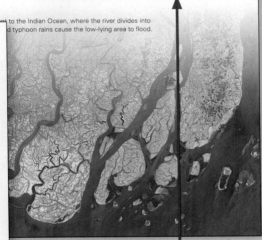

Review Questions

1. What role do streams have in the hydrologic cycle? Indicate various sources of water in streams.
2. Describe the different types of drainage networks.
3. What factors determine whether a stream is permanent or ephemeral?
4. How does discharge vary according to the stream's length, climate, and position along the stream course?
5. Describe how streams and running water erode the Earth's surface.
6. What are the components of sediment load in a stream?
7. Distinguish between a stream's competence and its capacity.
8. What factors determine the position of the base level? What is the difference between a local base level and the ultimate base level?
9. Why do canyons form in some places and valleys in others?
10. How does a braided stream differ from a meandering stream?
11. Describe how meanders form and are cut off and the landforms that result from the process.
12. Describe how deltas grow and develop. How do they differ from alluvial fans?
13. How does a stream-eroded landscape evolve as time passes?
14. How are superposed and antecedent drainages similar? How are they different?
15. What is the difference between a seasonal flood and a flash flood?
16. How do people try to prevent flood damage? Are the methods always successful? Why?
17. What is the recurrence interval of a flood? Why can't someone say that "the hundred-year flood happened last year, so I'm safe for another hundred years"?

On Further Thought

1. The northeastern two-thirds of Illinois, in the midwestern United States, was last covered by glaciers only 14,000 years ago. The rest of the state was last covered by glaciers over 100,000 years ago. Until the advent of modern agriculture, the recently glaciated area was a broad, grassy swamp, cut by very few stream channels. In contrast, the area that was glaciated overt 100,000 years ago is not swampy and has been cut by numerous stream valley. Why?
2. Records indicate that flood crests for a given amount of discharge along the Mississippi River have been getting higher since 1927, when a system of levees began to block off portions of the floodplain. Why?
3. The Ganges River carries an immense amount of sediment load, which has been building a huge delta in the Bay of Bengal. Look at the region using an atlas or *Google Earth*™, think about the nature of the watershed supplying water to the drainage network that feeds the Ganges, and explain why this river carries so much sediment.

THE VIEW FROM SPACE. The Ganges River carries sediment to the Indian Ocean, where the river divides into ... and typhoon rains cause the low-lying area to flood.

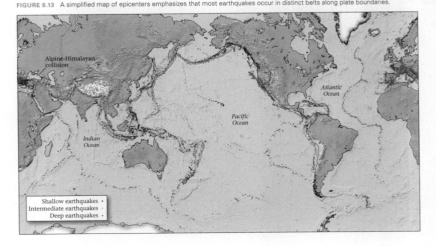

CHAPTER 8 A Violent Pulse: Earthquakes

Fortunately, such large earthquakes occur much less frequently than small earthquakes. There are about 100,000 magnitude 3 earthquakes every year, but a magnitude 8 earthquake happens only about once every three to five years (Fig. 8.12b).

TAKE-HOME MESSAGE

Seismographs measure earthquake vibrations and allow seismologists to determine the location and size of earthquakes. We specify earthquake size by an intensity value (damage caused) or a magnitude (the amplitude of ground motion). Magnitudes are logarithmic, meaning that ground shaking by a magnitude 8 event is 10 times larger than from a magnitude 7 event.

8.5 WHERE AND WHY DO EARTHQUAKES OCCUR?

Earthquakes do not take place everywhere on the globe. By plotting the distribution of earthquake epicenters on a map, seismologists find that most, but not all, earthquakes occur in fairly narrow **seismic belts**, or seismic zones. Most seismic belts cor-respond with plate boundaries. Earthquakes within these belts are plate-boundary earthquakes; ones that occur away from plate boundaries are intraplate earthquakes (the prefix *intra* means within) (Fig. 8.13). Notably, 80% of the earthquake energy released on Earth comes from the plate-boundary earthquakes in the belts surrounding the Pacific Ocean.

Notably, earthquakes do not occur at random depths in the Earth. Seismologists distinguish three classes of earthquakes based on hypocenter (focus) depth: shallow earthquakes occur in the top 20 km of the Earth, intermediate earthquakes take place between 20 and 300 km, and deep earthquakes occur down to a depth of about 660 km. Earthquakes cannot happen deeper in the Earth, because rock at great depth cannot rupture or change in a manner that produces shock waves. In this section, we look at the characteristics of earthquakes in various geologic settings and learn why earthquakes take place where they do.

Earthquakes at Plate Boundaries

The majority of earthquakes happen at faults along plate boundaries, for the relative motion between plates is accommodated primarily by slip on these faults. We find different kinds of faulting at different types of plate boundaries.

FIGURE 8.13 A simplified map of epicenters emphasizes that most earthquakes occur in distinct belts along plate boundaries.

On Further Thought

Essentials of Geology, Third Edition, helps students to go beyond the basic facts and begin to think about content and context. Each chapter provides not only a set of basic questions about content, but also a selection of critical thinking questions called On Further Thought. The questions challenge students to use the content of the chapter to address issues in the real world. They can be assigned as homework or can serve as discussion points in class.

New 3-D, Hollywood-Quality Animations

Earth: Videos of a Planet DVD Version 3.0 provides six spectacular new animations that illuminate geologic processes. These augment the existing 65 2-D animations available on StudySpace and the Instructor's CD-ROM. The new animations employ Hollywood-quality visuals that will amaze as they instruct. These state-of-the-art animations focus on the hardest-to-understand geologic processes, including *Formation of the Solar System*, *Plate Tectonics*, *Continental Collision*, *Meandering River*, *Glacier Erosion*, and more.

Outstanding Narrative

Essentials of Geology, Third Edition, contains an abundance of art, but it does not neglect the writing. Students delight in the book's comprehensive and complete narrative. Experience shows that the clear, organized explanation of a topic—in words—helps students understand the context of a concept and the relationships among concepts.

Glowing waves rise and flow, burning all life on their way, and freeze into black, crusty rock which adds to the height of the mountain and builds the land, thereby adding another day to the geologic past. . . . I became a geologist forever, by seeing with my own eyes: the Earth is alive!

—Hans Cloos (1886–1951),
on seeing the eruption of Mt. Vesuvius (Italy)

5.1 INTRODUCTION

Every few hundred years, one of the hills on Vulcano, an island in the Mediterranean Sea off the western coast of Italy, rumbles and spews out molten rock, glassy cinders, and dense "smoke" (actually a mixture of various gases, fine ash, and very tiny liquid droplets). Ancient Romans thought that such eruptions happened when Vulcan, the god of fire, fueled his forges beneath the island to manufacture weapons for the other gods. Geologic study suggests, instead, that eruptions take place when hot magma, formed by melting inside the Earth, rises through the crust and emerges at the surface. No one believes the Roman myth anymore, but the island's name evolved into the English word **volcano**, which geologists use to designate either an erupting vent through which molten rock reaches the Earth's surface or a mountain built from the products of eruption.

On the main peninsula of Italy, not far from Vulcano, another volcano, Mt. Vesuvius, towers over the Bay of Naples. Two thousand years ago, a prosperous Roman resort and trading town named Pompeii sprawled at the foot of Vesuvius. One morning in 79 C.E., earthquakes signaled the mountain's awakening. At 1:00 P.M. on August 24, a dark mottled cloud boiled up above Mt. Vesuvius's summit to a height of 27 km. As lightning sparked in its crown, the cloud drifted over Pompeii, turning day into night. Blocks and pellets of rock fell like hail, while fine ash and choking fumes enveloped the town (Fig. 5.1a). Frantic people rushed to escape, but for many it was too late. As the growing weight of volcanic debris began to crush buildings, an avalanche of hot ash swept over Pompeii, and by

FIGURE 5.1 The destruction of Pompeii.

(a) This 1817 painting by British artist J. M. W. Turner depicts the cataclysmic eruption of 79 C.E. that destroyed Pompeii.

(c) A plaster cast of a Pompeii resident who was buried in ash as he crouched in a corner of his house.

(b) The remains of Mt. Vesuvius lie in the distance, behind the excavated ruins of Pompeii. The town was buried and preserved beneath meters of ash and debris.

An Integrated and Innovative emedia program

Essentials of Geology, Third Edition, comes with an outstanding emedia package designed to create visually stunning lectures and promote learning.

Geotours

Using *Google Earth*'s 3-D satellite and aerial imagery, **Geotours** take students on virtual field trips to geologic sites around the world and can be used as lecture or homework/lab assignments. A **Geotours Workbook** is available online and in print.

Earth: Videos of a Planet DVD

For use in lecture, 3-D cinema quality animations help students visualize difficult concepts and 24 short film clips from the USGS Archive show them fascinating examples of geology in action.

Film clips include: Aspects of Eruption, Kilauea • Gas, and Lava Eruption, Kilauea • Lava Tubes, Kilauea • Lava Flow along the Coast • Sampling a Lava Flow • Basalt Lava Flows • Lava Crossing a Highway • Lava Eruption at Night • Lava Fountains • Mt. St. Helens 1980 • Hot Springs, Yellowstone • Mudpots, Yellowstone • Eruption of Old Faithful • Volcanic Hazards, Mt Rainier • Glacial Sculpting, Mt. Rainier • Exposing an Igneous Pluton • Debris Flows, Jiangjia Ravine • Grand Canyon: Bird's Eye View • Rift Basin Formation • Fault Development • Pleistocene Ice Sheet • Developing a Hurricane • Hurricane Winds in Action • Hurricane Storm Surge.

Animations

Over 60 original animations developed by Professor Marshak to illustrate dynamic Earth processes, emphasize plate tectonics, geologic hazards, and Earth Systems science concepts. These high-quality animations can be enlarged to full-screen view for lecture display and include VCR-like controls that allow instructors to control the pace of the animation. The animations can be easily accessed from the free StudySpace student website. Offline versions of the animations are available on the Instructor's DVD-ROM, linked from PowerPoint slides for easy classroom display.

Instructor's Resource DVD-ROM

Important Features on the Instructors' DVD-ROM and Instructor Website include:

- **Narrative Power Points**: This new source, developed by Stephen Marshak, presents images from the text and beyond with the appropriate sizing and spacing to show how they relate to one another. Each slide is fully customizable, with editable figures, labels, and captions.
- **Lecture PowerPoints**: These illustrated lecture outlines emphasize key points from each chapter
- **Animations**: Running locally on the instructor's computer, these classroom-ready animations require no plug-ins.
- **Zoomable Art**: Explore Gary Hincks's spectacular synoptic paintings in vivid detail using the zoomable art feature. Ideal for lecture use, these figures run locally from within PowerPoint—no plug-ins required.
- **GeoQuiz Clicker Questions**: These multiple-choice questions are perfect for use in classrooms equipped with "clicker" systems.

Instructor's Website

wwnorton.com/instructors

Online access to a rich array of resources: test bank, clicker questions, instructor's manual, art from the text, WebCT and Blackboard-ready content.

Blackboard and WebCT Coursepacks

Lecture and review resources—study plans, animations, Geotours, chapter quizzes, and more—in course-ready format.

Instructor's Manual and Test Bank

Prepared by John Werner of Seminole Community College, Terry Engelder of Pennsylvania State University, and Stephen Marshak, this manual offers material to help instructors as they prepare their lectures and includes over 1,200 multiple-choice and true-false questions. The test bank is available in printed form and electronically in *ExamView® Assessment Suite*, WebCT, and BlackBoard-ready formats.

StudySpace

wwnorton.com/studyspace
This innovative student website offers a wealth of resources for study and review, including:

- **Review materials**: Chapter overviews, vocabulary flashcards, and multiple-choice quizzes and a gradebook
- **Geotours**: 19 virtual field trips harness the power of *Google Earth™* and aid students in applying lessons.

- **Animations**: Over 60 original animations developed by Professor Marshak to illustrate dynamic Earth processes, emphasizing plate tectonics, geologic hazards, and Earth Systems science concepts.
- **Earth Science News**: Regular updates keep you up to date with the latest news about geologic events, recent advances, and contemporary environmental issues.

Ebook/Same great book, half the price!

nortonebooks.com
Essentials of Geology, Third Edition, is also available in an ebook format that replicates actual book pages and provides links to animations on StudySpace. Students can highlight text, use sticky notes, and work with zoomable images from the book.

Essentials of Geology

THIRD EDITION

Essentials of Geology

THIRD EDITION

Stephen Marshak
UNIVERSITY OF ILLINOIS

W. W. NORTON & COMPANY

NEW YORK   LONDON

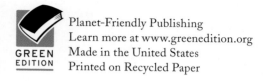
Planet-Friendly Publishing
Learn more at www.greenedition.org
Made in the United States
Printed on Recycled Paper

At W. W. Norton, we're committed to producing books in an earth-friendly manner and to helping our customers make greener choices.

Manufacturing books in the United States ensures compliance with strict environmental laws and eliminates the need for international freight shipping, a major contributor to global air pollution. And printing on recycled paper helps to minimize our consumption of trees, water, and fossil fuels. The text of *Essentials of Geology*, 3rd Edition, was printed on paper made with 30% post-consumer waste, and the cover was printed on paper made with 10% post-consumer waste.

According to Environmental Defense's Paper Calculator:
    Trees Saved: 140 • Air Emissions Eliminated: 15,577 pounds
    Water Saved: 51,062 gallons • Solid Waste Eliminated: 8,448 pounds

W. W. Norton & Company has been independent since its founding in 1923, when William Warder Norton and Mary D. Herter Norton first published lectures delivered at the People's Institute, the adult education division of New York City's Cooper Union. The Firm soon expanded its program beyond the Institute, publishing books by celebrated academics from America and abroad. By mid-century, the two major pillars of Norton's publishing program—trade books and college texts—were firmly established. In the 1950s, the Norton family transferred control of the company to its employees, and today—with a staff of four hundred and a comparable number of trade, college, and professional titles published each year—W. W. Norton & Company stands as the largest and oldest publishing house owned wholly by its employees.

Editor: Jack Repcheck
Senior project editor: Thomas Foley
Production manager: Christopher Granville
Copy editor: JoAnn Simony
Marketing manager: Betsy Twitchell
Managing editor, college: Marian Johnson
Science media supervisor: Sally Whiting
Photography editor: Kelly Mitchell
Editorial assistant: Jason Spears

Geotour spreads designed by Precision Graphics
Composition by Precision Graphics
Manufacturing by Courier Companies, Inc.
Illustrations for the Second and Third Editions by Precision Graphics

ISBN: 978-0-393-93238-6 (pbt.)

W. W. Norton & Company, Inc., 500 Fifth Avenue, New York, N.Y. 10110
www.wwnorton.com
W. W. Norton & Company Ltd., Castle House, 75/76 Wells Street, London W1T 3QT

1 2 3 4 5 6 7 8 9 0

Dedication

To Kathy, David, and Emma

Brief Contents

Special Features

What a Geologist Sees

Geology at a Glance

Contents

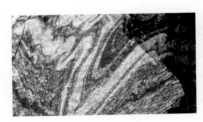

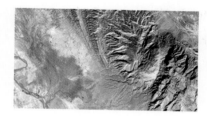

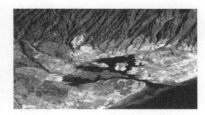

NARRATIVE THEMES

Why do earthquakes, volcanoes, floods, and landslides happen? What causes mountains to rise? How do beautiful landscapes develop? How have climate and life changed through time? When did the Earth form and by what process? Where do we dig to find valuable metals and where do we drill to find oil? Does sea level change? Do continents move? The study of geology addresses these important questions and many more. But from the birth of the discipline, in the late eighteenth century, until the mid-twentieth century, geologists considered each question largely in isolation, without pondering its relation to others. This approach changed, beginning in the 1960s, in response to the formulation of two "paradigm-shifting" ideas that have unified thinking about the Earth and its features. The first idea, called the *theory of plate tectonics*, states that the Earth's surface shell, rather than being static, consists of discrete plates that slowly move relative to each other so that the map of our planet continuously changes. Plate interactions cause earthquakes and volcanoes, build mountains, provide gases that make up the atmosphere, and affect the distribution of life on Earth. The second idea, the concept of *Earth System science*, emphasizes that our planet's water, land, atmosphere, and living inhabitants are dynamically interconnected. Materials constantly cycle among various living and nonliving reservoirs on, above, and within the planet, and the history of life is intimately linked to the history of the physical Earth.

Essentials of Geology, Third Edition, is an introduction to the study of our planet that uses the theory of plate tectonics and the concept of Earth System science throughout to weave together a number of narrative themes, including:

1. The solid Earth, the oceans, the atmosphere, and life are interconnected in complex ways, yielding a planet that is unique in the Solar System.
2. Most geologic processes reflect the interactions of plates.
3. The Earth is a planet, formed like other planets from dust and gas. But, in contrast to other planets, the Earth is a dynamic place where new geologic features continue to form and old ones continue to be destroyed.
4. The Earth is very old—about 4.57 billion years have passed since its birth. During this time, the map of the planet and its surface features have changed, and life has evolved.
5. Internal processes (driven by Earth's internal heat) and external processes (driven by heat from the Sun) interact at the Earth's surface to produce our landscapes.
6. Natural hazards—earthquakes, volcanoes, landslides, floods—can be studied and understood. In some cases, people can avoid their damaging effects.
7. Energy and mineral resources come from the Earth and are formed by geologic phenomena. Geologic study can help locate these resources.
8. Physical features of the Earth are linked to life processes, and vice versa.
9. Science comes from observation; people make scientific discoveries.
10. The study of geology provides an excellent means to improve science literacy.

These narrative themes serve as the *take-home message* of the book, a message that students will remember long after they finish their introductory geology course. In effect, they provide a mental "pegboard" on which students can organize and connect ideas, and develop a modern, coherent image of our planet.

PEDAGOGICAL APPROACH

Educational research demonstrates that students learn best when they actively engage with a combination of narrative text and narrative art. Some students respond more to the words of a textbook, which help to organize information, to provide answers to questions, to fill in the essential steps that link ideas together, and to help a student develop a personal context for understanding information. Some students respond more to narrative art—art designed to tell a story—for visual images help students comprehend and remember processes. *Essentials of Geology*, Third Edition, provides both of these learning tools.

The text has been crafted to be engaging, the art has been configured to tell a story, and the chapters are laid out to help students identify key principles.

Presentation alone, however, is not sufficient to make a book work. To be effective, books must also provide active-learning tools that challenge students to affirm their comprehension as they proceed through the text and to help them remember and internalize ideas. *Essentials of Geology*, Third Edition, provides active-learning experiences in several ways. Each chapter starts with a question—a *Geopuzzle*—that prompts students to think about what they already know, and prepares them to search for additional information as they read the chapter. *Take-home messages* at the end of each section help students identify key themes in context. Questions at the end of each chapter not only test basic knowledge, but also stimulate critical thinking. Finally, *Geotours* linked to each chapter guide students to spectacular geologic localities around the globe where they can apply their knowledge to reality.

ORGANIZATION

The topics covered in this book have been arranged so that students can build their knowledge of geology on a foundation of overarching principles. Thus, the book starts with cosmology and the formation of the Earth, and then introduces the architecture of our planet, from surface to center. With this background, students are prepared to delve into plate tectonics theory. Plate tectonics appears early in the book, so that students can relate the content of subsequent chapters to the theory. Knowledge of plate tectonics, for example, helps students understand the suite of chapters on minerals, rocks, and the rock cycle. Knowledge of plate tectonics and rocks together, in turn, provides a basis for studying volcanoes, earthquakes, and mountains. And with this background, students are prepared to see how the map of the Earth has changed through the vast expanse of geologic time, and how energy and mineral resources have developed. The book's final chapters address processes and problems occurring at or near the Earth's surface, from the unstable slopes of hills, down the course of rivers, to the shores of the sea and beyond. This section concludes with a topic of growing concern in society—global change, particularly climate change.

Although the sequence of chapters was chosen for a reason, this book is designed to be flexible enough for instructors to choose their own strategies for teaching geology. Geology is a nonlinear subject, in that individual topics are so interrelated that there is not always a single best way to order them. Thus, each chapter is self-contained, reiterating relevant material where necessary.

SPECIAL FEATURES

Essentials of Geology, Third Edition, contains a number of features that distinguish it from competing texts. Several of which are new to this edition.

Narrative Art

It's hard to understand features of the Earth System without being able to see them. To help students visualize topics, this book is lavishly illustrated, with figures that attempt to give a realistic context for a geologic feature without overwhelming students with extraneous detail. The talented artists who worked on the book have used the latest computer graphics software, resulting in the most sophisticated pedagogical art ever provided by a geoscience text. In addition to artist-produced drawings, the book provides abundant photographs from around the world, many of which were taken by the author. Where appropriate, photographs are accompanied by annotated sketches labeled "*What a geologist sees*," which help students to be certain that they see what the photo was intended to show.

In this edition, drawings and photographs have been integrated into *narrative art* that has been laid out, labeled, and annotated to tell a story—the figures are drawn to teach! Subcaptions are positioned adjacent to the relevant parts of a figure, labels point out key features, and balloons provide important annotation. The arrangement of subparts has been designed to convey time progression, where relevant. Color schemes in drawings have been tied to those of relevant photos, so that students can easily visualize the relationships between drawings and photos.

Geotours and the Geotour Workbook

Through the magic of *Google Earth*™, *Essentials of Geology*, Third Edition, provides a comprehensive set of active-learning exercises that take students on virtual field trips to see outstanding examples of geology from around the world. These *Geotours*, compiled at the end of the book, provide a narrative description of hundreds of field sites. The publisher provides instructors and students with a *.kmz* file that automatically flies to all the sites at the click of a mouse, and includes many incredible add-ons (developed by Scott Wilkerson), including map overlays, outcrop photos, and geologic cross sections. A *Geotours Workbook*, prepared by Scott Wilkerson and Stephen Marshak, provides active-learning exercises for each site, challenging students to fly around sites and make specific observations and measurements in context.

Featured Paintings: Geology at a Glance

In addition to individual figures, British artist Gary Hincks has created spectacular two-page annotated paintings for most chapters, called *Geology at a Glance*. These paintings integrate key concepts introduced in the chapters and visually emphasize the relationships between components of the Earth System. They provide students a way to review a subject . . . *at a glance.*

Take-Home Messages, Geopuzzles, and "On Further Thought" Questions

Each chapter begins with a question, a *Geopuzzle*, that challenges students to think about what they might already know or not know about the theme of the chapter. Such prompting helps students to be on the lookout for key ideas as they read. Every chapter section ends with a *Take-Home Message*, a brief summary that helps students identify and remember the highlight of the section in context. And each chapter concludes with *On Further Thought* questions that invite students to think beyond the basics.

Broad, but Concise, Coverage

Essentials of Geology, Third Edition, provides concise coverage of topics used in an introductory geology course. It has been shortened (by 300 pages), relative to its progenitor (*Earth: Portrait of a Planet 3e*), by removal of topics less frequently covered in an introductory geology course, and by streamlining the treatment of the remaining topics.

Interludes

The book contains several interludes. These, in effect, are "mini-chapters" which focus on topics that are self-contained but are not broad enough to require an entire chapter. The feature helps keep chapters reasonable in length, and provides flexibility in sequencing topics within a course.

Societal Issues

Geology's practical applications are addressed in several chapters. Students will learn about such topics as energy resources, mineral resources, global change, and mass wasting. Further, chapters on earthquakes, volcanoes, and landscapes highlight geologic hazards. And students are encouraged to apply their geologic understanding to environmental issues, where relevant.

Boxed Inserts

Throughout the text, boxes expand on specific topics by giving further scientific background, additional detail, or related information that's just plain interesting.

CHANGES IN THE THIRD EDITION

The newest edition of *Essentials of Geology* is not simply a cosmetic modification of the Second Edition. The text has been intelligently modified and thoroughly updated. Key changes include:

- *Narrative art*: Most of the art in the book has been reconfigured to produce narrative pieces that integrate drawings, photos, labels, and annotations in order to tell a story. Thus, students have two complete learning options in the book—the narrative text (written to engage), and the narrative art (drawn to teach).

- *Geotours*: Through the magic of *Google Earth*™, the book provides a comprehensive set of active-learning exercises that take students on virtual field trips to see outstanding examples of geology around the world.

- *New pedagogical tools*: This edition provides new components to help students actively retain what they learn. Each chapter begins with a *Geopuzzle*, each chapter section ends with a concise *Take-Home Message*, and each chapter ends with critical thinking questions entitled *On Further Thought*.

- *Incorporation of the latest discoveries and examples*: Recent research has yielded fundamental new understanding that belongs even in introductory books. While basic geoscience remains the backbone of *Essentials of Geology*, Third Edition, we have incorporated discussions of these new discoveries, and many others, into the text. To keep the discussion current, we have updated the treatment to include events that have happened since publication of the previous edition.

- *The View from Space*: Satellites orbiting the Earth provide spectacular imagery of our planet's surface. We have added numerous satellite images to this edition.

- *New Geology at a Glance synoptic art*: Famed geologic artist Gary Hincks has provided new, spectacular two-page paintings that provide an overview of key concepts. The drama of these paintings appeals to students' intuition.

- *Clarifications*: To help ensure that the text is as accurate and up-to-date as possible, we comprehensively revised the text to ensure that it is as clear as possible and have made complex discussions more readable.
- *Reorganization*: The chapter on volcanoes was moved up, so that it directly follows the chapter on igneous rocks, as the two topics are commonly taught in this sequence. The discussion of weathering and soils was broken out of the sedimentary rocks chapter and was made into a separate interlude.

SUPPLEMENTARY MATERIALS

For Students and Instructors . . .

Geotours at the click of a mouse. With the click of a mouse, *Google Earth*™ flies students to see geology in the field. Students and instructors may download a *.kmz* file which automatically enters hundreds of Geotour sites into the "places" folder in their computers' version of *Google Earth*™.

Geotours Workbook. Each student receives a copy of the *Geotours Workbook*, prepared by Scott Wilkerson and Stephen Marshak, which provides active-learning questions based on Geotour sites.

Narrative PowerPoints. All of the *PowerPoint*™ presentations for the book have been recreated to be properly scaled and labeled for the *PowerPoint*™ environment. Not only are they designed for instant use in the classroom, but they can serve as an at-home study aid for students. Supplemental photographs, provided by the author, are integrated into the collection so they appear in context.

Animations and videos. The book's accompanying DVD provides a rich collection of over 60 flash animations to illustrate geologic processes. In addition, the book comes with a DVD called *Videos of a Planet*, which provides three *3-D cinema-quality animations*, along with 24 film clips from the U.S. Geological Survey.

Zoomable art. Explore Gary Hincks's spectacular synoptic paintings in vivid detail using the zoomable art feature. Ideal for lecture use, these figures are included among the resources on the Instructor's DVD.

Ebook—nortonebooks.com. *Essentials of Geology*, Third Edition, is also available in an ebook format that replicates actual book pages and provides links to the animations on StudySpace. Students can highlight text, use sticky notes, and work with zoomable images from the book.

 StudySpace—wwnorton.com/studyspace. Free and open for students, StudySpace offers students assignment-driven study plans for each chapter including access to Geotours, Geotour worksheets, review materials, multiple-choice quizzes, flashcards, animations, a geology in the news feed, and more.

For Instructors Only . . .

Norton Media Library Instructor's DVD-ROM. The instructor DVD offers a wealth of easy-to-use multi-media resources, all structured around the text and designed for use in lecture presentations, including:

- *Editable chapter PowerPoints*™: These narrative presentations, designed by Stephen Marshak, take the book's art and photographs and places them in ready-to-use *PowerPoints*™. Images are properly sized, related images are provided on the same slide or in sequenced slides, and labeling has been streamlined so it doesn't clutter the image. Each slide also includes a brief title, so it can be a stand-alone teaching aid or study aid. Significantly, each figure part can easily be copied and transferred to an instructor's personal *PowerPoint*™ presentation.
- *Editable PowerPoint*™ *lectures*: These presentations, provided by Ron Parker of Earlham College, integrate illustrations from the book, supplemental slides, and text, into a format that can provide a basis for developing complete lectures.
- *Easy-to-use JPEGS*: All of the art and photographs from the text in JPEG format.
- *Flash animations*: The book's DVD and website provide over 60 flash animations that illustrate geologic processes.
- *Cinema-quality animations*: Three animations (Formation of the Earth; Glacial Advance and Retreat; and Plate Motions) have been developed for this book. These animations are provided as *Quicktime*™ movies on the book's DVD.
- *Zoomable art*: The DVD and website provide zoomable versions of the synoptic paintings by Gary Hincks. These allow instructors to zoom in closely to parts of the figure, in the context of the whole figure, during class presentations.
- *GeoQuiz Clicker Questions* and more . . .

Instructor's Manual and Test Bank. Prepared by John Werner of Seminole Community College, Terry Engelder of Pennsylvania State University, and Stephen Marshak, this manual offers useful material to help instructors as they prepare their lectures. It includes over 1,200 multiple-choice and true-false test questions. The test bank is available both in printed form and electronically in Exam-view and WebCT- or Blackboard-ready formats.

Instructor's Website—wwnorton.com/instructors. Online access to a rich array of resources: test bank, clicker questions, instructor's manual, art from the text WebCT- and Blackboard-ready content.

Blackboard and WebCT Coursepacks. Lecture and review resources—study plans, animations, *Geotours*, chapter quizzes, and more—in course-ready format.

ACKNOWLEDGMENTS

I am very grateful for the assistance of many people in bringing this book from the concept stage to the shelf in the first place, and for helping to provide the momentum needed to make this revision take shape.

First and foremost, I wish to thank my family, who have adopted "the book" as a member of the household and have tolerated the overabundance of photo stops on family trips. My wife, Kathy, carried out the immense task of cutting content from *Earth: Portrait of a Planet*, Third Edition, to produce the template for *Essentials of Geology*, Third Edition. She also edited new text, rationalized the art list, and cross-checked the many sets of proofs. Without her efforts, creation of this edition would not have been possible. My daughter, Emma, helped develop the concept of narrative art used in the book, updated the supplemental slide set, served as scale in photos, and provided feedback about how the book works from a student's perspective. My son, David, highlighted places where the writing could be improved, helped me to keep the project in perspective, and also served as scale in photographs.

The staff of W. W. Norton & Company have been incredibly supportive of my work and have been very generous in their investment in this project. Jack Repcheck, the editor, continues to be a fountain of sage advice and an understanding friend throughout its evolution. He has provided numerous innovative ideas that have strengthened the book and brought it to the attention of the geologic community. Betsy Twitchell, the marketing manager, has brought new insight and perspectives into this endeavor, and has done an amazing job of conveying the book's goals to its potential audience. Thom Foley and Christopher Granville have expertly and efficiently handled the task of managing production for this edition. Thom calmly and carefully processed the manuscript and kept the book on schedule, and Chris coordinated communication with the compositor and the artists. Susan Gaustad was the outstanding copy editior of the First Edition. This tradition continued through the efforts of JoAnn Simony on the Third Edition. JoAnn's sharp eyes and probing questions have improved the text immensely. Kelly Mitchell has been an excellent photo editor for the Third Edition, bringing the management of the photo collection into the twenty-first century, sorting out the figure credits, and locating excellent images where needed.

Production of the illustrations has involved many people over many years. I am indebted to the staff of Precision Graphics who have helped to create the style of the figures and have accommodated countless changes and tweaks without complaint. Stan Maddock and Joanne Brummet have been at the core of this effort, and I am forever thankful for their hard work. Jon Prince, Simon Shak, and Dan Whitaker creatively programmed the animations under the careful supervision of Andrew Troutt. For the Third Edition, Kristina Seymour at Precision Graphics has done a wonderful job of coordinating artwork and composition, and Jennifer Wasson has creatively handled page layout. Andrew Troutt has done an excellent job of transforming my approach to *PowerPoint*™ presentations into reality.

It has been great fun to interact with Gary Hincks, who painted the incredible two-page spreads, in part using his own designs and geologic insights. Some of Gary's paintings originally appeared in *Earth Story* (BBC Worldwide, 1998) and were based on illustrations jointly conceived by Simon Lamb and Felicity Maxwell, working with Gary. Others were developed specifically for *Earth: Portrait of a Planet* or for *Essentials of Geology*. Some of the chapter-opening quotes were found in *Language of the Earth*, compiled by F. T. Rhodes and R. O. Stone (Pergamon, 1981). During the initial development of the First Edition, I greatly benefited from discussions with Philip Sandberg, and during later stages in the development of the First Edition, Donald Prothero contributed text, editorial comments, and end-of-chapter material.

The three editions of this book and of its cousin, *Earth: Portrait of a Planet*, have benefited greatly from input by expert reviewers for specific chapters, by new general reviewers of the entire book, and by comments from faculty and students who have used the book and were kind enough to contact me or the publisher. This list of reviewers includes:

Jack C. Allen, Bucknell University
David W. Anderson, San Jose State University
Martin Appold, University of Missouri-Columbia
Philip Astwood, University of South Carolina
Eric Baer, Highline University
Victor Baker, University of Arizona
Julie Baldwin, University of Montana
Sandra Barr, Acadia University
Keith Bell, Carleton University

Mary Lou Bevier, University of British Columbia

Jim Black, Tarrant County College

Daniel Blake, University of Illinois

Ted Bornhorst, Michigan Technological University

Michael Bradley, Eastern Michigan University

Sam Browning, Massachusetts Institute of Technology

Bill Buhay, University of Winnipeg

Rachel Burks, Towson University

Peter Burns, University of Notre Dame

Katherine Cashman, University of Oregon

George S. Clark, University of Manitoba

Kevin Cole, Grand Valley State University

Patrick M. Colgan, Northeastern University

John W. Creasy, Bates College

Norbert Cygan, Chevron Oil, retired

Michael Dalman, Blinn College

Peter DeCelles, University of Arizona

Carlos Dengo, ExxonMobil Exploration Company

John Dewey, University of California, Davis

Charles Dimmick, Central Connecticut State University

Robert T. Dodd, State University of New York
at Stony Brook

Missy Eppes, University of North Carolina, Charlotte

Eric Essene, University of Michigan

James E. Evans, Bowling Green State University

Susan Everett, University of Michigan-Dearborn

Dori Farthing, SUNY-Geneseo

Grant Ferguson, St. Francis Xavier University

Leon Follmer, Illinois Geological Survey

Nels Forman, University of North Dakota

Bruce Fouke, University of Illinois

David Furbish, Vanderbilt University

Steve Gao, University of Missouri

Grant Garvin, John Hopkins University

Christopher Geiss, Trinity College, Connecticut

Gayle Gleason, SUNY Cortland

Cyrena Goodrich, Kingsborough Community College

William D. Gosnold, University of North Dakota

Lisa Greer, William & Mary College

Henry Halls, University of Toronto at Mississuaga

Bryce M. Hand, Syracuse University

Tom Henyey, University of South Carolina

Paul Hoffman, Harvard University

Curtis Hollabaugh, University of West Georgia

Bernie Housen, Western Washington University

Mary Hubbard, Kansas State University

Paul Hudak, University of North Texas

Neal Iverson, Iowa State University

Donna M. Jurdy, Northwestern University

Thomas Juster, University of Southern Florida

H. Karlsson, Texas Tech

Dennis Kent, Lamont Doherty/Rutgers

Jeffrey Knott, California State University, Fullerton

Ulrich Kruse, University of Illinois

Robert S. Kuhlman, Montgomery County
Community College

Lee Kump, Pennsylvania State University

David R. Lageson, Montana State University

Robert Lawrence, Oregon State University

Leland Timothy Long, Georgia Tech

Craig Lundstrom, University of Illinois

John A. Madsen, University of Delaware

Jerry Magloughlin, Colorado State University

Jennifer McGuire, Texas A&M University

Paul Meijer, Utrecht University (NL)

Jamie Dustin Mitchem, California University
of Pennsylvania

Alan Mix, Oregon State University

Kathy Nagy, University of Illinios-Chicago

Pamela Nelson, Glendale Community College

Robert Nowack, Purdue University

Charlie Onasch, Bowling Green State University

David Osleger, University of California, Davis

Eric Peterson, Illinois State University

Ginny Peterson, Grand Valley State University

Stephen Piercey, Laurentian University

Lisa M. Pratt, Indiana University

Mark Ragan, University of Iowa

Bob Reynolds, Central Oregon Community College

Joshua J. Roering, University of Oregon

Eric Sandvol, University of Missouri

William E. Sanford, Colorado State University

Roy Schlische, Rutgers University

Sahlemedhin Sertsu, Bowie State University

Roger D. Shew, University of North Carolina-Wilmington

Doug Shakel, Pima Community College

Norma Small-Warren, Howard University

Donny Smoak, University of South Florida

David Sparks, Texas A&M University

Angela Speck, University of Missouri

Tim Stark, University of Illinois

David Stetty, Jacksonville State University

Kevin G. Stewart, University of North Carolina
    at Chapel Hill

Don Stierman, University of Toledo

Gina Marie Seegers Szablewski, University of
    Wisconsin- Milwaukee

Barbara Tewksbury, Hamilton College

Thomas M. Tharp, Purdue University

Kathryn Thornbjarnarson, San Diego State University

Basil Tikoff, University of Wisconsin

Spencer Titley, University of Arizona

Robert T. Todd, State University of New York
    at Stony Brook

Torbjörn Törnqvist, University of Illinois, Chicago

Jon Tso, Radford University

Stacey Verardo, George Mason University

Barry Weaver, University of Oklahoma

Alan Whittington, University of Missouri

John Wickham, University of Texas at Arlington

Lorraine Wolf, Auburn University

Christopher J. Woltemade, Shippensburg University

I apologize if I inadvertently left anyone off this list.

ABOUT THE AUTHOR

Stephen Marshak is Professor of Geology at the University of Illinois, Urbana-Champaign, where he is also the Director of the School of Earth, Society, and Environment. He holds an A.B. from Cornell University, an M.S. from the University of Arizona, and a Ph.D. from Columbia University. Steve's research interests lie in structural geology and tectonics. This work has taken him into the field on a number of continents. Steve loves teaching and has won his college's and university's highest teaching awards. In addition to research papers and *Essentials of Geology*, Steve has authored *Earth: Portrait of a Planet*; and has co-authored *Earth Structure: An Introduction to Structural Geology and Tectonics*; and *Basic Methods of Structural Geology*.

THANKS!

I am very grateful to the faculty who selected earlier editions for their classes and to the students who engaged so energetically with the book. I particularly appreciate the comments from readers that helped to make improvements in this Third Edition. I continue to welcome your comments and can be reached at smarshak@illinois.edu.

Stephen Marshak

*To see the world in a grain of sand
and heaven in a wild flower.
To hold infinity in the palm of the hand
and eternity in an hour.*

—William Blake (British poet, 1757–1827)

And Just What Is Geology?

Students can see the Earth System at a glance in the Wasatch Mountains, Utah. Here, sunlight, air, water, rock, and life all interact to produce a complex and fascinating landscape.

GEOPUZZLE

Tourists might look at this photo and see a beautiful view. What do geologists see in this landscape?

1

Civilization exists by geological consent, subject to change without notice.

—Will Durant (1885–1981)

P.1 IN SEARCH OF IDEAS

Our C-130 Hercules transport plane rose from a smooth ice runway on the frozen sea surface at McMurdo Station, Antarctica, and headed south. We were off to spend a month studying unusual rocks exposed on a cliff about 250 km away. We climbed past the smoking summit of Mt. Erebus, Earth's southernmost volcano, and for the next hour flew along the Transantarctic Mountains, a ridge of rock that divides the continent into two parts, East Antarctica and West Antarctica (Fig. P.1). A vast ice sheet, in places over 3 km thick, covers East Antarctica—the surface of this ice sheet forms a high plain known as the Polar Plateau. From the plane's windows, we admired solid rivers of ice cutting through the Transantarctic Mountains, carrying ice from the Polar Plateau to the Ross Sea Ice Shelf. (Ice sheets and ice rivers are types of glaciers.) Suddenly, we heard the engines slow.

As the plane descended, it lowered its ski-equipped landing gear. The loadmaster shouted an abbreviated reminder of the emergency alarm code: "If you hear three short blasts of the siren, hold on for dear life!" The plane touched the surface of our first choice for a landing spot, the ice at the base of the rock cliff we wanted to study. *Wham, wham, wham, wham!!!!* As the plane's skis slammed into frozen snowdrifts on the ice surface at about 180 km an hour, it seemed as though a fairytale giant was shaking the plane. Seconds later, the landing aborted, we were airborne again, looking for a softer runway above the cliff. Finally, we landed in a field of deep snow, unloaded, and bade farewell to the plane. When the plane passed beyond the horizon, the silence of Antarctica hit us— no trees rustled, no dogs barked, and no traffic rumbled in this stark land of black rock and white ice. It would take us a day and a half to haul our sleds of food and equipment down to our study site (Fig. P.1). All this to look at a few dumb rocks?

Geologists, scientists who study the Earth, explore remote regions like Antarctica almost routinely. Such efforts often strike people in other professions as a strange way to make a living, as illustrated by Scottish poet Walter Scott's (1771–1832) description of geologists at work: "Some rin uphill and down dale, knappin' the chucky stones to pieces like sa' many roadmakers run daft. They say it is to see how the warld was made!"

Indeed—to see how the world was made, to see how it continues to evolve, to find its valuable resources, to prevent con-

FIGURE P.1 Geologic field work in Antarctica unlocks the mysteries of an ice-bound continent. A map of Antarctica emphasizes that the Transantarctic Mountains separate West Antarctica from East Antarctica.

A plane dropping geologists on a snowfield.

Sledding to a field site with crates of supplies.

tamination of its waters and soils, and to predict its dangerous movements. That is why geologists spend months at sea drilling holes in the ocean floor, why they scale mountains, camp in rain-drenched jungles, and trudge through scorching desert winds (Fig. P.2). That is why geologists use electron microscopes to examine the atomic structure of minerals, use mass spectrometers to define the composition of rock and water, and use super-computers to model the paths of earthquake waves. For over two centuries, geologists have pored over the Earth, *in search of ideas* to explain the processes that form and change our planet.

P.2 THE NATURE OF GEOLOGY

Geology, or geoscience, is the study of the Earth. Not only do geologists address academic questions, such as the formation and composition of the Earth, the causes of earthquakes and ice ages, and the evolution of life, they also address practical problems, such as how to prevent groundwater contamination, how to find oil and minerals, and how to stabilize slopes. And in recent years, geologists have made significant contributions

FIGURE P.3   Human-made cities cannot withstand the vibrations of a large earthquake. These apartment buildings collapsed during an earthquake in Turkey.

FIGURE P.2   A geologist studying exposed rocks on a mountain slope on the desert island of Zabargad, in the Red Sea.

to the study of global climate change. When news reports begin with "Scientists say . . ." and then continue with "an earthquake occurred today off Japan," or "landslides will threaten the city," or "contaminants from the proposed toxic waste dump are destroying the town's water supply," or "there's only a limited supply of oil left," the scientists referred to are geologists.

The fascination of geology attracts many people to careers in this science. Thousands of geologists work for oil, mining, water, engineering, and environmental companies, while a smaller number work in universities, government geological surveys, and research laboratories. Nevertheless, since the majority of students reading this book will not become professional geologists, it's fair to ask the question, Why should people, in general, study geology?

First, geology may be one of the most practical subjects you can learn. Ask yourself the following questions, and you'll realize that geologic processes, phenomena, and materials play major roles in daily life:

- Do you live in a region threatened by landslides, volcanoes, earthquakes, or floods (Fig. P.3)?
- Are you worried about the price of energy or about whether there will be a war in an oil-supplying country?
- Do you ever wonder about where the copper in your home's wires come from?
- Have you seen fields of green crops surrounded by desert and wondered where the irrigation water comes from?
- Would you like to buy a dream house on a beach or in the mountains?

Clearly, all citizens of the twenty-first century, not just professional geologists, need to make decisions concerning Earth-related issues. People can make more reasoned decisions if they have a basic understanding of geology.

Second, the study of geology gives you a perspective of the planet that no other field can. As you will see, the Earth is a complicated system. Its living organisms, climate, and solid rock interact with one another in a great variety of ways. Geologic study reveals Earth's antiquity (it's about 4.57 *billion* years old) and demonstrates how the planet has changed profoundly during its existence. What our ancestors considered the center of the Universe has become, with the development of geologic perspective, our "island in space" today. And what was believed to be an unchanging orb originating at the same time as humanity has become a dynamic planet that existed long before people did.

Third, the study of geology puts the accomplishments and consequences of human civilization in a broader context. View the aftermath of a large earthquake, flood, or hurricane, and it's clear that the might of natural geologic phenomena greatly exceeds the strength of human-made structures. But watch a bulldozer clear a swath of forest, or a dynamite explosion remove the top of a hill, or a prairie field evolve into a housing development, and it's clear that people can change the face of the Earth at rates that often exceed those of natural geologic processes.

Finally, when you finish reading this book, your view of the world may be forever colored by geologic curiosity. When you walk in the mountains, you will think of the many forces that shape and reshape the Earth's surface. When you hear about a natural disaster, you will have insight into the processes that brought it about. And when you next go on a road trip, the rock exposures along the highway will no longer be gray, faceless cliffs, but will present complex puzzles of texture and color telling a story of Earth's history.

P.3 THEMES OF THIS BOOK

A number of narrative themes appear (and reappear) throughout this text. These themes, listed below, can be viewed as the book's take-home message.

- *The Earth is a unique, evolving system.* Geologists increasingly recognize that the Earth is a complicated system; its interior, solid surface, oceans, atmosphere, and life forms interact in many ways to yield the landscapes and environment in which we live. Within this **Earth System**, chemical elements pass in cycles between different types of rock, between rock and sea, between sea and air, and between all of these entities and life. Aside from material added to the Earth by the occasional impact of a meteorite, all the material involved in these cycles originates in the Earth itself. Our planet is truly an island in space.

- *Plate tectonics explains many Earth processes.* Earth is not a homogeneous ball, but rather consists of concentric layers—from center to surface, Earth has a core, mantle, and crust. We live on the surface of the crust, where it meets the atmosphere and the oceans. In the 1960s, geologists recognized that the crust, together with the uppermost part of the underlying mantle, forms a 100- to 150-km-thick semirigid shell. Large cracks separate this shell into discrete pieces, called **plates**, which move very slowly relative to one another (Fig. P.4). The theory that describes this movement and its consequences is called the **theory of plate tectonics**, and it is the foundation for understanding most geologic phenomena. Although plates move very

FIGURE P.4 A simplified map of the Earth's plates. The arrows indicate the direction each plate is moving, and the length of the arrow indicateds plate velocity (the longer the arrow, the faster the motion). We discuss the types of plate boundaries in Chapter 2.

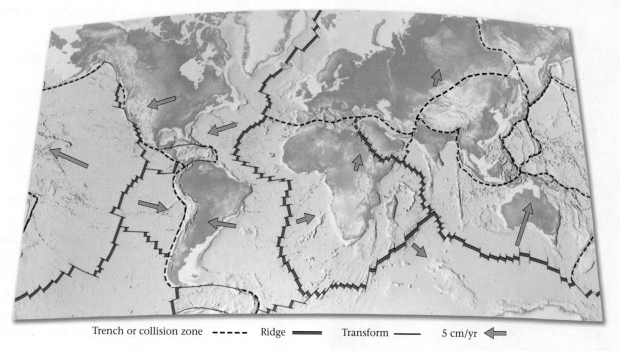

Trench or collision zone - - - - -   Ridge ━━━   Transform ━━━   5 cm/yr ⇐

slowly, generally less than 10 cm a year, their movements yield earthquakes, volcanoes, and mountain ranges, and cause the map of Earth's surface to change over time.

- *The Earth is a planet.* Despite the uniqueness of the Earth System, Earth can be viewed as a planet, formed like the other planets of the Solar System from dust and gas that encircled the newborn Sun.

- *The Earth is very old.* Geologic data indicate that the Earth formed 4.57 billion years ago—plenty of time for geologic processes to generate and destroy features of the Earth's surface, for life forms to evolve, and for the map of the planet to change. Plate-movement rates of only a few centimeters per year can move a continent thousands of kilometers if those movements continue for hundreds of millions of years. There is time enough to build mountains and time enough to grind them down, many times over. To define intervals of this time, geologists developed the **geologic time scale**. Figure P.5 provides the names of the major subdivisions of the geologic time scale. Chapter 10 discusses these in greater detail.

- *Internal and external processes drive geologic phenomena.* Internal processes are those phenomena driven by heat from inside the Earth. Plate movement is an example, and since plate movements cause mountain building, earthquakes, and volcanoes, we call all of these phenomena internal processes as well. External processes are those phenomena driven by heat supplied by radiation coming to the Earth from the Sun. This heat drives the movement of air and water, which grinds and sculpts the Earth's surface and transports the debris to new locations, where it accumulates. The interaction between internal and external processes forms the landscapes of our planet. As we'll see, gravity—the pull that one mass exerts on another—plays an important role in both internal and external processes.

- *Geologic phenomena affect our environment.* Volcanoes, earthquakes, landslides, floods, and even more subtle processes such as groundwater flow and contamination, or depletion of oil and gas reserves, are of vital interest to every inhabitant of this planet. Therefore, throughout this book we emphasize the linkages between geology and the environment.

- *Physical aspects of the Earth System are linked to life processes.* All life on this planet depends on such physical features as the minerals in soil; the temperature, humidity, and composition of the atmosphere; and the flow of surface and subsurface water. And life in turn affects and alters physical features. For example, the atmosphere's oxygen comes primarily from plant photosynthesis, a life activity.

- *Science comes from observation, and people make scientific discoveries.* Science does not consist of subjective guesses or arbitrary dogmas, but rather of a consistent set of objective statements resulting from the application of the **scientific method** (Box P.1). Every scientific idea must be tested thoroughly, and can be accepted only when supported by documented observations. Further, scientific ideas do not appear

FIGURE P.5 The geologic time scale.

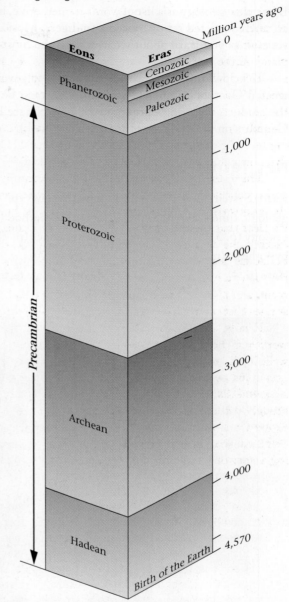

(a) The scale has been divided into eons and eras.

One thousand years ago = 1 Ka
(Ka stands for kila-annum)

One million years ago = 1 Ma
(Ma stands for mega-annum)

One billion years ago = 1 Ga
(Ga stands for giga-annum)

(b) Abbreviations for time units.

out of nowhere, but are the result of human efforts. Wherever possible, this book shows where geologic ideas came from, and tries to answer the question, "How do we know that?"

As you read this book, please keep these themes in mind. Don't view geology as a list of words to memorize, but rather as an interconnected set of concepts to digest. Most of all, enjoy yourself as you learn about what may be the most fascinating planet in the Universe.

To help illustrate the geology of our amazing world, we have created Geotours. Using *Google Earth*™, you'll be able to see the products of geologic processes for yourself (see Introducing Geotours on p. GT-1, near the end of the book).

Key Terms

Earth System (p. 4)	scientific laws (p. 7)
geologic time scale (p. 5)	scientific method (pp. 5–6)
geology (p. 3)	shatter cones (p. 7)
hypothesis (p. 7)	theory (p. 7)
plate (p. 4)	theory of plate tectonics (p. 4)

GEOPUZZLE REVISITED

A glance at the dirt (soil) beneath their feet makes geologists think of the processes that break rocks into a mass of loose grains in which plants can root. By studying the shape of the mountains, geologists imagine an earlier time when glaciers (rivers of ice) flowed slowly down the mountains, rasping and ripping at the underlying rock. By studying the rock itself, geologists picture an even earlier time when the present land surface lay kilometers underground, beneath a chain of erupting volcanoes.

BOX P.1

The Scientific Method

Sometime during the past 200 million years, a large block of rock or metal, which had been orbiting the Sun, slammed into our planet at a site in what is now the central United States, a landscape of flat cornfields. The impact of this block, a meteorite, released more energy than a nuclear bomb—a cloud of shattered rock and dust blasted skyward, and once-horizontal layers of rock from deep below the ground sprang upward and tilted on end beneath the gaping hole left by the impact. When the dust had settled, a huge crater surrounded by debris marked the surface of the Earth at the impact site. Later in Earth history, running water and blowing wind wore down this jagged scar. Some 15,000 years ago, sand, gravel, and mud carried by a vast glacier buried what remained, hiding it entirely from view (**Fig. 1a–c**). Wow! So much history beneath a cornfield. How do we know this? It takes scientific investigation.

The movies often portray science as a dangerous tool, capable of creating Frankenstein's monster, and scientists

as warped or nerdy characters with thick glasses and poor taste in clothes. In reality, science is simply the use of observation, experiment, and calculation to explain how nature operates, and scientists are people who study and try to understand natural phenomena. Scientists carry out their work using the **scientific method**, a sequence of steps for systematically analyzing scientific problems in a way that leads to verifiable results. Let's see how geologists employed the steps of the scientific method to come up with the meteorite-impact story.

1. *Recognizing the problem:* Any scientific project, like any detective story, begins by identifying a mystery. The cornfield mystery came to light when water drillers discovered limestone, a rock typically made of shell fragments, just below the 15,000-year-old glacial sediment. In surrounding regions, the rock at this depth consists of sandstone, made of cemented-together sand grains. Since limestone can be

used to build roads, make cement, and produce the agricultural lime used in treating soil, workers stripped off the glacial sediment and dug a quarry to excavate the limestone. They were amazed to find that rock layers exposed in the quarry tilted steeply and had been shattered by large cracks. In the surrounding regions, all rock layers are horizontal like the layers in a birthday cake, the limestone layer lies underneath the sandstone, and the rocks contain relatively few cracks. Curious geologists came to investigate, and soon realized that the geologic features of the land just beneath the cornfield presented a problem to be explained: What phenomena had brought limestone up close to the Earth's surface, had tilted the layering in the rocks, and had shattered the rocks?

2. *Collecting data:* The scientific method proceeds with the collection of observations or clues that point to an answer. Geologists studied the quarry and determined the age of its rocks, mea-

sured the orientation of the rock layers, and documented (made a written or photographic record of) the fractures that broke up the rocks.

3. *Proposing hypotheses:* A scientific **hypothesis** is merely a possible explanation, involving only naturally occurring processes, that can explain a set of observations. Scientists propose hypotheses during or after their initial data collection. The geologists working in the quarry came up with two alternative hypotheses. First, the features in this region could result from a volcanic explosion; and second, they could result from a meteorite impact.

4. *Testing hypotheses:* Since a hypothesis is no more than an idea that can be either right or wrong, scientists must put hypotheses through a series of tests to see if they work. The geologists at the quarry compared their field observations with published observations made at other sites of volcanic explosions and meteorite impacts, and studied the results of experiments designed to simulate such events. They learned that if the geologic features visible in the quarry were the result of volcanism, the quarry should contain rocks formed by the freezing of molten rock erupted by a volcano. But no such rocks were found. If, however, the features were the consequence of an impact, the rocks should contain **shatter cones**, small, cone-shaped cracks (Fig. 1). Shatter cones can easily be overlooked, so the geologists returned to the quarry specifically to search for them, and found them in abundance. The impact hypothesis passed the test!

Theories are scientific ideas supported by an abundance of evidence; they have passed many tests and have failed none. Scientists have much more confidence in a theory than they do in a hypothesis. Continued study in the quarry eventually yielded so much evidence for impact that the impact hypothesis came to be viewed as a theory. Scientists continue to test theories over a long time. Successful theories withstand these tests and are supported by so many observations that they come to be widely accepted. (As you will discover in Chapter 2, geologists consider the idea that continents drift around the surface of the Earth to be a theory, because so much evidence supports it.) However, some theories may eventually be disproven, to be replaced by better ones.

Some scientific ideas must be considered absolutely correct, for if they were violated, the natural Universe as we know it would not exist. Such ideas are called **scientific laws**, and examples include the law of gravity.

FIGURE 1   An ancient meteorite impact excavates a crater and permanantly changes rock beneath the surface.

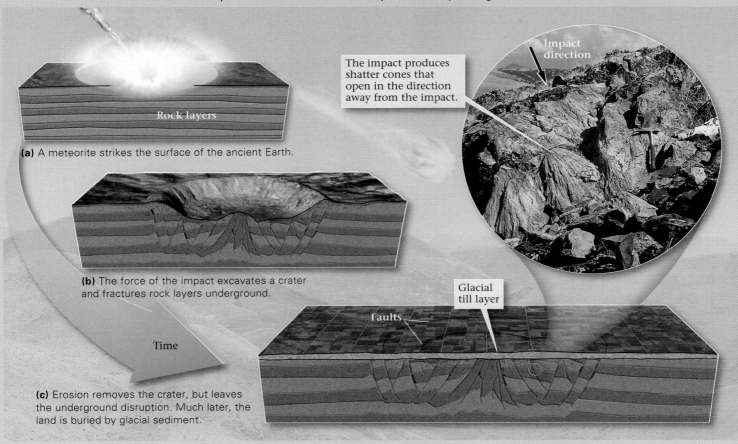

(a) A meteorite strikes the surface of the ancient Earth.

(b) The force of the impact excavates a crater and fractures rock layers underground.

(c) Erosion removes the crater, but leaves the underground disruption. Much later, the land is buried by glacial sediment.

The impact produces shatter cones that open in the direction away from the impact.

Impact direction

Rock layers

Time

Faults

Glacial till layer

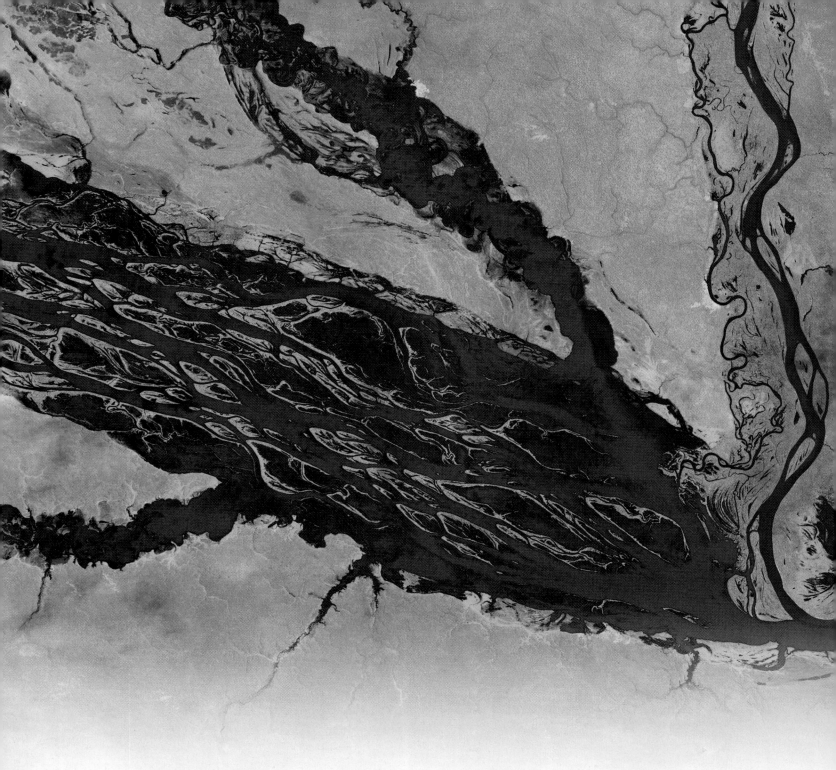

THE VIEW FROM SPACE   An image of the Earth taken by an orbiting satellite (Landsat 7) emphasizes that our planet is indeed a unique and evolving system. Here, a river draining the Amazon rainforest modifies the shape of vegetation-covered islands. No other planet hosts liquid water, or vegetation—no other planet looks like the Earth!

CHAPTER **1**

The Earth in Context

When the Hubble Space Telescope looks into what, to the naked eye, appears to be the black void of the night sky, it reveals a spectacle of disks and spirals of hazy light. Each of these is a distant galaxy, a cluster of as many as 300 billion stars. This is the fabric of space.

GEOPUZZLE

How did the Solar System (including the Earth) form, and what is the source of the material from which it formed?

1.1 INTRODUCTION

Sometime in the distant past, humans developed the capacity for complex, conscious thought. This amazing ability, which distinguishes our species from all others, brought with it the gift of curiosity, an innate desire to understand and explain the workings of ourselves and of all that surrounds us—our **Universe**. Questions that we ask about the Universe differ little from questions a child asks of a playmate: Where do you come from? How old are you? Such musings first spawned legends in which heroes, gods, and goddesses used supernatural powers to mold the planets and sculpt the landscape. Eventually, researchers began to apply scientific principles (see Box P.1) to the systematic study of the overall structure and history of the Universe, thereby establishing the modern discipline of scientific **cosmology**.

In this chapter, we begin with a brief introduction to the principles of scientific cosmology—we characterize the basic architecture of the Universe, introduce the big bang theory for the formation of the Universe, and discuss scientific ideas concerning the birth of the Earth. Then we outline the basic characteristics of our home planet by building an image of its surroundings, surface, and interior. Our high-speed tour of the Earth provides a reference frame for the remainder of this book.

TAKE-HOME MESSAGE

By the end of this chapter, you should know:

- the scientific explanation for the formation of the Universe and the Earth;
- the major components of the Solar System;
- the character of Earth's magnetic field, atmosphere, and surface;
- the basic materials that make up the Earth; and
- the nature of Earth's internal layering.

1.2 AN IMAGE OF THE UNIVERSE

What Is the Structure of the Universe?

Think about the mysterious spectacle of a clear night sky. What objects are up there? How big are they? How far away are they? How do they move? How are they arranged? In addressing such questions, ancient philosophers distinguished between stars (points of light whose locations relative to each other were fixed) and planets (tiny spots of light that move relative to the backdrop of stars). Over the centuries, two schools of thought developed concerning how to explain the configuration of stars and planets, and their relationships to the Earth, Sun, and Moon. The first school advocated a *geocentric* image (Fig. 1.1a), in which the

Earth sat without moving at the center of the Universe, while the Moon and the planets whirled around it within a globe of stars. The second school advocated a *heliocentric* image (Fig. 1.1b), in which the Sun lay at the center of the Universe, with the Earth and other planets orbiting around it. The geocentric image eventually gained the most followers, due to the influence of

FIGURE 1.1 Contrasting views of the Universe, as drawn by artists hundreds of years ago.

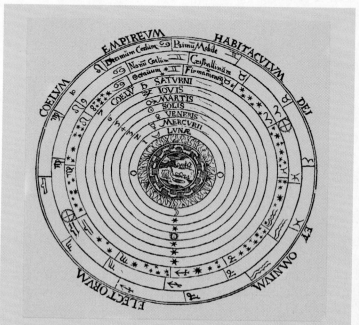

(a) The *geocentric* image of the Universe, showing the Earth at the center, surrounded by air, fire, and the other planets, all contained within the globe of the stars.

(b) The *heliocentric* image of the Universe, with the Sun at the center, as envisioned by Copernicus.

an Egyptian mathematician, Ptolemy (100–170 C.E.). Ptolemy developed equations that appeared to successfully predict the wanderings of the planets in the context of the geocentric model. During the Middle Ages (~476–1400 C.E.), church leaders in Europe adopted Ptolemy's geocentric image as dogma, because it justified the comforting thought that humanity's home occupies the most important place in the Universe. Anyone who disagreed with this view risked charges of heresy.

Then came the Renaissance. In fifteenth-century Europe, bold thinkers spawned a new age of exploration and scientific discovery. Thanks to the efforts of Nicolaus Copernicus (1473–1543) and Galileo Galilei (1564–1642), people gradually came to realize that the Earth and planets did indeed orbit the Sun and could not be at the center of the Universe. And when Isaac Newton (1643–1727) explained **gravity**, the attractive force that one object exerts on another (see Appendix), it finally became possible to explain the motions of these objects.

The Nature of Our Solar System

Eventually, astronomical study with powerful telescopes demonstrated that the Earth is but one of several planets orbiting the Sun, and that moons orbit most planets. The Sun, planets, moons, and countless other small objects held together by the "glue" of gravitational attraction comprise the **Solar System** (Fig. 1.2). Let's now discuss the components of the Solar System a bit more precisely.

The Sun accounts for 99.8% of the mass in the Solar System. The remaining 0.2% includes a great variety of objects, the largest of which are planets. Astronomers define a **planet** as an object that orbits a star, is spherical, and has "cleared its neighborhood of other objects." The last phrase in this definition sounds a bit strange at first, but merely implies that a planet's gravity has pulled in all particles of matter in its orbit. According to this definition, which was formalized

FIGURE 1.2 The relative sizes and positions of planets in the Solar System.

(a) The relative sizes of the planets. All are much smaller than the Sun, but the gas-giant planets are much larger than the terrestrial planets. Jupiter's diameter is about 11.2 times greater than that of Earth.

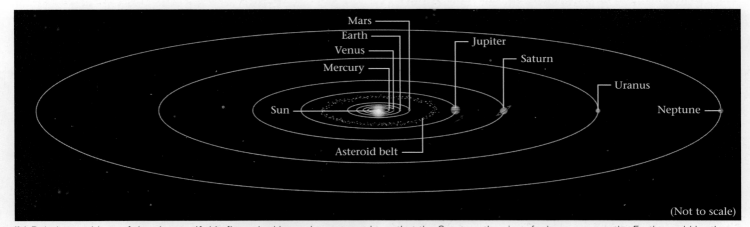

(b) Relative positions of the planets. If this figure had been drawn to scale so that the Sun was the size of a large orange, the Earth would be the size of a sesame seed 15 meters (49 feet) away. All planetary orbits lie roughly in the same plane.

in 2005, our Solar System includes eight planets—Mercury, Venus, Earth, Mars, Jupiter, Saturn, Uranus, and Neptune. (In 1930, astronomers discovered Pluto, a 2,390-km-diameter sphere of ice, whose orbit generally lies outside of Neptune's. Until 2005, Pluto was considered to be a planet, but it does not fit the modern definition and so has been dropped from the roster.) A **moon** is a sizeable body locked in orbit around a planet. All but two planets have moons—some, such as Earth's Moon, are large and spherical, but many are small and have irregular shapes. Our Solar System is not alone in hosting planets; in recent years, astronomers have found planets orbiting stars in many other systems.

Planets in our Solar System differ radically from one another both in size and composition. The inner planets (Mercury, Venus, Earth, and Mars), the ones closer to the Sun, are relatively small. Astronomers commonly refer to these as **terrestrial planets** because, like Earth, they consist of a shell of rock surrounding a ball of metallic iron alloy. The outer planets (Jupiter, Saturn, Uranus, and Neptune) are known as the **gas-giant planets**, or Jovian planets, because most of their mass consists of gas and "ice." (In this context, ice is a general term that includes the solid state of many materials that could be gaseous under Earth's surface conditions.) The adjective "giant" certainly seems appropriate, for these planets are huge—Jupiter, for example, has a mass 318 times larger than that of Earth and accounts for about 71% of the non-solar mass in the Solar System.

In addition to the planets, the Solar System contains many small objects. Millions of **asteroids** (chunks of rock and/or metal) comprise a belt between the orbits of Mars and Jupiter. Asteroids range in size from less than a centimeter to about 930 km in diameter. About a trillion bodies of ice lie in belts or clouds beyond the orbit of Neptune. Most of these icy fragments are tiny, but a few (including Pluto) have diameters of over 2,000 km and may be thought of as "dwarf planets." The gravitational pull of the planets has sent some of the icy objects on paths that take them into the inner part of the Solar System, where they begin to evaporate and form long tails—we call such objects **comets**.

Stars and Galaxies

Stars look like points of light. But in fact, **stars** are immense balls of incandescent gas in which nuclear reactions (see Appendix) produce intense heat and light. Our Sun is a medium-sized star. Stars are not randomly scattered through the Universe; gravity holds them together in immense groups, called **galaxies**. The Sun and over 300 billion stars together form the Milky Way galaxy. More than 100 billion galaxies constitute the visible Universe (see chapter opening photo). Galaxies are so far away that, to the naked eye, they look like stars in the night sky. The nearest galaxy to ours, Andromeda, lies over 2.2 million

light years away. A light year is the distance light travels in one year—about 10 trillion kilometers (6 trillion miles).

If we could view the Milky Way from a great distance, it would look like a flattened spiral, 100,000 light years across, with great curving arms slowly swirling around a glowing, disk-like center (Fig. 1.3). Presently, our Solar System lies near the outer edge of one of these arms and rotates around the center of the galaxy about once every 250 million years. We hurtle through space, relative to an observer standing outside the galaxy, at about 200 km per second.

TAKE-HOME MESSAGE

The Sun lies at the center of the Solar System and is but one star at the edge of one galaxy in a Universe of hundreds of billions of galaxies.

1.3 FORMING THE UNIVERSE

Do galaxies move with respect to other galaxies? Does the Universe become larger or smaller with time? Has the Universe always existed? Answers to these fundamental questions came from an understanding of a phenomenon called the Doppler effect, which we must introduce before proceeding.

Waves and the Doppler Effect

When a train whistle screams, the sound you hear has moved through the air from the whistle to your ear in the form of sound waves. **Waves** are disturbances that transmit energy from one point to another by causing periodic motions. As each sound wave passes, air alternately compresses, then expands. We refer to the distance between successive waves as the **wavelength**,

FIGURE 1.3 An image of what the Milky Way galaxy might look like if viewed from outside. Note that the galaxy consistsof spiral arms around a central cluster.

FIGURE 1.4 Manifestations of the Doppler effect.

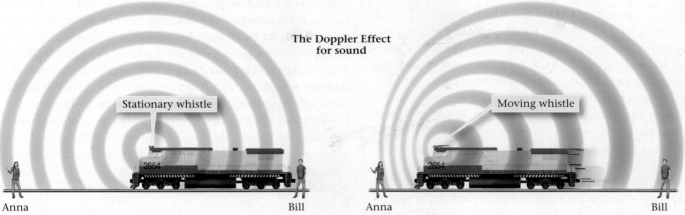

The Doppler Effect for sound

Stationary whistle

Anna

Bill

(a) The wavelength of sound waves emitted by a stationary train is the same in all directions.

Moving whistle

Anna

Bill

(b) Waves behind a moving train have a longer wavelength than those in front.

The Doppler Effect for light

Moving source of light

Waves that reach this observer are squeezed to shorter "blue-shifted" wavelengths.

Waves that reach this observer are spread out to longer "red-shifted" wavelengths.

V

Blue light (high frequency)

This observer sees no Doppler shift.

Red light (low frequency)

(c) The wavelength of blue light is less than that of red light. If a light source moves very fast, the Doppler effect results in a shifting of the wavelengths. The observed shift depends on the position of the observer.

and the number of waves that pass a point in a given time interval as the **frequency**. If the wavelength decreases, more waves pass a point in a given time interval, so the frequency increases. The pitch of a sound, meaning its note on the musical scale, depends on the frequency of the sound waves.

Imagine that you are standing on a station platform, and a train moves toward you. The train whistle's sound gets louder as the train approaches, but its pitch remains the same. Then, the instant the train passes, the pitch abruptly changes—it sounds like a lower note in the musical scale. Why? When the train moves toward you, the sound has a higher frequency (the waves are closer together so the wavelength is smaller), because the sound source, the whistle, has moved slightly closer to you between the instant that it emits one wave and the instant that it emits the next (Fig. 1.4a, b). When the train moves away from you, the sound has a lower frequency (the waves are farther apart), because the whistle has moved slightly farther from you between the instant it emits one

wave and the instant it emits the next. An Austrian physicist, C. J. Doppler (1803–1853), first interpreted this phenomenon, and thus the change in frequency that happens when a wave source moves is now known as the Doppler effect.

Light energy also moves in the form of waves. We can represent light waves symbolically by a periodic succession of crests and troughs (Fig. 1.4c). Visible light comes in many colors—the colors of the rainbow. The color you see depends on the frequency of the light waves, just as the pitch of a sound you hear depends on the frequency of sound waves. Red light has a longer wavelength (lower frequency) than does blue light. The Doppler effect also applies to light but can be noticed only if the light source moves very fast, at least a few percent of the speed of light. If a light source moves away from you, the light you see becomes redder, as the light shifts to longer wavelength or lower frequency. If the source moves toward you, the light you see becomes bluer, as the light shifts to higher frequency. We call these changes the **red shift** and the blue shift, respectively.

Does the Size of the Universe Change?

In the 1920s, astronomers around the world, including Edwin Hubble, after whom the Hubble Space Telescope was named, braved many a frosty night beneath the open dome of a mountaintop observatory in order to aim telescopes into deep space. These researchers were searching for distant galaxies. At first, they documented only the location and shape of newly discovered galaxies. But then, one astronomer began an additional project to study the wavelength of light produced by the distant galaxies. The results yielded a surprise that would forever change humanity's perception of the Universe.

Astronomers found, to their amazement, that the light of distant galaxies *displayed red shifts* relative to the light of nearby stars. Hubble pondered this mystery and, around 1929, realized that the red shifts must be a consequence of the Doppler effect, and thus that the distant galaxies must be moving away from Earth at an immense velocity. At the time, astronomers thought the Universe had a fixed size, so Hubble initially assumed that if some galaxies were moving away from Earth, others must be moving toward Earth. But this was not the case. On further examination, Hubble concluded that the light from all distant galaxies, regardless of their direction from Earth, exhibits a red shift. In other words, *all* distant galaxies are moving rapidly away from us.

How can all galaxies be moving away from us, regardless of which direction we look? Hubble puzzled over this question and finally recognized the solution: the whole Universe must be expanding! (To picture the expanding Universe, imagine a ball of bread dough with raisins scattered throughout. As the dough bakes and expands into a loaf, each raisin moves away from its neighbors, in every direction; Fig. 1.5a.) This idea came to be known as the **expanding Universe theory**.

Hubble's ideas marked a revolution in cosmological thinking. Now we picture the Universe as an expanding bubble, in which galaxies race away from each other at incredible speeds. This image immediately triggers the key question of cosmology: Did the expansion begin at some specific time in the past? If it did, then that instant would mark the beginning of the Universe, the beginning of space and time.

The Big Bang

Most astronomers have concluded that expansion did indeed begin at a specific time, with a cataclysmic explosion called the **big bang**. According to the big bang theory, all matter and energy—everything that now constitutes the Universe—was initially packed into an infinitesimally small point. The point exploded, according to current estimates, 13.7 (±1%) billion years ago.

FIGURE 1.5 The concept of the expanding Universe and the Big Bang.

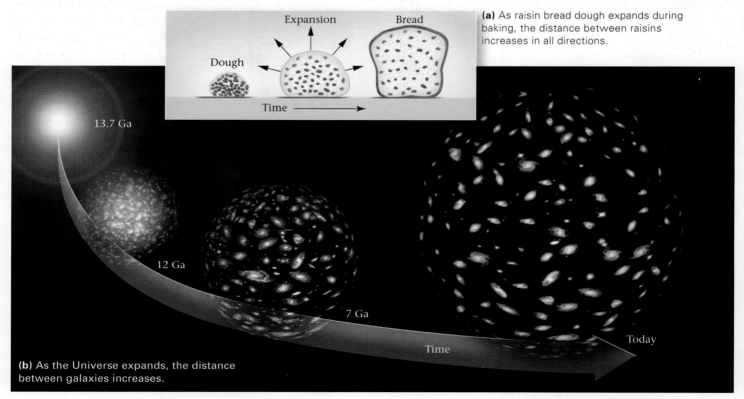

(a) As raisin bread dough expands during baking, the distance between raisins increases in all directions.

(b) As the Universe expands, the distance between galaxies increases.

Of course, no one was present at the instant of the big bang, so no one actually saw it happen. But by combining clever calculations with careful observations, researchers have developed a consistent model of how the Universe evolved, beginning an instant after the explosion (Fig. 1.5b). (Please see the Appendix for a review of the basic concepts of physics that you need to understand what follows.)

According to the contemporary model of the big bang, profound change happened at a fast and furious rate at the outset. During the first instant of existence, the Universe was so small, so dense, and so hot that it consisted entirely of energy—atoms, or even the smallest subatomic particles that make up atoms, could not even exist. Within a few seconds, however, hydrogen atoms could begin to form. And by the time the Universe reached an age of 3 minutes, when its temperature had fallen below 1 billion degrees, and its diameter had grown to about 53 million km (35 million miles), the collision and fusion (sticking together) of hydrogen atoms formed helium and other small atoms. Formation of new nuclei in the first few minutes of time is called big bang nucleosynthesis because it happened *before* any stars existed. This process could produce only small atoms, meaning ones containing a small number of protons, and it happened very rapidly. In fact, virtually all of the new atomic nuclei that would form by big bang nucleosynthesis existed by the end of 5 minutes.

Eventually, the Universe became cool enough for chemical bonds to bind atoms of certain elements together in molecules. Most notably, two hydrogen atoms could join to form molecules of H_2. As the Universe continued to expand and cool further, atoms and molecules slowed down and accumulated into patchy clouds called **nebulae**. The earliest nebulae of the Universe consisted almost entirely of hydrogen (74%, by volume) and helium (24%) gas.

Birth of the First Stars

When the Universe reached its 200 millionth birthday, it contained immense, slowly swirling, dark nebulae separated by vast voids of empty space (Fig. 1.6). The Universe could not remain this way forever, however, because of the invisible but persistent pull of gravity (see Appendix). Eventually, gravity began to remold the Universe pervasively and permanently.

All matter exerts gravitational pull—a type of force—on its surroundings, and as Isaac Newton first pointed out, the amount of pull depends on the amount of mass. Somewhere in the young Universe, the gravitational pull of an initially more massive region of a nebula began to suck in surrounding gas and, in a grand example of "the rich getting richer," grew in mass and, therefore, density (mass per unit volume). As this denser region attracted progressively more gas, the gas compacted into a smaller region, and the initial swirling move-

FIGURE 1.6 Nebulae in space look like clouds. In this Hubble Space Telescope picture, new stars are forming at the top of the nebula on the left. Stars that have already formed light up the nebulae from behind.

ment of gas transformed into a rotation around an axis. As gas continued to move inward, into a progressively smaller volume, the rotation rate became faster and faster. (A similar phenomenon happens when a spinning ice skater pulls her arms inward.) Because of its increased rotation, the nebula evolved into a disk shape. As more and more matter rained down onto the disk, it continued to grow, until eventually, gravity collapsed the inner portion of the disk into a dense ball. As the gas crammed into a smaller and smaller space, its temperature increased dramatically. Eventually, the central ball of the disk became hot enough to glow, and at this point it became a **protostar**.

A protostar continues to grow, by pulling in successively more mass, until its core becomes extremely dense and its temperature reaches about 10 million degrees. Under such conditions, hydrogen nuclei slam together so forcefully that they join, in a series of steps, to form helium nuclei. Such fusion reactions produce huge amounts of energy, and the mass becomes a fearsome furnace. When the first nuclear fusion reactions began in the first protostar, the body "ignited" and the first true star formed. When this happened, perhaps 800 million years after the big bang, the first starlight pierced the newborn Universe. This process would soon happen again and again, and many first-generation stars came into existence.

First-generation stars tended to be very massive, reaching 100 times the mass of the Sun. Astronomers have shown that the larger the star, the hotter it burns and the faster it runs out of fuel and dies. A huge star may survive only a few million years to a few tens of millions of years before it explodes to form a **supernova**. Thus, not long after the first generation of stars formed, the Universe began to be peppered with the first generation of supernova explosions.

TAKE-HOME MESSAGE

According to the big bang theory, the Universe started in a cataclysmic explosion about 13.7 Ga, and has been expanding ever since. Stars first formed when gravity caused nebulae of gases produced by the big bang to collapse inward, packing matter so tightly together that nuclear fusion reactions could begin.

1.4 **WE ARE ALL MADE OF STARDUST**

Where Do Elements Come From?

Nebulae from which the first-generation stars formed consisted entirely of small atoms (elements with atomic numbers smaller than 5; "atomic number" refers to the number of protons in the nucleus), because only these small atoms were generated by big bang nucleosynthesis. In contrast, the Universe of today contains ninety-two naturally occurring elements (see Appendix). Where do the other eighty-seven elements come from? In other words, how did elements such as carbon, sulfur, silicon, iron, gold, and uranium form? These elements, which are common on Earth, have larger atomic numbers—carbon has an atomic number of 6, and iron has an atomic number of 26. Physicists can demonstrate that these elements form during the life cycle of stars, by the process of stellar nucleosynthesis. Because of stellar nucleosynthesis we can consider stars to be element factories, constantly fashioning larger atoms out of smaller atoms.

What happens to the atoms formed in stars? Some escape into space during the star's lifetime, simply by moving fast enough to overcome the star's gravitational pull. The stream of atoms emitted from a star during its lifetime is a **stellar wind** (Fig. 1.7). Some escape only when a star dies. A low-mass star (like our Sun) releases a large shell of gas as it dies, whereas a high-mass star blasts matter into space during a supernova explosion (Fig. 1.8). Most very large atoms (those with atomic numbers greater than that of iron) require even more violent circumstances to form than can occur within a star. These atoms usually form as a result of supernova explosions. Once ejected into space, atoms from stars and supernova explosions form new nebulae or mix back into existing nebulae.

When the first generation of stars died, they left a legacy of new elements that mixed with residual gas from the big bang. A second generation of stars and associated planets formed out of the new, compositionally more diverse nebulae. Second-generation stars lived and died, and contributed elements to third-generation stars. Succeeding generations contain a greater proportion of heavier elements. Because different stars live for varied periods of time, at any given moment the Universe contains many different generations of stars. Our Sun may be a third-, fourth-, or fifth-generation star. Thus, the mix of elements we find on Earth includes relicts of primordial gas from the big bang as well as the disgorged guts of dead stars. Think of it—the elements that make up your body once resided inside a star!

The Nebula Theory for Forming the Solar System

Earlier in this chapter, we introduced scientific concepts of how stars form from nebulae. But we delayed our discussion of how the planets and other objects in our Solar System originated until we had discussed the production of heavier atoms such as carbon, silicon, iron, and uranium, because planets consist predominantly of these elements. Now that we've discussed stars as element factories, we return to the early history of the Solar System and introduce the nebula theory, an explanation for the origin of planets, moons, asteroids, and comets. According to the **nebula theory**, the Sun and all other objects in the Solar System formed from material that had been swirling about in a nebula. This process involved several stages. We've already discussed the stage of forming a star like the Sun from a nebula. Now, to complete the story of Solar System formation, we con-

FIGURE 1.7 In this image, a disk shades the Sun so we can see the stellar wind that the Sun produces.

FIGURE 1.8 Very heavy elements form during supernova explosions. Here we see the rapidly expanding shell of gas ejected into space from an explosion whose light reached the Earth in 1054 C.E. This shell is called the Crab Nebula.

sider the fate of the material in the flattened outer part of the disk, the material that did not become part of the star. This outer part is called the **protoplanetary disk**, because it is the source of planets (as well as of other moons, comets, and asteroids).

What did the protoplanetary disk consist of? The disk from which our Solar System formed contained all 92 elements, some as isolated atoms, and some bonded to others in molecules. Geologists divide the material formed from these atoms and molecules into two classes. *Volatile* materials—such as hydrogen, helium, methane, ammonia, water, and carbon monoxide—are materials that can exist as gas at the Earth's surface. In the pressure and temperature conditions of space, some volatile materials remain in gaseous form, but others freeze to form different kinds of "ice." *Refractory* materials are those that melt only at high temperatures, and they condense to form solid soot-sized particles of "dust" in the coldness of space. Initially, the protoplanetary disk may have been fairly homogeneous, meaning that it had much the same composition throughout. But as the proto-Sun began to form, the inner part of the disk became hotter, causing volatile elements to evaporate and drift to the outer portions of the disk. Thus, the inner part of the disk ended up with higher concentrations of dust, whereas the outer portions ended up with higher concentrations of ice. As this was happening, gravity caused the protoplanetary disk to evolve into a series of concentric rings.

How did the dusty, icy, and gassy rings transform into planets? Even before the proto-Sun ignited, the material of the surrounding rings began to clump and bind together. First, soot-sized particles merged to form sand-sized grains. Then, these grains stuck together to form grainy basketball-sized blocks, which in turn collided. If the collision was slow, blocks stuck together or simply bounced apart. If the collision was fast, one or both of the blocks shattered, producing smaller fragments that recombined later. Eventually, enough blocks coalesced to form **planetesimals**, bodies whose diameter exceeded about 1 km (Fig. 1.9). Because of their mass, the planetesimals exerted enough gravity to attract and pull in other objects that were nearby. Figuratively, planetesimals acted like vacuum cleaners, sucking in small pieces of dust and ice as well as smaller planetesimals that lay in their orbit, and in the process they grew progressively larger. Eventually, victors in the competition to attract mass grew into **protoplanets**, bodies approaching the size of today's planets. Once a protoplanet succeeded in incorporating virtually all the debris within its orbit, it became a full-fledged planet (see Geology at a Glance, pp. 18–19).

Early stages in the planet-forming process probably occurred very quickly—some computer models suggest that it may have taken only a few hundred thousand years to go from the dust and gas stage to the large planetesimal stage. Planets may have grown from planetesimals in 10 to 200 million years. In the inner orbits, where the protoplanetary disk consisted mostly of dust, small terrestrial planets composed of rock and metal formed. In the outer part of the Solar System, where significant amounts of ice existed, larger protoplanets grew, and these evolved into the gas-giant planets.

When did the planets form? Using techniques introduced in Chapter 10, geologists have concluded that meteorites thought to be leftover planetesimals formed at 4.57 Ga, and thus consider that date to be the birth date of the Solar System. If this date is correct, it means that the Solar System formed about 9 billion years after the big bang, and thus is only about a third as old as the Universe.

FIGURE 1.9 Cross section of a particular type of meteorite (a fragment of solid material that fell from space and landed on Earth) thought to display the texture of a small planetesimal.

Forming the Planets and the Earth-Moon System

1. Forming the solar system, according to the nebula hypothesis: A nebula forms from hydrogen and helium left over from the big bang, as well as from heavier elements that were produced by fusion reactions in stars or during explosions of stars.

2. Gravity pulls gas and dust inward to form an accretion disk. Eventually a glowing ball—the proto-Sun—forms at the center of the disk.

5. Forming the planets from planetesimals: Planetesimals grow by continuous collisions. Gradually, an irregularly shaped proto-Earth develops. The interior heats up and becomes soft.

6. Gravity reshapes the proto-Earth into a sphere. The interior of the Earth differentiates into a core and mantle.

8. The Moon forms from the ring of debris.

7. Soon after Earth forms, a small planet collides with it, blasting debris that forms a ring around the Earth.

3. "Dust" (particles of refractory materials) concentrates in the inner rings, while "ice" (particles of volatile materials) concentrates in the outer rings. Eventually, the dense ball of gas at the center of the disk becomes hot enough for fusion reactions to begin. When it ignites, it becomes the Sun.

4. Dust and ice particles collide and stick together, forming planetesimals.

9. Eventually, the atmosphere develops from volcanic gases. When the Earth becomes cool enough, moisture condenses and rains to create the oceans. Some gases may be added by passing comets.

The nebula theory successfully answers many questions about the Solar System. For example, why do we see such differences between the terrestrial planets and the gas-giant planets? As mentioned earlier, once the proto-Sun began to heat up, volatile materials in the inner part of the disk evaporated and migrated out, leaving refractory materials behind. When the Sun ignited and became a gigantic nuclear furnace, it emitted a strong solar wind that blew much of the remaining volatiles out of the inner Solar System. Thus, the inner rings ended up consisting mostly of dust, which coalesced to form rocky and metallic planets. Dust existed in the outer rings as well, and thus the gas-giant planets have rocky and metallic centers. But the outer rings also contained a huge amount of volatile material, which formed the ice and gas that makes up the bulk of the gas-giant planets.

Differentiation of the Earth and Formation of the Moon

When the larger planetesimals and protoplanets first formed, they had a fairly homogeneous distribution of material throughout, because the smaller pieces from which they formed all had much the same composition and collected together in no particular order. But large planetesimals did not stay homogeneous for long, because they began to heat up. The heat came primarily from two sources: the transformation of kinetic energy into thermal energy during collisions (see Appendix), and the decay of radioactive elements. In bodies whose temperature rose sufficiently to cause melting, denser iron alloy separated out and sank to the center of the body, whereas lighter rocky materials remained in a shell surrounding the center. By this process, called **differentiation**, protoplanets and large planetesimals developed internal layering early in their history. As we will see later, the central ball of iron alloy constitutes the body's core and the outer shell constitutes its mantle.

In the early days of the Solar System, planets continued to be bombarded by **meteorites** (solid objects falling from space that land on a planet) even after the Sun had ignited and differentiation had occurred (**Box 1.1**). Heavy bombardment in the early days of the Solar System pulverized the surfaces of planets and eventually left huge numbers of craters. It also contributed to heating the planets.

Based on the dating of Moon rocks, most geologists have concluded that at about 4.53 Ga, a Mars-sized protoplanet slammed into the newborn Earth. In the process, the colliding body disintegrated, along with a large part of the Earth's mantle. A ring of debris formed around the remaining, now molten Earth, and quickly coalesced to form the Moon. Of note, not all moons in the Solar System necessarily formed in this manner. Some may be protoplanets or comets captured by a larger planet's gravity.

Why Is the Earth Round?

Small planetesimals were jagged or irregular in shape, and asteroids today have irregular shapes. Planets, on the other hand, are more or less spherical. Why? Simply put, when a protoplanet gets big enough, gravity can change its shape. To picture how, imagine a block of cheese warming in an oven. As the cheese gets softer and softer, gravity causes it to spread out in a pancake-like blob. This model shows that gravitational force alone can cause material to change shape if the material is soft enough. Now let's apply this model to planetary growth.

The rock composing a small planetesimal is cool and strong enough so that the force of gravity is not sufficient to cause the rock to flow. But once a planetesimal grows beyond a certain critical size (about 1,000 km in diameter), its interior becomes warm and soft enough to flow in response to gravity. As a consequence, protrusions are pulled inward toward the center, and the planetesimal re-forms into a special shape that permits the force of gravity to be nearly the same at all points on its surface. This special shape is a sphere because in a sphere mass is evenly distributed around the center.

TAKE-HOME MESSAGE

The nebula theory states that stars and planets form when gravity pulls gas, dust, and ice together into a rotating disk. The center of the disk becomes a star. Rings around the star condense into solid planetesimals, which combine to form planets.

1.5 WELCOME TO THE EARTH SYSTEM AND ITS NEIGHBORHOOD

So far in this chapter, we've described scientific ideas about how the Universe, and then the Solar System, formed. Now, let's focus on our home planet and develop an image of Earth's overall architecture. To do this, imagine that we are explorers from outside our Solar System on a visit to the Earth for the first time. We will see that our planet consists of several components—the atmosphere (Earth's gaseous envelope), the hydrosphere (Earth's surface and near-surface water), the biosphere (Earth's great variety of life forms), the lithosphere (the outer, rigid shell of the Earth), and the interior (the softer material inside the Earth). These components, and the complex interactions among them, comprise the **Earth System**. Our planet remains a dynamic place, because heat inside the Earth, and the heat of the Sun, provide energy.

Our journey begins in interplanetary space, somewhere between Earth and Mars. Compared with the air we breathe at sea level, interplanetary space is a **vacuum**, meaning that it contains very little matter in a given volume. As our rocket approaches the Earth, and we see its beautiful bluish glow (Fig. 1.10), our instruments detect the planet's magnetic field, like a signpost shouting, "Approaching Earth!" A **magnetic field**, in a general sense, is the region affected by the force emanating from a magnet. This force, which grows progressively stronger as you approach the magnet, can attract or repel another magnet and can cause charged particles to move. Earth's magnetic field, like the familiar magnetic field around a bar magnet, is largely

BOX 1.1

THE REST OF THE STORY

Meteors and Meteorites

During the early days of the Solar System, the Earth collided with and incorporated countless planetesimals and smaller fragments of solid material lying in its path. Intense bombardment ceased about 3.9 Ga, but even today collisions with space objects continue, and over 1,000 tons of material (rock, metal, dust, and ice) fall to Earth, on average, every year. The vast majority of this material consists of fragments derived from comets and asteroids sent careening into the path of the Earth after billiard ball–like collisions with each other out in space, or because of the gravitational pull of a passing planet. Some of the material, however, consists of chips of the Moon or Mars, ejected into space when large objects collide with these bodies.

Astronomers refer to any object from space that enters the Earth's atmosphere as a meteoroid. Meteoroids move at speeds of up to 75 km/s, so fast that when they reach an altitude of about 150 km, friction with the atmosphere causes them to heat up and evaporate, leaving a streak of bright, glowing gas. The glowing streak, an atmospheric phenomenon, is a **meteor** (also known colloquially, though incorrectly, as a "falling star") (**Fig. 1**). Most visible meteors completely evaporate by an altitude of about 30 km. But dust-sized ones may slow down sufficiently to float to Earth, and larger ones (fist-sized or bigger) can survive the heat of entry to reach the surface of the planet. In some cases, meteoroids explode in brilliant fireballs.

Objects that strike the Earth are called **meteorites**. Most are asteroidal or planetary fragments, for the icy material of small cometary bodies is too fragile to survive the fall. Researchers recognize three basic classes of meteorites: iron (made of iron-nickel alloy), stony (made of rock), and stony iron (rock embedded in a matrix of metal). Of all known meteorites, about 93% are stony and 6% are iron. From their composition, researchers have concluded that some meteors (a special subcategory of stony meteorites called carbonaceous chondrites, because they contain carbon and small spherical nodules called chondrules) are asteroids derived from planetesimals that never underwent differentiation into a core and mantle. Other stony meteorites and all iron meteorites are asteroids derived from planetesimals that differentiated into a metallic core and a rocky mantle early in Solar System history but later shattered into fragments during collisions with other planetesimals. Most meteorites appear to be about 4.54 Ga, but carbonaceous chondrites are as old as 4.57 Ga, the oldest known material ever measured.

Although almost all meteorites are small and have not caused notable damage on Earth during human history, a very few have smashed through houses, dented cars, and bruised people. During the longer term of Earth history, however, there have been some catastrophic collisions that left huge craters. Similar craters pockmark the Moon (**Fig. 2**).

FIGURE 1   Meteors over Hong Kong in 2001.

FIGURE 2   Meteorite craters are found in abundance on the surface of the Moon.

FIGURE 1.10 A view of the Earth as seen from space. Oceans, clouds, and land stand out clearly.

a dipole, meaning it has a North Pole and a South Pole. We can portray the magnetic field by drawing magnetic field lines, the trajectories or paths along which needles (such as iron filings or compasses) would align if placed in the field (Fig. 1.11a, b).

The solar wind interacts with Earth's magnetic field, distorting it into a huge teardrop pointing away from the Sun. Fortunately, the magnetic field deflects most (but not all) of the wind, so that most of the particles in the wind do not reach Earth's surface. In this way, the magnetic field acts like a shield against the solar wind; the region inside this magnetic shield is called the magnetosphere (Fig. 1.11c).

Though it protects the Earth from most of the solar wind, the magnetic field does not stop our rocket ship, and we continue to speed toward the planet. At distances of about 3,000 km and 10,500 km out from the Earth, we encounter the Van Allen radiation belts, named for the physicist who first recognized them in 1959. These contain solar wind particles as well as cosmic rays (nuclei of atoms emitted from supernova explo-

FIGURE 1.11 A magnetic field permeates the space around the Earth. It can be symbolized by a bar magnet.

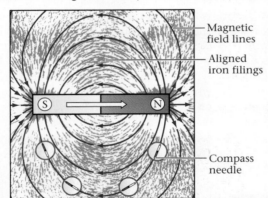

(a) A bar magnet produces a magnetic field. Magnetic field lines point into the "south pole" and out from the "north pole".

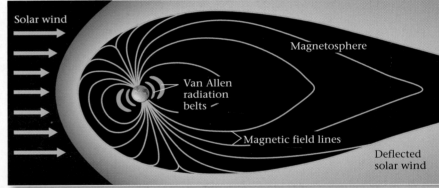

(c) Earth behaves like a magnetic dipole, but the field lines are distorted by the solar wind. The Van Allen radiation belts trap charged particles.

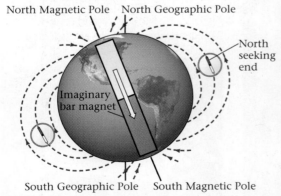

(b) We can represent the Earth's field by an imaginary bar magnet inside. Because of the way the magnetic flow lines point, the north end of this imaginary bar actually lies in the south, and the south end of this bar lies in the north, a rather confusing circumstance. By convention, the north magnetic pole is labeled as the one that lies near the north geographic pole of the Earth. This way, the north-seeking end of a compass points toward the north, since opposite ends of a magnet attract.

(d) Charged particles flow toward Earth's magnetic poles and cause gases in the atmosphere to glow, forming colorful aurorae in polar skies.

sions), that were moving so fast they were able to penetrate the weaker outer part of the magnetic field only to be trapped by the stronger magnetic field closer to the Earth. By trapping cosmic rays, the Van Allen belts protect life on Earth from dangerous radiation. Some charged particles make it past the Van Allen belts and are channeled along magnetic field lines to the polar regions of Earth. When these particles interact with gas atoms in the upper atmosphere they cause the gases to glow, like the gases in neon signs, creating spectacular aurorae (Fig. 1.11d).

TAKE-HOME MESSAGE

The Earth is a complex system consisting of many components. It has a magnetic field that shields our planet's surface from solar wind and cosmic rays.

1.6 THE ATMOSPHERE

As we descend further, we enter Earth's **atmosphere**, an envelope of gas consisting of 78% nitrogen (N_2) and 21% oxygen (O_2), with minor amounts (1% total) of argon, carbon dioxide (CO_2), neon, methane, ozone, carbon monoxide, and sulfur dioxide (Fig. 1.12a, b). Other terrestrial planets have atmospheres, but none of them are like Earth's.

The weight of overlying air squeezes on the air below, and thus pushes gas molecules in the air below closer together. Thus, both the density (mass per unit volume) of air and the air pressure (the amount of push that the air exerts on material beneath it) decrease with increasing elevation (Fig. 1.12c). Technically, we specify pressure in units of force, or push, per unit area. Such units include atmospheres (abbreviated atm) and bars, where 1 atm = 1.04 kilograms per square centimeter,

FIGURE 1.12 Characteristics of the atmosphere that envelops the Earth.

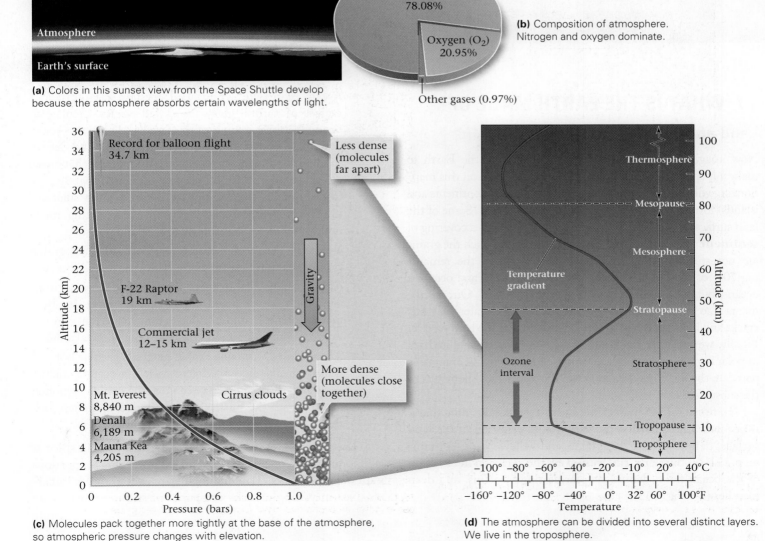

(a) Colors in this sunset view from the Space Shuttle develop because the atmosphere absorbs certain wavelengths of light.

(b) Composition of atmosphere. Nitrogen and oxygen dominate.

(c) Molecules pack together more tightly at the base of the atmosphere, so atmospheric pressure changes with elevation.

(d) The atmosphere can be divided into several distinct layers. We live in the troposphere.

or 14.7 pounds per square inch and 1 atm = 1.01 bars. At sea level, average air pressure is 1 atm, whereas on the peak of Mt. Everest, 8.85 km above sea level, air pressure is only 0.3 atm. In fact, 99% of atmospheric gas lies below 50 km. Humans cannot live for long at elevations greater than about 4.5 to 5.5 km.

The nature of the atmosphere changes with distance from the Earth's surface. Because of these changes, atmospheric scientists divide the atmosphere into layers. Most winds and clouds develop only in the lowest layer, the troposphere. The layers of the atmosphere that lie above the troposphere are named, in sequence from base to top: the stratosphere, the mesosphere, and the thermosphere (**Fig. 1.12d**). Boundaries between layers are called pauses. For example, the boundary between the troposphere and the overlying stratosphere is the tropopause. The pauses are defined as elevations at which temperature stops decreasing and starts increasing, or vice versa.

TAKE-HOME MESSAGE

Earth's atmosphere consists mostly of nitrogen and oxygen. Ninety-nine percent of the atmosphere's gas lies below an elevation of 50 km. So, relative to our planet's diameter, the atmosphere is very thin indeed.

1.7 WHAT IS THE EARTH MADE OF?

Land and Oceans

Now imagine that we've gone into orbit around the Earth to make a map of the planet. What features should go on this map? Starting with the most obvious, we note that land (continents and islands) forms about 30% of the surface (**Fig. 1.13**). Some of the land surface consists of solid rock, whereas some has a covering of **sediment** (materials such as sand and gravel, in which the grains are not stuck together). **Surface water** covers the remaining 70% of the Earth. Most surface water is salty and occupies oceans, but some is fresh and fills lakes and rivers. Our instruments also detect **groundwater**, which is the water that fills cracks and holes within rock and sediment under the land surface. Finally, we find that ice covers significant areas of land and sea in polar regions and at high elevations, and that living organisms populate the land, sea, air, and even the upper few kilometers of the subsurface.

To finish off our map of the Earth's surface, we note that it is not flat. **Topography**, the variation in elevation of the land surface, defines plains, mountains, and valleys (see Geotour 1 on p. GT-4). Similarly, **bathymetry**, the variation in elevation of the ocean floor, defines submarine ridges, plains, and deep trenches (see Fig. 1.13).

Categories of Earth Materials

At this point, we leave our fantasy space voyage and turn our attention to the materials that make up the solid Earth, because we need to be aware of these before we can discuss the architecture of the Earth's interior. We begin by reiterating that the Earth consists mostly of elements produced by fusion reactions in stars and supernova explosions. Only four elements (iron, oxygen, silicon, and magnesium) make up 90% of the Earth's mass; the remaining 10% consists of the other 88 elements (**Fig. 1.14**). The elements of the Earth comprise a great variety of materials. For reference in this chapter and the next, we introduce the basic categories of materials. All of these will be discussed in more detail later in the book.

- *Organic chemicals*: Carbon-containing compounds that either occur in living organisms, or have characteristics that resemble compounds in living organisms, are called **organic chemicals**.

- *Minerals*: A solid, natural substance in which atoms are arranged in an orderly pattern is called a **mineral**. A single coherent sample of a mineral that grew to its present shape is a crystal, whereas an irregularly shaped sample, or a fragment derived from a once-larger crystal or group of crystals, is a grain.

- *Glasses*: A solid in which atoms are not arranged in an orderly pattern is called **glass**.

- *Rocks*: Aggregates of mineral crystals or grains, and masses of natural glass, are called **rocks**. Geologists recognize three main groups of rocks. (1) Igneous rocks develop when hot molten (liquid) rock cools and freezes solid. (2) Sedimentary rocks form from grains that break off preexisting rock and become cemented together, or from minerals that precipitate out of a water solution. (3) Metamorphic rocks form when preexisting rocks undergo changes in response to heat and pressure.

- *Sediment*: An accumulation of loose mineral grains (grains that have not stuck together) is called sediment.

- *Metals*: Solids composed of metal atoms (such as iron, aluminum, copper, and tin) are called **metals**. In a metal, outer electrons are able to flow freely. An alloy is a mixture containing more than one type of metal atom.

- *Melts*: **Melts** form when solid materials become hot and transform into liquid. Molten rock is a type of melt—geologists distinguish between magma, which is molten rock beneath the Earth's surface, and lava, molten rock that has flowed out onto the Earth's surface.

The most common minerals in the Earth contain silica (a compound of silicon and oxygen) mixed in varying proportions with other elements. These minerals are called silicate minerals. Not surprisingly, rocks composed of silicate minerals are

FIGURE 1.13 This map of the Earth shows variations in elevation on both the land surface and the sea floor. Darker blues are deeper water in the ocean. Greens are lower elevation on land.

FIGURE 1.14 The proportions of major elements making up the mass of the whole Earth. Note that iron and oxygen account for most of the mass.

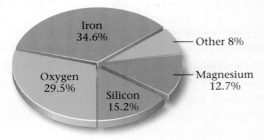

TAKE-HOME MESSAGE

The Earth consists mostly of silicate rock (such as granite, basalt, gabbro, peridotite) and iron alloy. Different types of silicate rock can be distinguished from each other by their composition (proportion of silicon to iron and magnesium) and by their grain size. Composition affects rock density.

silicate rocks. Geologists distinguish four classes of igneous silicate rocks based, in essence, on the proportion of silica to iron and magnesium. In order, from greatest to least proportion of silicon to iron and magnesium, these classes are: felsic (or silicic), intermediate, mafic, and ultramafic. As the proportion of silica in a rock increases, the density (mass per unit volume) decreases. Thus, felsic rocks are less dense than mafic rocks. Many different rock types occur in each class, as will be discussed in detail in Chapters 4 through 6. For now, we introduce the four rock types whose names are used in the discussion of the Earth's layers that follows. These are (1) *granite*, a felsic rock with large grains; (2) *gabbro*, a mafic rock with large grains; (3) *basalt*, a mafic rock with small grains; and (4) *peridotite*, an ultramafic rock with large grains.

1.8 HOW DO WE KNOW THAT THE EARTH HAS LAYERS?

The world's deepest mine shaft penetrates gold-bearing rock that lies about 3.5 km (2 miles) beneath South Africa. Though miners seeking this gold must begin their workday by plummeting straight down a vertical shaft for almost ten minutes aboard the world's fastest elevator, the shaft represents little more than a pinprick on Earth's surface, when compared with the planet's radius of 6,371 km. Even the deepest well ever drilled, a 12-km-deep hole in northern Russia, penetrates only the upper 0.02% of the Earth. We literally live on the thin skin of our planet, its interior forever inaccessible to our direct observation.

People have speculated about the Earth's interior since ancient times. What is the source of incandescent lavas spewed from volcanoes, of precious gems and metals, of sparkling spring water, and of the mysterious powers that shake the ground and topple buildings? Without the ability to observe the Earth's interior firsthand, pre-twentieth-century authors dreamed up fanciful images of it. For example, the English poet John Milton (1608–1674) described the underworld as a "dungeon horrible, on all sides round, as one great furnace flamed" (Fig. 1.15). Perhaps his image was inspired by volcanoes in the Mediterranean. In the eighteenth and nineteenth centuries, some European writers thought that the Earth's interior resembled a sponge, containing open caverns variously filled with molten rock, water, or air. In fact, in the popular 1864 novel *Journey to the Center of the Earth*, by the French author Jules Verne, three explorers find a route through interconnected caverns to the Earth's center.

Our present image of the Earth's interior, one made up of distinct layers, is the end product of many discoveries made during the past two hundred years. The image began to take form when researchers measured the mass of the whole Earth, and from this information calculated its average density. They discovered that our planet's average density far exceeds the density of common rocks found at its surface. The interior of the Earth, therefore, must contain denser material than its outermost layer. Eventually, researchers determined that most of the Earth's mass had to be concentrated closer to the center of the planet. Thus, they recognized that the Earth resembled a

FIGURE 1.16 An early image of Earth's internal layers.

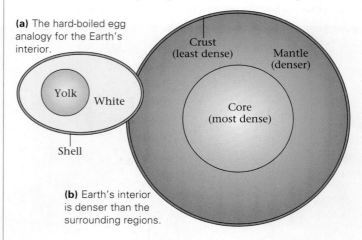

(a) The hard-boiled egg analogy for the Earth's interior.

Yolk White Shell

Crust (least dense) Mantle (denser) Core (most dense)

(b) Earth's interior is denser than the surrounding regions.

hard-boiled egg, in that it had three principal layers: a not-so-dense **crust** (like an eggshell, composed of rocks such as granite, basalt, and gabbro), a denser solid **mantle** in the middle (the "white," composed of a then-unknown material), and a very dense **core** (the "yolk," composed of a different unknown material) (Fig. 1.16). Clearly, many questions remained. How thick are the layers? Are the boundaries between layers sharp or gradational? And what exactly are the layers composed of?

Clues from the Study of Earthquakes: Refining the Image

One day in 1889, a physicist in Germany noticed that the pendulum in his lab began to move without having been touched. He reasoned that the pendulum was actually standing still while the Earth moved under it. A few days later, he read in a newspaper that a large **earthquake** (ground shaking due to the sudden breaking of rocks in the Earth) had taken place in Japan minutes before the movement of his pendulum began. The physicist deduced that vibrations due to the earthquake had traveled through the Earth from Japan and had jiggled his laboratory in Germany. The energy in such vibrations moves in the form of waves, called either seismic waves or earthquake waves, that resemble the shock waves you feel with your hands when you snap a stick (Fig. 1.17). The breaking of rock during an earthquake either produces a new fracture on which sliding occurs or causes sliding on a preexisting fracture. A fracture on which sliding occurs is called a **fault**.

Geologists immediately realized that the study of seismic waves traveling through the Earth might provide a tool for exploring the Earth's insides, much as ultrasound helps doctors study a patient's insides. Specifically, laboratory measurements demonstrated that earthquake waves travel at different velocities (speeds) through different materials. Thus, by detecting

FIGURE 1.15 A literary image of the Earth's insides: *The Fallen Angels Entering Pandemonium, from* [Milton's] *"Paradise Lost," Book 1,* by the English painter John Martin (1789–1854).

FIGURE 1.17 Faulting and earthquakes.

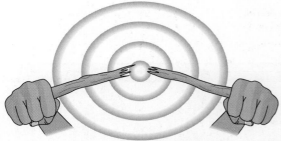

(a) Snapping a stick generates vibrations that pass through the stick to your hands.

Earthquake wave

Fault plane

(not to scale)

(b) Similarly, when the rock inside the Earth suddenly breaks and slips, forming a fracture called a fault, it generates shock waves that pass through the Earth and shake the surface.

depths at which velocities suddenly change, geoscientists pinpointed the boundaries between layers and even recognized subtler boundaries within layers. (Chapter 8 provides further details about earthquakes, and Interlude D shows *how* the study of earthquake waves defines the Earth's layers.)

Pressure and Temperature Inside the Earth

In order to keep underground tunnels from collapsing under the pressure created by the weight of overlying rock, mining engineers must design sturdy support structures. It is no surprise that deeper tunnels require stronger supports: the downward push from the weight of overlying rock increases with depth, simply because the mass of the overlying rock layer increases with depth. At the Earth's center, pressure probably reaches about 3,600,000 atm.

Temperature also increases with depth in the Earth. Even on a cool winter's day, miners who chisel away at gold veins exposed in tunnels 3.5 km below the surface swelter in temperatures of about 53°C (127°F). We refer to the rate of change in temperature with depth as the **geothermal gradient**. In the upper part of the crust, the geothermal gradient averages between 15° and 50°C per km. At greater depths, the rate

decreases (to 10°C per km or less). Thus, 35 km below the surface of a continent, the temperature reaches 400° to 700°C. No one has ever directly measured the temperature at the Earth's center, but calculations suggest it may exceed 4,700°C.

TAKE-HOME MESSAGE

Researchers have used a variety of techniques to understand the Earth's interior. They conclude that Earth has distinct layers—the crust, mantle, and core—and that both temperature and pressure increase with increasing depth. Earth's center is almost as hot as the Sun's surface.

1.9 WHAT ARE THE LAYERS MADE OF?

As a result of studies during the past century, geologists have a pretty clear sense of what the layers inside the Earth are made of. Let's now look at the properties of individual layers in detail, starting from the Earth's surface (Fig. 1.18a, b).

The Crust

When you stand on the surface of the Earth, you are standing on top of its outermost layer, the crust. The crust is our home and the source of all our resources. How thick is this all-important layer? Or, in other words, what is the depth to the crust-mantle boundary? An answer came from the studies of Andrija Mohorovičić, a researcher working in Zagreb, Croatia. In 1909, he discovered that the velocity of earthquake waves suddenly increased at a depth of tens of kilometers beneath the Earth's surface, and he suggested that this increase was caused by an abrupt change in the properties of rock (see Interlude D for further detail). Later studies showed that this change can be found most everywhere around our planet, though it occurs at different depths in different locations. Specifically, it's deeper beneath continents than beneath oceans. Geologists now consider the change to be the crust-mantle boundary, and they refer to it as the **Moho** in Mohorovičić's honor. The relatively shallow depth of the Moho (7 to 70 km, depending on location) as compared to the radius of the Earth (6,371 km) emphasizes that the crust is very thin indeed. In fact, the crust is only about 0.1% to 1.0% of the Earth's radius, so if the Earth were the size of a balloon, the crust would be about the thickness of the balloon's skin.

Geologists distinguish between two fundamentally different types of crust—oceanic crust, which underlies the sea floor, and continental crust, which underlies continents. The crust is not simply cooled mantle, like the skin on chocolate pudding, but rather consists of a variety of rocks that differ in composition (chemical makeup) from mantle rock.

FIGURE 1.18 A modern view of Earth's interior layers.

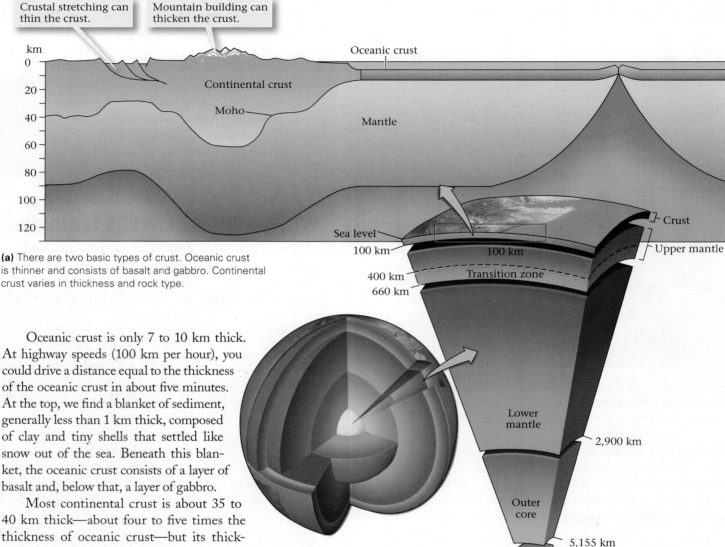

(a) There are two basic types of crust. Oceanic crust is thinner and consists of basalt and gabbro. Continental crust varies in thickness and rock type.

(b) By studying earthquake waves, geologists produced a refined image of Earth's interior, in which the mantle and core are subdivided.

Oceanic crust is only 7 to 10 km thick. At highway speeds (100 km per hour), you could drive a distance equal to the thickness of the oceanic crust in about five minutes. At the top, we find a blanket of sediment, generally less than 1 km thick, composed of clay and tiny shells that settled like snow out of the sea. Beneath this blanket, the oceanic crust consists of a layer of basalt and, below that, a layer of gabbro.

Most continental crust is about 35 to 40 km thick—about four to five times the thickness of oceanic crust—but its thickness varies significantly. In some places, continental crust has been stretched and thinned so it's only 25 km from the surface to the Moho. And in some places, the crust has been crumpled and thickened so that the depth to the Moho may be up to 70 km. In contrast to oceanic crust, continental crust contains a great variety of rock types, ranging from mafic to felsic in composition. On average, continental crust is less mafic than oceanic crust—it has a felsic (granite-like) to intermediate composition—so continental crust overall is less dense than oceanic crust. Notably, oxygen is the most abundant element in the crust (Table 1.1).

The Mantle

The mantle of the Earth forms a 2,885-km-thick layer surrounding the core. In terms of volume, it is the largest part of the Earth. In contrast to the crust, the mantle consists entirely of an ultramafic rock called peridotite. This means that peridotite, though rare at the Earth's surface, is actually the most abundant rock in our planet! On the basis of the occurrence of changes in the velocity of earthquake waves, geoscientists divide the mantle into two sublayers (see Fig. 1.18b): the **upper mantle**, down to a depth of 660 km, and the **lower mantle**, from 660 km down to 2,900 km. The **transition zone** is the section between 400 km and 660 km deep (see Interlude D).

Almost all of the mantle is solid rock. But even though it's solid, mantle rock below a depth of 100 to 150 km is so hot that it's soft enough to flow. This flow, however, takes place

TABLE 1.1 The abundance of elements in Earth's crust

Element	Symbol	Percentage by weight	Percentage by volume	Percentage by atoms
Oxygen	O	46.6	93.8	60.5
Silicon	Si	27.7	0.9	20.5
Aluminum	Al	8.1	0.8	6.2
Iron	Fe	5	0.5	1.9
Calcium	Ca	3.6	1	1.9
Sodium	Na	2.8	1.2	2.5
Potassium	K	2.6	1.5	1.8
Magnesium	Mg	2.1	0.3	1.4
All others	—	1.5	0.01	3.3

extremely slowly—at a rate of less than 15 cm a year. "Soft" here does not mean liquid; it simply means that over long periods of time mantle rock can change shape, like soft wax, without breaking. Note that we said *almost* all of the mantle is solid—in fact, up to a few percent of the mantle has melted. This melt occurs in films or bubbles between grains in the mantle between 100 and 200 km depth beneath the ocean floor.

Though, overall, the temperature of the mantle increases with depth, temperature also varies significantly with location even at the same depth. The warmer regions are less dense, while the cooler regions are denser. The distribution of warmer and cooler mantle indicates that the mantle convects like water in a simmering pot (see Appendix). Warm mantle gradually flows upward, while cooler, denser mantle sinks.

The Core

Early calculations suggested that the core had the same density as gold, so for many years people held the fanciful hope that vast riches lay at the heart of our planet. Alas, geologists eventually concluded that the core consists of a far less glamorous material, iron alloy (iron mixed with tiny amounts of other elements).

Studies of earthquake waves (see Interlude D), led geoscientists to divide the core into two parts, the outer core (between 2,900 and 5,155 km deep) and the inner core (from a depth of 5,155 km down to the Earth's center at 6,371 km). The outer core consists of *liquid* iron alloy. It can exist as a liquid because the temperature in the outer core is so high that even the great pressures squeezing the region cannot lock atoms into a solid framework. The iron alloy of the outer core can flow; this flow generates Earth's magnetic field (Fig. 1.19).

The inner core, with a radius of about 1,220 km, is a *solid* iron-nickel alloy that may reach a temperature of over 4,700°C. Even though it is hotter than the outer core, the inner core is a solid because it is deeper and is subjected to even greater pressure. The pressure keeps atoms from wandering freely, so they pack

together tightly in very dense materials. Recent data suggest that the inner core rotates slightly faster than the rest of the Earth.

TAKE-HOME MESSAGE

Continental crust is 25 to 70 km thick and, on average, resembles granite in composition, whereas oceanic crust is about 7 km thick, and consists of basalt. The base of the crust is called the Moho. Earth's mantle is much thicker than the crust and consists of very dense rock. The core consists of iron alloy. Its outer part is liquid, and its inner part is solid.

FIGURE 1.19 In 3-D, Earth's magnetic field can be visualized as invisible curtains of energy, generated by flow in the outer core.

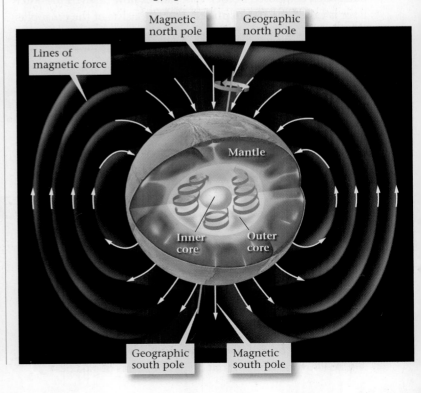

Magnetic north pole

Geographic north pole

Lines of magnetic force

Mantle

Inner core

Outer core

Geographic south pole

Magnetic south pole

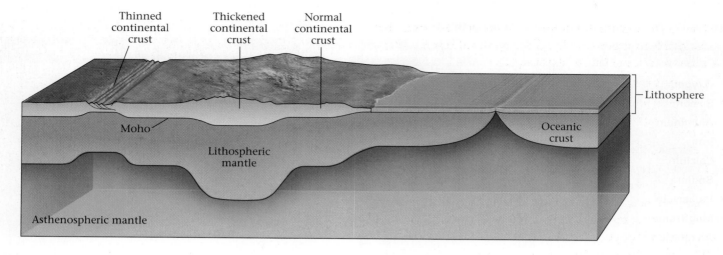

1.10 THE LITHOSPHERE AND THE ASTHENOSPHERE

So far, we have identified three major layers (crust, mantle, and core) inside the Earth that differ compositionally from each other. Earthquake waves travel at different velocities through these layers. An alternate way of thinking about Earth layers comes from studying the degree to which the material making up a layer can flow. In this context we distinguish between *rigid* materials, which can bend or break but cannot flow, and *plastic* materials, which are relatively soft and can flow without breaking.

Let's apply this concept to the outer portion of the Earth's shell. Geologists have determined that the outer 100 to 150 km of the Earth is relatively rigid. In other words, the Earth has an outer shell composed of rock that cannot flow easily. This outer layer is called the **lithosphere**, and it consists of the crust plus the uppermost part of the mantle. We refer to the portion of the mantle within the lithosphere as the lithospheric mantle. Note that the terms *lithosphere* and *crust* are not synonymous—the crust is just the upper part of the lithosphere. The lithosphere lies on top of the **asthenosphere**, which is the portion of the mantle in which rock can flow. The boundary between the lithosphere and asthenosphere occurs where the temperature reaches about 1,280°C, for at this temperature mantle rock becomes soft enough to flow. One final point: even though the asthenosphere can flow, do not think of it as a liquid. As we mentioned earlier, it is not. Rather, the asthenosphere is largely solid—a small amount of melt occurs only in the upper part. Asthenosphere flows at rates of up to 15 cm per year.

The two types of lithosphere are somewhat different (**Fig. 1.20**). Oceanic lithosphere, topped by oceanic crust, generally has a thickness of about 100 km. In contrast, continental lithosphere, topped by continental crust, generally has a thickness of about 150 km. Notice that the asthenosphere is entirely in the mantle and generally lies below a depth of 100 to 150 km. We can't assign a specific depth to the base of the asthenosphere because all of the mantle below 150 km can flow, but for convenience, some geologists place the base of the asthenosphere at the upper mantle/transition zone boundary.

Now, with an understanding of Earth's overall architecture at hand, we can discuss geology's grand unifying theory—plate tectonics. The next chapter introduces this key topic.

TAKE-HOME MESSAGE

The crust and outermost part of the mantle make up a 100- to 150-km-thick layer called the lithosphere that behaves rigidly and cannot flow. It overlies the asthenosphere, the region of the mantle that is soft enough to flow (though very slowly).

Chapter Summary

- The geocentric Universe concept places the Earth at the center of the Universe. The heliocentric (Sun-centered) Universe concept was not widely accepted until the Renaissance.
- The Earth is one of eight planets orbiting the Sun. Our Solar System lies on the outer edge of a slowly revolving galaxy. The Universe contains hundreds of billions of galaxies.
- The observed red shift of light from distant galaxies led to the expanding Universe theory. Astronomers propose that this expansion began after the big bang 13.7 billion years ago.
- The first atoms (hydrogen and helium) in the Universe developed less than 1 million years after the big bang. These atoms accumulated in vast gas clouds called nebulae.

- The Earth and the life forms on it contain elements that could only have been produced during the life cycle of stars. Thus, we are all made of stardust.

- Gravity caused clumps of gas in the nebulae to coalesce into flattened disks with bulbous centers. The protostars at the center of these disks eventually became dense and hot enough that fusion reactions began in them. When this happened, they became true stars.

- Planets developed from rings of gas and dust that surrounded protostars. The gas condensed into planetesimals that then merged to form protoplanets, and finally true planets. The inner rings became the terrestrial planets. The outer rings grew into gas-giant planets.

- The Moon formed from debris blasted free from Earth when a Mars-sized protoplanet collided with our planet.

- Planets assume a near-spherical shape because the warm rock inside is so soft that gravity can smooth out irregularities.

- The Earth has a magnetic field, which shields it from solar wind and cosmic rays.

- A layer of gas surrounds the Earth. This atmosphere (78% nitrogen, 21% oxygen, 1% other) can be subdivided into distinct layers. Air pressure decreases with elevation.

- The surface of the Earth can be divided into land (30%) and ocean (70%).

- The Earth consists of organic chemicals, minerals, glasses, rocks, metals, and melts. Most rocks on Earth contain silica (SiO_2). We distinguish among rock types based on the proportion of silica.

- The Earth's interior can be divided into three distinct layers: the crust, the mantle, and the core.

- Pressure and temperature both increase with depth in the Earth. The rate at which temperature increases as depth increases is the geothermal gradient.

- Studies of seismic waves have revealed the existence of sublayers in the core (outer core and inner core) and mantle (upper mantle and lower mantle).

- The crust is a thin skin that varies in thickness from 7 to 10 km beneath the oceans to 25 to 70 km beneath the continents. Oceanic crust is mafic, while average continental crust is felsic to intermediate. The mantle is made of ultramafic rock. The core consists of iron alloy—the outer core is liquid, and the inner core is solid. Flow in the outer core generates the Earth's magnetic field.

- The crust plus the upper part of the mantle constitute the lithosphere, a relatively rigid shell. The lithosphere lies over the asthenosphere, mantle that is capable of flowing.

GEOPUZZLE REVISITED

Geologists conclude that the Solar System formed from atoms generated by the big bang, and from atoms produced in stars or during the explosion of stars. Gravity pulled all this material together into a bulbous disk whose central ball became the Sun. The remainder of the disk condensed into planetesimals, which in turn coalesced to form planets.

Key Terms

asteroid (p. 12)	meteorite (p. 20)
asthenosphere (p. 30)	mineral (p. 24)
atmosphere (p. 23)	Moho (p. 27)
bathymetry (p. 24)	moon (p. 12)
big bang (p. 14)	nebula (p. 15)
comet (p. 12)	nebula theory (p. 16)
core (p. 26)	organic chemical (p. 24)
cosmology (p. 10)	planet (p. 11)
crust (p. 26)	planetesimal (p. 17)
differentiation (p. 20)	protoplanet (p. 17)
earthquake (p. 26)	protoplanetary disk (p. 17)
Earth System (p. 20)	protostar (p. 15)
expanding Universe theory (p. 14)	red shift (p. 13)
	rock (p. 24)
fault (p. 26)	sediment (p. 24)
frequency (p. 13)	silicate rocks (p. 25)
galaxy (p. 12)	Solar System (p. 11)
gas-giant planet (p. 12)	star (p. 12)
geothermal gradient (p. 27)	stellar wind (p. 16)
glass (p. 24)	supernova (p. 16)
gravity (p. 11)	surface water (p. 24)
groundwater (p. 24)	terrestrial planet (p. 12)
lithosphere (p. 30)	topography (p. 24)
lower mantle (p. 28)	transition zone (p. 28)
magnetic field (p. 21)	Universe (p. 10)
mantle (p. 26)	upper mantle (p. 28)
melt (p. 24)	vacuum (p. 21)
metal (p. 24)	wave (p. 12)
meteor (p. 21)	wavelength (p. 12)

Review Questions

1. Contrast the geocentric and heliocentric concepts of the Universe.
2. Describe how the Doppler effect works.
3. What does the red shift of distant galaxies tell us about their motion with respect to the Earth?
4. Briefly describe the steps in the formation of the Universe according to the big bang theory.
5. How does the composition of the Solar System, in terms of the elements making up the Sun and the planets, differ from that of the nebulae that existed a million years after the birth of the Universe? Explain the difference.
6. Why isn't the Earth homogeneous?
7. Describe how the Moon was formed.
8. Why is the Earth round?
9. What is the Earth's magnetic field? Draw a representation of the field on a piece of paper.
10. What is the Earth's atmosphere composed of? Why would you die of suffocation if you were to eject from a plane at 12 km without taking an oxygen tank with you?
11. Describe the major categories of materials constituting the Earth.
12. What are the principal layers of the Earth?
13. How do temperature and pressure change with increasing depth in the Earth?
14. What is the Moho? Describe the differences between continental crust and oceanic crust.
15. What is the mantle composed of? Is there any melt within the mantle?
16. What is the core composed of? How do the inner core and the outer core differ from each other?
17. What is the difference between the lithosphere and the asthenosphere? At what depth does the lithosphere/asthenosphere boundary occur? Is this above or below the Moho?

On Further Thought

1. Recent observations suggest that the Moon has a very small, solid core that is less than 3% of its mass. In comparison, Earth's core is about 33% of its mass. Explain why this difference might exist.
2. There is hardly any hydrogen or helium in the Earth's atmosphere, yet most of the nebulae from which the Solar System formed consisted of hydrogen and helium. Where did all this gas go?

3. The popular media sometimes imply that the crust floats on a "sea of magma." Is this a correct image of the mantle just below the Moho? Explain your answer.
4. As you will see later in this book, emplacement of a huge weight (e.g., a continental ice sheet) causes the surface of lithosphere to sink, just as your weight causes the surface of a trampoline to sink. Emplacement of such a weight does not, however, cause a change in the thickness of the lithosphere. How is this possible? (*Hint:* Think about the nature of the asthenosphere.)

THE VIEW FROM SPACE   A cluster of newborn stars makes the gas and dust of the nebula surrounding them glow, as seen in this image taken by the Hubble Space Telescope.

The Way the Earth Works: Plate Tectonics

Fossil leaves of *Glossopteris* from an exposure in Australia. The presence of this fossil on many continents was one of the observations that led to the proposal of continental drift.

GEOPUZZLE

At first glance, it looks like the continents on either side of the Atlantic Ocean once fit together quite nicely, like the pieces of a jigsaw puzzle. Do continents really move? Or, to put it another way, does the map of Earth's surface change over time?

2.1 INTRODUCTION

In September 1930, fifteen explorers led by a German meteorologist, Alfred Wegener, set out across the endless snowfields of Greenland to resupply two weather observers stranded at a remote camp. The observers were planning to spend the long polar night recording wind speeds and temperatures on Greenland's polar plateau. At the time, Wegener was well known, not only to researchers studying climate but also to geologists. Some fifteen years earlier, he had published a small book, *The Origin of the Continents and Oceans*, in which he had dared to challenge geologists' long-held assumption that the continents had remained fixed in position through all of Earth history. Wegener proposed, instead, that the continents once fit together like pieces of a giant jigsaw puzzle, to make one vast supercontinent. He suggested that this supercontinent, which he named **Pangaea** (pronounced Pan-jee-ah; Greek for "all land"), later fragmented into separate continents that drifted apart, moving slowly to their present positions (Fig. 2.1). This process came to be known as **continental drift**.

Wegener presented many observations in favor of continental drift, but he met with strong resistance. At a widely publicized 1926 geology conference in New York City, a crowd of celebrated American professors scoffed: "What force could possibly be great enough to move the immense mass of a continent?" Wegener's writings didn't provide a good answer, so despite all the supporting observations he had provided, most of the meeting's participants rejected continental drift.

Now, four years later, Wegener faced his greatest challenge. On October 30, 1930, Wegener reached the observers and dropped off enough supplies to last the winter. Wegener and one companion set out on the return trip the next day, but they never made it home.

Had Wegener survived to old age, he would have seen his hypothesis become the foundation of a scientific revolution. Today, geologists accept Wegener's ideas and take for granted that the map of the Earth constantly changes; continents waltz around its surface, variously combining and breaking apart through geologic time. The revolution began in 1960, when Harry Hess, of Princeton University, proposed that as continents drift apart, new ocean floor forms between them by a process that his contemporary Robert Dietz also described and named **sea-floor spreading**. Hess and others also suggested that continents move toward each other when the old ocean floor between them sinks back down into the Earth's interior, a process now called **subduction**. By 1968, geologists had developed a fairly complete model describing continental drift, sea-floor spreading, and subduction. In this model, Earth's lithosphere, its outer, relatively rigid shell, consists of about twenty distinct pieces, or **plates**, that slowly move relative to each other. Because we can confirm this model by many observations, it has gained the status of a theory, which we now call the theory of **plate tectonics**, from the Greek word *tekton*, which means builder; plate movements "build" regional geologic features. Geologists view plate tectonics as the grand unifying theory of geology, because it can successfully explain a great many geologic phenomena, as we will see.

In this chapter, we introduce the observations that led Wegener to propose his continental drift hypothesis. Then we look at paleomagnetism, the record of Earth's magnetic field in the past, which provides a key proof of continental drift. Next, we learn how observations about the sea floor made by geologists during the mid-twentieth century led Harry Hess to propose the concept of sea-floor spreading. We conclude by describing the many facets of modern plate tectonics theory.

FIGURE 2.1 Wegener's image of Pangaea, and his proposal of its subsequent breakup and dispersal.

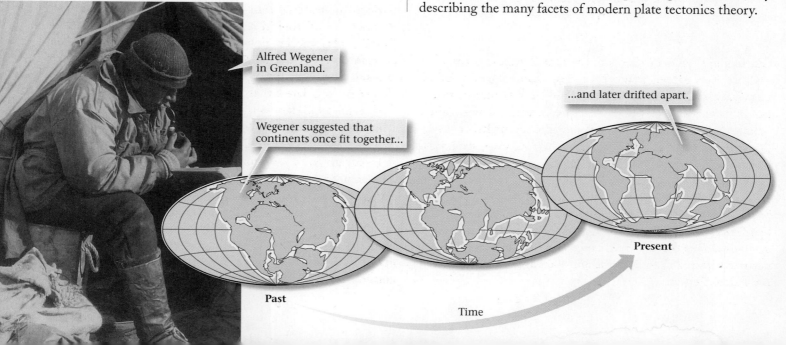

Alfred Wegener in Greenland.

Wegener suggested that continents once fit together...

...and later drifted apart.

Past

Time

Present

2.2 WEGENER'S EVIDENCE FOR CONTINENTAL DRIFT

Wegener suggested that a vast supercontinent, Pangaea, existed until near the end of the Mesozoic Era (the interval of geologic time that lasted from 251 to 65 million years ago). He suggested that Pangaea then broke apart, and the land masses drifted away from each other to form the continents we see today. Let's look at some of Wegener's arguments and see why he came to this conclusion.

The Fit of the Continents

Almost as soon as maps of the Atlantic coastlines became available in the 1500s, scholars noticed the fit of the continents. The northwestern coast of Africa could tuck in against the eastern coast of North America, and the bulge of eastern South America could nestle cozily into the indentation of southwestern Africa. Australia, Antarctica, and India could all connect to the southeast of Africa, while Greenland, Europe, and Asia could pack against the northeastern margin of North America. In fact, all the continents could be joined, with remarkably few overlaps or gaps, to create Pangaea. Wegener concluded that the fit was too good to be coincidence.

Locations of Past Glaciations

Glaciers are rivers or sheets of ice that flow across the land surface. As a glacier flows, it carries sediment grains of all sizes (clay, silt, sand, pebbles, and boulders). Grains protruding from the base of the moving ice carve scratches, called striations, into the substrate. When the ice melts, it leaves the sediment in a deposit called till and exposes striations. Thus, the occurrence of till and striations at a location serve as evidence that the region was covered by a glacier in the past. By studying the age of glacial till deposits, geologists have determined that large areas of the land were covered by glaciers during time intervals of Earth history called ice ages (discussed further in Chapter 18). One of these ice ages occurred from 280 to 260 million years ago, near the end of the Paleozoic Era.

Wegener was an Arctic climate scientist by training, so it's no surprise that he had a strong interest in glaciers. He knew that glaciers form mostly at high latitudes today. So he suspected that if he plotted a map of the locations of late Paleozoic glacial till and striations, he might gain insight into the locations of continents at that time. When he plotted these locations, he found that glaciers of this time interval occurred in southern South America, southern Africa, southern India, Antarctica, and southern Australia. These places are now widely separated and, with the exception of Antarctica, do not currently lie in cold polar regions (Fig. 2.2a). To Wegener's amazement, all late Paleozoic glaciated areas lie *adjacent* to each other on his map of Pangaea (Fig. 2.2b). Furthermore, when he plotted the orientation of glacial striations, they all pointed roughly outward from a location in southeastern Africa. In other words, Wegener determined that the distribution of glaciations at the end of the Paleozoic Era could easily be explained if the continents had been united in Pangaea, with the southern part of Pangaea lying beneath the center of a huge ice cap. This distribution of glaciation could not be explained if the continents had always been in their present positions.

The Distribution of Climatic Belts

If the southern part of Pangaea had straddled the South Pole at the end of the Paleozoic Era, then during this same time interval, southern North America, southern Europe, and northwestern Africa would have straddled the equator and would have had tropical or subtropical climates. Wegener searched for evidence that this was so by studying sedimentary rocks that were formed at this time, for the material making up these rocks can reveal clues to the past climate. For example, in the swamps and jungles of tropical regions, thick deposits of plant material accumulate, and when deeply buried, this material transforms into coal. And, in the clear, shallow seas of tropical regions, large reefs develop offshore. Finally, subtropical regions, on either side of the tropical belt, contain deserts, an environment in which sand dunes form and salt from evaporating seawater or salt lakes accumulates. Wegener speculated that the distribution of late Paleozoic coal, sand-dune deposits, and salt deposits could define climate belts on Pangaea.

Sure enough, in the belt of Pangaea that Wegener expected to be equatorial, late Paleozoic sedimentary rock layers include abundant coal and the relics of reefs; and in the portions of Pangaea that Wegener predicted would be subtropical, late Paleozoic sedimentary rock layers include relics of desert dunes and of salt (Fig. 2.2c). On a present-day map of our planet, these deposits are scattered around the globe at a variety of latitudes. In Wegener's Pangaea, the deposits align in continuous bands that occupy appropriate latitudes.

FIGURE 2.2 Wegener's evidence for continental drift came from analyzing the geologic record.

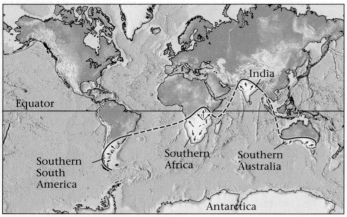

(a) The distribution of late Paleozoic glacial deposits on present-day Earth. Arrows indicate glacial striations.

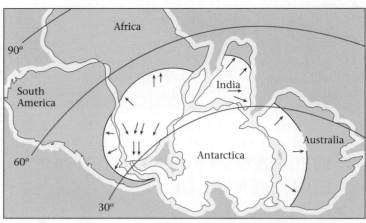

(b) On Pangaea, areas with glacial deposits fit together in a southern polar cap.

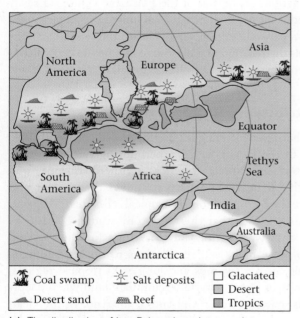

Coal swamp   Salt deposits   ☐ Glaciated
Desert sand   Reef   ☐ Desert
                                     ☐ Tropics

(c) The distribution of late Paleozoic rock types plots sensibly in the climate belts of Pangaea.

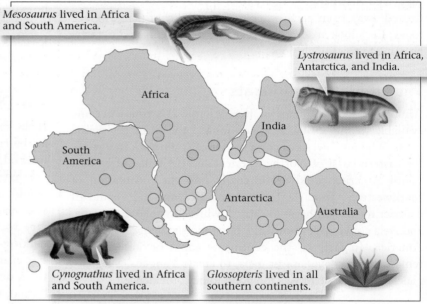

Mesosaurus lived in Africa and South America.

Lystrosaurus lived in Africa, Antarctica, and India.

Cynognathus lived in Africa and South America.

Glossopteris lived in all southern continents.

(d) Fossils of Mesozoic land-dwelling organisms occur on multiple continents. This would be hard to explain if continents were always separated by oceans.

The Distribution of Fossils

Today, different continents provide homes for different species. Kangaroos, for example, live only in Australia. Similarly, many kinds of plants grow only on one continent and not on others. Why? Because land-dwelling species of animals and plants cannot swim across vast oceans, and thus evolved independently on different continents. During a period of Earth history when all continents were in contact, however, land animals and plants could have migrated among many continents.

With this concept in mind, Wegener plotted fossil occurrences of land-dwelling species that existed during the late Paleozoic and early Mesozoic Eras (between about 300 and

210 million years ago) and found that these species had indeed existed on several continents (Fig. 2.2d). Wegener argued that the distribution of fossil species required the continents to have been adjacent to one another in the late Paleozoic and early Mesozoic Eras.

Matching Geologic Units

In the same way that an art historian can identify a Picasso painting and an architect a Victorian design, a geoscientist can identify a distinctive group of rocks. Wegener found that the same distinctive Precambrian (the interval of geologic time before 542 million years ago) rock assemblages occurred on the eastern coast of South America and the western coast of Africa, regions now separated by an ocean (Fig. 2.3a). If the continents had been joined to create Pangaea in the past, then these matching rock groups would have been adjacent to each other, and thus could have composed continuous blocks or belts. Wegener also noted that features of the Appalachian Mountains of the United States and Canada closely resemble mountain belts in southern Greenland, Great Britain, Scandinavia, and northwestern Africa (Fig. 2.3b), regions that would have lain adjacent to each other in Pangaea. Wegener thus demonstrated that not only did the coastlines of continents match, so did the rocks adjacent to the coastlines.

Criticism of Wegener's Ideas

Wegener's model of a supercontinent that later broke apart explained the distribution of late Paleozoic glaciers, coal, sand dunes, distinctive rock assemblages, and fossils. Clearly, he had compiled a strong case for continental drift. But as noted earlier, he could not adequately explain how or why continents drifted. He left on his final expedition to Greenland having failed to convince his peers, and he died without knowing that his ideas, after lying dormant for decades, would be reborn as the basis of the broader theory of plate tectonics.

In effect, Wegener was ahead of his time. It would take three more decades of research before geologists obtained sufficient data to test his hypotheses properly. Collecting this data required instruments and techniques that did not exist in Wegener's day. Of the many geologic discoveries that ultimately opened the door to plate tectonics, perhaps the most important came from the discovery of a phenomenon called paleomagnetism.

TAKE-HOME MESSAGE

In support of his proposal of continental drift, Wegener noted that coastlines on opposite sides of oceans matched, and that the observed distribution of ancient glaciations, climate belts, fossils, and rock units makes better sense if Pangaea existed.

2.3 PALEOMAGNETISM AND THE PROOF OF CONTINENTAL DRIFT

More than 1,500 years ago, Chinese sailors discovered that a piece of lodestone, when suspended from a thread, points in a northerly direction and can help guide a voyage. Lodestone exhibits this behavior because it consists of magnetite, an iron-rich mineral that, like a compass needle, aligns with Earth's magnetic field lines. Some rock types contain tiny crystals of magnetite, or other magnetic minerals, and thus behave overall like weak magnets. In this section, we explain how the study of such magnetic behavior led to the realization that rocks preserve **paleomagnetism**, a record of Earth's magnetic field

FIGURE 2.3 Further evidence of drift: rocks on different sides of the ocean match.

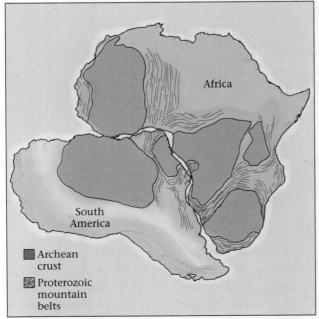

(a) Distinctive belts of rock in South America would align with similar ones in Africa, without the Atlantic.

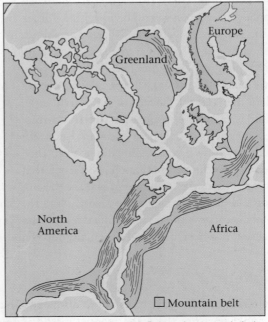

(b) If the Atlantic didn't exist, Paleozoic mountain belts on both coasts would be adjacent.

FIGURE 2.4 Features of Earth's magnetic field.

in the past. An understanding of paleomagnetism provided proof of continental drift and, as we'll see later in this chapter, contributed to the development of plate tectonics theory. As a foundation for introducing paleomagnetism, we first review the nature of the Earth's magnetic field.

Earth's Magnetic Field

Circulation of liquid iron alloy in the outer core of the Earth generates a magnetic field. (As you may learn in a physics course, a similar phenomenon happens in an electrical dynamo.) Earth's magnetic field resembles the field produced by a bar magnet, in that it has two ends of opposite polarity. Thus, we can represent Earth's field by a **magnetic dipole**, an imaginary arrow (Fig. 2.4a). This dipole intersects the surface of the planet at two points, known as the **magnetic poles**. By convention, the north magnetic pole is at the end of the Earth nearest the north geographic pole (the point where the northern end of the spin axis intersects the surface). The north-seeking (red) end of a compass needle points to the north magnetic pole.

Earth's magnetic poles move constantly, so, averaged over thousands of years, they roughly coincide with Earth's geographic poles (Fig. 2.4b). At present, however, the magnetic poles lie hundreds of kilometers away from the geographic poles, so the magnetic dipole tilts at about 11° relative to the Earth's spin axis. Because of this difference, a compass today does not point exactly to geographic north. The angle between the direction that a compass needle points and a line of longitude is the **magnetic declination** (Fig. 2.4c).

Invisible field lines curve through space between the magnetic poles. In a cross-section view, these lines lie parallel to the surface of the Earth (that is, are horizontal) at the equator, tilt at an angle to the surface in mid-latitudes, and plunge perpendicular to the surface at the magnetic poles (Fig. 2.4a). The angle between a magnetic field line and the surface of the Earth, at a given location, is called the **magnetic inclination**. If you place a magnetic needle on a horizontal axis so it can pivot up and down, and then carry it from the magnetic equator to the magnetic pole, you'll see that the inclination varies with latitude—it is 0° at the magnetic equator and 90° at the magnetic poles. (Note that the compass you may carry with you on a hike does not show inclination, because it has been balanced to remain horizontal.)

Paleomagnetism

In the early twentieth century, researchers developed instruments that could measure the weak magnetic field produced by rocks and made a surprising discovery. In rocks that formed

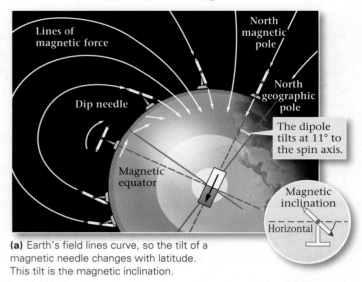

(a) Earth's field lines curve, so the tilt of a magnetic needle changes with latitude. This tilt is the magnetic inclination.

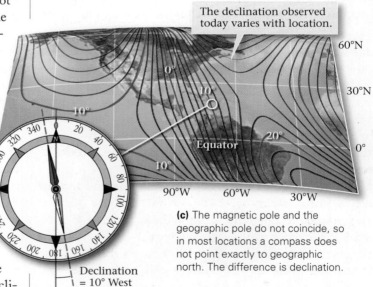

(b) A map of the magnetic pole position during the past 1800 years shows that the pole moves, but stays within high latitudes.

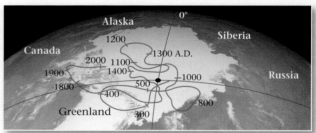

(c) The magnetic pole and the geographic pole do not coincide, so in most locations a compass does not point exactly to geographic north. The difference is declination.

millions of years ago, the orientation of the dipole representing the magnetic field of the rocks *is not the same* as that of present-day Earth (Fig. 2.5a). To understand this statement, consider an example. Imagine traveling to a location near the coast on the equator in South America where the inclination and declination are presently 0°. If you measure the weak magnetic field

FIGURE 2.5 Paleomagnetism and how it can form during cooling of lava.

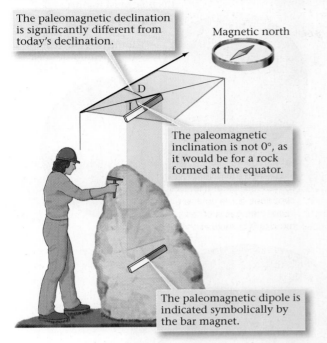

The paleomagnetic declination is significantly different from today's declination.

Magnetic north

The paleomagnetic inclination is not 0°, as it would be for a rock formed at the equator.

The paleomagnetic dipole is indicated symbolically by the bar magnet.

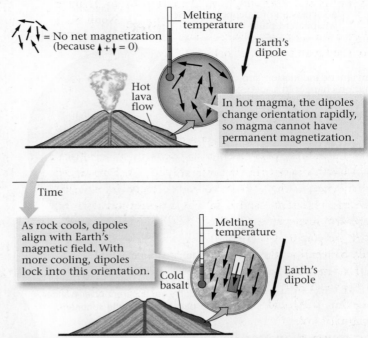

= No net magnetization (because ↑ + ↓ = 0)

Melting temperature

Earth's dipole

Hot lava flow

In hot magma, the dipoles change orientation rapidly, so magma cannot have permanent magnetization.

Time

As rock cools, dipoles align with Earth's magnetic field. With more cooling, dipoles lock into this orientation.

Melting temperature

Cold basalt

Earth's dipole

(a) A geologist finds an ancient rock sample. The orientation of the rock's paleomagnetism is different from that of today's field.

(b) Paleomagnetism can form when lava cools and becomes solid rock.

produced by, say, a 90-million-year-old rock, and represent the orientation of this field by an imaginary bar magnet, you'll find that this imaginary bar magnet does *not* point to the present-day north magnetic pole, because its declination differs from the declination that a compass at the location would display today, and its inclination is not 0°. Geologists have made similar measurements at many locations around the globe and have found that the magnetic fields of ancient rocks indicate the orientation of the magnetic field, relative to the rock, at the time the rock formed. This record, preserved in rock, is **paleomagnetism**.

How does paleomagnetism develop? One process happens when lava (molten rock) cools to form basalt (**Fig. 2.5b**). Lava is a very hot liquid containing no crystals. As it starts to cool and solidify into rock, tiny magnetite crystals begin to grow. At first, thermal energy (see Appendix) causes the magnetic dipole associated with each crystal to wobble and tumble chaotically. Thus, at any given instant, the dipoles of the magnetite specks are randomly oriented and the magnetic forces they produce cancel each other out. Eventually, the rock cools sufficiently that the dipoles slow down and, like tiny compass needles, align with the Earth's magnetic field. As the rock cools still more, these tiny compass needles lock into permanent parallelism with the Earth's magnetic field at the time the cooling takes place.

Apparent Polar Wander— A Proof of Continental Drift

Why doesn't the paleomagnetic dipole in rocks point to the present-day magnetic field? When geologists first attempted to answer this question, they assumed that continents were fixed in position, and thus concluded that the position of Earth's magnetic poles in the past were different than they are today. They introduced the term **paleopole** to refer to the supposed position of the Earth's magnetic north pole in the past. With this concept in mind, they set out to track what they thought was the change in position of the paleopole over time. To do this, they measured the paleomagnetism in a succession of rocks of different ages, from the same general location on a continent, and plotted the position of the associated succession of paleopole positions on a map (**Fig. 2.6a**). The successive positions of dated paleopoles trace out a curving line that came to be known as an **apparent polar-wander path**. At first, geologists assumed that the position of the continent from which they collected the samples was fixed, and interpreted the apparent polar-wander path to represent how the position of Earth's magnetic pole migrated through time. But were they in for a surprise! When they obtained polar-wander paths from many different localities, they found that *each continent has a different apparent polar-wander path*. The hypothesis that continents are fixed in position cannot

FIGURE 2.6   Apparent polar-wander paths, and their interpretation.

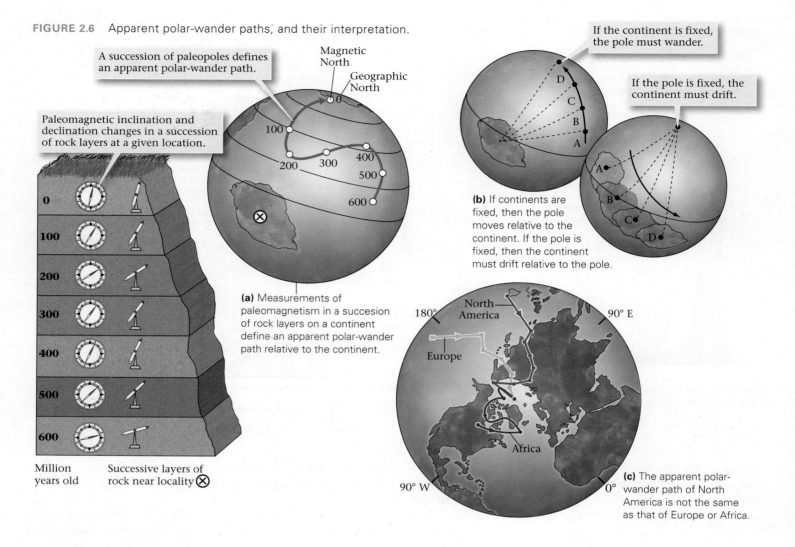

A succession of paleopoles defines an apparent polar-wander path.

Paleomagnetic inclination and declination changes in a succession of rock layers at a given location.

Magnetic North

Geographic North

(a) Measurements of paleomagnetism in a succesion of rock layers on a continent define an apparent polar-wander path relative to the continent.

Million years old

Successive layers of rock near locality ⊗

If the continent is fixed, the pole must wander.

If the pole is fixed, the continent must drift.

(b) If continents are fixed, then the pole moves relative to the continent. If the pole is fixed, then the continent must drift relative to the pole.

North America

Europe

Africa

180°

90° E

90° W

0°

(c) The apparent polar-wander path of North America is not the same as that of Europe or Africa.

explain this observation, for if the magnetic pole moved while all the continents stayed fixed, measurements from all continents should produce the same apparent polar-wander paths.

Geologists suddenly realized that they were looking at apparent polar-wander paths in the wrong way. *It's not the pole that moves relative to fixed continents, but rather the continents that move relative to a fixed pole* (Fig. 2.6b). Since each continent has its own unique polar-wander path (Fig. 2.6c), the continents must move with respect to each other. The discovery proved that Wegener was right all along—continents do drift!

TAKE-HOME MESSAGE

Rocks can preserve a record of the Earth's magnetic field in the past. The study of paleomagnetism indicates that continents have moved relative to Earth's magnetic poles, and thus proves that continental drift occurred.

2.4   SETTING THE STAGE FOR THE DISCOVERY OF SEA-FLOOR SPREADING

New Images of Sea-Floor Bathymetry

Military needs during World War II gave a boost to sea-floor exploration, for as submarine fleets grew, navies required detailed information about **bathymetry**, or depth variations. The invention of echo sounding (sonar) permitted such information to be gathered quickly. As a ship equipped with sonar travels, observers can obtain a continuous record of the depth of the sea floor. The resulting cross section showing depth plotted against location is called a bathymetric profile (Fig. 2.7a, b). By cruising back and forth across the ocean many times, investigators obtained a series of bathymetric profiles and from these constructed maps of the sea floor. (Geologists can now produce such maps much more rapidly using satellite data.) Bathymetric maps reveal several important features.

FIGURE 2.7 Bathymetry of mid-ocean ridges.

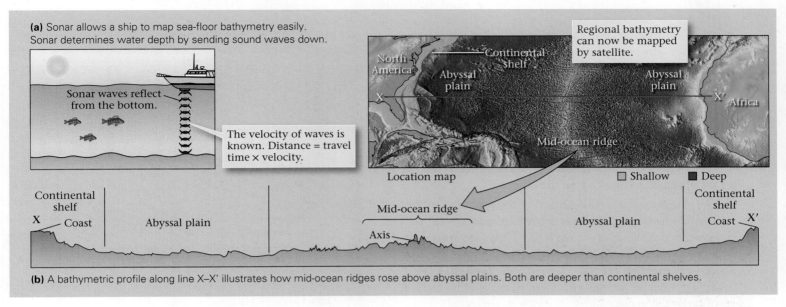

(a) Sonar allows a ship to map sea-floor bathymetry easily. Sonar determines water depth by sending sound waves down.

Sonar waves reflect from the bottom.

The velocity of waves is known. Distance = travel time × velocity.

Regional bathymetry can now be mapped by satellite.

North America

Continental shelf

Abyssal plain

Abyssal plain

X

X'

Mid-ocean ridge

Africa

Location map

☐ Shallow ■ Deep

Continental shelf

X Coast

Abyssal plain

Mid-ocean ridge

Axis

Abyssal plain

Continental shelf

Coast X'

(b) A bathymetric profile along line X–X' illustrates how mid-ocean ridges rose above abyssal plains. Both are deeper than continental shelves.

FIGURE 2.8 Other bathymetric features of the ocean floor, and the pattern of heat flow at mid-ocean ridges.

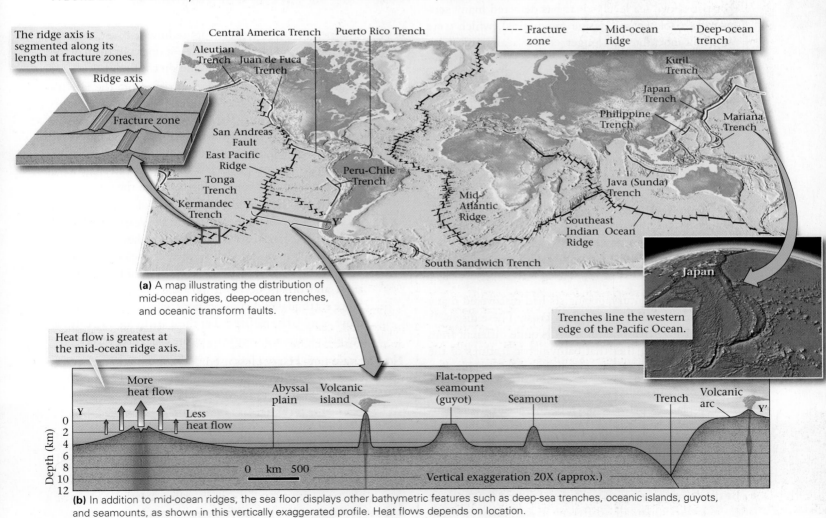

The ridge axis is segmented along its length at fracture zones.

Ridge axis

Fracture zone

---- Fracture zone   — Mid-ocean ridge   — Deep-ocean trench

Central America Trench   Puerto Rico Trench

Aleutian Trench Juan de Fuca Trench

Kuril Trench

Japan Trench

Philippine Trench

Mariana Trench

San Andreas Fault

East Pacific Ridge

Tonga Trench

Kermandec Trench

Peru-Chile Trench

Y

Y'

Mid-Atlantic Ridge

Java (Sunda) Trench

Southeast Indian Ocean Ridge

South Sandwich Trench

Japan

(a) A map illustrating the distribution of mid-ocean ridges, deep-ocean trenches, and oceanic transform faults.

Trenches line the western edge of the Pacific Ocean.

Heat flow is greatest at the mid-ocean ridge axis.

More heat flow

Less heat flow

Abyssal plain

Volcanic island

Flat-topped seamount (guyot)

Seamount

Trench

Volcanic arc

Y

Y'

Depth (km): 0 2 4 6 8 10 12

0 km 500

Vertical exaggeration 20X (approx.)

(b) In addition to mid-ocean ridges, the sea floor displays other bathymetric features such as deep-sea trenches, oceanic islands, guyots, and seamounts, as shown in this vertically exaggerated profile. Heat flows depends on location.

- *Mid-ocean ridges*: The floor beneath all major oceans includes **abyssal plains**, which are broad, relatively flat regions of the ocean that lie at a depth of about 4 to 5 km below sea level; and **mid-ocean ridges**, elongate submarine mountain ranges whose peaks lie only about 2 to 2.5 km below sea level (Fig. 2.7b; Fig. 2.8a). Geologists call the crest of the mid-ocean ridge the ridge axis. All mid-ocean ridges are roughly symmetrical—bathymetry on one side of the axis is nearly a mirror image of bathymetry on the other side.

- *Deep-ocean trenches*: Along much of the perimeter of the Pacific Ocean, and in a few other localities as well, the ocean floor reaches depths of 8 to 12 km—deep enough to swallow Mt. Everest. These deep areas occur in elongate troughs that are now referred to as **trenches** (Fig. 2.8b). Trenches border **volcanic arcs**, curving chains of active volcanoes.

- *Seamount chains*: Numerous volcanic islands poke up from the ocean floor: for example, the Hawaiian Islands lie in the middle of the Pacific. In addition to islands that rise above sea level, sonar has detected many **seamounts** (isolated submarine mountains), which were once volcanoes but no longer erupt. Seamounts with flat tops, due to ancient reef growth, are called guyots. Volcanic islands and seamounts typically occur in chains, but in contrast to the volcanic arcs that border deep-ocean trenches, only one island at the end of a seamount chain remains capable of erupting today.

- *Fracture zones*: Surveys reveal that the ocean floor is diced up by narrow bands of vertical cracks and broken-up rock. These **fracture zones** lie roughly at right angles to mid-ocean ridges, effectively segmenting the ridges into small pieces (Fig. 2.8a).

New Observations on the Nature of Oceanic Crust

By the mid-twentieth century, geologists had discovered many important characteristics of the sea-floor crust. These discoveries led them to realize that oceanic crust is quite different from continental crust, and further that bathymetric features of the ocean floor provide clues to the origin of the crust. Specifically:

- A layer of sediment composed of clay and the tiny shells of dead plankton covers much of the ocean floor. This layer *becomes progressively thicker* away from the mid-ocean ridge axis. But even at its thickest, the sediment layer is too thin to have been accumulating for the entirety of Earth history.

- By dredging up samples, geologists learned that oceanic crust is fundamentally different in composition from

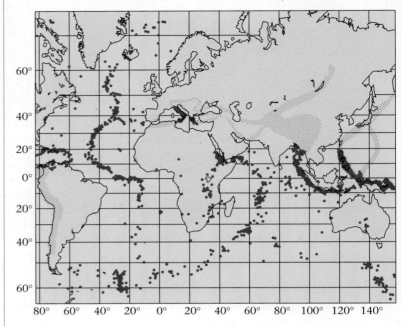

FIGURE 2.9 A 1953 map showing the distribution of earthquake locations in the ocean basins. Note that earthquakes occur in belts.

continental crust. Beneath its sediment cover, oceanic crust bedrock consists of basalt—it does not display the great variety of rock types found on continents.

- Heat flow, the rate at which heat rises from the Earth's interior up through the crust, is not the same everywhere in the oceans. Rather, more heat rises beneath mid-ocean ridges than elsewhere (Fig. 2.8b). This observation led geologists to speculate that magma might be rising into the crust just below the mid-ocean ridge axis, for hot molten rock could bring heat into the crust.

- When maps showing the distribution of earthquakes in oceanic regions became available in the years after World War II, geologists realized that earthquakes do not occur randomly, but rather define distinct belts (Fig. 2.9). Some belts follow trenches, some follow mid-ocean ridge axes, and others lie along portions of fracture zones. Since earthquakes define locations where rocks break and move, geologists realized that these bathymetric features are places where movements of the crust take place.

Now let's see how Harry Hess used these observations to come up with the hypothesis of sea-floor spreading.

TAKE-HOME MESSAGE

Discovery of bathymetric features (ridges, trenches, and fracture zones) and other characteristics of the sea floor, along with a compilation of maps showing the distribution of earthquakes, set the stage for the discovery of sea-floor spreading.

2.5 HARRY HESS AND HIS "ESSAY IN GEOPOETRY"

In the late 1950s, Harry Hess, after studying the observations described above, realized that because the sediment layer on the ocean floor was thin overall, the ocean floor might be much younger than the continents. Also, because the sediment thickened progressively away from mid-ocean ridges, the ridges themselves likely were younger than the deeper parts of the ocean floor. If this was so, then somehow *new ocean floor must be forming at the ridges*, and thus an ocean could be getting wider with time. But how? The association of earthquakes with mid-ocean ridges suggested to him that the sea floor was cracking and splitting apart at the ridge. The discovery of high heat flow along mid-ocean ridge axes provided the final piece of the puzzle, for it suggested the presence of molten rock beneath the ridges. In 1960, Hess suggested that molten rock rose upward beneath mid-ocean ridges and that this material solidified to form oceanic crust (Fig. 2.10). The new sea floor then moved away from the ridge, a process we now call sea-floor spreading. Hess realized that old ocean floor must be consumed somewhere, or the Earth would have to expand. He suggested that deep-ocean trenches might be places where the sea floor sank back into the mantle, and that earthquakes at trenches were evidence of this movement (Fig. 2.10), but he didn't understand how the movement took place.

Hess and his contemporaries realized that the sea-floor-spreading hypothesis instantly provided the long-sought explanation of how continental drift occurs. Continents passively move apart as the sea floor between them spreads at mid-ocean ridges, and they passively move together as the sea floor between them sinks back into the mantle at trenches. (Geologists now realize that it is the lithosphere that moves, not just the crust.) Thus, sea-floor spreading proved to be an important step on the route to plate tectonics—the idea seemed so good that Hess referred to his description of it as "an essay in geopoetry." But other key discoveries would have to take place before the whole theory of plate tectonics could come together.

TAKE-HOME MESSAGE

Ocean basins get wider with time due to the process of sea-floor spreading, and old ocean floor can sink back into the mantle by the process of subduction. As ocean basins get wider, continents drift apart. Subduction allows continents to move toward each other.

2.6 EVIDENCE FOR SEA-FLOOR SPREADING

For a hypothesis to become a theory (see Box P.1), it must be tested—scientists must demonstrate that the idea really works. During the 1960s, geologists found that sea-floor spreading successfully explains several previously baffling observations. Here we discuss two: (1) the existence of orderly variations in the strength of the magnetic field over the sea floor, producing a pattern of stripes called marine-magnetic anomalies; and (2) the variation in sediment thickness on the ocean crust, as measured by drilling.

Marine Magnetic Anomalies

Geologists can measure the strength of Earth's magnetic field with an instrument called a magnetometer. At any given location on the surface of the Earth, the magnetic field that you measure includes two parts: one produced by the main dipole of the Earth, due to circulation of molten iron in the outer core, and another produced by the magnetism of near-surface rock. A **magnetic anomaly** is the difference between the expected strength of the Earth's main dipole field at a certain location and the *actual* measured strength of the magnetic field at that location. Places where the field strength is stronger than expected are positive anomalies, and places where the field strength is weaker than expected are negative anomalies.

Geologists towed magnetometers back and forth across the ocean to map variations in magnetic field strength (Fig. 2.11a). As a ship cruised along its course, the magnetometer's gauge might first detect an interval of strong signal (a positive anomaly) and then an interval of weak signal (a negative anomaly). A graph of signal strength versus distance along the traverse, therefore, has a sawtooth shape (Fig. 2.11b). When geologists compiled data from many cruises on a map, these **marine magnetic anomalies** defined distinctive,

FIGURE 2.10 Harry Hess's basic concept of sea-floor spreading. Hess implied, incorrectly, that only the crust moved. We will see that this sketch is an oversimplification.

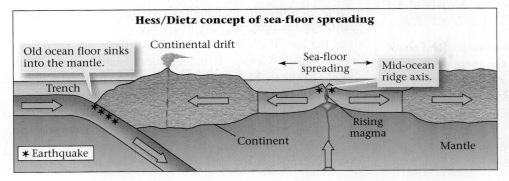

FIGURE 2.11 The discovery of marine magnetic anomalies.

(a) A ship towing a magnetometer detects changes in the strength of the magnetic field.

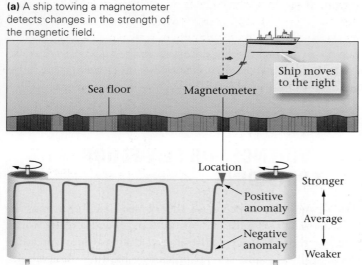

(b) On a paper record, intervals of stronger magnetism (positive anomalies) alternate with intervals of weaker magnetism (negtive anomalies).

The pattern of anomalies is symmetrical, relative to mid-ocean ridges.

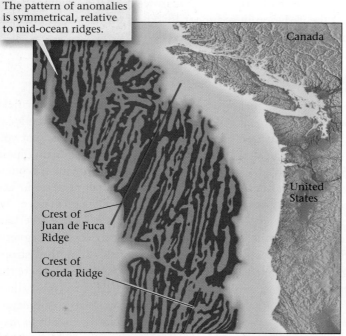

(c) After making many traverses, geologists make a map showing areas of positive anomalies (dark) and negative anomalies (light). These define alternating bands or stripes.

alternating bands. If we color positive anomalies dark and negative anomalies light, the pattern made by the anomalies resembles the stripes on a candy cane (Fig. 2.11c). The mystery of this marine magnetic anomaly pattern, however, remained unsolved until geologists recognized the existence of magnetic reversals.

Magnetic Reversals.   Recall that Earth's magnetic field can be represented by an arrow, representing the dipole, that pres-

ently points from the north magnetic pole to the south magnetic pole. When geologists measured the paleomagnetism of a succession of rock layers that had accumulated over a long period of time, they found that the polarity (which end of a magnet points north and which end points south) of the paleomagnetic field preserved in some layers was the same as that of Earth's present magnetic field, whereas in other layers it was the opposite (Fig. 2.12a, b). At first, observations of reversed polarity were largely ignored, thought to be the result of lightning strikes or of local chemical reactions between rock and water. But when repeated measurements from around the world revealed a systematic pattern of alternating normal and reversed polarity in rock layers, geologists realized that reversals were a global, not a local, phenomenon. They reached the unavoidable conclusion that, at various times during Earth history, *the polarity of Earth's magnetic field has suddenly reversed!* In other words, sometimes the Earth has normal polarity, as it does today, and sometimes it has reversed polarity (Fig. 2.12c). Times when the Earth's field flips from normal to reversed polarity, or vice versa, are called **magnetic reversals**. Reversals may be the result of changes in circulation patterns in the outer core. When the Earth has reversed polarity, the south magnetic pole lies near the north geographic pole, and the north magnetic pole lies near the south geographic pole. Thus, if you were to use a compass during periods when the Earth's magnetic field was reversed, the north-seeking end of the needle would point to the south geographic pole.

In the 1950s, about the same time geologists discovered polarity reversals, they developed a technique that permitted them to define the age of a rock in years, by measuring the decay of radioactive elements in the rock. The technique, called isotopic dating, will be discussed in detail in Chapter 10. Geologists applied the technique to determine the ages of rock layers in which they obtained their paleomagnetic measurements, and thus determined *when* the magnetic field of the Earth reversed. With this information, they constructed a history of magnetic reversals for the past 4.5 million years; this history is now called the **magnetic-reversal chronology**. A diagram representing the Earth's magnetic-reversal chronology (Fig. 2.12d) shows that reversals do not occur regularly, so the lengths of different polarity "chrons," the time intervals between reversals, are different. For example, we have had a normal-polarity chron for about the last 700,000 years. Before that, there was a reversed-polarity chron. Geologists named the youngest four polarity chrons (Brunhes, Matuyama, Gauss, and Gilbert) after scientists who had made important contributions to the study of rock magnetism. As more measurements became available, investigators realized that there were some short-duration reversals (less than 200,000 years long) within the chrons, and called these shorter durations polarity subchrons.

The Interpretation of Marine Anomalies. Why do marine magnetic anomalies exist? A graduate student in England, Fred Vine, working with his adviser, Drummond Mathews, and a Canadian geologist, Lawrence Morley (working independently), discovered a solution to this riddle. Simply put, a positive anomaly occurs over areas of sea floor where basalt has normal polarity. In these areas, the magnetic force produced by the basalt *adds* to the force produced by Earth's dipole and creates a stronger magnetic signal than expected, as measured by the magnetometer (Fig. 2.13a). A negative anomaly occurs over regions of sea floor where basalt has reversed polarity. Here, the magnetic force of the basalt *subtracts* from the force produced by the Earth's dipole and results in a weaker magnetic signal. Sea floor yielding positive anomalies formed at times when the Earth had normal polarity, whereas sea floor yielding negative anomalies formed when the Earth had reversed polarity.

The sea-floor-spreading model easily explains why the magnetic anomaly pattern on one side of a mid-ocean ridge is

FIGURE 2.12 Magnetic polarity reversals, and the chronology of reversals.

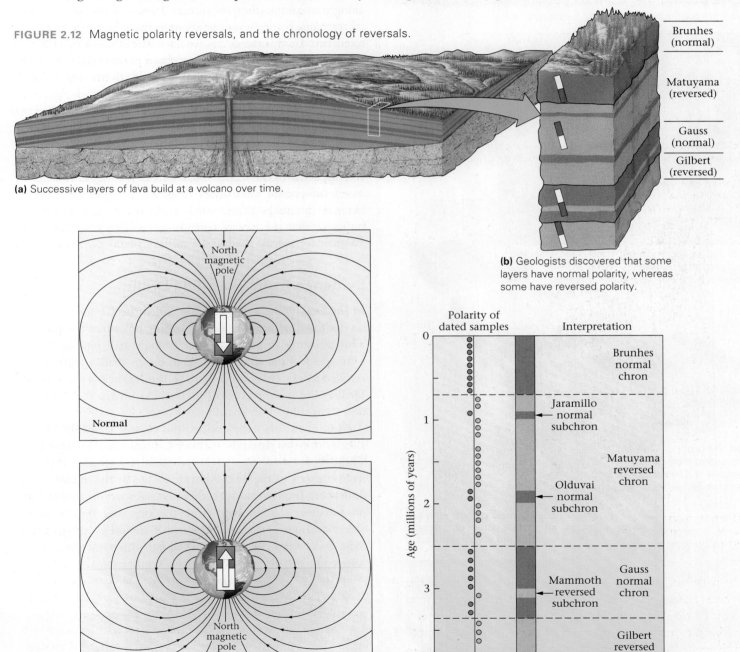

(a) Successive layers of lava build at a volcano over time.

(b) Geologists discovered that some layers have normal polarity, whereas some have reversed polarity.

(c) Geologists proposed that the Earth's magnetic field reverses polarity every now and then.

(d) Observations led to the production of a reversal chronology, with named polarity intervals.

FIGURE 2.13 The progressive development of magnetic anomalies, and the long-term reversal chronology.

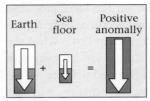

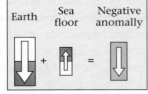

(a) Positive anomalies form when sea-floor rock has the same polarity as the present magnetic field. Negative anomalies form when sea-floor rock has polarity that is opposite to the present field.

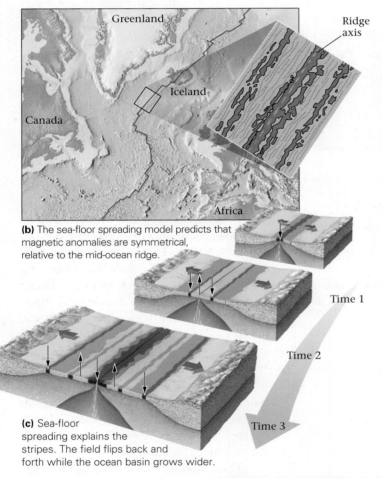

Greenland

Iceland

Canada

Africa

Ridge axis

(b) The sea-floor spreading model predicts that magnetic anomalies are symmetrical, relative to the mid-ocean ridge.

Time 1

Time 2

Time 3

(c) Sea-floor spreading explains the stripes. The field flips back and forth while the ocean basin grows wider.

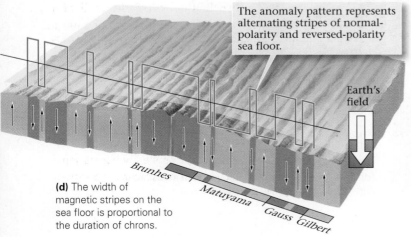

The anomaly pattern represents alternating stripes of normal-polarity and reversed-polarity sea floor.

Earth's field

Brunhes

Matuyama

Gauss

Gilbert

(d) The width of magnetic stripes on the sea floor is proportional to the duration of chrons.

a mirror image of the anomaly pattern on the other side (Fig. 2.13b). At Time 1 (sometime in the past), a time of normal polarity, the dark stripe of sea floor forms (Fig. 2.13c). The tiny dipoles of magnetite grains in basalt making up this stripe align with the Earth's field. As it forms, the rock in this stripe migrates away from the ridge axis, half to the right and half to the left. Later, at Time 2, the field has reversed, and the light-gray stripe forms with reversed polarity. As it forms, it too moves away from the axis, and still younger crust begins to develop along the axis. As the process continues over millions of years, many stripes form. A positive anomaly exists along the ridge axis today because this represents sea floor that has developed during the most recent interval of time, a chron of normal polarity. The magnetism of the rock along the ridge adds to the magnetism of the Earth's field.

By relating the stripes on the Atlantic sea floor to magnetic reversals found in basalt layers on land, geologists dated the sea floor back to an age of 4.5 million years, and they found that the relative widths of anomaly stripes on the sea floor exactly corresponded to the relative durations of polarity chrons in the magnetic-reversal chronology (Fig. 2.13d). The relationship between anomaly-stripe width and polarity-chron duration indicates that the rate of spreading has been constant in the Atlantic for at least the last 4.5 million years.

Evidence from Deep-Sea Drilling

In the late 1960s, a drilling ship called the *Glomar Challenger* set out to sail around the ocean drilling holes into the sea floor. This amazing ship could lower enough drill pipe to drill in 5-km-deep water and could continue to drill until the hole reached a depth of about 1.7 km (1.1 miles) below the sea floor. Drillers brought up cores of rock and sediment that geoscientists then studied on board.

On one of its early cruises, the *Glomar Challenger* drilled a series of holes through sea-floor sediment to the basalt layer. These holes were spaced at progressively greater distances from the axis of the Mid-Atlantic Ridge. If the model of sea-floor spreading was correct, then the sediment layer should be progressively thicker away from the axis, and the age of the oldest sediment just above the basalt should be progressively older away from the axis. When the drilling and the analyses were complete, the predictions were confirmed.

TAKE-HOME MESSAGE

Marine magnetic anomalies form because reversals of the Earth's magnetic polarity take place while sea-floor spreading occurs. The discovery and interpretation of these anomalies proved the sea-floor-spreading hypothesis.

2.7 WHAT DO WE MEAN BY PLATE TECTONICS?

The discovery of sea-floor spreading and the paleomagnetic proof of continental drift set off a scientific revolution in geology in the 1960s and 70s. Geologists realized that many of their existing interpretations of global geology, based on the premise that the positions of continents and oceans remain fixed in position through time, were simply wrong! Researchers dropped what they were doing and turned their attention to studying the broader implications of sea-floor spreading and continental drift. New studies clarified the meaning of a plate, defined the types of plate boundaries, constrained plate motions, related plate motions to earthquakes and volcanoes, showed how plate motions generate mountain belts and seamount chains, and outlined the history of past plate motions. From these, the modern theory of plate tectonics evolved. Below, we describe lithosphere plates and their boundaries, and then outline the basic principles of plate tectonics theory.

The Concept of a Lithosphere Plate

We learned earlier that geoscientists divide the outer part of the Earth into two layers. The **lithosphere** consists of the crust plus the top (cooler) part of the upper mantle. It behaves relatively rigidly, meaning that when a force pushes or pulls on it, it does not flow but rather bends or breaks (Fig. 2.14a). The lithosphere floats on a relatively soft, or "plastic," layer called the **asthenosphere**, composed of warmer (>1,280°C) mantle that can flow slowly when acted on by force. The asthenosphere can undergo convection, like water in a pot, though much more slowly, but the lithosphere cannot.

Continental lithosphere and oceanic lithosphere differ markedly in their thickness. On average, continental lithosphere has a thickness of 150 km, whereas old oceanic lithosphere has a thickness of about 100 km (Fig. 2.14b). (For reasons discussed later in this chapter, new oceanic lithosphere at a mid-ocean ridge is much thinner.) Recall that the crustal part of continental lithosphere ranges from 25 to 70 km thick and consists of relatively low-density felsic and intermediate rock (see Chapter 1). In contrast, the crustal part of oceanic lithosphere is only 7 to 10 km thick and consists of relatively high-density mafic rock. Because of these differences, the continental lithosphere "floats" at a higher level than does the oceanic lithosphere.

The lithosphere forms the Earth's relatively rigid shell. But unlike the shell of a hen's egg, the lithosphere shell contains a number of major breaks, which separate it into distinct pieces. As noted earlier, we call the pieces **lithosphere plates**, or simply plates. The breaks between plates are known as **plate boundaries**. Geoscientists distinguish twelve major plates and several microplates. Note that some plates consist entirely of oceanic lithosphere whereas some plates consist of both oceanic *and* continental lithosphere.

FIGURE 2.14 Nature of the lithosphere.

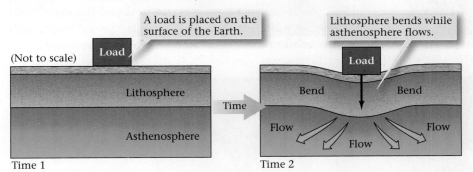

(a) The lithosphere is fairly rigid, but when a glacier or volcano is placed on its surface, it bends. This can happen because the underlying "plastic" asthenosphere can flow out of the way.

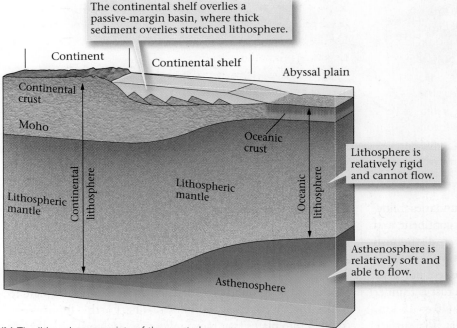

(b) The lithosphere consists of the crust plus the uppermost mantle. It is thicker beneath continents than beneath oceans.

FIGURE 2.15 The locations of plate boundaries and the distribution of earthquakes.

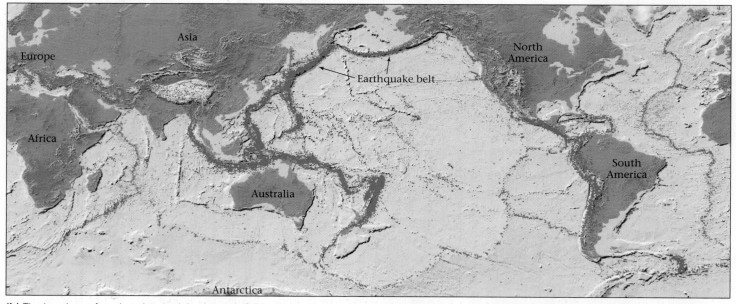

(a) A map of major plates shows that some consist entirely of oceanic lithosphere, whereas some consist of both continental and oceanic lithosphere. Active continental margins lie along plate boundaries; passive margins do not.

(b) The locations of earthquakes (red dots) mostly fall in distinct bands that correspond to plate boundaries. Relatively few earthquakes occur in the stabler plate interiors.

As illustrated in **Figure 2.15a**, some plate boundaries follow continental margins, the boundary between a continent and an ocean, but others do not. For this reason, we distinguish between two types of continental margins: **active margins** are plate boundaries, whereas **passive margins** are not plate boundaries. Along passive margins, continental crust is thinner than in continental interiors (Fig. 2.14b). Thick (10 to 15 km) accumulations of sediment cover this thinned crust. The surface of this sediment layer is a broad, shallow (less than 500 m deep)

region called the continental shelf, home to the major fisheries of the world.

The Basic Principles of Plate Tectonics

With the background provided above, we can restate plate tectonics theory concisely as follows. The Earth's lithosphere is divided into plates that move relative to each other and relative to the underlying asthenosphere. Plate movement occurs at rates of about 1 to 15 cm per year. As a plate moves, its internal area remains largely rigid and intact, but rock along plate boundaries undergoes deformation (cracking, sliding, bending, stretching, and squashing) as the plate grinds or scrapes against its neighbors or pulls away from its neighbors. As plates move, so do the continents that form part of the plates, resulting in continental drift. Because of plate tectonics, the map of Earth's surface constantly changes.

Identifying Plate Boundaries

How do we recognize the location of a plate boundary? The answer becomes clear from looking at a map showing the locations of earthquakes (Fig. 2.15b). Recall from Chapter 1 that earthquakes are vibrations caused by shock waves that are generated where rock breaks and suddenly shears (slides) along a fault (a fracture on which sliding occurs). The focus of the earthquake is the spot where the fault slips, and the epicenter marks the point on the surface of the Earth directly above the focus. Earthquake epicenters do not speckle the Earth's surface randomly, like buckshot on a target. Rather, the majority occur in relatively narrow, distinct belts. These earthquake belts define the position of plate boundaries, because the fracturing and slipping that occur along plate boundaries as plates move generate earthquakes. Plate interiors, regions away from the plate boundaries, remain relatively earthquake-free because they do not accommodate much movement. While earthquakes serve as the most definitive indicator of a plate boundary, other prominent geologic features also develop along plate boundaries, as you will learn by the end of this chapter.

Geologists define three types of plate boundaries, based simply on the relative motions of the plates on either side of the boundary (Fig. 2.16a–c). A boundary at which two plates move apart from each other is a **divergent boundary**. A boundary at which two plates move toward each other so that one plate sinks beneath the other is a **convergent boundary**. And a boundary at which one plate slips along the side of another plate is a **transform boundary**. Each type of boundary looks and behaves differently from the others, as we will now see.

TAKE-HOME MESSAGE

The Earth's lithosphere, its rigid shell, consists of about twenty plates that move relative to each other. The distribution of earthquakes delineates the boundaries between plates. Geologists recognize three distinct types of plate boundaries.

2.8 DIVERGENT PLATE BOUNDARIES AND SEA-FLOOR SPREADING

At divergent boundaries, or spreading boundaries, two oceanic plates move apart by the process of sea-floor spreading. Note that an *open space does not develop* between diverging plates. Rather, as the plates move apart, new oceanic lithosphere forms along the divergent boundary (Fig. 2.17a). This process takes place at a submarine mountain range called a mid-ocean ridge that rises 2 km above the adjacent abyssal plains of the ocean. Thus, geologists also commonly refer to a divergent boundary as a mid-ocean ridge, or simply a ridge.

To characterize a divergent boundary more completely, let's look at one mid-ocean ridge in more detail (Fig. 2.17b). The Mid-Atlantic Ridge extends from the waters between northern Greenland and northern Scandinavia southward across

FIGURE 2.16 The three types of plate boundaries differ based on the nature of relative movement.

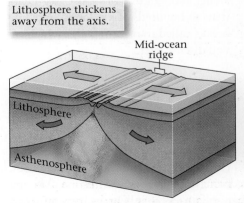

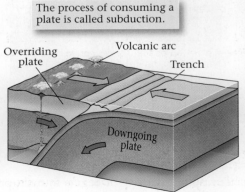

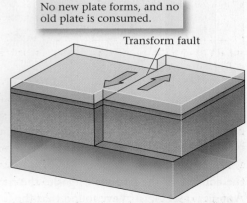

(a) At a divergent boundary, two plates move away from the axis of a mid-ocean ridge. New oceanic lithosphere forms.

(b) At a convergent boundary, two plates move toward each other; the downgoing plate sinks beneath the overriding plate.

(c) At a transform boundary, two plates slide past each other on a vertical fault surface.

FIGURE 2.17 The process of sea-floor spreading.

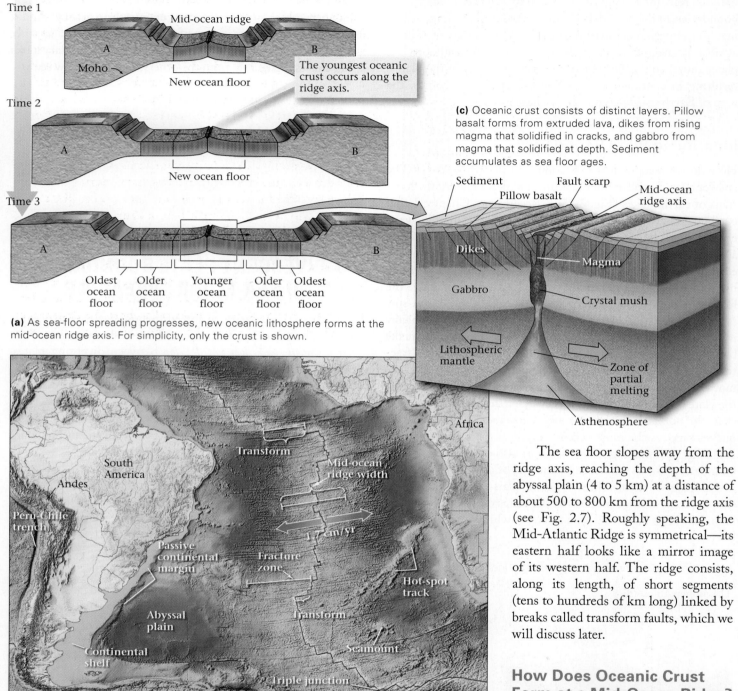

Time 1

Mid-ocean ridge

A   B

Moho

New ocean floor

The youngest oceanic crust occurs along the ridge axis.

Time 2

A   B

New ocean floor

Time 3

A   B

Oldest ocean floor | Older ocean floor | Younger ocean floor | Older ocean floor | Oldest ocean floor

(a) As sea-floor spreading progresses, new oceanic lithosphere forms at the mid-ocean ridge axis. For simplicity, only the crust is shown.

(c) Oceanic crust consists of distinct layers. Pillow basalt forms from extruded lava, dikes from rising magma that solidified in cracks, and gabbro from magma that solidified at depth. Sediment accumulates as sea floor ages.

Sediment   Fault scarp   Mid-ocean ridge axis

Pillow basalt

Dikes   Magma

Gabbro

Crystal mush

Lithospheric mantle

Zone of partial melting

Asthenosphere

(b) The bathymetry of the Mid-Atlantic Ridge in the Southern Atlantic Ocean. The lighter colors are shallower depths.

South America

Andes

Peru-Chile trench

Transform

Mid-ocean ridge width

Passive continental margin

Fracture zone

1.7 cm/yr

Africa

Hot-spot track

Abyssal plain

Transform

Continental shelf

Seamount

Triple junction

The sea floor slopes away from the ridge axis, reaching the depth of the abyssal plain (4 to 5 km) at a distance of about 500 to 800 km from the ridge axis (see Fig. 2.7). Roughly speaking, the Mid-Atlantic Ridge is symmetrical—its eastern half looks like a mirror image of its western half. The ridge consists, along its length, of short segments (tens to hundreds of km long) linked by breaks called transform faults, which we will discuss later.

How Does Oceanic Crust Form at a Mid-Ocean Ridge?

As sea-floor spreading takes place, hot asthenosphere rises beneath the ridge (Fig. 2.17c). As this asthenosphere rises,

the equator to the latitude opposite the southern tip of South America. Geologists have found that the formation of new sea floor takes place only along the axis (centerline) of the ridge. The axis lies at water depths of about 2 to 2.5 km.

it begins to melt, producing molten rock, or magma. Magma has a lower density than solid rock, so it behaves buoyantly and rises, as oil rises above vinegar in salad dressing. It eventually accumulates in the crust below the ridge axis, filling a region

FIGURE 2.18 A column of superhot water gushing from a vent (known as a "black smoker") along the mid-ocean ridge. The cloud of "smoke" actually consistes of tiny mineral grains.

volcanoes. The resulting lava cools to form a layer of basalt. This basalt occurs in small blobs called pillows. We can't easily see the submarine volcanoes because they occur at depths of more than 2 km beneath sea level, but they have been observed by the research submarine *Alvin*. Observers in the submarine have also detected chimneys spewing hot, mineralized water that rose through cracks in the sea floor, after being heated by magma below the surface. These chimneys are called **black smokers** because the water they emit looks like dark smoke; the color comes from a suspension of tiny mineral grains that precipitate in the water as the water cools (Fig. 2.18).

called a magma chamber. As the magma cools, it turns into a mush of crystals. Some of the magma solidifies completely along the side of the chamber to make the coarse-grained, mafic igneous rock called gabbro. The rest rises still higher to fill vertical cracks, where it solidifies and forms wall-like sheets, or dikes, of basalt. Finally, some magma rises all the way to the surface of the sea floor at the ridge axis and spills out of small submarine

As soon as it forms, new oceanic crust moves away from the ridge axis, and when this happens, more magma rises from below, so still more crust forms. In other words, like a vast, continuously moving conveyor belt, magma from the mantle rises to the Earth's surface at the ridge, solidifies to form oceanic crust, and then moves laterally away from the ridge. Because all sea floor forms at mid-ocean ridges, the youngest sea floor occurs on either side of the ridge axis, and sea floor becomes progressively older away from the ridge. In the Atlantic Ocean, the oldest sea floor lies adjacent to the passive continental margins on either side of the ocean (Fig. 2.19). The

FIGURE 2.19 This map of the world shows the age of the sea floor. Note how the sea floor grows older with increasing distance from the ridge axis. (Ma = million years ago)

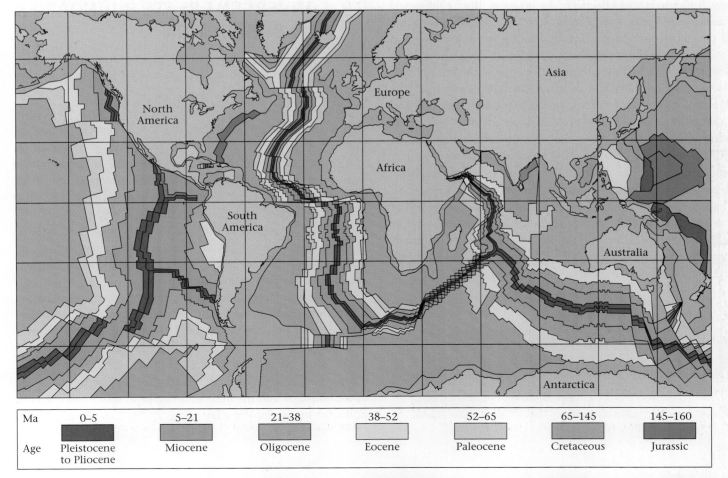

Ma	0–5	5–21	21–38	38–52	52–65	65–145	145–160
Age	Pleistocene to Pliocene	Miocene	Oligocene	Eocene	Paleocene	Cretaceous	Jurassic

oldest ocean floor on our planet is in the western Pacific Ocean; this crust formed 200 million years ago.

The tension (stretching force) applied to newly formed solid crust as spreading takes place breaks this new crust, resulting in the formation of faults. Movement (slip) on the faults causes divergent-boundary earthquakes and produces numerous cliffs, or scarps, that lie parallel to the ridge axis.

How Does the Lithospheric Mantle Form at a Mid-Ocean Ridge?

So far, we've seen how oceanic crust forms at mid-ocean ridges. What about the formation of the mantle part of the oceanic lithosphere? This part consists of the cooler uppermost area of the mantle, in which temperatures are less than about 1,280°C. At the ridge axis, such temperatures occur almost at the base of the crust, because of the presence of rising hot asthenosphere and hot magma, so the lithospheric mantle beneath the ridge axis effectively doesn't exist. But as the newly formed oceanic crust moves away from the ridge axis, the crust and the uppermost mantle directly beneath it gradually cool as they lose heat to the ocean above. As soon as mantle rock cools below 1,280°C, it becomes, by definition, part of the lithosphere.

As oceanic lithosphere continues to move away from the ridge axis, it continues to cool, so the lithospheric mantle, and therefore the oceanic lithosphere as a whole, grows progressively thicker (Fig. 2.20a, b). Note that this process doesn't change the thickness of the oceanic crust, for the crust formed entirely at the ridge axis. The rate at which cooling and lithospheric thickening occur decreases with distance from the ridge axis. In fact, by the time the lithosphere is about 80 million years old it has just about reached its maximum thickness. As lithosphere thickens and gets cooler and denser, it sinks into the asthenosphere, like a ship taking on ballast. Thus, the ocean is deeper over older ocean floor than over younger ocean floor.

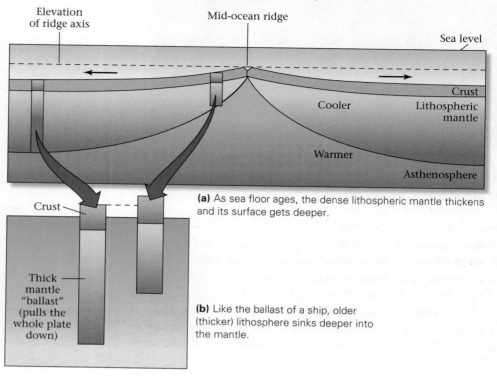

FIGURE 2.20 Changes accompanying the aging of lithosphere.

(a) As sea floor ages, the dense lithospheric mantle thickens and its surface gets deeper.

(b) Like the ballast of a ship, older (thicker) lithosphere sinks deeper into the mantle.

TAKE-HOME MESSAGE

Sea-floor spreading occurs at divergent plate boundaries. Through this process, new oceanic plates form and move apart. These plate boundaries are delineated by mid-ocean ridges, along which submarine volcanoes erupt.

2.9 CONVERGENT PLATE BOUNDARIES AND SUBDUCTION

At convergent plate boundaries, two plates, at least one of which is oceanic, move toward one another. But rather than butting each other like angry rams, one oceanic plate bends and sinks down into the asthenosphere beneath the other plate. Geologists refer to the sinking process as subduction, so convergent boundaries are also known as subduction zones. Because subduction at a convergent boundary consumes old ocean lithosphere and thus closes (or "consumes") oceanic basins, geologists also refer to convergent boundaries as consuming boundaries, and because they are delineated by deep-ocean trenches, they are sometimes simply called trenches (see Fig. 2.8). The amount of oceanic plate consumption worldwide, averaged over time, equals the amount of sea-floor spreading worldwide, so the surface area of the Earth remains constant through time.

Subduction occurs for a simple reason: oceanic lithosphere, once it has aged at least 10 million years, is denser than asthenosphere and thus can sink through the asthenosphere. When it lies flat on the surface of the asthenosphere, oceanic lithosphere doesn't sink, because the resistance of the asthenosphere to flow is too great. However, once the end of the convergent plate bends down and slips into the mantle, it begins to sink like an anchor falling to the bottom of a lake (Fig. 2.21a). As the lithosphere

FIGURE 2.21 During the process of subduction, oceanic lithosphere sinks back into the deeper mantle.

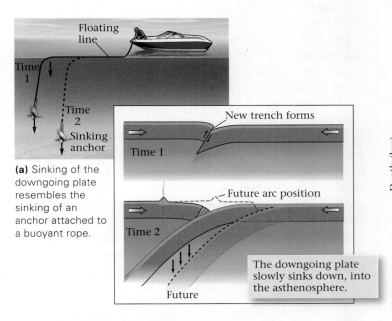

(a) Sinking of the downgoing plate resembles the sinking of an anchor attached to a buoyant rope.

The downgoing plate slowly sinks down, into the asthenosphere.

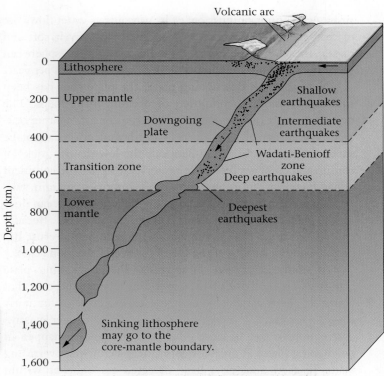

(b) A belt of earthquakes (dots) defines the position of the downgoing plate in the region above a depth of about 660 km.

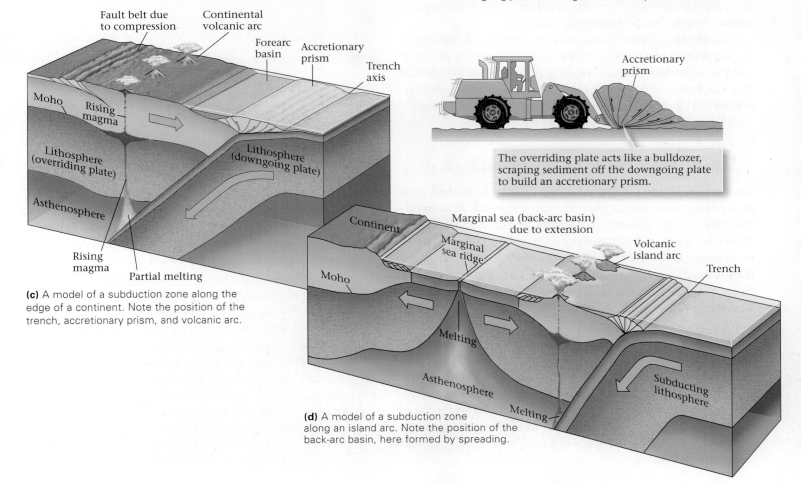

(c) A model of a subduction zone along the edge of a continent. Note the position of the trench, accretionary prism, and volcanic arc.

The overriding plate acts like a bulldozer, scraping sediment off the downgoing plate to build an accretionary prism.

(d) A model of a subduction zone along an island arc. Note the position of the back-arc basin, here formed by spreading.

sinks, asthenosphere flows out of its way, just as water flows out of the way of an anchor. Even though it is relatively soft and plastic, the asthenosphere resists flow, so oceanic lithosphere can sink only very slowly, at a rate of less than about 15 cm per year.

Note that the downgoing plate, the plate that has been subducted, *must* be composed of oceanic lithosphere. The overriding plate, which does not sink, can consist of either oceanic or continental lithosphere. Continental crust cannot be subducted because it is too buoyant; the low-density rocks of continental crust act like a life preserver keeping the continent afloat. If continental crust moves into a convergent margin, subduction eventually stops. Because of subduction, all ocean floor on the planet is less than about 200 million years old. Because continental crust cannot subduct, some continental crust has persisted at the surface of the Earth for over 3.8 billion years.

At convergent plate boundaries, the downgoing plate grinds along the base of the overriding plate, a process that generates large earthquakes. These earthquakes occur fairly close to the Earth's surface, so some of them trigger massive destruction in coastal cities. But earthquakes also happen in downgoing plates at greater depths, deep below the overriding plate. In fact, geologists have detected earthquakes within downgoing plates to a depth of 660 km. The belt of earthquakes in a downgoing plate is called a **Wadati-Benioff zone**, after its two discoverers (Fig. 2.21b).

At depths greater than 660 km, conditions leading to earthquakes evidently do not occur. Recent observations, however, indicates that some downgoing plates do continue to sink *below* a depth of 660 km—they just do so without generating earthquakes. In fact, studies suggest that the lower mantle may be a graveyard for old subducted plates.

Geologic Features of a Convergent Boundary

To become familiar with the various geologic features that occur along a convergent plate boundary, let's look at an example, the boundary between the western coast of the South American Plate and the eastern edge of the Nazca Plate (a portion of the Pacific Ocean floor). A deep-ocean trench, the Peru-Chile Trench, delineates this boundary (see Fig. 2.17b). Such trenches form where the plate bends as it starts to sink into the asthenosphere.

In the Peru-Chile Trench, as the downgoing plate slides under the overriding plate, sediment (clay and plankton) that had settled on the surface of the downgoing plate, as well as sand that fell into the trench from the shores of South America, gets scraped up and incorporated in a wedge-shaped mass known as an accretionary prism (Fig. 2.21c). An accretionary prism forms in basically the same way as a pile of snow in front of a plow, and like the snow, the sediment tends to be squashed and contorted during the formation of the prism.

A chain of volcanoes known as a volcanic arc develops behind the accretionary prism (see Geotour 2 on p. GT-6 to see a volcanic arc and other features of plate boundaries). As we will see in Chapter 4, the magma that feeds these volcanoes forms at or just above the surface of the downgoing plate when the plate reaches a depth of about 150 km below the Earth's surface. If the volcanic arc forms where an oceanic plate subducts beneath continental lithosphere, the resulting chain of volcanoes grows on the continent and forms a continental volcanic arc (see Fig. 2.21c). In some cases, the plates compress together, causing a belt of faults to form behind the arc. If, however, the volcanic arc forms where one oceanic plate subducts beneath another oceanic plate, the resulting volcanoes form a chain of islands known as a volcanic island arc (Fig. 2.21d). A marginal sea, or back-arc basin, the small ocean basin between an island arc and the continent, forms either in cases where subduction happens to begin offshore, trapping ocean lithosphere behind the arc, or where stretching of the lithosphere behind the arc leads to the formation of a small spreading ridge between the arc and the continent.

TAKE-HOME MESSAGE

At a convergent plate boundary, an oceanic plate sinks into the mantle beneath the edge of another plate. This process allows the two plates to move toward each other and ocean basins to close. Trenches and volcanic arcs delineate convergent boundaries.

2.10 TRANSFORM PLATE BOUNDARIES

When researchers began to explore the bathymetry of mid-ocean ridges in detail, they discovered that mid-ocean ridges are not smooth, uninterrupted lines, but rather consist of short segments that appear to be offset from each other (see Fig. 2.17b). Narrow belts of broken and irregular sea floor lie roughly at right angles to the ridge segments, intersect the ends of the segments, and extend beyond the ends of the segments. These belts came to be known as fracture zones. Originally, researchers incorrectly assumed that the entire length of each fracture zone was a fault, and that slip on a fracture zone had displaced segments of the mid-ocean ridge sideways, relative to each other. In other words, they imagined that a mid-ocean ridge initiated as a continuous, fence-like line that only later was broken up by faulting. But when information about the distribution of earthquakes along mid-ocean ridges became available, it was clear that this model could not be correct. Earthquakes, and therefore active fault slip, occur *only* on the segment of a fracture zone that lies between two ridge segments (Fig. 2.22a). The portions of fracture zones that extend beyond the edges of ridge segments, out into the abyssal plain, are not active.

FIGURE 2.22 The concept of transform faulting.

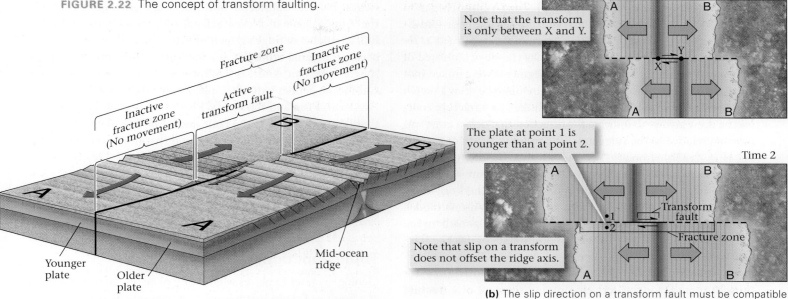

Note that the transform is only between X and Y.

Time 1

The plate at point 1 is younger than at point 2.

Time 2

Transform fault

Fracture zone

Note that slip on a transform does not offset the ridge axis.

(b) The slip direction on a transform fault must be compatible with the configuration of spreading. The length of a transform can stay constant over time.

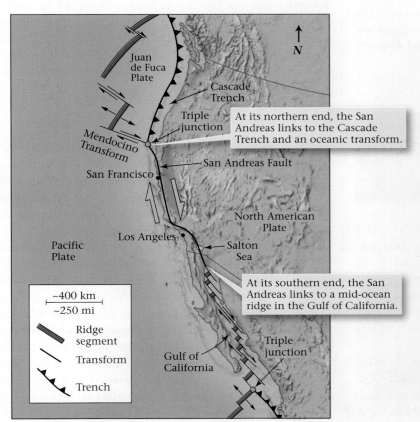

Fracture zone

Inactive fracture zone (No movement)

Active transform fault

Inactive fracture zone (No movement)

Mid-ocean ridge

Younger plate

Older plate

(a) Transform faults on the ocean floor link ridge segments. Only the fault between the two ridge axes is the active transform.

Juan de Fuca Plate

Cascade Trench

Triple junction

Mendocino Transform

At its northern end, the San Andreas links to the Cascade Trench and an oceanic transform.

San Andreas Fault

San Francisco

North American Plate

Los Angeles

Salton Sea

Pacific Plate

At its southern end, the San Andreas links to a mid-ocean ridge in the Gulf of California.

~400 km
~250 mi

Ridge segment

Transform

Trench

Gulf of California

Triple junction

N

(c) The San Andreas Fault is a transform plate boundary between the North American and Pacific Plates. The Pacific is moving northwest, relative to North America.

Fault trace

(d) In southern California, the San Andreas Fault cuts a dry landscape. The fault trace is in the narrow valley. The land has been pushed up slightly along the fault.

The distribution of movement along fracture zones remained a mystery until a Canadian researcher, J. Tuzo Wilson, began to think about fracture zones in the context of the sea-floor spreading concept. Wilson proposed that fracture zones formed *at the same time* as the ridge axis itself, and thus the ridge consisted of separate segments to start with. These segments were linked (not offset) by fracture zones. With this idea in mind, he drew a sketch map showing two ridge-axis segments linked by a fracture zone, and he drew arrows to indicate the direction that ocean crust was moving, relative to the ridge axis, as a result of sea-floor spreading (Fig. 2.22b). Look at Wilson's arrows. Clearly, the movement direction on the active portion of the fracture zone must be opposite to the movement direction that researchers originally thought occurred on the structure. Further, in Wilson's model, slip occurs only along the segment of the fracture zone between the two ridge segments. Plates on opposite sides of inactive fracture zones move together, as one plate.

Wilson introduced the term **transform boundary** (or transform fault) for the actively slipping segment of a fracture zone between two ridge segments, and he pointed out that these are a third type of plate boundary. At a transform boundary, one plate slides sideways past another, but no new plate forms and no old plate is consumed. Transform boundaries are, therefore, defined by a vertical fault on which the slip direction parallels the Earth's surface.

TAKE-HOME MESSAGE

At transform plate boundaries, one plate slips sideways past another. Most transform boundaries link segments of mid-ocean ridges. But some, such as the San Andreas Fault, cut across continental crust.

FIGURE 2.23 Examples of triple junctions. The triple junctions are marked by dots.

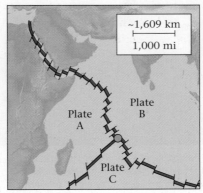

(a) A ridge-ridge-ridge triple junction occurs in the Indian Ocean.

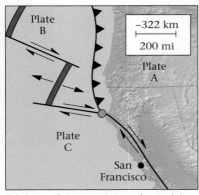

(b) A transform-trench-transform triple junction occurs at the north end of the San Andreas Fault.

So far we've discussed only transforms along mid-ocean ridges. Not all transforms link ridge segments. Some, such as the Alpine Fault of New Zealand, link trenches, while others link a trench to a ridge segment. Further, not all transform faults occur in oceanic lithosphere; a few cut across continental lithosphere. The San Andreas Fault, for example, which cuts across California, defines part of the plate boundary between the North American Plate and the Pacific Plate—the portion of California that lies to the west of the fault (including Los Angeles) is part of the Pacific Plate, while the portion that lies to the east of the fault is part of the North American Plate (Fig. 2.22c, d).

2.11 SPECIAL LOCATIONS IN THE PLATE MOSAIC

Triple Junctions

Geologists refer to places where *three* plate boundaries intersect at a point as **triple junctions**, and name them after the types of boundaries that intersect. For example, the triple junction formed where the Southwest Indian Ocean Ridge intersects two arms of the Mid–Indian Ocean Ridge (this is the triple junction of the African, Antarctic, and Australian Plates) is a ridge-ridge-ridge triple junction (Fig. 2.23a). The triple junction north of San Francisco is a trench-transform-transform triple junction (Fig. 2.23b).

Hot Spots

Most subaerial (above sea level) volcanoes are situated in the volcanic arcs that border trenches. Volcanoes also lie along mid-ocean ridges, but ocean water hides most of them. The volcanoes of volcanic arcs and mid-ocean ridges are plate-boundary volcanoes, in that they formed as a consequence of movement along the boundary. Not all volcanoes on Earth are plate-boundary volcanoes, however. For example, the big island of Hawaii, a large volcano, lies in the middle of the Pacific Plate, and Yellowstone National Park, site of recent volcanic activity, lies in the interior of the North American Plate. Worldwide, geoscientists have identified about one hundred volcanoes that exist as isolated points and are not a consequence of movement at a plate boundary. These are called hot-spot volcanoes, or simply **hot spots** (Fig. 2.24). Most hot spots are located in the interiors of plates, away from the boundaries, but a few grew on mid-ocean ridges.

What causes hot-spot volcanoes? In the early 1960s, J. Tuzo Wilson noted that *active* hot-spot volcanoes (examples that are erupting or may erupt in the future) build an island at the end of a chain of *dead* volcanic islands and seamounts (formerly active volca-

FIGURE 2.24 The dots represent the locations of selected hot-spot volcanoes. The red lines represent hot-spot tracks. The most recent volcano (dot) is at one end of this track. Some of these are extinct, indicating that the plume no longer exists. Some hot spots are fairly recent and do not have tracks. Dashed tracks indicate places where a track was broken by sea-floor spreading.

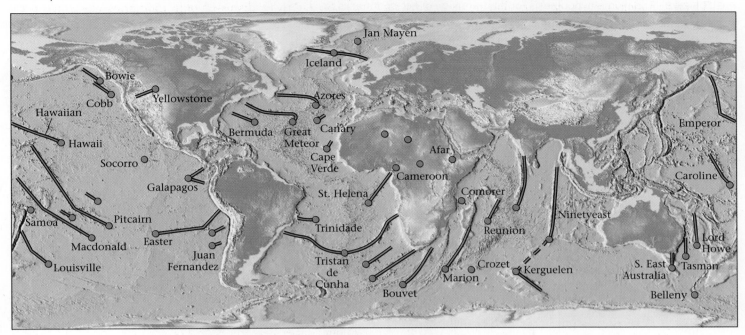

noes that will never erupt again). This configuration is different from that of volcanic arcs along convergent plate boundaries—at volcanic arcs, all of the volcanoes are active. With this image in mind, Wilson suggested that the position of the heat source causing a hot-spot volcano is fixed, relative to the moving plate. In Wilson's model, the active volcano represents the present-day location of the heat source, whereas the chain of dead volcanic islands represents locations on the plate that were once over the heat source but progressively moved off.

A few years later, Jason Morgan suggested that the heat source for hot spots is a **mantle plume**, a column of very hot rock rising up through the mantle to the base of the lithosphere (Fig. 2.25a–d). In Morgan's model, plumes originate deep in the mantle. Rock in the plume, though solid, is soft enough to flow, and rises buoyantly because it is less dense than surrounding cooler rock. When the hot rock of the plume reaches the base of the lithosphere, it partially melts (for reasons discussed in Chapter 4) and produces magma that seeps up through the lithosphere to the Earth's surface. The chain of extinct volcanoes, or **hot-spot track**, forms when the overlying plate moves over a fixed plume. This movement slowly carries the volcano off the top of the plume, so that it becomes extinct. A new, younger volcano grows over the plume. While the deep-mantle plume model is easy to visualize, it is not the only explanation for hot-spot tracks, and it remains controversial. Some geologists argue that plumes originate at shallow depths in the mantle, or that hot spots are not due to plumes at all, but to other phenomena.

The Hawaiian chain provides a clear example of the volcanism associated with a hot-spot track. Volcanic eruptions occur today only on the big island of Hawaii. Other islands to the northwest are remnants of dead volcanoes, the oldest of which is Kauai. To the northwest of Kauai, still older volcanic remnants are found. To the northwest of Midway Island, the track bends in a more northerly direction, and the volcanic remnants no longer poke above sea level; we refer to this segment as the Emperor seamount chain. The bend is due to a change in the direction of the Pacific Plate about 40 Ma.

Some hot spots lie within continents. For example, several have been active in the interior of Africa, and one now underlies Yellowstone National Park. The famous geysers (natural steam and hot-water fountains) of Yellowstone exist because hot magma, formed above the Yellowstone hot spot, lies not far below the surface of the park. A few hot spots lie on mid-ocean ridges. Where this happens, a volcanic island may protrude above sea level, because the hot spot produces far more magma than does a normal mid-ocean ridge. Iceland, for example, formed where a hot spot underlies the Mid-Atlantic Ridge. The extra volcanism of the hot spot built up the island of Iceland so that it rises almost 3 km above other places on the Mid-Atlantic Ridge.

TAKE-HOME MESSAGE

Three plate boundaries join at a triple junction. Hot spots are places where volcanoes exist as isolated points that are not necessarily a direct consequence of movement at plate boundaries. The magma at hot spots may form by melting at the top of mantle plumes.

FIGURE 2.25 The plume hyppothesis for the formation of hot-spot tracks.

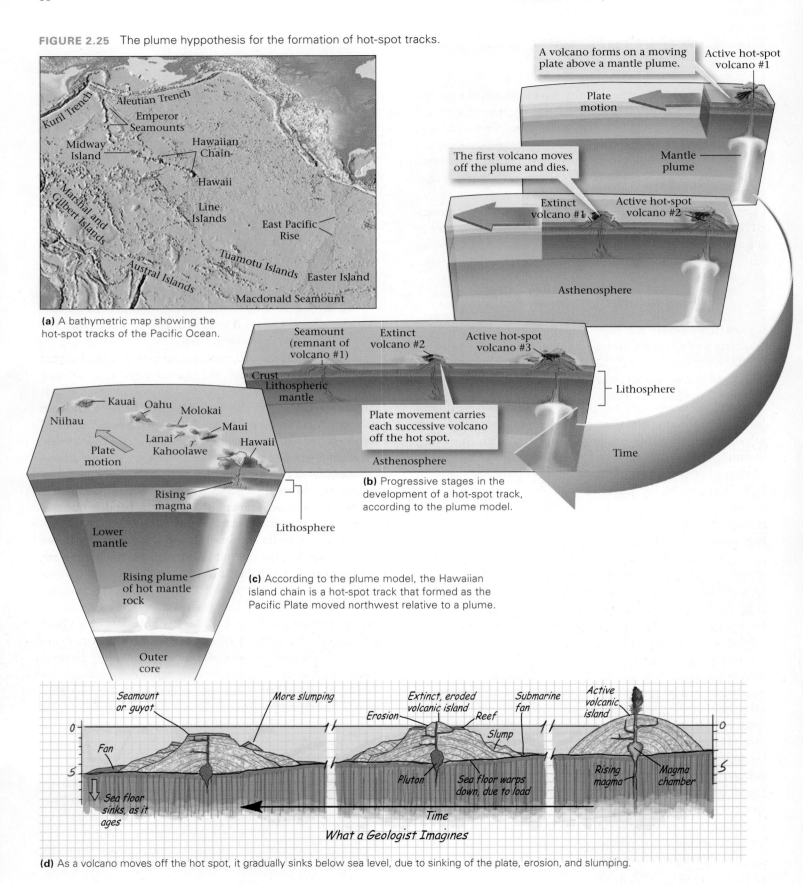

(a) A bathymetric map showing the hot-spot tracks of the Pacific Ocean.

(b) Progressive stages in the development of a hot-spot track, according to the plume model.

(c) According to the plume model, the Hawaiian island chain is a hot-spot track that formed as the Pacific Plate moved northwest relative to a plume.

(d) As a volcano moves off the hot spot, it gradually sinks below sea level, due to sinking of the plate, erosion, and slumping.

2.12 HOW DO PLATE BOUNDARIES FORM AND DIE?

The configuration of plates and plate boundaries visible on our planet today has not existed for all of geologic history, and will not exist indefinitely into the future. Because of plate motion, oceanic plates form and are later consumed, while continents merge and later split apart. How does a new divergent boundary come into existence, and how does a convergent boundary cease to exist? Most new divergent boundaries form when a continent splits and separates into two continents. We call this process **rifting**. A convergent boundary ceases to exist when a piece of buoyant lithosphere, such as a continent or an island arc, moves into the subduction zone. We call this process **collision**.

Continental Rifting

A **continental rift** is a linear belt in which continental lithosphere undergoes rifting, or pulls apart (**Fig. 2.26a**). The lithosphere stretches horizontally, so it thins vertically, much like a piece of taffy you pull between your fingers. Near the surface of the continent, where the crust is cold and brittle, stretching causes rock to break and faults to develop. As a consequence of faulting, blocks of rock slide down, leading to the formation of a low area that gradually becomes buried by sediment. Deeper in the crust, and in the underlying lithospheric mantle, rock is warmer and softer, so stretching takes place in a plastic manner without breaking the rock. The whole region that stretches is the rift.

As continental lithosphere thins, hot asthenosphere rises beneath the rift and starts to melt. Molten rock erupts at volcanoes along the rift. If rifting continues for a long enough time, the continent breaks in two, a new mid-ocean ridge forms, and sea-floor spreading begins. The relict of the rift evolves into a passive margin (see Fig. 2.14b). In some cases, however, rifting stops before the continent splits in two. Then, the rift remains as a permanent scar in the crust, defined by a belt of faults, volcanic rocks, and a thick layer of sediment.

Perhaps the most spectacular example of an active rift is located in eastern Africa; geoscientists aptly refer to it as the East African Rift (**Fig. 2.26b, c**). To astronauts in orbit, the rift looks like a giant gash in the crust. On the ground, it consists of a deep trough bordered on both sides by high cliffs formed by faulting. Along the length of the rift, several major volcanoes smoke and fume; these include the snow-crested Mt. Kilimanjaro, towering over 6 km above the savannah. At its north end, the rift joins the Red Sea Ridge and the Gulf of Aden Ridge at a triple junction. Another major rift, known as the Basin and Range Province, breaks up the landscape of the western United States between Salt Lake City, Utah, and Reno, Nevada (**Fig. 2.26d**). Here, movement on numerous faults tilted blocks of crust to form narrow mountain ranges,

while sediment that eroded from the blocks filled the adjacent basins (the low areas between the ranges).

Collision

India was once a small, separate continent that lay far to the south of Asia. But subduction consumed the ocean between India and Asia, and India moved northward, finally slamming into the southern margin of Asia 40 to 50 million years ago. Continental crust, unlike oceanic crust, is too buoyant to subduct. So when India collided with Asia, the attached oceanic plate broke off and sank down into the deep mantle, while India pushed hard into Asia, squeezing the rocks and sediment that once lay between the two continents into the 8-km-high welt that we now know as the Himalayan Mountains. During this process, not only did the surface of the Earth rise, but the crust became thicker. The crust beneath a collisional mountain range can be up to 60 to 70 km thick, about twice the thickness of normal continental crust.

Geoscientists refer to the process during which two buoyant pieces of lithosphere converge and squeeze together as collision (**Fig. 2.27**). Some collisions involve two continents, whereas some involve continents and an island arc. When a collision is complete, the convergent plate boundary that once existed between the two colliding pieces ceases to exist. Collisions yield some of the most spectacular mountains on the planet, such as the Himalayas and the Alps. They also yielded major mountain ranges in the past, which subsequently eroded away so that today we see only their relicts. For example, the Appalachian Mountains in the eastern United States formed as a consequence of three collisions. After the last one, a collision between Africa and North America around 280 million years ago, North America became part of the Pangaea supercontinent.

TAKE-HOME MESSAGE

Rifting can split a continent in two and can lead to the formation of a new divergent plate boundary. When two continents come together at a convergent plate boundary, they collide, a mountain belt forms, and subduction ceases.

2.13 WHAT DRIVES PLATE MOTION?

We've now discussed the many facets of plate tectonics theory (see **Geology at a Glance**, pp. 62–63). But what drives plate motion? When geoscientists first proposed plate tectonics, they thought the process occurred simply because convective flow in the asthenosphere actively dragged plates along, as if the plates were rafts on a flowing river. Thus, early images depicting plate motion showed simple convection cells—elliptical flow paths

FIGURE 2.26 During the process of rifting, lithosphere stretches.

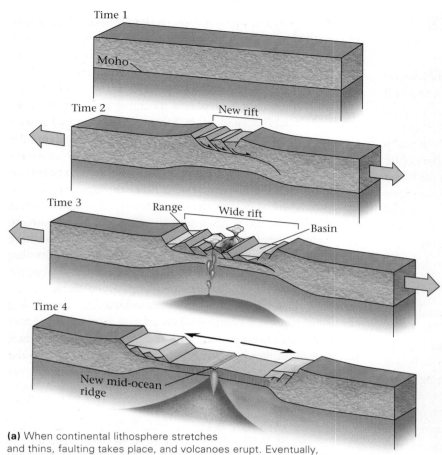

Time 1

Moho

Time 2

New rift

Time 3

Range   Wide rift   Basin

Time 4

New mid-ocean
ridge

(a) When continental lithosphere stretches
and thins, faulting takes place, and volcanoes erupt. Eventually,
the continent splits in two and a new ocean basin forms.

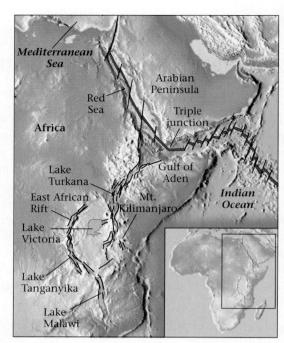

Mediterranean
Sea

Arabian
Peninsula

Red
Sea

Triple
junction

Africa

Gulf of
Aden

Lake
Turkana

East African
Rift

Mt.
Kilimanjaro

Indian
Ocean

Lake
Victoria

Lake
Tanganyika

Lake
Malawi

(b) The East African Rift is growing today. The Red Sea
started as a rift.

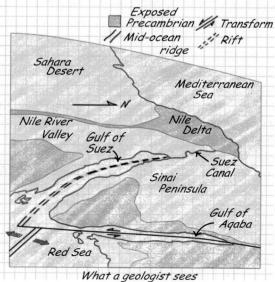

Exposed
Precambrian   Transform

Mid-ocean
ridge   Rift

Sahara
Desert

Mediterranean
Sea

N

Nile River
Valley

Nile
Delta

Gulf of
Suez

Suez
Canal

Sinai
Peninsula

Gulf of
Aqaba

Red Sea

What a geologist sees

(c) Astronauts can see how rifting has opened
up gulfs on either side of the Sinai Peninsula.

Snake River Plain

Reno

Salt Lake City

Basin
and
Range

Sierra Nevada

N

San
Andreas
fault

Colorado
Plateau

Rio
Grande
rift

Basin
and
Range

250 km

(d) The Basin and Range Province is a rift. Faulting bounds the narrow
north-south-trending mountains, separated by basins. The arrows
indicate the direction of stretching.

FIGURE 2.27 Continental collision.

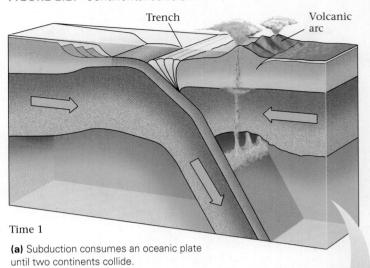

Trench

Volcanic arc

Time 1

(a) Subduction consumes an oceanic plate until two continents collide.

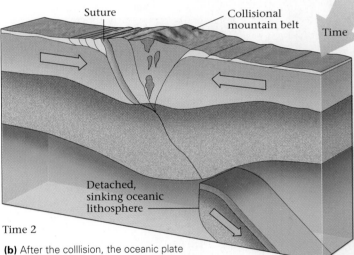

Suture

Collisional mountain belt

Time

Detached, sinking oceanic lithosphere

Time 2

(b) After the colllision, the oceanic plate detaches and sinks into the mantle. Rock caught in the collision zone gets broken, bent, and squashed, and forms a mountain range.

of convecting asthenosphere—beneath mid-ocean ridges. At first glance, this hypothesis looked pretty good, but on closer examination, it failed. Among other reasons, it is impossible to draw a global arrangement of convection cells that can explain the complex geometry of plate boundaries on Earth. Gradually, geoscientists came to the conclusion that convective flow within the asthenosphere occurs, but does not directly drive motion. In other words, hot asthenosphere does rise in some places and sink in others, probably because of temperature contrasts, and plate motion overall is a manifestation of this convection, but the local directions of this convective flow do not necessarily define the local directions of plate motion. Today, geoscientists favor the hypothesis that two forces—ridge-push force and slab-pull force—strongly influence the motion of individual plates.

Ridge-push force develops because mid-ocean ridges lie at a higher elevation than the adjacent abyssal plains of the ocean (Fig. 2.28a). To understand ridge-push force, imagine you have a glass containing a layer of water over a layer of honey. By tilting the glass momentarily and then returning it to its upright position, you can create a temporary slope in the boundary between these substances. While the boundary has this slope, gravity causes the elevated honey to push against the glass adjacent to the side where the honey surface lies at lower elevation. The geometry of a mid-ocean ridge resembles this situation, for the surface of the sea floor is higher along a mid-ocean ridge axis than in adjacent abyssal plains. Gravity causes the elevated lithosphere at the ridge axis to push on the lithosphere that lies farther from the axis, making it move away. As lithosphere moves away from the ridge axis, new hot asthenosphere rises to fill the gap. Note that the upward movement of asthenosphere beneath a mid-ocean ridge is a *consequence* of sea-floor spreading, not the cause.

Slab-pull force, the force that subducting, downgoing plates apply to oceanic lithosphere at a convergent margin, arises simply because lithosphere that was formed more than 10 million years ago is denser than asthenosphere, so it can sink into the

FIGURE 2.28 Forces driving plate motions. Both ridge push and slab pull make plates move.

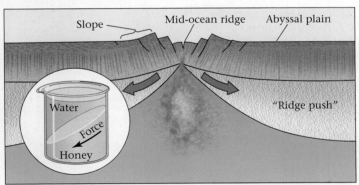

Slope Mid-ocean ridge Abyssal plain

Water Force Honey

"Ridge push"

(a) Ridge push develops because the region of a rift is elevated. Like a wedge of honey with a sloping surface, the mass of the ridge pushes sideways.

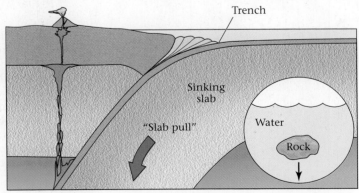

Trench

Sinking slab

"Slab pull" Water Rock

(b) Slab pull develops because lithosphere is denser than the underlying asthenosphere, and sinks like a stone in water (though much more slowly).

The Theory of Plate Tectonics

Hot-spot
volcano

Transform
plate
boundary

Volcanic arc

Trench

Continental rift

Convergent plate
boundary

Subducting oceanic
lithosphere

Collisional mountain belt

Continental crust

Continental lithosphere

Lithospheric mantle

Asthenosphere

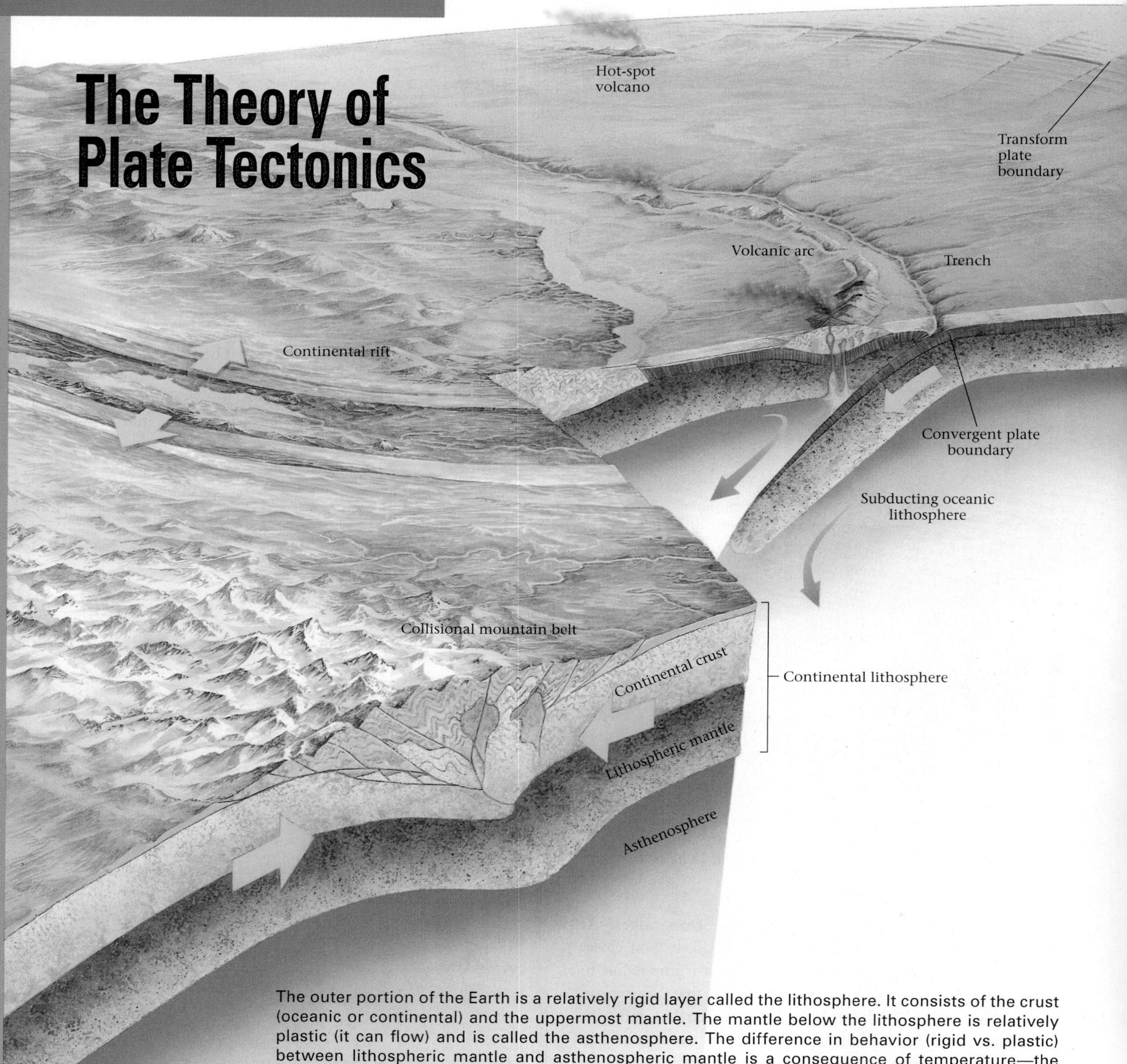

The outer portion of the Earth is a relatively rigid layer called the lithosphere. It consists of the crust (oceanic or continental) and the uppermost mantle. The mantle below the lithosphere is relatively plastic (it can flow) and is called the asthenosphere. The difference in behavior (rigid vs. plastic) between lithospheric mantle and asthenospheric mantle is a consequence of temperature—the former is cooler than the latter. Continental lithosphere is typically about 150 km thick, while oceanic lithosphere is about 100 km thick. (Note: they are not drawn to scale in this image.)

According to the theory of plate tectonics, the lithosphere is broken into about twenty plates that move relative to each other. Most of the motion is accommodated by sliding along plate boundaries (the edges of plates); plate interiors stay relatively unaffected by this motion. There are three kinds of plate boundaries.

1. Divergent boundaries: Here, two plates move apart by a process called sea-floor spreading. A mid-ocean ridge delineates a divergent boundary. Asthenospheric mantle rises beneath a mid-ocean

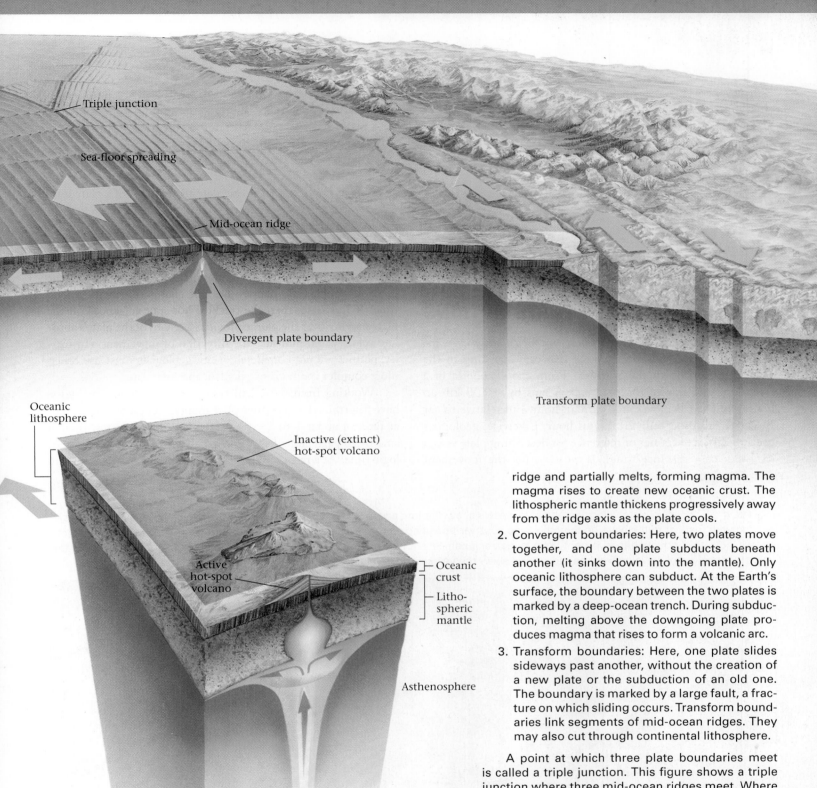

Triple junction

Sea-floor spreading

Mid-ocean ridge

Divergent plate boundary

Transform plate boundary

Oceanic
lithosphere

Inactive (extinct)
hot-spot volcano

Active
hot-spot
volcano

Oceanic
crust

Litho-
spheric
mantle

Asthenosphere

Mantle
plume

ridge and partially melts, forming magma. The magma rises to create new oceanic crust. The lithospheric mantle thickens progressively away from the ridge axis as the plate cools.

2. Convergent boundaries: Here, two plates move together, and one plate subducts beneath another (it sinks down into the mantle). Only oceanic lithosphere can subduct. At the Earth's surface, the boundary between the two plates is marked by a deep-ocean trench. During subduction, melting above the downgoing plate produces magma that rises to form a volcanic arc.

3. Transform boundaries: Here, one plate slides sideways past another, without the creation of a new plate or the subduction of an old one. The boundary is marked by a large fault, a fracture on which sliding occurs. Transform boundaries link segments of mid-ocean ridges. They may also cut through continental lithosphere.

A point at which three plate boundaries meet is called a triple junction. This figure shows a triple junction where three mid-ocean ridges meet. Where two continents collide, a collisional mountain belt forms. This happens because continental crust is too buoyant to be subducted. At a continental rift, a continent stretches and may break in two. Rifts are marked by the existence of many faults. If a continent breaks apart, a new mid-ocean ridge develops.

Hot-spot volcanoes form above plumes of hot mantle rock that rise from near the core-mantle boundary. As a plate drifts over a hot spot, it leaves a chain of extinct volcanoes.

asthenosphere (Fig. 2.28b). Thus, once an oceanic plate starts to sink, it gradually pulls the rest of the plate along behind it, like an anchor pulling down the anchor line. This "pull" is the slab-pull force.

TAKE-HOME MESSAGE

Though convective movement in the mantle may contribute to plate motion, it probably isn't the dominant force acting on a plate. The details of plate motions appear to be related to ridge-push and slab-pull forces.

2.14 THE VELOCITY OF PLATE MOTIONS

How fast do plates move? *It depends on your frame of reference.* To illustrate this concept, imagine two cars speeding in the same direction down the highway. From the viewpoint of a tree along the side of the road, car A zips by at 100 km an hour, while car B moves at 80 km an hour. But relative to car B, car A moves at only 20 km an hour. Likewise, geologists use two different frames of reference for describing plate velocity (velocity = distance/time). If we describe the movement

of plate A with respect to plate B, then we are talking about **relative plate velocity**. But if we describe the movement of both plates relative to a fixed point in the mantle, then we are speaking of **absolute plate velocity** (Fig. 2.29).

To determine relative plate motions, geoscientists measure the distance of a known magnetic anomaly from the axis of a mid-ocean ridge, and then calculate the velocity of a plate relative to the ridge axis by applying this equation: plate velocity = distance from the anomaly to the ridge axis divided by the age of the anomaly. The velocity of the plate on one side of the ridge relative to the plate on the other is twice this value.

To estimate absolute plate motions, we can *assume* that the position of a mantle plume does not change much for a long time. If this is so, then the track of hot-spot volcanoes on the plate moving over the plume provides a record of the plate's absolute velocity and indicates the direction of movement. (In reality, plumes are not completely fixed; geologists must use other, more complex methods to calculate absolute plate motions.)

Working from the calculations described above, geologists have determined that relative plate motions on Earth today occur at rates of about 1 to 15 cm per year. But these rates, though small, can yield large displacements given the immensity of geologic time. At a rate of 10 cm per year, a plate can move 100 km

FIGURE 2.29 Relative plate velocities: the blue arrows show the rate and direction at which the plate on one side of the boundary is moving with respect to the plate on the other side. Outward-pointing arrows indicate spreading (divergent boundaries), inward-pointing arrows indicate subduction (convergent boundaries), and parallel arrows show transform motion. The length of an arrow represents the velocity. Absolute plate velocities: the red arrows show the velocity of the plates with respect to a fixed point in the mantle.

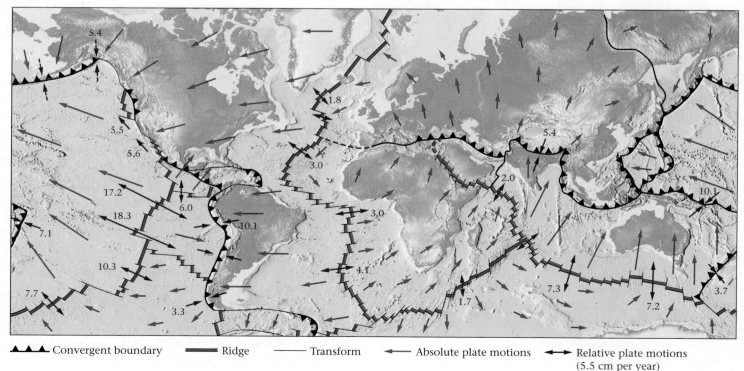

▲▲▲ Convergent boundary ━━━ Ridge ━━━ Transform ◄── Absolute plate motions ◄──► Relative plate motions (5.5 cm per year)

FIGURE 2.30 Due to plate tectonics, the map of Earth's surface slowly changes. Here we see the assembly, and later the breakup, of Pangaea during the past 400 million years.

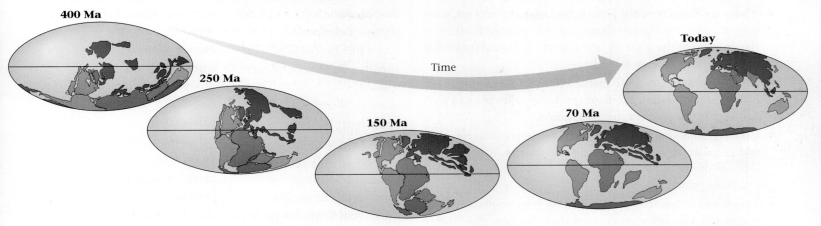

in a million years. Can we detect such slow rates? Until the last decade, the answer was no. Now the answer is yes, because of satellites orbiting the Earth with **global positioning system (GPS)** technology. Automobile drivers use GPS receivers to find their destinations, and geologists use them to monitor plate motions. If we calculate carefully enough, we can detect displacements of millimeters per year. In other words, we can now see the plates move!

Taking into account many data sources that define the motion of plates, geologists have greatly refined the image of continental drift that Wegener tried so hard to prove nearly a

century ago. We can now see how the map of our planet's surface has evolved during the past 400 million years (Fig. 2.30), and even before.

TAKE-HOME MESSAGE

Plates move at 1 to 15 cm/y, about the rate that your fingernails grow. We can describe the "relative motion" of one plate with respect to another, or the "absolute motion" of a plate with respect to the underlying asthenosphere. GPS can now detect plate motion.

Chapter Summary

- Alfred Wegener proposed that continents had once been joined together to form a single huge supercontinent (Pangaea) and had subsequently drifted apart. This idea is called continental drift.

- Wegener drew from several different sources of data to support his hypothesis: (1) the correlation of coastlines; (2) the distribution of late Paleozoic glaciers; (3) the distribution of late Paleozoic equatorial climatic belts; (4) the distribution of fossil species; (5) the match-up of distinctive rock assemblages that are now on opposite sides of the ocean were adjacent on Pangaea.

- Rocks retain a record of the Earth's magnetic field as it existed at the time the rocks formed. This record is called paleomagnetism. By measuring paleomagnetism in successively older rocks, geologists discovered apparent polar-wander paths.

- Apparent polar-wander paths are different for different continents. This observation can be explained by movement of

continents with respect to each other, while the Earth's magnetic poles remain roughly fixed.

- Around 1960, Harry Hess proposed the hypothesis of sea-floor spreading. According to this hypothesis, new sea floor forms at mid-ocean ridges, above a band of upwelling mantle, then spreads symmetrically away from the ridge axis. Eventually, the ocean floor sinks back into the mantle at deep-ocean trenches.

- Geologists documented that the Earth's magnetic field reverses polarity every now and then. The record of reversals is called the magnetic-reversal chronology.

- The proof of sea-floor spreading came from the interpretation of marine magnetic anomalies and drilling of the sea floor.

- The lithosphere, the rigid outer layer of the Earth, is broken into discrete plates that move relative to each other. Plates consist of the crust and the uppermost (cooler) mantle. Lithosphere plates effectively float on the underlying soft asthenosphere. Continental drift and sea-floor spreading are manifestations of plate movement.

- Most earthquakes and volcanoes occur along plate boundaries; the interiors of plates remain relatively rigid and intact.

- There are three types of plate boundaries—divergent, convergent, and transform—distinguished from each other by the movement the plate on one side of the boundary makes relative to the plate on the other side.

- Divergent boundaries are marked by mid-ocean ridges. At divergent boundaries, sea-floor spreading takes place, a process that forms new oceanic lithosphere.

- Convergent boundaries are marked by deep-ocean trenches and volcanic arcs. At convergent boundaries, oceanic lithosphere of the downgoing plate is subducted beneath an overriding plate.

- Transform boundaries are marked by large faults at which one plate slides sideways past another. No new plate forms and no old plate is consumed at a transform boundary.

- Triple junctions are points where three plate boundaries intersect.

- Hot spots are places where volcanism occurs at an isolated volcano. As a plate moves over the hot spot, the volcano moves off and dies, and a new volcano forms over the hot spot. Hot spots may be caused by mantle plumes.

- A large continent can split into two smaller ones by the process of rifting. During rifting, continental lithosphere stretches and thins. If it finally breaks apart, a new mid-ocean ridge forms and sea-floor spreading begins.

- Convergent plate boundaries cease to exist when a buoyant piece of crust (a continent or an island arc) moves into the subduction zone. When that happens, collision occurs.

- Ridge-push force and slab-pull force contribute to driving plate motions. Plates move at rates of about 1 to 15 cm per year. Modern satellite measurements can detect these motions.

Key Terms

absolute plate velocity (p. 64)	magnetic-reversal chronology (p. 44)
abyssal plain (p. 42)	
active margin (p. 48)	mantle plume (p. 57)
apparent polar-wander path (p. 39)	marine magnetic anomaly (p. 43)
asthenosphere (p. 47)	mid-ocean ridge (p. 42)
bathymetry (p. 40)	paleomagnetism (pp. 37, 39)
black smoker (p. 51)	paleopole (p. 39)
collision (p. 59)	Pangaea (p. 34)
continental drift (p. 34)	passive margin (p. 48)
continental rift (p. 59)	plate (p. 34)
convergent boundary (p. 49)	plate boundary (p. 47)
divergent boundary (p. 49)	plate tectonics (p. 34)
fracture zone (p. 42)	relative plate velocity (p. 64)
global positioning system (GPS) (p. 65)	ridge-push force (p. 61)
	rifting (p. 59)
hot spot (p. 56)	sea-floor spreading (p. 34)
hot-spot track (p. 57)	seamount (p. 42)
lithosphere (p. 47)	slab-pull force (p. 61)
lithosphere plate (p. 47)	subduction (p. 34)
magnetic anomaly (p. 43)	transform boundary (pp. 49, 56)
magnetic declination (p. 38)	
magnetic dipole (p. 38)	trench (p. 42)
magnetic inclination (p. 38)	triple junction (p. 56)
magnetic pole (p. 38)	volcanic arc (p. 42)
magnetic reversal (p. 44)	Wadati-Benioff zone (p. 54)

GEOPUZZLE REVISITED

Many lines of evidence indicate that the position of continents indeed changes over time, and thus that the map of the Earth's surface is not fixed. Not only do coastlines match, but the observed distribution of rock units, fossils, and climate belts, and evidence of past glaciations all point to the occurrence of continental drift. Drift occurs because ocean basins grow wider by the process of sea-floor spreading, or get narrower by the process of subduction. The documentation and interpretation of marine magnetic anomalies proved that sea-floor spreading does happen.

Review Questions

1. What was Wegener's continental drift hypothesis? What was his evidence?

2. How do apparent polar-wander paths show that the continents, rather than the poles, have moved?

3. Describe the basic characteristics of mid-ocean ridges, deep-ocean trenches, and seamount chains.

4. Describe the hypothesis of sea-floor spreading.

5. How did the observations of heat flow and seismicity support the hypothesis of sea-floor spreading?

6. What is a marine magnetic anomaly? How is it detected?

7. Describe the pattern of marine magnetic anomalies across a mid-ocean ridge. How is this pattern explained?

8. Did drilling into the sea floor contribute further proof of sea-floor spreading? If so, how?

9. What are the characteristics of a lithosphere plate?

10. How does oceanic lithosphere differ from continental lithosphere in thickness, composition, and density?

11. What are the basic premises of plate tectonics?

12. How do we identify a plate boundary?

13. Describe the three types of plate boundaries.

14. How does crust form along a mid-ocean ridge?

15. Why is subduction necessary on a nonexpanding Earth with spreading ridges?

16. Describe the major features of a convergent boundary.

17. Why are transform plate boundaries required on an Earth with spreading and subducting plate boundaries?

18. What is a triple junction?

19. How is a hot-spot track produced, and how can hot-spot tracks be used to track the past motions of a plate?

20. Describe the characteristics of a contintental rift, and give examples of where this process is occurring today.

21. Describe the process of continental collision, and give examples of where this process has occurred.

22. Discuss the major forces that move lithosphere plates.

23. Explain the difference between relative plate velocity and absolute plate velocity.

On Further Thought

1. Why are the marine magnetic anomalies bordering the East Pacific Rise in the southeastern Pacific Ocean wider than those bordering the Mid-Atlantic Ridge in the South Atlantic Ocean?

2. The Pacific Plate moves north relative to the North American Plate at a rate of 6 cm per year. How long will it take Los Angeles (a city on the Pacific Plate) to move northwards by 480 km, the present distance between Los Angeles and San Francisco?

3. Look at a map of the western Pacific Ocean, and examine the position of Japan with respect to mainland Asia. Japan's older crust contains rocks similar to those of eastern Asia. Presently, there are many active volcanoes along the length of Japan. With these facts in mind, explain how the Japan Sea (the region between Japan and the mainland) formed.

THE VIEW FROM SPACE This image was produced by Christoph Hormann, using computer rendering techniques. It shows the Caucasus Mountains between the Black Sea and the Caspian Sea. This range is forming due to the collision between two continental masses.

CHAPTER 3

Patterns in Nature: Minerals

This photo is real, not a computer collage! We're seeing the world's largest known mineral crystals jutting from the walls of a cave in Chihuahua, Mexico. The crystals are of the mineral gypsum; they formed by precipitation from a water solution.

GEOPUZZLE

In the game of *Twenty Questions,* you try to guess the identity of an object that your friend is thinking about and start by asking, "Is it animal, vegetable, or mineral?" Do geologists consider everything on Earth that is not "animal" or "vegetable" to be "mineral?"

3.1 INTRODUCTION

Zabargad Island rises barren and brown above the Red Sea, about 70 km off the coast of southern Egypt. Nothing grows on Zabargad, except for scruffy grass and a few shrubs, so no one lives there now. But in ancient times many workers toiled on this 5-square-km patch of desert, gradually chipping their way into the side of its highest hill. They were searching for glassy green, pea-sized pieces of peridot, a prized gem. Carefully polished peridots were worn as jewelry by ancient Egyptians. Eventually, some of the gems appeared in Europe, set into crowns and scepters (Fig. 3.1). These peridots now glitter behind glass cases in museums, millennia after first being pried free from the Earth, and perhaps 10 million years after first being formed by the bonding together of still more ancient atoms.

Peridot is one of about 4,000 minerals that have been identified on Earth so far, and fascinate collectors and geologists alike. Mineralogists, people who specialize in the study of minerals, discover fifty to one hundred new minerals every year. Each different mineral has a name. Some names come from Latin, Greek, German, or English words describing a certain characteristic; some honor a person; some indicate the place where the mineral was first recognized; and some reflect a particular element in the mineral. Several minerals have more than one name—for example, peridot is the gem-quality version of

FIGURE 3.1 A royal crown studded with jewels. The gemstone near the base is a green peridot.

FIGURE 3.2 Copper ore is a useful mineral that serves as a source of copper metal.

(a) Malachite is a type of copper ore ($Cu_2[CO_3][OH]_2$); it contains copper plus other chemicals.

(b) The copper for pots is produced by processing ore minerals.

olivine, a common mineral. Although the vast majority of mineral types are rare, forming only under special conditions, many are quite common and are found in a variety of rock types.

Why study minerals? Without exaggeration, we can say that *minerals are the building blocks of our planet*. To a geologist, almost any study of Earth materials depends on an understanding of minerals, for minerals make up the rocks and sediments that make up the Earth and its landscapes. Minerals are also important from a practical standpoint (see Chapter 12). Industrial minerals serve as the raw materials for manufacturing chemicals, concrete, and wallboard. Ore minerals are the source of valuable metals like copper and gold and provide energy resources like uranium (Fig. 3.2a, b). And certain forms of minerals—gems—delight the eye as jewelry. Unfortunately, though, some minerals pose environmental hazards. No wonder **mineralogy**, the study of minerals, fascinates professionals and amateurs alike.

In this chapter, we begin by presenting the geologic definition of a mineral, then look at how minerals form and at the main characteristics that enable us to identify them. Finally, we note the basic scheme that geologists use to classify minerals. This chapter assumes that you understand the fundamental concepts of matter and energy, especially the nature of atoms, molecules, and chemical bonds. If you are rusty on these topics, please review the Appendix. We summarize basic terms from chemistry in **Box 3.1**, for convenience.

TAKE-HOME MESSAGE

When you finish this chapter, you should understand what minerals are, how they form, and how they are classified. You should also know how to identify a mineral, and recognize the difference between gems and ordinary mineral specimens.

3.2 WHAT IS A MINERAL?

To a geologist, a **mineral** is a naurally occurring solid, formed by geologic processes, that has a crystalline structure and a definable chemical composition. Almost all minerals are inorganic. Let's pull apart this mouthful of a definition and examine its meaning in detail.

- *Naturally occurring*: Minerals are produced in nature, not in factories. We need to emphasize this point because in recent decades, industrial chemists have learned how to synthesize materials that have characteristics virtually identical to those of real minerals. These materials are not minerals in a geologic sense, though they are referred to in the commercial world as "synthetic minerals."

- *Solid*: A solid is a state of matter that can maintain its shape indefinitely, and thus will not conform to the shape of its container (see Appendix). Liquids (such as oil or water) and gases (such as air) cannot be minerals.

- *Crystalline structure*: The atoms that make up a mineral are not distributed randomly and cannot move around easily. Rather, they are fixed in a specific, orderly pattern. A material in which atoms are fixed in an orderly pattern is called a crystalline solid. Mineralogists refer to the pattern itself (the imaginary framework representing the arrangement of atoms) as a **crystal lattice** (Fig. 3.3a).

- *Definable chemical composition*: This simply means that it is possible to write a chemical formula for a mineral (Box 3.1). Some minerals contain only one element, but most

FIGURE 3.3 The nature of crystalline and noncrystalline materials.

A crystal face

Quartz crystal

Scaffolding

Washington Monument

(a) This quartz crystal contains an orderly arrangement of atoms. The arrangement resembles scaffolding.

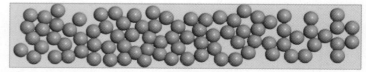

(b) Atoms in noncrystalline solids, such as glass, are not orderly.

are compounds of two or more elements. For example, diamond and graphite have the formula C, because they consist entirely of carbon. Quartz has the formula SiO_2—it contains the elements silicon and oxygen in the proportion of one silicon atom for every two oxygen atoms. Some formulas are more complicated: for example, the formula for biotite is $K(Mg,Fe)_3(AlSi_3O_{10})(OH)_2$.

- *Inorganic*: We must first distinguish between organic and inorganic chemicals. Organic chemicals are molecules containing carbon-hydrogen bonds. Although some organic chemicals contain only carbon and hydrogen, others also contain oxygen, nitrogen, and other elements in varying quantities. Sugar ($C_{12}H_{22}O_{11}$), for example, is an organic chemical. *Almost all minerals are inorganic.* Thus, sugar and protein are not minerals. But, we have to add the qualifier "almost all" because mineralogists do consider about thirty organic substances formed by "the action of geologic processes on organic materials" to be minerals. Examples include the crystals that grow in ancient deposits of bat guano.

With these definitions in mind, we can make an important distinction between minerals and glasses. Both minerals and glasses are solids, in that they can retain their shape indefinitely. But a mineral is crystalline, and glass is not. Whereas atoms, ions, or molecules in a mineral are ordered into a crystal lattice, like soldiers standing in formation, those in a glass are arranged in a semichaotic way, like people at a party, in small clusters or chains that are neither oriented in the same way nor spaced at regular intervals (Fig. 3.3b).

If you ever need to figure out whether a substance is a mineral or not, just check it against the criteria listed above. Is motor oil a mineral? No—it's a liquid. Is table salt a mineral? Yes—it's a solid crystalline compound with the formula NaCl.

TAKE-HOME MESSAGE

For a substance to be a mineral, it must meet several criteria: it must have an orderly arrangement of atoms inside, it must have a definable chemical formula, it must be solid, it must occur in nature, and it must have been formed by geologic processes.

BOX 3.1

SCIENCE TOOLBOX

Some Basic Definitions from Chemistry

To describe minerals, we need to use several terms from chemistry (for a more in-depth discussion, see the Appendix). To avoid confusion, terms are listed in an order that permits each successive term to utilize previous terms.

- **Element:** a pure substance that cannot be separated into other elements.
- **Atom:** the smallest piece of an element that retains the characteristics of the element. An atom consists of a nucleus surrounded by a cloud of orbiting electrons; the nucleus is made up of protons and neutrons (except in hydrogen, whose nucleus contains only one proton and no neutrons). Electrons have a negative charge, protons have a positive charge, and neutrons have a neutral charge. An atom that has the same number of electrons as protons is said to be neutral, in that it does not have an overall electrical charge.
- **Atomic number:** the number of protons in an atom of an element.
- **Atomic weight:** approximately the number of protons plus neutrons in an atom of an element.
- **Ion:** an atom that is not neutral. An ion that has an excess negative charge (because it has more electrons than protons) is an anion, whereas an ion

that has an excess positive charge (because it has more protons than electrons) is a cation. We indicate the charge with a superscript. For example, Cl^- has a single excess electron; Fe^{2+} is missing two electrons.

- **Chemical bond:** an attractive force that holds two or more atoms together. For example, *covalent bonds* form when atoms share electrons. *Ionic bonds* form when a cation and anion (ions with opposite charges) get close together and attract each other. In materials with metallic bonds, some of the electrons can move freely.
- **Molecule:** two or more atoms bonded together. The atoms may be of the same element or of different elements.
- **Compound:** a pure substance that can be subdivided into two or more elements. The smallest piece of a compound that retains the characteristics of the compound is a molecule.
- **Chemical:** a general name used for a pure substance (either an element or a compound).
- **Chemical formula:** a shorthand recipe that itemizes the various elements in a chemical and specifies their relative proportions. For example, the formula for water, H_2O, indicates that water consists of molecules in which two hydrogens bond to one oxygen.

- **Chemical reaction:** a process that involves the breaking or forming of chemical bonds. Chemical reactions can break molecules apart or create new molecules and/or isolated atoms.
- **Mixture:** a combination of two or more elements or compounds that can be separated without a chemical reaction. For example, a cereal composed of bran flakes and raisins is a mixture—you can separate the raisins from the flakes without destroying either.
- **Solution:** a type of material in which one chemical (the solute) dissolves (becomes completely incorporated) in another (the solvent). In solutions, a solute may separate into ions during the process. For example, when salt (NaCl) dissolves in water, it separates into sodium (Na^+) and chloride (Cl^-) ions. In a solution, atoms or molecules of the solvent surround atoms, ions, or molecules of the solute.
- **Precipitate:** (noun) a compound that forms when ions in liquid solution join together to create a solid that settles out of the solution; (verb) the process of forming solid grains by separation and settling from a solution. For example, when saltwater evaporates, solid salt crystals precipitate and settle to the bottom of the remaining water.

3.3 BEAUTY IN PATTERNS: CRYSTALS AND THEIR STRUCTURE

What Is a Crystal?

The word crystal brings to mind sparkling chandeliers, elegant wine goblets, and shiny jewels. But, as is the case with the word mineral, geologists have a more precise definition. A **crystal** is a single, continuous (that is, uninterrupted) piece of a crystalline solid, typically bounded by flat surfaces called **crystal faces** that grow naturally as the mineral forms. The word comes from the Greek *krystallos*, meaning ice. Many crystals have beautiful shapes that look like they belong in the pages of a geometry book. The angle between two adjacent crystal faces of one specimen is identical to the angle between the corresponding faces of another specimen. For example, a perfectly formed quartz crystal looks like an obelisk (Fig. 3.4a). The angle between the faces of the columnar part of a quartz crystal is exactly 120°. This rule holds regardless of whether the whole crystal is big or small and regardless of whether all of the faces are the same size (Fig. 3.4a). Crystals come in a great variety of shapes, including cubes, trapezoids, pyramids, octahedrons, hexagonal columns, blades, needles, columns, and obelisks (Fig. 3.4b).

Because crystals have a regular geometric form, people have always considered crystals to be special, perhaps even a source of magical powers. For example, shamans of some cultures relied on talismans or amulets made of crystals, which supposedly brought power to their wearer or warded off evil spirits. Scientists have concluded, however, that crystals have no effect on health or mood. For millennia, crystals have inspired awe because of the way they sparkle, but such behavior is simply a consequence of how crystal structures interact with light.

What's Inside a Crystal?

What do the insides of a mineral actually look like? We can picture atoms in minerals as tiny balls packed together tightly and held in place by chemical bonds. The way in which atoms are packed defines the **crystal structure** of the mineral. To illustrate crystal structures, we look at a few examples. Halite (rock salt) consists of oppositely charged ions that stick together because opposite charges attract. In halite, six chloride (Cl^-) ions surround each sodium (Na^+) ion, producing an overall arrangement of atoms that defines the shape of a cube (Fig. 3.5a, b). Diamond, by contrast, is a mineral made entirely of carbon. In diamond, each atom bonds to four neighbors arranged in the form of a tetrahedron, so some naturally formed diamond crystals have the shape of a double tetrahedron (Fig. 3.5c). Graphite, another mineral composed entirely of carbon, behaves very differently from diamond. In contrast to diamond, graphite is so soft that it can be used as the "lead"

FIGURE 3.4 Some characteristics of crystals.

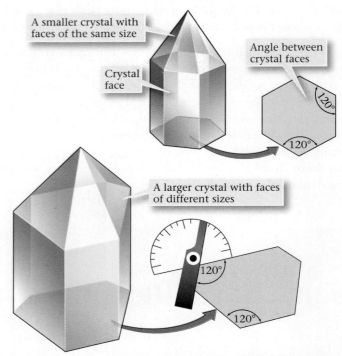

A smaller crystal with faces of the same size

Crystal face

Angle between crystal faces

A larger crystal with faces of different sizes

(a) For a given mineral, the angle between adjacent crystal faces in one specimen is the same as the angle between corresponding faces in another, regardless of specimen size.

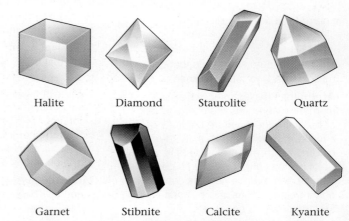

Halite   Diamond   Staurolite   Quartz

Garnet   Stibnite   Calcite   Kyanite

(b) Crystals come in a variety of shapes, including cubes, prisms, blades, and pyramids. Some terminate at a point and some terminate with flat surfaces.

in a pencil; as you move a pencil across paper, tiny flakes of graphite peel off the pencil point and adhere to the paper. This behavior occurs because the carbon atoms in graphite are not arranged in tetrahedra, but rather occur in sheets (Fig. 3.5d). The sheets are bonded to each other by weak bonds and thus can separate from each other easily. Two different minerals (such as diamond and graphite) that have the same composition but different crystal structures are **polymorphs**.

The orderly arrangement of atoms inside a crystal—its crystal structure—provides one of nature's most spectacular examples of a pattern. The pattern on wallpaper may be defined by the regular spacing of clumps of flowers. Similarly, the pattern in a crystal is defined by the regular spacing of atoms (**Fig. 3.6a, b**). If the crystal contains more than one type of atom, the atoms alternate in a regular way. The orderly arrangement controls the outward shape of crystals. For example, if the atoms are packed into the shape of a cube, a crystal of the mineral will have faces that intersect at 90° angles.

FIGURE 3.5 The nature of crystalline structure in minerals.

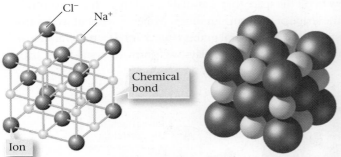

(a) In a ball-and-stick model of halite, the balls are ions, and the sticks are chemical bonds.

(b) This ball model gives a better sense of how ions pack together in crystal.

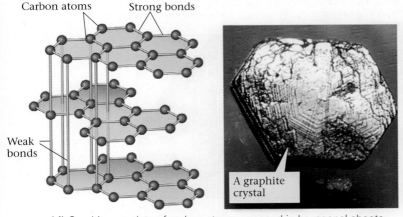

(c) In a diamond, carbon atoms are arranged in tetrahedra. All of the bonds are strong .

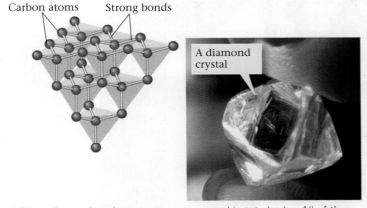

(d) Graphite consists of carbon atoms arranged in hexagonal sheets. The sheets are connected by weak bonds.

FIGURE 3.6 Patterns and symmetry in minerals.

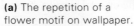

(a) The repetition of a flower motif on wallpaper.

(b) The repetition of alternating sulfur and lead atoms in the mineral galena (PbS).

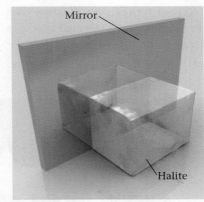

(c) Minerals display symmetry. One half of a halite crystal is a mirror image of the other. Snowflakes are symmetrical hexagons.

The pattern of atoms or ions in a mineral displays **symmetry**, meaning that the shape of one part of a mineral is a mirror image of the shape of another part. For example, if you were to cut a halite crystal in half and place one half against a mirror, the crystal would look whole again (**Fig. 3.6c**).

The Formation and Destruction of Minerals

New mineral crystals can form in five ways. First, they can form by the solidification of a melt, meaning the freezing of a liquid. For example, ice crystals, a type of mineral, are made by freezing water. Second, they can form by precipitation from a solution, meaning that atoms, molecules, or ions dissolved in water bond together and separate out of the water. Salt crystals, for example, develop when you evaporate saltwater. Third, they can form by solid-state diffusion, the movement of atoms or ions through a solid to arrange into a new crystal structure, a process that takes place very slowly. For example, garnets grow by diffusion in solid rock. Fourth, minerals can form at interfaces between the physical and biological components of the Earth System by a process called biomineralization. This occurs when living organisms cause minerals to precipitate either within or on their bodies, or immediately adjacent to their bodies. For example, clams and other shelled organisms extract ions from water to produce mineral shells. Fifth, minerals can precipitate directly from a gas. This process typically occurs around volcanic vents or around geysers, for at such locations volcanic gases or steam enter the atmosphere and cool abruptly. Some of the bright yellow sulfur deposits found in volcanic regions form in this way.

The first step in forming a crystal is the chance formation of a seed, or an extremely small crystal (**Fig. 3.7a**). Once the seed exists, other atoms in the surrounding material attach themselves to the face of the seed. As the crystal grows, crystal faces move outward but maintain the same orientation (**Fig. 3.7b**). The youngest part of the crystal is at its outer edge.

In the case of crystals formed by the solidification of a melt, atoms begin to attach to the seed when the melt becomes so cool that thermal vibrations can no longer break apart the attraction between the seed and the atoms in the melt. Crystals formed by precipitation from a solution develop when the solution becomes saturated, meaning the number of dissolved ions per unit volume of solution becomes so great that they can get close enough to each other to bond together.

As crystals grow, they develop their particular crystal shape, based on the geometry of their internal structure. The shape is defined by the relative dimensions of the crystal (needle-like, sheet-like, etc.) and the angles between crystal faces. Typically, the growth of minerals is restricted in one or more directions, because existing crystals act as obstacles. In such cases, minerals grow to fill the space that is available, and their shape is controlled by the shape of their surroundings. Minerals without well-formed crystal faces are anhedral grains (**Fig. 3.7c**). If a mineral's growth is uninhibited so that it displays well-formed crystal faces, then it

FIGURE 3.7 The growth of crystals.

Ions attach to the crystal face.

(a) New crystals nucleate and begin to precipitate out of a water solution. As time progresses, they grow into the open space.

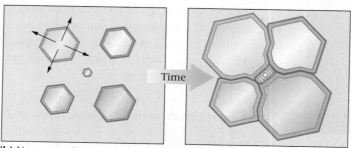

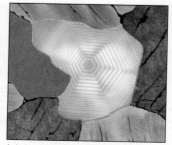

(b) New crystals grow outward from the central seed. As time passes, they maintain their shape until they interfere with each other.

(c) A crystal growing in a confined space will be anhedral.

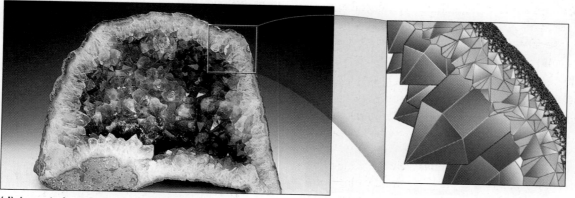

(d) A geode from Brazil consists of purple quartz crystals (amethyst) that grew from the wall into the center. The enlargement sketch indicates that the crystals are euhedral.

is a euhedral crystal. The surface crystals of a **geode**, a mineral-lined cavity in rock, may be euhedral (Fig. 3.7d).

A mineral can be destroyed by melting, dissolving, or some other chemical reaction. Melting involves heating a mineral to a temperature at which thermal vibration of the atoms or ions in the lattice break the chemical bonds holding them to the lattice. The atoms or ions then separate, either individually or in small groups, to move around again freely. Dissolution occurs when you immerse a mineral in a solvent, such as water. Atoms or ions then separate from the crystal face and are surrounded by solvent molecules. Chemical reactions can destroy a mineral when it comes in contact with reactive materials. For example, iron-bearing minerals react with air and water to form rust. The action of microbes in the environment can also destroy minerals. In effect, microbes "eat" certain minerals, using the energy stored in the chemical bonds that hold the atoms of the mineral together as their source of energy for metabolism.

TAKE-HOME MESSAGE

In a crystalline material, atoms are arranged in a very regular pattern. Minerals are crystalline materials that form by several geographic processes, including solidification from a melt, precipitation from a solution, and diffusion in solids.

3.4 HOW CAN YOU TELL ONE MINERAL FROM ANOTHER?

Amateur and professional geologists alike get a kick out of recognizing minerals. They might hover around a display case in a museum and name specimens without bothering to look at the labels. How do they do it? The trick lies in learning to recognize the basic physical properties (visual and material characteristics) that distinguish one mineral from another. Some physical properties, such as shape and color, can be seen from a distance. Others, such as hardness and magnetization, can be determined only by handling the specimen or by performing an identification test on it. Identification tests include scratching the mineral against another object, placing it near a magnet, weighing it, tasting it, or placing a drop of acid on it. Let's examine some of the physical properties most commonly used in mineral identification.

- *Color*: Color results from the way a mineral interacts with light. Sunlight contains the whole spectrum of colors; each color has a different wavelength. A mineral absorbs certain wavelengths, so the color you see when looking at a specimen represents the wavelengths the mineral does not absorb. Certain minerals always have the same color, but many show a range of colors (Fig. 3.8a). Color variations in a mineral reflect the presence of impurities.

- *Streak*: The **streak** of a mineral refers to the color of a powder produced by pulverizing the mineral. You can obtain a streak by scraping the mineral against an unglazed ceramic plate (Fig. 3.8b). The color of a mineral powder tends to be less variable than the color of a whole crystal, and thus provides a fairly reliable clue to a mineral's identity. Calcite, for example, always yields a white streak even though pieces of calcite may be white, pink, or clear.

- *Luster*: **Luster** refers to the way a mineral surface scatters light. Geoscientists describe luster simply by comparing the appearance of the mineral with the appearance of a familiar substance. For example, minerals that look like metal have a metallic luster, whereas those that do not have a nonmetallic luster—the adjectives are self-explanatory (Fig. 3.8c, d). Terms used for types of nonmetallic luster include silky, glassy, satiny, resinous, pearly, or earthy.

- *Hardness*: **Hardness** is a measure of the relative ability of a mineral to resist scratching, and therefore represents the resistance of bonds in the crystal structure to being broken. The atoms or ions in crystals of a hard mineral are more strongly bonded than those in a soft mineral. Hard minerals can scratch soft minerals, but soft minerals cannot scratch hard ones. Diamond, the hardest mineral known, can scratch anything, which is why it is used to cut glass. In the early 1800s, a mineralogist named Friedrich Mohs listed some minerals in sequence of relative hardness; a mineral with a hardness of 5 can scratch all minerals with a hardness of 5 or less. This list, the **Mohs hardness scale**, helps in mineral identification. When you use the scale (Table 3.1), it helps to compare the hardness of a mineral with a common item such as your fingernail, a penny, or a glass plate.

- *Specific gravity*: **Specific gravity** represents the density of a mineral, as specified by the ratio between the weight of a volume of the mineral and the weight of an equal volume of water at 4°C. For example, one cubic centimeter of quartz has a weight of 2.65 grams, whereas one cubic centimeter of water has a weight of 1.00 gram. Thus, the specific gravity of quartz is 2.65. In practice, you can develop a feel for specific gravity by hefting minerals in your hands. A piece of galena (lead ore) "feels" heavier than a similar-sized piece of quartz.

- *Crystal habit*: The **crystal habit** of a mineral refers to the shape of a single crystal with well-formed crystal faces, or to the character of an aggregate of many well-formed crystals that grew together as a group (Fig. 3.8e). The habit depends on the internal arrangement of atoms in the crystal. A description of habit generally includes adjectives that define relative dimensions of the crystal and the geometric shape of the crystal.

FIGURE 3.8 Physical characteristics of minerals.

(a) Color is diagnostic of some minerals, but not all. For example, quartz can come in many colors.

(b) To obtain the streak of a mineral, rub it against a porcelain plate. The streak consists of mineral powder.

(c) Pyrite has a metallic luster because it gleams like metal.

(d) Feldspar has a nonmetallic luster.

(e) Crystal habit refers to the shape or character of the crystal. The specimen on the left is bladed, and the one on the right is fibrous.

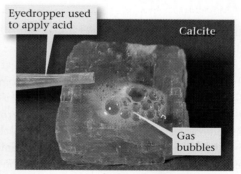

(f) Calcite reacts with hydrochloric acid to produce carbon dioxide gas.

(g) Magnetite is magnetic.

TABLE 3.1 Mohs hardness scale. Mohs' numbers are relative—in reality, diamond is 3.5x harder than corundum, as the graph shows.

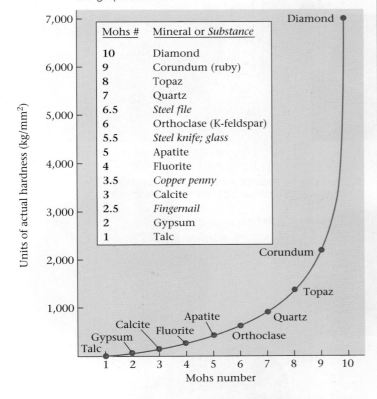

Mohs #	Mineral or *Substance*
10	Diamond
9	Corundum (ruby)
8	Topaz
7	Quartz
6.5	*Steel file*
6	Orthoclase (K-feldspar)
5.5	*Steel knife; glass*
5	Apatite
4	Fluorite
3.5	*Copper penny*
3	Calcite
2.5	*Fingernail*
2	Gypsum
1	Talc

For example, crystals that are roughly the same length in all directions are called equant or blocky, those that are much longer in one dimension than in others are columnar or needle-like, those shaped like sheets of paper are platy, and those shaped like knives are bladed.

- *Special properties:* Some minerals have distinctive properties that readily distinguish them from other minerals. For example, calcite ($CaCO_3$) reacts with dilute hydrochloric acid (HCl) to produce carbon dioxide (CO_2) gas (**Fig. 3.8f**). Dolomite ($CaMg[CO_3]_2$) also reacts with acid, but not as strongly. Graphite makes a gray mark on paper (it's the "lead" in pencils), magnetite attracts a magnet (**Fig. 3.8g**), halite tastes salty, and plagioclase has striations (thin parallel corrugations or stripes) on its surface.

- *Fracture and cleavage:* Different minerals fracture (break) in different ways, depending on the internal arrangement of atoms. If a mineral breaks to form distinct planar surfaces that have a specific orientation in relation to the crystal structure, then we say that the mineral has **cleavage** and we refer to each surface as a cleavage plane. Cleavage forms in directions where the bonds holding atoms together in the crystal are the weakest (**Fig. 3.9a–e**). Some minerals have one direction of cleavage. For exam-

ple, mica has very weak bonds in one direction but strong bonds in the other two directions. Thus, it easily splits into parallel sheets; the surface of each sheet is a cleavage plane. Other minerals have two or three directions of cleavage that intersect at a specific angle. For example, halite has three sets of cleavage planes that intersect at right angles, so halite crystals break into little cubes. Materials that have no cleavage at all (because bonding is equally strong in all directions) break either by forming irregular fractures or by forming conchoidal fractures (**Fig. 3.9f**). **Conchoidal fractures** are smoothly curving, clamshell-shaped surfaces; they typically form in glass. Cleavage planes are sometimes hard to distinguish from crystal faces (**Fig. 3.9g**).

TAKE-HOME MESSAGE

The physical characteristics of minerals (such as color, crystal shape, hardness, cleavage, and luster) are a manifestation of the crystal structure and chemical composition of minerals. You can identify a mineral by examining these characteristics.

3.5 ORGANIZING OUR KNOWLEDGE: MINERAL CLASSIFICATION

Minerals can be separated into a small number of groups, or mineral classes. You may think, "Why bother?" Classification schemes are useful because they help organize information and streamline discussion. Biologists, for example, classify animals into groups based on how they feed their young and on the architecture of their skeletons, and botanists classify plants according to the way they reproduce and by the shape of their leaves. In the case of minerals, a good means of classification eluded researchers until it became possible to determine the chemical makeup of minerals. A Swedish chemist, Baron Jöns Jacob Berzelius (1779–1848), analyzed minerals and noted chemical similarities among many of them. Berzelius, along with his students, then established that most minerals can be *classified by specifying the principal anion* (negative ion) or anionic group (negative molecule) within the mineral. We now take a look at these mineral classes, focusing especially on silicates, the class that constitutes most of the rock in the Earth.

The Mineral Classes

Mineralogists distinguish several principal classes of minerals. Here are some of the major ones.

- *Silicates:* The fundamental component of most **silicates** in the Earth's crust is the SiO_4^{4-} anionic group. In the

FIGURE 3.9 The nature of mineral cleavage and fracture.

(a) Mica has one strong plane of cleavage and splits into sheets.

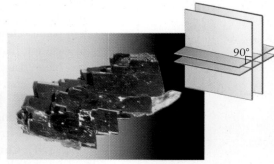

(b) Pyroxene has two planes of cleavage that intersect at 90°.

(c) Amphibole has two planes that intersect at 60°.

Halite breaks into cubes.

(d) Halite has three mutually perpendicular planes of cleavage.

Calcite breaks into rhombs.

(e) Calcite has three planes of cleavage, one of which is inclined.

Irregular fracture

Garnet

Quartz

Conchoidal fracture

(f) Minerals without cleavage can develop irregular or conchoidal fractures.

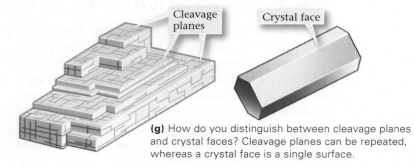

Cleavage planes

Crystal face

(g) How do you distinguish between cleavage planes and crystal faces? Cleavage planes can be repeated, whereas a crystal face is a single surface.

SiO_4^{4-} group, four oxygen atoms surround a single silicon atom. We will learn more about silicates in the next section.

- *Oxides*: Oxides consist of metal cations bonded to oxygen anions. Typical oxide minerals include hematite (Fe_2O_3; Fig. 3.8b) and magnetite (Fe_3O_4; Fig. 3.8g).

- *Sulfides*: Sulfides consist of a metal cation bonded to a sulfide anion (S^{2-}). Examples include galena (PbS) and pyrite (FeS_2; Fig. 3.8c).

- *Sulfates*: Sulfates consist of a metal cation bonded to the SO_4^{2-} anionic group. Many sulfates form by precipitation out of water at or near the Earth's surface. An example is gypsum ($CaSO_4 \cdot 2H_2O$).

- *Halides*: The anion in a halide is a halogen ion (such as chloride [Cl^-] or fluoride [F^-]), an element from the second column from the right in the periodic table (see Appendix). Halite, or rock salt (NaCl; Fig. 3.9d), and fluorite (CaF_2), a source of fluoride, are common examples.

- *Carbonates*: In **carbonates**, the molecule CO_3^{2-} serves as the anionic group. Elements such as calcium or magnesium bond to this group. The two most common carbonates are calcite ($CaCO_3$; Fig. 3.9e) and dolomite ($CaMg[CO_3]_2$).

- *Native metals*: Native metals consist of pure masses of a single metal. The metal atoms are bonded by metallic bonds. Copper and gold, for example, may occur as native metals. A gold nugget is a mass of native gold that has been broken out of a rock.

Silicates: The Major Rock-Forming Minerals

Silicate minerals make up over 95% of the continental crust. Rocks of the oceanic crust and the Earth's mantle consist almost entirely of silicate minerals. Thus, silicate minerals are the most common minerals on Earth.

Most silicate minerals in the crust consist of combinations of a fundamental building block, the SiO_4^{4-} anionic group, better known as the **silicon-oxygen tetrahedron**, in which oxygen atoms are positioned in the corners of a tetrahedron, a pyramid-like shape with four triangular faces (**Fig. 3.10a**).

FIGURE 3.10 The structure of silicate minerals.

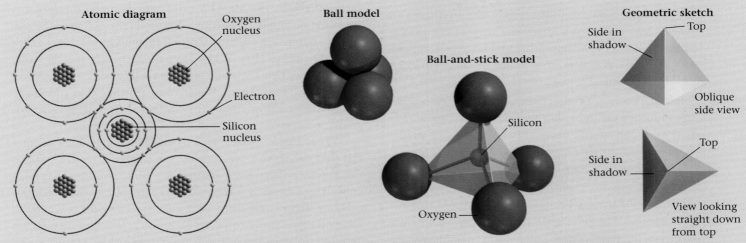

(a) The fundamental building block of a silicate mineral is the silicon-oxygen tetrahedron. Oxygens occupy the corners of the tetrahedron, and silicon lies at the center. Geologists portray the tetrahedron in a number of different ways.

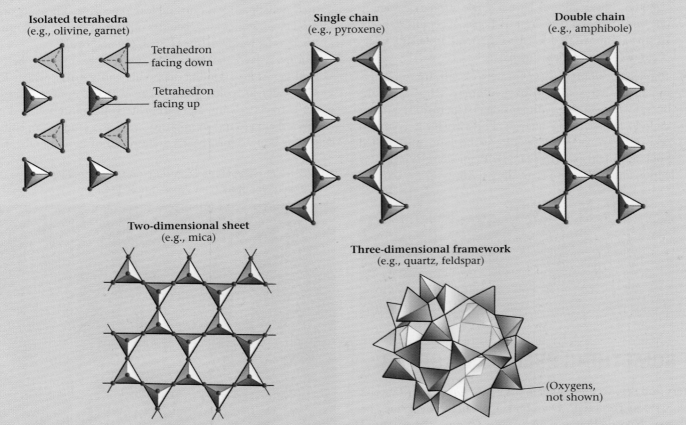

(b) The classes of silicate minerals differ from one another by the way in which the silicon-oxygen tetrahedra are linked. Where the tetrahedra link, they share an oxygen atom. Oxygen atoms are shown in red. Positive ions (not shown) occupy spaces between tetrahedra.

Silica tetrahedra can link together, forming larger molecules, by sharing oxygen atoms. Silicate minerals are divided into seven groups, of which five are described below. Groups are distinguished from each other on the basis of how silicon-oxygen tetrahedra in the mineral link together (**Fig. 3.10b**). The number of links determines how many oxygen atoms are shared between tetrahedra and, therefore, the ratio of silicon (Si) to oxygen (O) in the mineral.

- *Independent tetrahedra*: In this group, the tetrahedra are independent and do not share any oxygen atoms. The attraction between the tetrahedra and positive ions holds such minerals together. This group includes olivine, a glassy green mineral, and garnet (Fig. 3.1; Fig. 3.9f).
- *Single chains*: In a single-chain silicate, the tetrahedra link to form a chain by sharing two oxygen atoms. The most common of the many different types of single-chain silicates are pyroxenes (Fig. 3.9b).
- *Double chains*: In a double-chain silicate, the tetrahedra link to form a double chain by sharing two or three oxygen atoms. Amphiboles are the most common type (Fig. 3.9c).
- *Sheet silicates*: The tetrahedra in this group share three oxygen atoms and therefore link to form two-dimensional sheets. Other ions and, in some cases, water molecules fit between the sheets in some sheet silicates. Because of their structure, sheet silicates have a single strong cleavage in one direction, and they occur in books of very thin sheets. In this group we find micas (Fig. 3.9a) and clays. Clays occur only in extremely tiny flakes.
- *Framework silicates*: In a framework silicate, each tetrahedron shares all four oxygen atoms with its neighbors, forming a three-dimensional structure. Examples include feldspar and quartz. The two most common feldspars are plagioclase and orthoclase, also called potassium feldspar (Fig. 3.8d). Plagioclase typically is white, and orthoclase typically is pink.

TAKE-HOME MESSAGE

Mineralogists classify minerals into a number of different classes (such as silicates, carbonates, oxides, and sulfides) on the basis of their chemical composition. Silicates are the most abundant. Inside, they consist of arrangements of silicon-oxygen tetrahedra.

3.6 SOMETHING PRECIOUS—GEMS!

Mystery and romance follow famous gems. Consider the stone now known as the Hope Diamond, recognized by name the world over. No one knows who first dug it out of the ground (**Box 3.2**). Was it mined in the 1600s, or was it stolen off an ancient religious monument? What we do know is that in the 1600s, a French trader named Jean Baptiste Tavernier obtained a large (112.5 carats, where 1 carat = 200 milligrams), rare blue diamond in India, perhaps from a Hindu statue, and carried it back to France. King Louis XIV bought the diamond and had it fashioned into a jewel of 68 carats. This jewel vanished in 1762 during a burglary. Perhaps it was lost forever—perhaps not. In 1830, a 44.5-carat blue diamond mysteriously appeared on the jewel market for sale. Henry Hope, a British banker, purchased the stone, which then became known as the Hope Diamond (**Fig. 3.11**). It changed hands several times until 1958, when the famous New York jeweler Harry Winston donated it to the Smithsonian Institution in Washington, D.C., where it now sits behind bulletproof glass in a heavily guarded display.

What makes stones such as the Hope Diamond so special that people risk life and fortune to obtain them? What is the difference between a gemstone, a gem, and any other mineral? A gemstone is a mineral that has special value because it is rare and people consider it beautiful. A **gem** is a cut and finished stone ready to be set in jewelry. Jewelers distinguish between precious stones (such as diamond, ruby, sapphire, and emerald), which are particularly rare and expensive, and semiprecious stones (such as topaz, tourmaline, aquamarine, and garnet), which are less rare and less expensive. The category of semiprecious stones also includes opaque or translucent minerals such as lapis, malachite (Fig. 3.2a), and opal.

In everyday language, pearls and amber may also be considered gemstones. Unlike diamonds and garnets, which form inorganically in rocks, pearls form in living oysters when the oyster extracts calcium and carbonate ions from water and precipitates them around an impurity, such as a sand grain, embedded in its body. Thus, pearls are a result of biomineralization.

FIGURE 3.11 The Hope Diamond, now on display in the Smithsonian Institution, in Washington, D.C.

BOX 3.2

THE REST OF THE STORY

Where Do Diamonds Come From?

Diamonds consist of carbon, which typically accumulates only at or near Earth's surface. Experiments demonstrate that the temperatures and pressures needed to form diamond are so extreme that, in nature, they generally occur only at depths of around 150 km below the Earth's surface. Under these conditions, the carbon atoms that were arranged in hexagonal sheets in graphite rearrange to form the much stronger and more compact structure of diamond. (Of note, engineers can duplicate these conditions in the laboratory; corporations manufacture several tons of diamonds a year.)

How does carbon get down into the mantle, where it transforms into diamond? Geologists speculate that the process of subduction provides the means. Carbon-containing rocks and sediments in oceanic lithosphere plates at the Earth's surface can be carried down to depth at a convergent plate boundary. This carbon transforms

into diamond, some of which becomes trapped in the lithospheric mantle beneath continents.

But if diamonds form in the mantle, then how do they return to the surface? One possibility is that the process of rifting cracks the continental crust and causes a small part of the underlying lithospheric mantle to melt. Magma generated during this process rises to the surface, bringing the diamonds with it. Near the surface, the magma cools and solidifies to form a special kind of igneous rock called kimberlite (named for Kimberley, South Africa, where it was first found). Diamonds brought up with the magma are embedded in the kimberlite (**Fig. 1**). Kimberlite magma contains a lot of dissolved gas and thus froths to the surface very rapidly. Kimberlite rock commonly occurs in carrot-shaped bodies called kimberlite pipes that are 50 to 200 m across and at least 1 km deep.

Controversial measurements suggest that many of the diamonds that sparkle on engagement rings today were formed 3.2 billion years ago. The diamonds sat at depths of 150 km in the Earth until two rifting events, one of which took place in the late Precambrian and the other during the late Mesozoic, released them to the surface, like genies out of a bottle. The Mesozoic rifting event led to the breakup of Pangaea.

In places where diamonds occur in solid kimberlite, they can

be obtained only by digging up the kimberlite and crushing it, to separate out the diamonds. But nature can also break diamonds free from the Earth. In places where kimberlite has been exposed at the ground surface for a long time, the rock chemically reacts with water and air (a process called weathering; see Interlude B). These reactions cause most minerals in kimberlite to disintegrate, creating sediment that washes away in rivers. Diamonds are so strong that they remain as solid grains in river gravel. Thus, many diamonds have been obtained simply by separating them from recent or ancient river gravel.

Diamond-bearing kimberlite pipes occur in many places around the world, particularly where very old continental lithosphere exists. Southern and central Africa, Siberia, northwestern Canada, India, Brazil, Borneo, Australia, and the U.S. Rocky Mountains all have pipes (see **Geotour 3**). Rivers and glaciers, however, have transported diamond-bearing sediments great distances from their original sources. In fact, diamonds have even been found in farm fields of the midwestern United States. Not all natural diamonds are valuable; value depends on color and clarity. Diamonds that contain imperfections (cracks, or specks of other material), or are dark gray in color, are not used for jewelry. These stones, called industrial diamonds, are used instead as abrasives, for diamond powder is so hard (10 on the Mohs hardness scale) that it can be used to grind away any other substance.

Gem-quality diamonds come in a range of sizes. Jewelers measure diamond size in carats, where one carat equals 200 milligrams (0.2 grams)—one ounce equals 142 carats. (Note that a carat measures gemstone weight, whereas a karat specifies the purity of gold.) The largest diamond ever found, a stone called the Cullinan Diamond, was discovered in South Africa in 1905. It weighed 3,106 carats (621 grams) before being cut. By comparison, the diamond on a typical engagement ring weighs less than one carat. Diamonds are rare, but not as rare as their price suggests. A worldwide consortium of diamond producers stockpile the stones so as not to flood the market and drive the price down.

FIGURE 1 Diamond occurrences.

A diamond mine pit.

A diamond embedded in solid kimberlite.

Most pearls used in jewelry today are "cultured" pearls, made by artificially introducing round sand grains into oysters in order to stimulate pearl production. Amber is also formed by organic processes—it consists of fossilized tree sap. But because amber consists of organic compounds that are not arranged in a crystal structure, it does not meet the definition of a mineral.

In some cases, gemstones are merely pretty and rare versions of more common minerals. For example, ruby is a special version of the common mineral corundum, and emerald is a special version of the common mineral beryl (Fig. 3.12a). As for the beauty of a gemstone, this quality lies basically in its color and, in the case of transparent gems, its "fire"—the way the stone bends and internally reflects the light passing through it, and disperses the light into a spectrum. Fire makes a diamond sparkle more than a similarly cut piece of glass.

Gemstones form in many ways. Some solidify from a melt, some form by diffusion, some precipitate out of a water solution in cracks, and some are a consequence of the chemical interaction of rock with water near the Earth's surface. Many gems come from pegmatites, particularly coarse-grained igneous rocks formed by the solidification of steamy melt.

Most gems used in jewelry are "cut" stones. The smooth **facets** on a gem are ground and polished surfaces made with a faceting machine (Fig. 3.12b). Facets are not the natural crystal faces of the mineral, nor are they cleavage planes, though gem cutters sometimes make the facets parallel to cleavage directions and will try to break a large gemstone into smaller pieces by splitting it on a cleavage plane. A faceting machine consists of a doping arm, a device that holds a stone in a specific orientation; and a lap, a rotating disk covered with a wet paste of grinding powder and water. The gem cutter fixes a gemstone to the end of the doping arm and positions the arm so that it holds the stone against the moving lap. The movement of the lap grinds a facet. When the facet is complete, the gem cutter rotates the arm by a specific angle, lowers the stone, and grinds another facet. The geometry of the facets defines the cut of the stone. Different cuts have names, such as "brilliant," "French," "star," and "pear." Grinding facets is a lot of work—a typical engagement-ring diamond with a brilliant cut has fifty-seven facets (Fig. 3.12c)!

FIGURE 3.12 Cutting gemstones.

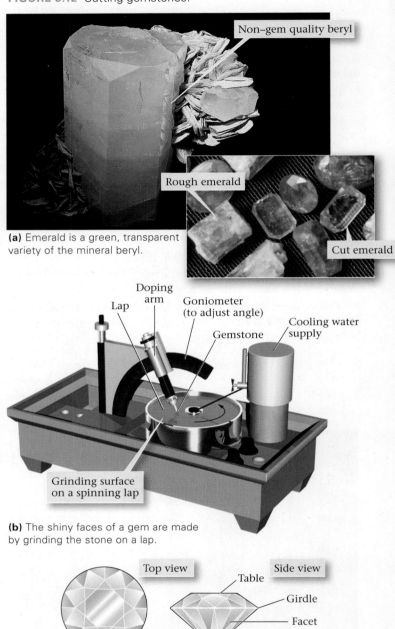

(a) Emerald is a green, transparent variety of the mineral beryl.

(b) The shiny faces of a gem are made by grinding the stone on a lap.

(c) There are many different "cuts" for a gem. Here we see the top and side views of a brilliant-cut diamond.

Chapter Summary

- Minerals are naturally occurring, solid substances with a definable chemical composition and an internal structure characterized by an orderly arrangement of atoms, ions, or molecules in a crystalline lattice. Most minerals are inorganic.

- In the crystalline lattice of minerals, atoms occur in a specific pattern—one of nature's finest examples of ordering.

- Minerals can form by the solidification of a melt, precipitation from a water solution, diffusion through a solid, the metabolism of organisms, and precipitation from a gas.

- Close to 4,000 different types of minerals are known, each with a name and distinctive physical properties (such as color, streak, luster, hardness, specific gravity, crystal habit, and cleavage).

- The unique physical properties of a mineral reflect its chemical composition and crystal structure. By observing these physical properties, you can identify minerals.

- The most convenient way for classifying minerals is to group them according to their chemical composition. Mineral classes include: silicates, oxides, sulfides, sulfates, halides, carbonates, and native metals.

- The silicate minerals are the most common on Earth. The silicon-oxygen tetrahedron, a silicon atom surrounded by four oxygen atoms, is the fundamental building block of silicate minerals.

- Groups of silicate minerals are distinguished from each other by the ways in which the silicon-oxygen tetrahedra that constitute them are linked.

- Gem stones are minerals known for their beauty and rarity. The facets on cut gems used in jewelry are made by grinding and polishing the stones with a faceting machine.

GEOPUZZLE REVISITED

The geologic definition of a mineral is much narrower than the definition used in everyday conversation. Just because something is not or was not alive doesn't mean that it's a mineral. Minerals have an orderly internal crystalline structure and must have formed by geologic processes.

Key Terms

carbonates (p. 78)
cleavage (p. 77)
conchoidal fracture (p. 77)
crystal (p. 72)
crystal face (p. 72)
crystal habit (p. 75)
crystal lattice (p. 70)
crystal structure (p. 72)
facet (p. 82)
gem (p. 80)
geode (p. 75)
hardness (p. 75)

luster (p. 75)
mineral (p. 70)
mineralogy (p. 69)
Mohs hardness scale (p. 75)
polymorph (p. 72)
silicate (p. 77)
silicon-oxygen tetrahedron (p. 79)
specific gravity (p. 75)
streak (p. 75)
symmetry (p. 73)

Review Questions

1. What is a mineral, as geologists understand the term? How is this definition different from the everyday usage of the word?

2. Why is glass not a mineral?

3. Salt is a mineral, but the plastic making up an inexpensive pen is not. Why not?

4. Describe the several ways that mineral crystals can form.

5. Why do some minerals occur as euhedral crystals, whereas others occur as anhedral grains?

6. List and define the principal physical properties used to identify a mineral.

7. How can you determine the hardness of a mineral? What is the Mohs hardness scale?

8. How do you distinguish cleavage surfaces from crystal faces on a mineral? How does each type of surface form?

9. What is the prime characteristic that geologists use to separate minerals into classes?

10. On what basis do mineralogists organize silicate minerals into distinct groups?

11. What is the relationship between the way in which silicon-oxygen tetrahedra bond in micas and the characteristic cleavage of micas?

12. Why are some minerals considered gemstones? How do you make the facets on a gem?

On Further Thought

1. Compare the chemical formula of magnetite with that of biotite. Why is magnetite mined as iron ore, but biotite is not?

2. Imagine that you are given two milky white crystals, each about 2 cm across. You are told that one of the crystals is composed of plagioclase and the other of quartz. How can you determine which is which?

3. Could you use crushed calcite to grind and form facets on a diamond? Why or why not?

ANOTHER VIEW   Sapphires come in many colors, and can be cut into many shapes.

Rock Groups

It took years of back-breaking labor for nineteenth-century workers to chisel and chip ledges and tunnels through hard rock to run a rail line across a mountain range. In the process, the workers became very familiar with the nature of rock.

A.1 INTRODUCTION

During the 1849 gold rush in the Sierra Nevada of California, only a few lucky individuals actually became rich. The rest of the "forty-niners" either slunk home in debt or took up less glamorous jobs in new towns such as San Francisco. These towns grew rapidly, and soon the American west coast was demanding large quantities of manufactured goods from east coast factories. Making the goods was no problem, but getting them to California meant either a stormy voyage around the southern tip of South America or a trek with stubborn mule teams through the deserts of Nevada and Utah. The time was ripe to build a railroad linking the east and west coasts of North America, and, with much fanfare, the Central Pacific line decided to punch one right across the peaks of the Sierras. In 1863, while the Civil War raged elsewhere in the United States, the company transported six thousand Chinese laborers across the Pacific in the squalor of unventilated cargo holds and set them to work chipping ledges and blasting tunnels. Along the way, untold numbers of laborers died of frostbite, exhaustion, mistimed blasts, landslides, or avalanches.

Through their efforts, the railroad laborers certainly gained an intimate knowledge of how rock feels and behaves—it's solid, heavy, and hard! They also found that some rocks seemed to break easily into layers but others did not, and some rocks were dark colored while others were light colored. They realized, like anyone who looks closely at rock exposures, that rocks are not just gray, featureless masses, but rather come in a great variety of colors and textures.

Why are there so many distinct types of rocks? The answer is simple: rocks can form in many different ways and from many different materials. Because of the relationship between rock type and the process of formation, *rocks provide a historical record of geologic events and give insight into interactions among components of the Earth System.*

The next few chapters are devoted to a discussion of rocks and a description of how rocks form; this interlude serves as a general introduction. Here, we learn what the term *rock* means to geologists, what rocks are made of, and how to distinguish the three principal groups of rocks. We also look at how geologists study rocks.

A.2 WHAT IS ROCK?

To geologists, **rock** is a coherent, naturally occurring solid, consisting of an aggregate of minerals or, less commonly, a mass of glass. Now let's take this definition apart.

- *Coherent*: A rock holds together, and thus must be broken to be separated into pieces. As a result of its coherence, rock can form cliffs or can be carved into sculptures. A pile of unattached grains does not constitute a rock.
- *Naturally occurring*: Geologists consider only naturally occurring materials to be rocks, so manufactured materials, such as concrete and brick, do not qualify.
- *An aggregate of minerals or a mass of glass*: The vast majority of rocks consist of an aggregate (a collection) of many mineral grains, and/or crystals, stuck or grown together. (*Note*: a grain is any fragment or piece of mineral, rock, or glass.) Technically, a single mineral crystal is a "mineral specimen," not a rock, even if it is meters long. Some rocks contain only one kind of mineral, whereas others contain several different kinds. A few of the rock types that form at volcanoes (see Chapter 4) consist of glass, which may occur either as a homogeneous mass or as an accumulation of tiny glass shards.

What holds rock together? Grains in rock stick together to form a coherent mass either because they are bonded by natural **cement**, mineral material that precipitates from water and fills the space between grains (**Fig. A.1a**), or because they interlock with one another like pieces in a jigsaw puzzle (**Fig. A.1b**). Rocks whose grains are stuck together by cement are called **clastic**, whereas rocks whose crystals interlock with one another are called **crystalline**. Glassy rocks hold together either because they originate as a continuous mass (that is, they have no separate grains) or because the separate glassy grains welded together while still hot.

A.3 ROCK OCCURRENCES

At the surface of the Earth, rock occurs either as broken chunks (pebbles, cobbles, or boulders; see Chapter 5) that have moved from their point of origin by falling down a slope or by being transported in ice, water, or wind, or as **bedrock** which is still attached to the Earth's crust. Geologists refer to an exposure of bedrock as an **outcrop**. An outcrop may appear as a rounded knob out in a field, as a ledge forming a cliff or ridge, on the face of a stream cut (where running water cut down into bedrock), or along human-made road cuts and excavations (**Fig. A.2a–c**).

To people who live in cities or forests or on farmland, outcrops of bedrock may be unfamiliar, since the outcrops may be covered by vegetation, sand, mud, gravel, soil, water, asphalt, concrete, or buildings. Outcrops are particularly rare in regions such as the midwestern United States, where, during the past

FIGURE A.1   Rocks, aggregates of mineral grains and/or crystals, can be clastic or crystalline.

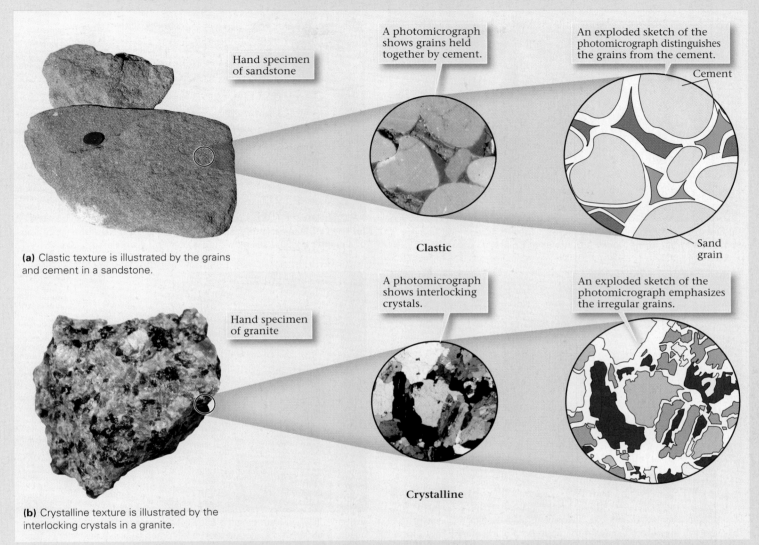

Hand specimen of sandstone

A photomicrograph shows grains held together by cement.

An exploded sketch of the photomicrograph distinguishes the grains from the cement.

Cement

Sand grain

Clastic

(a) Clastic texture is illustrated by the grains and cement in a sandstone.

Hand specimen of granite

A photomicrograph shows interlocking crystals.

An exploded sketch of the photomicrograph emphasizes the irregular grains.

Crystalline

(b) Crystalline texture is illustrated by the interlocking crystals in a granite.

million years, ice-age glaciers melted and left behind thick deposits of debris (see Chapter 18).

A.4 THE BASIS OF ROCK CLASSIFICATION

Beginning in the eighteenth century, geologists struggled to develop a sensible way to classify rocks, for they realized, as did miners from centuries past, that not all rocks are the same. Classification schemes help us organize information and remember significant details about materials or objects, and they help us recognize similarities and differences among them. One of the earliest classification schemes divided rocks into three groups—so-called primary, secondary, and ter-

tiary—on the basis of the incorrect perception that the groups had formed in a time succession.

By the end of the eighteenth century, many geologists had realized that a *genetic scheme* for classifying rocks—a scheme that focuses on the origin (genesis) of rocks—would be the best approach for classifying rocks, and this is the approach that we continue to use today. Using this approach, geologists recognize three basic groups: (1) **igneous rocks**, which form by the freezing (solidification) of molten rock, or melt (Fig. A.3a); (2) **sedimentary rocks**, which form either by the cementing together of fragments (grains) broken off preexisting rocks or by the precipitation of mineral crystals out of water solutions at or near the Earth's surface (Fig. A.3b); and (3) **metamorphic rocks**, which form when preexisting rocks change character in response to a change in pressure and temperature conditions,

FIGURE A.2 Types of rock exposures.

(a) Outcrops are natural rock exposures. These outcrops rise as cliffs above the forest of the Rocky Mountains in Colorado.

(b) Highway engineers excavate road cuts to provide a level route for a highway, as in this example near Kingston, New York.

(c) Stream cuts form when flowing water grinds down into the land surface, as in this example from Arizona.

FIGURE A.3 Examples of the three major rock groups.

Igneous

(a) Lava (molten rock) freezes to form igneous rock. Here, the molten tip of a brand-new flow still glows red. Older flows are already solid.

Sedimentary

(b) Sand, formed from grains eroded off the rock cliffs, collects on the beach. If buried and turned to rock, it becomes layers of sandstone, such as those making up the cliffs.

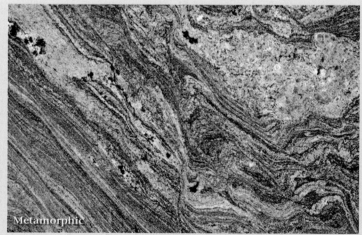

Metamorphic

(c) Metamorphic rock forms when preexisting rocks endure changes in temperature and pressure, and/or are subjected to shearing, stretching, or shortening. New minerals and textures form.

and/or as a result of squashing, stretching, or shear (Fig. A.3c). Metamorphic change occurs in the solid state, which means that it does not require melting. Each of the three groups contains many different individual rock types, distinguished from one another by physical characteristics such as:

- *grain size*: The dimensions of individual grains in a rock may be measured in millimeters or centimeters. Some grains are so small that they can't be seen without a microscope, whereas others are as big as a fist or larger. Some grains are **equant**, meaning that they have the same dimensions in all directions; some are **inequant**, meaning that the dimensions are not the same in all directions (Fig. A.4a, b). In some rocks, all the grains are the same size, whereas other rocks contain a variety of different-sized grains.

- *composition*: A rock is a mass of chemicals. The term "rock composition" refers to the proportions of different chemicals making up the rock. The proportion of chemicals, in turn, affects the proportion of different minerals constituting the rock.

- *texture*: This term refers to the arrangement of grains in a rock, that is, the way grains connect to one another and whether or not inequant grains are aligned parallel to each other. The concept of rock texture will become easier to grasp as we look at different examples of rocks in the following chapters.

- *layering*: Some rock bodies appear to contain distinct layering, defined either by bands of different composi-

tions or textures, or by the alignment of inequant grains so that they trend parallel to each other. Different types of layering occur in different kinds of rocks. For example, the layering in sedimentary rocks is called **bedding**, whereas the layering in metamorphic rocks is called **metamorphic foliation** (Fig. A.5a, b).

FIGURE A.4 Describing grains in rock.

Magnification reveals a variety of grains.

Equant Inequant

This rock is an aggregate of mineral grains.

1 millimeter

Inequant grains align to form foliation.

1 meter

(a) Grains in rock come in a variety of shapes. Some are equant, whereas some are inequant. In this example of metamorphic rock, inequant grains align to define a foliation.

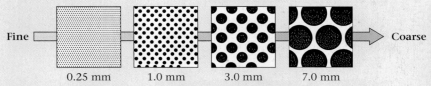

Fine Coarse

0.25 mm   1.0 mm   3.0 mm   7.0 mm

(b) Geologists define grain size by using this comparison chart.

FIGURE A.5 Layering in rock.

Horizontal bedding

Younger beds

Older beds

Tilted bedding

(a) Bedding in a sedimentary rock, here defined by alternating layers of coarser and finer grains, as exposed on a cliff along an Oregon beach. Older beds were tilted before younger ones were deposited.

Vertical foliation plane

(b) Foliation in metamorphic rock, here defined by alternating layers of light minerals and dark minerals.

FIGURE A.6   Studying rocks in thin section.

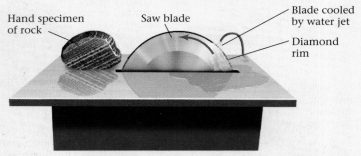

(a) Using a special saw, a geologist cuts a thin chip of a rock specimen.

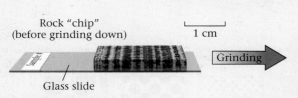

Rock "chip" (before grinding down)

1 cm

Glass slide

Grinding

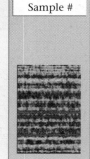

Sample #

(b) The geologist glues the chip to a glass slide and grinds it down until it is so thin that light can pass through it.

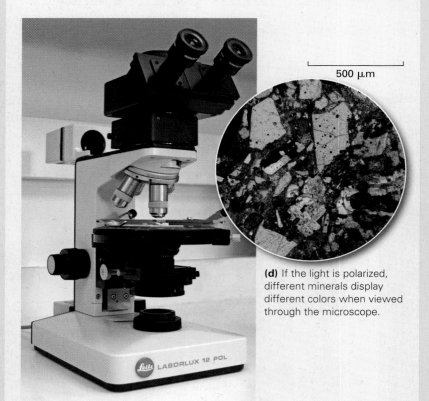

500 μm

(d) If the light is polarized, different minerals display different colors when viewed through the microscope.

(c) With a petrographic microscope, it's possible to view thin sections with light that shines through the sample from below.

Each individual rock type has a name. Names come from a variety of sources. Some come from the dominant component making up the rock, some from the region where the rock was first discovered or is particularly abundant, some from a root word of Latin origin, and some from a traditional name used by people in an area where the rock is found. All told, there are hundreds of different rock names, though in this book we will introduce only about thirty.

A.5 STUDYING ROCK

Outcrop Observations

The study of rocks begins by examining a rock in an outcrop. If the outcrop is big enough, such an examination will reveal relationships between the rock you're interested in and the rocks around it, and will allow you to detect layering. Geologists carefully record observations about an outcrop, then break off a **hand specimen**, a fist-sized piece, which they can examine more closely with a hand lens (magnifying glass). Observation with a hand lens enables geologists to identify sand-sized or larger mineral grains, and may enable them to describe the texture of the rock.

Thin-Section Study

Geologists often must examine rock composition and texture in minute detail in order to identify a rock and develop a hypothesis for how it formed. To do this, they take a specimen back to the lab, make a very thin slice (about 0.03 mm thick, the thickness of a human hair) mounted on a glass slide. They study the resulting **thin section** (Fig. A.6a–c) with a petrographic microscope (*petro* comes from the Greek word for rock). A petrographic microscope differs from an ordinary microscope in that it illuminates the thin section with transmitted polarized light. This means that the illuminating light beam first passes through a special filter that makes all the light waves in the beam vibrate in the same plane, and then the light passes up through the thin section. An observer looks through the thin section as if it were a window. When illuminated with transmitted polarized light, each type of mineral grain displays a unique suite of colors (Fig. A.6d). The specific color the observer sees depends on both the identity of the grain and its orientation with respect to the waves of polarized light, for a crystal interferes with polarized light and allows only certain wavelengths to pass through.

The brilliant colors and strange shapes in a thin section rival the beauty of an abstract painting or stained glass. By examining a thin section with a petrographic microscope, geologists can identify most of the minerals constituting the rock and can describe the way in which the grains connect to each other. A photograph taken through a petrographic microscope is called a **photomicrograph**.

High-Tech Analytical Equipment

Beginning in the 1950s, high-tech electronic instruments became available that enabled geologists to examine rocks on an even finer scale than is possible with a petrographic microscope. Modern research laboratories typically boast instruments such as electron microprobes, which can focus a beam of electrons on a small part of a grain to create a signal that defines the chemical composition of the mineral (Fig. A.7); mass spectrometers, which analyze the proportions of atoms with different atomic weights contained in a rock; and X-ray diffractometers, which identify minerals by looking at the way X-ray beams pass through crystals in a rock. Such instruments, in conjunction with optical examination, can provide geolo-

FIGURE A.7 An electron microprobe uses a beam of electrons to analyze the chemical composition of minerals.

Image   Electron-beam source   Detector

gists with highly detailed characterizations of rocks, which in turn help them understand how the rocks formed and where the rocks came from. This information enables geologists to use the study of rocks as a basis for deciphering Earth history.

Key Terms

bedding (p. 89)

bedrock (p. 86)

cement (p. 86)

clastic (p. 86)

crystalline (p. 86)

equant (p. 89)

hand specimen (p. 90)

igneous rock (p. 87)

inequant (p. 89)

metamorphic foliation (p. 89)

metamorphic rock (p. 87)

outcrop (p. 86)

photomicrograph (p. 91)

rock (p. 86)

sedimentary rock (p. 87)

thin section (p. 90)

ANOTHER VIEW This quarry, in northwestern Italy, provides blocks of pure white marble, some of which were carved into beautiful sculptures. It also provides a view into the bedrock that lies beneath rugged peaks.

CHAPTER **4**

Up from the Inferno: Magma and Igneous Rocks

The light-colored rock of the Torres del Paines in Chile is an intrusion of granite. This granite formed when melt froze deep in the crust. It has been exposed due to mountain building and erosion.

GEOPUZZLE

Where does the red-hot molten rock that spills and/or blasts out of a volcano come from, and what does it turn into?

4.1 INTRODUCTION

Every now and then, an incandescent liquid—hot molten rock, or melt—fountains from a crater or crack on the big island of Hawaii. Hawaii is a **volcano**, a vent at which melt from inside the Earth spews onto the planet's surface. The transfer of melt from inside the Earth onto its surface is a volcanic eruption. Some of the melt, which is called **lava** once it has reached the Earth's surface, pools around the vent, while the rest runs down

the mountainside as a viscous (syrupy) red-yellow stream called a **lava flow**. Near its source, this lava has a temperature of 1,100 to 1,200°C and moves swiftly, cascading over escarpments at speeds of up to 60 km per hour (Fig. 4.1a). At the base of the mountain, the lava flow slows but advances nonetheless, engulfing roads, houses, or vegetation in its path (Fig. 4.1b). As the flow cools, it slows down. Its surface darkens and crusts over, occasionally breaking to reveal the hot, sticky mass that continues to ooze within. Finally, the flow stops moving entirely, and within days or weeks the once red-hot melt has become a hard, black solid through and through (Fig. 4.1c). This process can repeat again and again over geologic time (Fig. 4.1d). A new **igneous rock**, made by the freezing of a melt, has formed.

FIGURE 4.1 Formation and evolution of lava flows.

(a) Lava erupts as a fountain from a volcanic vent on Hawaii. A fast-moving river of lava then flows downslope.

(b) At a distance from the vent, the lava has completely crusted over with new rock, but the interior of the flow remains molten.

(c) Eventually, the flow cools completely and becomes a layer of new rock. This flow engulfed a road on Hawaii.

(d) Over time, many lava flows can accumulate one on top of another to build a large volcano. A canyon cut into the volcano exposes ancient flows.

Considering the fiery heat of the melt from which igneous rocks develop, the name igneous—from the Latin *ignis*, meaning fire—makes sense.

Although some igneous rocks solidify at the surface during volcanic eruptions such as those on Hawaii, a vastly greater volume results from solidification of melt underground, out of sight. Geologists refer to melt that exists below the Earth's surface as **magma**. Rock made by the freezing of magma underground, after it has pushed its way (intruded) into preexisting rock of the crust, is **intrusive igneous rock**. Some intrusive rock freezes within a relatively large, irregular volume called a magma chamber, whereas some freezes in tabular cracks or chimney-like columns. Rock that forms by the freezing of lava above ground, after it spills out (extrudes) onto the surface of the Earth and comes into contact with the atmosphere or ocean, is **extrusive igneous rock** (**Fig. 4.2a**). Extrusive igneous rock includes both solid lava flows, formed when streams or mounds of lava solidify on the surface of the Earth, and deposits of pyroclastic debris (from the Greek word *pyro*, meaning fire). Such debris includes **volcanic ash**, consisting of fine particles of glass that form when a spray of lava erupts into the air and freezes instantly. Some ash billows up several kilometers into the sky above a volcano, and eventually drifts down in a snow-like ash fall. But some ash rushes down the side of the volcano in a scalding avalanche called an ash flow. Pyroclastic debris also includes larger fragments formed when clots of lava freeze in the air, or when the force of the eruption blasts apart preexisting rock of a volcano (**Fig. 4.2b**). We'll provide further detail about pyroclastic debris in Chapter 5.

A great variety of igneous rocks exist on Earth (see **Geotour 4** on p. GT-10); they make up all of the oceanic crust, and most of the continental crust. To understand why and how these rocks form, and why there are so many different kinds, we first discuss why magma forms, why it rises, how it flows, and how it freezes in intrusive and extrusive environments. We then look at the scheme that geologists use to classify igneous rocks.

TAKE-HOME MESSAGE

By the end of this chapter, you should understand why melts form and rise into the crust, how the nature of an igneous rock reflects the conditions under which it solidifies, and how geologists classify the great variety of igneous rocks.

4.2 WHY DOES MAGMA FORM?

The Earth is hot inside, due to old heat left over from the planet's formation (see Chapter 1), as well as new heat produced by the decay of radioactive elements. But the popular image that the solid crust of the Earth floats on a sea of molten rock is *not* correct. Magma forms only in special places where preexisting solid rock melts. Below, we describe conditions that lead to melting.

Melting due to a decrease in pressure (decompression). We can portray the variation in temperature with depth in the Earth on a graph by a curving line called the **geotherm**. Beneath typical oceanic crust, temperatures comparable to those of lava occur in the upper mantle (**Fig. 4.3a**). But even though the upper mantle is very hot, its rock stays solid because it is also under great pressure from the weight of overlying rock. Basically, pressure at great depth prevents atoms from breaking free of solid mineral crystals. Because pressure prevents melting, a decrease in pressure can permit melting. Specifically, if the pressure affecting hot mantle rock decreases while the temperature remains unchanged, magma forms. This kind of melting, called decompression melting, occurs where hot mantle rock rises to shallower depths in the Earth (**Fig. 4.3b**).

FIGURE 4.2   The intrusive and extrusive realms.

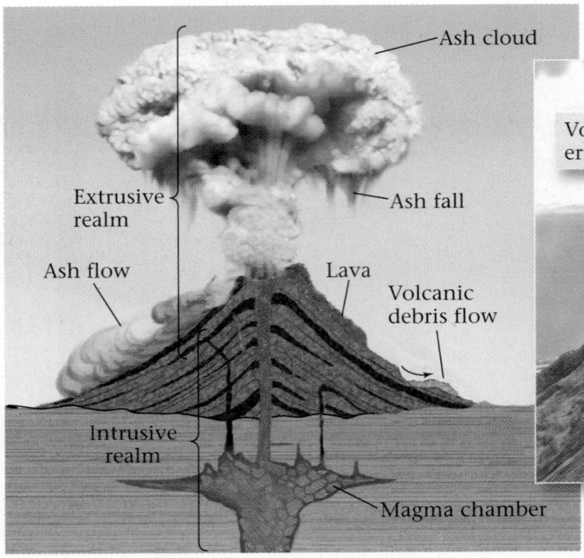

(a) The intrusive realm lies underground. Lava flows, as well as various types of ash eruptions, produce extrusive rocks.

(b) The extrusive realm is at the surface. This Alaskan volcano consists of lava and ash. Some of this material breaks into blocks, due to the force of eruptions, or due to slip downslope.

FIGURE 4.3 Decompression melting.

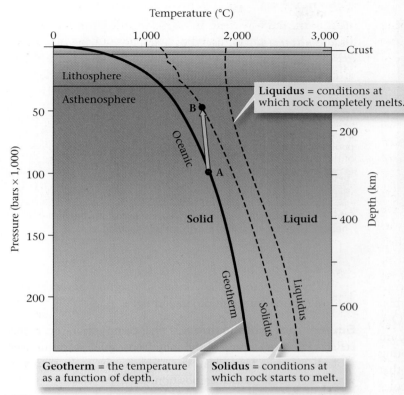

Geotherm = the temperature as a function of depth.

Liquidus = conditions at which rock completely melts.

Solidus = conditions at which rock starts to melt.

(a) Decompression melting takes place when the pressure acting on hot rock decreases. As this graph of pressure and temperature conditions in the Earth shows, when rock rises from point A to point B, the pressure decreases a lot, but the rock cools only a little, so the rock begins to melt.

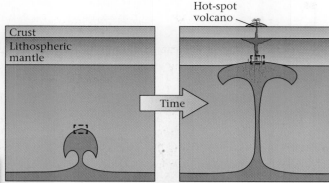

Decompression melting in a mantle plume

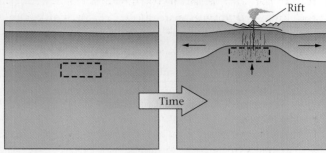

Decompression melting beneath a rift

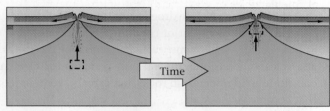

Decompression melting beneath a mid-ocean ridge

(b) The conditions leading to decompression melting occur in several different geologic environments. In each case, a volume of hot asthenosphere (outlined by dashed lines) rises to a shallower depth.

Melting as a result of the addition of volatiles. Magma also forms at locations where chemicals called volatiles mix with hot mantle rock. Volatiles are substances such as water (H_2O) and carbon dioxide (CO_2) that evaporate easily and can exist in gaseous forms at the Earth's surface. When volatiles mix with hot rock, they help break chemical bonds. So if you add volatiles to a solid, hot dry rock, the rock begins to melt (**Fig. 4.4a**). In effect, adding volatiles decreases a rock's melting temperature. Melting due to addition of volatiles is sometimes called flux melting.

Melting as a result of heat transfer from rising magma. When magma from the mantle rises up into the crust, it brings heat with it. This heat raises the temperature of the surrounding crustal rock and, in some cases, the rise in temperature may be sufficient to melt part of the crustal rock (**Fig. 4.4b**). To picture the process, imagine injecting hot fudge into ice cream; the fudge transfers heat to the ice cream, raises its temperature, and causes it to melt. We call such melting heat-transfer melting, because it results from the transfer of heat from a hotter material to a cooler one.

4.3 WHAT IS MAGMA MADE OF?

All magmas contain silicon and oxygen, which bond to form the silicon-oxygen tetrahedron. But magmas also contain varying proportions of other elements such as aluminum (Al), calcium (Ca), sodium (Na), potassium (K), iron (Fe), and magnesium (Mg). Because magma is a liquid, its atoms do not lie in an orderly crystalline lattice but are grouped instead in clusters or short chains, relatively free to move with respect to one another.

FIGURE 4.4 Flux melting and heat-transfer melting.

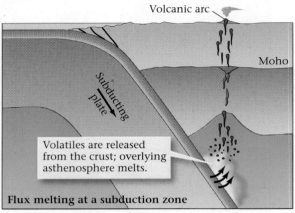

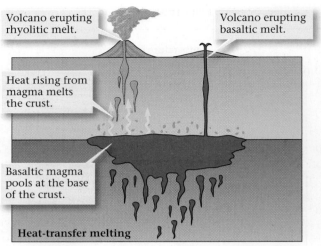

(a) Flux melting occurs where volatiles enter hot mantle; this happens at subduction zones.

(b) Heat-transfer melting occurs when rising magma brings heat up with it and melts overlying or surrounding rock.

"Dry" magmas contain no volatiles. "Wet" magmas, in contrast, include up to 15% dissolved volatiles such as water, carbon dioxide, nitrogen (N_2), hydrogen (H_2), and sulfur dioxide (SO_2). These volatiles come out of the Earth at volcanoes in the form of gas. Usually water constitutes about half of the gas erupting at a volcano. Thus, magma not only contains the elements that constitute solid minerals in rocks, but can also contain molecules that become water or air.

The Major Types of Magma

Geologists distinguish four major types of magma by measuring the percentage of silica (SiO_2) within the magma (Table 4.1). As we will see later in this chapter and in Chapter 5, these varieties of magma differ in their physical characteristics, such as how easily they flow and freeze to form different kinds of rocks. The names of the magma types reflect the elements within the magma. For example, mafic magma gets its name because it has a relatively high proportion of iron oxide (FeO, or Fe_2O_3) and magnesium oxide (MgO)—"ma" stands for magnesium and "-fic" stands for iron, from the Latin *ferric*. Silicic magma contains a high proportion of silica. (Note that, at an elementary level, geologists use the terms *felsic* and *silicic* interchangeably.) Intermediate magma gets its name simply because this type has a composition between felsic and mafic.

TABLE 4.1 **The Four Categories of Magma**

Felsic (or silicic) magma	66–76% silica
Intermediate magma	52–66% silica
Mafic magma	45–52% silica
Ultramafic magma	38–45% silica

Why are there so many kinds of magma? Several factors control magma composition, including:

Source rock composition. The composition of a melt reflects the composition of the solid from which it was derived. Not all magmas form from the same source rock, so not all magmas have the same composition.

Partial melting. Under the temperature and pressure conditions that occur in the Earth, only about 2 to 30% of an original rock melts to produce magma at a given location—the temperature at sites of magma production simply never gets high enough to melt the entire original rock before the magma has a chance to migrate away from its source. **Partial melting** refers to the process by which only part of an original rock melts to produce magma (Fig. 4.5a). Magmas formed by partial melting tend to be more felsic than the original rock from which they were derived—for example, partial melting of an ultramafic rock produces a mafic magma. Since magma generally migrates away from the site of melting, the unmelted rock left behind tends to be more mafic than the original rock.

Assimilation. As magma sits in a magma chamber before completely solidifying, it may incorporate chemicals dissolved from the wall rocks of the chamber or from blocks that detached from the wall and sank into the magma (Fig. 4.5b). This process is called contamination or **assimilation**.

Magma mixing. Different magmas formed in different locations from different sources may enter a magma chamber. In some cases, the originally distinct magmas mix to create a new, different magma. Thoroughly mixing a felsic magma with a mafic magma in equal proportions produces an intermediate magma.

FIGURE 4.5 Phenomena that can affect the composition of magma.

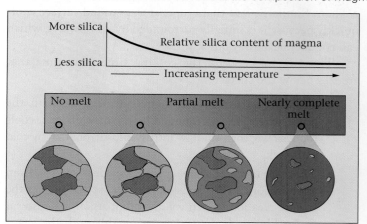

(a) Partial melting: The first-formed melt will be richer in silica than the original rock. As melting continues, magma becomes more mafic.

(b) Mixing and assimilation: Heat provided by deep magma partially melts wall rock; the new magma then mixes with deep magma. Also, blocks of wall rock can dissolve (assimilate) in the deep magma.

TAKE-HOME MESSAGE

Magma is a very hot (650°C to 1,280°C) liquid made up of varying proportions of Si, O, Fe, Mg, Ca, and other chemicals. The composition of a magma depends on its source, on how it interacts with its surroundings, and on whether crystals sink as they form.

4.4 MOVEMENT OF MOLTEN ROCK

If magma stayed put once it formed, new igneous rocks would not develop in or on the crust. But it doesn't stay put; magma tends to move upward, away from where it formed. In some cases, it reaches the Earth's surface and erupts at a volcano. This movement is a key component of the Earth System, because it transfers material from deeper parts of the Earth upwards, and provides the raw material from which new rocks and the atmosphere and ocean form.

Why Does Magma Rise?

Magma rises for two reasons. First, magma rises because it is less dense than surrounding rock; buoyancy drives magma upward just as it drives a wooden block up through water. Second, magma rises because the weight of overlying rock creates pressure at depth that literally squeezes magma upward. The same process happens when you step into a puddle barefoot and mud squeezes up between your toes.

What Controls the Speed of Flow?

Viscosity, or resistance to flow, affects the speed with which magmas or lavas move. Magmas with low viscosity flow more

easily than those with high viscosity, just as water flows more easily than molasses. Viscosity depends on temperature, volatile content, and silica content. Hotter magma is less viscous than cooler magma, just as hot tar is less viscous than cool tar, because thermal energy breaks bonds and allows atoms to move more easily. Similarly, magmas or lavas containing more volatiles are less viscous than dry (volatile-free) magmas, because the volatile atoms also tend to break apart bonds and may accumulate to form bubbles. Mafic magmas are less viscous than felsic magmas, because silicon-oxygen tetrahedra tend to link together in the magma to create long chains that can't move past each other easily. Thus, hotter mafic lavas have relatively low viscosity and flow in thin sheets over wide regions, but cooler felsic lavas are highly viscous and clump at the volcanic vent (Fig. 4.6a, b).

TAKE-HOME MESSAGE

Magma rises from its source because it is buoyant and because of pressure due to overlying rocks. As it moves, magma pushes into existing cracks or forces open new cracks. Its resistance to flow (viscosity) depends on its composition, temperature, and gas content.

4.5 TRANSFORMING MAGMA AND LAVA INTO ROCK

It may seem strange to speak of freezing in the context of forming rock, when most people think of freezing as the transformation of liquid water to solid ice as the temperature drops below 0°C (32°F). Nevertheless, the freezing of liquid melt to form solid igneous rock represents the same phenomenon, except that igneous rocks freeze at high temperatures—

FIGURE 4.6 Viscosity affects lava behavior.

(a) Felsic to intermediate lava is very viscous. When it erupts, it may form a mound-like lava dome around the volcano's vent.

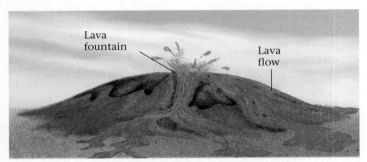

(b) Mafic lava has relatively low viscosity. It can erupt in fountains, move long distances, and form thin lava flows.

between 650°C and 1,100°C. To put such temperatures in perspective, remember that home ovens attain a maximum temperature of only 260°C.

What makes magma freeze? In most cases, magma freezes simply when it cools below its freezing temperature and crystals start to grow. Because temperatures decrease toward the Earth's surface, cooling automatically happens when magma rises. The cooler environment may be cool wall rock (meaning the rock surrounding the intrusion) if the magma intrudes underground, or it may be the atmosphere or the ocean if the magma extrudes as lava at the Earth's surface. Let's now consider factors controlling the rate of cooling, and consider how magma changes as it cools.

How Fast Does Magma Cool?

The time it takes for a magma to cool depends on how fast it is able to transfer heat into its surroundings. To see why, think about the process of cooling coffee. If you pour hot coffee into a thermos bottle and seal it, the coffee stays hot for hours; because of insulation, the coffee in the thermos loses heat to the air outside only very slowly. Like the thermos bottle, surrounding rock acts as an insulator in that it transports heat away from a magma very slowly, so magma underground (in an intrusive environment) cools slowly. In contrast, if you spill

coffee on a table, it cools quickly because it loses heat to the cold air. Similarly, lava that erupts at the ground surface cools quickly because it is initially surrounded by air or water, which can conduct heat away quickly.

Three factors control the cooling time of magma that freezes below the surface.

- *The depth of intrusion*: Magma intruded deep in the crust, where it is surrounded by warm wall rock, cools more slowly than shallow intrusions, surrounded by cold wall rock.

- *The shape and size of a magma body*: Heat escapes from an intrusion at the intrusion's surface, so the greater the surface area for a given volume of intrusion, the faster it cools. Thus, a body of magma roughly with the shape of a pancake cools faster than one with the shape of a melon. And since the ratio of surface area to volume increases as size decreases, a body of magma the size of a car cools faster than one the size of a ship (Fig. 4.7a, b).

- *The presence of circulating groundwater*: Water passing through magma absorbs and carries away heat, much like the coolant that flows around an automobile engine.

Changes in Magma During Cooling: Fractional Crystallization

Most people are familiar with the process of forming ice out of liquid water—cool the water to a temperature of 0°C and crystals of ice start to form. Keep the temperature cold enough for long enough and all the water becomes solid, composed entirely of one type of mineral—water ice. The process of freezing magma or lava is much more complex, because molten rock contains many different compounds, not just water, so *during freezing of molten rock, many different minerals form.* Further, not all of these minerals form at the same time (Box 4.1). To get a sense of this complexity, let's look at an example.

When a mafic magma starts to freeze, mafic (iron- and magnesium-rich) minerals (such as olivine and pyroxene) start to crystallize first. These solid crystals are denser than the remaining magma, so they start to sink (Fig. 4.7c). Some react chemically with the remaining magma as they sink, but some reach the floor of the magma chamber and become isolated from the magma. This process of progressive crystal formation and settling is called **fractional crystallization**—it progressively extracts iron and magnesium from the magma, so the remaining magma becomes more felsic. If a magma freezes completely before a lot of fractional crystallization has occurred, the magma becomes mafic igneous rock. But freezing of a magma that has been left over after a lot of fractional crystallization has occurred produces felsic igneous rock.

BOX 4.1

THE REST OF THE STORY

Bowen's Reaction Series

In the 1920s, Norman L. Bowen began a series of laboratory experiments designed to determine the sequence in which silicate minerals crystallize from a melt. First, Bowen melted powdered mafic igneous rock by raising its temperature to about 1,280°C. Then he cooled the melt just enough to cause part of it to solidify. Finally, he "quenched" the remaining melt by submerging it quickly in cold mercury. Quenching, which means sudden cooling to form a solid, transformed any remaining liquid into glass, trapping the early-formed crystals. Bowen identified mineral crystals formed before quenching with a microscope, and he analyzed the chemical composition of the remaining glass.

After experiments at different temperatures, Bowen found that as new crystals form, they extract certain chemicals preferentially from the melt (**Fig. 1a**). Thus, the chemical composition of the remaining melt progressively changes as the melt cools. Bowen described the specific sequence of mineral-producing reactions that take place in a cooling, initially mafic, magma. This sequence is now called **Bowen's reaction series** in his honor.

Let's examine the sequence more closely. In a cooling melt, olivine and calcium-rich plagioclase form first. The Ca-plagioclase reacts with the melt to form more plagioclase, which contains more sodium (Na). Meanwhile, some olivine crystals react with the remaining melt to produce pyroxene, which may encase olivine crystals or even replace them. However, some of the olivine and Ca-plagioclase crystals settle out of the melt, taking iron, magnesium, and calcium atoms with them. By this process, the remaining melt becomes enriched with silica. As the melt continues to cool, plagioclase continues to form, with later-formed plagioclase having progressively more sodium (Na) than earlier-formed plagioclase. Pyroxene crystals react with melt to form amphibole, and then amphibole reacts with the remaining melt to form biotite. All the while, crystals continue to settle out, so the remaining melt continues to become more felsic. At temperatures of between 650°C and 850°C, only about 10% melt remains, and this melt has a high silica content. At this stage, the final melt freezes, yielding quartz, K-feldspar, and muscovite.

On the basis of his observations, Bowen realized that there are two tracks to the reaction series. The "discontinuous reaction series" refers to the sequence olivine, pyroxene, amphibole, biotite, K-feldspar/muscovite/quartz: each step yields a different class of silicate mineral. The "continuous reaction series" refers to the progressive change from calcium-rich to sodium-rich plagioclase: the steps yield different versions of the same mineral (**Fig. 1b**). It's important to note that not all minerals listed in the series appear in all igneous rock. For example, a mafic magma may completely crystallize before felsic minerals such as quartz or K-feldspar have a chance to form.

FIGURE 1 Bowen's reaction series indicates the succession of crystallization in cooling magma.

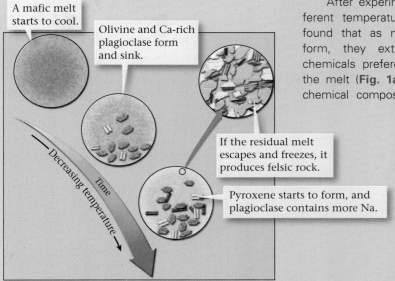

(a) With decreasing temperature, fractional crystallization begins and the composition of the remaining magma becomes more felsic.

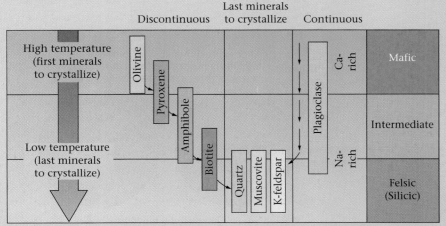

(b) This chart displays the discontinuous and continuous reaction series. Rocks formed from minerals at the top of the series are mafic, whereas rocks formed from the bottom of the series are felsic.

4.6 HOW DO EXTRUSIVE AND INTRUSIVE ENVIRONMENTS DIFFER?

With a background on how melts form and freeze, we can now introduce key features of the two settings—intrusive and extrusive—in which igneous rocks form.

FIGURE 4.7 Factors that affect the freezing of molten rock.

(a) The larger the ratio of its surface area to volume, the faster an intrusion cools. This ratio depends on both size and shape.

Extrusive Igneous Settings

Not all volcanic eruptions are the same, so not all extrusive rocks are the same. Some volcanoes erupt streams of low-viscosity lava that run down the flanks of the volcano and then spread over the countryside. When this lava freezes, it forms a relatively thin lava flow. Such flows may cool in days to months. In contrast, some volcanoes erupt viscous masses of lava that pile into domes, and still others erupt explosively, sending clouds of volcanic ash and debris skyward, and avalanches of ash tumbling down the sides of the volcano.

Which type of eruption occurs depends largely on a magma's composition and volatile content. As noted earlier, volatile-rich felsic lavas tend to erupt explosively and form thick ash and debris deposits (**Fig. 4.8a, b**). Mafic lavas tend to have low viscosity and spread in broad, thin flows (**Fig. 4.8c, d**). We discuss the products of extrusive settings in more detail in Chapter 5.

Intrusive Igneous Settings

Magma rises and intrudes into preexisting rock by slowly percolating upward between grains and/or by forcing open cracks. The magma that doesn't make it to the surface freezes solid underground in contact with preexisting rock and becomes intrusive igneous rock. Geologists commonly refer to the preexisting rock into which magma intrudes as wall rock, and the boundary between it and an intrusive igneous rock as an intrusive contact.

Geologists distinguish among different types of intrusions on the basis of their shape. Tabular intrusions, or sheet intrusions, are planar and are of roughly uniform thickness. Most are in the range of centimeters to tens of meters thick, and tens of meters to tens of kilometers long. A **dike** is a tabular intrusion that cuts across preexisting layering (bedding or foliation),

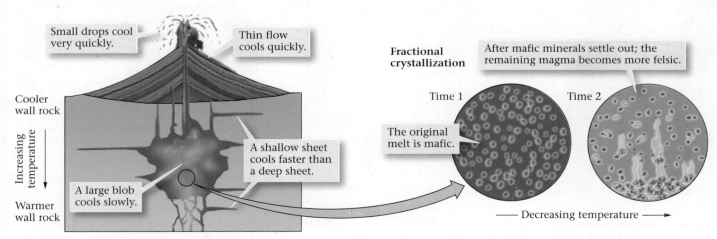

(b) The cooling rate of molten rock depends on the size and shape of the magma or lava body, and on its depth.

(c) The process of fractional crystallization results in a progressive change in magma composition during freezing.

FIGURE 4.8 Examples of extrusive igneous settings.

(a) This volcanic explosion produced two styles of ash eruption.

Ash cloud

Pyroclastic flow

(b) Layers of volcanic ash and debris, Hawaii.

20.cm

Some of the lava freezes in mid-air.

Lava flow

(c) This volcano is producing lava flows and fountains.

(d) A stack of over 50 lava flows, capped by debris, Mt. Vesuvius.

whereas a **sill** is a tabular intrusion that injects parallel to layering (Fig. 4.9a–d). In places where tabular intrusions cut across rock that does not have layering, a nearly vertical, wall-like tabular intrusion is a dike, whereas a nearly horizontal, tabletop-shaped tabular intrusion is a sill. Some intrusions start to inject between layers but then dome upward, creating a blister-shaped intrusion known as a **laccolith**.

Plutons are irregular or blob-shaped intrusions that range in size from tens of meters across to tens of kilometers across (Fig. 4.10a–d). The intrusion of numerous plutons in a

FIGURE 4.9   Igneous sills and dikes, examples of tabular intrusions.

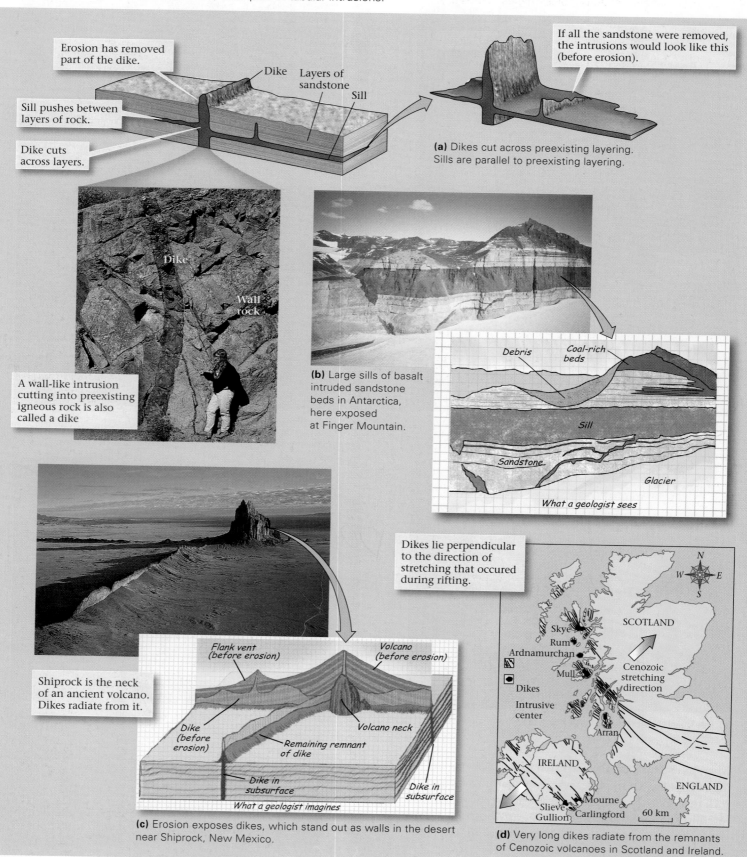

Erosion has removed part of the dike.

Sill pushes between layers of rock.

Dike cuts across layers.

Dike

Layers of sandstone

Sill

If all the sandstone were removed, the intrusions would look like this (before erosion).

(a) Dikes cut across preexisting layering. Sills are parallel to preexisting layering.

Dike

Wall rock

A wall-like intrusion cutting into preexisting igneous rock is also called a dike

(b) Large sills of basalt intruded sandstone beds in Antarctica, here exposed at Finger Mountain.

Debris

Coal-rich beds

Sill

Sandstone

Glacier

What a geologist sees

Dikes lie perpendicular to the direction of stretching that occured during rifting.

Shiprock is the neck of an ancient volcano. Dikes radiate from it.

Flank vent (before erosion)

Volcano (before erosion)

Dike (before erosion)

Volcano neck

Remaining remnant of dike

Dike in subsurface

Dike in subsurface

What a geologist imagines

(c) Erosion exposes dikes, which stand out as walls in the desert near Shiprock, New Mexico.

SCOTLAND

Skye

Rum

Ardnamurchan

Mull

Dikes

Intrusive center

Cenozoic stretching direction

Arran

IRELAND

ENGLAND

Slieve Gullion

Mourne Carlingford

60 km

(d) Very long dikes radiate from the remnants of Cenozoic volcanoes in Scotland and Ireland.

FIGURE 4.10 Igneous plutons, "blob-shaped" intrusions.

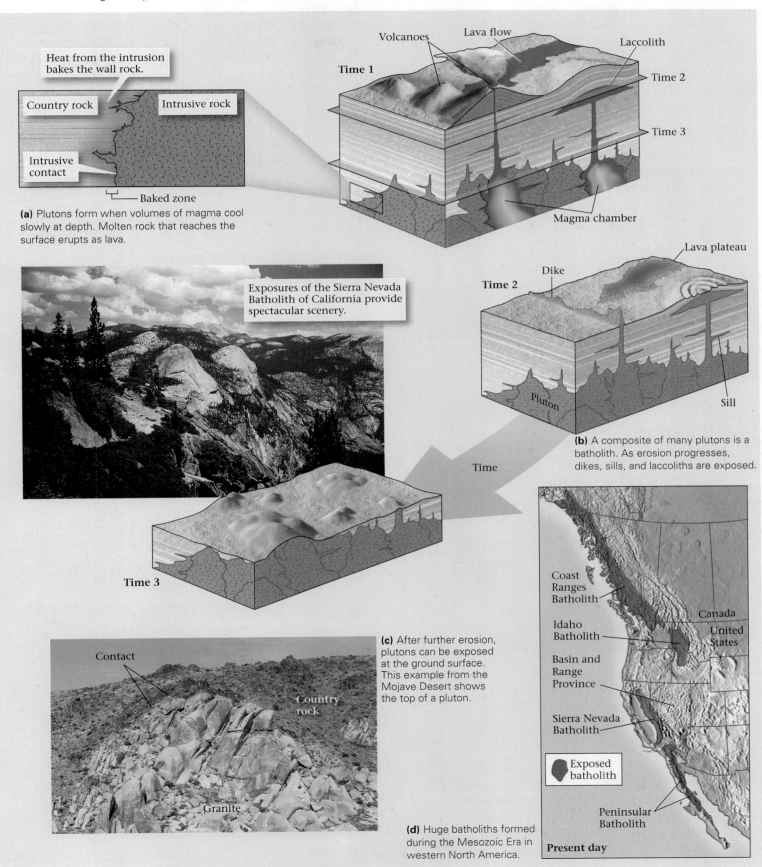

Heat from the intrusion bakes the wall rock.

Country rock

Intrusive rock

Intrusive contact

Baked zone

(a) Plutons form when volumes of magma cool slowly at depth. Molten rock that reaches the surface erupts as lava.

Volcanoes   Lava flow   Laccolith

Time 1   Time 2

Time 3

Magma chamber

Exposures of the Sierra Nevada Batholith of California provide spectacular scenery.

Lava plateau

Dike

Time 2

Pluton   Sill

(b) A composite of many plutons is a batholith. As erosion progresses, dikes, sills, and laccoliths are exposed.

Time

Time 3

(c) After further erosion, plutons can be exposed at the ground surface. This example from the Mojave Desert shows the top of a pluton.

Contact

Country rock

Granite

Coast Ranges Batholith

Idaho Batholith

Basin and Range Province

Sierra Nevada Batholith

Canada

United States

Exposed batholith

Peninsular Batholith

(d) Huge batholiths formed during the Mesozoic Era in western North America.

Present day

region creates a vast composite body that may be several hundred kilometers long and over 100 km wide; such immense masses of igneous rock are called **batholiths**. The rock making up the Sierra Nevada mountains of California is a batholith formed from plutons that intruded between 145 and 80 million years ago.

Where does the space for intrusions come from? Dikes form in regions where the crust is being stretched horizontally, such as in a rift. Thus, as the magma that makes a dike forces its way up into a crack, the crust opens up sideways (**Fig. 4.11a**). Intrusion of sills occurs near the surface of the Earth, so the pressure of the magma effectively pushes up the rock above the sill, leading to uplift of the Earth's surface (**Fig. 4.11b**).

How does the space for a pluton develop? Some geologists propose that a pluton is a frozen "diapir," meaning a light-bulb-shaped blob of magma that pierced overlying rock and pushed

it aside as it rose (**Fig. 4.11c**). Another explanation involves **stoping**, a process during which magma assimilates wall rock, and blocks of wall rock break off and sink into the magma (**Fig. 4.11d**). If a stoped block does not melt entirely, but rather becomes surrounded by new igneous rock, it is a **xenolith**, after the Greek word *xeno*, meaning foreign. More recently, geologists have proposed that plutons form by injection of several superimposed dikes or sills, which coalesce and recrystallize to become a single, massive body.

If intrusive igneous rocks form beneath the Earth's surface, why can we see them exposed today? Over long periods of geologic time, mountain building slowly uplifts huge masses of rock. Moving water, wind, and ice eventually strip away great thicknesses of overlying rock and expose the intrusive rock that has formed below.

TAKE-HOME MESSAGE

Volcanoes erupt lava and ash. But not all magma erupts. Some solidifies underground either in tabular intrusions (dikes and sills) or blob-like intrusions (plutons). In places, immense volumes of intrusive igneous rock form, producing batholiths.

4.7 HOW DO YOU DESCRIBE AN IGNEOUS ROCK?

If you wander around a city admiring building facades, you'll find that many facades consist of igneous rock, for such rocks tend to be very durable. If you had to describe one of these rocks to a friend, what adjectives might you use?

You would probably start by noting the rock's color. Overall, is the rock dark or light? More specifically, is it gray, pink, white, or black? Describing color may not be easy, because

FIGURE 4.11 Making room for an igneous intrusion.

Dike

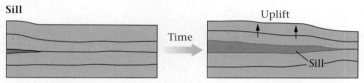

(a) Dikes fill space formed when crust undergoes horizontal stretching.

Sill

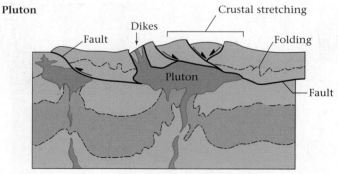

(b) Sills intrude between layers and may cause the uplift of the land surface.

Pluton

(c) The magma comprising a pluton may rise as a buoyant blob, or may intrude as a succession of sheets. Some plutons fill space produced by crustal stretching.

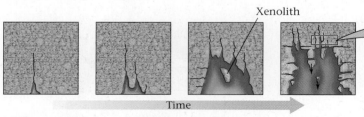

(d) Plutons may intrude by stoping. When blocks of wall rock fall into the magma, some dissolve but others may remain as xenoliths.

The white rock (granite) is intruding into the dark wall rock.

Formation of Igneous Rocks

Igneous rock forms by cooling of magma underground, or lava at the surface. Those that solidify underground are intrusive, whereas those that solidify at the surface are extrusive. The type of igneous rock that forms depends on the composition of the melt and the environment of cooling.

Stratified volcanic tuff

In the extrusive environment, melt may cool quickly and have a glassy texture. Melt that explodes into the air forms ash and other debris with fragmental texture.

Increasing silica content →

↑ Fast cooling

↓ Slow cooling

MAFIC	FELSIC
Scoria (glassy)	Obsidian (glassy)
Basalt (fine grained)	Rhyolite (fine grained)
Gabbro (coarse grained)	Granite (coarse grained)

In the intrusive environment crystals grow together to form an interlocking texture. Slower cooling makes coarser grains.

Minerals in an igneous rock crystallize in succession as the melt cools.

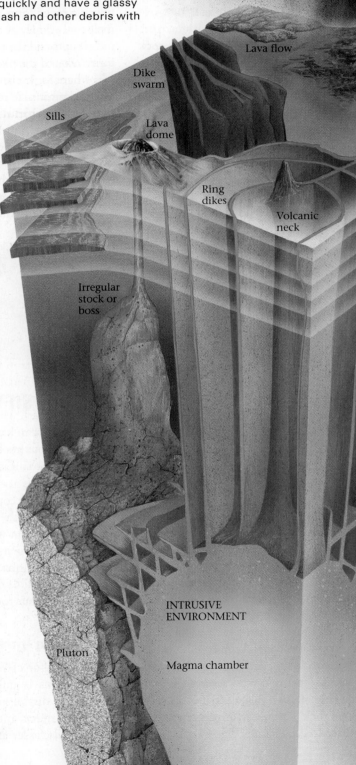

EXTRUSIVE ENVIRONMENT

Pyroclastic flow

Lava flow

Dike swarm

Sills

Lava dome

Laccolith

Ring dikes

Volcanic neck

Irregular stock or boss

INTRUSIVE ENVIRONMENT

Pluton

Magma chamber

some igneous rocks contain many visible mineral grains, each with a different color, but even so, you'll probably be able to characterize the overall hue of the rock. What physical factors control the color of igneous rock? Generally, the color reflects the rock's composition, but it isn't always so simple, because color may also be influenced by grain size and by the presence of trace amounts of impurities. (For example, the presence of a small amount of iron oxide gives rock a reddish tint.)

Next, you would probably characterize the rock's texture. A description of igneous texture indicates whether the rock consists of glass, crystals, or fragments. If the rock consists of crystals or fragments, a description of texture also specifies the grain size. Here are the common terms for defining texture:

- *Interlocking texture*: Rocks that consist of mineral crystals that intergrow when the melt solidifies, and thus fit together like pieces of a jigsaw puzzle, have an interlocking texture (**Fig. 4.12a**). Rocks with an interlocking texture are **crystalline igneous rocks**. The interlocking of crystals in these rocks occurs because once some grains have developed, they interfere with the growth of later-formed grains. The last grains to form end up filling irregular spaces between already existing grains.

  Geologists distinguish subcategories of crystalline igneous rocks according to the size of the crystals. Coarse-grained (phaneritic) rocks have crystals large enough to be identified with the naked eye. Fine-grained (aphanitic) rocks have crystals too small to be identified with the naked eye. Porphyritic rocks have larger crystals surrounded by a mass of fine crystals. In a porphyritic rock, the larger crystals are called phenocrysts, while the mass of finer crystals is called groundmass.

- *Fragmental texture*: Rocks consisting of igneous chunks and/or shards that are packed together, welded together, or cemented together after having solidified, are **fragmental igneous rocks** (Fig. 4.12a).

- *Glassy texture*: Rocks made of a solid mass of glass, or of tiny crystals surrounded by glass are **glassy igneous rocks** (Fig. 4.12a). Glassy rocks fracture conchoidally (Fig. 4.12b).

What factors control the texture of igneous rocks? In the case of nonfragmental rocks, *texture largely reflects cooling rate*. The presence of glass indicates that cooling happened so quickly that the atoms within a lava didn't have time to arrange into crystal lattices. Crystalline rocks form when a melt cools more slowly. In crystalline rocks, grain size depends on cooling time. A melt that cools rapidly, but not rapidly enough to make glass, forms fine-grained rock, because many crystal seeds form but none has time to grow large (**Fig. 4.12c**). A melt that cools very slowly forms a coarse-grained rock, because relatively few seeds form and each crystal has time to grow large (Fig. 4.12c).

Because of the relationship between cooling rate and texture, lava flows, dikes, and sills tend to be composed of fine-grained igneous rock. In contrast, plutons tend to be composed of coarse-grained rock. Specifically, plutons that intrude into hot wall rock at great depth cool very slowly and thus tend to have larger crystals than plutons that intrude into cool country rock at shallow depth, where they cool relatively rapidly. Porphyritic rocks form when a melt cools in two stages. First, the melt cools slowly at depth, so that phenocrysts form. Then, the melt erupts and the remainder cools quickly, so that groundmass forms around the phenocrysts.

There is, however, an exception to the standard cooling rate and grain size relationship. A very coarse-grained igneous rock called **pegmatite** doesn't necessarily cool slowly. Pegmatite contains crystals up to tens of centimeters across and occurs in dikes. Because pegmatite occurs in dikes, which generally cool quickly, the coarseness of the rock may seem surprising. Researchers have shown that pegmatites are coarse because they form from water-rich melts in which atoms can move around so rapidly that large crystals can grow very quickly.

TAKE-HOME MESSAGE

Igneous rocks come in a variety of textures—crystalline (made of interlocking crystals), glassy (made of a solid mass of glass), and fragmental (made of broken-up fragments of rock and/or ash). The grain size of crystalline rocks depends on the rate of cooling.

4.8 CLASSIFYING IGNEOUS ROCKS

Because melts can have a variety of compositions and can freeze to form igneous rocks in many different environments above and below the surface of the Earth, we observe a wide spectrum of igneous rock types. We classify these according to their texture and composition. Studying a rock's texture tells us about the rate at which it cooled, as we've seen, and therefore the environment in which it formed (see **Geology at a Glance**, p. 105). Studying its composition tells us about the original source of the magma and the way in which the magma evolved before finally solidifying. Below, we introduce some of the most important igneous rock types.

Crystalline Igneous Rocks

The scheme for classifying the principal types of crystalline igneous rocks is quite simple. The different compositional classes are distinguished on the basis of silica content—ultramafic, mafic, intermediate, or felsic—whereas the different textural classes are distinguished according to whether or

FIGURE 4.12 Textures and types of igneous rocks.

Crystalline   Fragmental   Glassy

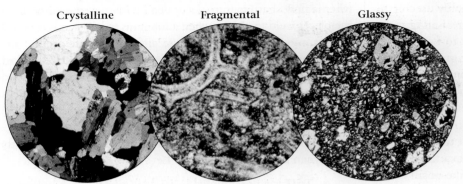

Obsidian

(a) Photomicrographs of thin sections reveal the different textures of igneous rocks.

(b) Obsidian fractures conchoidally.

Fine grained   Coarse grained

Felsic

Increasing silica content

Mafic

Rhyolite

Granite, cut by pegmatite

Andesite

Diorite

Basalt

Gabbro

(c) Examples of igneous rocks, arranged by grain size and composition. Different environments yield different rock types.

not the grains are coarse or fine. The chart in **Figure 4.13** gives the texture and composition of the most commonly used crystalline rock names. As a rough guide, the color of an igneous rock reflects its composition: mafic rocks tend to be black or dark gray, intermediate rocks tend to be lighter gray or greenish gray, and felsic rocks tend to be light tan to pink or maroon. Figure 4.12 provides images of some of these rocks.

Note that, according to Figure 4.13, rhyolite and granite have the same chemical composition but differ in grain size. Which of these two rocks develops from a melt of felsic composition depends on the cooling rate. A felsic lava that solidifies quickly at the Earth's surface or in a thin dike or sill turns into fine-grained rhyolite, but the same magma, if solidifying slowly at depth in a pluton, turns into coarse-grained granite. A similar situation holds for mafic lavas—a mafic lava that cools quickly in a lava flow forms basalt, but a mafic magma that cools slowly forms gabbro.

Glassy Igneous Rocks

Glassy texture develops more commonly in felsic igneous rocks because the high concentration of silica inhibits the easy growth of crystals. But basaltic and intermediate lavas can form glass if they cool rapidly enough. In some cases, a rapidly cooling lava freezes while it still contains a high concentration of gas bubbles—these bubbles remain as open holes known as **vesicles**. Geologists distinguish among several different kinds of glassy rocks.

- **Obsidian** is a mass of solid, felsic glass. It tends to be black or brown (Fig. 4.12b). Because it breaks conchoidally, sharp-edged pieces split off its surface when you hit a sample with a hammer. Pre-industrial people worldwide used such pieces for arrowheads, scrapers, and knife blades.
- **Pumice** is a felsic volcanic rock that contains abundant vesicles, giving it the appearance of a sponge. Pumice forms by the quick cooling of frothy lava that resembles the head of foam in a glass of beer. In some cases, pumice contains so many air-filled pores that it can actually float on water, like styrofoam (Fig. 4.14a).
- **Scoria** is a mafic volcanic rock that contains abundant vesicles (more than about 30%). Generally, the bubbles in scoria are bigger than those in pumice, and the rock, overall, looks darker (Fig. 4.14b).

Fragmental Igneous Rocks

Geologists distinguish among different kinds of fragmental igneous rocks according to the size of the fragments and the way

in which the fragments stick together. Fragmental rocks form when a flow shatters into pieces during its movement and then the pieces weld together; when a fountain of lava sends droplets of lava into the air, which freeze and accumulate around the vent; or when ash, crystals, and preexisting volcanic rock blast from a volcano during an eruption. Fragmental rocks composed of debris that has been blasted out of a volcano or thrown out in a fountain are commonly called **pyroclastic rocks** (from the Greek *pyro*, meaning fire, and from the word clastic, which signifies that the rocks are composed of grains that were already solid when they stuck together). Let's consider two examples.

- **Tuff** is a fine-grained pyroclastic igneous rock composed of volcanic ash. It may contain fragments of pumice.
- **Volcanic breccia** consists of larger fragments of volcanic debris that either fall through the air and accumulate, or form when a lava flow breaks into pieces.

FIGURE 4.13 Igneous rocks are classified based on composition and texture.

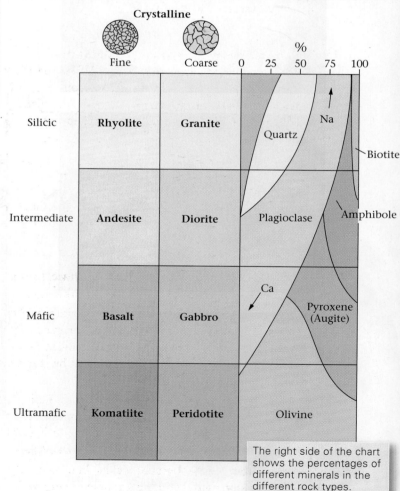

The right side of the chart shows the percentages of different minerals in the different rock types.

FIGURE 4.14 Vesicle-filled volcanic rocks.

(a) Pumice is so light that paper can hold it up.

(b) Scoria looks like a dark sponge, though very hard.

In this chapter, we've focused on the diversity of igneous rocks, and on why and where they form. We've seen that extrusive rocks erupt at volcanoes. In the next chapter, we will examine volcanic eruptions in greater detail and will discuss the geologic settings at which igneous activity takes place.

TAKE-HOME MESSAGE

Geologists classify igneous rocks on the basis of texture and composition. For a given rock composition (proportion of silica), there is one name for a fine-grained version and another for a coarse-grained version. Other names are used for glassy and fragmental rocks.

Chapter Summary

- Magma is liquid rock (melt) under the Earth's surface. Lava is melt that has erupted from a volcano at the Earth's surface.

- Magma forms when hot rock in the Earth partially melts. This process only occurs under certain circumstances—when the pressure decreases, when volatiles are added to hot rock, and when heat is transferred into the crust by magma rising from the mantle into the crust.

- Magma occurs in a range of compositions: felsic (silicic), intermediate, mafic, and ultramafic. The composition of magma reflects the original composition of the rock from which the magma formed and the way the magma evolves.

- Magma rises from depth because of its buoyancy and because of pressure caused by the weight of overlying rock.

- Magma viscosity depends on composition. Felsic magma is more viscous than mafic magma.

- The rate at which intrusive magma cools depends on the depth at which it intrudes, the size and shape of the magma body, and whether circulating groundwater is present. The cooling time influences the texture of an igneous rock.

- Extrusive igneous rocks form from lava that erupts out of a volcano. Intrusive igneous rocks develop from magma that freezes inside the Earth.

- Lava may solidify to form flows, or it may explode into the air to form ash.

- Intrusive igneous rocks form when magma intrudes into preexisting rock below Earth's surface. Blob-shaped intrusions are called plutons. Tabular intrusions that cut across layering are dikes, and those that form parallel to layering are sills. Huge intrusions, made up of many plutons, are known as batholiths.

- Igneous rocks are classified according to texture and composition.

GEOPUZZLE REVISITED

Molten rock, or magma, forms in the upper mantle and lower crust, but only at special localities—along convergent and divergent plate boundaries, at hot spots, and in rifts. Most magma freezes underground, but some erupts as lava or ash at volcanoes. When melt cools and solidifies, it becomes igneous rock.

Key Terms

assimilation (p. 96)

batholith (p. 104)

Bowen's reaction series (p. 99)

crystalline igneous rock (p. 106)

dike (p. 100)

extrusive igneous rock (p. 94)

fractional crystallization (p. 98)

fragmental igneous rock (p. 106)

geotherm (p. 94)

glassy igneous rock (p. 106)

igneous rock (p. 93)

intrusive igneous rock (p. 94)

laccolith (p. 101)

lava (p. 93)

lava flow (p. 93)

magma (p. 94)

obsidian (p. 108)

partial melting (p. 96)

pegmatite (p. 106)

pluton (p. 101)

pumice (p. 108)

pyroclastic rock (p. 108)

scoria (p. 108)

sill (p. 101)

stoping (p. 104)

tuff (p. 108)

vesicle (p. 108)

volcanic ash (p. 94)

volcanic breccia (p. 108)

volcano (p. 93)

xenolith (p. 104)

Review Questions

1. How is the process of freezing magma similar to that of freezing water? How is it different?

2. What are the sources of heat in the Earth? How did the first igneous rocks on the planet form?

3. Describe the three processes that lead to the formation of magmas.

4. Why are there so many different types of magmas?

5. Why do magmas rise from depth to the surface of the Earth?

6. What factors control the viscosity of a melt?

7. What factors control the cooling time of a magma within the crust?

8. How does grain size reflect the cooling time of a magma?

9. What does the mixture of grain sizes in a porphyritic igneous rock indicate about its cooling history?

On Further Thought

1. If you look at the Moon, even without a telescope, you see broad areas where its surface appears relatively darker and smoother. These areas are called *mare* (plural: *maria*), from the Latin word for "sea." The term is misleading, for they are not bodies of water, but rather plains of igneous rock formed after huge meteors struck the Moon and formed very deep craters. These impacts occurred early in the history of the Moon, when its insides were warmer. With this background information in mind, propose a cause for the igneous activity, and suggest the type of igneous rock that fills the mare. (*Hint*: Think about how the presence of a deep crater affects pressure in the region below the crater, and think about the viscosity of a magma that could spread over such a broad area.)

THE VIEW FROM SPACE The lighter colored elliptical areas in this satellite image of a region called the Pilbara block, in northwestern Australia, are complex batholiths consisting of Archean granite. The darker rock surrounding the batholiths consists of basalt, komatiite, and sedimentary rock. The field of view (side-to-side) is about 300 km.

CHAPTER 5

The Wrath of Vulcan: Volcanic Eruptions

On the coast of Hawaii, glowing red streams of lava spill into the sea, instantly vaporizing the water. Each flow adds a bit more land to the island.

GEOPUZZLE

Why do volcanoes exist? Why do they occur where they do? Are all eruptions the same?

Glowing waves rise and flow, burning all life on their way, and freeze into black, crusty rock which adds to the height of the mountain and builds the land, thereby adding another day to the geologic past. . . . I became a geologist forever, by seeing with my own eyes: the Earth is alive!

—Hans Cloos (1886–1951),
on seeing the eruption of Mt. Vesuvius (Italy)

5.1 INTRODUCTION

Every few hundred years, one of the hills on Vulcano, an island in the Mediterranean Sea off the western coast of Italy, rumbles and spews out molten rock, glassy cinders, and dense "smoke" (actually a mixture of various gases, fine ash, and very tiny liquid droplets). Ancient Romans thought that such eruptions happened when Vulcan, the god of fire, fueled his forges beneath the island to manufacture weapons for the other gods. Geologic study suggests, instead, that eruptions take place when

hot magma, formed by melting inside the Earth, rises through the crust and emerges at the surface. No one believes the Roman myth anymore, but the island's name evolved into the English word **volcano**, which geologists use to designate either an erupting vent through which molten rock reaches the Earth's surface or a mountain built from the products of eruption.

On the main peninsula of Italy, not far from Vulcano, another volcano, Mt. Vesuvius, towers over the Bay of Naples. Two thousand years ago, a prosperous Roman resort and trading town named Pompeii sprawled at the foot of Vesuvius. One morning in 79 C.E., earthquakes signaled the mountain's awakening. At 1:00 P.M. on August 24, a dark mottled cloud boiled up above Mt. Vesuvius's summit to a height of 27 km. As lightning sparked in its crown, the cloud drifted over Pompeii, turning day into night. Blocks and pellets of rock fell like hail, while fine ash and choking fumes enveloped the town (**Fig. 5.1a**). Frantic people rushed to escape, but for many it was too late. As the growing weight of volcanic debris began to crush buildings, an avalanche of hot ash swept over Pompeii, and by

FIGURE 5.1 The destruction of Pompeii.

(a) This 1817 painting by British artist J. M. W. Turner depicts the cataclysmic eruption of 79 C.E. that destroyed Pompeii.

(b) The remains of Mt. Vesuvius lie in the distance, behind the excavated ruins of Pompeii. The town was buried and preserved beneath meters of ash and debris.

(c) A plaster cast of a Pompeii resident who was buried in ash as he crouched in a corner of his house.

the next day the town had vanished beneath a 6-m-thick gray-black blanket of pyroclastic debris. This covering protected the ruins of Pompeii so well that when archaeologists excavated the town 1,800 years later, they found an amazingly complete record of Roman daily life (Fig. 5.1b). In addition, they discovered open spaces in the debris. Out of curiosity, they filled the spaces with plaster, and then dug away the surrounding ash. The spaces turned out to be fossil casts of Pompeii's unfortunate inhabitants, their bodies forever twisted in agony or huddled in despair (Fig. 5.1c).

Clearly, volcanoes are unpredictable and dangerous. Volcanic activity can build a towering, snow-crested mountain or can blast one apart. It can provide the fertile soil that enables a civilization to thrive, or a rain of destruction that can snuff one out. Because of the diversity of volcanic activity and its consequences, this chapter sets out ambitious goals. We first look more closely at the products of volcanic eruptions and the basic characteristics of volcanoes. Then we examine the different kinds of volcanic eruptions on Earth and why they occur where they do. Finally, we examine the hazards posed by volcanoes, efforts by geoscientists to predict eruptions and help minimize the damage they cause, and the possible influence of eruptions on climate and civilization.

TAKE-HOME MESSAGE

By the end of this chapter, you should be aware of the diverse materials produced by eruptions, how eruptions can differ from one another, and why volcanoes form where they do. Finally, you should recognize the nature of volcanic hazards.

5.2 THE PRODUCTS OF VOLCANIC ERUPTIONS

The drama of a volcanic eruption transfers materials from inside the Earth to our planet's surface. Products of an eruption come in three forms—lava flows, pyroclastic debris, and gas.

Lava Flows

Sometimes it races down the side of a volcano like a fast-moving, incandescent stream, sometimes it builds into a rubble-covered mound at a volcano's summit, and sometimes it oozes like a sticky but scalding paste. Clearly, not all lava behaves in the same way when it rises out of a volcano. Therefore, not all lava flows look or behave the same. Why? The character of a lava flow primarily reflects its viscosity, and not all lavas have the same viscosity. Differences in viscosity depend, in turn, on

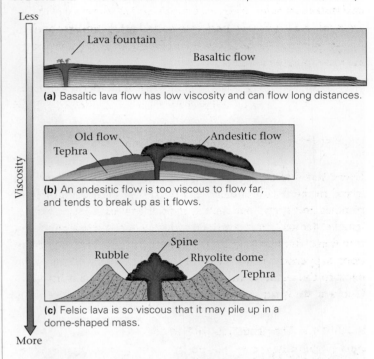

FIGURE 5.2 The character of a lava flow depends on its viscosity.

(a) Basaltic lava flow has low viscosity and can flow long distances.

(b) An andesitic flow is too viscous to flow far, and tends to break up as it flows.

(c) Felsic lava is so viscous that it may pile up in a dome-shaped mass.

a variety of factors including chemical composition, temperature, gas content, and crystal content. Silica content plays a particularly key role in controlling viscosity—silica-poor lava (basaltic) is less viscous, and thus flows farther, than does silica-rich (rhyolitic) lava (Fig. 5.2). To illustrate the different ways in which lava behaves, we now examine flows of different compositions.

Basaltic lava flows.   Basaltic (mafic) lava has very low viscosity when it first emerges from a volcano because it contains relatively little silica and is very hot. Thus, on the steep slopes near the summit of a volcano, it flows very quickly, sometimes at speeds of over 30 km per hour. The lava slows down to less-than-walking pace after it has traveled several kilometers and has started to cool. Most flows measure less than 10 km long, but the ends of some flows are as much as 500 km from the source.

How can lava travel so far? Although all the lava in a flow moves when it first emerges, rapid cooling causes the surface of the flow to crust over after the flow has moved a distance from the source. The solid crust serves as insulation, allowing the hot interior of the flow to remain liquid and continue to move. An insulated, tunnel-like conduit within a flow, through which lava moves, is a **lava tube**. In some cases, lava tubes drain and eventually become empty tunnels. Once a lava flow has covered the land and freezes, it becomes a solid blanket of rock (Fig. 5.3a, b).

The surface texture of a basaltic lava flow when it finally freezes reflects the timing of freezing relative to its movement.

FIGURE 5.3 Features of basaltic lava flows. They have low viscosity, and thus can flow long distances. Their surface and interior can be complex.

(a) A fast-moving flow coming from Mt. Etna, Sicily.

(b) A basaltic lava flow covers a highway in Hawaii.

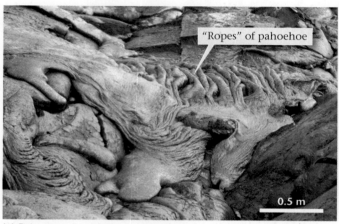

(c) Pahoehoe developing on a new lava flow in Hawaii.

(d) The rubbly surface of an a'a' flow, Sunset Crater, Arizona

(e) Columnar jointing develops when the interior of a flow cools and cracks. Such jointing develops in dikes and sills. This example is Devil's Postpile in California.

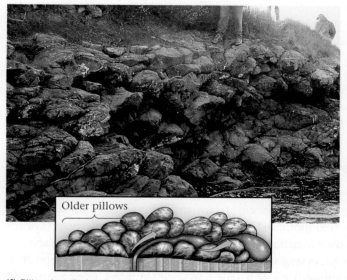

(f) Pillow basalt develops when lava erupts underwater. Later uplift may expose them above sea level, as in this Oregon outcrop.

Flows with warm, pasty surfaces wrinkle into smooth, glassy, rope-like ridges; geologists have adopted the Hawaiian word **pahoehoe** (pronounced pa-hoy-hoy) for such flows (Fig. 5.3c). If the surface layer of the lava freezes and then breaks up due to the continued movement of lava underneath, it becomes a jumble of sharp, angular fragments, creating a rubbly flow also called by its Hawaiian name, **a'a'** (pronounced ah-ah) (Fig. 5.3d). Footpaths made by people living in basaltic volcanic regions follow the smooth surface of pahoehoe flows rather than the rough, foot-slashing surface of a'a' flows.

During the final stages of cooling, lava flows contract and may fracture into roughly hexagonal columns. This type of fracturing is called **columnar jointing** (Fig. 5.3e).

Basaltic flows that erupt under water look different from those that erupt on land, because the lava cools so much more quickly. Because of rapid cooling, submarine basaltic lava forms a glass-encrusted blob, or pillow, on freezing (Fig. 5.3f). The rind of a pillow momentarily stops the flow's advance, but within minutes the pressure of the lava squeezing into the pillow breaks the rind, and a new blob of lava squirts out, and then itself freezes into a pillow.

Andesitic and rhyolitic lava flows. Because of its higher silica content and thus its greater viscosity, andesitic lava cannot flow as easily as basaltic lava. When erupted, andesitic lava first forms a large mound above the vent. This mound then advances slowly down the volcano's flank at only about 1 to 5 m a day, in a lumpy flow with a bulbous snout. Typically, andesitic flows are less than 10 km long. Because the lava moves so slowly, the outside of the flow has time to solidify; so as it moves, the surface breaks up into angular blocks, and the whole flow looks like a jumble of rubble.

Rhyolitic lava is the most viscous of all lavas because it is the most silicic and the coolest. Therefore, it tends to accumulate either above the vent in a dome-like mass, called a lava dome (Fig. 5.4), or in short and bulbous flows rarely more than 1 to 2 km long. Sometimes rhyolitic lava freezes while still in the vent, and then pushes upward as a column-like spire or spine up to 100 m above the vent. Rhyolitic flows, where they do form, have broken and blocky surfaces.

Volcaniclastic Deposits

On a mild day in February 1943, as Dionisio Pulido prepared to sow the fertile soil of his field in the Trans-Mexican volcanic belt 330 km (200 miles) west of Mexico City, an earthquake jolted the ground again, as it had dozens of times in the previous days. But this time, to Dionisio's amazement, the surface of his field visibly bulged upward by a few meters, and then cracked. Ash and sulfurous fumes filled the air, and Dionisio fled. When he returned the following morning, his rich

FIGURE 5.4 This rhyolite dome formed around 1350 C.E. in Panum Crater, California.

Tephra cone

Rhyolite dome

soil lay buried beneath a 40-m-high mound of gray cinders—Dionisio had witnessed the birth of Paricutín, a new volcano. During the next several months, Paricutín erupted continuously, at times blasting clots of lava into the sky like fireworks. By the following year, it had become a steep-sided cone 330 m high. Nine years later, when the volcano ceased all activity, its lava and debris covered 25 square km, and Dionisio's farm and those of his neighbors were gone.

This description of Paricutín's eruption, and that of Vesuvius described at the beginning of this chapter, emphasizes that volcanoes can erupt large quantities of fragmental igneous material. Geologists use the general term **volcaniclastic deposits** for accumulations of this material. Volcaniclastic deposits include **pyroclastic debris** (from the Greek *pyro*, meaning fire), which forms from lava that flies into the air and freezes. It also includes the debris formed when an eruption blasts apart preexisting volcanic rock that surrounds the volcano's vent, and the debris that accumulates after tumbling down the volcano in landslides or after being transported in water-rich slurries. Let's look at these components in more detail—you'll see that different types form in association with different kinds of eruptions.

Pyroclastic debris from basaltic eruptions. Basaltic magma rising in a volcano may contain dissolved volatiles. As such magma approaches the surface, the volatiles form bubbles. When the bubbles reach the surface, they burst and eject clots and drops of molten magma upward to form dramatic fountains (Fig. 5.5a). To picture this process, think of the droplets that spray from a newly opened bottle of soda. Pea- to plum-sized fragments of glassy lava and scoria are called cinders, or **lapilli**, from the Latin word for little stones. Flying droplets may trail thin strands of lava, which freeze into filaments of glass known as Pelé's hair, after the Hawaiian goddess of volcanoes, and the droplets themselves freeze into streamlined glassy beads known

FIGURE 5.5 Pyroclastic debris from basaltic eruptions.

(a) Fountains of lava may erupt from basaltic volcanoes.

Blocks and bombs litter the slope of a Hawaiian volcano.

Bombs may become streamlined as they fly through the air.

(b) Clots of lava that freeze in the air form bombs. Fragments of rock blasted off of the volcano form angular blocks. Droplets freeze into cinders.

as Pelé's tears. Apple- to refrigerator-sized fragments are called **bombs**, and these too may become streamlined during flight (Fig. 5.5b).

Pyroclastic debris from explosive eruptions. Intermediate and felsic lava is more viscous than basalt, and may be more gas-rich. As a result, eruptions of these magmas tend to be explosive (Fig. 5.6a). Debris ejected from explosive eruptions includes **ash**, composed of shards of glass formed when extremely tiny droplets of lava spray into the air and freeze instantly (Fig. 5.6b), and pumice fragments, formed when frothy lava freezes and then breaks apart. Lapilli produced during explosive eruptions consist of marble-sized pumice fragments or snowball-like balls of ash that form when ash mixes with water in the air and clumps together (Fig. 5.6c). Gravity pulls part of the rising column of ash down the side of the volcano in a fast-moving **pyroclastic flow** (Fig. 5.6d). This deadly cloud is also called a glowing avalanche, or in French, a *nuée ardente*. Ash from an explosive eruption may also rise as a cloud high above the volcano and be carried away by the wind to fall like snow elsewhere (Fig. 5.6e).

In April 1902, a devastating pyroclastic flow erupted from Mt. Pelée on the Caribbean island of Martinique. On the morning of May 8, a rhyolite dome that had obstructed the throat of the volcano broke. The immense pressure that had been building beneath the obstruction was suddenly released, and in the same way champagne bursts out of a bottle when the cork is pulled, a cloud of ash spewed out of Mt. Pelée. Col-

lapse of the ash column formed a pyroclastic flow—at a temperature of 200°C to 450°C—that swept down Pelée's flank. This flow rode on a cushion of air and reached speeds of up to 300 km per hour before slamming into the busy port town of St. Pierre. Within moments, the town's buildings were flattened and its 28,000 inhabitants were dead of incineration or asphyxiation. Only two people survived—one was a prisoner who was protected by the stout walls of his underground cell. Similar eruptions have happened more recently on the nearby island of Montserrat, but with a much smaller death toll because of timely evacuation.

Pyroclastic deposits. Unconsolidated deposits of pyroclastic grains, regardless of size, constitute **tephra**. Ash, or ash mixed with lapilli, becomes **tuff** when buried and transformed into coherent rock. Tuff that formed from ash that fell like snow from the sky is called air-fall tuff. A sheet of tuff that formed from a pyroclastic flow is an **ignimbrite**. Ash in an ignimbrite is so hot that it may weld together to form a hard mass.

Other volcaniclastic deposits. As noted earlier, the force of volcanic eruptions can fragment and disperse preexisting volcanic rock that has built up around the volcano. Chunks of fragmented preexisting volcanic rock are called **blocks**, as distinct from bombs (see Fig. 5.5b). Unconsolidated tephra and accumulations of blocks are quite weak, and prone to sliding down the flanks of volcanoes to form chaotic jumbles downslope (see Fig. 4.2b). When mixed with 10 to 25% water, from rain

FIGURE 5.6 Types of pyroclastic debris from felsic eruptions.

(a) An explosive volcanic eruption.

Wind

Stratospheric haze

Falling ash

Rising column

Falling lapilli

Collapsing column

Nuée ardente

(b) Ash consists of tiny specks of glass.

(c) Damp ash may clump in mid-air to form marble-sized lapilli.

A pyroclastic flow on a volcano in Japan.

Ash from the eruption of Montserrat blew skyward and avalanched seaward.

(d) A pyroclastic flow is an avalanche of hot ash.

Air-fall ash near Montserrat.

(e) Air-fall ash drops from clouds of ash that rise high in the atmosphere.

A lahar from the Mt. St. Helens eruption.

(f) A lahar is a slurry of ash mixed with water.

or melting snow, the material forms a **volcanic debris flow** that moves slowly downslope. Very wet mixtures of ash, coarser debris, and water form a slurry known as a **lahar**. A lahar can rush down valleys at speeds of up to 50 km per hour, and can travel tens of kilometers (Fig. 5.6f). When debris flows and lahars settle and drain, they yield a layer consisting of volcanic blocks suspended in an ashy mud.

Volcanic Gas

Most magma contains dissolved gases, including water, carbon dioxide, sulfur dioxide, and hydrogen sulfide (H_2O, CO_2, SO_2, and H_2S). In fact, up to 9% of a magma may consist of gaseous components. Generally, lavas with more silica contain a greater proportion of gas. Volcanic gases come out of solution when the magma approaches the Earth's surface and pressure decreases, just as bubbles come out of solution in a soda when you pop the bottle top off. In low-viscosity magma, gas bubbles can rise faster than the magma moves, and thus most reach the surface of the magma and enter the atmosphere before the lava does. Thus, some volcanoes may, for a while, produce large quantities of steam, without much lava (Fig. 5.7a). The last bubbles to form, however, freeze into the lava and become holes called **vesicles** (Fig. 5.7b). In high-viscosity magmas, the gas has trouble escaping because bubbles can't push through the sticky lava. When this happens, explosive pressures build inside or beneath the volcano.

TAKE-HOME MESSAGE

Volcanoes erupt lava (ranging in composition from mafic to felsic), pyroclastic debris (ranging in size from tiny ash shards to large blocks), and gas. Some ash falls like snow, whereas some rushes down the flanks of volcanoes in glowing avalanches.

5.3 THE ARCHITECTURE AND SHAPE OF VOLCANOES

As we saw in Chapter 4, melting in the upper mantle and lower crust produces magma, which rises into the upper crust. Typically, this magma accumulates underground in a **magma chamber**, an open space or a zone of highly fractured rock that can contain a large quantity of magma. Some of the magma freezes in the magma chamber and transforms into intrusive igneous rock, but some rises through an opening, or conduit, to the Earth's surface and erupts to form a volcano. In some volcanoes, the conduit has the shape of a vertical pipe, while

FIGURE 5.7 The gas component of volcanic eruptions.

(a) A volcano in Alaska erupting large quantities of steam.

(b) Gas bubbles frozen in lava produce vesicles, as in this block from Sunset Crater, Arizona.

in others the conduit is a crack called a **fissure** (Fig. 5.8a, b). Initially, a fissure may erupt a curtain of lava.

With time, the solid products of eruption (lava and/or pyroclastic debris) accumulate around a conduit to form a mound or cone. At the top of the mound, a circular depression called a **crater** (shaped like a bowl, up to 500 m across and 200 m deep) develops, either during eruption as material accumulates around the summit vent or just after eruption as the summit collapses into the drained conduit. Eruptions that happen in the summit crater are summit eruptions. In some volcanoes, a secondary conduit or fissure breaks through along the sides, or flanks, of the volcano, causing a flank eruption (Fig. 5.9a).

After major eruptions, the center of the volcano may collapse into the large, drained magma chamber below, creating a **caldera**, a big circular depression up to thousands of meters across and up to several hundred meters deep. Typically, a caldera has steep walls and a fairly flat floor (Fig. 5.9b–d). Some

FIGURE 5.8 Crater eruptions and fissure eruptions come from conduits of different shapes.

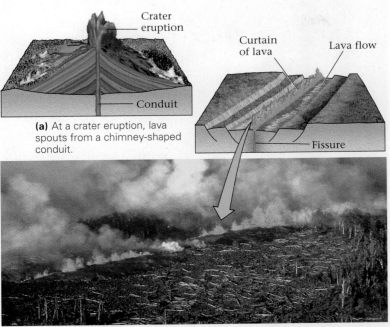

(a) At a crater eruption, lava spouts from a chimney-shaped conduit.

(b) At a fissure eruption, lava erupts as a curtain from a long crack.

FIGURE 5.9 The formation of volcanic calderas.

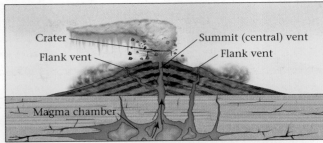

(a) As an eruption begins, the magma chamber inflates with magma. There can be a central vent and one or more flank vents.

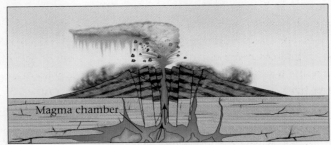

(b) During an eruption, the magma chamber drains, and the central portion of the volcano collapses downward.

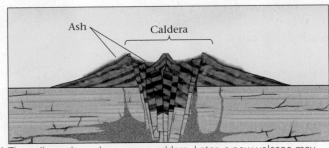

(c) The collapsed area becomes a caldera. Later, a new volcano may begin to grow within the caldera.

calderas form when a volcano explodes, blasting so much material into the air that only a portion of the volcano remains. Note that calderas differ from craters in terms of size, shape, and mode of formation.

Geologists distinguish among three different shapes of subaerial (above sea level) volcanoes. **Shield volcanoes**, so named because they resemble a soldier's shield lying on the ground, are broad, gentle domes (Fig. 5.10a). Shields generally form from low-viscosity basaltic lava flows. **Cinder cones** consist of cone-shaped piles of tephra (Fig. 5.10b). Typically, cinder cones are symmetrical and have deep craters at their summit. **Stratovolcanoes**, also called composite volcanoes, are large and cone-shaped (steeper near the summit), and consist of alternating layers of lava, tephra, and debris (Fig. 5.10c). Their shape, exemplified by Japan's Mt. Fuji, supplies the classic image most people have of a volcano. Explosions or landslides may destroy the classic shape at times.

TAKE-HOME MESSAGE

Volcanoes erupt from chimney-like conduits or from fissures. Basaltic eruptions build shield volcanoes, fountaining lava builds cinder cones, and alternating ash and lava eruptions produce stratovolcanoes. Explosions may produce large calderas.

(d) This caldera in Oregon formed about 7,700 years ago. Afterwards, it filled with water to become Crater Lake.

FIGURE 5.10 Different shapes of volcanoes.

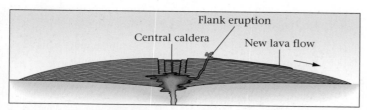

(b) A cinder cone on the flank of a larger volcano in Arizona. The pile of cinders has assumed the angle of repose. A lava flow covers the land surface in the distance.

(a) A shield volcano, made from successive flows of low-viscosity basalt, has very gentle slopes.

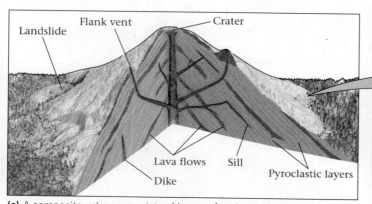

Mt. Fuji, a composite volcano in Japan, last erupted in 1707.

(c) A composite volcano consists of layers of tephra and lava. Volcanic debris flows and ash avalanches modify slopes and contribute to the development of a classic cone-like shape.

5.4 ERUPTIVE STYLES: WILL IT FLOW, OR WILL IT BLOW?

The 1990 eruption of Kilauea in Hawaii produced lakes and rivers of lava that cascaded down the volcano's flanks. In contrast, the 1980 eruption of Mt. St. Helens in Washington climaxed with a tremendous explosion that blanketed the surrounding countryside with tephra. Clearly, different volcanoes erupt in different ways. In fact, successive eruptions from the same volcano may even differ from one another. Geologists refer to the character of an eruption as the eruptive style, and make the following distinctions.

- **Effusive eruptions**: These eruptions produce mainly lava flows. Most yield low-viscosity basaltic lavas, which can stream tens to hundreds of kilometers. During effu-

sive eruptions, lava may pool in lakes around the vent, or spray up in fountains.

- **Explosive eruptions**: These eruptions produce clouds and avalanches of pyroclastic debris (Box 5.1). Pyroclastic eruptions happen when gas expands in the rising magma but cannot escape. Eventually, the pressure becomes so great that it blasts the lava, along with previously solidified volcanic rock, out of the volcano. In some cases, an explosive eruption blasts the volcano apart and leaves behind a large caldera. Such explosions, awesome in their power and catastrophic in their consequences, eject cubic kilometers of igneous materials. The resulting plume of debris resembles the mushroom cloud above a nuclear explosion. Coarse-grained ash and lapilli settle from the cloud close to the volcano, whereas finer ash settles farther away.

Note that the type of volcano (shield, cinder cone, or composite) depends on its eruptive style. Volcanoes that have only effusive eruptions become shield volcanoes, those that generate small pyroclastic eruptions due to fountaining basaltic lava yield cinder cones, and those that alternate between effusive and large pyroclastic eruptions become composite volcanoes. Large explosions yield calderas and blanket the surrounding countryside with ash. If the ash is hot enough, as happens in a glowing avalanche, the ash welds to form sheets of ignimbrite.

Why are there such contrasts in eruptive style? Eruptive style depends on the viscosity and gas content of the magma in the volcano. These characteristics, in turn, depend on the composition and temperature of the magma and on the environment (subaerial or submarine) in which the eruption occurs. Traditionally, geologists have classified volcanoes according to their eruptive style, each style named after a well-known example (see **Geology at a Glance**, pp. 122–123).

TAKE-HOME MESSAGE

Effusive eruptions produce lava, which can flow for tens of kilometers, sometimes through lava tunnels. Explosive eruptions occur when pressure builds up in the volcano or if water enters the volcano and turns to steam. Eruptive style reflects lava composition.

5.5 GEOLOGIC SETTINGS OF VOLCANISM

Different styles of volcanism occur at different locations on earth (see **Geotour 5** on p. GT-12). Most eruptions occur along plate boundaries, but major eruptions also occur at hot spots (Fig. 5.11). We'll now look at the settings in which eruptions occur, in the context of plate tectonics theory.

FIGURE 5.11 A map showing the distribution of volcanoes around the world, and the basic geologic settings in which volcanoes form, in the context of plate-tectonics theory.

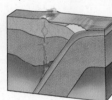  

(I) = Island arc   (C) = Continental arc   (R) = Rift   (H) = Hot spot   (M) = Mid-ocean ridge

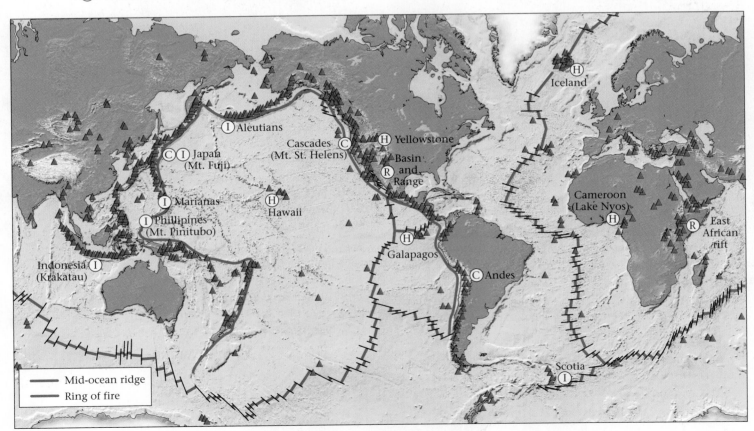

Volcanoes

Volcanic eruptions are a sight to behold, and in some cases a hazard to fear. Beneath a volcano, magma formed in the upper mantle or the lower crust rises to fill a magma chamber near the Earth's surface. When the pressure in this magma chamber becomes great enough, magma forces upward through a conduit, or crack, to the ground surface, and erupts.

Once molten rock has erupted at the surface, it is lava. Some lava spills down the side of the volcano to make a lava flow. In some cases, lava spatters or fountains out of the volcanic vent in little blobs or drops that cool quickly in the air to create fragmental igneous rock called tephra, or cinders. Larger blobs ejected by a volcano become volcanic bombs, which attain a streamlined shape as they fall.

Cinder Cones

Some volcanic eruptions produce fountains of small lava clots. These clots quickly freeze into glassy cinders, which accumulate rapidly into conical piles, or cinder cones, around the vent.

Caldera Formulation

When an explosive eruption suddenly drains a magma chamber, the overlying volcano collapses to produce a large circular depression called a caldera. Eventually, a new volcano may grow within the caldera. Some calderas fill with water and become lakes.

Vulcanian eruptions occur when a build up of gas and magma explode.

Strombolian crater explosions frequently burst through thinly crusted lava.

Hawaiian fountain explosions are caused by escaping gas.

Plinean explosions shoot a huge column of pumice fragments up to 50 km into the atmosphere. The ash fall rains down and the column collapses back around the vent traveling overland as a pyroclastic flow.

Side ve

Eroded cone

Lava cone

Lava flow

Sills

Dikes

Cinde

Lava pavement (cracked/broken)

| Lava flow 50 km | Mud flow 150 km | Pyroclastic flow 200 km | Ash fall 2500 km |

The distance volcanic hazards can travel from an eruption.

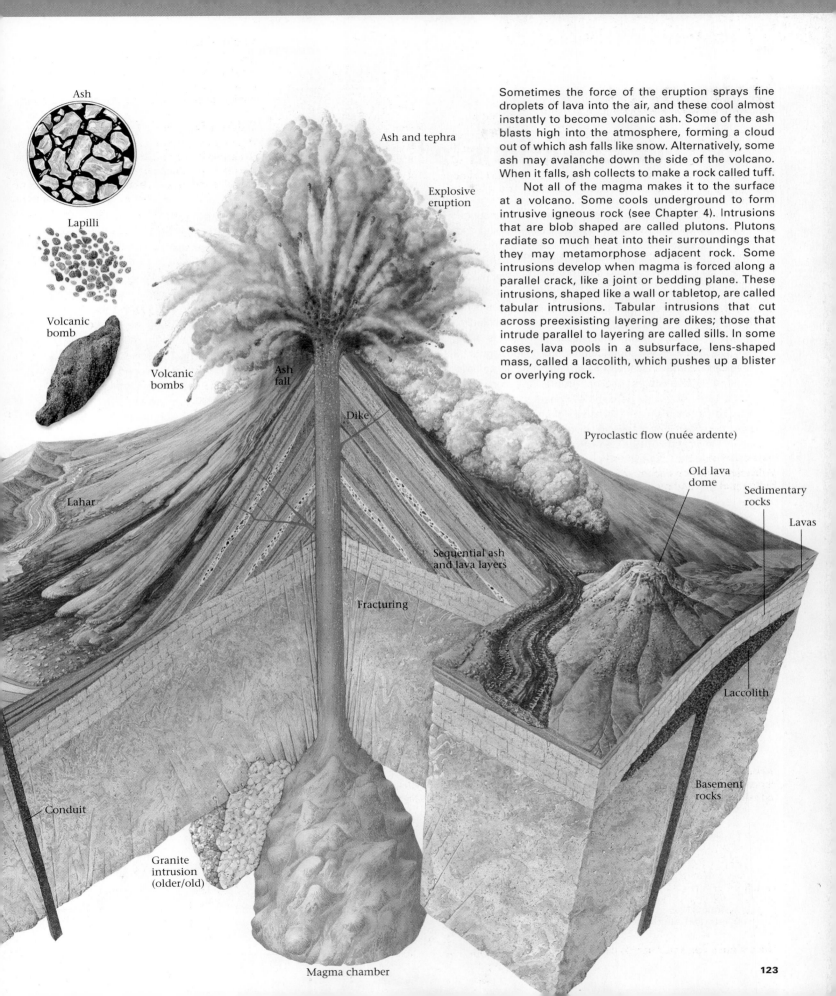

Ash

Lapilli

Volcanic bomb

Ash and tephra

Explosive eruption

Volcanic bombs

Ash fall

Dike

Lahar

Conduit

Granite intrusion (older/old)

Magma chamber

Sequential ash and lava layers

Fracturing

Pyroclastic flow (nuée ardente)

Old lava dome

Sedimentary rocks

Lavas

Laccolith

Basement rocks

Sometimes the force of the eruption sprays fine droplets of lava into the air, and these cool almost instantly to become volcanic ash. Some of the ash blasts high into the atmosphere, forming a cloud out of which ash falls like snow. Alternatively, some ash may avalanche down the side of the volcano. When it falls, ash collects to make a rock called tuff.

Not all of the magma makes it to the surface at a volcano. Some cools underground to form intrusive igneous rock (see Chapter 4). Intrusions that are blob shaped are called plutons. Plutons radiate so much heat into their surroundings that they may metamorphose adjacent rock. Some intrusions develop when magma is forced along a parallel crack, like a joint or bedding plane. These intrusions, shaped like a wall or tabletop, are called tabular intrusions. Tabular intrusions that cut across preexisisting layering are dikes; those that intrude parallel to layering are called sills. In some cases, lava pools in a subsurface, lens-shaped mass, called a laccolith, which pushes up a blister or overlying rock.

THE HUMAN ANGLE

BOX 5.1

Volcanic Explosions to Remember

Explosions of volcanoes generate enduring images of destruction. The historical record shows a vast range in the volume of debris erupted, even though the largest *observed* eruption (Tambora in 1815) was small compared to an explosion that took place over 600,000 years ago in what is now Yellowstone National Park, Wyoming (**Fig. 1**). Let's look at two notable cases.

Mt. St. Helens, a snow-crested stratovolcano in the Cascades of the northwestern United States, had not erupted since 1857. However, geologic evidence suggested that the mountain had a violent past, punctuated by many explosive eruptions. On March 20, 1980, an earthquake

announced that the volcano was awakening once again. A week later, a crater 80 m in diameter burst open at the summit and began emitting gas and pyroclastic debris. Geologists who set up monitoring stations to observe the volcano noted that its north side was beginning to bulge markedly, suggesting that the volcano was filling with magma and that the magma was making the volcano expand like a balloon. Their concern that an eruption was imminent led local authorities to evacuate people in the area.

The climactic eruption came suddenly. At 8:32 A.M. on May 18, a geologist, David Johnston, monitoring the volcano

from a distance of 10 km, shouted over his two-way radio, "Vancouver, Vancouver, this is it!" An earthquake had triggered a huge landslide that caused 3 cubic km of the volcano's weakened north side to slide away. The sudden landslide released pressure on the magma in the volcano, causing a sudden and violent expansion of gases that blasted through the side of the volcano (**Fig. 2**). Rock, steam, and ash screamed north at the speed of sound and flattened a forest and everything in it over an area of 600 square km (**Fig. 3**). Tragically, Johnston, along with sixty others, vanished forever.

FIGURE 2   The eruption of Mt. St. Helens, 1980.

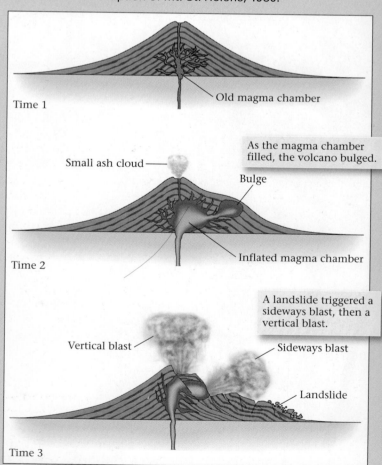

FIGURE 1   The relative amounts of pyroclastic debris (in cubic km) ejected during major historic eruptions of the past two centuries. Notice that the 1815 Tambora eruption was over five times bigger than the 1883 Krakatau eruption, which in turn was over five times larger than the 1980 Mt. St. Helens eruption.

Mt. St. Helens, 1980 C.E.,
1 km³ (0.24 cubic miles)

Krakatau, 1883 C.E.,
18 km³ (4.3 cubic miles)

Crater Lake, 7,600 B.C.E,
75 km³ (18 cubic miles)

Phlegrean Fields, 40,000 B.C.E.,
200 km³ (48 cubic miles)

Yellowstone, 630,000 B.C.E.,
1,000 km³ (240 cubic miles)

Yellowstone 2 Ma.
2,500 km³ (600 cubic miles)

Mt. Pinatubo, 1991 C.E.,
10 km³ (2.4 cubic miles)

Vesuvius, 79 C.E.,
25 km³ (6 cubic miles)

Tambora, 1815 C.E.,
145 km³ (35 cubic miles)

Yellowstone, 1.3 Ma,
250 km³ (62 cubic miles)

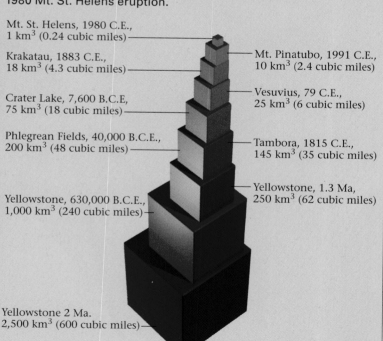

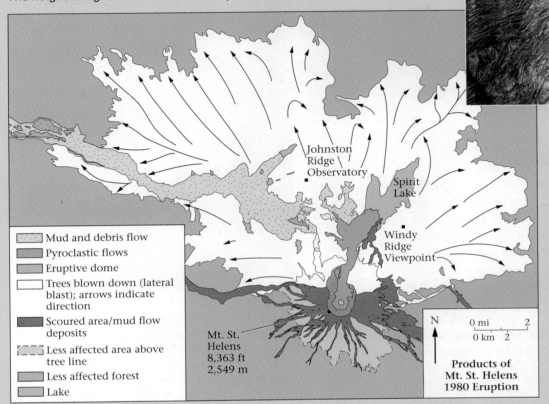

FIGURE 3   A map shows the dimensions of the region destroyed by the eruption of Mt. St. Helens. The arrows indicate the blast direction. The neighboring forest was flattened by a blast of rock, steam, and ash.

Johnston Ridge Observatory

Spirit Lake

Windy Ridge Viewpoint

Mud and debris flow
Pyroclastic flows
Eruptive dome
Trees blown down (lateral blast); arrows indicate direction
Scoured area/mud flow deposits
Less affected area above tree line
Less affected forest
Lake

Mt. St. Helens
8,363 ft
2,549 m

N

0 mi   2
0 km   2

Products of Mt. St. Helens 1980 Eruption

The blast knocked trees down, as if they were toothpicks.

Seconds after the sideways blast, a vertical column carried about 540 million tons of ash (about 1 cubic km) 25 km into the sky, where the jet stream carried it away so that it was able to circle the globe. In towns near the volcano, a blizzard of ash choked roads and buried fields. Water-saturated ash formed viscous slurries, or lahars, that flooded river valleys, carrying away everything in their path.

When the eruption was finally over, the once cone-like peak of Mt. St. Helens had disappeared—the summit now lay 440 m lower, and the once snow-covered mountain was a gray mound with a large gouge in one side. The volcano came alive again in 2004, but did not explode.

An even greater explosion happened in 1883. Krakatau, a volcano in the sea between Indonesia and Sumatra, where the Indian Ocean floor subducts beneath Southeast Asia, had grown to become a 9-km-long island rising 800 m (2,600 feet)

above the sea. On May 20, the island began to erupt with a series of large explosions, yielding ash that settled as far as 500 km away. Smaller explosions continued through June and July, and steam and ash rose from the island, forming a huge black cloud that rained ash into the surrounding straits. Ships sailing by couldn't see where they were going, and their crews had to shovel ash off the decks.

Krakatau's demise came at 10 A.M. on August 27, perhaps when the volcano cracked and the magma chamber suddenly flooded with seawater. The resulting blast, five thousand times greater than the Hiroshima atomic bomb explosion, could be heard as far as 4,800 km away, and sub-audible sound waves traveled around the globe seven times. Giant waves pushed out by the explosion slammed into coastal towns, killing over 36,000 people. Near the volcano, a layer of ash up to 40 m thick accumulated. When the air finally cleared,

Krakatau was gone, replaced by a submarine caldera some 300 m deep (**Fig. 4**). All told, the eruption shot 20 cubic km of rock into the sky. Some ash reached elevations of 27 km. Because of this ash, people around the world could view spectacular sunsets during the next several years.

FIGURE 4   Profile of Krakatau, before and after the eruption. Note that a new volcano (Anak Krakatau) has formed.

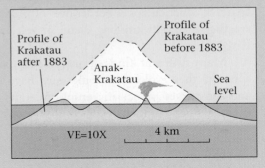

Profile of Krakatau after 1883

Profile of Krakatau before 1883

Anak-Krakatau

Sea level

VE=10X   4 km

FIGURE 5.12 The interior of an oceanic hot-spot volcano is complicated. Initially, eruption produces pillow basalts. When the volcano emerges above sea level, it becomes a shield volcano. The margins of the island frequently undergo slumping, and the weight of the volcano pushes down the surface of the lithosphere. The Hawaiian islands (inset) exemplify this architecture.

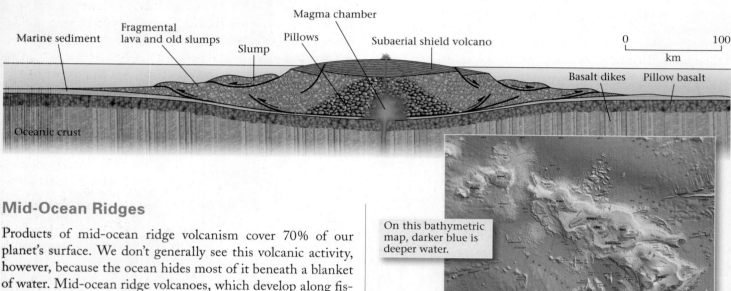

On this bathymetric map, darker blue is deeper water.

Mid-Ocean Ridges

Products of mid-ocean ridge volcanism cover 70% of our planet's surface. We don't generally see this volcanic activity, however, because the ocean hides most of it beneath a blanket of water. Mid-ocean ridge volcanoes, which develop along fissures parallel to the ridge axis, are not all continuously active. Each one turns on and off in a time scale measured in tens to hundreds of years. They erupt basalt, which, because it cools so quickly underwater, forms pillow-lava mounds. Water that heats up as it circulates through the crust near the magma chamber bursts out of hydrothermal (hot-water) vents along these mounds. As mentioned in Chapter 2, these vents are called black smokers.

Convergent Boundaries

Most of the subaerial volcanoes on Earth lie along convergent plate boundaries (subduction zones). The volcanoes form when volatile compounds such as water and carbon dioxide are released from the subducting plate and rise into the overlying hot mantle, causing melting and producing magma, which then rises through the lithosphere and erupts. Some of these volcanoes grow on oceanic crust and become volcanic island arcs, such as the Marianas of the western Pacific. Others grow on continental crust, building continental volcanic arcs such as the Cascade volcanic chain of Washington and Oregon. Typically, individual volcanoes in volcanic arcs lie about 50 to 100 km apart. Subduction zones border over 60% of the Pacific Ocean, creating a 20,000-km-long chain of volcanoes known as the Ring of Fire.

Many different kinds of magma form at volcanic arcs. As a result, these volcanoes sometimes have effusive eruptions and sometimes pyroclastic eruptions—and occasionally they explode. Such eruptions yield composite volcanoes, such as the elegant symmetrical cone of Mt. Fuji (see Fig. 5.10c) and the blasted-apart hulk of Mt. St. Helens (see Box 5.1).

Continental Rifts

The rifting of continental crust yields a wide array of different types of volcanoes, because (as in the case of continental hot spots) the magma that feeds these volcanoes comes both from the partial melting of the mantle and from the partial melting of the crust. Rifts host basaltic fissure eruptions, in which curtains of lava fountain up or linear chains of cinder cones develop. They also host explosive rhyolitic volcanoes, and in some places even stratovolcanoes.

Oceanic Hot-Spot Volcanoes (Hawaii)

When a hot-spot volcano first forms on oceanic lithosphere, basaltic magma erupts at the surface of the sea floor. At first, such submarine eruptions yield an irregular mound of pillow lava. With time, the volcano grows up above the sea surface and becomes an island. When the volcano emerges from the sea, the basalt lava that erupts no longer freezes so quickly, and thus flows as a thin sheet over a great distance. Thousands of thin basalt flows pile up, layer upon layer, to build a broad, dome-shaped shield volcano with gentle slopes (Fig. 5.12). As the volcano grows, portions of it can't resist the pull of gravity and slip seaward, creating large submarine slumps.

Continental Hot-Spot Volcanoes (Yellowstone National Park)

Yellowstone National Park lies at the northeast end of a track of volcanism marked by a string of calderas. The oldest of these calderas, at the southwest end of the track, erupted 16 million years ago (Fig. 5.13a, b). Recent and ongoing activity beneath Yellowstone has yielded fascinating landforms, volcanic rock deposits, and geysers. Eruptions at the Yellowstone hot spot differ from those in Hawaii in an important way: unlike Hawaii, the Yellowstone hot spot erupts both basaltic lava and rhyolitic pyroclastic debris. This happens because basaltic magma rising from the asthenosphere heats up and partially

FIGURE 5.13 Hot-spot volcanic activity in Yellowstone National Park.

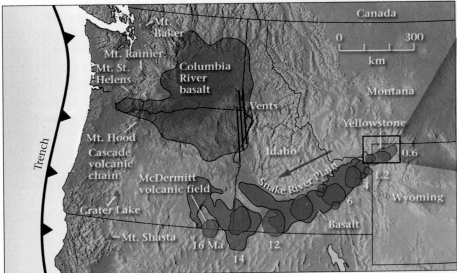

(a) Yellowstone lies at the end of a continental hot-spot track. Progressively older calderas follow the Snake River Plain. The blue arrow indicates plate motion.

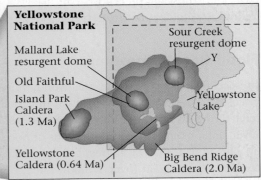

(b) Yellowstone overlies a caldera that exploded twice during the past 2 Ma.

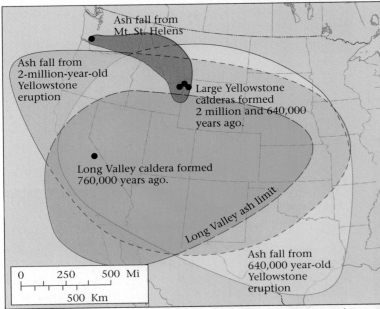

(c) The ash produced by explosions of the Yellowstone calderas covered vast areas—much more than Mt. St. Helens.

(d) Felsic tuffs form the colorful walls of Yellowstone Canyon.

melts the continental crust. Rhyolitic eruptions produce thick tuffs that now crop out as yellow and red rocks in the canyon of the Yellowstone River (Fig. 5.13c, d).

About 640,000 years ago, an immense pyroclastic flow, as well as a cloud of ash, blasted out of the Yellowstone region. Close to the eruption, ignimbrites up to tens of meters thick formed, and ash from the giant cloud sifted down over the United States as far east as the Mississippi River. The eruption, 1,000 times more powerful than that of Mt. St. Helens, left a huge caldera, almost 100 km across, which now dominates the landscape of Yellowstone. Magma remains beneath Yellowstone today, causing geyser activity.

Flood-Basalt Eruptions

In several locations around the world, huge sheets of low-viscosity lava erupted and spread out in vast sheets. Geologists refer to the lava of these sheets as **flood basalt** (Fig. 5.14a). Over time, many successive eruptions of flood basalt can build up a broad plateau. The aggregate volume of rock in such a plateau may be so great (over 175,000 km³), that geologists also refer to the plateau as a **large igneous province** (**LIP**).

An example of a LIP, the Columbia River Plateau, occurs in Washington and Oregon (see Fig. 5.13). The basalt here, which erupted around 15 million years ago, reaches a thickness of 3.5 km. Geologists have identified about 300 individual flows in the Columbia River Plateau. Lava in some of these flows traveled great distances—up to 600 km—from its source. Even larger flood-basalt provinces occur in eastern Siberia (an occurrence known as the "Siberian Traps"), the Deccan Plateau of India, the Paraná region of Brazil, and the Karroo Plateau of south Africa.

What causes flood-basalt eruptions? A popular hypothesis suggests that flood basalts form when a mantle plume first rises beneath a region that is undergoing rifting (Fig. 5.14b). As the plume reaches the base of the lithosphere, it has a bulbous head containing a huge amount of partially molten rock. Stretching and thinning of the overlying lithosphere results in further decompression of the plume head, and causes even more melt to form. The melt rises along fissures that form in the rift, and erupts spectacularly at the surface.

Iceland—a Hot Spot on a Ridge

Iceland is one of the few places on Earth where mid-ocean ridge volcanism has built a mound of basalt that protrudes above the sea. The island formed where a hot spot lies beneath the Mid-Atlantic Ridge—the presence of this hot spot (probably due to an underlying mantle plume) means that far more magma erupted here than beneath other places along the ridge. Because Iceland straddles a divergent plate boundary, it is being stretched apart, with faults forming as a consequence. Indeed, the central part of the island is a narrow rift, in which the youngest volcanic rocks of the island have erupted (Fig. 5.15a, b). This rift *is* the trace of the Mid-Atlantic Ridge. Faulting cracks the crust and so provides a conduit to a magma chamber. Thus, eruptions on Iceland tend to be fissure eruptions, yielding either curtains of lava that are many kilometers long or linear chains of small cinder cones.

TAKE-HOME MESSAGE

Most volcanic activity takes place on plate boundaries. The sea generally hides divergent-boundary volcanism—Iceland is an exception. Subduction produces island arcs and continental arcs. The latter include stratovolcanoes. Volcanism also occurs along rifts and at hot spots.

FIGURE 5.14 One hypothesis that explains flood basalts is a plume bringing very hot mantle to the base of rifting lithosphere.

(a) Flood basalt layers exposed on the wall of a canyon in Idaho.

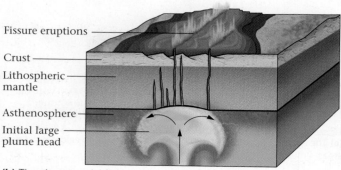

(b) The plume model for forming flood basalts.

FIGURE 5.15 Iceland's oceanic hot-spot plateau.

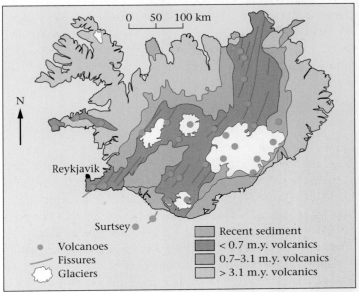

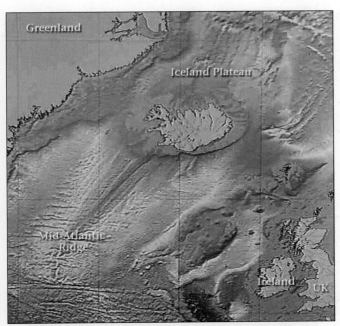

(a) A geologic map of Iceland shows how the youngest volcanics occur in the central rift, effectively the on-land portion of the Mid-Atlantic Ridge.

(b) A bathymetric map shows that Iceland sits atop a huge plateau straddling the Mid-Atlantic Ridge. Red is shallower water, blue is deeper.

5.6 BEWARE: VOLCANOES ARE HAZARDS!

Like earthquakes, volcanoes are natural hazards that have the potential to cause great destruction to humanity, in both the short term and the long term. According to one estimate, volcanic eruptions in the last two thousand years have caused about a quarter of a million deaths—much fewer than those caused by earthquakes, but nevertheless a sizable number. Considering the rapid expansion of cities, far more people live in dangerous proximity to volcanoes today than ever before, so if anything, the hazard posed by volcanoes has gotten worse—imagine if a Krakatau-like explosion were to occur next to a major city today. Let's now look at the different kinds of threats posed by volcanic eruptions.

Hazards Due to Eruptive Materials

Threat of lava flows. When you think of an eruption, perhaps the first threat that comes to mind is the lava that flows from a volcano. Indeed, on many occasions lava has overwhelmed towns (Fig. 5.16a–c). Basaltic lava from effusive eruptions is the greatest threat, because it can flow quickly and spread over a broad area. In Hawaii, recent lava flows have buried roads, housing developments, and vehicles (see Fig. 5.3b). Usually people have time to get out of the way of such flows, but not necessarily with their possessions. All they can do is watch helplessly from a distance as an advancing flow engulfs their homes. Before the

lava even touches it, the building may burst into flames from the intense heat. The most disastrous lava flow in recent times came from the 2002 eruption of Mt. Nyiragongo in the Congo. Lava flows traveled almost 50 km and flooded the streets of Goma, encasing the streets with a 2-m-thick layer of basalt. The flows destroyed almost half the city.

Threat of pyroclastic flows. Pyroclastic flows race down the flanks of a volcano at speeds of 100 to 300 km per hour (Fig. 5.16d). The largest can travel tens to hundreds of kilometers. The volume of ash contained in such glowing avalanches is not necessarily great—St. Pierre on Martinique was covered only by a thin layer of dust after the pyroclastic flow from Mt. Pelée had passed—but the cloud can be so hot and poisonous that it means instant death to anyone caught in its path, and because it moves so fast, the force of its impact can flatten buildings and forests.

Threat of ash and lapilli. During a pyroclastic eruption, large quantities of ash erupt into the air, later to fall back to Earth. Close to the volcano, pumice and lapilli tumble out of the sky, smashing through or crushing roofs of nearby buildings. These materials can accumulate into a blanket up to several meters thick. Winds can carry fine ash over a broad region. In the Philippines, for example, a typhoon spread heavy air-fall ash from the 1991 eruption of Mt. Pinatubo so that it covered a 4,000-square-km area (Fig. 5.16e, f). Because of heavy rains, the ash became soggy and heavy, and it was particularly damaging to roofs. Ash buries crops, may spread toxic chemicals that

FIGURE 5.16 Hazards due to lava and ash from volcanic eruptions.

Lava Flows

(a) A lava flow reaches a house in Hawaii and sets it on fire.

(b) Lava from Mt. Etna threatens a town.

(c) Residents rescue household goods after a lava flow filled the streets of Goma, along the East African Rift.

(d) This empty school bus was engulfed by lava in Hawaii.

Pyroclastic Debris

(e) A pyroclastic flow chases fleeing firefighters during the eruption of Mt. Unzen in Japan.

(f) This 1989-1990 eruption of Redoubt Volcano, Alaska, produced immense clouds of ash.

(g) A blizzard of ash fell from the cloud erupted by Mt. Pinatubo in the Philippines.

poison the soil, and insidiously infiltrates machinery, causing moving parts to wear out.

Fine ash from an eruption can also present a hazard to airplanes. Like a sandblaster, the sharp, angular ash abrades turbine blades, greatly reducing engine efficiency. The ash, along with sulfuric acid formed from the volcanic gas, scores windows and damages the fuselage. Also, when heated inside a jet engine, the ash melts, creating a liquid that sprays around the turbine and freezes; the resulting glassy coating restricts the air flow and causes the engine to flame out. In 1982, a British Airways 747 flew through the ash cloud over a volcano on Java. Corrosion turned the windshield opaque, and ingested ash caused all four engines to fail. For thirteen minutes, the plane glided earthward, dropping from 11.5 km (37,000 feet) to 3.7 km (12,000 feet) above the black ocean below. As passengers assumed a brace position for ditching at sea, the pilots tried repeatedly to restart the engines. Suddenly, in the oxygen-rich air of lower elevations, the engines roared back to life. The plane swooped back into the sky and headed for an emergency landing in Jakarta, where, without functioning instruments and with an opaque windshield, the pilot brought the 263 passengers and crew back to the ground safely. To land, he had to squint out an open side window, with only his toes touching the controls.

Other Hazards Related to Eruptions

Threat of the blast. Most exploding volcanoes direct their fury upward. But some, like Mt. St. Helens, explode sideways. The forcefully ejected gas and ash, like the blast of a bomb, flattens everything in its path. In the case of Mt. St. Helens, the region around the volcano had been a beautiful pine forest; but after the eruption, the once-towering trees were stripped of bark and needles and lay scattered over the hill slopes like matchsticks (see Box 5.1).

Threat of landslides. Eruptions commonly trigger large landslides along a volcano's flanks. The debris, composed of ash and solidified lava that erupted earlier, can move quite fast (250 km per hour) and far. During the eruption of Mt. St. Helens, 8 billion tons of debris took off down the mountainside, careened over a 360-m-high ridge, and tumbled down a river valley, until the last of it finally came to rest over 20 km from the volcano.

Threat of lahars. When volcanic ash and other debris mix with water, the result is a slurry that resembles freshly mixed concrete. A flow of this slurry, known as a lahar, can move downslope at speeds of over 50 km per hour. Because lahars are denser and more viscous than water, they pack more force than flowing water and can literally carry away everything in their path. The lahars of Mt. St. Helens traveled more than 40 km from the volcano, following existing drainages (**Fig. 5.17**). When they had passed, they left a gray and barren wake of mud, boulders, broken bridges, and crumpled houses, as if a giant knife had scraped across the landscape.

Lahars may develop in regions where snow and ice cover an erupting volcano, for the eruption melts the snow and ice, thereby creating an instant supply of water. Perhaps the most destructive lahar of recent times accompanied the eruption of the snow-crested Nevado del Ruiz in Colombia on the night of November 13, 1985. The lahar surged down a valley like a 40-m-high wave, hitting the sleeping town of Armero, 60 km from the volcano. Ninety percent of the buildings in the town were gone, replaced by a 5-m-thick layer of mud, which now entombs the bodies of 25,000 people.

Threat of earthquakes. Earthquakes accompany almost all major volcanic eruptions, for the movement of magma breaks rocks underground. Such earthquakes may trigger landslides on the volcano's flanks and can cause buildings to collapse and dams to rupture, even before the eruption itself begins.

Threat of tsunamis (giant waves). Where explosive eruptions occur in the sea, the blast and the underwater collapse of

FIGURE 5.17 Hazards due to lahars and gas.

A lahar buried the town of Amero, Colombia.

Cattle killed by a CO_2 cloud in Cameroon.

a caldera generate huge sea waves, or tsunamis, tens of meters high. Most of the 36,000 deaths attributed to the 1883 eruption of Krakatau were due not to ash or lava, but rather to tsunamis that slammed into nearby coastal towns (see Box 5.1).

Threat of gas. We have already seen that volcanoes erupt not only solid material, but also large quantities of gases such as water vapor, carbon dioxide, sulfur dioxide, and hydrogen sulfide. Usually the gas eruption accompanies the lava and ash eruption, with the gas contributing only a minor part of the calamity. But occasionally gas erupts alone and snuffs out life in its path without causing any other damage. Such an event occurred in 1986 near Lake Nyos in western Africa.

Lake Nyos is a small but deep lake filling the crater of an active volcano in Cameroon. Though only 1 km across, the lake reaches a depth of over 200 m. Because of its depth, the cool bottom water of the lake does not mix with warm surface water. Carbon dioxide gas slowly bubbles out of cracks in the floor of the crater and dissolves in the cool bottom water. Apparently, by August 21, 1986, the bottom water had become supersaturated in carbon dioxide. On that day, perhaps triggered by a landslide or wind, the lake "burped" and expelled a forceful froth of CO_2 bubbles. Because it is denser than air, this invisible gas flowed down the flank of the volcano and spread out over the countryside for about 23 km before dispersing. Although not toxic, carbon dioxide cannot provide oxygen for metabolism or oxidation. When the gas cloud engulfed the village of Nyos, it quietly put out the cooking fires and suffocated the sleeping inhabitants. The next morning, the landscape looked exactly as it had the day before, except for the lifeless bodies of 1,742 people and about 6,000 head of cattle (see Fig. 5.17).

TAKE-HOME MESSAGE

Volcanoes can be dangerous! The lava flows, pyroclastic debris, explosions, mud flows (lahars), landslides, earthquakes, and gas clouds produced during eruptions can destroy cities and farmland. Ash flows move very fast and incinerate everything in their path.

5.7 PROTECTION FROM VULCAN'S WRATH

Taken together, the various damaging phenomena accompanying a volcanic eruption can devastate a society. For example, archeological research suggests that the Minoan culture, which thrived in the eastern Mediterranean during the Bronze Age (beginning 3000 B.C.E.), disappeared during the century following explosive eruptions of the Santorini volcano in 1645 B.C.E. All that remains of Santorini today is a huge caldera whose rim projects above sea level as the Greek island of Thera. Ash clouds, tsunamis, and earthquakes generated by Santorini may have so disrupted their daily life that the Minoan people moved elsewhere, leaving behind their elaborately decorated palaces.

Clearly, volcanic eruptions are a natural hazard of extreme danger. Can anything be done to protect lives and property from this danger? The answer is, yes. Below, we first examine the evidence that geologists use to determine if a volcano has the potential to erupt, and then we consider the suite of observations that may allow geologists to predict the timing of an eruption.

Active, Dormant, and Extinct Volcanoes

Geologists refer to volcanoes that are erupting, have erupted recently, or are likely to erupt soon as **active volcanoes** and distinguish them from **dormant volcanoes**, which have not erupted for hundreds to thousands of years but do have the potential to erupt again in the future. Volcanoes that were active in the past but have shut off entirely and will never erupt in the future are called **extinct volcanoes**. As examples, geologists consider Hawaii's Kilauea to be active, for it currently erupts and has erupted frequently during recorded history. In contrast, Mt. Rainier in the Cascades last erupted centuries to millennia ago, but since subduction continues along the western edge of Oregon and Washington, the volcano could erupt in the future, and so it is considered dormant. Devil's Tower, in Wyoming, developed in association with volcanism millions of years ago but this could not happen again because the geologic cause for volcanism in the area no longer exists. Thus, volcanism in the Devil's Tower area is extinct.

How do you determine if a volcano is active, dormant, or extinct? One way is to examine the historical record. Another is to determine the age of erupted rocks, and search for evidence that the volcano is still in a tectonically active area. Finally, you can examine the landscape character of the volcano. Specifically, the shape and extent of erosion of a volcano depends on whether it has been erupting recently or ceased erupting long ago. Specifically, the shape (shield, stratovolcano, or cinder cone) of an erupting volcano depends primarily on the eruptive style, because at an erupting volcano, the process of construction happens faster than the process of erosion.

Once a volcano stops erupting, erosion attacks. The rate at which erosion destroys a volcano depends on whether it's composed of pyroclastic debris or lava. Cinder cones and ash piles can wash away quickly. In contrast, composite or shield volcanoes, which have been armor plated by lava flows, can withstand the attack of water and ice for quite some time. In the end, however, erosion wins out, and you can tell an old volcano that has not erupted for a long time from a volcano that has erupted recently by the extent to which river or glacial valleys have been carved into its flanks (**Fig. 5.18a**). In some

FIGURE 5.18 The shape of a volcano changes as it is eroded.

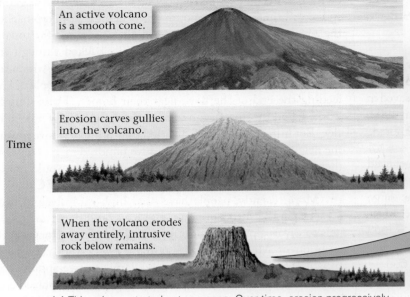

An active volcano is a smooth cone.

Time

Erosion carves gullies into the volcano.

When the volcano erodes away entirely, intrusive rock below remains.

(a) This volcano started out as a cone. Over time, erosion progressively removed weaker rock and exposed intrusive rock once a few km beneath the volcano.

(b) Devil's Tower in Wyoming is 260 m high. It formed when magma cooled deep beneath a volcano at about 40 Ma. Huge columnar joints formed when the magma cooled.

cases, the softer exterior of a volcano completely erodes away, leaving behind the plug of harder frozen magma that once lay within or deep beneath the volcano, as well as the network of dikes that radiate from this plug. You can see good examples of such landforms at Shiprock, New Mexico (Fig. 4.9c), and at Devil's Tower, Wyoming (**Fig. 5.18b**).

Predicting Eruptions

Little can be done to predict an eruption at a given active or dormant volcano beyond a few months or years, except to define the recurrence interval, the average time between eruptions. But short-term (weeks to months) predictions of impending volcanic activity, unlike short-term predictions of earthquakes, *are* actually feasible. Some volcanoes send out distinct warning signals announcing that an eruption may take place very soon, for as magma squeezes into the magma chamber, it causes a number of changes that geologists can measure.

- *Earthquake activity*: Movement of magma generates vibrations in the Earth. When magma flows into a volcano, rocks surrounding the magma chamber crack, and blocks slip with respect to each other. Such cracking and shifting also causes earthquakes. Thus, in the days or weeks preceding an earthquake, the region between 1 and 7 km beneath a volcano becomes seismically active.
- *Changes in heat flow*: The presence of hot magma increases the local heat flow, the amount of heat passing through

rock. In some cases, the increase in the heat flow melts snow or ice on the volcano, triggering floods and lahars even before an eruption occurs.

- *Changes in shape*: As magma fills the magma chamber inside a volcano, it pushes outward and can cause the surface of the volcano to bulge; the same effect happens when you blow into a balloon.
- *Increases in gas emission and steam*: Even though magma remains below the surface, gases bubbling out of the magma, or steam formed by the heating of groundwater by the volcano, percolate upward through cracks in the Earth and rise from the volcanic vent. So an increase in the volume of gas emission, or of new hot springs, indicates that magma has entered the ground below.

Because geologists can determine when magma has moved into the magma chamber of a volcano, government agencies now send monitoring teams to a volcano at the first sign of activity. These teams set up instruments to record earthquakes, measure the heat flow, determine changes in the volcano's shape, and analyze emissions.

Controlling Volcanic Hazards

Danger-assessment maps. Let's say that a given active volcano has the potential to erupt in the near future. What can we do to prevent the loss of life and property? Since we can't prevent the eruption, the first and most effective precaution is to define the regions that can be directly affected by the eruption—to compile a volcanic danger-assessment map (Fig. 5.19). These maps delineate areas that lie in the path of potential lava flows, lahars, debris flows, or pyroclastic flows.

Evacuation. Unfortunately, because of the uncertainty of prediction, the decision about whether or not to evacuate is a hard one. In the case of Mt. St. Helens, hundreds of lives were saved in 1980 by timely evacuation, but in the case of Mt. Pelée in 1902, thousands of lives were lost because warning signs were ignored.

Diverting flows. In traditional cultures, people believed that gods or goddesses controlled volcanic eruptions, so when a volcano rumbled, they provided offerings to appease the deity. Sometimes, people have used direct force to change the direction of a flow or even to stop it. For example, during a 1669

eruption of Mt. Etna, a lively volcano on the Italian island of Sicily, basaltic lava formed a glowing orange river that began to spill down the side of the mountain. When the flow approached the town of Catania, 16 km from the summit, fifty townspeople protected by wet cowhides boldly hacked through the chilled side of the flow to create an opening through which the lava could exit. They hoped thereby to cut off the supply of lava feeding the end of the flow, near their homes. Their strategy worked, and the flow began to ooze through the new hole in its side. But unfortunately, the diverted flow began to move toward the neighboring town of Paterno. Five hundred men of Paterno then chased away the Catanians so that the hole would not be kept open, and eventually the flow swallowed part of Catania.

More recently, people have used high explosives to blast breaches in the flanks of flows, and have built dams and channels to divert flows. Major efforts to divert flows from a 1983 eruption of Mt. Etna, and again in 1992, were successful. Inhabitants of Iceland used a particularly creative approach in 1973 to stop a flow before it overran a town—they sprayed cold seawater onto the flow to freeze it in its tracks (Fig. 5.20). The flow did stop short of the town, but whether this was a consequence of the cold shower it received remains unknown.

TAKE-HOME MESSAGE

Volcanoes don't erupt continuously and don't last forever, so we distinguish among active, dormant, and extinct volcanoes. It is possible to predict eruptions and take precautions. Once a volcano ceases to erupt, erosion destroys its eruptive shape.

FIGURE 5.19 A volcanic hazard-assesment map for the Mt. Rainier area in Washington (courtesy of the U.S. Geological Survey). The different colors on the map indicate different kinds of hazards.

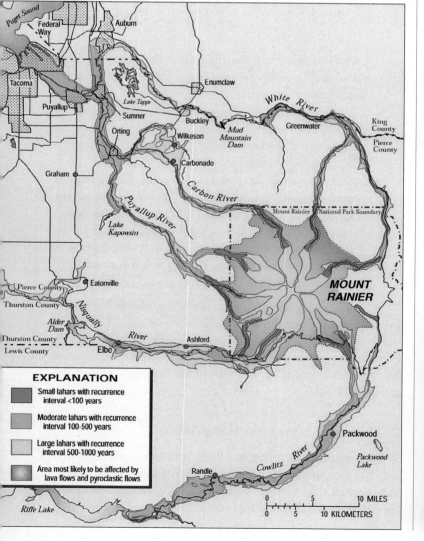

EXPLANATION

- Small lahars with recurrence interval <100 years
- Moderate lahars with recurrence interval 100-500 years
- Large lahars with recurrence interval 500-1000 years
- Area most likely to be affected by lava flows and pyroclastic flows

FIGURE 5.20 As a basalt flow encroached on this town in Iceland, firefighters used forty-three pumps to dump over 6 million cubic m of seawater on the lava to freeze it and stop the flow.

5.8 THE EFFECT OF VOLCANOES ON CLIMATE

In 1783, Benjamin Franklin was living in Europe, serving as the American ambassador to France. The summer of that year seemed to be unusually cool and hazy. Franklin, who was an accomplished scientist as well as a statesman, couldn't resist seeking an explanation for this phenomenon, and learned that in June of 1783, a huge volcanic eruption had taken place in Iceland. He wondered if the "smoke" from the eruption had prevented sunlight from reaching the Earth, thus causing the cooler temperatures. Franklin reported this idea at a meeting, and by doing so, may well have been the first scientist ever to suggest a link between eruptions and climate.

Franklin's idea seemed to be confirmed in 1815, when Mt. Tambora in Indonesia exploded. Tambora's explosion ejected over 100 cubic km of ash and pumice into the air. The sky became so hazy that stars dimmed by a full magnitude. Temperatures dipped so low in the northern hemisphere that 1816 became known as "the year without a summer." The unusual weather of that year inspired writers and artists. For example, Byron's 1816 poem "Darkness" contains the gloomy lines "The bright Sun was extinguish'd, and the stars / Did wander darkling in the eternal space . . . Morn came and went—and came, and brought no day."

Geoscientists have witnessed other examples of eruption-triggered coolness more recently. In the months following the 1883 eruption of Krakatau and the 1991 eruption of Pinatubo, global temperatures noticeably dipped. To study the effect of volcanic activity on climate even further in the past, geologists have studied ice from the glaciers of Greenland and Antarctica. Glacial ice has layers, each of which represents the snow that fell in a single year. Some layers contain concentrations of sulfuric acid, formed when sulfur dioxide from volcanic gas dissolves in the water from which snow forms. These layers indicate years in which major eruptions occurred. Years in which ice contains acid correspond to years during which the thinness of tree rings elsewhere in the world indicates a cool growing season.

How can a volcanic eruption create these cooling effects? When a large explosive eruption takes place, fine ash and aerosols (tiny liquid droplets) enter the stratosphere. It takes only about two weeks for the ash and aerosols to circle the planet. They stay suspended in the stratosphere for many months to years, because they are above the weather and do not get washed away by rainfall. The haze they produce causes cooler average temperatures because it absorbs incoming visible solar radiation during the day but does not absorb the infrared radiation that rises from the Earth's surface at night.

TAKE-HOME MESSAGE

The ash, gases, and aerosols produced by an explosive eruption can be blown around the globe, and this material can cause significant global cooling. Climatic effects, as well as other consequences of eruptions, may have hastened the end of some civilizations.

5.9 VOLCANOES ON OTHER PLANETS

We conclude this chapter by looking beyond the Earth, for our planet is not the only one in the Solar System to have hosted volcanic eruptions. We can see the effects of volcanic activity on our nearest neighbor, the Moon, just by looking up on a clear night. The broad darker areas of the Moon, the maria (singular mare) (after the Latin word for "sea"), consist of flood basalts that erupted over 3 billion years ago (Fig. 5.21a). Geologists propose that the flood basalts formed when huge meteors collided with the Moon, blasting out giant craters. Crater formation decreased the pressure in the Moon's mantle so that it underwent partial melting, producing basaltic magma that rose to the surface and filled the craters.

On Venus, about 22,000 volcanic edifices have been identified. Some of these even have caldera structures at their crests (Fig. 5.21b). Though no volcanoes currently erupt on Mars, the planet's surface displays a record of a spectacular volcanic past. The largest known mountain in the Solar System, Olympus Mons (Fig. 5.21c), is an extinct shield volcano on Mars. The base of Olympus Mons is 600 km across, and its peak rises 25 km above the surrounding plains.

Active volcanism currently occurs on Io, one of the many moons of Jupiter. Cameras in the *Galileo* spacecraft have recorded huge volcanoes on Io in the act of spraying plumes of sulfur gas into space (Fig. 5.21d) and have tracked immense, moving lava flows. Different colors of erupted material make the surface of this moon resemble a pizza. Researchers have proposed that the volcanic activity is due to tidal power: the gravitational pull exerted by Jupiter and by other moons alternately stretches and then squeezes Io, generating sufficient friction to keep Io's mantle hot. Geologists have also detected eruptions from moons of Saturn (Fig. 5.21e).

TAKE-HOME MESSAGE

Space exploration reveals that volcanism not only occurs on Earth, but has also left its mark on other terrestrial planets. Satellites have detected active eruptions on the icy moons of Jupiter and Saturn.

FIGURE 5.21   Volcanism on other planets and moons in the Solar System.

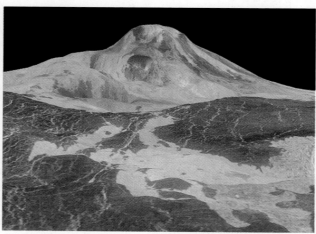

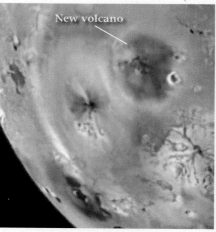

(a) Maria of the Moon were seas of basaltic lava.

(b) A volcano rises above the plains of Venus.

(d) An active volcano erupts sulfur on Io, a moon of Jupiter.

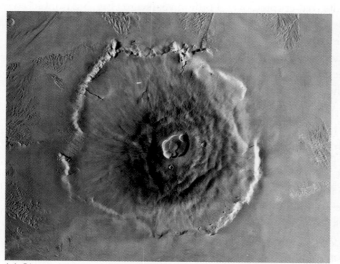

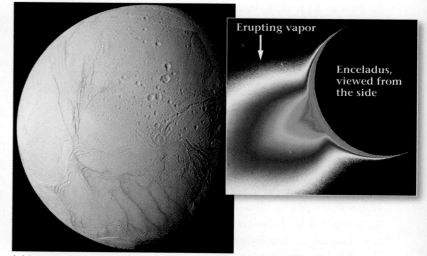

(c) Olympus Mons rises 22 km above the surface of Mars. It's the largest volcano in the solar system.

(e) Large cracks at the southern end of Enceladus, a moon of Saturn, erupt water vapor.

Chapter Summary

- Volcanoes are vents at which molten rock (lava), pyroclastic debris, gas, and aerosols erupt at the Earth's surface. A hill or mountain created from the products of an eruption is also called a volcano.

- The characteristics of a lava flow depend on its viscosity, which in turn depends on its temperature and composition.

- Basaltic lavas can flow great distances. Pahoehoe flows have smooth, ropy surfaces, whereas a'a' flows have rough, rubbly surfaces. Andesitic and rhyolitic lava flows tend to pile into mounds at the vent.

- Pyroclastic debris includes powder-sized ash, marble-sized lapilli, and apple- to refrigerator-sized blocks and bombs. Some falls from the air, whereas some forms glowing avalanches that rush down the side of the volcano.

- Eruptions may occur at a volcano's summit or from fissures on its flanks. The summit of an erupting volcano may collapse to form a bowl-shaped depression called a caldera.

- A volcano's shape depends on the type of eruption. Shield volcanoes are broad, gentle domes. Cinder cones are steep-sided, symmetrical hills composed of tephra. Composite volcanoes can become quite large and consist of alternating layers of pyroclastic debris and lava.

- The type of eruption depends on several factors, including the lava's viscosity and gas content. Effusive eruptions produce only flows of lava, whereas explosive eruptions produce clouds and flows of pyroclastic debris.
- Different kinds of volcanoes form in different geologic settings as defined by plate tectonics theory.
- Volcanic eruptions pose many hazards: lava flows overrun roads and towns, ash falls blanket the landscape, pyroclastic flows incinerate towns and fields, landslides and lahars bury the land surface, earthquakes topple structures and rupture dams, tsunamis wash away coastal towns, and invisible gases suffocate nearby people and animals.
- Eruptions can be predicted by earthquake activity, changes in heat flow, changes in shape of the volcano, and the emission of gas and steam.
- We can minimize the consequences of an eruption by avoiding construction in danger zones and by drawing up evacuation plans. In a few cases, it may be possible to divert flows.
- Immense flood basalts cover portions of the Moon. The largest known volcano, Olympus Mons, towers over the surface of Mars. Satellites have documented evidence for eruptions on moons of Jupiter and Saturn.

GEOPUZZLE REVISITED

Earth's volcanoes develop because there are places in the upper mantle and crust where partial melting takes place and magma forms. Though some magma freezes underground, some rises through conduits to the surface and erupts as lava or pyroclastic debris. The special places where melting takes place can be understood in the context of plate tectonics theory. Not all eruptions are the same, in part because not all lava has the same composition. Some eruptions spew out lava that flows in fast-moving streams, whereas others end in a cataclysmic explosion that blankets the countryside in ash.

Key Terms

a'a' (p. 115)

active volcano (p. 132)

ash (p. 116)

blocks (p. 116)

bombs (p. 116)

caldera (p. 118)

cinder cone (p. 119)

columnar jointing (p. 115)

crater (p. 118)

dormant volcano (p. 132)

effusive eruption (p. 120)

explosive eruption (p. 120)

extinct volcano (p. 132)

fissure (p. 118)

flood basalt (p. 128)

ignimbrite (p. 116)

lahar (p. 118)

lapilli (p. 115)

large igneous province (LIP) (p. 128)

lava tube (p. 113)

magma chamber (p. 118)

pahoehoe (p. 115)

pyroclastic debris (p. 115)

pyroclastic flow (p. 116)

shield volcano (p. 119)

stratovolcano (p. 119)

tephra (p. 116)

tuff (p. 116)

vesicles (p. 118)

volcanic debris flow (p. 118)

volcaniclastic deposits (p. 115)

volcano (p. 112)

Review Questions

1. Describe the three different kinds of material that can erupt from a volcano.
2. Describe the differences between a pyroclastic flow and a lahar.
3. Describe the differences among shield volcanoes, stratovolcanoes, and cinder cones. How are these differences explained by the composition of their lavas and other factors?
4. Why do some volcanic eruptions consist mostly of lava flows, while others are explosive and do not produce flows?
5. Describe the activity in the mantle that leads to hot-spot eruptions.
6. How do continental-rift eruptions form flood basalts?
7. Contrast an island volcanic arc with a continental volcanic arc. What is a hot-spot volcano?
8. Identify some of the major volcanic hazards, and explain how they develop.
9. How do geologists predict volcanic eruptions?
10. Explain how steps can be taken to protect people from the effects of eruptions.

On Further Thought

1. The Long Valley Caldera, near the Sierra Nevada Mountains, exploded about 700,000 years ago and produced an immense ash fall called the Bishop Tuff. About 30 km to the northwest lies Mono Lake, with an island in the middle and a string of craters extending south from its south shore. Hot springs and tufa deposits can be found along the lake. You can see the lake on *Google Earth*™ at Lat 37° 59' 56.58" N Long 119° 2' 18.20" W. Explain the origin of Mono Lake. Do you think that it represents a volcanic hazard?

INTERLUDE B

A Surface Veneer:
Sediments and Soils

Students traversing a field in Ireland step on grass, rooted in soil, derived from light-brown unconsolidated sediment. Bedrock, here consisting of gray sandstone and shale, lies still further down.

B.1 INTRODUCTION

In the 1950s, the government of Egypt decided to build the Aswan High Dam to trap water of the Nile River in a huge reservoir before the water could reach the Mediterranean Sea. In the process of identifying a good site for the dam's foundation, geologists discovered that the present-day Nile River flows on the surface of a 1.5-km-thick layer of gravel, sand, and mud that fills what was once a canyon as large as the Grand Canyon (Fig. B.1). How could the river once have carved a canyon 1.5 km deep, and why did this canyon later fill with sediment?

The origin of the pre-Nile canyon remained a mystery until the summer of 1970, when geologists began to study the material that forms the floor of the Mediterranean Sea. They expected to find layers consisting of the shells of plankton (tiny floating organisms) that had settled out of the water, or of clay that rivers had carried to the sea. To their surprise, however, they also found a 2-km-thick layer of halite and gypsum. Such salts form when seawater dries up. The researchers proposed that to yield a salt layer that is 2 km thick, the entire Mediterranean would have had to dry up completely several times, with the sea refilling after each drying event. They realized that this discovery solved the mystery of the pre-Nile canyon.

FIGURE B.1 The present Nile River flows on sediment that filled a deep canyon, cut when the Mediterranean basin was dry.

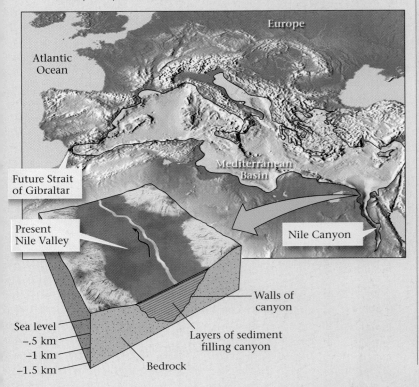

When the Mediterranean Sea dried up, the Nile River could dig down and cut a canyon. And when the sea refilled with water, this canyon flooded and filled with sand and gravel.

Why did the Mediterranean Sea dry up? Only 10% of the water in the Mediterranean comes from rivers, and since the sea lies in a hot, dry region, 10 times that amount of water evaporates from its surface each year. Thus, most of the water in the Mediterranean enters through the Strait of Gibraltar from the Atlantic Ocean. If this flow stops, the Mediterranean Sea evaporates. About 6 million years ago, the northward-drifting African Plate collided with the European Plate, forming a natural dam separating the Mediterranean from the Atlantic. When global sea level dropped, water stopped flowing in from the Atlantic and the Mediterranean evaporated. All the salt that had been dissolved in its water accumulated as a solid deposit of halite and gypsum on the floor of the resulting basin, and the pre-Nile canyon formed. When sea level rose, a gigantic flood rushed from the Atlantic into the Mediterranean, filling the basin again. This process repeated many times. About 5.5 million years ago, the Mediterranean rose to its present level, and gravel, sand, and mud carried by the Nile River filled the Nile canyon to its present level.

Geologists refer to the kinds of deposits just described—sand, mud, gravel, halite and gypsum layers, and shell fragments—as sediment. **Sediment**, broadly defined, consists of loose fragments of rocks or minerals broken off bedrock, mineral crystals that precipitate directly out of water, and shells or shell fragments.

Sediments form by the weathering (physical and chemical breakdown) of preexisting rock. They form a surface veneer, or "cover," on bedrock (Fig. B.2). This veneer ranges in thickness from nonexistent, in places where bedrock crops out at the Earth's surface, to a few kilometers. Some sediments transform into soil, essential for life. Let's now look at how weathering produces sediment, and how soils form and evolve.

B.2 WEATHERING: FORMING SEDIMENT

Weathering refers to the combination of processes that break up and corrode solid rock, eventually transforming it into sediment. We can say that rock that has not undergone weathering is unweathered or "fresh" (Fig. B.3). Rock exposed at the Earth's surface sooner or later crumbles away because of weathering. Just as a plumber can unclog a drain by using physical force (with a plumber's snake) or by causing a chemical reaction (with a dose of liquid drain opener), nature can attack rocks in two ways, so geologists distinguish between two types of weathering: physical and chemical.

FIGURE B.2 A layer of unconsolidated sediment (sand, clay, and cobbles), topped by dark soil, overlies bedrock in this outcrop along the coast of western Ireland.

FIGURE B.3 This outcrop shows the contrast between fresh and weathered granite. The rock below the notebook is fresh—the outcrop face is a fairly smooth fracture. The rock above the notebook is weathered—the outcrop face is crumbly, breaking into grains that have fallen and collected on the ledge.

Weathered granite

Unweathered (fresh) granite

Physical Weathering

Physical weathering, sometimes also referred to as mechanical weathering, breaks intact rock into unconnected clasts (grains or chunks), collectively called debris or detritus. Each size range of clasts has a name (Table B.1). Many different phenomena contribute to physical weathering:

Jointing. Rocks buried deep in the Earth's crust endure enormous pressure due to the weight of overlying rock or overburden. Rocks at depth are also warmer than rocks nearer the surface because of the Earth's geothermal gradient. Over long periods of time, moving water, air, and ice at the Earth's surface grind away and remove overburden, so rock formerly at depth rises closer to the Earth's surface. As a result, the pressure squeezing this rock decreases, and the rock becomes cooler. A change in pressure and temperature causes rock to change shape slightly. Such changes cause hard rock to break into pieces. Natural cracks that form in rocks due to removal of overburden or due to cooling (and for other reasons as well; see Chapter 9) are known as **joints**.

Almost all rock outcrops contain joints. Some joints are fairly planar, some curving, and some irregular. For example, large granite plutons split into onion-like sheets along joints that lie parallel to the mountain face; this process is called exfoliation. Sedimentary rock beds, however, tend to break into rectangular blocks (Fig. B.4a). Regardless of their orientation, the formation of joints turns formerly intact bedrock into loose blocks. Eventually, these blocks fall from the outcrop at which they formed. After a while, they may collect in an apron of talus, the rock rubble at the base of a slope (Fig. B.4b).

Frost wedging. Freezing water bursts pipes and shatters bottles because water expands when it freezes and pushes the walls of the container apart. The same phenomenon happens in rock. When the water trapped in a joint freezes, it forces the joint open and may cause the joint to grow. Such frost wedging helps break blocks free from intact bedrock (Fig. B.5a).

Root wedging. Have you ever noticed how the roots of an old tree can break up a sidewalk? Even though the wood of roots

TABLE B.1 Clasts are classified by grain diameter

Boulders	More than 256 millimeters (mm)
Cobbles	Between 64 mm and 256 mm
Pebbles	Between 2 mm and 64 mm
Sand	Between 1/16 mm and 2 mm
Silt	Between 1/256 mm and 1/16 mm
Mud	Less than 1/256 mm

FIGURE B.4 Joints (natural cracks) break bedrock into blocks and sheets, which can tumble down a slope.

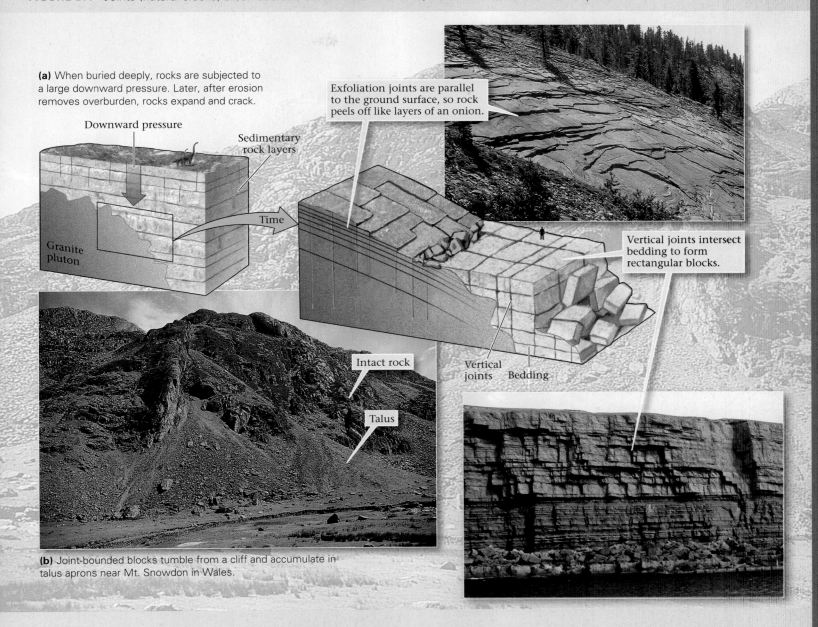

(a) When buried deeply, rocks are subjected to a large downward pressure. Later, after erosion removes overburden, rocks expand and crack.

Downward pressure

Sedimentary rock layers

Granite pluton

Time

Exfoliation joints are parallel to the ground surface, so rock peels off like layers of an onion.

Vertical joints intersect bedding to form rectangular blocks.

Intact rock

Talus

Vertical joints   Bedding

(b) Joint-bounded blocks tumble from a cliff and accumulate in talus aprons near Mt. Snowdon in Wales.

doesn't seem very strong, as roots expand they apply pressure to their surroundings. Tree roots that grow into joints can push joints open in a process known as root wedging (Fig. B.5b).

Salt wedging. In arid climates, dissolved salt in groundwater precipitates and grows as crystals in open pore spaces in rocks. This process, called salt wedging, pushes apart the surrounding grains and so weakens the rock that when exposed to wind and rain, the rock disintegrates into separate grains. The same phenomenon happens in coastal areas, where salt spray percolates into rock and then dries (Fig. B.5c).

Thermal expansion. When the heat of an intense forest fire bakes a rock, the outer layer of the rock expands. On cooling, the layer contracts. This change creates forces in the rock sufficient to make the outer part of the rock break off in sheet-like pieces.

Animal attack. Animal life also contributes to physical weathering: burrowing creatures, from earthworms to gophers, push open cracks and move rock fragments. And in the past century, humans have become perhaps the most energetic agent of physical weathering on the planet. When we excavate quarries,

FIGURE B.5 Wedging is one type of physical (mechanical) weathering.

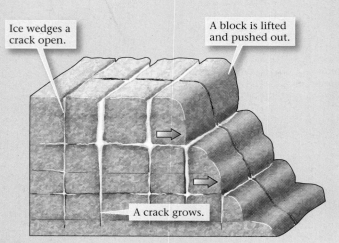

Ice wedges a crack open.

A block is lifted and pushed out.

A crack grows.

(a) When the water that fills cracks freezes, it expands and wedges the cracks open.

(b) Tree roots can wedge cracks open.

(c) Salt wedging led to disintegraton of gravestones in Whitby, England.

foundations, mines, or roadbeds by digging and blasting, we shatter and displace rock that might otherwise have remained intact for millions of years more.

Chemical Weathering

Up to now we've taken the plumber's-snake approach to breaking up rock. Now let's look at the liquid-drain-opener approach. **Chemical weathering** refers to the chemical reactions that alter or destroy minerals when rock comes in contact with water solutions or air. Common reactions involved in chemical weathering include the following:

- *Dissolution.* Chemical weathering during which minerals dissolve into water is called dissolution. Dissolution primarily affects salts and carbonate minerals, but even quartz dissolves slightly (Fig. B.6a, b).
- *Hydrolysis.* During hydrolysis, water chemically reacts with minerals and breaks them down (*lysis* means loosen in Greek) to form other minerals. For example, hydrolysis reactions in feldspar produce clay.
- *Oxidation.* Oxidation reactions in rocks transform iron-bearing minerals (such as biotite and pyrite) into a rusty-brown mixture of various iron-oxide and iron-hydroxide minerals.
- *Hydration.* Hydration, the absorption of water into the crystal structure of minerals, causes some minerals, such as certain types of clay, to expand. Such expansion weakens rock.

Not all minerals undergo chemical weathering at the same rates. Some weather in a matter of months or years, whereas others remain unweathered for millions of years. For example,

when a granite (which contains quartz, mica, and feldspar) undergoes chemical weathering, everything but quartz transforms to clay. That's why beaches typically consist of quartz sand; quartz is the most common mineral left after the other minerals have turned to clay and washed away.

Until fairly recently, geoscientists tended to think of chemical weathering as a strictly inorganic chemical reaction, occurring entirely independently of life forms. But we now realize that organisms play a major role in the chemical-weathering process. For example, the roots of plants, fungi, and lichens secrete organic acids that help dissolve minerals in rocks; these organisms extract nutrients from the minerals. Microbes, such as bacteria, are amazing in that they literally eat minerals for lunch. Bacteria pluck off molecules from minerals and use the energy from the molecules' chemical bonds to supply their own life force.

Physical and Chemical Weathering Working Together

So far we've looked at the processes of chemical and physical weathering separately, but in the real world they happen together, aiding one another in disintegrating rock to form sediment (see Geology at a Glance, pp. 144–145).

Physical weathering speeds up chemical weathering. To understand why, keep in mind that chemical-weathering reactions take place at the surface of a material. Thus, the overall rate at which chemical weathering occurs depends on the ratio of surface area to volume—the greater the surface area, the faster the volume as a whole can chemically weather. When jointing (physical weathering) breaks a large block of rock into smaller pieces, the surface area increases, so chemical weathering happens faster (Fig. B.7a).

FIGURE B.6 Dissolution is one type of chemical weathering.

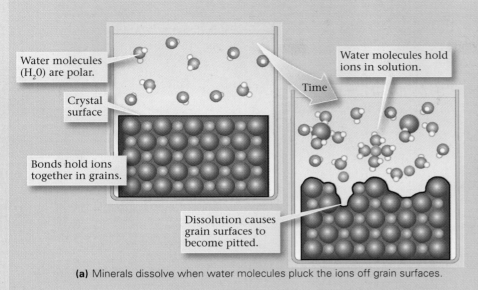

Water molecules (H₂0) are polar.

Crystal surface

Bonds hold ions together in grains.

Time

Water molecules hold ions in solution.

Dissolution causes grain surfaces to become pitted.

(a) Minerals dissolve when water molecules pluck the ions off grain surfaces.

(b) Dissolution along joints intersecting the surface of limestone bedrock in Ireland produced troughs.

FIGURE B.7 Physical and chemical processes work together during the weathering process.

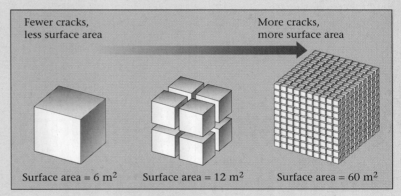

Fewer cracks, less surface area

More cracks, more surface area

Surface area = 6 m²   Surface area = 12 m²   Surface area = 60 m²

(a) As rock breaks apart due to physical weathering, the surface area increases relative to volume.

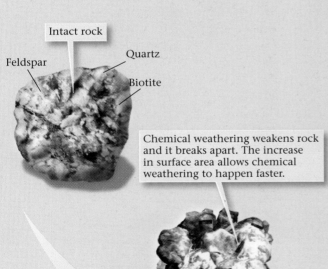

Intact rock

Feldspar

Quartz

Biotite

Chemical weathering weakens rock and it breaks apart. The increase in surface area allows chemical weathering to happen faster.

Rock has broken into loose grains; feldspar has turned into clay.

Quartz   Clay

A current washes clay away. Tumbling quartz grains become rounder.

Time

(b) Chemical weathering weakens rock so it breaks apart. As this happens, the surface area increases, so chemical weathering happens still faster. Eventually, the rock completely disaggregates to form sediment. Weathering of granite produces quartz sand and clay.

Glacial erosion

River erosion

Weathered granite

Cliff retreat

Glacial deposition

Limestone dissolution

Weathering, Sediment, and Soil Production

Similarly, chemical weathering speeds up physical weathering by dissolving away grains or cements that hold a rock together, transforming hard minerals (like feldspar) into soft minerals (like clay), and causing minerals to absorb water and expand. These phenomena make rock weaker, so it can disintegrate more easily (Fig. B.7b).

Note that weathering happens faster at edges, and even faster at the corners of broken blocks. This is because weathering attacks a flat face from only one direction, an edge from

two directions, and a corner from three directions. Thus, with time, edges of blocks become blunt and corners become rounded (Fig. B.8a). In rocks such as granite, which do not contain layering that can affect weathering rates, rectangular blocks transform into a spheroidal shape (Fig. B.8b).

When different rocks in an outcrop undergo weathering at different rates, we say that the outcrop has undergone differential weathering. As a result of differential weathering, cliffs composed of a variety of rock layers take on a stair-step or saw-

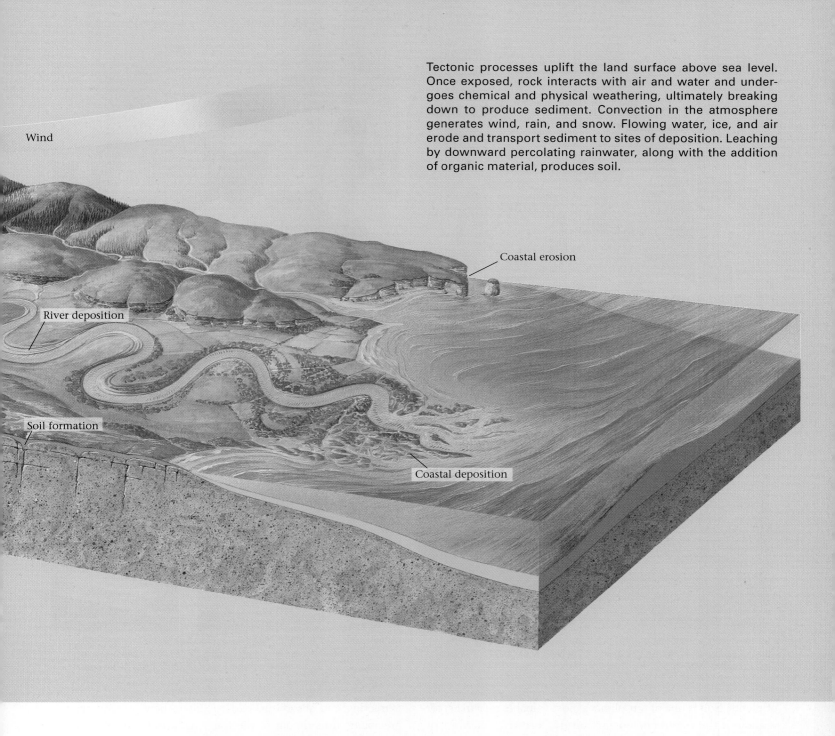

Wind

Coastal erosion

River deposition

Soil formation

Coastal deposition

Tectonic processes uplift the land surface above sea level. Once exposed, rock interacts with air and water and undergoes chemical and physical weathering, ultimately breaking down to produce sediment. Convection in the atmosphere generates wind, rain, and snow. Flowing water, ice, and air erode and transport sediment to sites of deposition. Leaching by downward percolating rainwater, along with the addition of organic material, produces soil.

tooth shape (Fig. B.8c). You can easily see the consequences of differential weathering if you walk through a graveyard. The inscriptions on some headstones are sharp and clear, whereas those on other stones have become blunted or have even disappeared (Fig. B.8d). That's because the minerals in these different stones have different resistances to weathering. Granite, an igneous rock with a high quartz content, retains inscriptions the longest. But marble, a metamorphic rock composed of calcite, dissolves away relatively rapidly in acidic rain.

B.3 SOIL

If you've ever had the chance to dig in a garden, you've seen firsthand that the material in which flowers grow looks and feels different from beach sand or from potter's clay. We call the material in a garden dirt or, more technically, soil. **Soil** consists of rock or sediment that has been modified by physical and chemical interaction with organic material and rainwater, over time, to produce a substrate that can support the growth

FIGURE B.8   Differential weathering.

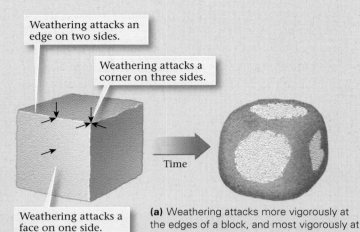

Weathering attacks an edge on two sides.

Weathering attacks a corner on three sides.

Weathering attacks a face on one side.

Time

(a) Weathering attacks more vigorously at the edges of a block, and most vigorously at the corners. Thus, homogeneous rocks tend to weather into rounded blocks.

(b) Weathering along cracks in granite of the Mojave Desert led to the formation of rounded blocks. This style is called spheroidal weathering.

Weak shale

Strong sandstone

(c) Sawtooth weathering profiles develop in sequences of alternating strong and weak layers on this exposure in New Mexico. Weak layers are indented, whereas strong layers protrude.

(d) Inscriptions on a granite headstone (left) last for centuries, but those on a marble headstone (right) may weather away in decades. These gravestones are in the same cemetery and are about the same age.

of plants. Soil is one of our planet's most valuable resources, for without it there could be no agriculture, forestry, ranching, or home gardening.

Three processes taking place at or just below the surface of the Earth contribute to soil formation. First, chemical and physical weathering produces loose debris, new mineral grains (such as clay), and ions in solution. Second, rainwater percolates through the debris and carries dissolved ions and clay flakes downward. The region in which this downward transport occurs is called the **zone of leaching**, because leaching means

extracting and absorbing. Farther down, new mineral crystals precipitate directly out of the water or form by reaction of the water with debris, and the water leaves behind its load of fine clay. The region in which new minerals and clay collect is the **zone of accumulation** (Fig. B.9a). Third, microbes, fungi, plants, and animals interact with sediment, by producing acids that weather grains, by absorbing nutrient atoms, and by leaving behind waste and remains. Plant roots and burrowing animals (insects, worms, and gophers) churn and break up the soil, and microbes metabolize mineral grains and release chemicals.

Because different soil-forming processes operate at different depths, soils typically develop distinct zones, known as **soil horizons**, arranged in a vertical sequence called a **soil profile** (Fig. B.9b). Let's look at an idealized soil profile, from top to bottom, using a soil formed in a temperate forest as our example. The highest horizon is the O-horizon (the prefix stands for organic), so called because it consists almost entirely of organic matter and contains barely any mineral matter. Below the O-horizon we find the A-horizon, in which humus (organic debris) has decayed further and has mixed with mineral grains (clay, silt, and sand). Water percolating through the A-horizon causes chemical weathering reactions to occur and produces ions in solution and new clay materials. The downward-moving water eventually carries soluble chemicals and fine clay deeper into the subsurface. The O- and A-horizons constitute dark-gray to blackish-brown topsoil, the fertile portion of soil that farmers till for planting crops. In some places, the A-horizon grades downward into the E-horizon, a soil level that has undergone substantial leaching but has not yet mixed with organic material. Ions and clay accumulate in the B-horizon, or subsoil. Note from our description that the O-, A-, and E-horizons make up the zone of leaching, whereas the B-horizon makes up the zone of accumulation.

Finally, at the base of a soil profile we find the C-horizon, which consists of material derived from the substrate that's been chemically weathered and broken apart, but has not yet undergone leaching or accumulation. The C-horizon grades downward into unweathered bedrock, or into unweathered sediment.

As farmers, foresters, and ranchers well know, the soil in one locality can differ greatly from the soil in another, in terms of composition, thickness, and texture. And crops that grow well in one type of soil may wither and die in another. Such diversity exists because the makeup of a soil depends on several soil-forming factors (Fig. B.10a, b).

- *Climate*: Large amounts of rainfall and warm temperatures accelerate chemical weathering and cause most of the soluble elements to be leached. In regions with small amounts of rainfall and cooler temperatures, soils take a long time to develop and can retain unweathered minerals and soluble components. Climate seems to be the single most important factor in determining the nature of soils that develop.

- *Substrate composition*: Some soils form on basalt, some on granite, some on volcanic ash, some on recently deposited quartz silt. These different substrates consist of different materials, so the soils formed on them end up with different chemical compositions.

- *Slope steepness*: A thick soil can accumulate under land that lies flat. But on a steep slope, regolith may wash

FIGURE B.9 Formation of soil horizons.

Rain enters ground.

Plant debris accumulates.

Worms churn.

Microbes and fungi metabolize.

Roots weather minerals.

Downward percolating water transports ions and clay.

Ions and clay accumulate

Zone of leaching

Zone of accumulation

Time

O

A

E

B

C

Topsoil

Transition

Subsoil

Weathered bedrock

Solid bedrock

(a) Soil character depends on climate, for climate controls rainfall and vegetation.

(b) Eventually, distinct soil horizons develop, each with a characteristic composition and texture.

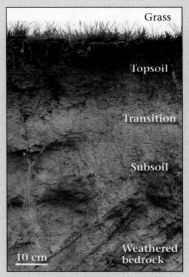

Grass

Topsoil

Transition

Subsoil

10 cm

Weathered bedrock

(c) Soil horizons exposed on the wall of a gully in eastern Brazil.

FIGURE B.10   Factors that control the character of soil.

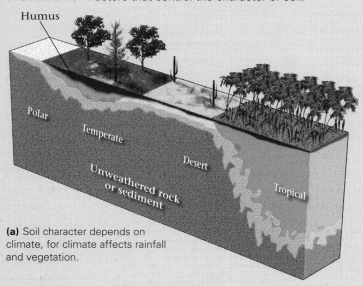

(a) Soil character depends on climate, for climate affects rainfall and vegetation.

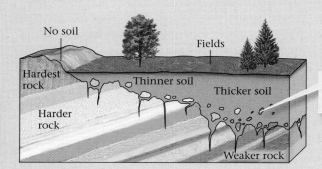

Thicker soil develops over a weaker substrate.

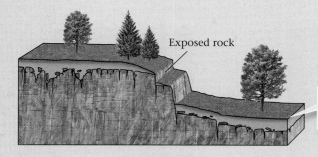

Thicker soil develops over gentler slopes.

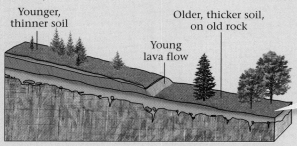

An older soil is thicker than a younger soil.

(b) Soil character also depends on the strength of the substrate, the steepness of the slope, and the length of time soil has been forming.

away before it can evolve into a soil. Thus, all other factors being equal, soil thickness increases as the slope angle decreases.

- *Wetness*: Depending on the details of local topography and on the depth below the surface at which groundwater occurs, some soil is wetter than other soil in the same region. Wet soils tend to contain more organic material than do dry soils.

- *Time*: Because soil formation is an evolutionary process, a young soil tends to be thinner and less developed than an old soil. The rate of soil formation varies greatly with environment.

- *Vegetation type*: Different kinds of plants extract or add different nutrients and quantities of organic matter to a soil. Also, some plants have deeper root systems than others and help prevent soil from washing away.

Soil scientists worldwide have struggled mightily to develop a rational scheme for classifying soils. Not all schemes utilize the same criteria, and even today there is not worldwide agreement on which works best. In the United States, a country that includes many climates at mid-latitudes, many soil scientists use the U.S. Comprehensive Soil Classification System, which distinguishes among twelve orders of soil based on the physical characteristics and environment of soil formation (see **Table B.2**). Rainfall and vegetation play a key role in determining the type of soil that forms. For example, in deserts, where there is very little rainfall and sparse vegetation, an aridisol forms (**Fig. B.11a**). Aridisols have no O-horizon (because there is so little organic material), and the A-horizon is thin. Soluble minerals, specifically calcite, that would be washed away entirely if there were more rainfall, instead accumulate in the B-horizon. In fact, capillary action may bring calcite *up* from deeper down as water evaporates at the ground surface. Calcite cements clasts together in the B-horizon to form a rock-like mass called calcrete. In temperate environments, an alfisol forms—this soil has an O-horizon, and because of moderate amounts of rainfall, materials leached from the A-horizon accumulate in the B-horizon (**Fig. B.11b**). In a tropical climate, oxisols develop. Here, so much rainfall percolates down into the ground that all reactive minerals in the soil undergo chemical weathering, producing ions and clay that flush downward. This process leaves an A-horizon that contains substantial amounts of stable iron-oxide, aluminum-oxide, and aluminum-hydroxide residues (**Fig. B.11c**). The resulting brick-red material is commonly called laterite.

TABLE B.2 Soil orders: U.S. Comprehensive Soil Classification System (see also the map on p. 151)

Alfisol	Gray/brown, has subsurface clay accumulation and abundant plant nutrients. Forms in humid forests.
Andisol	Forms in volcanic ash.
Aridisol	Low in organic matter, has carbonate horizons. Forms in arid environments.
Entisol	Has no horizons. Formed very recently.
Gelisol	Underlain with permanently frozen ground.
Histosol	Very rich in organic debris. Forms in swamps and marshes.
Inceptisol	Moist, has poorly developed horizons. Formed recently.
Mollisol	Soft, black, and rich in nutrients. Forms in subhumid to subarid grasslands.
Oxisol	Very weathered, rich in aluminum oxide and iron oxide, low in plant nutrients. Forms in tropical regions.
Spodosol	Acidic, low in plant nutrients, ashy, has accumulations of iron and aluminum. Forms in humid forests.
Ultisol	Very mature, strongly weathered soils, low in plant nutrients.
Vertisol	Clay-rich soils capable of swelling when wet, and shrinking and cracking when dry.

Canadians use a different scheme focusing only on soils that develop north of the 40th parallel. This scheme works well for cooler, high-latitude climates.

As we have seen, soils take time to form, so soils capable of supporting crops or forests should be considered a natural resource worthy of protection. However, agriculture, overgrazing, and clear-cutting have led to the destruction of soil. Crops rapidly remove nutrients from soil, so if they are not replaced, the soil will not contain sufficient nutrients to maintain plant life. When the natural plant cover disappears, the surface of the soil becomes exposed to wind and water. Actions such as the impact of falling raindrops or the rasping of a plow break up the soil at the surface, with the result that it can wash away in water or blow away as dust. When this happens, **soil erosion**, the removal of soil by running water or by wind, takes place (Fig. B.12). In some cases, almost six tons of soil may be lost from an acre of land per year. Human activities increase rates of soil erosion by 10 to 100 times, so that it far exceeds the rate of soil formation. Droughts exacerbate the situation. For example, during the 1930s a succession of droughts killed off so much vegetation in the American plains that wind stripped the land of soil and caused devastating dust storms. Large numbers of people were forced to migrate away from the Dust Bowl of Oklahoma and adjacent areas.

FIGURE B.11 Examples of soil classification.

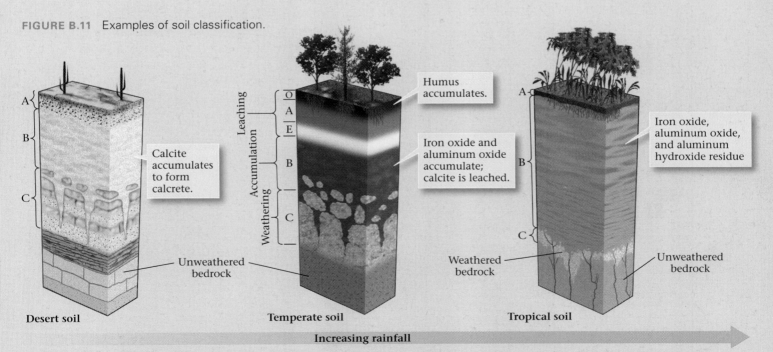

Humus accumulates.

Iron oxide and aluminum oxide accumulate; calcite is leached.

Iron oxide, aluminum oxide, and aluminum hydroxide residue

Calcite accumulates to form calcrete.

Unweathered bedrock

Weathered bedrock

Unweathered bedrock

Desert soil   Temperate soil   Tropical soil

Increasing rainfall

(a) Aridisol forms in deserts. Rainfall is so low that no O-horizon forms and soluble minerals accumulate in the B-horizon.

(b) Alfisol forms in temperate climates. An O-horizon forms, and less soluble materials accumulate in the B-horizon.

(c) Oxisol forms in tropical climates where percolating rainwater leaches all soluble minerals, leaving only iron- and aluminum-rich residues.

The consequences of rainforest destruction on soil are particularly profound. In an established rainforest, lush growth provides sufficient organic debris so that trees can grow. But if the forest is logged, or cleared for agriculture, the humus rapidly disappears, leaving laterite that contains few nutrients. Crop plants consume whatever nutrients there are so rapidly that the soil becomes infertile after only a year or two, useless for agriculture and unsuitable for regrowth of rainforest trees.

Key Terms

chemical weathering (p. 142)

joints (p. 140)

physical weathering (p. 140)

sediment (p. 139)

soil (p. 145)

soil erosion (p. 149)

soil horizon (p. 147)

soil profile (p. 147)

weathering (p. 139)

zone of accumulation (p. 146)

zone of leaching (p. 146)

FIGURE B.12 In this image, we can see that the lack of natural plant coverage has led to severe soil erosion by wind. Similar conditions produced the Dust Bowl of the 1930s.

ANOTHER VIEW U.S. Department of Agriculture map of soil types around the world.

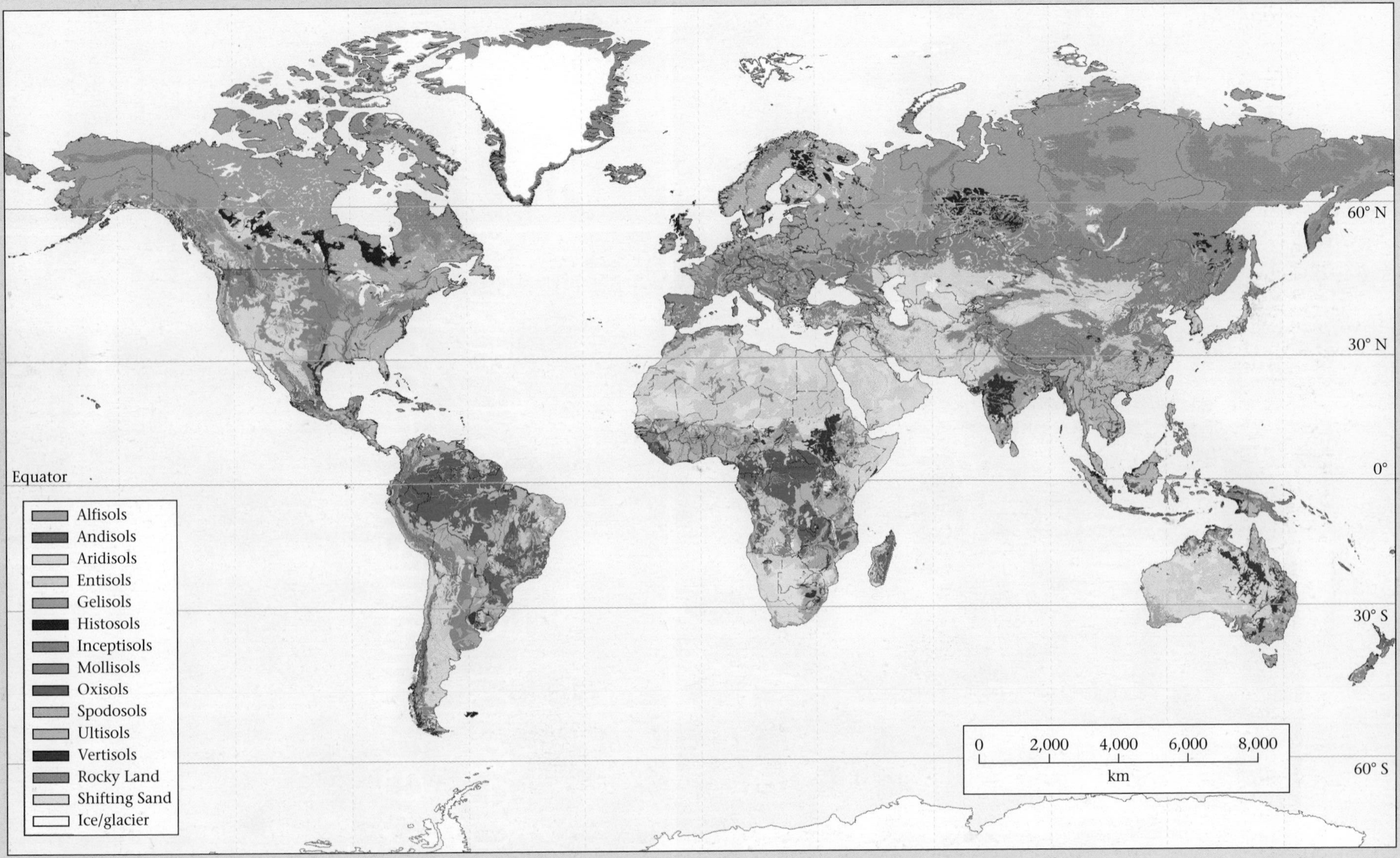

Equator

Alfisols
Andisols
Aridisols
Entisols
Gelisols
Histosols
Inceptisols
Mollisols
Oxisols
Spodosols
Ultisols
Vertisols
Rocky Land
Shifting Sand
Ice/glacier

60° N

30° N

0°

30° S

60° S

0   2,000   4,000   6,000   8,000
km

CHAPTER **6**

Pages of Earth's Past: Sedimentary Rocks

These cliffs, in Bryce Canyon (Utah), expose beds of sedimentary rock deposited in lakes, and by streams, over 40 million years ago.

Why do the walls of the Grand Canyon display spectacular layers of different-colored rock? Why do some layers form vertical cliffs while others do not?

6.1 INTRODUCTION

On an isolated, windswept drilling platform in the North Sea off the coast of Scotland, a group of roughnecks ready a multimillion-dollar drill—their plan is to penetrate the sea floor and see what lies beneath. The North Sea formed as a consequence of rifting tens of millions of years ago. During rifting, what was once dry land between Great Britain and continental Europe slowly sank or, in geologic parlance, "subsided." Rivers carried sediments from the surrounding land into the newborn North Sea, and these sediments collected in layers. At certain stages in the process, salts precipitated from seawater, and the shells of sea creatures settled and collected on the sea floor. As the drilling begins, a geologist stationed on the deck of the platform examines the material flushed out of the lengthening drill hole by high-pressure water. At first, drilling brings up soft mud and loose sand, silt, pebbles, and shell fragments. But as the hole goes deeper, the material coming up holds together in soft but coherent clumps. Eventually, when the hole has entered layers that now lie almost a kilometer below the sea floor, the drilling fluid flushes out chips and chunks of solid rock. When the geologist studies these fragments, she finds that they consist of grains of sand or silt cemented together, tightly packed clay that is harder than pottery, masses of shell fragments, or solid aggregates of salt crystals. The geologist has observed the transition of loose sediment into solid layers of sedimentary rock as burial depth progressively increases.

Formally defined, **sedimentary rock** is rock that forms at or near the surface of the Earth by the cementing together of loose grains that had been produced by physical or chemical weathering of preexisting rock, by the precipitation of minerals from water solutions, by the cementing together of shells and shell fragments, or by the growth of masses of shell-producing organisms. Layers of sedimentary rock are like the pages of a book, recording tales of ancient events and ancient environments on the ever-changing face of the Earth. They occur only in the upper part of the crust, and form a "cover" that buries the underlying "basement" of igneous and/or metamorphic rock (Fig. 6.1).

In Interlude B, we introduced the concept of weathering, and showed how it attacks bedrock, breaking it down into ions and loose sediment grains. What happens next? Some of the sediment may become incorporated in soil, as we have seen. But some becomes buried and transformed into sedimentary rock. In this chapter, we discuss the various kinds of sedimentary rock and the ways in which they form, and we consider what sedimentary rocks can tell us about the history of the Earth System.

TAKE-HOME MESSAGE

By the end of this chapter, you should understand the various processes that produce sedimentary rocks, be able to describe and classify such rocks, and be able to recognize the clues they contain about the past history of the Earth.

FIGURE 6.1 Sedimentary rocks form a cover over older basement rock.

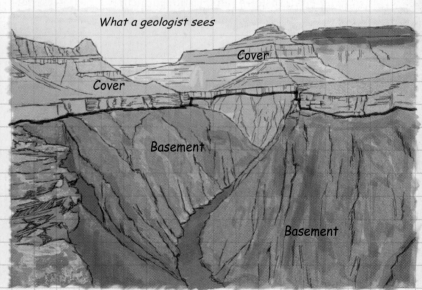

(a) In the Grand Canyon, we see a contact between cover (here, horizontal sedimentary layers) and basement (metamorphic rock).

(b) A geologist's sketch emphasizes the contact between the sedimentary cover and basement.

6.2 CLASSES OF SEDIMENTARY ROCKS

Geologists divide sedimentary rocks into four major classes, based on their mode of origin. (1) **Clastic sedimentary rock** consists of cemented-together solid fragments and grains derived from preexisting rocks (clastic comes from the Greek *klastos*, meaning broken). (2) **Biochemical sedimentary rock** consists of shells. (3) **Organic sedimentary rock** consists of carbon-rich relics of plants. And (4) **chemical sedimentary rock** is made up of minerals that precipitate directly from water solutions. Let's now look at these classes in more detail.

Clastic Sedimentary Rocks

Formation. Nine hundred years ago, a thriving community of Native Americans inhabited the high plateau of Mesa Verde, Colorado. In hollows beneath huge overhanging ledges, they built multistory stone-block buildings that have survived to this day. Clearly, the blocks are solid and durable—they are, after all, rock. But if you were to rub your thumb along one, it would feel gritty, and small grains of quartz would break free and roll under your thumb, for the block consists of quartz sand grains cemented together. Geologists call such rock a **sandstone**.

Sandstone is an example of clastic, or detrital, sedimentary rock. It consists of solid grains (**clasts**, or detritus) stuck together to form a solid mass. The grains can consist of individual minerals (such as grains of quartz or flakes of clay) or fragments of rock (such as pebbles of granite). The loose grains of sediment transform into clastic sedimentary rock by the following five steps (Fig. 6.2a, b).

- *Weathering*: Detritus forms by disintegration of bedrock in response to physical and chemical weathering.
- *Erosion*: **Erosion** refers to the combination of processes that separate rock or regolith from its substrate and carry it away.
- *Transportation*: Wind, water, or ice move sediment. The ability of a medium to carry sediment depends on its viscosity and velocity. Solid ice can carry sediment of any size, regardless of how slowly the ice moves. Very fast-moving, turbulent water can transport coarse fragments (cobbles and boulders), moderately fast-moving water can carry only sand and gravel, and slowly moving water carries only silt and mud. Strong winds can move sand and dust, but gentle breezes carry only dust.
- *Deposition*: **Deposition** is the process by which sediment settles out of the transporting material. Sediment settles out of wind or moving water when these fluids slow, because as the velocity decreases, the fluid no longer has the ability to move sediment. Sediment is deposited by ice when the ice melts.
- *Lithification*: Geologists refer to the transformation of loose sediment into solid rock as **lithification**. The lithification of clastic sediment involves two steps. First, when the sediment has been buried, pressure caused by the weight of overlying material squeezes out water and air that had been trapped between clasts, and clasts compact together tightly. Compacted sediment may then be bound together to make coherent sedimentary rock by the process of **cementation**. Cement consists of minerals (commonly quartz or calcite) that precipitate from groundwater, and fill the spaces between clasts. Cement acts like glue and holds detritus together.

FIGURE 6.2 The five steps in clastic sedimentary rock formation.

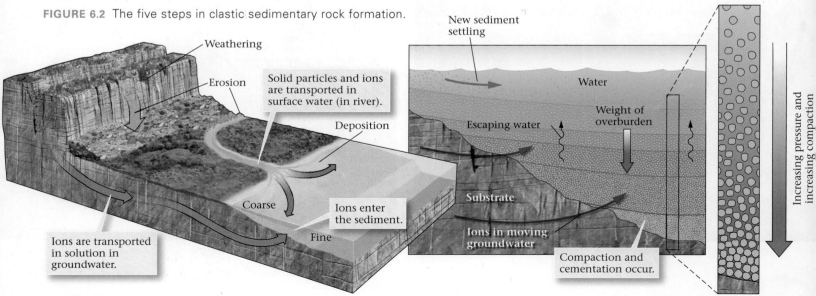

(a) Clasts produced by weathering undergo erosion, transportation, and deposition.

(b) The process of lithification takes place during progressive burial.

Classifying clastic sedimentary rocks.
Say that you pick up a clastic sedimentary rock and want to describe it sufficiently so that, from your words alone, another person can picture the rock. What characteristics should you mention? Geologists find the following characteristics most useful.

FIGURE 6.3 Grain characteristics used in describing sedimentary rock.

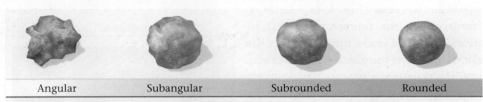

| Angular | Subangular | Subrounded | Rounded |

(a) Individual grains may become progressively more round as the edges and corners break off.

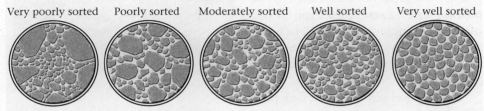

Very poorly sorted   Poorly sorted   Moderately sorted   Well sorted   Very well sorted

(b) Sorting refers to the distribution of grain sizes. Sediment tends to become better sorted as it moves away from the source.

- *Clast size.* Size refers to the diameter of clasts making up a rock. Names used for clast size, listed in order from coarsest to finest, are: boulder, cobble, pebble, sand, silt, and clay (see Table B.1).

- *Clast composition.* Composition refers to the makeup of clasts in sedimentary rock. Clasts may be composed of individual minerals or of rock fragments.

- *Angularity and sphericity.* Angularity indicates the degree to which clasts have smooth or angular corners and edges (Fig. 6.3a). Sphericity, in contrast, refers to the degree to which a clast is equidimensional, or resembles a sphere.

- *Sorting.* **Sorting** of clasts indicates the degree to which the clasts in a rock are all the same size or include a variety of sizes (Fig. 6.3b). Well-sorted sediment consists entirely of sediment of the same size, whereas poorly sorted sediment contains a mixture of more than one grain size.

- *Character of cement.* Not all clastic sedimentary rocks have the same kind of cement. In some, the cement consists predominantly of quartz, whereas in others, it consists predominantly of calcite.

With these characteristics in mind, we can distinguish among several common types of clastic sedimentary rocks, listed in Table 6.1. This table provides *common* rock names—specialists sometimes use other, more precise names based on more complex classification schemes.

Characteristics of a sedimentary rock provide clues to the source of the sediment, and to the environment of deposition. To see how, let's follow the fate of rock fragments as they gradually move from a cliff face in the mountains via a river to the seashore. Different kinds of sediment develop along the route, and overall, sediment changes character as it moves further from the source. Each of these types of sediment, if buried and lithified, would yield a different kind of sedimentary rock.

To start, imagine that some large blocks of rock tumble off a cliff and slam into other blocks already at the bottom. The impact shatters the blocks, producing a pile of angular fragments with sharp edges. If these fragments were to be

TABLE 6.1 Classifying clastic sedimentary rocks

Clast Size*	Clast Character	Rock Name (Alternate Name)
Coarse to very coarse	Rounded pebbles and cobbles	Conglomerate
	Angular clasts	Breccia
	Large clasts in muddy matrix	Diamictite
Medium to coarse	Sand-sized grains	Sandstone
	• quartz grains only	• quartz sandstone (quartz arenite)
	• quartz and feldspar sand	• arkose
	• sand-sized rock fragments	• lithic sandstone
	• sand and rock fragments in a clay-rich matrix	• wacke (informally called graywacke)
Fine	Silt-sized clasts	Siltstone
Very fine	Clay and/or very fine silt	Shale (if it breaks into platy sheets)
		Mudstone (if it doesn't break into platy sheets)

*For precise diameters, see Table B.1.

cemented together, the resulting rock would be **breccia** (Fig. 6.4a). Later, a storm causes the fragments (clasts) to slide downslope into a turbulent river. In the water, clasts bang into each other and into the riverbed, a process that shatters them into still smaller pieces and breaks off their sharp edges. Angular clasts gradually become rounded clasts. When the river water slows, pebbles and cobbles stop moving and form a mound or bar of gravel. Burial and lithification of these rounded clasts produces **conglomerate** (Fig. 6.4b).

If the gravel stays put for a long time, it undergoes chemical weathering. As a consequence, cobbles and pebbles break apart into individual mineral grains, eventually producing a sand-sized mixture of quartz, feldspar, and clay. Clay is so fine that it can be carried far downstream, so in sandbars not far from the source, we find a mixture of quartz and some feldspar grains—this sediment, if buried and lithified, becomes **arkose** (Fig. 6.4c). Over time, feldspar grains in sand continue to weather into clay so that gradually, during successive events that wash the sediment farther downstream, the sand loses feldspar and ends up being composed almost entirely of durable quartz grains. Some of the sand may make it to the sea, where waves carry it to beaches. This sediment, when buried and lithified, becomes quartz sandstone (Fig. 6.4d). Meanwhile, silt and clay may accumulate in the flat areas bordering streams, regions called floodplains (see Chapter 14) that become inundated only during floods, or in a wedge of sediment, called a delta, that accumulates in the sea at the mouth of the river. Some of the silt and mud may be collected in lagoons or mudflats along the shore. The silt, when lithified, becomes **siltstone**, and the mud, when lithified, becomes **shale** or **mudstone** (Fig. 6.4e).

FIGURE 6.4 Different kinds of sediments lithify into different kinds of sedimentary rocks.

Sediment ⟶ Lithification ⟶ Sedimentary rock

(a) Lithification of an accumulation of angular clasts yields breccia.

(b) Layers of river gravel lithify into conglomerate.

(c) Sediment deposited in an alluvial fan, close to its source, can be feldspar rich. Lithification of this sediment yields arkose.

Alluvial fan

Biochemical and Organic Sedimentary Rocks: Byproducts of Life

The Earth System involves interactions between living organisms and the physical planet. Numerous organisms have developed the ability to extract dissolved ions from seawater to make solid shells. When the organisms die, the solid material in their shells becomes incorporated in biochemical sedimentary rock. The living cells of plants, algae, bacteria, and plankton also provide materials that can be incorporated in sedimentary rocks. The rocks formed from this material contain organic chemicals, so they are called organic sedimentary rocks. Geologists recognize several different types of biochemical and organic sedimentary rocks, which we now describe.

Limestone (biochemical).   A snorkeler gliding above a reef sees an incredibly diverse community of coral and algae, around which creatures such as clams, oysters, snails (gastropods), and lampshells (brachiopods) live, and above which plankton float

FIGURE 6.4　*(continued)*

(d) Layers of beach or dune sand lithify into sandstone.

Sandstone

Shale

(e) Layers of mud, here exposed beneath marsh grass, lithify to form shale. Here, the thin-bedded shale is interbedded with sandstone.

Since the principal compound making up limestone is $CaCO_3$, geologists consider limestone to be a type of carbonate rock.

Limestone comes in a variety of textures, because the material that forms it accumulates in a variety of ways. For example, limestone can originate from reef builders (such as coral) that grew in place, from shell debris that was transported, or from carbonate mud that settled like snow out of water. Because of this variety, geologists distinguish among fossiliferous limestone, consisting of visible fossil shells or shell fragments (Fig. 6.5b); micrite, consisting of very fine carbonate mud; and chalk, consisting of plankton shells. Experts recognize many other types as well.

Typically, ancient limestone is a massive light-gray to dark-bluish-gray rock that breaks into chunky blocks—it doesn't look much like a pile of shell fragments (Fig. 6.5c). That's because several processes take place that change the texture of the rock over time. For example, water

(Fig. 6.5a). Though they all look so different from each other, many of these organisms share an important characteristic: they make solid shells of calcium carbonate ($CaCO_3$), which occurs either as the minerals calcite or aragonite. When the organisms die, the shells remain and may accumulate. Rocks formed dominantly from this material are the biochemical version of **limestone**.

passing through the rock precipitates new cement, and also dissolves some carbonate grains and causes new ones to grow.

Chert (biochemical). If you walk beneath the northern end of the Golden Gate Bridge in California, you will find outcrops of reddish, almost porcelain-like rock occurring in 3- to

FIGURE 6.5　The formation of carbonate rocks (limestone).

(b) Fossiliferous limestone consists of fossil shells and shell fragments that have been cemented together.

Relict of a small reef

(a) In this modern coral reef, corals produce shells. If buried and preserved, these become limestone.

(c) A Vermont quarry shows the gray color of 400 Ma limestone. The white mounds are relicts of small reefs.

FIGURE 6.6 Examples of biochemical and organic sedimentary rocks.

(a) This bedded chert developed on the deep sea floor by the deposition of plankton that secrete silica shells.

Part of a folded chert layer is outlined in white.

Sandstone and shale

Coal layer

(b) Coal is deposited in layers (beds), just like other kinds of sedimentary rocks.

15-cm-thick layers (Fig. 6.6a). Hit it with a hammer, and the rock cracks, almost like glass, creating smooth, spoon-shaped (conchoidal) fractures. Geologists call this rock biochemical chert; it's made from cryptocrystalline quartz (*crypto* is Greek for hidden), meaning quartz grains that are too small to be seen without the extreme magnification of an electron microscope. The chert beneath the Golden Gate Bridge formed from the shells of silica-secreting plankton that accumulated on the sea floor. Gradually, after burial, the shells dissolved, forming a silica-rich solution. Chert then precipitated from this solution.

Organic (carbonaceous) rocks: coal and oil shale. The Industrial Revolution of the nineteenth century depended on power provided by steam engines. After decimating forests to provide fuel for these engines, industrialists turned to coal for fuel. Coal is a black, combustible rock consisting of over 50% carbon, and thus differs markedly from the other sedimen-

tary rocks discussed so far—the carbon of coal occurs either as pure carbon or in organic chemicals. We consider coal to be a sedimentary rock because it is made up of detritus deposited in layers (Fig. 6.6b). We'll look more at coal formation in Chapter 12. Here, we simply need to know that the carbon and the organic chemicals making coal come from the remains of plant material that died and accumulated on the floor of a forest or swamp. Coal forms when the remains become deeply buried, where heat and pressure compact the plant material and drive off volatiles (hydrogen, water, carbon dioxide, ammonia), leaving a concentration of carbon.

Chemical Sedimentary Rocks

The colorful terraces, or mounds, around the vents of hot-water springs; the immense layers of salt that underlie the floor of the Mediterranean Sea; the smooth, sharp point of an ancient arrowhead—these materials all have something in common. They all consist of rock formed primarily by the precipitation of minerals out of water solutions. We call such rocks chemical sedimentary rocks. They typically have a crystalline texture, partly formed during their original precipitation and partly when, at a later time, new crystals grow at the expense of old ones.

Evaporites: the products of saltwater evaporation. In 1965, two daredevil drivers in jet-powered cars battled to be the first to set the land-speed record of 600 mph. On November 7, Art Arfons, in the *Green Monster*, peaked at 576.127 mph, but eight days later Craig Breedlove, driving the *Spirit of America*, reached 600.601 mph. Traveling at such speeds, a driver must maintain an absolutely straight line; any turn will catapult the vehicle out of control. Thus, high-speed trials take place on extremely long and flat racecourses. Not many places can

FIGURE 6.7   The formation of evaporite deposits.

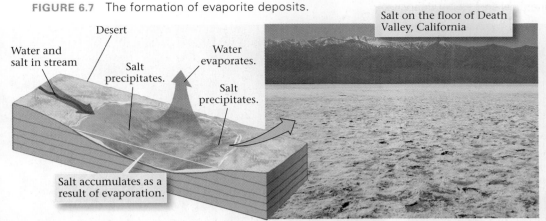

(a) In lakes with no outlet, tiny amounts of salt brought in by streams stays behind as the water evaporates. When the water evaporates entirely, a white crust of salt remains.

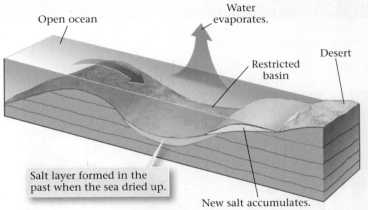

(b) Salt precipitation can also occur along the margins of a restricted marine basin, if saltwater evaporates faster than it can be resupplied.

provide such conditions—the Bonneville Salt Flats, near the Great Salt Lake of central Utah, do. The salt flats formed by the evaporation of an ancient salt lake. Under the heat of the Sun, the water turned to gas and drifted up into the atmosphere. The salt that had been dissolved in the water stayed behind.

Salt precipitation occurs wherever there is saturated salt-water. This happens along desert lakes with no outlet and along margins of restricted seas (Fig. 6.7a, b). For thick deposits of salt to form, large volumes of water must evaporate. Because salt deposits form as a consequence of evaporation, geologists refer to them as **evaporites**. The specific type of salt consti-tuting an evaporite depends on the amount of evaporation. When 80% of the water evaporates, gypsum forms; and when 90% of the water evaporates, halite precipitates.

Travertine (chemical limestone).   **Travertine** is a rock composed of crystalline calcium carbonate formed by chemi-cal precipitation from groundwater that has seeped out at the ground surface either in hot- or cold-water springs, or on the walls of caves. What causes this precipitation? It happens, in part, when the groundwater degas-ses, meaning that some of the carbon dioxide that had been dissolved in the groundwater bubbles out of solution. Removal of carbon dioxide decreases the ability of the water to hold dissolved carbonate. Precipita-tion also occurs when water evapo-rates, thereby increasing the concen-tration of carbonate. Various kinds of microbes live in the environments in which travertine accumulates, so biologic activity may also contribute to the precipitation process. Traver-tine produced at springs forms ter-races and mounds that are meters or even hundreds of meters thick (Fig. 6.8a). In cave settings, travertine builds up beautiful and complex growth forms called speleothems (Fig. 6.8b).

Dolostone: replacing calcite with dolomite.   **Dolostone**, also a carbonate rock, differs from limestone in that it contains the mineral dolomite ($CaMg[CO_3]_2$). Where does the mag-nesium come from? Most dolostone forms by a chemical reac-tion between solid calcite and magnesium-bearing groundwater. Much of the dolostone you may find in an outcrop actually originated as limestone but later changed as dolomite replaced calcite. This change may take place beneath lagoons along a shore soon after the limestone formed, or a long time later, after the limestone has been buried deeply.

Chert (replacement).   A tribe of Native Americans, the Onondaga, once lived off the land in eastern New York State. Here, outcrops of limestone contain layers of a black chert (Fig. 6.9a). Because of the way it breaks, the tribe's artisans could fashion sharp-edged tools (arrowheads and scrapers) from this chert, so the Onondaga collected it for their own toolmaking industry and for use in trade with other people. Unlike the deep-sea (biochemical) chert described earlier, the chert collected by the Onondaga formed when crypto-crystalline quartz gradually *replaced* calcite crystals within a body of limestone long after the limestone was depos-ited; geologists call such material replacement chert. Chert comes in many colors (black, white, red, brown, green, gray), depending on the impurities it contains. Petrified wood is chert that forms when silica-rich sediment, such as ash from a volcanic eruption, buries a forest. Dissolved silica pre-cipitates as cryptocrystalline quartz within wood, gradually replacing the wood's cellulose. The chert deposit retains the shape of the wood and even its growth rings. Some chert, known as agate, precipitates in concentric rings inside hol-lows in a rock (Fig. 6.9b).

FIGURE 6.8 Examples of travertine (chemical limestone) deposits.

Terrace of new travertine

2 m

(a) Travertine accumulates in terraces at Mammoth Hot Springs in Yellowstone Park, Wyoming.

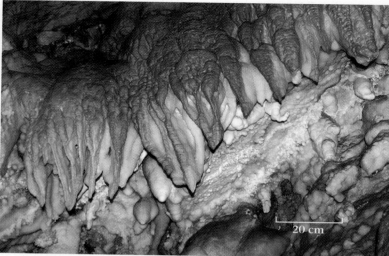

20 cm

(b) Travertine speleothems form as calcite-rich water drips from the ceiling of Timpanogos Cave in Utah.

FIGURE 6.9 Examples of chert that precipitated in place.

Layer of black chert

Tilted limestone beds

(a) Replacement chert forms as layers of nodules between tilted limestone beds in New York.

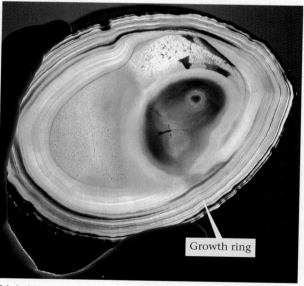

Growth ring

(b) A thin slice of Brazillian agate, lit from the back, shows growth rings.

TAKE-HOME MESSAGE

The many kinds of sedimentary rocks differ from one another by their mode of origin and/or composition. For example, clastic rocks form from cemented-together grains, limestones from shells or chemical precipitates, evaporates from saline solutions, and organic rocks from plant debris.

6.3 SEDIMENTARY STRUCTURES

Geologists use the term **sedimentary structure** for the layering of sedimentary rocks, for surface features on layers formed during deposition, and for the arrangement of grains within layers. Here, we examine some of the more important types.

Bedding and Stratification

Let's start by introducing the jargon for discussing sedimentary layers. A single layer of sediment or sedimentary rock with a recognizable top and bottom is called a **bed**; the boundary

between two beds is a bedding plane; several beds together constitute **strata**; and the overall arrangement of sediment into a
sequence of beds is bedding, or stratification.

Why does bedding form? To find the answer, we need to
think about how sediment is deposited. Changes in the climate,
water depth, current velocity, or the sediment source control
the type of sediment deposited at a location at a given time.
For example, on a normal day a slow-moving river may carry
only silt, which collects on the river bed (Fig. 6.10). During a
flood, the river flows faster and carries sand and pebbles, so a
layer of sandy gravel forms over the silt layer. Then, when the
flooding stops, more silt buries the gravel. If this succession of
sediments become lithified and exposed for you to see, they
appear as alternating beds of siltstone and sandy conglomerate.
Bedding can look like stripes on an outcrop.

During geologic time, long-term changes in a depositional
environment can take place. Thus, a given sequence of strata
may differ markedly from sequences of strata above or below.
A sequence of strata that is distinctive enough to be traced
across a fairly large region is called a **stratigraphic formation**, or simply a formation (Fig. 6.11a). For example, a region
may contain a succession of alternating sandstone and shale
beds deposited by rivers, overlain by beds of marine limestone deposited later when the region was submerged by the
sea. A stratigrapher might identify the sequence of sandstone
and shale beds as one formation and the sequence of limestone
beds as another. Formations are often named after the locality
where they were first found and studied. A map that portrays
the distribution of stratigraphic formations is called a geologic
map (Fig. 6.11b).

FIGURE 6.10 The formation of bedding during deposition of sediment.
In this case, the bedding reflects current strength.

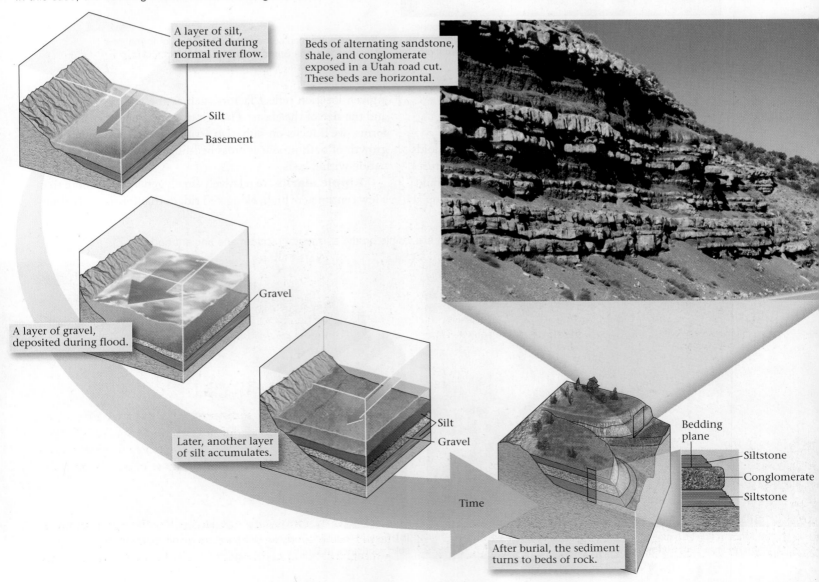

A layer of silt,
deposited during
normal river flow.

Beds of alternating sandstone,
shale, and conglomerate
exposed in a Utah road cut.
These beds are horizontal.

Silt

Basement

A layer of gravel,
deposited during flood.

Gravel

Later, another layer
of silt accumulates.

Silt

Gravel

Time

Bedding
plane

Siltstone

Conglomerate

Siltstone

After burial, the sediment
turns to beds of rock.

FIGURE 6.11 The concept of a stratigraphic formation.

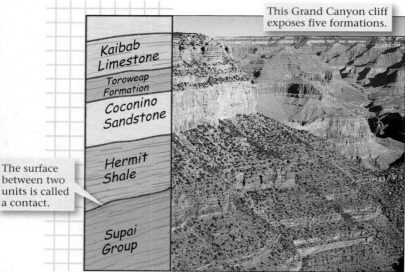

This Grand Canyon cliff exposes five formations.

Kaibab Limestone

Toroweap Formation

Coconino Sandstone

Hermit Shale

The surface between two units is called a contact.

Supai Group

(a) The names of formations consisting of one rock type may indicate the rock type (e.g., Kaibab Limestone). The name of a formation including more than one rock type includes the word "formation" (Toroweap Formation). Several related formations comprise a group (Supai Group).

(b) A geologic map portrays the distribution of formations in a portion of the Grand Canyon. Each color band is a specific formation.

Ripple Marks, Dunes, and Cross Bedding: A Consequence of Deposition in a Current

Many clastic sedimentary rocks accumulate in moving fluids (wind, rivers, or waves). Fascinating sedimentary structures develop at the interface between the sediment and the fluid. These structures are called bedforms. Bedforms that develop at a given location reflect factors such as the velocity of the flow and the size of the clasts. Though there are many types of bedforms, we'll focus on only two—ripple marks and dunes. The growth of both produces cross bedding, a special type of lamination within beds.

Ripple marks are relatively small, generally no more than a few centimeters high, elongated ridges that form on a bed sur-

FIGURE 6.12 Ripple marks, a type of sedimentary structure, are visible on the surface of modern and ancient beds.

(a) Modern ripples exposed at low tide along a sandy beach on the shore of Cape Cod, Massachusetts.

(b) These 145-million-year-old ripples are preserved on a tilted bed of solid sandstone at Dinosaur Ridge, Colorado.

FIGURE 6.13 Cross bedding, a type of sedimentary structure within a bed.

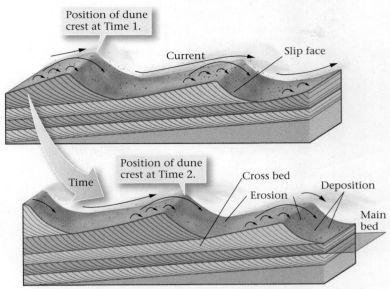

(a) Cross beds form as sand blows up the windward side of a dune or ripple, and then accumulates on the slip face. With time, the dune crest moves.

(b) A cliff face in Zion National Park, Utah, displays large cross beds formed between 200 and 180 million years ago, when the region was a desert with large sand dunes.

face at right angles to the direction of current flow. You can find ripples on modern beaches and preserved on bedding planes of ancient rocks (Fig. 6.12a, b). Dunes look like ripples, only they are larger. For example, dunes on the bed of a stream may be tens of centimeters high, and wind-formed dunes formed in deserts may be tens of to over 100 meters high.

If you examine a vertical slice cut into a ripple or dune, you will find distinct internal laminations that are inclined at an angle to the boundary of the main sedimentary layer. Such laminations are called **cross beds**. To see how cross beds develop, imagine a current of air or water moving uniformly in one direction (Fig. 6.13a). The current erodes and picks up clasts from the upstream part of the bedform and deposits them on the downstream or leeward part. Sediment builds up on the leeward side until gravity causes it to slip down. With time, the leeward side of the bedform builds in the downstream direction. The curving surface of the slip face establishes the shape of the cross beds. Eventually, a new cross-bedded layer builds out over a preexisting one. The boundary between two successive layers is called the main bedding, and the internal curving surfaces within the layer constitute the cross bedding (Fig. 6.13b).

Turbidity Currents and Graded Beds

Sediment deposited on a submarine slope might not stay in place forever. For example, an earthquake or storm might disturb this sediment and cause it to slip downslope. If the sediment is loose enough, it mixes with water to create a murky, turbulent cloud. This cloud is denser than clear water, and thus flows downslope like an underwater avalanche (Fig. 6.14a–c).

We call this moving submarine suspension of sediment a **turbidity current**. Downslope, the turbidity current slows. When this happens, the sediment that it has carried starts to settle out. Larger grains sink faster through a fluid than do finer grains, so the coarsest sediment settles out first. Progressively finer grains accumulate on top, with the finest sediment (clay) settling out last. This process forms a **graded bed**—that is, a layer of sediment in which grain size varies from coarse at the bottom to fine at the top. Geologists refer to a deposit from a turbidity current as a turbidite.

Bed-Surface Markings

A number of features appear on the surface of a bed as a consequence of events that happen during deposition or soon after, while the sediment layer remains soft. These bed-surface markings include the following.

- *Mud cracks*: If a mud layer dries up after deposition, it cracks into roughly hexagonal plates that typically curl up at their edges. We refer to the openings between the plates as mud cracks. Later, these fill with sediment and can be preserved (Fig. 6.15a, b).

- *Scour marks*: As currents flow over a sediment surface, they may scour out small troughs called scour marks parallel to the current flow. These indentations can be buried and preserved.

- *Fossils*: Fossils are relics of past life. Some fossils are shell imprints or footprints on a bedding surface (see Interlude D).

FIGURE 6.14 The development of graded bedding from turbidity currents.

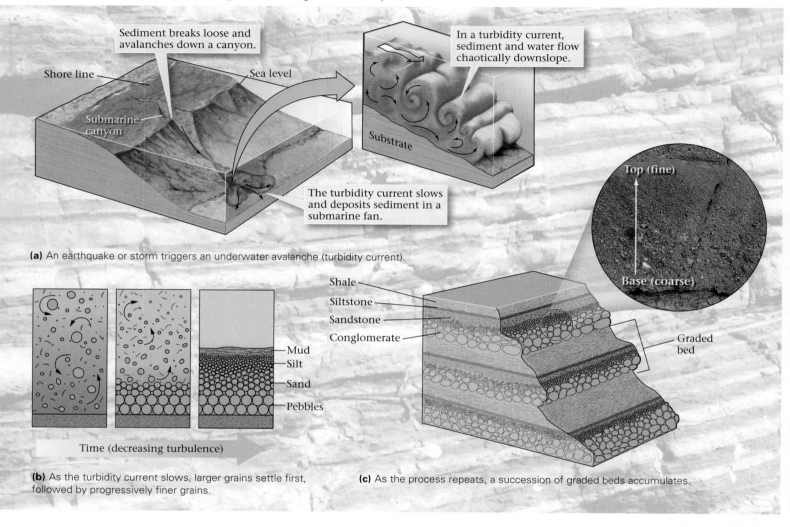

Sediment breaks loose and avalanches down a canyon.

Shore line

Sea level

Submarine canyon

In a turbidity current, sediment and water flow chaotically downslope.

Substrate

The turbidity current slows and deposits sediment in a submarine fan.

Top (fine)

Base (coarse)

(a) An earthquake or storm triggers an underwater avalanche (turbidity current).

Mud
Silt
Sand
Pebbles

Time (decreasing turbulence)

(b) As the turbidity current slows, larger grains settle first, followed by progressively finer grains.

Shale
Siltstone
Sandstone
Conglomerate

Graded bed

(c) As the process repeats, a succession of graded beds accumulates.

FIGURE 6.15 Mud cracks, a sedimentary structure formed by drying out of mud.

(a) Mud cracks in red mud at Bryce Canyon, Utah. Note how the edges of the mud plates curl up.

20 cm

(b) Mud cracks preserved in a 410-million-year-old bed exposed on the base of a cliff in New York.

Why Study Sedimentary Structures?

Sedimentary structures are not just a curiosity, but rather provide clues that help geologists understand the environment in which clastic sedimentary beds were deposited. For example, the presence of ripple marks and cross bedding indicates that layers were deposited in a current. The presence of mud cracks indicates that the sediment layer was exposed to the air and dried out on occasion. Graded beds indicate deposition by turbidity currents. And fossil types can tell us whether sediment was deposited along a river or in the deep sea, for different species of organisms live in different environments. In the next section of this chapter, we examine these environments in greater detail.

TAKE-HOME MESSAGE

Sedimentary rocks occur in beds, because deposition takes place in discrete episodes. A sequence of beds makes up a stratigraphic formation. Sedimentary structures, such as ripple marks, cross beds, and graded beds, give clues to the environment of deposition.

6.4 HOW DO WE RECOGNIZE DEPOSITIONAL ENVIRONMENTS?

Geologists refer to the conditions in which sediment was deposited as the depositional environment. Examples include beach, glacial, and river environments. To identify these environments, geologists, like detectives, look for such clues as grain size, composition, sorting, and roundness of clasts, which can tell us how far the sediment has traveled from its source and whether it was deposited from the wind, from a fast-moving current, or from a stagnant body of water. Clues such as fossil content and sedimentary structures can help us decide whether the sediments were deposited subaerially, just off the coast, or in the deep sea. Now let's look at some examples of different depositional environments and the sediments deposited in them, by imagining that we are taking a journey from the mountains to the sea, examining sediments as we go (see **Geology at a Glance** on pp. 168–169 and **Geotour 6** on p. GT-14).

Terrestrial (Nonmarine) Sedimentary Environments

Terrestrial sedimentary environments are those formed on dry land or in fresh water. Here, oxygen-rich surface or subsurface water may react with iron in the sediment to produce iron oxide. Thus, some terrestrial sedimentary strata have a reddish hue—strata with this color are informally called redbeds.

Glacial environments. We begin high in the mountains, where it's so cold that more snow collects in the winter than melts away, so glaciers—rivers of ice—develop and slowly flow downslope. Because ice is a solid, it can move sediment of any size. So as a glacier moves down a valley in the mountains, it carries along *all* the sediment that falls on its surface from adjacent cliffs or gets plucked from the ground at its base. At the end of the glacier, where the ice finally melts away, it drops its sedimentary load and makes a pile of glacial till (Fig. 6.16a). Till is unsorted and unstratified—it contains clasts ranging from clay size to boulder size all mixed together.

Mountain stream environments. As we walk down beyond the end of the glacier, we enter a realm where turbulent streams rush downslope in mountain valleys. This fast-moving water has the power to carry large clasts; in fact, during floods, boulders and cobbles tumble down the stream bed. Between floods, when water flow slows, the largest clasts settle out to form gravel and boulder beds, while the stream carries finer sediments like sand and mud further downstream (Fig. 6.16b). Sedimentary deposits of a mountain stream would, therefore, include conglomerate.

Alluvial fan environments. Our journey now takes us to the mountain front, where the fast-moving stream empties onto a plain. In arid regions, where there is not enough water for the stream to flow continuously, the stream deposits its load of sediment near the mountain front, producing a wedge-shaped apron of gravel and sand called an alluvial fan (Fig. 6.16c). Deposition takes place here because when the stream pours from a canyon mouth and spreads out over a broader region, friction with the ground causes the water to slow down, and slow-moving water does not have the power to move coarse sediment. The sand here still contains feldspar grains, for these have not yet weathered into clay. Alluvial-fan sediments become arkose and conglomerate.

Sand dune environments. If the climate is very dry, few plants can grow and the ground surface lies exposed. Strong winds can move dust and sand. The dust gets carried away and the resulting well sorted sand can accumulate in dunes. Thus, thick layers of well sorted sandstone, in which we see large cross beds, are relics of desert sand-dune environments (Fig. 6.16d).

River environments. In climates where streams flow, we find several distinctive depositional environments. Rivers transport sand, silt, and mud. The coarser sediments tumble along the bed in the river's channel, while the finer sediments drift along, suspended in the water. This fine sediment settles out along the banks of the river, or on the floodplain, the flat region on either side of the river that is covered with water only during floods. On the floodplain, mud layers dry out between floods, leading to the

FIGURE 6.16 Examples of nonmarine depositional environments.

(a) Glacial till at the end of a glacier in France.

(b) Boulders and cobbles deposited by a mountain stream in Colorado.

(c) An alluvial fan in Death Valley, California.

(d) Sand dunes in Brazil.

(e) Deposits of an ancient river channel in Indiana. Note how the floor of the channel cuts across older strata. The geologist's sketch emphasizes the relationship.

Edge of photo

What a geologist sees

Younger floodplain deposits

Channel fill

Older floodplain deposits

(Talus)

(f) Laminated mud from a lake bed.

formation of mud cracks. River sediments lithify to form sandstone, siltstone, and shale. Typically, channels of coarser sediment are surrounded by layers of fine-grained floodplain deposits; in cross section, the channel has a lens-like shape (Fig. 6.16e).

Lake environments. In temperate climates, where water remains at the surface throughout the year, lakes form. In lakes, the relatively quiet water can't move coarse sediment; any coarse sediment brought into the lake by a stream settles out at the stream's outlet. Only fine clay makes it out into the center of the lake, where it settles to form mud on the lake bed. Thus, lake sediments typically consist of finely laminated shale (Fig. 6.16f).

In 1885, an American geologist named G. K. Gilbert studied small deltas that formed where small streams (carrying gravel, sand, and silt) emptied into lakes. He showed that such deltas contain three components (Fig. 6.17): topset beds composed of gravel, foreset beds of gravel and sand, and silty bottomset beds.

Marine Sedimentary Environments

Marine environments, as the name implies, form along the coasts, or at depth on the sea floor. The type of sediment at a location depends on the environment, water depth, and whether or not clastic grains are available.

FIGURE 6.17 A simple "Gilbert-type" delta formed where a stream enters a lake.

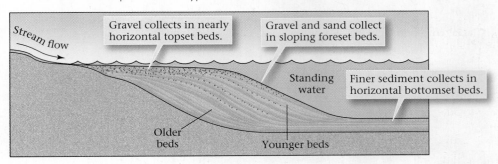

Stream flow

Gravel collects in nearly horizontal topset beds.

Gravel and sand collect in sloping foreset beds.

Standing water

Finer sediment collects in horizontal bottomset beds.

Older beds

Younger beds

Marine delta deposits. After following the river downstream for a long distance, we reach its mouth, where it empties into the sea. Here, the river builds a delta of sediment out into the sea. Deltas were so named because the map shape of some deltas (such as the Nile Delta of Egypt) resembles the Greek letter *delta* (Δ), as we discuss further in Chapter 14. River water stops flowing when it enters the sea, so sediment settles out.

Large deltas are much more complex than the type that Gilbert studied; they include many different sedimentary environments. Sea-level changes may cause the position of the different environments to move with time. Deposits of an ocean-margin delta produce a great variety of sedimentary rock types (Fig. 6.18a).

Coastal beach sands. Now we leave the delta and wander along the coast. Oceanic currents transport sand along the coastline. The sand washes back and forth in the surf, so it becomes well sorted (waves winnow out mud and silt) and well rounded, and because of the back-and-forth movement of ocean water over the sand, the sand surface may become rippled (Fig. 6.18b). Thus, if you find well-sorted, medium-grained sandstone, perhaps with ripple marks, you may be looking at the remnants of a beach environment.

Shallow-marine clastic deposits. From the beach, we proceed offshore. In deeper water, where wave energy does not stir the sea floor, finer sediment accumulates. Because the water here may be only meters to a few tens of meters deep, geologists refer to this depositional setting as a shallow-marine environment. Clastic sediments that accumulate in this environment tend to be fine-grained, well-sorted, well-rounded silt, and they are inhabited by a great variety of organisms such as mollusks and worms. Thus, if you see smooth beds of siltstone and mudstone containing marine fossils, you may be looking at shallow-marine clastic deposits.

FIGURE 6.18 Examples of coastal depositional environments.

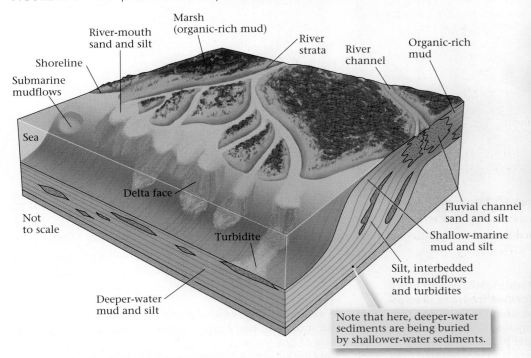

Marsh (organic-rich mud)

River-mouth sand and silt

River strata

River channel

Organic-rich mud

Shoreline

Submarine mudflows

Sea

Delta face

Not to scale

Turbidite

Deeper-water mud and silt

Fluvial channel sand and silt

Shallow-marine mud and silt

Silt, interbedded with mudflows and turbidites

Note that here, deeper-water sediments are being buried by shallower-water sediments.

(a) A major river delta along an ocean coast is a complex depositional environment. Sea-level changes affect locations of depositional settings.

(b) Waves on this California beach wash and sort the sand.

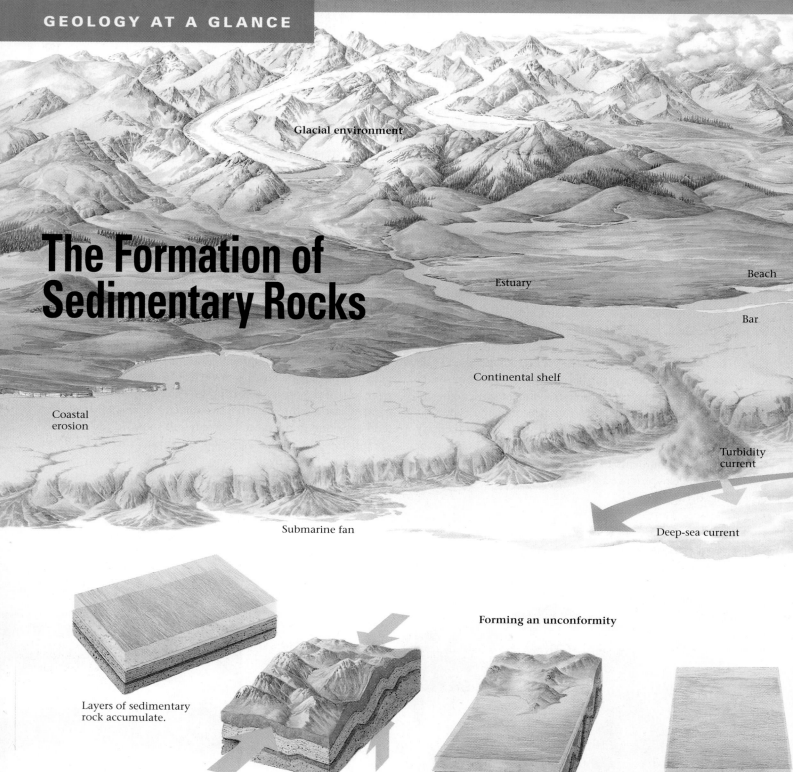

The Formation of Sedimentary Rocks

Glacial environment

Estuary

Beach

Bar

Continental shelf

Coastal erosion

Turbidity current

Submarine fan

Deep-sea current

Layers of sedimentary rock accumulate.

Mountain building folds the rock layers.

Forming an unconformity

The mountains are eroded; the folded layers are submerged.

New sedimentary layers accumulate.

Categories of sedimentary rocks include clastic sedimentary rocks, chemical sedimentary rocks (formed from the precipitation of minerals out of water), and biochemical sedimentary rocks (formed from the shells of organisms). Clastic sedimentary rocks develop when grains (clasts) break off preexisting rock by weathering and erosion and are transported to a new location by wind, water, or ice; the grains are deposited to create sediment layers, which are then cemented together. We distinguish among types of clastic sedimentary rocks on the basis of grain size.

The character of a sedimentary rock depends on the composition of the sediment and on the environment in which

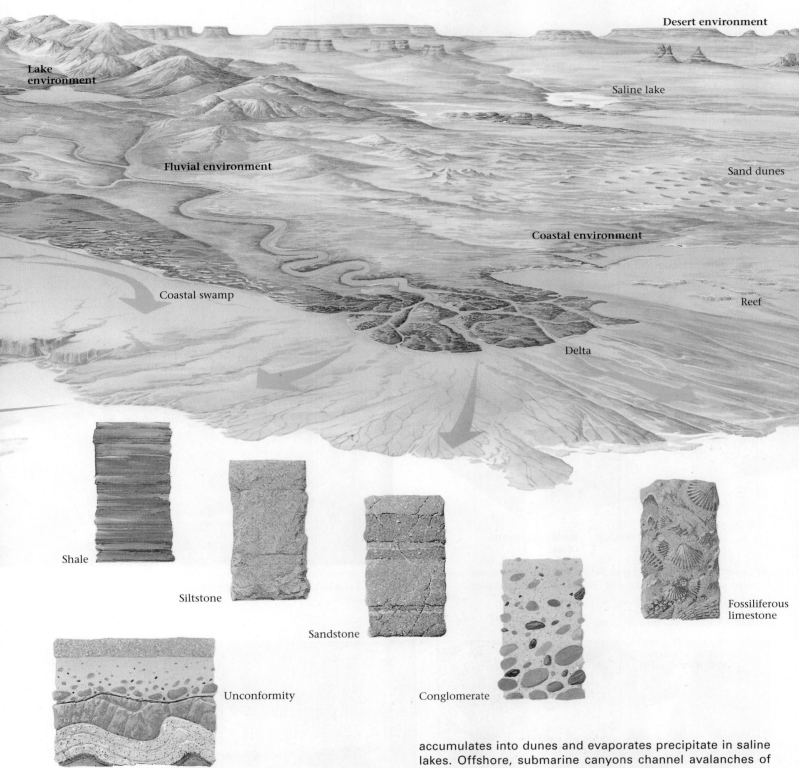

Lake environment

Desert environment

Saline lake

Fluvial environment

Sand dunes

Coastal environment

Coastal swamp

Reef

Delta

Shale

Siltstone

Sandstone

Conglomerate

Fossiliferous limestone

Unconformity

it accumulated. For example, glaciers carry sediment of all sizes, so they leave deposits of poorly sorted (different-sized) till; streams deposit coarser grains in their channels and finer ones on floodplains; a river slows down at its mouth and deposits an immense pile of silt in a delta. Fossiliferous limestone develops on coral reefs. In desert environments, sand accumulates into dunes and evaporates precipitate in saline lakes. Offshore, submarine canyons channel avalanches of sediment, or turbidity currents, out to the deep-sea floor.

Sedimentary rocks tell the history of the Earth. For example, the layering, or bedding, of sedimentary rocks is initially horizontal. So where we see layers bent or folded, we can conclude that the layers were deformed during mountain building. Where horizontal layers overlie folded layers, we have an unconformity: for a time, sediment was not deposited, and/or older rocks were eroded away.

Shallow-water carbonate environments. In shallow-marine settings far from the mouth of a river, where relatively little clastic sediment (sand and mud) enters the water, warm, clear, nutrient-rich water hosts an abundance of organisms. Their shells, which consist of carbonate minerals, make up most of the sediment that accumulates (Fig. 6.19a, b). The nature of carbonate sediment depends on the water depth. Beaches collect sand composed of shell fragments; lagoons (protected bodies of quiet water) are sites where carbonate mud accumulates; and reefs consist of coral and coral debris. Farther offshore of a reef, we can find a sloping apron of reef fragments. Shallow-water carbonate environments transform into sequences of limestone.

Deep-marine deposits. We conclude our journey by sailing offshore. Along the transition between coastal regions and the deep ocean, turbidity currents deposit graded beds. Farther offshore, in the deep-ocean realm, only fine clay and plankton provide a source for sediment. The clay eventually settles out onto the deep sea floor, forming deposits of finely laminated mudstones, and plankton shells settle to form chalk (from calcite shells; Fig. 6.20a, b) or chert (from siliceous shells). Thus, deposits of mudstone, chalk, or bedded chert indicate a deep-marine origin.

TAKE-HOME MESSAGE

Different types of sedimentary rocks accumulate in different terrestrial and marine depositional environments. By examining rock types and sedimentary structures, geologists can deduce the depositional environment in which the sediment accumulated.

FIGURE 6.19 Reef environments for the deposition of carbonate rocks.

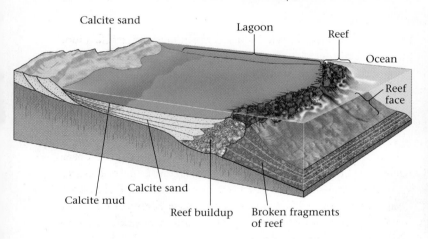

(a) Carbonate reefs form along shorelines in warm-water environments. In detail, reefs include many distinct depositional environments.

(b) A dramatic reef surrounds an island in the tropical Pacific. Deeper water is darker. Note the surf along the edge of the reef.

FIGURE 6.20 Examples of deep-marine sediment.

(a) These plankton shells, which make up some kinds of deep-marine sediment, are so small that they could pass through the eye of a needle.

(b) The chalk cliffs of southeastern England consist of plankton shells deposited on the sea floor tens of millions of years ago.

6.5 **SEDIMENTARY BASINS**

The sedimentary veneer on the Earth's surface varies greatly in thickness. If you stand in central Siberia or south-central Canada, you will find yourself on igneous and metamorphic basement rocks that are over a billion years old—there are no sedimentary rocks anywhere in sight. Yet if you stand along the southern coast of Texas, you would have to drill through over 15 km of sedimentary beds before reaching igneous and metamorphic basement. Thick accumulations of sediment form only in special regions where the surface of the Earth's lithosphere sinks, as sediment collects. Geologists use the term **subsidence** to refer to the sinking of lithosphere, and the term **sedimentary basin** for the sediment-filled depression. In what geologic settings do sedimentary basins form? An understanding of plate tectonics theory provides the answers.

Types of Sedimentary Basins in the Context of Plate Tectonics Theory

Geologists distinguish among different kinds of sedimentary basins on the basis of the region of a lithosphere plate in which they formed. Let's consider a few examples.

- *Rift basins*: These form in continental rifts, regions where the lithosphere has been stretched. As the rift grows, slip on faults drops blocks of crust down, creating low areas bordered by narrow mountain ridges. These troughs fill with sediment.

- *Passive-margin basins*: These form along the edges of continents that are not plate boundaries. They are underlain by stretched lithosphere, the remnants of a rift whose evolution ultimately led to the formation of a mid-ocean ridge. Passive-margin basins form because subsidence of stretched lithosphere continues long after rifting ceases and sea-floor spreading begins. They fill with sediment carried to the sea by rivers and with carbonate rocks formed in coastal reefs.

- *Intracontinental basins*: These develop in the interiors of continents, initially because of subsidence over a rift. They may continue to subside even hundreds of millions of years after they first formed, for reasons that are not well understood.

- *Foreland basins*: These form on the continent side of a mountain belt because as the mountain belt grows, large slices of rock are pushed up and onto the surface of the continent. The weight of these slices pushes down on the surface of the lithosphere, creating a wedge-shaped depression adjacent to the mountain range that fills with sediment eroded from the range.

Transgression and Regression

Sea-level changes control the succession of sediments that we see in a sedimentary basin. At times during Earth history, sea level has risen by as much as a couple of hundred meters, creating shallow seas that submerge the interiors of continents. At other times sea level has fallen by a couple of hundred meters, exposing even the continental shelves to air. Sea-level changes may be due to a number of factors, including climate changes, which control the amount of ice stored in polar ice caps, and changes in the volume of ocean basins. When sea level rises, the coast migrates inland. We call this process **transgression**. As the coast migrates, it gets progressively buried by deeper-water sediment. When sea level falls, the coast migrates seaward. We call this process **regression** (Fig. 6.21). The process of transgression and regression leads to the formation of broad blankets of sediment.

Diagenesis

Earlier in this chapter we discussed the process of lithification, by which sediment hardens into rock. Lithification is an aspect of a broader phenomenon called diagenesis. Geologists use the term **diagenesis** for all the physical, chemical, and biological processes that transform sediment into sedimentary rock and that alter characteristics of sedimentary rock after the rock has formed.

In sedimentary basins, sedimentary rocks may become buried very deeply. As a result, the rocks endure higher pressures and temperatures and come in contact with warm groundwater. Diagenesis under such conditions can cause chemical reactions in the rock that produce new minerals, and can cause cement to dissolve or precipitate.

As temperature and pressure increase still deeper in the subsurface, the changes that take place in rocks become more profound. At sufficiently high temperature and pressure, a new assemblage of minerals forms, and/or mineral grains become aligned parallel to each other. Geologists consider such changes to be examples of metamorphism. The transition between diagenesis and metamorphism in sedimentary rocks is gradational and occurs between temperatures of 150°C and 300°C. In the next chapter, we enter the realm of true metamorphism.

TAKE-HOME MESSAGE

In order for a thick deposit of sediment to accumulate, the surface of the Earth must sink (subside) and form a depression called a sedimentary basin. Sediment fills a basin as it subsides. Changes take place in the sediment as it becomes buried more deeply and reacts with water.

FIGURE 6.21 The concept of transgression and regression, during deposition of sedimentary sequence.

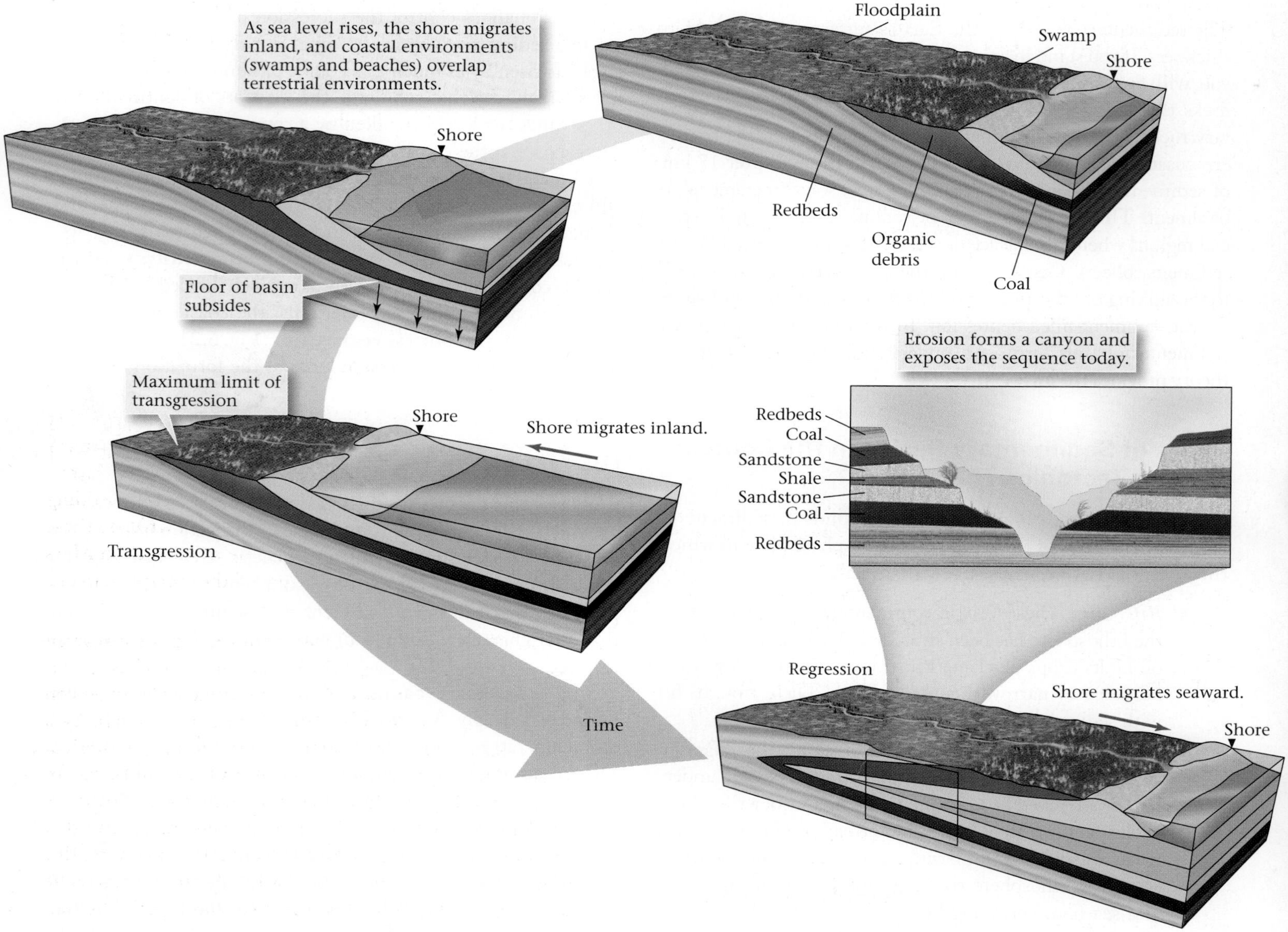

As sea level rises, the shore migrates inland, and coastal environments (swamps and beaches) overlap terrestrial environments.

Shore

Floor of basin subsides

Maximum limit of transgression

Shore

Shore migrates inland.

Transgression

Floodplain

Swamp

Shore

Redbeds

Organic debris

Coal

Erosion forms a canyon and exposes the sequence today.

Redbeds
Coal
Sandstone
Shale
Sandstone
Coal
Redbeds

Regression

Shore migrates seaward.

Shore

Time

Chapter Summary

- Geologists recognize four major classes of sedimentary rocks. Clastic rocks form from cemented-together grains that were first produced by weathering, then were transported, deposited, and lithified. Biochemical rocks develop from the shells of organisms. Organic rocks consist of plant debris or of altered plankton remains. Chemical rocks precipitate directly from water.

- Sedimentary structures include bedding, cross bedding, graded bedding, ripple marks, dunes, and mud cracks. They serve as clues to depositional settings.

- Glaciers, streams, alluvial fans, deserts, rivers, lakes, deltas, beaches, shallow seas, and deep seas each accumulate a different assemblage of sedimentary strata.

- Thick piles of sedimentary rocks accumulate in sedimentary basins, regions where the lithosphere sinks.

- Transgressions occur when sea level rises and the coastline migrates inland. Regressions occur when sea level falls and the coastline migrates seaward.

- Diagenesis involves processes leading to lithification and processes that alter sedimentary rock once it has formed.

GEOPUZZLE REVISITED

The Grand Canyon cuts down through an over 1.5 km-thick succession of sedimentary strata, recording a long history of deposition in a variety of environments during successive transgressions and regressions of the sea. Contrasting layers consist of contrasting rock types—each rock type formed in a different depositional environment. Not only are different layers different colors (in part due to the amount of oxidized iron in the rock), but they also have different grain sizes and bedding thicknesses. Stronger rock units (sandstone and limestone) form steep cliffs, whereas weaker units (shale) form gentler slopes.

Key Terms

arkose (p. 156)
bed (p. 160)
biochemical sedimentary rock (p. 154)
breccia (p. 156)
cementation (p. 154)
chemical sedimentary rock (p. 154)
clastic sedimentary rock (p. 154)
clasts (p. 154)
conglomerate (p. 156)
cross bed (p. 163)
deposition (p. 154)
diagenesis (p. 171)
dolostone (p. 159)
erosion (p. 154)
evaporite (p. 159)
graded bed (p. 163)
limestone (p. 157)

lithification (p. 154)
mudstone (p. 156)
organic sedimentary rock (p. 154)
regression (p. 171)
ripple marks (p. 162)
sandstone (p. 154)
sedimentary basin (p. 171)
sedimentary rock (p. 153)
sedimentary structure (p. 160)
shale (p. 156)
siltstone (p. 156)
sorting (p. 155)
strata (p. 161)
stratigraphic formation (p. 161)
subsidence (p. 171)
transgression (p. 171)
travertine (p. 159)
turbidity current (p. 163)

Review Questions

1. Describe how a clastic sedimentary rock forms from its unweathered parent rock.
2. Explain how biochemical sedimentary rocks form.
3. Describe how grain size and shape, sorting, sphericity, and angularity change as sediments move downstream.
4. Describe the two different kinds of chert. How are they similar? How are they different?

5. What kinds of conditions produce evaporites?
6. How does dolostone differ from limestone?
7. Describe how cross beds form. How can you read the current direction from cross beds?
8. Describe how a turbidity current forms and moves. How does it produce graded bedding?
9. Compare deposits of an alluvial fan with those of a deep-marine deposit.
10. Why don't sediments accumulate everywhere? What types of tectonic conditions are required to create basins?
11. What kinds of changes may take place during diagenesis?

On Further Thought

1. Recent exploration of Mars by robotic vehicles suggests that layers of sedimentary rock cover portions of the planet's surface. On the basis of examining images of these layers, some researchers claim that the layers contain cross bedding and relicts of gypsum crystals. At face value, what do these features suggest about depositional environments on Mars in the past? (*Note*: Interpretation of the images remains controversial.)

2. The Gulf Coast of the United States is a passive-margin basin that contains a very thick accumulation of sediment. Drilling reveals that the base of the sedimentary succession in this basin consists of redbeds. These are overlain by a thick layer of evaporite. The evaporite, in turn, is overlain by deposits composed dominantly of sandstone and shale. In some intervals, sandstone occurs in channels and contains ripple marks, and the shale contains mud cracks. In other intervals, the sandstone and shale contains fossils of marine organisms. The sequence contains hardly any conglomerate or arkose. Be a sedimentary detective, and explain the succession of sediment in the basin.

3. Examine the Bahamas with *Google Earth*™ or NASA World Wind. (You can find a high-resolution image at Lat 23°58'40.98"N Long 77°30'20.37"W). Note that broad expanses of very shallow water surround the islands, that white sand beaches occur along the coast of the islands, and that small reefs occur offshore. What does the sand consist of, and what rock will it become if it eventually becomes buried and lithified? Compare the area of shallow water in the Bahamas area with the area of Florida. The bedrock of Florida consists mostly of shallow marine limestone. What does this observation suggest about the nature of the Florida peninsula in the past? Keep in mind that sea level on Earth changes over time. Presently, most of the land surface of Florida lies at less than 50 m (164 feet) above sea level.

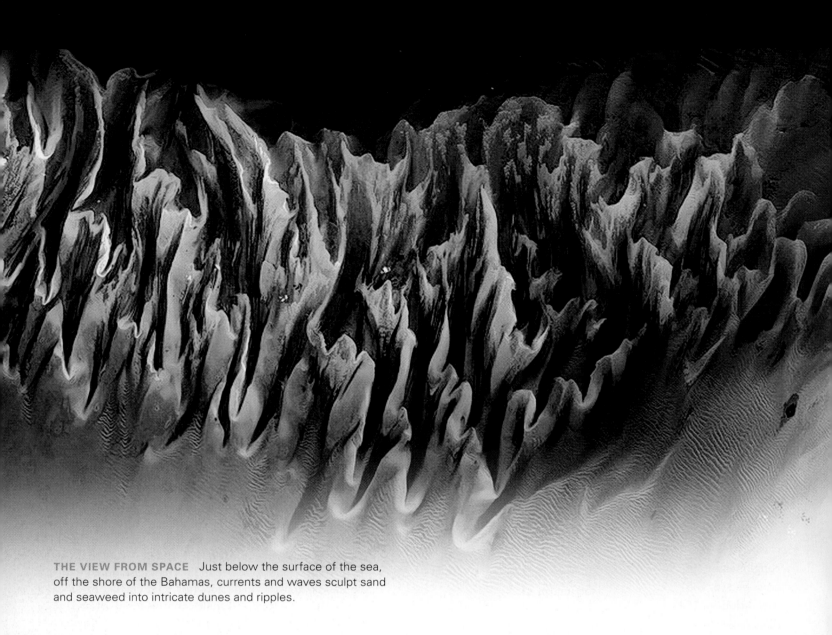

THE VIEW FROM SPACE Just below the surface of the sea, off the shore of the Bahamas, currents and waves sculpt sand and seaweed into intricate dunes and ripples.

CHAPTER **7**

Metamorphism: A Process of Change

A geologist examines an outcrop of 2.7 billion-year-old metamorphic rock in the Canadian Shield of Ontario. The curved layering in this rock gives the impression that the rock flowed like plastic. The layering is metamorphic foliation.

GEOPUZZLE

Marble, the rock from which Michelangelo carved his sculptures, contains the same chemicals as limestone, a sedimentary rock. But grains in marble interlock, and the form of layering in marble suggests that the rock once flowed. The rock can't be igneous because its composition is unlike that of any known magma. So, how do rocks like marble form?

Nothing in the world lasts, save eternal change.

—Honorat de Bueil (1589–1650)

7.1 INTRODUCTION

Cool winds sweep across Scotland for much of the year. In this blustery climate, vegetation has a hard time taking hold, so the landscape provides countless outcrops of barren rock. During the latter half of the eighteenth century, James Hutton became fascinated with the Earth and examined these outcrops, hoping to learn how rock formed. Hutton found that many features in the outcrops resembled the products of present-day sediment deposition and volcanic activity, and soon he came to an understanding of how sedimentary and igneous rock form. But Hutton also found rock that contained minerals and textures quite different from those in sedimentary and igneous samples. He described this puzzling rock as "a mass of matter which had evidently formed originally in the ordinary manner . . . but which is now extremely distorted in its structure . . . and variously changed in its composition."

The rock that so puzzled Hutton is now known as metamorphic rock, from the Greek words *meta*, meaning change, and *morphe*, meaning form. In modern terms, a **metamorphic rock** is one that forms when a preexisting rock, or **protolith**, undergoes a solid-state change in response to the modification of its environment. This process of change is called **metamorphism**. Let's look at the components of our definition more closely. By "solid-state," we mean that a metamorphic rock does *not* form by solidification of magma—remember that geologists consider rocks that solidified from magma to be igneous. By "change," we mean that metamorphism produces new minerals that did not occur in the protolith, and/or produces a new texture (arrangement of mineral grains) that is distinct from that of the protolith. And by "modification of environment," we mean that metamorphism takes place when a protolith endures a rise or fall in temperature and/or pressure, undergoes compression and shear, or reacts with very hot water.

Hutton did more than just note the existence of metamorphic rock—he also tried to understand why metamorphism takes place. Because he found metamorphic rocks adjacent to igneous intrusions, he concluded that metamorphism can take place when heat from an intrusion "cooks" the rock into which it intrudes. And because he found that metamorphic rocks can occur over broad regions in the absence of intrusions, he speculated that metamorphism can also take place when rock becomes deeply buried, as occurs during mountain building.

From Hutton's day to the present, geologists have undertaken field studies, laboratory experiments, and theoretical calculations to better characterize metamorphism. In this chapter, we present the results of their work. We begin by explaining the causes of metamorphism and the basis for classifying metamorphic rocks. We conclude by discussing the geologic settings in which these rocks form. As you will see, Hutton's speculations on the origin of metamorphic rock were basically correct, but they represented only part of the story—the rest of the story could not take shape until the theory of plate tectonics was proposed.

TAKE-HOME MESSAGE

By the end of this chapter, you should understand that rocks may be modified in response to changes in temperature, pressure, and other environmental conditions. This process, called metamorphism, alters the texture of rock and/or mineral assemblage in rock.

7.2 CONSEQUENCES AND CAUSES OF METAMORPHISM

What Is a Metamorphic Rock?

If someone were to put a rock on a table in front of you, how would you know that it is metamorphic? First, metamorphic rocks can have a **metamorphic texture**, meaning that the grains in the rock have grown in place within the solid rock. Second, metamorphic rocks can possess **metamorphic minerals**, new minerals that grow only under metamorphic temperatures and pressures. In fact, metamorphism can produce a group of minerals which together make up a metamorphic mineral assemblage. And third, metamorphic rocks can have **metamorphic foliation**, defined by the parallel alignment of platy minerals (such as mica) and/or the presence of alternating light-colored and dark-colored layers. When these characteristics develop, a metamorphic rock becomes as different from its protolith as a butterfly is from a caterpillar. For example, metamorphism of red shale can yield a metamorphic rock consisting of aligned mica flakes and brilliant garnet crystals (Fig. 7.1a). Metamorphism of limestone composed of cemented-together fossil fragments can yield a metamorphic rock consisting of large interlocking crystals of calcite (Fig. 7.1b).

The formation of metamorphic textures and minerals takes place very slowly—it may take thousands to millions of years—and it involves several processes, which sometimes occur alone and sometimes together. The most common processes are:

- *Recrystallization*, which changes the shape and size of grains without changing the identity of the mineral constituting the grains (Fig. 7.2a).
- *Phase change*, which transforms one mineral into another mineral with the same composition but a different crystal structure. On an atomic scale, phase change involves the rearrangement of atoms.

■ *Metamorphic reaction,* or *neocrystallization* (from the Greek *neos,* for new), which results in the growth of new mineral crystals that differ from those of the protolith (Fig. 7.2b). During neocrystallization, chemical reactions digest minerals of the protolith to produce new minerals of the metamorphic rock. For this to take place, atoms migrate (diffuse) through solid crystals, a very slow process, and/or dissolve and reprecipitate at grain boundaries.

■ *Pressure solution,* which happens when a rock is squeezed more strongly in one direction than in others in the presence of water. Mineral grains dissolve where their surfaces are pressed against other grains, producing ions that migrate through the water to precipitate elsewhere (Fig. 7.2c).

■ *Plastic deformation,* which happens when a rock is squeezed or sheared at elevated temperatures. Under these conditions, minerals behave like soft plastic and change shape without breaking (Fig. 7.2d).

Caterpillars undergo metamorph*osis* because of hormonal changes in their bodies. Rocks undergo metamorph*ism* when they are subjected to heat, pressure, compression and shear, and/or very hot water. Let's now consider the details of how these agents of metamorphism operate.

FIGURE 7.1 Metamorphism causes changes in mineral makeup and texture.

(a) A specimen of red shale (left) contains clay, quartz, and iron oxide. When this rock undergoes intense metamorphism, it may change into a gneiss containing different minerals. This gneiss sample (right) contains biotite, quartz, feldspar, and purple garnet.

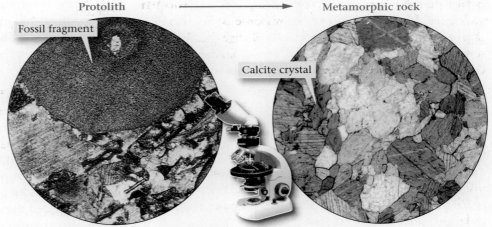

(b) A photomicrograph of a limestone in thin section (left) shows tiny fossil fragments and lime mud. After metamorphism, the texture of the rock changes completely and we see large, interlocking crystals of calcite (right). Note that the color seen here depends on the orientation of the crystals.

Metamorphism Due to Heating

When you heat cake batter, the batter transforms into a new material—cake. Similarly, when you heat a rock, its ingredients transform into a new material—metamorphic rock. Why? Think about what happens to atoms in a mineral grain as the grain warms. Heat causes the atoms to vibrate rapidly, stretching and bending chemical bonds that lock atoms to their neighbors. If bonds stretch too far and break, atoms detach from their original neighbors, move slightly, and form new bonds with other atoms. Repetition of this process leads to rearrangement of atoms within grains, or to migration of atoms into and out of grains. As a consequence, recrystallization and/or neocrystallization take place, enabling a metamorphic mineral assemblage to grow in solid rock.

Metamorphism takes place at temperatures between those at which diagenesis occurs and those that cause melting. Roughly speaking, this means that most metamorphic rocks you find in outcrops on continents formed at temperatures of between 200°C and 850°C.

FIGURE 7.2   Metamorphic processes, as seen through a microscope.

Protolith ⟶ Metamorphic rock

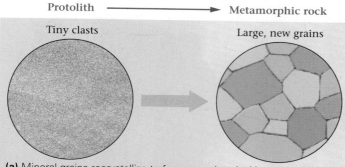

(a) Mineral grains recrystallize to form new, interlocking grains of the same mineral. Typically, grains get larger.

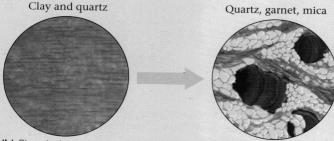

(b) Chemical reactions change the original assemblage of minerals into a new, metamorphic assemblage of minerals.

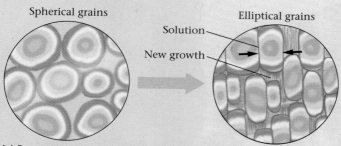

(c) Pressure solution dissolves grains on the sides undergoing more pressure, and precipitates new mineral material where the pressure is lower. Arrows indicate the squeezing direction.

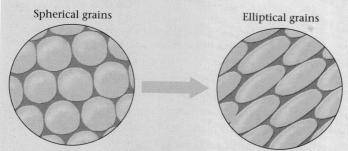

(d) Plastic deformation changes the shape of grains—without breaking them—as a result of squeezing or shear at high temperatures.

Metamorphism Due to Pressure

As you swim underwater in a swimming pool, water squeezes against you equally from all sides—in other words, your body feels pressure. Pressure can cause a material to collapse inward. For example, if you pull an air-filled balloon down to a depth of 10 m in a lake, the balloon becomes significantly smaller. Pressure can have the same effect on minerals. Near the Earth's surface, minerals with relatively open crystal structures can be stable. However, if you subject these minerals to extreme pressure, the atoms pack more closely together and denser minerals tend to form. Such transformations involve phase changes and/or neocrystallization.

Changing Pressure and Temperature Together

So far, we've considered changes in pressure and temperature as separate phenomena. But in the Earth, pressure and temperature change together with increasing depth. For example, at a depth of 8 km, temperature in the crust reaches about 200°C and pressure reaches about 2.3 kbar. If a rock slowly becomes buried to a depth of 20 km, as can happen during mountain building, temperature in the rock increases to 500°C, and pressure to 5.5 kbar. Experiments and calculations show that the "stability" of certain minerals (the ability of a mineral to survive a long time) and mineral assemblages depend on *both* pressure and temperature. Thus, a metamorphic rock formed at 8 km does not contain the same minerals as one formed at 20 km.

Compression, Shear, and Development of Preferred Orientation

Imagine that you have just built a house of cards and, being in a destructive mood, you step on it. The structure collapses because the downward push you apply with your foot exceeds the push provided by air in other directions. We can say that we have subjected the cards to compression (Fig. 7.3a). Compression flattens a material (Fig. 7.3b). Shear, in contrast, moves one part of a material sideways, relative to another. If, for example, you place a deck of cards on a table, then set your hand on top of the deck and move your hand parallel to the table, you shear the deck (Fig. 7.3c).

When rocks are subjected to compression and shear at elevated temperatures and pressures, they can change shape without breaking. As it changes shape, the internal texture of a rock also changes. For example, platy (pancake-shaped) grains become parallel to one another, and elongate (cigar-shaped) grains align in the same direction. Both platy and elongate grains are inequant grains, meaning that the dimension of a grain is not the same in all directions; in contrast, equant grains

FIGURE 7.3 Compression and shear changes rock grains during metamorphism.

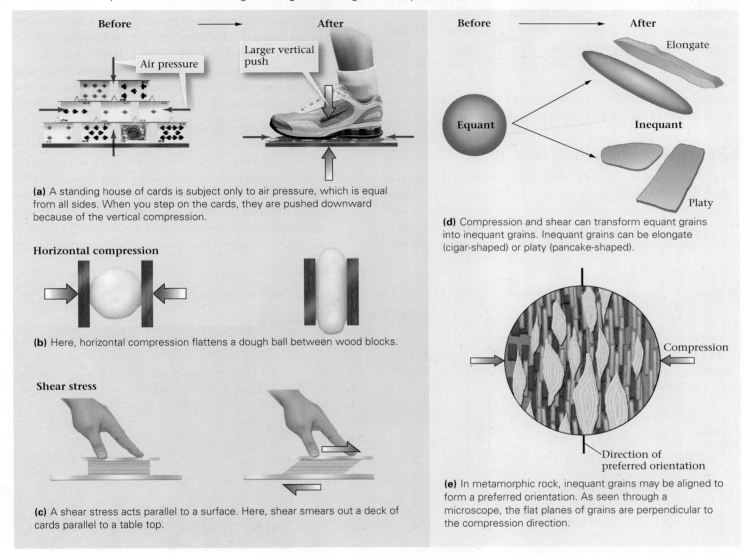

(a) A standing house of cards is subject only to air pressure, which is equal from all sides. When you step on the cards, they are pushed downward because of the vertical compression.

Horizontal compression

(b) Here, horizontal compression flattens a dough ball between wood blocks.

Shear stress

(c) A shear stress acts parallel to a surface. Here, shear smears out a deck of cards parallel to a table top.

(d) Compression and shear can transform equant grains into inequant grains. Inequant grains can be elongate (cigar-shaped) or platy (pancake-shaped).

(e) In metamorphic rock, inequant grains may be aligned to form a preferred orientation. As seen through a microscope, the flat planes of grains are perpendicular to the compression direction.

have roughly the same dimensions in all directions (Fig. 7.3d). The alignment of inequant minerals in a rock results in a **preferred orientation** (Fig. 7.3e).

The Role of Hot Water

Metamorphic reactions commonly take place in the presence of very hot water, because water permeates the crust. Such hydrothermal fluids chemically react with rock. For example, hydrothermal fluids accelerate metamorphic reactions, because atoms involved in the reactions can migrate faster through a fluid than they can through a solid, and hydrothermal fluids provide water that can be absorbed by minerals during metamorphic reac-

tions. Finally, fluids passing through a rock may pick up some dissolved ions and drop off others, and thus can change the overall chemical composition of a rock during metamorphism. The process of changing a rock's chemical composition by reactions with hydrothermal fluids is called **metasomatism**.

TAKE-HOME MESSAGE

Metamorphism takes place in response to changes in temperature, pressure, application of differential stress (squashing, stretching, or shearing), and/or reaction with hydrothermal fluids. Application of compression and shear during metamorphism aligns mineral grains.

7.3 TYPES OF METAMORPHIC ROCKS

Coming up with a way to classify and name the great variety of metamorphic rocks on Earth hasn't been easy. After decades of debate, geologists have found it most convenient to divide metamorphic rocks into two fundamental classes: foliated rocks and nonfoliated rocks. Each class contains several rock types. We distinguish nonfoliated rocks from each other primarily by their component minerals, whereas foliated rocks are distinguished partly by their component minerals and partly by the nature of their foliation.

Foliated Metamorphic Rocks

To understand this class of rocks, we first need to discuss the nature of **foliation** in more detail. The word comes from the Latin *folium*, for leaf. Geologists use *foliation* to refer to the repetition of planar surfaces or layers in a metamorphic rock (Fig. 7.4a). Some layers are indeed as thin as a leaf, or thinner, but some may be over a meter thick. Foliation can give metamorphic rocks a striped or streaked appearance in an outcrop, and/or give them the ability to split into thin sheets. A foliated metamorphic rock has foliation because it contains inequant mineral crystals that are aligned parallel to one another, defining preferred mineral orientation, and/or because the rock has alternating dark-colored and light-colored layers.

Foliated metamorphic rocks can be distinguished from one another according to their composition, their grain size, and the nature of their foliation. The most common types include:

- *Slate*: The finest-grained foliated metamorphic rock, **slate**, forms by metamorphism of shale or mudstone—rocks composed of clay—under relatively low pressures and temperatures. Slate contains a strong foliation, called slaty cleavage, which allows it to split into thin sheets that make excellent roofing shingles (see Fig. 7.4a). The slaty cleavage develops when compression causes clay flakes to reorient and regrow into an orientation perpendicular to

FIGURE 7.4 Slate is a foliated metamorphic rock that forms at relatively low temperature and pressure.

(a) A block of slate splits easily along cleavage planes, which may be at a high angle to the bedding planes. Workers split slate to produce roof shingles that, when overlapped, make a watertight surface (see insert).

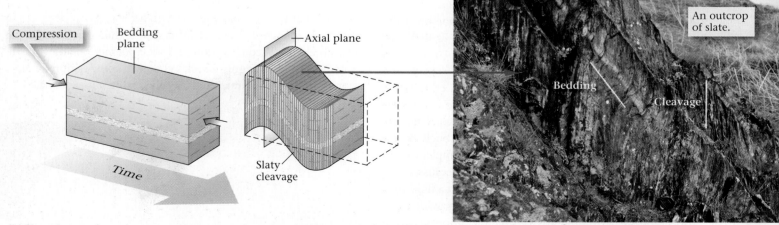

(b) Slaty cleavage forms in response to compressive stress. In this example, layers also bend to form folds, as slaty cleavage develops. The cleavage tends to be oriented parallel to the axial plane, an imaginary surface that, simplistically, divides the fold in half.

the direction of compression. Because of this relation-ship, end-on compression of shale beds produces vertical slaty cleavage. Commonly, such compression also causes the layers to bend and become folded (Fig. 7.4b).

- *Phyllite*: **Phyllite** is a fine-grained metamorphic rock with a foliation caused by the preferred orientation of very fine-grained white mica. The word comes from the Greek word *phyllon*, meaning leaf. The parallelism of translucent fine-grained mica gives phyllite a silky sheen known as phyllitic luster (Fig. 7.5a). Phyllite forms by the metamorphism of slate at a temperature high enough to cause neocrystallization of white mica.

- *Metaconglomerate*: Under the metamorphic conditions that produce slate or phyllite, a protolith of conglom-erate becomes **metaconglomerate**. Specifically, pres-sure solution and plastic deformation flatten pebbles and cobbles into pancake-like shapes. The alignment of inequant clasts defines a foliation (Fig. 7.5b).

- *Schist*: **Schist** is a medium- to coarse-grained meta-morphic rock that possesses a type of foliation, called schistosity, defined by the preferred orientation of *large* mica flakes (muscovite and/or biotite; Fig. 7.5c). Schist forms at a higher temperature than does phyllite.

- *Gneiss*: **Gneiss** is a compositionally layered metamor-phic rock, typically composed of alternating dark-colored and light-colored layers that range in thickness from millimeters to meters. This compositional layering, or gneissic banding, gives gneiss a striped appearance (Fig. 7.6a).

    How does the banding in gneiss form? Banding in some examples of gneiss evolved directly from the original bedding in a rock. For example, metamor-phism of a protolith consisting of alternating beds of sandstone and shale produces a gneiss consisting of alternating beds of quartzite and mica. It is more common for gneissic banding to form when the pro-tolith undergoes an extreme amount of shearing under conditions in which the rock can flow like soft plastic (Fig. 7.6b). Such flow stretches, folds, and smears out preexisting compositional contrasts in the rock and transforms them into aligned sheets. Finally, banding in some gneisses develops by an incompletely under-stood process called metamorphic differentiation. During this process, chemical reactions segregate dif-ferent minerals into different layers (Fig. 7.6c).

- *Migmatite*: Under certain conditions, gneiss may *begin* to melt, producing felsic magma. If the melt freezes again before flowing out of the source area, a mixture of igne-ous rock and relict metamorphic rock forms. This mix-ture is called **migmatite**.

FIGURE 7.5 Examples of foliated metamorphic rocks formed at higher temperatures and pressures.

(a) During formation of phyllite, clay recrystallizes to form tiny mica flakes that reflect light, giving the rock a sheen.

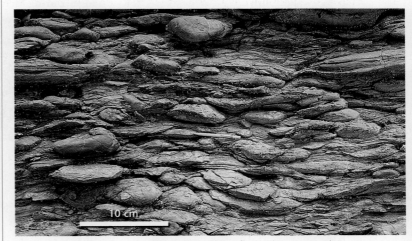

(b) In metaconglomerate, pebbles and cobbles flatten into a pancake shape without cracking.

(c) A schist contains coarse mica flakes, along with other metamorphic minerals.

FIGURE 7.6 The formation of gneiss, which takes place at very high temperatures and pressures.

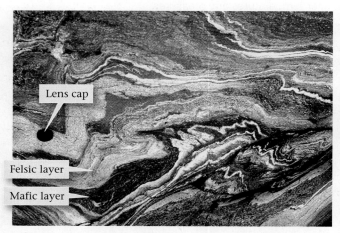

(a) Gneiss displays compositional banding (alternating felsic and mafic layers). The layers can become contorted.

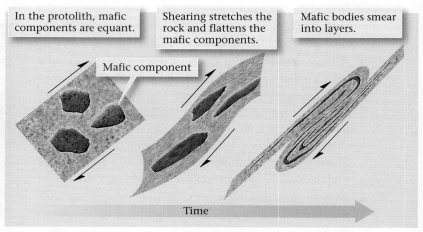

In the protolith, mafic components are equant.

Shearing stretches the rock and flattens the mafic components.

Mafic bodies smear into layers.

Mafic component

Time

(b) Formation of gneiss, in some cases, involves extreme shear. Original contrasting rock types are smeared into parallel layers.

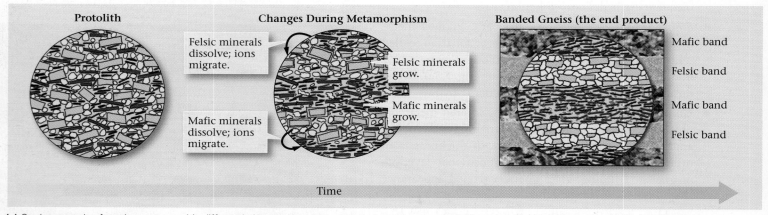

Protolith

Changes During Metamorphism

Banded Gneiss (the end product)

Felsic minerals dissolve; ions migrate.

Felsic minerals grow.

Mafic minerals grow.

Mafic minerals dissolve; ions migrate.

Mafic band

Felsic band

Mafic band

Felsic band

Time

(c) Gneiss may also form by metamorphic differentiation, during which chemical reactions cause felsic and mafic minerals to grow in distinct, separate layers.

Nonfoliated Metamorphic Rocks

Nonfoliated metamorphic rocks contain minerals that recrystallized or grew during metamorphism, but have no foliation. The lack of foliation means either that metamorphism occurred in the absence of compression and shear, or that most of the new crystals are equant. We list below some of the rock types that can occur without foliation.

- *Hornfels*: **Hornfels** is a nonfoliated rock that contains a variety of metamorphic minerals. The specific mineral assemblage in a hornfels depends on the composition of the protolith and on the temperature and pressure of metamorphism.

- *Quartzite*: **Quartzite** forms by the metamorphism of pure quartz sandstone. During metamorphism, preexisting quartz grains recrystallize, creating new, larger grains. In the process, the distinction between cement and grains disappears, open pore space disappears, and the grains become interlocking. When quartzite cracks, the fracture cuts across grain boundaries—in contrast, fractures in sandstone curve *around* grains. Quartzite looks glassier than sandstone and does not have the grainy, sandpaper-like surface characteristic of sandstone (Fig. 7.7a). Depending on the impurities contained in the quartz, quartzite can vary in color from white to gray, purple, or green.

- *Marble*: The metamorphism of limestone yields **marble**. During the formation of marble, calcite composing the protolith recrystallizes, so fossil shells, pore space, and the distinction between grains and cement disappear. Thus, marble typically consists of a fairly uniform mass of interlocking calcite crystals.

  Sculptors love to work with marble because the rock is relatively soft and has a uniform texture that gives it the cohesiveness and homogeneity needed to fashion large, smooth, highly detailed sculptures. Marble comes in a variety of colors—white, pink, green, and black—depending on the impurities it contains. Michelangelo, one of the great Italian Renaissance artists, sought large, unbroken blocks of creamy white marble from quarries in the Italian Alps for his masterpieces (Fig. 7.7b). If the original protolith contained layers of other rock types, such as shale, the resulting marble has color banding that makes it a prized decorative stone (Fig. 7.7c).

TAKE-HOME MESSAGE

Geologists divide metamorphic rock into two general classes based on whether or not the rock contains foliation. Types of foliated rocks (slate, phyllite, schist, and gneiss) differ from each other in terms of the nature of foliation, which in turn reflect metamorphic conditions.

7.4 HOW DO WE DEFINE THE INTENSITY OF METAMORPHISM?

Not all metamorphism takes place under the same physical conditions. For example, rocks carried to a great depth beneath a mountain range undergo more intense metamorphism than do rocks closer to the surface. Geologists use the term **metamorphic grade** in a somewhat informal way to indicate the intensity of metamorphism, meaning the amount or degree of metamorphic change. (To provide a more complete indication of the intensity of metamorphism, geologists use the concept of metamorphic facies; see Box 7.1.) Classification of metamorphic grade depends primarily on temperature, because temperature plays the dominant role in determining the extent of recrystallization and neocrystallization during metamorphism. Metamorphic rocks that form at relatively low temperatures (between about 250°C and 400°C) are low-grade rocks, and metamorphic rocks that form at relatively high temperatures (over 600°C) are high-grade rocks. Intermediate-grade rocks form at temperatures between these two extremes (Fig. 7.8a).

Different grades of metamorphism yield different metamorphic mineral assemblages. As grade increases, recrystallization and neocrystallization produce coarser grains and new mineral assemblages that are stable at higher temperatures and pressures (Fig. 7.8b).

Geologists discovered that the presence of certain minerals, known as index minerals, in a rock indicates the approx-imate metamorphic grade of the rock. The line on a map along which an index mineral first appears is called an isograd (from the Greek *iso*, meaning equal). All points along an isograd have approximately the same metamorphic grade. **Metamorphic zones** are regions between two isograds; zones are named after an index mineral that was not present in the previous, lower-grade zone. To compare rocks of different grades, you could take a hike from central New York State eastward into central Massachusetts in the eastern United States. Your path starts in a region where rocks were not metamorphosed, and it takes you into the internal part of the Appalachian Mountain belt, where rocks were intensely metamorphosed. As a consequence, you cross several metamorphic zones (Fig. 7.8c).

TAKE-HOME MESSAGE

The type of metamorphic rock (low grade vs. high grade) that forms at a location depends on the conditions of metamorphism. For example, a given metamorphic mineral assemblage forms under a definable range of temperature and pressure.

FIGURE 7.7 Examples of nonfoliated metamorphic rocks.

(a) This unfoliated maroon quartzite from Wisconsin breaks on smooth fractures that cut across grains.

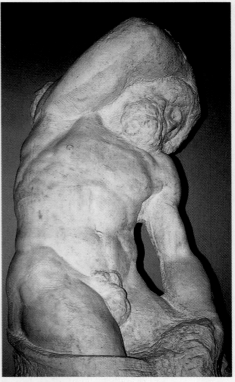

(b) Michelangelo used nonfoliated white marble for his spectacular sculptures.

(c) The marble in this floor tile has color banding inherited from original bedding. The layers became contorted during metamorphism.

BOX 7.1

Metamorphic Facies

In the early years of the twentieth century, geologists working in Scandinavia, where erosion by glaciers has left beautiful, nearly unweathered exposures, came to realize that metamorphic rocks, in general, do not consist of a hodgepodge of minerals formed at different times and in different places, but rather consist of a distinct assemblage of minerals that grew in association with each other at a certain pressure and temperature. It seemed that the mineral assemblage present in a rock more or less represented a condition of chemical equilibrium, meaning that the chemicals making up the rock had organized into a group of mineral grains that were—to anthropomorphize a bit—comfortable with each other and their surroundings, and thus did not feel the need to change further. These geologists also realized that the specific mineral assemblage in a rock depends on pressure and temperature conditions, *and* on the composition of the protolith.

This discovery led the geologists to propose the concept of metamorphic facies. A **metamorphic facies** is a set of metamorphic mineral assemblages indicative of a certain range of pressure and temperature. Each specific assemblage in a facies reflects a specific protolith composition. According to this definition, a given metamorphic facies includes several different kinds of rocks that differ from each other in terms of chemical composition and, therefore, mineral content—but all the rocks of a given facies formed under roughly the same temperature and pressure conditions. Geologists recognize several facies, of which the major ones are zeolite, hornfels, greenschist, amphibolite, blueschist, eclogite, and granulite. The names of the different facies are based on a distinctive feature or mineral found in some of the rocks of the facies.

We can represent the approximate conditions under which metamorphic facies formed by using a pressure-temperature graph (**Fig. 1**). Each area on the graph, labeled with a facies name, represents the approximate range of temperatures and pressures in which mineral assemblages characteristic of that particular facies form. For example, a rock subjected to the pressure and temperature at Point A (4.5 kbar and 400°C) develops a mineral assemblage characteristic of the greenschist facies. As the

graph implies, the boundaries between facies cannot be precisely defined, and there are likely broad transitions between facies. We can also portray the geothermal gradients of different crustal regions on the graph. Beneath mountain ranges, for example, the geothermal gradient passes through the zeolite, greenschist, amphibolite, and granulite facies. In contrast, in the accretionary prism that forms at a subduction zone, temperature increases slowly with increasing depth, so blueschist assemblages can form.

FIGURE 1 The common metamorphic facies. The boundaries between the facies are depicted as wide bands because they are gradational and approximate. Note that some amphibolite-facies rocks and all granulite-facies rocks form at pressure-temperature conditions to the right of the melting curve for wet granite. Thus, such metamorphic rocks develop *only* if the protolith is dry. One of the facies depicted on the graph is not mentioned in the text: specifically, P-P ("prehnite-pumpellyite") facies, named for two metamorphic minerals.

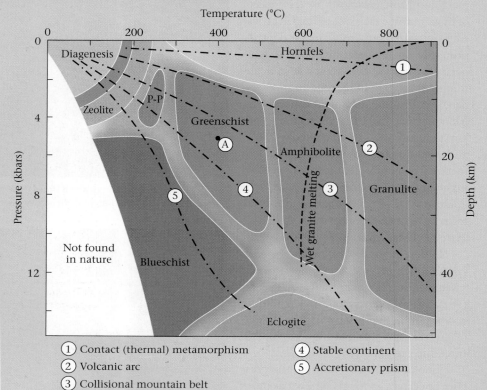

① Contact (thermal) metamorphism   ④ Stable continent
② Volcanic arc   ⑤ Accretionary prism
③ Collisional mountain belt

FIGURE 7.8   Intensity of metamorphism is indicated by metamorphic grade.

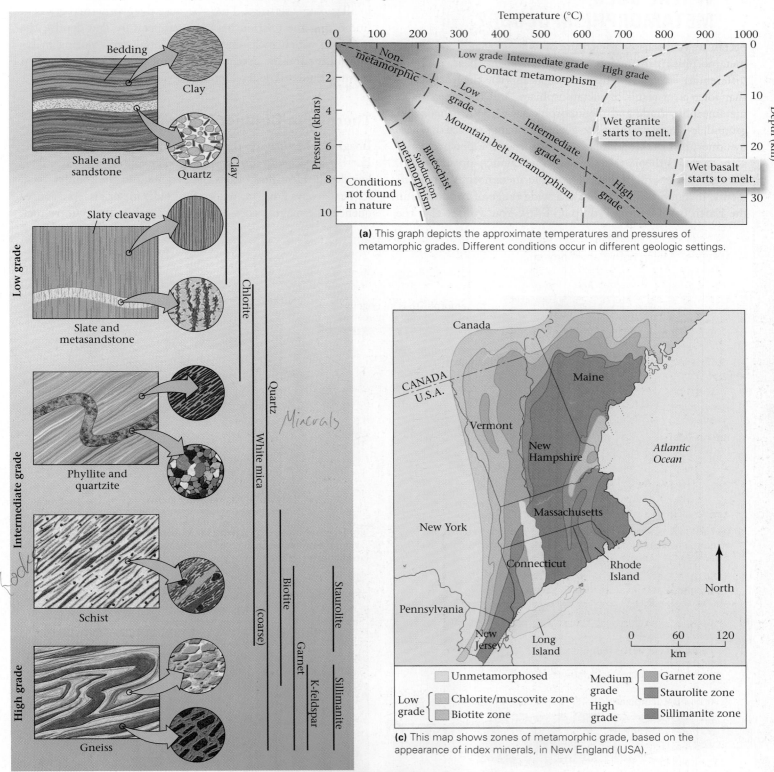

(a) This graph depicts the approximate temperatures and pressures of metamorphic grades. Different conditions occur in different geologic settings.

(b) Here, we see the consequences of the progressive metamorphism of shale and sandstone from low grade to high grade during mountain building. Lines on the right indicate the range of grade in which key metamorphic minerals form.

(c) This map shows zones of metamorphic grade, based on the appearance of index minerals, in New England (USA).

7.5 WHERE DOES METAMORPHISM OCCUR?

So far, we've discussed the nature of changes that occur during metamorphism, the agents of metamorphism (heat, pressure, differential stress, and hydrothermal fluids), the rock types that form as a result of metamorphism, and the concepts of metamorphic grade and metamorphic facies. With this background, let's now examine the geologic settings on Earth where metamorphism takes place, as viewed from the perspective of plate tectonics theory (see **Geology at a Glance**, pp. 188–189). Because of the wide range of possible metamorphic environments, metamorphism occurs at a wide range of conditions in the Earth (see **Geotour 7** on p. GT-16). You will see that the conditions under which metamorphism occurs are not the same in all geologic settings. That's because the geothermal gradient (indicating the relation between temperature and depth), the extent to which rocks endure differential stress during metamorphism, and the extent to which rocks interact with hydrothermal fluids all depend on the geologic environment.

Thermal or Contact Metamorphism

Imagine a hot magma that rises from great depth beneath the Earth's surface and intrudes into cooler country rock at a shallow depth. Heat flows from the magma into the country rock, for heat always flows from hotter to colder materials. As a consequence, the magma cools and solidifies while the country rock

FIGURE 7.9 Metamorphic environments.

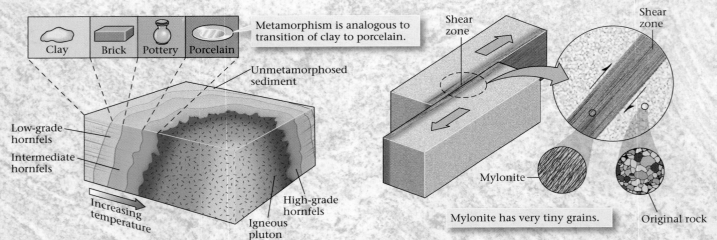

(a) Heat radiated from a large pluton can produce a metamorphic aureole, in which hornfels develops. Grade decreases progressively away from the pluton contact.

(b) Dynamic metamorphism occurs in response to shear of a rock under metamorphic conditions. The shear changes the texture, but might not affect the mineral content.

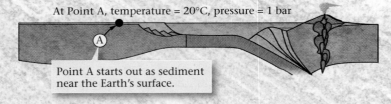

(c) Dynamothermal metamorphism occurs during mountain building, when part of the crust is shoved over another part. Pressure and temperature increase while shearing takes place.

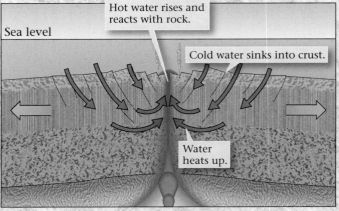

(d) Hydrothermal metamorphism occurs when hot water circulates through rock. This process is common along mid-ocean ridges.

BOX 7.2

THE HUMAN ANGLE

Pottery Making— An Analog for Thermal Metamorphism

A brick for the wall of an adobe house, an earthenware pot, a stoneware bowl, or a translucent porcelain teacup may all be formed from the same lump of soft clay, scooped from the surface of the Earth and shaped by human hands. This pliable and slimy muck is a mixture of very fine clay minerals and quartz grains formed during the chemical weathering of rock and water. Fine potter's clay for making white china contains a particular clay mineral called kaolinite, named after the locality in China (called *Kauling*, meaning high ridge) where it was originally discovered.

People in arid climates make adobe bricks by forming damp clay into blocks, which they then dry in the sun. Such bricks can be used for construction *only* in arid climates, because if it rains heavily, the bricks will rehydrate and turn back into sticky muck—drying clay in the sun does not change the structure of the clay minerals.

To make a more durable material, brick makers place clay blocks in a kiln and bake ("fire") them at high temperatures. This process makes the bricks hard and impervious to water. Potters use the same process to make jugs. In fact, fired clay jugs that were used for storing wine and olive oil have been found intact in sunken Greek and Phoenician ships that have rested on the floor of the Mediterranean Sea for thousands of years! Clearly, the firing of a clay pot fundamentally and permanently changes clay in a way that makes it physically different (see Fig. 7.9a). In other words, firing causes a thermal metamorphic change in the mineral assemblage that composes pottery. The extent of the transformation depends on the

kiln temperature, just as the grade of metamorphic rock depends on temperature. Potters usually fire earthenware at about 1,100°C and stoneware (which is harder than a knife or fork) at about 1,250°C. To produce porcelain—fine china—the clay must partially melt at even higher temperatures. Just as it begins to melt, the potter cools it quickly. Such quenching of the melt creates glass, which gives porcelain its translucent, vitreous (glassy) appearance.

heats up. In addition, hydrothermal fluids circulate through both the intrusion and the country rock. As a consequence of the heat and hydrothermal fluids, the country rock undergoes metamorphism, with the highest-grade rocks forming immediately adjacent to the pluton, where the temperatures were highest, and progressively lower-grade rocks forming farther away. The distinct belt of metamorphic rock that forms around an igneous intrusion is called a **metamorphic aureole** or contact aureole (Fig. 7.9a). The width of an aureole depends on the size and shape of the intrusion, and on the amount of hydrothermal circulation—larger intrusions produce wider aureoles.

The local metamorphism caused by igneous intrusion can be called either **thermal metamorphism** (see Box 7.2), to emphasize that it develops in response to heat without a change in pressure and without differential stress, or **contact metamorphism**, to emphasize that it develops adjacent to

an intrusion. Because this metamorphism takes place without application of compression or shear, aureoles contain hornfels, a nonfoliated metamorphic rock.

Contact metamorphism occurs anywhere that the intrusion of plutons occurs. According to plate tectonics theory, plutons intrude into the crust at convergent plate boundaries, in rifts, and during the mountain building that takes place when continents collide.

Burial Metamorphism

As sediment gets buried in a subsiding sedimentary basin, the pressure increases due to the weight of overburden, and the temperature increases due to the geothermal gradient. In the upper few kilometers, the temperatures and pressures are low enough that the changes taking place in the sediment can be

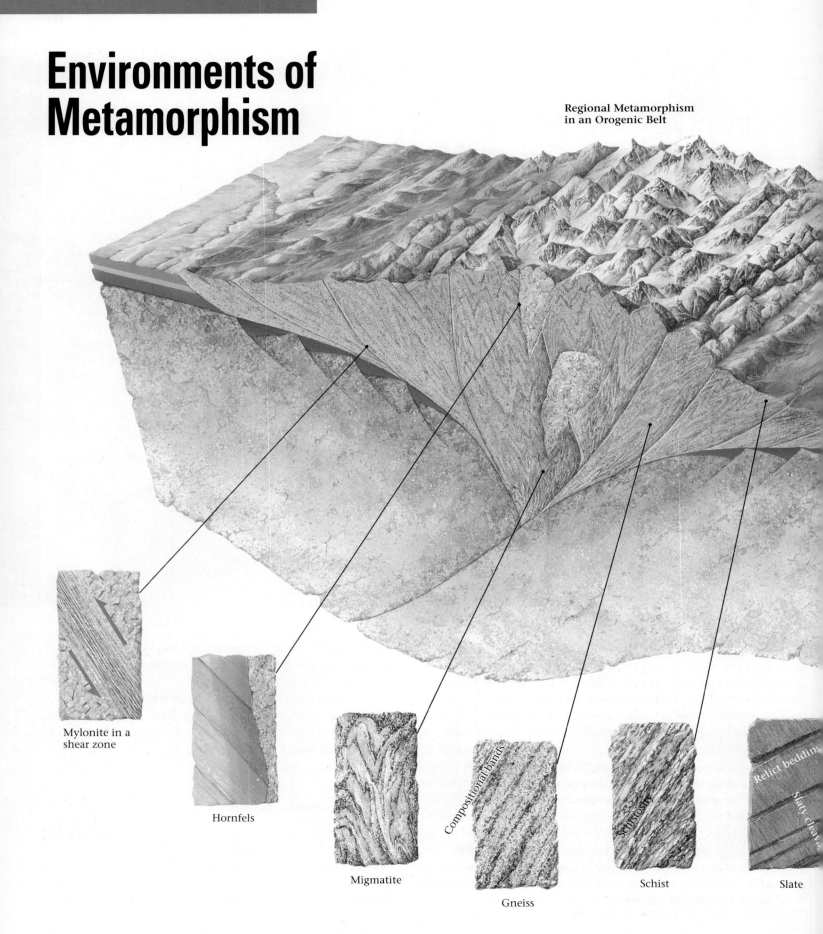

Environments of Metamorphism

Regional Metamorphism
in an Orogenic Belt

Mylonite in a
shear zone

Hornfels

Migmatite

Compositional bands

Gneiss

Schistosity

Schist

Relict bedding

Slaty cleavage

Slate

Unmetamorphosed shale

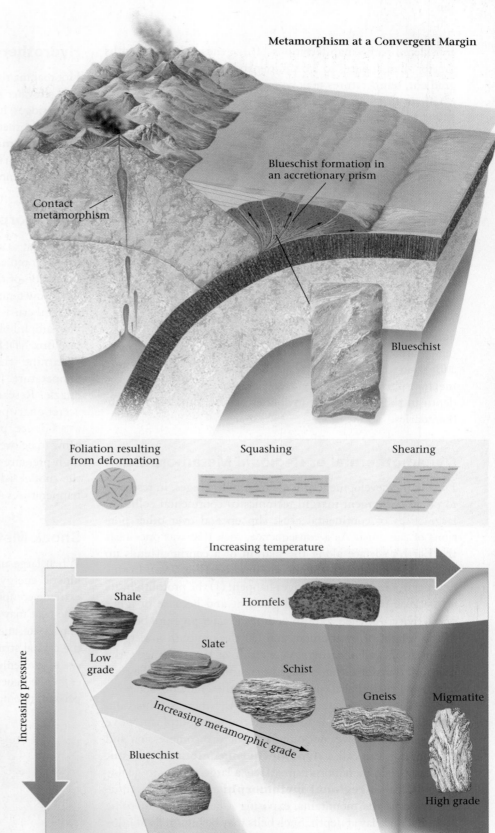

Contact metamorphism

Blueschist formation in an accretionary prism

Blueschist

Foliation resulting from deformation

Squashing

Shearing

Increasing temperature

Increasing pressure

Shale

Low grade

Slate

Schist

Hornfels

Increasing metamorphic grade

Gneiss

Migmatite

Blueschist

High grade

Metamorphic rocks form when a preexisting rock (a protolith) undergoes changes in texture and/or mineral content in the solid state, in response to changes in temperature, pressure, or differential stress or in response to interactions with hydrothermal fluids. Some metamorphic rocks are nonfoliated (they do not have metamorphic layering), whereas others are foliated (they do have metamorphic layering). Foliation results when rock is compressed or sheared during metamorphism, causing minerals to grow or rotate into parallelism with each other. Dynamothermal (regional) metamorphism occurs during mountain building and during subduction, when sea-floor sediment is carried to the base of an accretionary prism. Contact metamorphism takes place around an igneous intrusion, or pluton, caused by the heat released by the pluton.

Geologists distinguish among metamorphic rocks according to the type of foliation and the mineral assemblage a rock contains. Hornfels is unfoliated and forms as a result of contact metamorphism. Mylonite develops when shearing creates a foliation but not necessarily a change in types of minerals. Slate, which forms from shale, contains slaty cleavage; clay flakes are typically aligned at an angle to bedding. Schist contains coarse grains of mica (muscovite and/or biotite) aligned parallel to each other. Gneiss has compositional banding. (Migmatite forms when part of the rock melts, and thus it is a mixture of metamorphic and igneous rock.) Quartzite is composed predominantly of quartz (it is metamorphosed sandstone), whereas marble is composed predominantly of calcite or dolomite (it is metamorphosed limestone or dolostone). Quartzite and marble are usually unfoliated.

The types of minerals and foliation in a metamorphic rock indicate the rock's grade. High-grade rocks, such as gneiss, form at higher temperatures and pressures, whereas low-grade rocks, such as schist, form at lower pressures and temperatures. Blueschist is an unusual metamorphic rock that develops under relatively high pressures but relatively low temperatures—the environment of an accretionary prism.

considered to be manifestations of diagenesis. But at depths greater than about 8 to 15 km, depending on the geothermal gradient, temperatures may be great enough for metamorphic reactions to begin, and low-grade metamorphic rocks form. Metamorphism due only to the consequences of very deep burial is called **burial metamorphism**.

Dynamic Metamorphism

Faults are surfaces on which one piece of crust slides, or shears, past another. Near the Earth's surface (in the upper 10 to 15 km) this movement can fracture rock, breaking it into angular fragments or even crushing it to a powder. But at greater depths, rock is so warm that it behaves like soft plastic as shear along the fault takes place. During this process, the minerals in the rock recrystallize. We call this process **dynamic metamorphism**, because it occurs as a *consequence of shearing alone*, under metamorphic conditions, without requiring a change in temperature or pressure. The resulting rock, a **mylonite**, has a foliation that roughly parallels the fault (Fig. 7.9b). Dynamic metamorphism takes place anywhere that faulting occurs at depth in the crust. Thus, mylonites can be found at all plate boundaries.

Dynamothermal or Regional Metamorphism

During the development of large mountain ranges, in response to either convergent-margin tectonics or continental collision, large slices of continental crust slip up and over other portions of the crust. As a consequence, rock that was once near the Earth's surface along the margin of a continent ends up at great depth beneath the mountain range (Fig. 7.9c). In this new environment, three changes happen: (1) the protolith heats up because of the geothermal gradient and because of igneous activity; (2) the protolith endures greater pressure because of the weight of overburden; and (3) the protolith undergoes compression and shearing. As a result of these changes, the protolith transforms into foliated metamorphic rock. The type of foliated rock that forms depends on the grade of metamorphism—slate forms at shallower depths, whereas schist and gneiss form at greater depths. Since the metamorphism we've just described involves not only heat but also compression and shearing, we can call it **dynamothermal metamorphism**. Typically, such metamorphism affects a large region, so geologists also call it **regional metamorphism**. Erosion eventually removes the mountains, exposing a belt of metamorphic rock that once lay at depth. Such belts may be hundreds of kilometers wide and thousands of kilometers long.

Hydrothermal Metamorphism

Hot magma rises beneath the axis of mid-ocean ridges, so when cold seawater sinks through cracks down into the oceanic crust along ridges, it heats up and transforms into hydrothermal fluid. This fluid then rises through the crust, near the ridge, causing **hydrothermal metamorphism** of ocean-floor basalt (Fig. 7.9d). Eventually, the fluid escapes through vents back into the sea; these vents are called black smokers (see Chapter 2).

Metamorphism in Subduction Zones

Blueschist is a relatively rare rock that contains an unusual blue-colored amphibole. Laboratory experiments indicate that formation of this mineral requires very high pressure but relatively low temperature. Such conditions do not develop in continental crust—usually, at the high pressure needed to produce blue amphibole, temperature in continental crust is also high (see Box 7.1). So to figure out where blueschist forms, we must determine where high pressure can develop at relatively low temperature. Plate tectonics theory provides the answer to this puzzle. Researchers found that blueschist occurs only in the accretionary prisms that form at subduction zones (see Geology at a Glance, pp. 188–189). They realized that because prisms grow to be over 20 km thick, rock at the base of the prism feels high pressure (due to the weight of overburden). But because the subducted oceanic lithosphere beneath the prism is cool, temperatures at the base of the prism remain relatively low.

Shock Metamorphism

When large meteorites slam into the Earth, a vast amount of kinetic energy instantly transforms into heat, and a pulse of extreme compression (a shock wave) propagates into the Earth. The heat may be sufficient to melt or even vaporize rock at the impact site, and the extreme compression of the shock wave causes the crystal structure of quartz grains in rocks below the impact site to suddenly undergo a phase change to a more compact mineral called coesite. Such changes are called **shock metamorphism**, to emphasize their relationship to a large impact.

TAKE-HOME MESSAGE

Contact metamorphism develops around igneous intrusions. Regional metamorphism occurs beneath mountain ranges, where rock undergoes compression and shear at high temperatures and pressures. Impact shock and fluid circulation also cause metamorphism.

7.6 WHERE DO YOU FIND METAMORPHIC ROCKS?

When you stand on an outcrop of metamorphic rock, you are standing on material that once lay many kilometers beneath the surface of the Earth. How does metamorphic rock return to the Earth's surface? Geologists refer to the overall process by which deeply buried rocks end up back at the surface as **exhumation**.

Exhumation results from several processes in the Earth System that happen simultaneously. Let's look at the specific processes that contribute to bringing high-grade metamorphic rocks from below a collisional mountain range back to the surface (Fig. 7.10). First, as two continents progressively push together, the rock caught between them squeezes upward, or is uplifted, much like dough pressed in a vise; the upward movement takes place by slip on faults and by plastic-like flow of rock. Second, as the mountain range grows, the crust at depth beneath it warms up and becomes softer. Eventually, the range starts to collapse under its own weight, much like a block of soft cheese placed in the hot sun. As a result of this collapse, the upper crust spreads out laterally. This movement stretches the upper part of the crust in the horizontal direction and causes it to become thinner in the vertical direction. As the upper part of the crust becomes thinner, the deeper crust ends up closer to the surface. Third, erosion takes place at the surface; weathering, landslides, river flow, and glacial flow together play the role of a giant rasp, stripping away rock at the surface and exposing rock that was once below the surface.

Keeping in mind the processes that form metamorphic rock and cause exhumation, let's ask the question, Where are metamorphic rocks presently exposed? You can start your quest to find metamorphic rock outcrops by hiking into a mountain range. As we've seen, the process of mountain building produces and eventually exhumes metamorphic rocks. The towering cliffs in the interior of a mountain range typically reveal schist, gneiss, and quartzite (Fig. 7.11a). Even after the peaks have eroded away, the record of mountain building remains in the form of a belt of metamorphic rock at the ground surface.

Vast expanses of metamorphic rock crop out in continental shields. A **shield** is a broad region of long-lived stable continental crust where Phanerozoic sedimentary cover either was not deposited or has been eroded away so that Precambrian rocks are exposed (Fig. 7.11b, c). These rocks were metamorphosed during a succession of Precambrian mountain-building events that led to the original growth of continents.

TAKE-HOME MESSAGE

Metamorphic rocks can be found in mountain ranges and in shields. Their exposure at the Earth's surface requires exhumation, which involves uplift, erosion, and in some cases, the thinning of upper crust by faulting.

FIGURE 7.10 Processes that bring metamorphic rock back to Earth's surface. Three phenomena contribute to exhumation of rocks at depth. Here, the red dot (representing metamorphic rocks formed at the base of a mountain range) gets progressively closer to the surface over time.

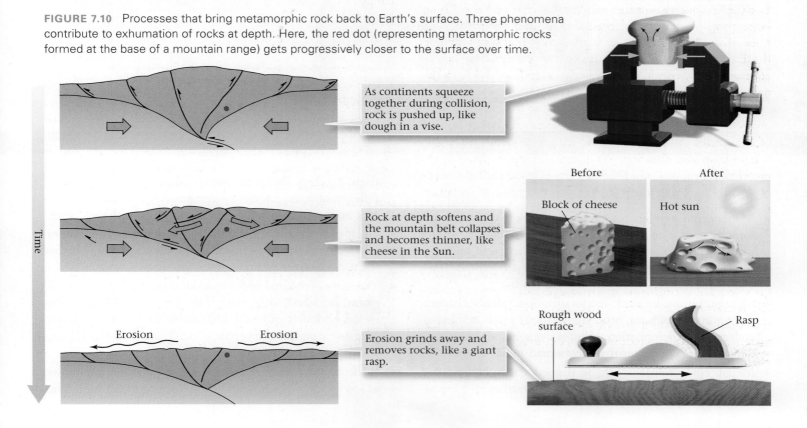

Time

As continents squeeze together during collision, rock is pushed up, like dough in a vise.

Rock at depth softens and the mountain belt collapses and becomes thinner, like cheese in the Sun.

Before　　After

Block of cheese　　Hot sun

Erosion　　Erosion

Erosion grinds away and removes rocks, like a giant rasp.

Rough wood surface　　Rasp

(a) The Gunnison River carved a deep canyon into the Precambrian basement rock in Colorado.

(b) A photograph from an airplane window of the flat landscape of the eastern Canadian shield.

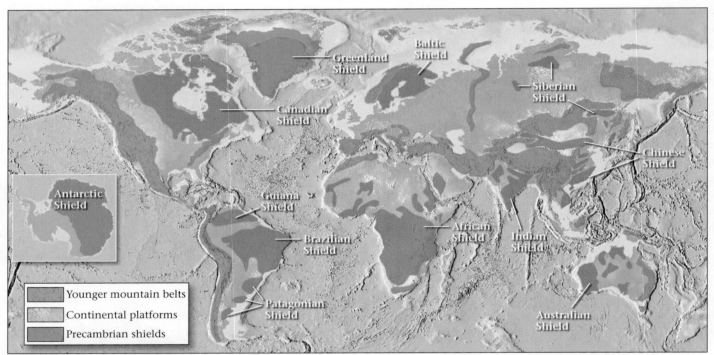

(c) A map showing the distribution of shields, areas where broad expanses of Precambrian crust, including Precambrian metamorphic rocks, crop out.

FIGURE 7.11 Examples of rock exposures consisting of Precambrian metamorphic rocks.

Chapter Summary

- Metamorphism refers to changes in a rock that result in the formation of a metamorphic mineral assemblage, and/or metamorphic foliation, in response to change in temperature and/or pressure, to the application of differential stress, and to interaction with hydrothermal fluids.

- Metamorphism involves recrystallization, phase changes, metamorphic reactions (neocrystallization), pressure solution, and/or plastic deformation. If hot-water solutions bring in or remove elements, we say that metasomatism has occurred.

- Metamorphic foliation can be defined either by preferred mineral orientation (aligned inequant crystals) or by composi-

tional banding. Preferred mineral orientation develops where differential stress causes the compression and shearing of a rock so that its inequant grains align parallel with each other.

- Geologists separate metamorphic rocks into two classes, foliated rocks and nonfoliated rocks, depending on whether the rocks contain foliation.

- The class of foliated rocks includes slate, phyllite, metaconglomerate, schist, and gneiss. The class of nonfoliated rocks includes hornfels, quartzite, and marble. Migmatite, a mixture of igneous and metamorphic rock, forms under conditions where partial melting begins.

- Rocks formed under relatively low temperatures are known as low-grade rocks, whereas those formed under high temperatures are known as high-grade rocks. Intermediate-grade rocks develop between these two extremes. Different metamorphic mineral assemblages form at different grades.

- Geologists track the distribution of different grades of rock by looking for index minerals. Isograds indicate the location at which index minerals first appear. A metamorphic zone is the region between two isograds.

- A metamorphic facies is a group of metamorphic mineral assemblages that develop under a specified range of temperature and pressure conditions.

- Thermal metamorphism (also called contact metamorphism) occurs in an aureole surrounding an igneous intrusion. Burial metamorphism occurs at depth in a sedimentary basin. Dynamically metamorphosed rocks form along faults, where rocks undergo plastic shearing. Dynamothermal metamorphism (also called regional metamorphism) results when rocks undergo heating and shearing during mountain building. Hydrothermal metamorphism takes place due to the circulation of hot water in oceanic crust at mid-ocean ridges. Shock metamorphism happens during the impact of a meteorite.

- We find extensive areas of metamorphic rocks in mountain ranges. Blueschist forms in accretionary prisms. Shields expose Precambrian metamorphic rocks.

GEOPUZZLE REVISITED

Marble is one of many different types of metamorphic rocks, formed due to changes in mineral content and/or rock texture, which happen when a rock is subjected to changes in temperature, pressure, and/or differential stress. The marble in Michelangelo's sculptures started as fossiliferous limestone (in the Jurassic). It was buried deeply, heated, and sheared during the formation of the Apennine Mountains. Exhumation, due partly to extensional collapse and partly to erosion, has returned the rock to the surface.

Key Terms

burial metamorphism (p. 190)

contact metamorphism (p. 187)

dynamic metamorphism (p. 190)

dynamothermal metamorphism (p. 190)

exhumation (p. 191)

foliation (p. 180)

gneiss (p. 181)

hornfels (p. 182)

hydrothermal metamorphism (p. 190)

marble (p. 182)

metaconglomerate (p. 181)

metamorphic aureole (p. 187)

metamorphic facies (p. 184)

metamorphic foliation (p. 176)

metamorphic grade (p. 183)

metamorphic mineral (p. 176)

metamorphic rock (p. 176)

metamorphic texture (p. 176)

metamorphic zone (p. 183)

metamorphism (p. 176)

metasomatism (p. 179)

migmatite (p. 181)

mylonite (p. 190)

phyllite (p. 181)

preferred mineral orientation (p. 179)

protolith (p. 176)

quartzite (p. 182)

regional metamorphism (p. 190)

schist (p. 181)

shield (p. 191)

shock metamorphism (p. 190)

slate (p. 180)

thermal metamorphism (p. 187)

Review Questions

1. How are metamorphic rocks different from igneous and sedimentary rocks?

2. What two features characterize most metamorphic rocks?

3. What phenomena cause metamorphism?

4. What is metamorphic foliation, and how does it form?

5. How does slate differ from a phyllite? How does phyllite differ from a schist? How does schist differ from a gneiss?

6. Why is hornfels nonfoliated?

7. What is a metamorphic grade, and how can it be determined? How does grade differ from facies?

8. Describe the geologic settings where thermal, dynamic, and dynamothermal metamorphism take place.

9. Why does metamorphism happen at the site of meteor impacts and along mid-ocean ridges?

10. How does plate tectonics explain the combination of low-temperature but high-pressure minerals found in a blueschist?

11. Where would you go if you wanted to find exposed metamorphic rocks, and how did such rocks return to the surface of the Earth after being at depth in the crust?

On Further Thought

1. Do you think that you would be likely to find a broad region (hundreds of km across and hundreds of km long) in which the outcrop consists of high-grade hornfels? Why or why not? (*Hint*: Think about the causes of metamorphism and the conditions under which a hornfels forms.)

2. Would we likely find broad regions of gneiss and schist on the Moon? Why or why not?

The Rock Cycle

The photos below show three major rock groups of the Earth System. Over time, the atoms in one rock type may be incorporated in another—this transfer is called the rock cycle.

Lava, spilling out of a Hawaiian volcano, becomes igneous rock when it freezes.

Beds of sedimentary rock form the mesas of Monument Valley, Arizona.

Metamorphic rock, as illustrated in a Brazilian quarry wall, may contain complex foliation.

C.1 INTRODUCTION

"Stable as a rock." This familiar expression implies that a rock is permanent, unchanging over time. But it isn't. In the time frame of Earth history, a span of over 4.5 billion years, atoms making up one rock type may be rearranged or moved elsewhere, eventually becoming part of another rock type. Later, the atoms may move again to form a third rock type, and so on. Geologists call the progressive transformation of Earth materials from one rock type to another the **rock cycle** (Fig. C.1), one of many examples of cycles acting in or on the Earth. A discussion of the rock cycle illustrates the relationships among the rock types described in the previous three chapters.

By following the arrows in Figure C.1, you can see that there are many paths around or through the rock cycle. For example, igneous rock may weather and erode to produce sediment, which lithifies to form sedimentary rock. The new sedimentary rock may become buried so deeply that it transforms into metamorphic rock, which then could partially melt and produce magma. This magma later solidifies to form new igneous rock. We can symbolize this path as igneous → sedimentary → metamorphic → igneous. But alternatively, the metamorphic rock could be uplifted and eroded to form new sediment and, later, new sedimentary rock, without melting. This path takes a shortcut through the cycle that we can symbolize as igneous → sedimentary → metamorphic → sedimentary. Likewise, the igneous rock could be metamorphosed directly, without first turning to sediment. This metamorphic rock could then be eroded to produce sediment that becomes sedimentary rock, defining another shortcut path: igneous → metamorphic → sedimentary. To get a clearer sense of how the rock cycle works, let's look at one example in the context of plate tectonics.

FIGURE C.1 The stages of the rock cycle, showing various alternative pathways.

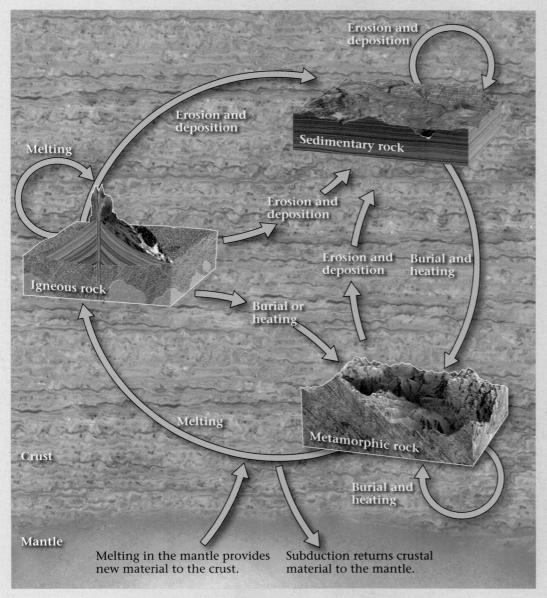

Erosion and deposition

Erosion and deposition

Melting

Sedimentary rock

Erosion and deposition

Erosion and deposition

Burial and heating

Igneous rock

Burial or heating

Melting

Metamorphic rock

Crust

Burial and heating

Mantle

Melting in the mantle provides new material to the crust.

Subduction returns crustal material to the mantle.

C.2 A CASE STUDY OF THE ROCK CYCLE

Material can enter the rock cycle when magma rises from the mantle. Suppose the magma erupts and forms basalt (an igneous rock) at a continental hot-spot volcano (Fig. C.2a). Interaction with wind, rain, and vegetation gradually weathers the basalt, physically breaking it into smaller fragments and chemically altering it to yield clay. Water washes the newly formed clay away and transports it downstream—if you've ever seen a brown-colored river, you've seen clay traveling to a site of deposition. Eventually the river reaches the sea, where the water slows down and the clay settles out.

Let's imagine, for this example, that the clay settles out along

FIGURE C.2 An example of the rock cycle in the context of plate tectonics.

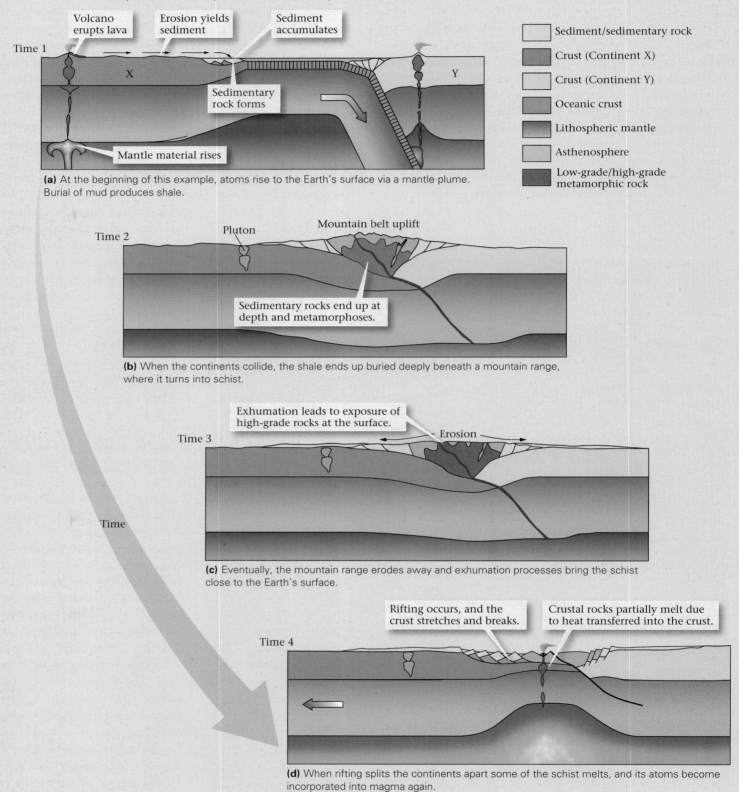

(a) At the beginning of this example, atoms rise to the Earth's surface via a mantle plume. Burial of mud produces shale.

(b) When the continents collide, the shale ends up buried deeply beneath a mountain range, where it turns into schist.

(c) Eventually, the mountain range erodes away and exhumation processes bring the schist close to the Earth's surface.

(d) When rifting splits the continents apart some of the schist melts, and its atoms become incorporated into magma again.

the margin of continent X and forms a deposit of mud. Gradually, through time, the mud becomes buried and the clay flakes pack tightly together, resulting in a new sedimentary rock, shale. The shale resides 6 km below the continental shelf for millions of years, until the adjacent oceanic plate subducts and another continent, Y, collides with X. The edge of the encroaching continent pushes over the shale and buries it very deeply. As the mountains grow, the shale that had once been 6 km below the surface ends up 20 km below the surface, and under the pressure and temperature conditions present at this depth, it metamorphoses into schist (Fig. C.2b).

The story's not over. Once mountain building stops, erosion grinds away the mountain range, and exhumation brings some of the schist to the ground surface. Some of this schist erodes to form sediment, which is carried off and deposited elsewhere to form new sedimentary rock. But other schist remains preserved below the surface (Fig. C.2c). Eventually, continental rifting takes place at the site of the former mountain range, and the crust containing the schist begins to split apart. When this happens, some of the schist partially melts and a new felsic magma forms. This felsic magma rises to the surface of the crust and freezes into rhyolite, a new igneous rock (Fig. C.2d). In terms of the rock cycle, we're back at the beginning, having once again made igneous rock.

C.3 RATES OF MOVEMENT THROUGH THE ROCK CYCLE

We have seen that not all atoms pass through the rock cycle in the same way. Similarly, not all atoms pass through the rock cycle at the same rate, and for that reason we find rocks of many different ages at the surface of the Earth. Some rocks remain in one form for less than a few million years, while others stay unchanged for most of Earth history. A rock in the Appalachian Mountains has passed through stages of the rock cycle many times during the past few hundred million years, because the eastern margin of North America has been subjected to multiple events of basin formation, mountain building, and rifting since the end of the Precambrian. In contrast, rocks exposed on Precambrian shields have remained unchanged for over a billion years.

Most atoms that comprise continental rocks never return to the mantle, because continental crust is buoyant and does not subduct. However, a small amount of sediment that erodes off a continent ends up in deep-ocean trenches, and some of this gets carried back into the mantle by subduction. In fact, recent research suggests that metamorphic and igneous rocks at the base of the continental crust may be scraped off and transported down into the mantle by subduction.

Our tour of the rock cycle has focused on continental rocks. What about the oceans? Oceanic crust consists of igneous rock (basalt and gabbro) overlain by sediment. Because a layer of water blankets the crust, erosion does not affect it, so oceanic crustal rock generally does not follow the path into the sedimentary loop of the rock cycle. But sooner or later, oceanic crust subducts. When this happens, the rock of the crust first undergoes metamorphism, for as it sinks, it is subjected to progressively higher temperatures and pressures.

C.4 WHAT DRIVES THE ROCK CYCLE IN THE EARTH SYSTEM?

The rock cycle occurs because the Earth is a dynamic planet and there are many rock-forming environments (see Geology at a Glance, pp. 198–199). The planet's internal heat and gravitational field drive plate movements. Plate interactions cause the uplift of mountain ranges, a process that exposes rock to weathering, erosion, and sediment production. Plate interactions also generate the geologic settings in which metamorphism occurs, where rock melts to provide magma, and where sedimentary basins develop.

At the surface of the Earth, the gases initially released by volcanism collect to form the ocean and atmosphere. Heat (from the Sun) and gravity drive convection in the atmosphere and oceans, leading to wind, rain, ice, and currents—the agents of weathering and erosion. Weathering and erosion grind away at the surface of the Earth and send material into the sedimentary loop of the cycle. In the Earth System, life also plays a key role by adding corrosive oxygen to the atmosphere, and by directly contributing to weathering. In sum, external energy (solar heat), internal energy (Earth's internal heat), gravity, and life all play a role in driving the rock cycle by keeping some of the mantle, crust, atmosphere, and oceans in motion.

Key Term

rock cycle (p. 195)

Rock-Forming Environments and the Rock Cycle

Drainage networks collect surface water that can transport sediment to the ocean.

Sand dunes form from grains carried by the wind.

In a desert environment, rock weathers and fragments. Debris falls in landslides.

Flash floods carry sediment out of canyons to form an alluvial fan.

Volcanic eruptions emit lava and ash, which form new igneous rock at Earth's surface.

Sedimentary rocks make a cover on the surface of continents.

The crust and lithospheric mantle stretch and thin in a rift.

Magma rises from the mantle. Heat from this magma causes contact metamorphism.

Deep levels of continents consist of ancient metamorphic and igneous rocks. This is the basement of the continents.

Continental margins slowly sink and are buried by new sediment.

Partial melting occurs in the asthenosphere to produce new magma.

km
0
10
20
30
40
50
60
70
80
90
100

Rocks form in many different environments. Igneous rocks develop where melt rises from depth and cools. Intrusive igneous rocks form where magma cools underground; extrusive igneous rocks form where lava and ash erupt at the surface.

Weathering and erosion break up existing rock and produce sediment. Different kinds of sediments develop in different places, reflecting both the composition of the source and the setting in which the sediment accumulates. We distinguish among sediment that collects in alluvial fans, desert dunes, river channels and floodplains, deltas, coral reefs, coastlines, the continental shelf, the deep sea, and the toe of a glacier. When this sediment eventually gets buried and undergoes lithification, new sedimentary rocks form.

Under certain conditions, preexisting rocks can undergo change in the solid state—metamorphism—which produces

Glaciers erode rock and can transport sediment of all sizes.

In a region of continental collision, rocks that were near the surface are deeply buried and metamorphosed.

In humid climates, thick soils develop.

Magma that cools and solidifies underground forms igneous intrusions.

Along coastal plains, rivers meander. Sediment collects in the channel and floodplain.

Where a river enters the sea, sediment settles out to form a delta.

Reefs grow from calcite-secreting organisms. These will eventually turn into limestone.

Many different kinds of sediment accumulate along coastlines, building out a continental shelf.

Underwater avalanches carry a cloud of sediment that settles to form a submarine fan.

Fine clay and plankton shells settle on the oceanic crust.

The oceanic crust consists of igneous rocks formed at a mid-ocean ridge.

metamorphic rocks. Contact metamorphism is due to heat released by an intrusion of magma. Regional metamorphism occurs where tectonic processes cause rocks from the surface to be buried very deeply.

Because the Earth is dynamic, environments change through time. Tectonic processes cause new igneous rocks to form. When exposed at the surface, these rocks weather to make sediment. The slow sinking of some regions creates sedimentary basins in which sediment accumulates and new sedimentary rocks form. Later, these rocks may be buried deeply and metamorphosed. Uplift as a result of mountain building exposes the rocks to the surface, where they may once again be transformed into sediment. This progressive transformation is called the rock cycle.

CHAPTER 8

A Violent Pulse: Earthquakes

In the tragic aftermath of an earthquake that rocked India, Pakistan, and Afghanistan in October of 2005, rescue workers struggle to free victims trapped beneath the rubble of this apartment building in Islamabad, Pakistan. When the ground shakes, walls may tumble.

GEOPUZZLE

Why do earthquakes happen where they do? Can people predict when an earthquake will occur?

8.1 **INTRODUCTION**

As the morning of January 17, 1994, approached, residents of Northridge, a suburb near Los Angeles, slept peacefully in anticipation of the Martin Luther King Day holiday. But beneath the quiet landscape, a disaster was in the making. For many years, the imperceptibly slow movement of the Pacific Plate relative to the North American Plate had been bending the rocks making up the California crust. But like a stick that you flex with your hands, rock can bend only so far before it snaps (Fig. 8.1a, b). Under California, the "snap" happened at 4:31 A.M., 10 km down. It sent pulses of energy racing through the crust at an average speed of 11,000 km (7,000 miles) per hour—ten times the speed of sound.

When the energy pulses, or shocks, reached the Earth's surface, the ground bucked up and down and swayed from side to side. Sleepers bounced out of bed, homes slipped off their foundations, and freeway bridges disconnected from their supports. As more and more shocks arrived, walls swayed and toppled, roofs collapsed, and steep hill slopes bordering the coast gave way. Ruptured gas lines fed fires that had been ignited in the rubble by water heaters and sparking wires. Then, forty seconds after it started, the motion stopped, and the shouts and sirens of rescuers replaced the crash and clatter of breaking masonry and glass. A major **earthquake**—an episode of ground shaking—had occurred.

Earthquakes have affected the Earth since the solid lithosphere first formed. Most are a consequence of lithosphere-plate movement; they punctuate each step in the growth of mountains, the drift of continents, and the opening and closing of ocean basins. And, perhaps of more relevance to us, earthquakes have afflicted human civilization since the construction of the first village, and they have directly caused the deaths of over 3.5 million people during the past two millennia. Ground shaking, giant waves, landslides, and fires during earthquakes turn cities to rubble. The destruction caused by some earthquakes may even have changed the course of civilization.

Earthquakes are a fact of life on planet Earth: almost 1 million detectable earthquakes happen every year. Fortunately, most cause no damage or casualties, because they are too small or they occur in unpopulated areas. But a few hundred earthquakes per year rattle the ground sufficiently to damage buildings and injure their occupants, and every five to twenty years, on average, a great earthquake triggers a horrific calamity. What geologic phenomena generate earthquakes? Why do earthquakes take place where they do? How do they cause damage? Can we predict when earthquakes will happen, or even prevent them from happening? These questions have puzzled seismologists (from the Greek word *seismos*, for shock or earthquake), the geoscientists who study earthquakes, for decades. In this chapter, we seek some of the answers.

FIGURE 8.1 What happens during an earthquake?

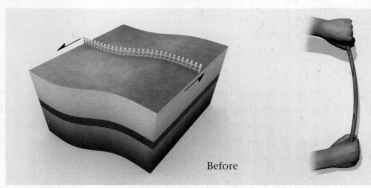

(a) Before an earthquake, rock bends elastically, like a stick that you arch between your hands. The drawing exaggerates the amount of bending.

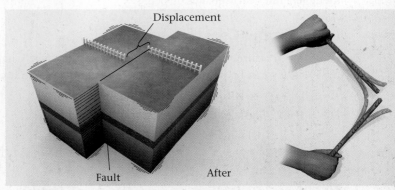

(b) Eventually, the rock breaks, and sliding suddenly occurs on a fault. This break generates vibrations. You feel such vibrations when you break a stick.

TAKE-HOME MESSAGE

By the end of this chapter, you should understand what causes earthquakes, why earthquakes occur where they do, how earthquakes are measured, the meaning of earthquake magnitude, the ways that earthquakes cause damage, and the limitations of earthquake prediction.

8.2 **WHAT CAUSES EARTHQUAKES?**

Ancient cultures offered a variety of explanations for **seismicity** (earthquake activity), most of which involved the action or mood of a giant animal or god. Scientific study suggests that seismicity instead occurs for several reasons, including

- the sudden formation of a new **fault** (a fracture on which sliding occurs),
- sudden slip on an existing fault,
- a sudden change in the arrangement of atoms in the minerals of rock,

- movement of magma in a volcano,
- the explosion of a volcano,
- a giant landslide,
- a meteorite impact, or
- an underground nuclear-bomb test.

As we learned in Chapter 1, the place within the Earth where rock ruptures and slips, or the place where an explosion occurs, is the **hypocenter** (or focus) of the earthquake. Energy radiates from the hypocenter. The point on the surface of the Earth that lies directly above the hypocenter is the **epicenter** (Fig. 8.2a, b).

The formation and movement of faults cause the vast majority of destructive earthquakes, so typically the hypocenter of an earthquake is associated with a fault surface. Thus, we'll focus our investigation on how faults develop and why their movement generates earthquakes.

Faults in the Crust

Faults are fractures on which slip or sliding occurs (see Chapter 9). They can be pictured simplistically as surfaces that cut

FIGURE 8.2 Earthquake hypocenters and epicenters.

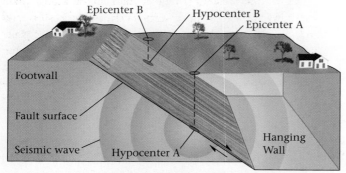

(a) The hypocenter, or focus, is the place underground where rock breaks and generates earthquake waves. The epicenter is the point on the surface of the Earth directly above the hypocenter.

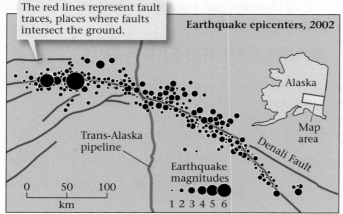

(b) Geologists represent epicenters as dots on a map; dot size indicates the size of the earthquake. Epicenters typically lie on or near a fault trace.

the crust. Nineteenth-century miners who encountered faults in mine tunnels referred to the rock mass above a sloping fault plane as the hanging wall, because it hung over their heads, and the rock mass below the fault plane as the footwall, because it lay beneath their feet. The miners described the direction in which rock masses slipped on a fault by specifying the direction that the hanging wall moved in relation to the footwall, and we still use these terms today (see **Geology at a Glance**, pp. 204–205). When the hanging wall slips down the slope of the fault, it's a normal fault. When the hanging wall slips up the slope, it's a reverse fault if steep and a thrust fault if shallowly sloping (Fig. 8.3a–c). Strike-slip faults have near-vertical planes on which slip occurs parallel to an imaginary horizontal line, called a strike line, on the fault plane—no up or down motion takes place here (Fig. 8.3d). In Chapter 9, we will introduce other types of faults.

Normal faults form in response to stretching or extension of the crust. Reverse or thrust faults develop in response to squeezing (compression) and shortening of the crust, and strike-slip faults form where one block of crust slides past another laterally. By measuring the distance between the two ends of a fence, a distinctive sedimentary bed, or an igneous dike that's been offset by a fault, geologists define the **displacement**, the amount of slip, on the fault (Fig. 8.4a, b).

Faults are found almost everywhere—but don't panic! Not all of them are likely to be the source of earthquakes. Faults that have moved recently or are likely to move in the near future are called *active* faults (and if they generate earthquakes, news media sometimes refer to them as earthquake faults). Faults that last moved in the distant past and probably won't move again in the near future are called *inactive* faults. Some faults have been inactive for billions of years. The intersection between a fault and the ground surface is the fault trace, or fault line. In places where an active normal or reverse fault intersects the ground, movement on the fault displaces the ground surface and generates a small step called a **fault scarp** (see Fig. 8.3a).

Faulting Generates Earthquake Energy

What is the relationship between faulting and earthquakes? Earthquakes can happen either when rock breaks and a new fault forms, or when a preexisting fault suddenly slips again. Let's look more closely at these two causes.

- *Earthquakes due to fault formation*: Imagine that you grip each side of a brick-shaped block of rock with a clamp. Apply an upward push on one of the clamps and a downward push on the other. By doing so, you have applied a "stress" to the rock. (**Stress** refers to a push, pull, or shear.) At first, the rock bends slightly but doesn't break (Fig. 8.5a). In fact, if you were to stop applying stress

FIGURE 8.3 The basic types of faults. Fault types are distinguished from one another by the nature of slip.

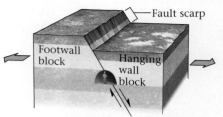

(a) Normal faults form during extension of the crust. The hanging wall moves down.

(b) Reverse faults form during shortening of the crust. The hanging wall moves up and the fault is steep.

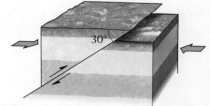

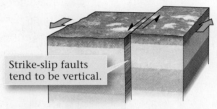

(c) Thrust faults also form during shortening. The fault's slope is gentle (less than 30°).

(d) On a strike-slip fault, one block slides laterally past another, so no vertical displacement takes place.

fracture forms, the block breaks in two. The rock on one side suddenly slides past the rock on the other side, and any elastic bending that had built up is released so the rock straightens out (Fig. 8.5c). Because sliding occurs, the fracture at this stage has become a fault. Such fault formation and slip, like the snap of a stick, generate vibrations—an earthquake.

A fault can't slip forever, for friction eventually slows and stops the movement, just as friction slows and stops a book sliding across a table. Friction, defined as the force that resists sliding on a surface, is caused by the existence of bumps or protrusions on fault surfaces that act like tiny anchors.

at this stage, the rock would return to its original shape. Geologists refer to such a phenomenon as elastic behavior—the same phenomenon happens when a rubber band returns to its original shape or a bent stick straightens out after you let go. Now repeat the experiment, but bend the rock even more. If you bend the rock far enough, a number of small cracks or breaks start to form (Fig. 8.5b). Eventually the cracks connect to one another to form a fracture that cuts across the entire block of rock. The instant that such a throughgoing

■ *Earthquakes due to slip on a preexisting fault*: Once a fault comes into being, it can act like a scar in the Earth's crust, in that it remains weaker than surrounding, intact crust. When stress builds sufficiently, the fault slips, and this movement takes place before stress becomes great enough to fracture surrounding intact rock. Note that after each slip event, no earthquakes happen on the fault until stress builds again. Geologists refer to such alternation between stress buildup, when no slip occurs, and slip events (earthquakes) as **stick-slip behavior**.

FIGURE 8.4 Examples of fault displacement on the San Andreas Fault in California.

(a) A wooden fence built across the fault was offset during the 1906 San Francisco earthquake. The displacement indicates that the motion was strike-slip.

(b) An aerial photograph shows offset of a dirt road by slip on the fault in 1999.

Faulting in the Crust

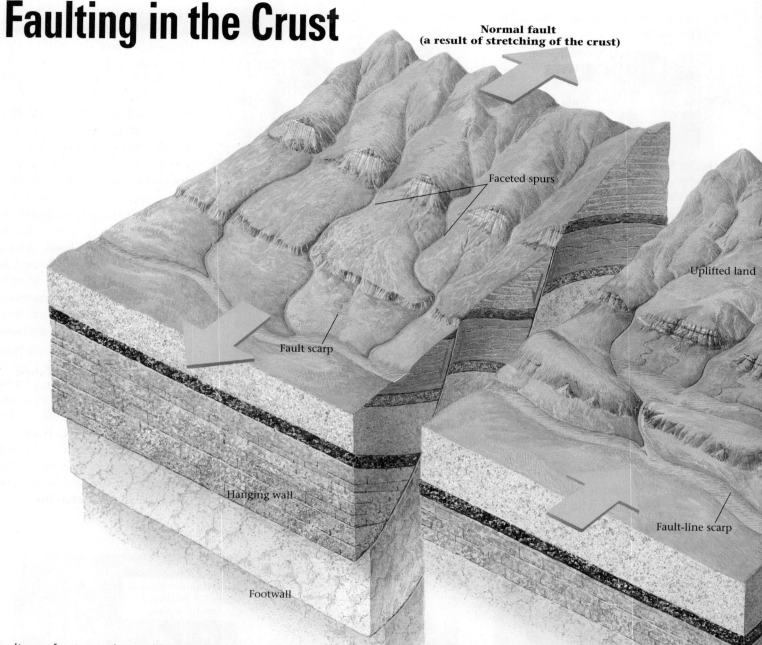

Normal fault
(a result of stretching of the crust)

Faceted spurs

Uplifted land

Fault scarp

Hanging wall

Footwall

Fault-line scarp

Faults are fractures along which one block of crust slides past another block. Sometimes movement takes place slowly and smoothly, without earthquakes, but other times the movement is sudden, and rocks break as a consequence. The sudden breaking of rock sends shock waves, called seismic waves, through the crust, creating vibrations at the Earth's surface—an earthquake.

Geologists recognize three types of faults. If the hanging-wall block (the rock above a fault plane) slides down the fault's slope relative to the footwall block (the rock below the fault plane), the fault is a normal fault. (Normal faults form where the crust is being stretched apart, as in a continental rift.) If the hanging-wall block is being pushed up the slope of the fault relative to the footwall block, then the fault is a reverse fault. (Reverse faults develop where the crust is being com-

pressed or squashed, as in a collisional mountain belt.) If one block of rock slides past another and there is no up or down motion, the fault is a strike-slip fault. Strike-slip fault planes tend to be nearly vertical.

If a fault displaces the ground surface, it creates a ledge called a fault scarp. Sometimes we can identify the trace (or line) of a fault on the land surface because the rock of the hanging wall has a different resistance to erosion than the rock of the footwall; a ledge formed along this line due to erosion is a fault-line scarp. Where fault scarps cut a system of rivers and valleys, the ridges are truncated to make triangular facets. Strike-slip faults may offset ridges, streams, and orchards sideways. If there is a slight extension along the fault, the land surface sinks, and a sag pond develops.

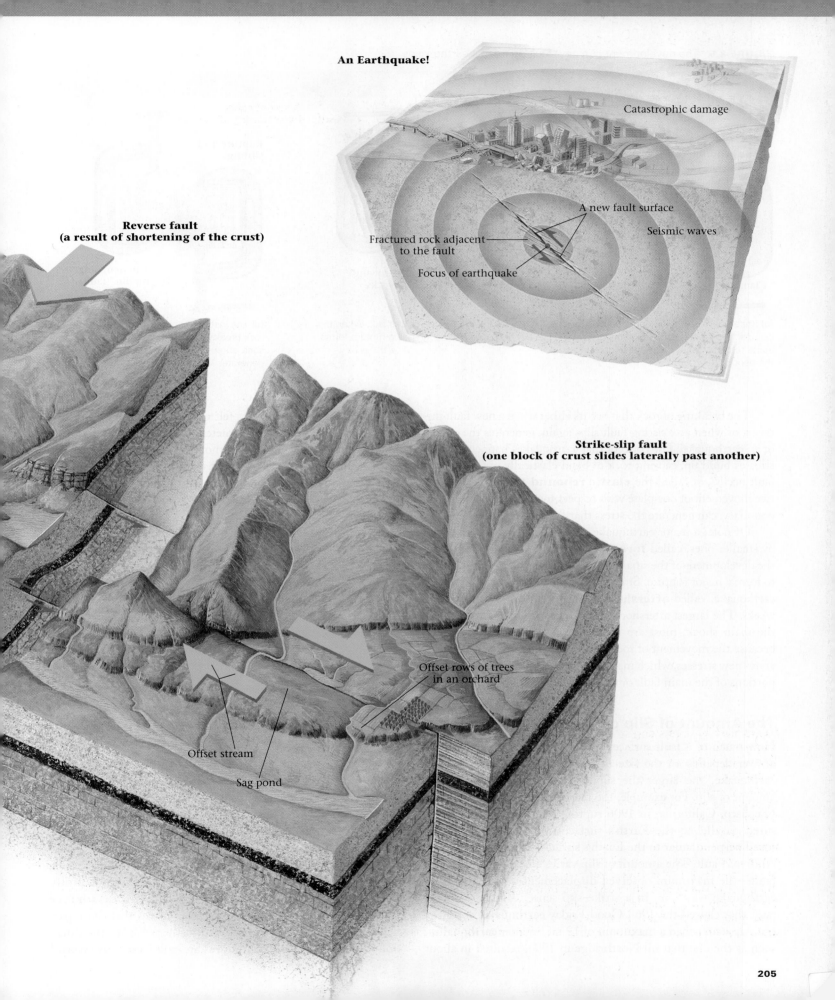

An Earthquake!

Catastrophic damage

A new fault surface

Seismic waves

Fractured rock adjacent to the fault

Focus of earthquake

**Reverse fault
(a result of shortening of the crust)**

**Strike-slip fault
(one block of crust slides laterally past another)**

Offset rows of trees
in an orchard

Offset stream

Sag pond

FIGURE 8.5 The development of a new fault generates earthquakes.

Time

Elastic bending

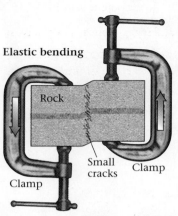

Rock

Small cracks

Clamp

Clamp

(a) Imagine a block of rock gripped by two clamps. Move one clamp up, and the rock starts to bend. Small cracks develop along the bend.

Cracking

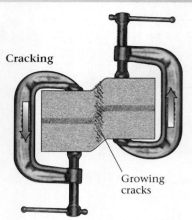

Growing cracks

(b) Eventually, the cracks link. When this happens, a through-going rupture forms.

Rupture and sliding

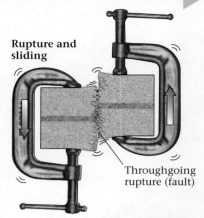

Throughgoing rupture (fault)

(c) The instant that the rupture forms, the rock breaks into two pieces that slide past each other. The energy that is released generates vibrations (earthquake energy).

The breaking of rock that occurs either when a new fault initiates, or when an existing fault slips again, generates the energy of an earthquake. The concept that earthquakes happen because stresses build up, causing rock to bend elastically until slip on a fault occurs, is called the **elastic-rebound theory**. The relative movement of one plate with respect to another, along a plate boundary, can generate the stress that drives this process.

Of note, a major earthquake along a fault may be preceded by smaller ones, called **foreshocks**, which possibly result from the development of the smaller cracks that will eventually link up to form a major rupture. Smaller earthquakes that follow a major earthquake, called **aftershocks**, may occur for days to several weeks. The largest aftershock tends to be ten times smaller than the main shock; most are much smaller. Aftershocks happen because the movement of rock during the main earthquake generates new stresses, which may be large enough to reactivate small portions of the main fault or to activate small, nearby faults.

The Amount of Slip on Faults

How much of a fault surface slips during an earthquake? The answer depends on the size of the earthquake: the larger the earthquake, the larger the slipped area and the greater the amount of slip. For example, the major earthquake that hit San Francisco, California, in 1906 ruptured a 430-km-long (measured parallel to the Earth's surface) by 15-km-high (measured perpendicular to the Earth's surface) segment of the San Andreas Fault. The amount of slip varies along the length of a fault—the maximum observed displacement during the 1906 earthquake was 7 m, in a strike-slip sense. Slip on a thrust fault that caused the 1964 Good Friday earthquake in southern Alaska reached a maximum of 12 m. Smaller earthquakes, such as the one that hit Northridge in 1994, resulted in about

0.5 m slip, and the smallest-felt earthquakes result from displacements measured in millimeters to centimeters.

Although the cumulative movement on a fault during a human life span may not amount to much, over geologic time the cumulative movement becomes significant. For example, if earthquakes occurring on a strike-slip fault cause 1 cm of displacement per year, on average, the fault's movement will yield 10 km of displacement after 1 million years.

TAKE-HOME MESSAGE

Most earthquakes happen due to the sudden rupture of rock accompanying the formation or reactivation of a fault. A hypocenter is the location in the Earth where an earthquake occurs, and an epicenter is the point on the surface of the Earth directly above. Friction stops slip on a fault. But stress then builds until it can trigger another slip event.

8.3 SEISMIC WAVES

How does the energy produced at the hypocenter of an earthquake travel to the surface or even pass through the entire Earth? Earthquake energy travels through rock and sediment in the form of waves. We call these waves **seismic waves** (or earthquake waves). You feel such waves when you touch one end of a brick and tap the other end with a hammer.

Seismologists distinguish among different types of seismic waves on the basis of where and how the waves move. **Body waves** pass through the interior of the Earth, whereas **surface waves** travel along the Earth's surface. Waves that cause particles of material to move back and forth *parallel* to the direction in which the wave itself moves are called **compressional**

waves. As a compressional wave passes, the material first compresses (or squeezes together), then dilates (or expands). To see this kind of motion in action, push on the end of a spring and watch as the little pulse of compression moves along the length of the spring. Waves that cause particles of material to move back and forth *perpendicular* to the direction in which the wave itself moves are called **shear waves**. To see shear-wave motion, jerk the end of a rope up and down and watch how the up-and-down motion travels along the rope. With these concepts in mind, we can define four basic types of seismic waves (Fig. 8.6a–c):

- P-waves (P stands for primary) are compressional body waves.
- S-waves (S stands for secondary) are shear body waves.
- L-waves (L stands for Love, the name of a seismologist) are surface waves that cause the ground to ripple back and forth, producing a snake-like movement.
- R-waves (R stands for Rayleigh, the name of a physicist) are surface waves that cause the ground to ripple up and down.

FIGURE 8.6 Different types of earthquake waves.

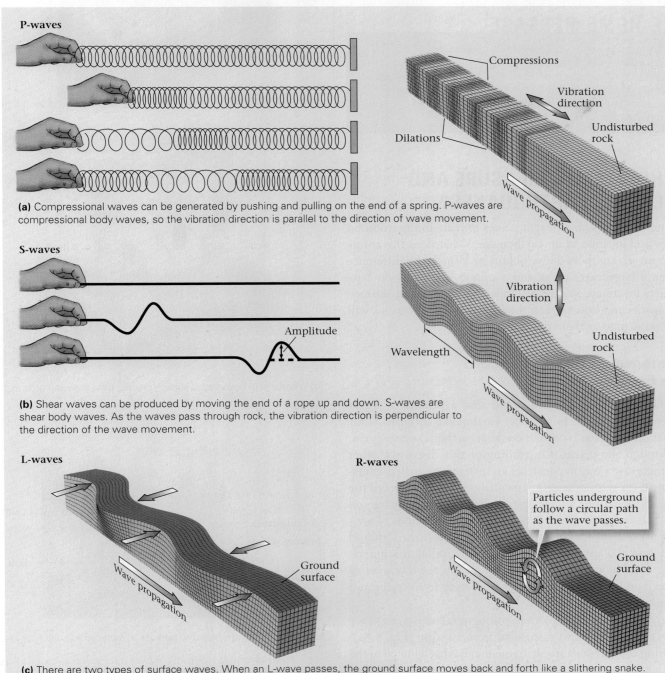

P-waves

(a) Compressional waves can be generated by pushing and pulling on the end of a spring. P-waves are compressional body waves, so the vibration direction is parallel to the direction of wave movement.

Compressions

Vibration direction

Dilations

Undisturbed rock

Wave propagation

S-waves

Amplitude

Wavelength

Vibration direction

Undisturbed rock

Wave propagation

(b) Shear waves can be produced by moving the end of a rope up and down. S-waves are shear body waves. As the waves pass through rock, the vibration direction is perpendicular to the direction of the wave movement.

L-waves

Wave propagation

Ground surface

R-waves

Particles underground follow a circular path as the wave passes.

Wave propagation

Ground surface

(c) There are two types of surface waves. When an L-wave passes, the ground surface moves back and forth like a slithering snake. R-waves make the ground surface go up and down. Both types of waves die out with increasing depth.

The different types of seismic waves travel at different velocities. P-waves travel the fastest and thus arrive first. S-waves travel more slowly, at about 60% of the speed of P-waves, so they arrive later. Surface waves are the slowest of all.

Friction absorbs energy as waves pass through a material, and waves bounce off layers and obstacles in the Earth, so the amount of energy carried by seismic waves decreases the farther they travel. Thus, people near the epicenter may be thrown off their feet by a large earthquake, but those 100 km away barely feel it. Similarly, an earthquake caused by slip on a fault deep in the crust causes less damage than one caused by slip on a fault near the surface.

TAKE-HOME MESSAGE

Earthquake energy travels as seismic waves. Body waves (P-waves and S-waves) travel through the interior of the Earth, whereas surface waves travel along the surface. Ground shaking, due to arrival of waves, generally decreases with distance from the epicenter.

8.4 HOW DO WE MEASURE AND LOCATE EARTHQUAKES?

Most news reports about earthquakes provide information on the size and location of an earthquake. What does this information mean, and how do we obtain it? What's the difference between a large earthquake and a minor one? How do seismologists locate an epicenter? Understanding how a seismograph works and how to read the information it provides will allow you to answer these questions.

Seismographs and the Record of an Earthquake

In 1889, a German physicist noted that a pendulum in his lab moved subsequent to a deadly earthquake in Japan. This observation confirmed speculations that earthquake energy can pass through the planet. On reading of this discovery, other researchers saw a way to construct an instrument, called a **seismograph** (or seismometer), that can systematically record the ground motion from an earthquake happening anywhere on Earth. Seismologists now use two basic configurations of seismographs, one for measuring vertical (up-and-down) ground motion and the other for measuring horizontal (back-and-forth) ground motion (Fig. 8.7a, b). Typically, the instruments are placed on bedrock in sheltered areas, away from traffic and other urban noise (Fig. 8.7c).

A mechanical vertical-motion seismograph consists of a heavy weight (like a pendulum) suspended from a spring (Fig. 8.8a). The spring connects to a sturdy frame that has been bolted to the

FIGURE 8.7 The basic operation of a seismograph.

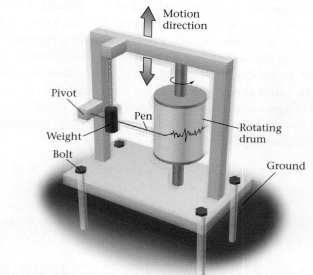

(a) A vertical-motion seismograph records up-and-down ground motion.

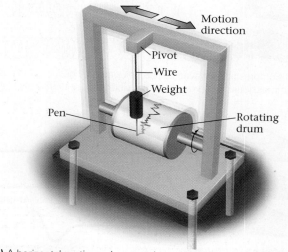

(b) A horizontal-motion seismograph records back-and-forth ground motion.

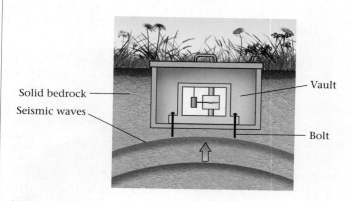

(c) Seismographs are bolted to bedrock in a protected shelter or vault.

FIGURE 8.8 How a vertical-motion seismograph works.

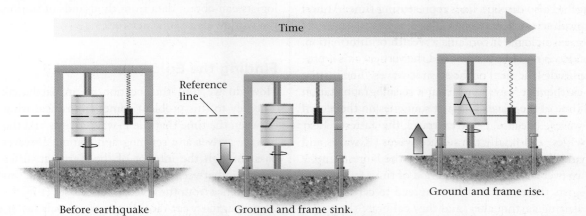

Time

Reference line

Before earthquake

Ground and frame sink.

Ground and frame rise.

(a) Before an earthquake, the pen traces a straight line. During an earthquake, the paper roll moves up and down while the pen stays in place.

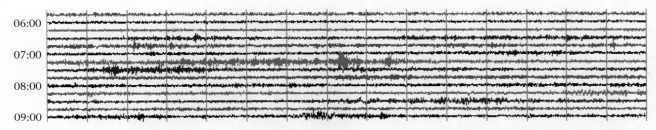

Paper

Surface waves

P-wave arrival

S-wave arrival

Surface-wave arrival

Aftershock

06:00
07:00
08:00
09:00

11:15
12:15
13:15
14:15

0   10   20   30   40   50   0   10
Minutes

(b) This closeup of a seismogram shows the signals generated by different kinds of seismic waves.

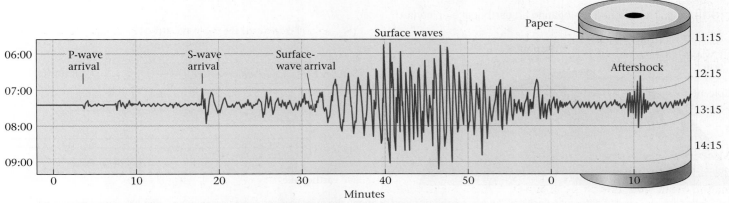

06:00
07:00
08:00
09:00

(c) A digital seismic record from a seismograph station in Arkansas. The space between vertical lines represents one minute. Colors have no meaning but make the figure more readable. Each color band represents the record of 15 minutes.

ground. A pen extends sideways from the weight and touches a vertical revolving cylinder of paper that has been connected to the seismograph frame. When an earthquake wave arrives and causes the ground surface to move up and down, it makes the seismograph frame also move up and down. The weight, however, because of its inertia (the tendency of an object at rest to remain at rest), remains fixed in space. As a consequence, the revolving paper roll moves up and down under the pen. The pen, therefore, traces out the waves representing the up-and-down movement. (If the paper cylinder did not revolve, the pen would move back and forth in the same place on the paper.) A

mechanical horizontal-motion seismograph works on the same principle, except that the paper cylinder is horizontal and the weight hangs from a wire. Sideways back-and-forth movement of this seismograph causes the pen to trace out waves. In sum, the key to a seismograph is the presence of a weight that stays fixed in space while everything else moves around it. Modern seismographs work on the same principle, except the weight is a magnet that moves relative to a wire coil, producing an electric signal that can be recorded digitally.

The waves traced by a seismograph provide a record of the earthquake called a **seismogram** (Fig. 8.8b, c). In order for seismologists to determine when a particular earthquake wave arrives, the record also displays lines representing time. At first glance, a typical seismogram looks like a messy squiggle of lines, but to a seismologist it contains a wealth of information. The horizontal axis represents time, and the vertical axis represents the amplitude (the size) of the seismic waves. The instant at which an earthquake wave appears at a seismograph station is the arrival time of the wave. The first squiggles on the record represent P-waves, because P-waves travel the fastest. Next come the S-waves, and finally the surface waves (R-waves and L-waves). Typically, the surface waves have the largest amplitude and arrive over a relatively long period of time.

Seismologists the world over have agreed to certain standards for measuring earthquakes, and they calibrate their seismographs to a precise time signal so that they can compare seismograms from different parts of the world. Today, seismologists can obtain data from thousands of stations constituting the worldwide seismic network.

Finding the Epicenter

How do we find the location of an earthquake's epicenter? The key to this problem comes from measuring the *difference* between the time that the P-wave arrives and the time that the S-wave arrives at a seismograph station. P-waves and S-waves pass through the interior of the Earth at different velocities. The delay between P-wave and S-wave arrival times increases as the distance from the epicenter increases (Fig. 8.9a). To picture why, imagine a car race. If one car travels faster than another, the distance between them increases as the race proceeds.

FIGURE 8.9 The method for locating an earthquake epicenter.

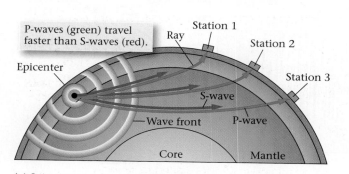

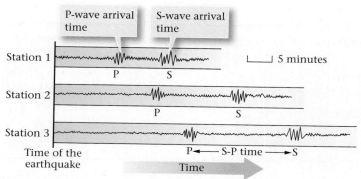

(a) Seismic waves travel at different velocities. The greater the distance between the epicenter and the seismograph station, the longer it takes for earthquake waves to arrive, and the greater the delay between the P-wave and S-wave arrival times.

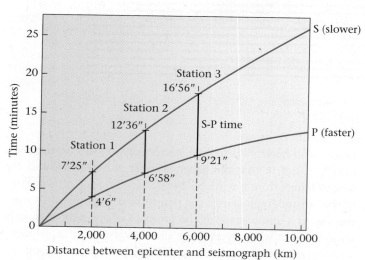

(b) We can represent the different arrival times of P-waves and S-waves on a travel-time curve.

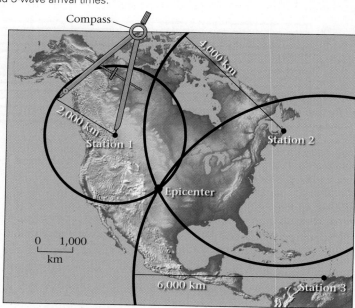

(c) If an earthquake epicenter lies 2,000 km from Station 1, we draw a circle with a radius of 2,000 km around the station at the scale of the map. We repeat for the other two stations. The intersection is the epicenter.

We can represent the time delay between P- and S-wave arrivals on a graph with a line called a travel-time curve, which plots the time since the earthquake waves began to propagate from below the epicenter on the vertical axis and the distance to the epicenter on the horizontal axis (**Figure 8.9b**). To use a travel-time curve for determining the distance to an epicenter, start by measuring the time difference between the P- and S-waves on your seismogram; this is called the S–P (pronounced S minus P) time. Then draw a line segment on a piece of tracing paper to represent this amount of time, at the scale used for the vertical axis of the graph. Orient the line segment parallel to the time axis and move it back and forth until one end lies on the P-wave curve and the other end lies on the S-wave curve (this gives the S–P time). Extend the line down to the horizontal axis, and simply read off the distance to the epicenter. You have now determined the distance at which the S–P time difference on the graph equals the time difference you observed.

The analysis of one seismogram tells you only the distance between the epicenter and the seismograph station; it does not tell you in which direction from the station the epicenter lies. To determine the map position of the epicenter, we can use a method called triangulation, by plotting the distance from the epicenter to *three* stations. For example, say you know that the epicenter lies 2,000 km from Station 1, 4,000 km from Station 2, and 6,000 km from Station 3. On a map, draw a circle around each station, such that the radius of the circle is the distance between the station and the epicenter, at the scale of the map. The epicenter lies at the intersection of the three circles, for this is the only point at which the epicenter has the appropriate measured distance from all three stations (**Fig. 8.9c**).

Defining the Size of an Earthquake

Some earthquakes shake the ground violently and cause extensive damage, whereas others can barely be felt. Seismologists have developed means to define size in a uniform way, so that they can make systematic comparisons among earthquakes. Seismologists use two distinct approaches for characterizing the relative sizes of earthquakes—the first yields a number called the intensity and the second yields a number called magnitude.

Mercalli intensity scale. In 1902, the Italian scientist Giuseppe Mercalli developed the first widely used scale for characterizing earthquake size. This scale, called the **Mercalli intensity scale**, defines the intensity of an earthquake by the amount of damage it causes—that is, by its destructiveness. Geologists now use the modified Mercalli scale (MMI), with Roman numerals, as shown in **Table 8.1**. Note that the specification of the intensity of an earthquake depends on a *subjective* assessment of the damage, not on a particular measurement

with an instrument. Note also that the Mercalli intensity value varies with distance from the epicenter, because earthquake energy dies out as the waves travel farther through the Earth. Seismologists draw lines, called contours, that delimit zones in which the earthquake has a specific intensity (Fig. 8.10).

TABLE 8.1 Modified Mercalli intensity scale

MMI	Destructiveness (Perceptions of the Extent of Damage)
I	Detected only by seismic instruments; causes no damage.
II	Felt by a few stationary people, especially in upper floors of buildings; suspended objects, such as lamps, may swing.
III	Felt indoors; standing automobiles sway on their suspensions; it seems as though a heavy truck is passing.
IV	Shaking awakens some sleepers; dishes and windows rattle.
V	Most people awaken; some dishes and windows break, unstable objects tip over; trees and poles sway.
VI	Shaking frightens some people; plaster walls crack, heavy furniture moves slightly, and a few chimneys crack, but overall little damage occurs.
VII	Most people are frightened and run outside; a lot of plaster cracks, windows break, some chimneys topple, and unstable furniture overturns; poorly built buildings sustain considerable damage.
VIII	Many chimneys and factory smokestacks topple; heavy furniture overturns; substantial buildings sustain some damage, and poorly built buildings suffer severe damage.
IX	Frame buildings separate from their foundations; most buildings sustain damage, and some buildings collapse; the ground cracks, underground pipes break, and rails bend; some landslides occur.
X	Most masonry structures and some well-built wooden structures are destroyed; the ground severely cracks in places; many landslides occur along steep slopes; some bridges collapse; some sediment liquifies; concrete dams may crack; facades on many buildings collapse; railways and roads suffer severe damage.
XI	Few masonry buildings remain standing; many bridges collapse; broad fissures form in the ground; most pipelines break; severe liquefaction of sediment occurs; some dams collapse; facades on most buildings collapse or are severely damaged.
XII	Earthquake waves cause visible undulations of the ground surface; objects are thrown up off the ground; there is complete destruction of buildings and bridges of all types.

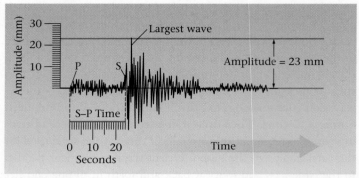

FIGURE 8.11 Using the Richter magnitude scale

(a) To calculate the Richter magnitude from a seismograph, first measure the S–P (S minus P) time, to determine the distance to the epicenter. Then measure the amplitude of the largest wave.

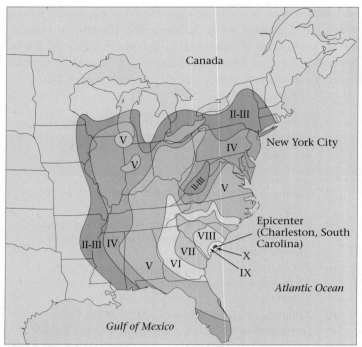

FIGURE 8.10 This map shows modified Mercalli intensity contours for the 1886 Charleston, South Carolina, earthquake. Note that near the epicenter, ground shaking reached MMI = X, and in New York City, ground shaking reached MMI = II to III.

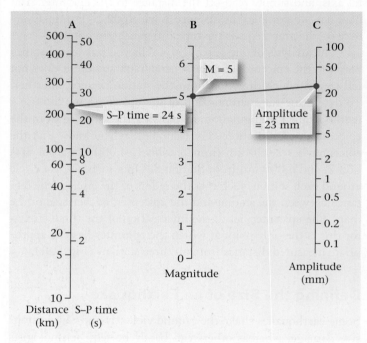

(b) Draw a line from the point on Column A representing the S–P time or the distance to the epicenter to the point on Column C representing the wave amplitude. Read the Richter magnitude off Column B.

Earthquake magnitude scales. When you read a report of an earthquake disaster in the news, you will likely come across a phrase that reads something like, "An earthquake with a magnitude of 7.2 struck the city yesterday at noon." What does this mean? The **magnitude** of an earthquake is a number that indicates its relative size as determined by measuring the maximum amplitude of ground motion recorded by a seismograph. By "amplitude of ground motion," we mean the amount of up-and-down or back-and-forth motion of the ground. The larger the ground motion, the greater the deflection of a seismograph pen or needle as it traces out a seismogram. When calculating a magnitude, seismologists take into account the distance between the epicenter and the seismograph, so a magnitude value does not depend on this distance.

The American seismologist Charles Richter developed the concept of defining and measuring earthquake magnitude in 1935. The scale he proposed came to be known as the **Richter scale**. A simple chart allows you to determine the magnitude if you know the amplitude of the largest deflection on a seismograph, and the distance from the epicenter (Fig. 8.11a, b).

Richter's scale became so widely used that news reports often include wording such as, "The earthquake registered a 7.2 on the Richter scale." These days, however, seismologists actually use several different magnitude scales, not just the Richter scale, because the original Richter scale works well only for shallow earthquakes that are close to the seismograph. Because

of the distance limitation, a number on the original Richter scale is now also called a local magnitude (M_L). The **moment-magnitude scale** now appears to provide the most accurate representation of an earthquake's size. To calculate the moment magnitude (M_W), seismologists must measure the amplitude of several different seismic waves, determine the area of the fault that slipped, and determine the amount of slip that occurred. Because of the way it is calculated, the M_W scale does a better job of distinguishing among very large earthquakes. For example, the largest recorded earthquake in history, the great 1960 Chilean quake, registered as an 8.5 on the M_L scale and as a 9.5 on the M_W scale. The larger number makes more sense, because this earthquake was indeed much larger than other known events for which $M_L = 8.5$.

TABLE 8.2 **Adjectives for describing earthquakes**

Adjective	Magnitude	Approximate maximum intensity at epicenter	Effects
Great	>8.0	X to XII	Major to total destruction
Major	7.0 to 7.9	IX to X	Great damage
Strong	6.0 to 6.9	VII to VIII	Moderate to serious damage
Moderate	5.0 to 5.9	VI to VII	Slight to moderate damage
Light	4.0 to 4.9	IV to V	Felt by most; slight damage
Minor	<3.9	III or smaller	Felt by some; hardly any damage

FIGURE 8.12 Measuring the energy released by earthquakes.

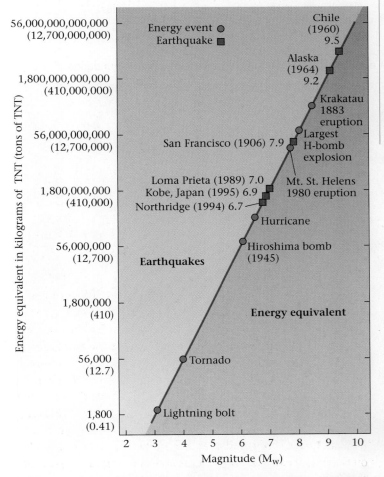

What magnitude is given in modern reports of earthquakes in the newspaper? It depends on how soon the article appears after an earthquake has taken place. For early reports, seismologists report a preliminary magnitude, such as an M_L, which can be calculated quickly. Later on, after they have had the chance to collect the necessary data, seismologists report an M_W, which becomes the number generally used for archival records.

All magnitude scales are logarithmic, meaning that an increase of one unit of magnitude represents a tenfold increase in the maximum ground motion. Thus, a magnitude 8 earthquake results in ground motion that is 10 times greater than that of a magnitude 7 earthquake, and 1,000 times greater than that of a magnitude 5 earthquake. To make life easier, seismologists use familiar adjectives to describe the size of an earthquake, as listed in Table 8.2.

As we pointed out earlier, earthquakes release energy. Seismologists can calculate the energy release from equations that relate moment magnitude to energy. Not all versions of this calculation yield the same result, so energy estimates must be taken as an approximation. According to some researchers, a magnitude 6 earthquake releases about as much energy as the atomic bomb that was dropped on Hiroshima in 1945, and the 1964 Alaska earthquake released significantly more energy than the largest hydrogen bomb ever detonated. Notably, an increase in magnitude by one integer represents approximately a thirty-two-fold increase in energy. Thus, a magnitude 8 earthquake releases about 1 million times more energy than a magnitude 4 earthquake (**Fig. 8.12a**). In fact, a single magnitude 8.9 earthquake releases as much energy as the entire *average* global annual release of seismic energy coming from all other earthquakes combined!

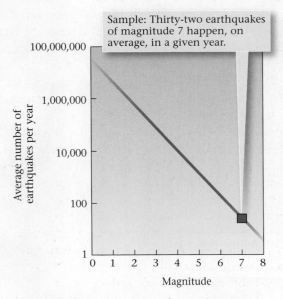

(a) Energy released by earthquakes increases dramatically with magnitude. Great earthquakes release vastly more energy than the largest bombs.

(b) The number of earthquakes per year of a given magnitude decreases with increasing magnitude.

Fortunately, such large earthquakes occur much less frequently than small earthquakes. There are about 100,000 magnitude 3 earthquakes every year, but a magnitude 8 earthquake happens only about once every three to five years (Fig. 8.12b).

TAKE-HOME MESSAGE

Seismographs measure earthquake vibrations and allow seismologists to determine the location and size of earthquakes. We specify earthquake size by an intensity value (damage caused) or a magnitude (the amplitude of ground motion). Magnitudes are logarithmic, meaning that ground shaking by a magnitude 8 event is 10 times larger than from a magnitude 7 event.

8.5 WHERE AND WHY DO EARTHQUAKES OCCUR?

Earthquakes do not take place everywhere on the globe. By plotting the distribution of earthquake epicenters on a map, seismologists find that most, but not all, earthquakes occur in fairly narrow **seismic belts**, or seismic zones. Most seismic belts correspond with plate boundaries. Earthquakes within these belts are plate-boundary earthquakes; ones that occur away from plate boundaries are intraplate earthquakes (the prefix *intra* means within) (Fig. 8.13). Notably, 80% of the earthquake energy released on Earth comes from the plate-boundary earthquakes in the belts surrounding the Pacific Ocean.

Notably, earthquakes do not occur at random depths in the Earth. Seismologists distinguish three classes of earthquakes based on hypocenter (focus) depth: shallow earthquakes occur in the top 20 km of the Earth, intermediate earthquakes take place between 20 and 300 km, and deep earthquakes occur down to a depth of about 660 km. Earthquakes cannot happen deeper in the Earth, because rock at great depth cannot rupture or change in a manner that produces shock waves. In this section, we look at the characteristics of earthquakes in various geologic settings and learn why earthquakes take place where they do.

Earthquakes at Plate Boundaries

The majority of earthquakes happen at faults along plate boundaries, for the relative motion between plates is accommodated primarily by slip on these faults. We find different kinds of faulting at different types of plate boundaries.

FIGURE 8.13 A simplified map of epicenters emphasizes that most earthquakes occur in distinct belts along plate boundaries.

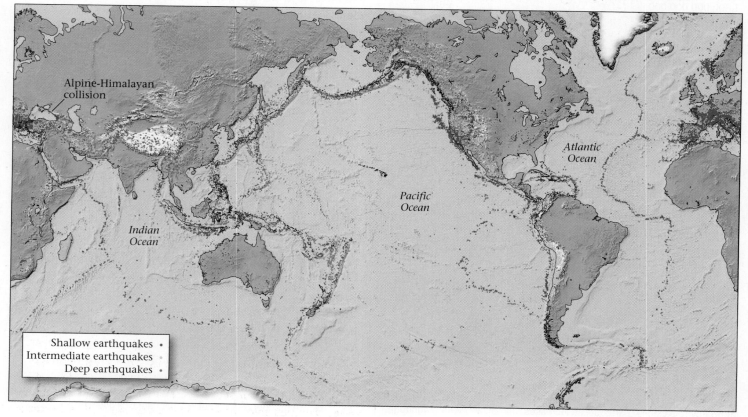

FIGURE 8.14 The distribution of earthquakes at a mid-ocean ridge. Note that normal faults occur along the ridge axis, and strike-slip faults occur along active transform faults. Earthquakes don't take place along inactive fracture zones.

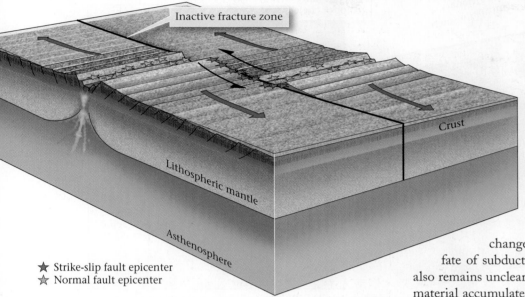

★ Strike-slip fault epicenter
☆ Normal fault epicenter

Divergent-plate-boundary seismicity.   At divergent plate boundaries (mid-ocean ridges), two oceanic plates form and move apart. Divergent boundaries are segmented, and spreading segments are linked by transform faults. Therefore, two kinds of faults develop at divergent boundaries. Along spreading segments, stretching generates normal faults, whereas along the transform faults that link spreading segments, strike-slip faulting occurs (Fig. 8.14). Seismicity along mid-ocean ridges takes place at shallow depths.

Convergent-plate-boundary seismicity.   Convergent plate boundaries are complicated regions at which several different kinds of earthquake take place. Specifically, as the downgoing plate begins to subduct, it bends and scrapes along the base of the overriding plate. Bending causes normal faults to develop in the downgoing plate, seaward of the trench. Large thrust faults develop along the contact between the downgoing and overriding plates, and shear on these faults can produce disastrous, shallow earthquakes. In some cases, the push applied by the downgoing plate compresses the overriding plate and triggers shallow faulting in the overriding plate.

In contrast to other types of plate boundaries, convergent plate boundaries also host intermediate and deep earthquakes. These occur in the downgoing slab as it sinks into the mantle, defining the sloping band of seismicity called a **Wadati-Benioff zone**, after the seismologists who first recognized it. Intermediate and deep earthquakes happen partly in response to stresses caused by shear between the downgoing plate and the mantle, and partly by the slab pull of the deeper part of the plate on the shallower part.

Considering the depth at which they occur, intermediate and deep earthquakes of a Wadati-Benioff zone present a problem: Shouldn't the Earth be too warm to undergo brittle deformation at such depths? Seismologists suggest that intermediate earthquakes can happen because the interior of the downgoing plate takes a long time to heat up, and the rock in the plate's interior remains cool and brittle enough to break even down to about 300 km. Deep earthquakes, however, remain a mystery. Some researchers suggest that they occur when minerals in the rock suddenly collapse as a result of great pressure, and the atoms rearrange into new minerals that take up less space—the sudden change in volume would generate a shock. The fate of subducting lithosphere below a depth of 660 km also remains unclear. Recent evidence suggests that some plate material accumulates at 660 km, while some sinks still deeper into the mantle but no longer generates earthquakes.

Earthquakes in southern Alaska, Japan, the western coast of South America, Mexico, and other places along the Pacific Rim serve as examples of convergent boundary earthquakes. One such event happened on Good Friday of 1964, near Anchorage, Alaska. This earthquake had a seismic moment magnitude (M_W) of 9.2, and its vibrations wrecked substantial parts of towns all along the coast. In 1995, an earthquake with a M_W of 6.9 struck Kobe, Japan, causing immense devastation and killing over five thousand people (Fig. 8.15a–c).

Transform-plate-boundary seismicity.   At transform plate boundaries, where one plate slides past another without the production or consumption of oceanic lithosphere, most faulting results in strike-slip motion. The majority of transform faults in the world link segments of oceanic ridges, but a few, such as the San Andreas Fault of California and the Alpine Fault of New Zealand, cut across continental lithosphere. All transform-fault earthquakes have a shallow focus, so the larger ones on land can cause disaster.

As an example of a transform-fault earthquake, consider the slip of the San Andreas Fault near San Francisco in 1906 (Fig. 8.16a). In the wake of the gold rush, San Francisco was a booming city with broad streets and numerous large buildings. But it was built on the transform boundary along which the Pacific Plate moves north at an average of 6 cm per year relative to North America. Because of the stick-slip behavior of the fault, this movement doesn't occur smoothly but happens in jerks, each of which causes an earthquake. At 5:12 A.M. on April 18, the fault slipped by as much as 7 m, and earthquake waves slammed into the city. Witnesses watched in horror as

FIGURE 8.15   An example of a convergent-margin earthquake.

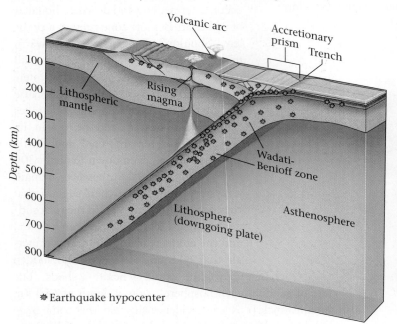

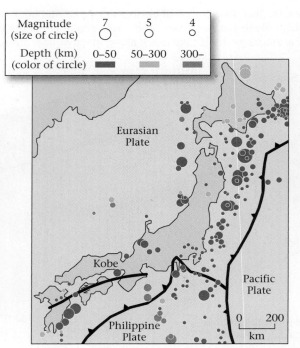

(a) At a convergent plate boundary, earthquakes occur along the contact between the two plates, as well as in the downgoing plate and overriding plate.

(b) Map of earthquake epicenters and subduction zones in and near Japan.

(c) A collapsed building after the Kobe earthquake in Japan.

the street undulated like ocean waves. Buildings swayed and banged together, laundry lines stretched and snapped, church bells rang, and then towers, facades, and houses toppled. Judging from the damage, seismologists estimate that the largest shock would have registered a seismic moment magnitude of 7.9. Most building collapse took place downtown. Fire followed soon after, consuming huge areas of the city, for most buildings were made of wood (Fig. 8.16b). In the end, about five hundred people died, and a quarter of a million were left homeless.

The San Francisco earthquake has not been the only one to strike along the San Andreas and nearby related faults. Over a

dozen major earthquakes have happened on these faults during the past two centuries, including the 1857 magnitude 8.7 earthquake just east of Los Angeles, and the 1989 magnitude 7.1 Loma Prieta earthquake, which occurred 100 km south of San Francisco but nevertheless shut down a World Series game and caused the collapse of a double-decker freeway. And as noted earlier, the Northridge quake struck near San Francisco in 1994 (Fig. 8.16c).

Earthquakes at Continental Rifts and Collision Zones

Continental rifts.   The stretching of continental crust at continental rifts generates normal faults (Fig. 8.17). Active rifts today include the East African Rift, the Basin and Range Province (mostly in Nevada, Utah, and Arizona), and the Rio Grande Rift (in New Mexico). In all these places, shallow earthquakes occur, similar in nature to the earthquakes at mid-ocean ridges. But in contrast to mid-ocean ridges, these seismic zones can be located under populated areas, and thus cause major damage.

Collision zones.   Two continents collide when the oceanic lithosphere that once separated them has been completely subducted. Such collisions produce great mountain ranges such as the Alpine-Himalayan chain. Though a variety of earthquakes happen in collision zones, the most common result from movement on thrust faults (see Fig. 8.17).

FIGURE 8.16 The San Andreas fault system, a continental transform.

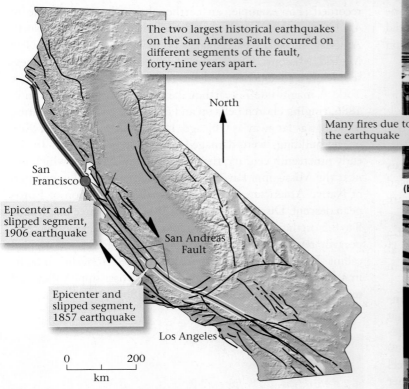

The two largest historical earthquakes on the San Andreas Fault occurred on different segments of the fault, forty-nine years apart.

North

San Francisco

Epicenter and slipped segment, 1906 earthquake

San Andreas Fault

Epicenter and slipped segment, 1857 earthquake

Los Angeles

0   200
km

(a) The San Andreas Fault system in California. Note that the system includes many faults in a 100-km-wide band.

Many fires due to the earthquake

(b) A street in San Francisco after the 1906 earthquake.

(c) The Northridge earthquake caused a double-decker bridge to collapse, as support columns gave way.

The complex collision between Africa and Europe generated the stress that triggered an earthquake near Lisbon, Portugal, on All Saints' Day (November 1), 1755. Aftershocks lasted for days. Innumerable artworks by great Renaissance masters were destroyed, along with a library that housed all the records of Portuguese exploration. Voltaire (1694–1778), the great French writer, immortalized the earthquake in his novel *Candide*. How, pondered Voltaire, could this world be the best of all possible worlds if such a calamity could happen and ruin good people?

Stress resulting from a collision between the Indian subcontinent and Asia caused the catastrophic magnitude 7.6 earthquake of October 2005 in the Kashmir region along the border of

FIGURE 8.17 The tectonic settings in which earthquakes occur in continental lithosphere. Subduction-related earthquakes in continental crust are not shown.

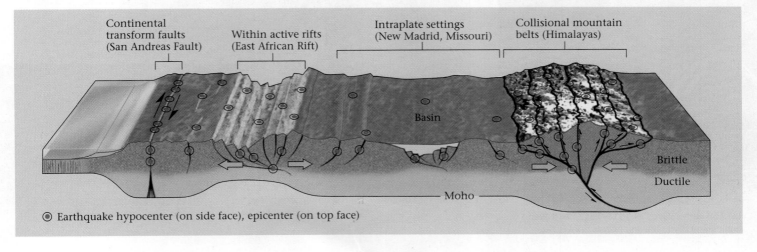

Continental transform faults (San Andreas Fault)

Within active rifts (East African Rift)

Intraplate settings (New Madrid, Missouri)

Collisional mountain belts (Himalayas)

Basin

Brittle

Ductile

Moho

◉ Earthquake hypocenter (on side face), epicenter (on top face)

India and Pakistan. In this region of poorly constructed homes, ground shaking resulted in the collapse of whole towns. Associated landslides ripped out roads, denying access to potential rescuers. At least 86,000 people perished, and another 4 million were left homeless.

Intraplate Earthquakes

Some earthquakes occur in the interiors of plates and are not associated with plate boundaries, active rifts, or collision zones (see Fig. 8.17). These **intraplate earthquakes** account for only about 5% of the earthquake energy released in a year. Almost all have a shallow focus. Seismologists are still trying to understand the causes of intraplate earthquakes. Most favor the idea that force applied to the boundary of a plate can cause the interior of the plate to break suddenly at weak, preexisting fault zones, some of which first formed in the Precambrian.

Because the crust in an intraplate setting tends to be composed of stronger rock than the crust along a plate boundary, it tends to transmit seismic waves more efficiently than does plate-boundary crust. As an analogy, if you were to bang a hammer on a block of solid rock and on a block of styrofoam, you would feel the impact on the other side of the rock, but not on the other side of the styrofoam because the impact would be absorbed by the styrofoam. Thus, intraplate earthquakes can be felt over a very broad region.

In Europe, a number of intraplate earthquakes have been recorded. For example, an earthquake with a magnitude of 4.8 hit central England, near Birmingham, in 2002. In North America, intraplate earthquakes occur in the vicinities of New Madrid, Missouri; Charleston, South Carolina; eastern Tennessee; Montréal, Quebec; and the Adirondack Mountains, New York. A magnitude 7.3 earthquake occurred near Charleston in 1886, ringing church bells up and down the coast and vibrating buildings as far away as Chicago. In Charleston itself, over 90% of the buildings were damaged, and sixty people died. In the early nineteenth century, the region of New Madrid, which lies near the Mississippi River, was inhabited by a small population of Native Americans and an even smaller population of European descent. During the winter of 1811–1812, three magnitude 8 to 8.5 earthquakes struck the region. The ground motion temporarily reversed the flow of the Mississippi River and toppled cabins (Fig. 8.18a). The earthquakes resulted from slip on faults in a Precambrian rift that underlies the Mississippi Valley and remains seismically active today (Fig. 8.18b).

TAKE-HOME MESSAGE

Most, but not all, earthquakes happen along plate boundaries. Events also occur in rifts and collision zones, and along weak faults within plate interiors. Most catastrophic earthquakes happen at convergent boundaries and along continental transforms.

FIGURE 8.18 New Madrid, Missouri, is an example of intraplate seismic activity.

(a) The earthquakes of 1811–1812 destroyed cabins and disrupted the Mississippi River.

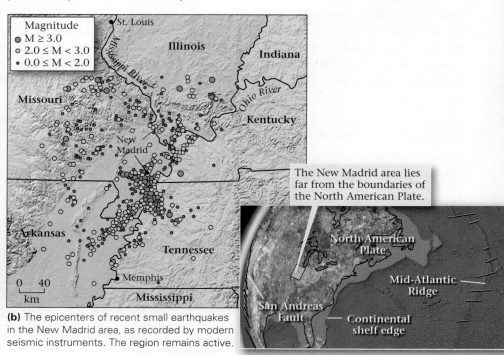

(b) The epicenters of recent small earthquakes in the New Madrid area, as recorded by modern seismic instruments. The region remains active.

The New Madrid area lies far from the boundaries of the North American Plate.

8.6 HOW DO EARTHQUAKES CAUSE DAMAGE?

An area ravaged by a major earthquake is a heartbreaking sight. The terror and sorrow etched on the faces of survivors mirror the inconceivable destruction. This destruction comes as a result of many processes.

Ground Shaking and Displacement

An earthquake starts suddenly and may last from a few seconds to a few tens of seconds. Different kinds of earthquake waves cause different kinds of ground motion (Fig. 8.19). If the motions are large enough, they may knock you down or even throw you in the air. In great earthquakes, the ground's movement can have an amplitude of as much as 1 m, but in moderate earthquakes, motions fall in the range of a few centimeters or less.

The nature and severity of the shaking at a given location depend on four factors: (1) the magnitude of the earthquake, because larger-magnitude events release more energy; (2) the distance from the hypocenter, because earthquake energy decreases as waves pass through the Earth; (3) the nature of the substrate at the location (that is, the character and thickness of different materials beneath the ground surface) because earthquake waves tend to be amplified in weaker substrate; and (4) the "frequency" of the earthquake waves (where frequency equals the number of oscillations that pass a point in a specified interval of time).

If you're out in an open field during an earthquake, ground motion alone won't kill you—you may be knocked off your feet and bounced around a bit, but your body is too flexible to break. Buildings and bridges aren't so lucky (Fig. 8.20a–d). When earthquake waves pass, they sway, twist back and forth, or lurch up and down, depending on the type of wave motion. As a result, connectors between the frame and façade of a building may separate, so the facade crashes to the ground. The flexing of walls shatters windows and makes roofs collapse. Floors or bridge decks may rise up and slam down on the columns that support them, thereby crushing the columns. Some buildings collapse with their floors piling on top of one another like pancakes in a stack. Some buildings and bridges simply tip over. The majority of earthquake-related deaths and injuries happen when people are hit by flying debris or are crushed beneath falling walls or roofs. Aftershocks worsen the problem, because they may topple already weakened buildings, trapping rescuers. During earthquakes, roads, rail lines, and pipelines may also buckle and rupture. If a building, fence, road, pipeline, or rail line straddles a fault trace, slippage on the fault can crack the structure and separate it into two pieces.

FIGURE 8.19 Types of ground motion during earthquakes. The ground can shake in many ways at once, causing surface structures to move.

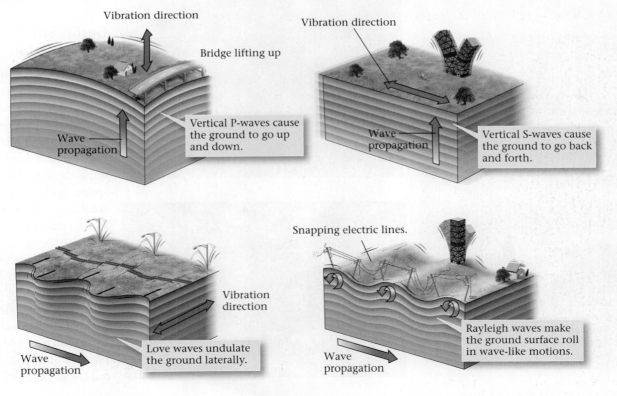

FIGURE 8.20 Examples of earthquake damage due to vibration.

(a) During a 1999 earthquake in Turkey, concrete buildings collaped when supports gave way and floors piled on each other like pancakes.

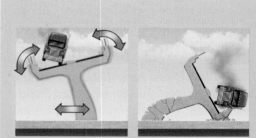

(b) An elevated bridge tipped over during the 1995 Kobe, Japan, earthquake.

(c) Concrete bridge supports were crushed during the 1994 Northridge earthquake when the overlying bridge bounced up and slammed back down.

(d) A neighborhood of masonry buildings in Armenia collapsed during a 1999 earthquake because the walls broke apart.

Landslides

The shaking of an earthquake can cause ground on steep slopes or ground underlain by weak sediment to give way. This movement results in a landslide, the tumbling and flow of soil and rock downslope. Landslides occur commonly along the coast of California, for movement on faults has rapidly uplifted this coastline in the past few million years, resulting in the development of steep cliffs. When earthquakes take place, the cliffs collapse, often carrying expensive homes down to the beach below (Fig. 8.21a, b). Such events lead to the misperception that "California will someday fall into the sea." Although small portions of the coastline do collapse, the state as a whole remains firmly attached to the continent, despite what Hollywood scriptwriters say.

Sediment Liquefaction

Places where the substrate contains wet sediment can be particularly hazardous during an earthquake. In beds of wet sand or silt, ground shaking causes the sediment grains to try and settle together. But because the spaces (pores) between grains are filled with water, water pressure in the pores increases and pushes the grains apart, and the wet silt or sand becomes a fluid-like slurry. The abrupt loss of strength of a wet sandy sediment in response to ground shaking is a phenomenon called **liquefaction**, and it can cause major damage during an earthquake. Buildings whose foun-

(a) Shaking triggers landslides, which caused a steep slope along the coast of California to collapse, carrying part of a home with it.

(b) During the 1964 Alaska earthquake, slumping caused the land to give way beneath parts of Anchorage.

(c) Liquefaction under their foundations caused these apartment buildings in Taiwan to tip over.

The dark area in this aerial photograph is the slump that was Turnagain Heights.

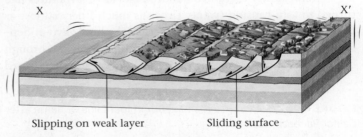

Slipping on weak layer   Sliding surface

(d) During the 1964 Turnagain Heights disaster, a landslide carried a neighborhood out to sea. Slip occured on a weak layer.

dations lie in liquefied material may sink or even tip over (Fig. 8.21c). In some cases, certain kinds of wet clay, when shaken, transform into a muddy ooze and can also give way.

If shaking weakens sediment, the ground above may give way and slide down the slope. Such an event happened during the 1964 Alaska earthquake, when the sediment beneath a housing development in the coastal suburb of Turnagain Heights liquefied (Fig. 8.21d). The land broke into a series of blocks that slid seaward on a weak sediment layer, carrying with it dozens of houses, which were transformed into a jumble of splintered wood and shattered windows.

In some cases, the liquefaction of sand layers below the ground surface makes the sand erupt through holes or cracks in overlying sediment, producing small mounds of sand called sand volcanoes or sand boils (Fig. 8.22a). Liquefaction may also cause bedding in unconsolidated sequences of sediment to break up, and ground settling due to underlying liquefaction may cause large fissures to develop in the overlying sediments (Fig. 8.22b).

Fire

The shaking during an earthquake can make lamps, stoves, or candles with open flames tip over, and it may break wires or topple power lines, generating sparks. As a consequence, areas already turned to rubble, and even areas not so badly damaged, may be consumed by fire. Ruptured gas pipelines and oil tanks feed the flames, sending columns of fire erupting skyward (Fig. 8.23). Firefighters might not even be able to reach the fires, because the doors to the firehouse won't open or rubble blocks the streets. Moreover, firefighters may find themselves without water, for ground shaking and landslides damage water lines.

FIGURE 8.22 The formation of sand volcanoes and ground fissures.

Before

— Compacted mud

— Unconsolidated sand

☐ Mud
☐ Sand

Time

Sand volcanoes

After

(a) Ground shaking may cause wet sand to squirt up through cracks in the overlying sediment to form a row of sand volcanoes.

(b) Liquefaction of a sand layer beneath the dry soil of this field caused the soil to crack and fissures to develop.

If a fire gets a good start, it can become an unstoppable inferno. Most of the destruction of the 1906 San Francisco earthquake, in fact, resulted from fire. For three days, the blaze spread through the city until firefighters contained it by blasting a firebreak. By then, five hundred blocks of structures had turned to ash, causing twenty times as much financial loss as the shaking itself. When a large earthquake hit Tokyo in September 1923, fires set by cooking stoves spread quickly through the wood-and-paper buildings, creating an inferno that heated the air above the city. The hot air rose like a balloon, and when cool air rushed in, creating wind gusts of over 100 mph, the wind stoked the blaze and incinerated 120,000 people.

Tsunamis

The azure waters and palm-fringed islands of the Indian Ocean's east coast hide one of the most seismically active plate boundaries on Earth—the Sunda Trench. Along this convergent boundary, the Indian Ocean floor subducts at about 6 cm per year, leading to slip on large thrust faults. Just before 8:00 A.M. on December 26, 2004, the crust above a 1300-km-long by 100-km-wide portion of one of these faults lurched westward by as much as 15 m. The rupture started at the hypocenter, and then propagated north at 2.8 km/s; thus, the rupturing process took nine minutes. This slip triggered a great earthquake ($M_W = 9.3$) and pushed the sea floor up by tens of centimeters. The rise of the sea floor, in turn, shoved up the overlying water. Because the area that rose was so broad, the volume of displaced water was immense. As a consequence, tragedy of an unimaginable extent was about to unfold. Water from above the upthrust sea floor began moving outward

FIGURE 8.23 Broken gas mains erupt in fountains of flame after the 1994 Northridge, California, earthquake.

FIGURE 8.24 Tsunamis can be generated when faulting displaces the sea floor over a broad area.

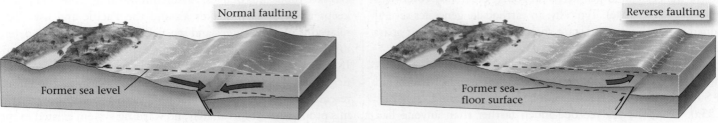

(a) Tsunamis are generated by sudden displacement of the sea floor during either normal or reverse faulting.

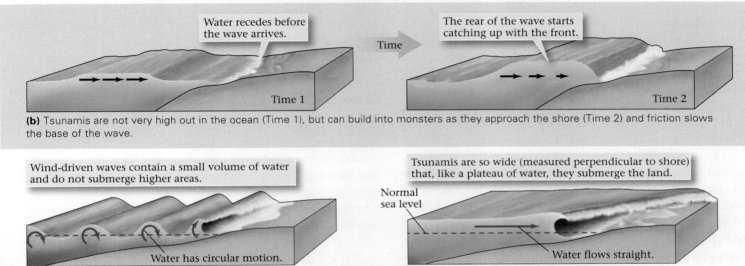

(b) Tsunamis are not very high out in the ocean (Time 1), but can build into monsters as they approach the shore (Time 2) and friction slows the base of the wave.

Wind-driven waves contain a small volume of water and do not submerge higher areas.

Tsunamis are so wide (measured perpendicular to shore) that, like a plateau of water, they submerge the land.

(c) Tsunamis are very different from storm-generated waves in that they contain much greater volumes of water.

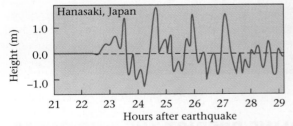

(d) A single earthquake can generate several tsunamis, as recorded here by a tidal gauge in Japan, twelve hours after the great 1960 earthquake in Chile.

from above the fault zone, a process that generated a series of giant waves, or **tsunamis**, traveling at speeds of about 800 km per hour (500 mph)—almost the speed of a jet plane (Fig. 8.24a). Tsunami is a Japanese word that translates literally as "harbor wave," an apt name because tsunamis can be particularly damaging to harbor towns. This name replaces the older term, "tidal wave." Tsunamis have no relationship to tides, and calling them tidal waves can be misleading. Tsunamis can be caused by undersea earthquakes but, as we will see in Chapter 13, they can also be set off by large undersea landslides.

Regardless of cause, tsunamis are very different from familiar, wind-driven storm waves. Large wind-driven waves can reach heights of 10 to 30 meters in the open ocean. But even such monsters have wavelengths of only tens of meters, and thus only involve a relatively small volume of water. In contrast, although a tsunami in deep water may cause a rise in sea level of at most only a few tens of centimeters—a ship crossing one wouldn't even notice—tsunamis have wavelengths of tens to hundreds of kilometers and thus involve a huge volume of water. In simpler terms, we can think of the width of a tsunami, in map view, as being more than 100 times the width of a wind-driven wave. Because of this difference, a storm wave and a tsunami have very different effects when they strike the shore.

When a water wave approaches the shore, friction between the base of the wave and the sea floor slows the bottom of the wave, so the back of the wave catches up to the front, and the added volume of water builds the wave higher (Fig. 8.24b). The top of the wave may fall over the front of the wave and cause a breaker. In the case of a wind-driven wave, the breaker may be tall when it washes onto the beach, but the wave is so narrow that it doesn't contain much water. Thus, the wave makes it only part way up the beach before it runs out of water, friction slows it to a stop, and gravity causes the water to spill back seaward. In the case of a tsunami, the wave is so wide that, as friction slows the wave, it builds into a plateau that can be tens of meters high and many kilometers wide. Thus, when a tsunami reaches shore, it contains so much water that it crosses the beach and just keeps on going, eventually covering a huge area (**Fig. 8.24c**). Typically, an earthquake generates several tsunamis that arrive on distant shores as much as an hour apart (**Fig. 8.24d**).

Tsunami damage can be catastrophic. The December 2004 wave struck Banda Aceh, a city at the north end of the island of Sumatra, as it was waking to a beautiful, cloudless day (**Fig. 8.25a**). First, the sea receded much farther than anyone had ever seen, exposing large areas of reefs that normally remained submerged even at low tide. People walked out onto the exposed reefs in wonder. But then, with a rumble that grew to a roar, a wall of frothing water began to build in the distance and approach land (**Fig. 8.25b**). Puzzled bathers first watched, then ran inland in panic when the threat became clear. As the tsunami approached shore, friction with the sea floor had slowed it to less than 30 km an hour, but it still moved faster than people could run. In places, the wave front reached heights of 15 to 30 m (45 to 100 feet) as it slammed into Banda Aceh (**Fig. 8.25c**).

The impact of the water ripped boats from their moorings, snapped trees, battered buildings into rubble, and tossed cars and trucks like toys. And the water just kept coming, eventually flooding low-lying land as far as about 7 km inland (**Fig. 8.25d**). It drenched forests and fields with salt water (deadly to plants) and buried fields and streets with up to a meter of sand and mud. Eventually the water slowed and then began to rush back to the shore, but at Banda Aceh, at least two more tsunamis struck before the first one had entirely receded, so the water remained high for some time. When the water level finally returned to normal, a jumble of flotsam, as well as the bodies of unfortunate victims, floated out to sea and drifted away.

Geologists refer to the tsunami that struck Banda Aceh as a near-field (or local) tsunami, because of its proximity to the earthquake. But the horror of Banda Aceh was merely a preamble to the devastation that would soon visit other stretches of Indian Ocean coast. Far-field (or distant) tsunamis crossed the ocean and struck Sri Lanka 2.5 hours after the earthquake, the coast of India half an hour after that, and the coast of Africa, on the west side of the Indian Ocean, 5.5 hours after the earthquake. Coastal towns vanished, fishing fleets sank, and beach resorts collapsed into rubble. In the end, perhaps more than 220,000 people died that day.

The 2004 Indian Ocean event remains etched in people's minds because of the immense death toll and the nonstop news coverage. But it is not unique. Tsunamis generated by the great Chilean earthquake of 1960 ($M_W = 9.5$) destroyed coastal towns of South America and crossed the Pacific, causing a 10.7-m-high wall of water to strike Hawaii 15 hours later. When, 21 hours after the earthquake, the tsunami reached Japan, it flattened coastal villages and left 50,000 people homeless. The tsunami following the 1964 Good Friday earthquake in Alaska destroyed ports at Valdez and Kodiak.

Because of the danger of tsunamis, predicting their arrival can save thousands of lives. A tsunami warning center in Hawaii keeps track of earthquakes around the Pacific and uses data relayed from tide gauges and sea-floor pressure gauges to determine whether a particular earthquake has generated a tsunami. If observers detect a tsunami, they flash warnings to authorities around the Pacific. In recent years, to help in the effort, researchers have used computer models that predict how tsunamis propagate. Unfortunately, no warning system existed in the Indian Ocean, and even though Hawaiian observers detected the earthquake and realized that a tsunami was likely, they did not have the means to contact many local authorities. Even if they had, affected towns had no evacuation plans in place. A multinational team has begun work to remedy this situation.

Disease

Once the ground shaking and fires have stopped, disease may still threaten lives in an earthquake-damaged region. Earthquakes cut water and sewer lines, destroying clean water supplies and exposing the public to bacteria, and they cut transportation lines, preventing food and medicine from reaching the area. The severity of such problems depends on the ability of emergency services to cope.

TAKE-HOME MESSAGE

Earthquakes cause devastation because ground shaking topples buildings. But destruction can also be caused by landslides, tsunamis, and sediment liquefaction triggered by the shaking. After an earthquake, survivors face the threats of fire and disease.

8.7 CAN WE PREDICT THE "BIG ONE"?

We have seen that large earthquakes occurring near population centers cause catastrophe (see **Geotour 8** on p. GT-18). Needless to say, many lives could be saved if only it were possible to know exactly when and where an earthquake will happen, so people could evacuate dangerous buildings, turn off gas and electricity, and, with more warning, build stronger structures.

Can seismologists predict earthquakes? The answer depends on the time frame of the prediction. With our present understanding of the distribution of seismic zones and the frequency at which earthquakes occur, we can make long-term predictions (on the time scale of decades to centuries). For example, with some certainty, we can say that a major earthquake will rattle California during the next 100 years, and that a major earthquake won't strike central Canada during the next ten years. But despite

FIGURE 8.25 The great Indian Ocean tsunami of 2004.

(a) A devastating tsunami was triggered by an earthquake off Sumatra. Three hours later, the leading wave struck the coasts of Sri Lanka and India. A computer model shows the wave.

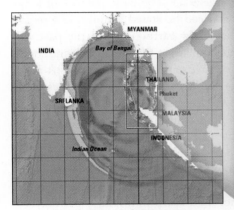

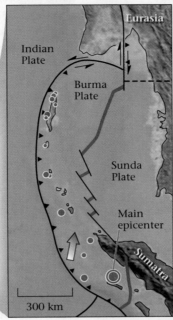

Colors represent wave height—yellow is highest. There were several waves.

The earthquake was caused by subduction at a trench. Red dots are epicenters.

(b) This snapshot shows the wave rushing toward the coast of Sumatra. Recession of water in advance of the wave exposed a reef.

(c) The wave blasts through a grove of palm trees as it strikes the coast of Thailand.

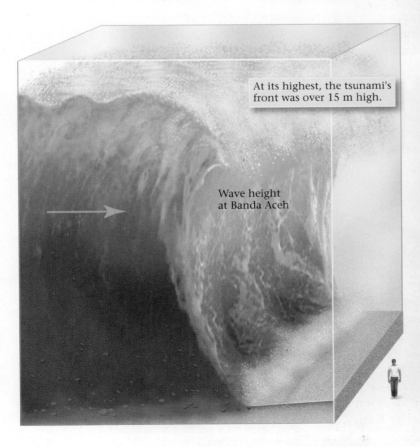

At its highest, the tsunami's front was over 15 m high.

Wave height at Banda Aceh

(d) Satellite photos of the Indonesian province of Aceh before and after the tsunami struck. Note that the city was washed away and the beach vanished.

extensive research, seismologists *cannot* make accurate short-term predictions (on the time scale of hours to weeks). Thus, we cannot say, for example, that an earthquake will happen in Montreal at 2:43 P.M. on January 17.

In this section, we look at the scientific basis of both long- and short-term predictions and consider the consequences of a prediction. Seismologists refer to studies leading to predictions as seismic-risk, or seismic-hazard assessment. On the basis of this work, they produce maps that assign levels of seismic risk to different regions.

Long-Term Predictions

When making a prediction, we use the word *probability*, because a prediction only gives the likelihood of an event. For example, a seismologist may say, "The probability of a major earthquake occurring in the next 20 years in this state is 20%." This sentence implies that there's a one-in-five chance that the earthquake will happen during a twenty-year period. Urban planners and civil engineers can use long-term predictions to help create building codes for a region—codes requiring stronger buildings make sense for regions with greater seismic risk. They may also use predictions to determine whether or not to build vulnerable structures such as nuclear power plants, hospitals, or dams in potentially seismic areas. Seismologists base long-term earthquake predictions on two pieces of information: the identification of seismic zones and the **recurrence interval** (the average time between successive events).

To identify a seismic zone, seismologists produce a map showing the epicenters of earthquakes that have happened during a set period of time (say, 30 years). Clusters or belts of epicenters define the seismic zone. The basic premise of long-term earthquake prediction can be stated as follows: a region in which there have been many earthquakes in the past will likely experience more earthquakes in the future. Seismic zones, therefore, are regions of greater seismic risk. This doesn't mean that a disastrous earthquake can't happen far from a seismic zone—they can and do—but the risk is less.

Epicenter maps can be produced with data from only the past fifty years or so, because before that time seismologists did not have enough seismographs to locate epicenters accurately. Fortunately, geologists can assess seismic risk by examining landforms for evidence of recent faulting. For example, the presence of a distinct fault scarp in a landscape indicates that faulting has happened so recently that erosion has not yet had time to grind away the evidence (Fig. 8.26a).

FIGURE 8.26 Identifying recent fault movement and recurrence interval.

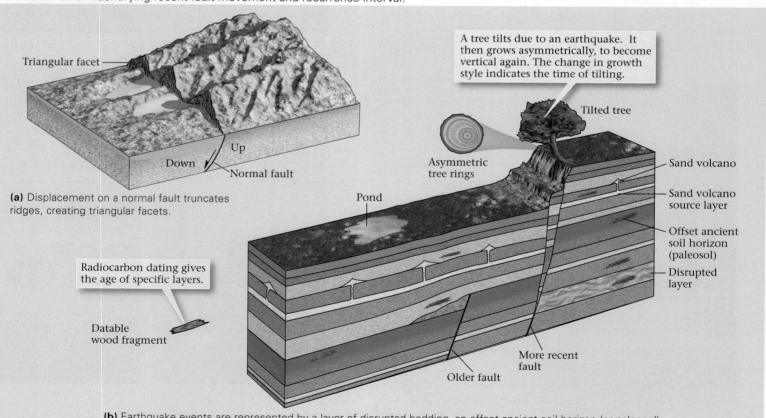

(a) Displacement on a normal fault truncates ridges, creating triangular facets.

(b) Earthquake events are represented by a layer of disrupted bedding, an offset ancient soil horizon (or paleosol), a layer of sand volcanoes, and a bent tree. Dating several events helps define the recurrence interval.

FIGURE 8.27 Examples of seismic-hazard maps.

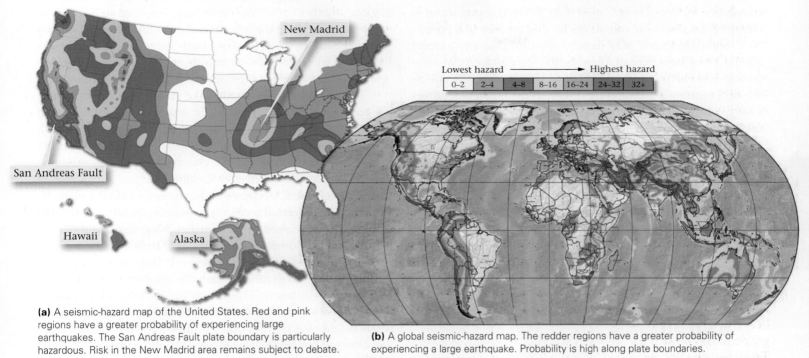

(a) A seismic-hazard map of the United States. Red and pink regions have a greater probability of experiencing large earthquakes. The San Andreas Fault plate boundary is particularly hazardous. Risk in the New Madrid area remains subject to debate.

(b) A global seismic-hazard map. The redder regions have a greater probability of experiencing a large earthquake. Probability is high along plate boundaries.

To determine the recurrence interval for large earthquakes in a seismic zone, seismologists must determine when large earthquakes happened there in the past. Since the historical record does not provide information far enough back in time, we must study geologic evidence for great earthquakes. For example, we can examine sedimentary strata near a fault to find layers of sand volcanoes and disrupted bedding in the stratigraphic record. Each layer, whose age can be determined by using radiocarbon dating of plant fragments, records the time of an earthquake. We can obtain additional information by dating offset soil horizons now buried beneath other sediment (Fig. 8.26b). With such information, seismologists are able to produce regional maps illustrating seismic risk (Fig. 8.27a, b).

Seismologists suspect that places called seismic gaps, where a known active fault has not slipped for a long time, may be particularly dangerous. In a seismic gap, either the fault must be moving nonseismically, or the stress must be building up to be released by a major earthquake in the future.

Short-Term Predictions

Short-term predictions, which could lead to such precautions as evacuating dangerous buildings, shutting off gas and electricity, and readying emergency services, are not reliable and may never be. Nevertheless, there are clues to imminent earthquakes, and seismologists have been working hard to understand them. The first clue comes from the detection of foreshocks. A swarm (a cluster of events during a short period of time) of foreshocks may indicate the cracking that precedes a major rupture along a fault zone. In this regard, foreshocks are analogous to the cracking noises you hear just before a tree limb breaks off and falls down. But foreshocks do not always occur, and even if they do, they usually can be identified only in hindsight, because they may be indistinguishable from other small earthquakes.

Another possible data source for short-term predictions comes from the precise laser surveying of the ground. Before an earthquake, a region of crust may undergo distortion, either in response to the buildup of elastic strain in the rock, or because of the development of small, open cracks that cause the crust to increase in volume. The land surface may bulge or sink, or straight lines on the ground may become bent. The detection of such movements by laser surveying can hint at an upcoming earthquake. More recently, researchers have been able to use satellite data to detect distortion of the land surface. Geologists also have begun to use computer models of stress to predict where stress buildups may lead to earthquakes.

Other changes that have been explored but have not been confirmed as precursors of earthquakes include the following: changes in the water level in wells; appearance of gases, such as radon or helium, in wells; changes in the electrical conductivity of rock underground; and unusual animal behavior. Believers in these proposed clues suggest that they all reflect the occurrence of cracking in the crust prior to an earthquake, but most investigators remain skeptical.

As long as short-term predictions remain questionable, emergency service planners must ask, What if a prediction is wrong? Should schools and offices be shut because of a prediction? Should millions of dollars be spent to evacuate people? Should a city be deserted, allowing for the possibility of looters? Should the public be notified, or should only officials be notified, creating a potential for rumor? If the prediction proves wrong, can seismologists be sued? No one really knows the answers to these questions.

TAKE-HOME MESSAGE

Researchers can determine regions where earthquakes are more likely, but not exactly when and where an event will occur. Seismic hazard is greater where seismicity has happened more frequently in the past and therefore has a shorter recurrence interval.

8.8 EARTHQUAKE ENGINEERING AND ZONING

The loss of human life from an earthquake of a given size varies widely. The loss depends on a number of factors, most notably the proximity of an epicenter to a population center, the depth of the hypocenter, the style of construction in the epicentral region, whether or not the earthquake occurred in a region of steep slopes or along the coast, whether building foundations are on solid bedrock or on weak substrate, whether the earthquake happened when people were outside or inside, and whether the government was able to provide emergency services promptly.

For example, a 1988 earthquake in Armenia was not much bigger than a 1971 earthquake in southern California, but it caused almost 500 times as many deaths (24,000 versus 50). The difference in death toll reflects differences in the style and quality of construction and the characteristics of the substrate. The unreinforced concrete-slab buildings and masonry houses of Armenia collapsed, whereas the structures in California had, by and large, been erected according to building codes that take into account stresses caused by earthquakes. Most flexed and twisted but did not fall down and crush people. The terrible 1976 earthquake in T'ang-shan, China, killed over a quarter of a million people because the ground beneath the epicenter had been weakened by coal mining and collapsed, and because buildings were poorly constructed. Mexico City's 1985 earthquake proved disastrous because the city lies over a sedimentary basin whose composition and bowl-like shape focused seismic energy. During the 1989 Loma Prieta quake in California, portions of Route 880 in Oakland that were built on a weak substrate collapsed, whereas portions built on bedrock or gravel remained standing.

FIGURE 8.28 Preventing damage and injury during an earthquake.

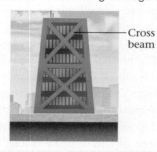
Cross beam

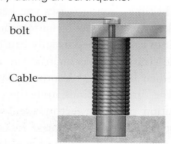
Anchor bolt

Cable

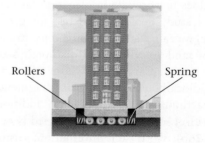
Rollers   Spring

Buildings are less likely to collapse if they are wider at the base and if cross beams are added for strength.

Wrapping a bridge's support columns in cable and bolting the span to the columns will prevent the bridge from collapsing so easily.

Placing buildings on rollers or shock absorbers lessens the severity of the vibrations.

(a) Damage can be prevented if buildings are designed to withstand vibration.

(c) If an earthquake strikes, take cover under a sturdy table near a wall.

(b) Buoys can detect a tsunami in the open ocean, so people on land can be warned.

We can mitigate or diminish their consequences by taking sensible precautions. Clearly, earthquake engineering (the designing of buildings that can withstand shaking) and earthquake zoning (the determination of where land is stable and where it might collapse) can help save lives and property. In regions prone to large earthquakes, buildings and bridges should be constructed so they are able to withstand vibrations without collapsing (Fig. 8.28a). They should be somewhat flexible so that ground motions can't crack them, and supports should be strong enough to maintain loads far in excess of the loads caused by their static (nonmoving) weight. Wrapping steel cables around bridge support columns makes them many times stronger. Bolting the bridge spans to the top of a column prevents the spans from bouncing off. Concrete-block, unreinforced concrete, and unreinforced brick buildings crack and tumble under conditions in which wood-frame, steel-girder, or reinforced concrete buildings remain standing. Traditional heavy, brittle tile roofs shatter and bury the inhabitants inside, whereas sheet-metal or asphalt shingle roofs do not. Loose decorative stone and huge open-span roofs do not fare well when vibrated and should be avoided.

Similarly, developers should avoid construction on land underlain by weak mud that could liquefy. They should not build on top of, on, or at the base of steep escarpments because the escarpments could fail and produce landslides, and they should avoid locating large population centers downstream of dams (which could crack and collapse, causing a flood). And they should avoid constructing vulnerable buildings directly over active faults, because fault movement could crack and destroy the buildings. Cities in seismic zones need to draw up emergency plans to deal with disaster. Communication centers should be situated in safe localities, and strategies need to be implemented for providing supplies under circumstances where roads may be impassable. And in coastal areas, tsunami warning systems need to be implemented (Fig. 8.28b).

Finally, communities and individuals should learn to protect themselves during an earthquake. In your home, keep emergency supplies, bolt bookshelves to walls, strap the water heater in place, install locking latches on cabinets, know how to shut off the gas and electricity, know how to find the exit, have a fire extinguisher handy, and know where to go to find family members. Schools and offices should have earthquake-preparedness drills. When an earthquake strikes, stay away from buildings. If you are trapped inside, a heavy table may provide some protection (Fig. 8.28c). As long as lithosphere plates continue to move, earthquakes will continue to shake. But we can learn to live with them.

TAKE-HOME MESSAGE

Earthquakes are a fact of life on this dynamic planet. People in regions facing high seismic risk should build on stable ground, avoid unstable slopes, and design construction that can survive shaking. Evacuation planning saves lives after an event.

Chapter Summary

- Earthquakes are episodes of ground shaking. Earthquake activity is called seismicity.

- Most earthquakes happen when rock breaks during faulting. A fault is a fracture on which sliding occurs. The place where rock breaks and earthquake energy is released is called the hypocenter (focus), and the point on the ground directly above the hypocenter is the epicenter.

- Active faults are faults on which movement is likely. Inactive faults ceased being active long ago, but can still be recognized because of the displacement across them. Displacement on active faults that intersect the ground surface may yield a fault scarp.

- According to elastic-rebound theory, during fault formation, rock elastically bends, then cracks. Eventually, cracks link to form a throughgoing rupture on which sliding occurs. When this happens, the rock breaks and vibrates, and this generates an earthquake. Faults exhibit stick-slip behavior, in that they move in sudden increments.

- Earthquake energy travels in the form of seismic waves. Body waves, which pass through the interior of the Earth, include P-waves (compressional waves) and S-waves (shear waves). Surface waves, which pass along the surface of the Earth, include R-waves (Rayleigh waves) and L-waves (Love waves).

- We can detect earthquake waves by using a seismograph.

- Seismograms demonstrate that different earthquake waves arrive at different times, because they travel at different velocities. Using the difference between P-wave and S-wave arrival times, seismologists can pinpoint the epicenter location.

- The Mercalli intensity scale is based on documenting the damage caused by an earthquake. Magnitude scales, such as the Richter scale, are based on measuring the amount of ground motion, as indicated by traces of waves on a seismogram. The moment-magnitude scale takes into account the amount of slip, the length and depth of the rupture, and the strength of the ruptured rock.

- A magnitude 8 earthquake yields about 10 times as much ground motion as a magnitude 7 earthquake and releases about 32 times as much energy.

- Most earthquakes occur in seismic belts, or zones, of which the majority lie along plate boundaries. Intraplate earthquakes happen in the interior of plates. Different kinds of earthquakes happen at different kinds of plate boundaries.

- Earthquake damage results from ground shaking (which can topple buildings), landslides (set loose by vibration), sediment liquefaction (the transformation of compacted clay into a muddy slurry), fire, and tsunamis (giant waves).

- Seismologists predict that earthquakes are more likely in seismic zones than elsewhere, and can determine the recurrence interval (the average time between successive events) for great earthquakes. But it may never be possible to pinpoint the exact time and place at which an earthquake will take place.

- Earthquake hazards can be reduced with better construction practices and zoning, and by educating people about what to do during an earthquake.

GEOPUZZLE REVISITED

Most earthquakes happen when rock abruptly breaks during the formation of a new fault, or during renewed slip on an existing fault. Stress drives the process, building up slowly until it exceeds the rock's strength. Major seismic belts (regions of frequent earthquake activity) coincide with plate boundaries, collision zones, and rifts, and thus delineate regions where relative plate motion is being accommodated. A few seismic belts, however, do occur along ancient, weak faults within plates. Seismic risk is clearly greater in seismic belts, for stress builds up more rapidly in these regions. But it is not possible to predict exactly where or when an earthquake will occur within a zone.

Key Terms

aftershocks (p. 206)
body waves (p. 206)
compressional waves (p. 206)
displacement (p. 202)
earthquake (p. 201)
elastic-rebound theory (p. 206)
epicenter (p. 202)
fault (p. 201)
fault scarp (p. 202)
foreshocks (p. 206)
hypocenter (p. 202)
intraplate earthquake (p. 218)
liquefaction (p. 220)
magnitude (p. 212)
Mercalli intensity scale (p. 211)
moment-magnitude scale (p. 212)
recurrence interval (p. 226)
Richter scale (p. 212)
seismic belt (p. 214)
seismicity (p. 201)
seismic waves (p. 206)
seismogram (p. 209)
seismograph (p. 208)
shear waves (p. 207)
stick-slip behavior (p. 203)
stress (p. 202)
surface waves (p. 206)
tsunami (p. 223)
Wadati-Benioff zone (p. 215)

Review Questions

1. Compare normal, reverse, and strike-slip faults.
2. Describe elastic-rebound theory and the concept of stick-slip behavior.
3. Describe the motions of the four types of seismic waves. Which are body waves, and which are surface waves?
4. Explain how the vertical and horizontal components of an earthquake are detected on a seismograph.
5. Explain the differences among the scales used to describe the size of an earthquake.
6. How does seismicity on mid-ocean ridges compare with seismicity at convergent or transform boundaries? Do all earthquakes occur at plate boundaries?
7. What is a Wadati-Benioff zone, and why was it important in understanding plate tectonics?
8. Describe the types of damage caused by earthquakes.
9. What is a tsunami, and why does it form?
10. Explain how liquefaction occurs in an earthquake, and how it can cause damage.
11. How are long-term and short-term earthquake predictions made? What is the basis for determining a recurrence interval, and what does a recurrence interval mean?
12. What types of structure are most prone to collapse in an earthquake? What types are most resistant to collapse?

On Further Thought

1. Is seismic risk greater in a town on the west coast of South America or in one on the east coast? Explain your answer.

2. The northeast-trending Ramapo Fault crops out north of New York City near the east coast of the United States. Precambrian gneiss forms the hills to the northwest of the fault, and Mesozoic sedimentary rock underlies the lowlands to the southeast. (You can see the fault on Google Earth by going to Lat 41° 10' 21.12" N Long 74° 5' 12.36" W. Once you're there, tilt the image and fly northeast along the fault.) Where the fault crosses the Hudson River, there is an abrupt bend in the river. A nuclear power plant was built near this bend. There are numerous bogs, containing sediments deposited over the past several thousand years, on the surface of the basin. Geologic studies suggest that the Ramapo Fault first formed during the Precambrian, was reactivated during the Paleozoic, and was the site of major displacement during the Mesozoic rifting that separated North America from Africa. Imagine that you are a geologist with the task of determining the seismic risk of the fault. What evidence of present-day or past seismic activity could you look for? What does the long-term geologic history of the fault imply about its strength?

3. On the seismogram of an earthquake recorded at a seismic station in Paris, France, the S-wave arrives 6 minutes after the P-wave. On the seismogram obtained by a station in Mumbai, India, for the same earthquake, the difference between the P-wave and S-wave arrival times is 4 minutes. Which station is closer to the epicenter? From the information provided, can you pinpoint the location of the epicenter? Explain.

THE VIEW FROM SPACE   A view of Scotland in winter shows the rugged highlands blanketed by snow. A narrow, northwest-southeast-trending valley slashes across the landscape. This valley delineates the trace of the Great Glen Fault, a structure that first formed over 200 million years ago. Fracturing made the rock along the fault more susceptible to weathering and erosion, leading to formation of the valley. In places, the valley has filled with water to form deep, elongate lakes. One of these is the infamous Loch Ness, supposedly home to the fictitious "Loch Ness monster." The Great Glen fault is not a present-day plate boundary, so it is not very active, seismically. Nevertheless, it is a crustal weakness, so occasionally small to moderate earthquakes happen along it.

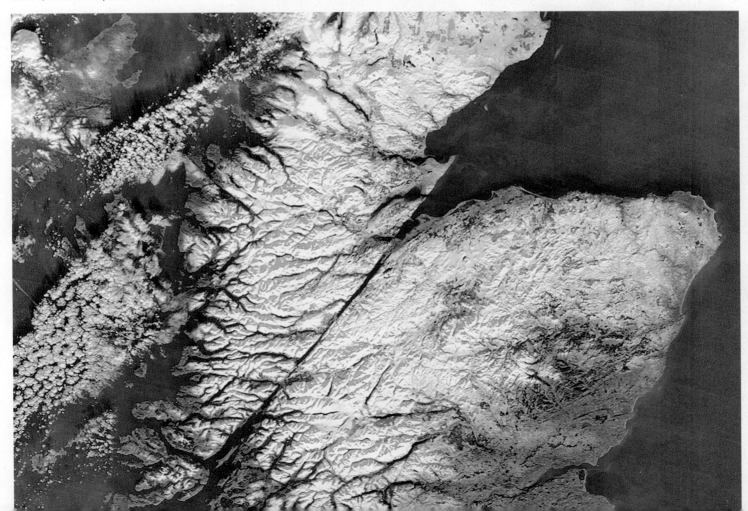

Seeing Inside the Earth

D.1 INTRODUCTION

We live on the Earth's skin and can see light years into space just by looking up. But when we look down—that's another story! We can't use our eyes to look through rock, because it is opaque. So how do we learn about what's inside this planet? Tunneling and drilling aren't much help because they do little more than prick the surface. Fortunately, as discussed in Chapter 1, nineteenth-century geologists realized that measurements of the Earth's mass and shape provide indirect clues to the mystery of what's inside and, from these clues, they could determine that the Earth is not homogeneous but rather consists of three concentric layers: a crust (of low density), a mantle (of intermediate density), and a core (of high density). Further study showed that the crust beneath continents differs from the crust beneath oceans—continental crust consists of a variety of felsic, intermediate, and mafic rock, whereas oceanic crust consists almost entirely of mafic rock (Fig. D.1). However, the determination of the depths of the boundaries between Earth's layers and the division of the layers into sublayers with distinct properties could not be made until the twentieth century, when it became possible to measure seismic waves passing through the Earth. By studying the speed and direction in which seismic waves travel, seismologists can effectively "see" details of our planet's internal layers.

In this interlude, we look at the behavior of seismic waves as they pass through our planet, and we learn how this behavior characterizes Earth's interior. We begin by highlighting a few key points about seismic waves, then move on to the phenomena of wave reflection and refraction. Finally, we witness the discoveries of the different layer boundaries in the Earth. This interlude, incorporating information about earthquakes and seismic waves provided in Chapter 8, completes the journey to the center of the Earth that we began in Chapter 1.

D.2 THE MOVEMENT OF SEISMIC WAVES THROUGH THE EARTH

Wave Fronts and Travel Times

Recall that a sudden rupture of intact rock or the frictional slip of rock on a fault produce seismic waves. These waves move outward from the point of rupture, the earthquake hypocenter, in all directions at once. The boundary between the rock through which a wave has passed and the rock through which it has not yet passed is called a **wave front**. A wave front expands outward from the earthquake focus like a growing bubble. We can represent a succession of waves in a drawing by a series of concentric wave fronts. A line connecting the changing position of an imaginary point on a wave front as the front moves through rock is called a **seismic ray**. Note that seismic rays are perpendicular to wave fronts, so each point on the wave front follows a slightly different ray (Fig. D.2a). The time it takes for a wave to travel from the focus to a seismograph station along a given ray is the **travel time** along that ray.

The ability of a seismic wave to travel through a certain material, and the velocity at which it travels, depend on the character of the material. Factors such as density (mass per unit volume), rigidity (how stiff or resistant to bending a material is), and compressibility (how easily a material's volume changes in response to

FIGURE D.1 The nineteenth-century three-layer image of the Earth.

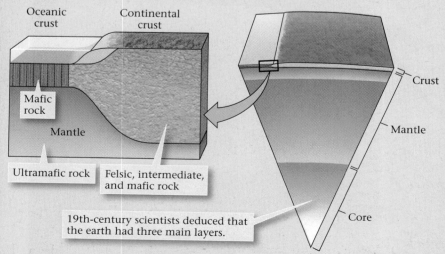

Oceanic crust

Continental crust

Mafic rock

Mantle

Ultramafic rock

Felsic, intermediate, and mafic rock

19th-century scientists deduced that the earth had three main layers.

Crust

Mantle

Core

FIGURE D.2   The propagation of earthquake waves.

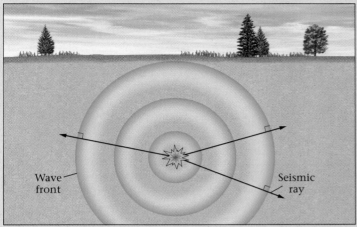

(a) An earthquake sends out waves in all directions.

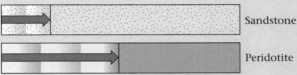

(b) Seismic waves travel at different velocities in different rock types. After a given time, the wave will have traveled further in peridotite than in sandstone.

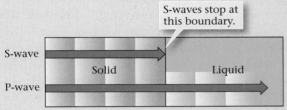

(c) Both P-waves and S-waves can travel through a solid, but only P-waves can travel through a liquid.

(d) P-waves travel faster in solid iron alloy than in liquid, such as molten iron alloy.

squashing) all affect seismic-wave movement. Studies of seismic waves reveal the following:

- Seismic waves travel at different velocities in different rock types (Fig. D.2b). For example, P-waves travel at 8 km per second in peridotite (an ultramafic igneous rock), but at only 3.5 km per second in sandstone (a porous sedimentary rock). Therefore, waves accelerate or slow down if they pass from one rock into another.
- Both P-waves and S-waves can travel through a solid, but only P-waves can travel through a liquid (Fig. D.2c).

- In general, seismic waves travel more slowly in a liquid than in a solid. For example, they travel more slowly in magma than in solid rock, and more slowly in molten iron alloy than in solid iron alloy (Fig. D.2d).

The Reflection and Refraction of Wave Energy

Shine a flashlight into a container of water so that the light ray hits the boundary (or interface) between water and air at an angle. Some of the ray bounces off the water surface and heads back up into the air, while some enters the water (Fig. D.3a). The light ray that enters the water bends at the air-water boundary, so that the angle between the ray and the boundary in the air is different from the angle between the ray and the boundary in the water. Physicists refer to the light ray that bounces off the air-water boundary and heads back into the air as the reflected ray, and the ray that bends at the boundary as the refracted ray. The phenomenon of bouncing off is **reflection**, and the phenomenon of bending is **refraction**. Wave reflection and refraction take place at the interface between two materials, if the wave travels at different velocities through the two materials.

The amount and direction of refraction at a boundary depend on the contrast in wave velocity across the boundary and on the angle at which a wave hits the interface. As a rule, if waves enter a layer through which they travel more slowly, the rays representing the waves bend down and away from the interface (Fig. D.3b). This relation makes sense if you picture a car driving from a paved surface diagonally onto a sandy beach—the wheel that rolls onto the sand first slows down relative to the wheel still on the pavement, causing the car to turn. Alternatively, if the ray were to pass from a layer in which it travels slowly into one in which it travels more rapidly, it would bend up and toward the interface (see Fig. D.3b).

Seismic energy travels in the form of waves, so seismic waves reflect and/or refract when reaching the interface between two rock layers if the waves travel at different velocities in the two layers. For example, imagine a layer of sandstone overlying a layer of basalt. Seismic velocities in sandstone are slower than in basalt, so as seismic waves reach the boundary, some reflect while some refract.

D.3 SEISMIC STUDY OF EARTH'S INTERIOR

Discovering the Crust-Mantle Boundary

The concept that seismic waves refract at boundaries between different layers allowed geologists to define the depth of the crust-mantle boundary. In 1909, Andrija Mohorovičić, a Croatian seismologist, noted that P-waves arriving at seismograph

FIGURE D.3 Refraction and reflection of waves.

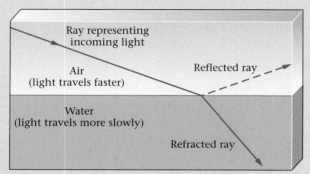

(a) A ray of light reflects and partly refracts when it crosses the boundary between two different materials.

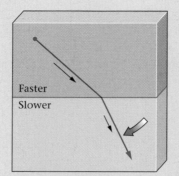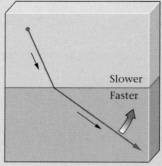

(b) A ray that enters a material through which it travels more slowly bends away from the boundary. A ray that enters a faster medium bends toward the boundary.

stations less than 200 km from the epicenter traveled at an average speed of 6 km per second, whereas P-waves arriving at seismographs more than 200 km from the epicenter traveled at an average speed of 8 km per second. To explain this observation, he suggested that the first P-waves reaching *nearby* seismographs followed a short, shallow path through the crust, a substance through which they travel more slowly. The first P-waves to reach *distant* seismographs, however, followed a deeper path through the mantle and refracted at the crust-mantle boundary; thus, for most of their journey, they traveled more rapidly (Fig. D.4a, b).

Calculations based on this observation require the crust-mantle boundary beneath continents to be at a depth of about 35 to 40 km. As we learned in Chapter 1, this boundary is now called the **Moho**, in honor of Mohorovičić.

Defining the Structure of the Mantle

By studying travel times, seismologists have determined that seismic waves travel at different velocities at different depths in the mantle. Between a depth of about 100 and 200 km in the mantle beneath oceanic lithosphere, seismic velocities are slower than in the overlying lithospheric mantle (Fig. D.5). This 100- to 200-km deep layer is called the **low-velocity zone**; here, the prevailing temperature and pressure conditions cause peri-

dotite to partially melt by up to 2%. The melt, a liquid, coats solid grains and fills voids between grains. Because seismic waves travel more slowly through liquids than through solids, the coatings of melt slow seismic waves down. In the context of plate tectonics theory, the low-velocity zone is the weak layer on which oceanic lithosphere plates move. Below the low-velocity zone, the mantle does not contain melt. (Seismologists do not find a well-developed low-velocity zone beneath continents.)

Below a depth of about 200 km, seismic-wave velocities in the mantle increase with depth. Seismologists interpret this increase to mean that mantle peridotite becomes progressively less compressible, more rigid, and denser with depth. This proposal makes sense, considering that the weight of overlying rock increases with depth, and as pressure increases, the atoms making up rock squeeze together more tightly and are not as free to move. Because of refraction, the progressive increase in seismic velocity with depth causes seismic rays to curve in the mantle. To understand the shape of a curved ray, look at Figure D.6a, which represents a portion of the mantle by a series of imaginary layers, each permitting a slightly greater seismic-wave velocity than the layer above. Every time a seismic ray crosses the boundary between adjacent layers, it refracts a little toward the boundary. After the ray has crossed several layers, it has bent so much that it begins to head back up toward the top of the stack. Now if we replace the stack of distinct layers with a single layer in which

FIGURE D.4 Discovery of the Moho.

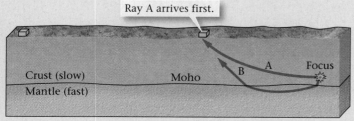

(a) Seismic waves traveling only in the crust reach a nearby seismograph first.

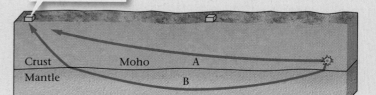

(b) Seismic waves traveling for most of their path in the mantle reach a distant seismograph first.

FIGURE D.5 The velocity of P-waves in the mantle changes with depth.

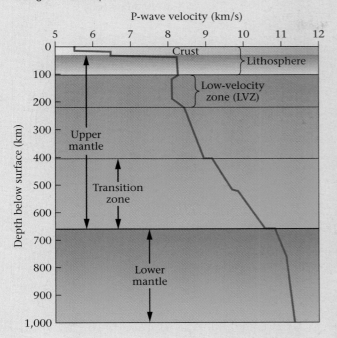

FIGURE D.6 Curving of seismic waves.

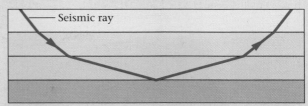

(a) In a stack of discrete layers, rays bend at each boundary. If the velocity is progressively faster in each lower layer, the ray eventually bends back and returns to the surface.

(b) In a material in which the velocity increases gradually with depth, rays curve smoothly and eventually return to the surface

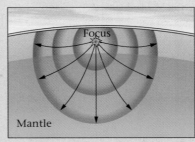

(c) The velocity of seismic waves increases with depth in the mantle, so rays curve and wave fronts are oblong.

velocity increases gradually with depth at a constant rate, the wave follows a smoothly curving path (**Fig. D.6b, c**).

At depths between 410 km and 660 km (see Fig. D.5), seismic velocity increases in a series of abrupt steps. Experiments suggest that these **seismic-velocity discontinuities** mark depths at which phase changes occur, meaning that pressure causes atoms in minerals to rearrange and pack together more tightly, thereby changing the rock's density, compressibility, and rigidity. Researchers are still trying to determine if chemical changes also occur at the discontinuities. Because of these seismic-velocity discontinuities, as we learned in Chapter 1 but now can see more clearly, seismologists subdivide the mantle into the **upper mantle** (above 410 km), the **transition zone** (between 410 and 660 km), and the **lower mantle** (below 660 km).

Discovering the Core-Mantle Boundary

During the first decade of the twentieth century, seismologists installed seismographs at many stations around the world, expecting to be able to record waves produced by a large earthquake anywhere on Earth. In 1914, one of these seismologists, Beno Gutenberg, discovered that P-waves from an earthquake do not arrive at seismographs lying in a band between 103° and 143° from the earthquake epicenter, as measured along the circumference of the Earth. This band is now called the **P-wave shadow zone** (**Fig. D.7a**). If the density of the Earth increased gradually with depth all the way to the center, the shadow zone would not exist, because rays passing into the

interior would curve up and reach every point on the surface. Thus, the presence of a shadow zone means that deep in the Earth a major interface exists where seismic waves *abruptly* refract down (implying that the velocity of seismic waves suddenly decreases). This interface, now called the **core-mantle boundary**, lies at a depth of about 2,900 km.

To see why the P-wave shadow zone exists, follow the two seismic rays labeled A and B in Figure D.7a. Ray A curves smoothly in the mantle (we are ignoring seismic-velocity discontinuities in the mantle) and passes just above the core-mantle boundary before returning to the surface. It reaches the surface 103° from the epicenter. In contrast, Ray B just penetrates the boundary and refracts down into the core. Ray B then curves through the core and refracts again when it crosses back into the mantle. As a consequence, Ray B intersects the surface at more than 143° from the epicenter.

The downward bending of seismic waves when they pass from the mantle down into the core indicates that seismic velocities in the core are slower than in the mantle. Thus, even though the core is deeper and denser than the mantle, the outer part of the core must be less rigid than the mantle.

FIGURE D.7 Shadow zones and the discovery of the Earth's core.

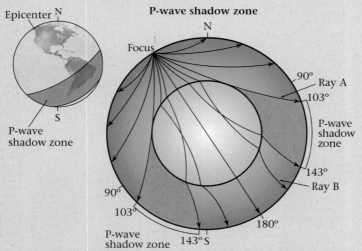

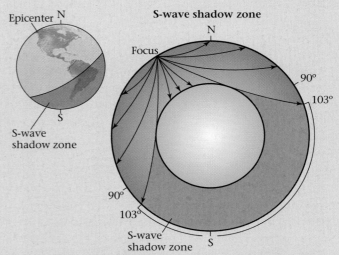

(a) P-waves do not arrive in the P-wave shadow zone, because they refract at the core-mantle boundary.

(b) S-waves do not arrive in the S-wave shadow zone, because they cannot pass through the liquid outer core.

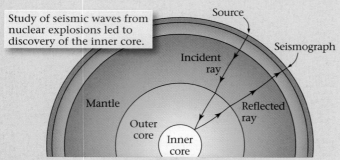

Study of seismic waves from nuclear explosions led to discovery of the inner core.

(c) Seismic waves reflect off the inner core–outer core boundary.

Discovering the Nature of the Core

Geologists had inferred, based on measurements of the Earth's density and on the study of meteorites, that the Earth's core consists of iron alloy. But it was not until the modern study of how seismic waves pass through the interior of the Earth that it became possible to answer a key question: Is the core, or at least part of it, solid or liquid? A study of S-waves gave seismologists the answer. They found that S-waves do not arrive at stations located between 103° and 180° from the epicenter (a band called the **S-wave shadow zone**). This means that S-waves cannot pass through the core at all—otherwise, an S-wave headed straight down through the Earth would appear on the other side. Remember that S-waves can travel only through solids. Thus, the fact that S-waves do not pass through the core means that the core, or at least part of it, consists of liquid (Fig. D.7b).

At first, seismologists thought that the entire core might be liquid iron alloy. But in 1936, a Danish seismologist, Inge Lehmann, discovered that P-waves passing through the core reflected off a boundary within the core. She then proposed that the core is made up of two parts: an **outer core** consisting of liquid iron alloy and an **inner core** consisting of solid

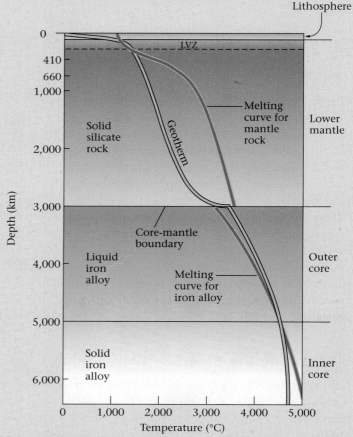

(d) A graph of the geotherm and melting curve for the Earth. Note that the melting temperature is less than the Earth's temperature in the outer core, so the outer core is molten.

iron alloy. Lehmann's work defined the existence of the inner core but could not locate the depth at which the inner core–outer core interface occurs. This depth was eventually located by measuring the exact time it took for seismic waves generated by nuclear explosions to penetrate the Earth, bounce off the inner core–outer core boundary, and return to the surface (Fig. D.7c). The measurements showed that the inner core–outer core boundary occurs at a depth of 5,155 km.

Why does the core have two layers—a liquid outer layer and a solid inner one? The answer lies in a graph that plots the temperature of the Earth, and the melting temperature of Earth materials, against increasing depth (Fig. D.7d). In the outer core, iron alloy is above its melting temperature, whereas in the inner core, iron alloy is below its melting temperature. Pressure is so great in the inner core that iron atoms lock into solid crystals, even though the temperature is very high. Of note, recent studies of seismic waves passing through the inner core have shown that the inner core rotates slightly faster than the rest of the Earth—it makes an extra revolution about once every twenty-five years.

A Modern Image of Earth's Layers

Through painstaking effort, seismologists have developed a graph, known as a **velocity-versus-depth curve**, that shows the depths at which seismic velocity suddenly changes and thus helped to establish the principal layers and sublayers in the Earth (Fig. D.8). This graph is an average for the whole Earth. Modern research now focuses on developing refined graphs for specific locations.

D.4 FINE-TUNING OUR IMAGE OF THE EARTH'S INTERIOR

Seismic Tomography

In recent years, sophisticated computers have been able to use global seismic data to create a three-dimensional image of seismic-wave velocities within the Earth. This type of analysis, known as **seismic tomography**, resembles the method employed by a medical CAT-scan machine (*CAT* stands for computer-aided tomography). Tomography (from the Greek word for slice) allows us to picture variations in seismic velocities visible on slices through the Earth. In seismic tomography studies, researchers compare travel times for seismic waves that follow different paths through the Earth, and they distinguish regions in which waves travel unexpectedly fast from regions where waves travel unexpectedly slowly as compared with the average model of Figure D.8. Overall, tomographic studies show that a layered image of the Earth like that depicted in Figure D.8, though reasonable as a rough approximation, is not accurate in detail. Rather, it appears that there can be different seismic velocities at a given depth in the Earth (Fig. D.9a).

Why are there distinct zones of lower-than-expected velocities and distinct zones of higher-than-expected velocities

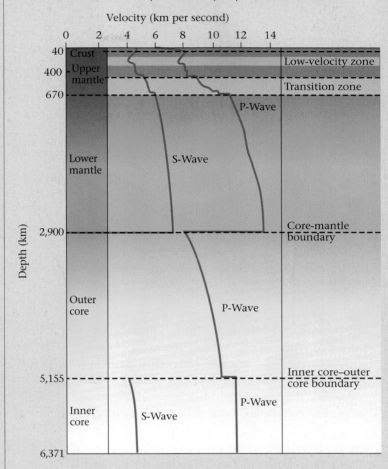

FIGURE D.8 The velocity-versus-depth profile of the Earth.

at the same depth in the mantle? The velocity contrasts indicate that there are blotchy variations in the nature of materials within the mantle. Seismologists suggest that these contrasts reflect temperature variations (hotter rocks are less rigid than cooler rocks) and/or compositional variations. If they represent temperature variations, then the distribution of hotter and cooler zones in the mantle supports the theory that convection occurs in the mantle—the hotter zones are rising, and the cooler zones are sinking. Seismic-tomography studies can also detect remnants of subducted oceanic plates (Fig. D.9b). Overall, tomographic images suggest that the interior of the Earth should be thought of as a dynamic system (Fig. D.9c).

Seismic-Reflection Profiling

Seismic techniques are also letting us fine-tune our image of the crust. During the past half century, geologists have found that by using dynamite, by banging large weights against the Earth's surface, or by releasing bursts of compressed air into the ocean, they can create artificial seismic waves that propagate down into the Earth and reflect off the boundaries between different layers of rock in the crust. By recording the time at which these

FIGURE D.9   Tomographic images of the Earth's interior, and their interpretation.

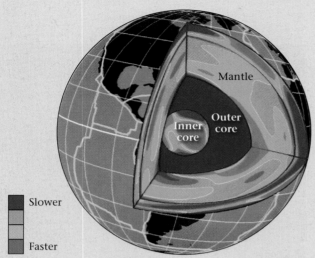

Slower

Faster

(a) Tomographic image of the Earth. Slower areas may be relatively cooler than their surroundings, and faster areas may be relatively warmer.

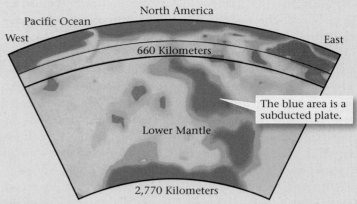

The blue area is a subducted plate.

(b) Closeup tomographic image of the mantle beneath North America and the Pacific Ocean.

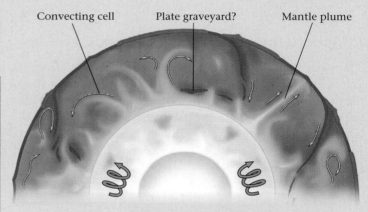

Convecting cell   Plate graveyard?   Mantle plume

(c) The modern view of the Earth's complex and dynamic interior.

reflected waves return to the surface, geologists can calculate the depth to the layers, and can produce a cross-sectional view of the crust called a **seismic-reflection profile** (Fig. D.10a, b). This image defines the depths at which specific strata occur and reveals the presence of subsurface folds (bends in layers) and faults. Oil companies must obtain seismic-reflection profiles, despite their high cost, because they allow geologists to identify likely locations for oil and gas.

Recently, computers have become so sophisticated that geologists can produce three-dimensional seismic-reflection images of the crust. These provide so much detail that geologists can trace out a ribbon of sand representing the channel of an ancient stream even where the sand lies buried kilometers below the Earth's surface.

New Research Directions

The deep interior of the Earth has become one of the most active research areas of recent decades, as geologists have developed exotic new laboratory techniques to study the nature of materials under very high pressures and temperatures. And new data from satellites allow researchers to link the strength of gravitational pull at the Earth's surface to features in the mantle below (Box D.1).

A major new research initiative, called *EarthScope*, involves placing hundreds of seismographs in a grid across swaths of the United States. Seismologists hope that data collected from these instruments will enable them to greatly refine images of the interior and test current models of it.

Key Terms

core-mantle boundary (p. 235)
geoid (p. 239)
inner core (p. 236)
lower mantle (p. 235)
low-velocity zone (p. 234)
Moho (p. 234)
outer core (p. 236)
P-wave shadow zone (p. 235)
reflection (p. 233)
refraction (p. 233)
seismic ray (p. 232)

seismic-reflection profile (p. 238)
seismic tomography (p. 237)
seismic-velocity discontinuities (p. 235)
S-wave shadow zone (p. 236)
transition zone (p. 235)
travel time (p. 232)
upper mantle (p. 235)
velocity-versus-depth curve (p. 237)
wave front (p. 232)

FIGURE D.10 Seismic-reflection profiling.

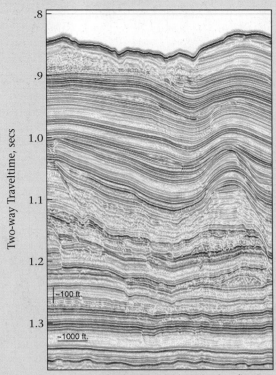

(a) Trucks thumping on the ground to generate the signal needed for making a seismic-reflection profile.

(b) After computer analysis, the data yield a seismic-reflection profile. Color bands represent horizons in the stratigraphic sequence.

Is the Earth Really Round?

In Chapter 1, we noted that as planetesimals grow into planets they eventually get large enough, warm enough, and soft enough for gravitational force to transform them from irregular shapes into spheres. On a perfect sphere, the pull of gravity is exactly the same at all points on the surface.

But is the Earth really a *perfect* sphere? The answer is No! To start with, Earth's rotation produces centrifugal force, which flattens the planet and causes the radius from the equator to the center (6,378 km, or 3,963 miles) to exceed the radius from the pole to the center (6,357 km, or 3,950 miles). But flattening due to centrifugal force isn't the whole story. Satellite measurements show that the Earth's surface actually has broad bumps and dimples. If these irregularities were greatly exaggerated, our planet would look somewhat like a warped pear (**Fig. 1**). This shape, a more accurate representation of the Earth's surface, is called the **geoid**.

You can visualize the geoid as the shape that sea level would have if the Earth were covered by a global ocean and sea level depended on the pull of gravity. Where gravity is stronger, the surface sinks lower (for gravity pulls the surface down), and where gravity is weaker, the surface rises higher. The difference between the highest point on the geoid and the lowest point is about 210 m. High areas may lie over warmer asthenosphere, where mantle is upwelling, and lower layers over cooler asthenosphere, where mantle is sinking.

FIGURE 1 An exaggerated representation of the geoid showing how the Earth's surface is distorted by highs and lows in the gravity field.

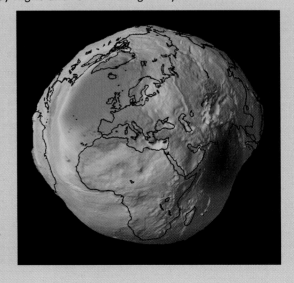

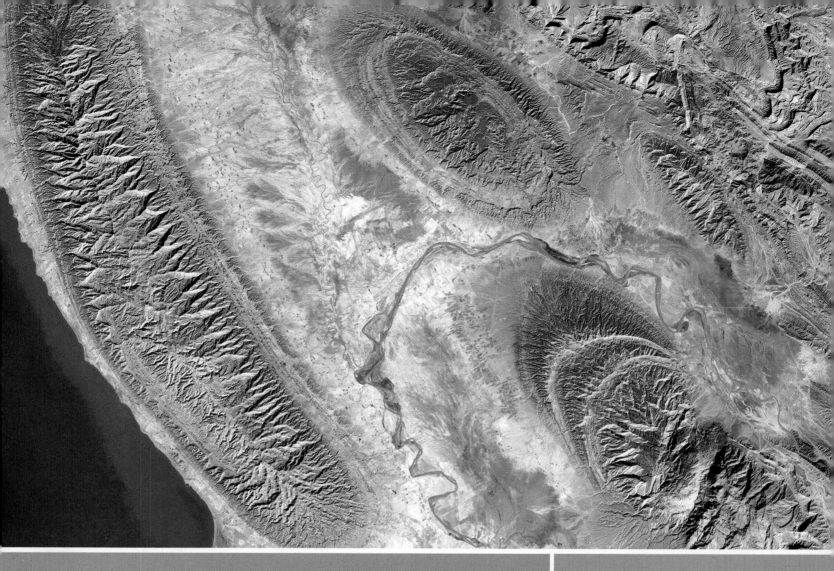

CHAPTER **9**

Crags, Cracks, and Crumples: Crustal Deformation and Mountain Building

Subduction of the Persian Gulf beneath the Asian continent, in the vicinity of Konari, Iran, produced large folds. In this satellite view, dipping strata outline the ellipsoidal shape of the folds. The large fold near the coast is about 10 km wide. Some of the folds are cored by salt. In the fold at the lower right, gray salt emerged at the surface.

GEOPUZZLE

Mountain belts include some of the most spectacular scenery on Earth. Why do mountain ranges form? Why are the rocks within these ranges bent and broken? And how long can a mountain range survive?

Innumerable peaks, black and sharp, rose grandly into the dark blue sky, their bases set in solid white, their sides streaked and splashed with snow, like ocean rocks with foam. . . . [Mountains] are nature's poems carved on tables of stone. . . . How quickly these old monuments excite and hold the imagination!

—John Muir, from *Wilderness Essays*

9.1 INTRODUCTION

Geographers call the peak of Mt. Everest "the top of the world," for this mountain, which lies in the Himalayas of south Asia, rises higher than any other on Earth. The cluster of flags on Mt. Everest's summit flap at 8.85 km (29,029 feet) above sea level—almost the cruising height of modern jets. No one can survive very long at the top, for the air there is too thin to breathe. In 1953, the British explorer Sir Edmund Hillary and Tenzing Norgay, a Nepalese guide, became the first to reach the summit. By 2008, more than 2,000 other people had also succeeded—but almost 200 people have died trying.

To geologists, mountains provide one of the most obvious indications of dynamic activity on Earth. To make a mountain, Earth forces must lift cubic kilometers of rock skyward against the pull of gravity. With the exception of the large volcanoes formed over hot spots, mountains do not occur in isolation, but rather as part of linear ranges variously called mountain belts or **orogens** (from the Greek words *oros*, meaning mountain, and *genesis*, meaning formation). Geographers define about a dozen major orogens and numerous smaller ones worldwide (**Fig. 9.1**).

The process of forming a mountain belt not only raises the surface of the crust, a process called **uplift**, but also causes rocks to undergo **deformation**, a process by which rocks bend or break in response to compression, tension, or shearing. Deformation produces geologic structures, including **joints** (cracks), **faults** (fractures on which one body of rock slides past another), **folds** (bends or wrinkles), and **foliation** (layering resulting from the alignment of mineral grains or the development of compositional bands). Mountain building may also involve metamorphism and igneous activity.

A mountain-building event, or orogeny, may last for tens of millions of years. As the land rises, erosion starts to grind it away, producing sediment and sculpting awesome, jagged topography. When the process of uplift ceases, erosion can bevel a range down to near sea level. At that point, only a low-lying belt of fractured, contorted, and metamorphosed rock remains—such crustal scars serve as a permanent monument to what had once been a region of high peaks.

In this chapter, we learn about geologic structures and other phenomena that happen during mountain building (see **Geotour 9** on p. GT-20). We will also learn why mountains and structures form, in the context of plate tectonics theory.

9.2 ROCK DEFORMATION IN THE EARTH'S CRUST

Deformation and Strain

As noted above, orogeny causes deformation (bending, breaking, shortening, stretching, or shearing), which in turn yields geologic structures. To get a visual sense of deformation, let's compare a road cut along a highway in the central Great Plains of North America, a region that has not undergone orogeny, with a cliff in the Alps (**Fig. 9.2a, b**).

The road cut, which lies at an elevation of only about 100 m above sea level, exposes nearly horizontal beds of sandstone, shale, and limestone—these beds have the same orientation that they had when first deposited. Sand grains in sandstone beds of this outcrop have a nearly spherical shape (the same shape they had when deposited), and clay flakes in the shale lie roughly parallel to the bedding, because of compaction. Rock of this outcrop is undeformed, meaning that it contains no geologic structures other than a few joints.

In an Alpine cliff, exposed at an elevation of 3 km, rocks look very different. Here, we find layers of quartzite, slate, and marble (the metamorphic equivalent of sandstone, shale, and limestone) in contorted beds whose shapes resemble the wrinkles in a rug that has been pushed across the floor. These wrinkles are folds. Grains in the quartzite may resemble flattened eggs, and the clay flakes in the slate are aligned at a steep angle to the bedding. In fact, the rock splits on planes called cleavage that parallel the clay flakes and cut across the bedding. Finally, if we try tracing the quartzite and slate layers along the outcrop face, we might find that they abruptly terminate at a sloping surface marked by broken-up rock. This surface is a fault. In this example, the quartzite and slate slid along the fault from where they first formed to get to their present location on top of marble layers.

FIGURE 9.1 Digital map of world topography, showing the locations of major mountain ranges.

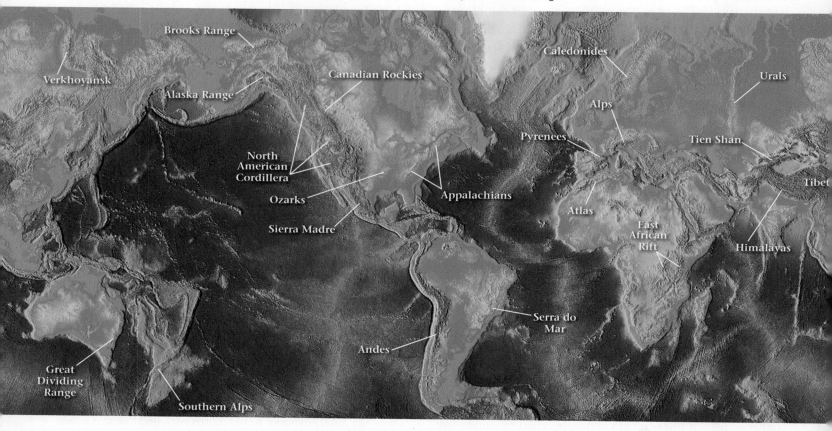

Clearly, the beds in the Alpine cliff have been deformed, and as a result the cliff exposes a variety of geologic structures. Beds no longer have the same shape and position that they had when first formed, and the shape and orientation of grains has changed. In sum, deformation includes one or more of the following (Fig. 9.3a–c): (1) a change in location (displacement); (2) a change in orientation (rotation); and (3) a change in shape (distortion). Deformation can be fairly obvious when observed in an outcrop.

Geologists refer to the *change in shape* that deformation causes as **strain**. We distinguish among different kinds of strain according to how the rock changes shape. If a layer of rock becomes longer, it has undergone stretching, but if the layer becomes shorter, it has undergone shortening (Fig. 9.4a–c). If

FIGURE 9.2 Deformation changes the character and configuration of rocks.

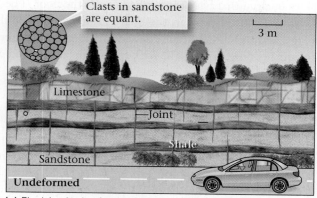

(a) Flat-lying beds of strata along a highway in the Great Plains of North America are essentially undeformed. A few joints, formed when overlying rock eroded away, are visible.

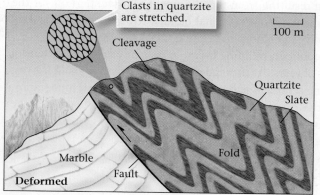

(b) In a mountain belt, deformation may cause layers to undergo folding and faulting. In addition, foliation (such as slaty cleavage) and stretched clasts may develop.

a change in shape involves the movement of one part of a rock body past another so that angles between features in the rock change, the result is called shear strain (Fig. 9.4d, e).

Brittle versus Ductile Deformation

Imagine that a plate tumbles off a table and lands on a hard floor—the plate breaks and shatters into pieces. Similarly, if you strike a glass window with a ball, the window cracks and may even shatter. All such phenomena serve as familiar examples of **brittle deformation** (Fig. 9.5a, b). Now, imagine that you squeeze a ball of soft dough between a book and a table top—the dough flattens into a pancake. Similarly, if you bend a stick of chewing gum, it changes from a plane into a curve. During such **ductile deformation**, objects change shape without visibly breaking (Fig. 9.5c, d).

What actually happens within mineral grains during these two different kinds of deformation? Recall that the atoms that make up mineral grains are connected by chemical bonds. During brittle deformation, many of the bonds break and stay broken, leading to the formation of a permanent crack across which material no longer connects. During ductile deformation, simplistically, some bonds break but new ones quickly form. In this way, the atoms within grains rearrange, and the grains change shape without permanent cracks forming.

Why do rocks inside the Earth sometimes deform brittlely and sometimes ductilely? The behavior of a rock depends on:

- *Temperature*: Warm rocks tend to deform ductilely, whereas cold rocks tend to deform brittlely. Heat makes materials softer.
- *Pressure*: Under great pressures deep in the Earth, rock behaves more ductilely than it does under low pressures near the surface. Pressure effectively prevents rock from separating into fragments.

FIGURE 9.3 The components of deformation include displacement, rotation, and distortion.

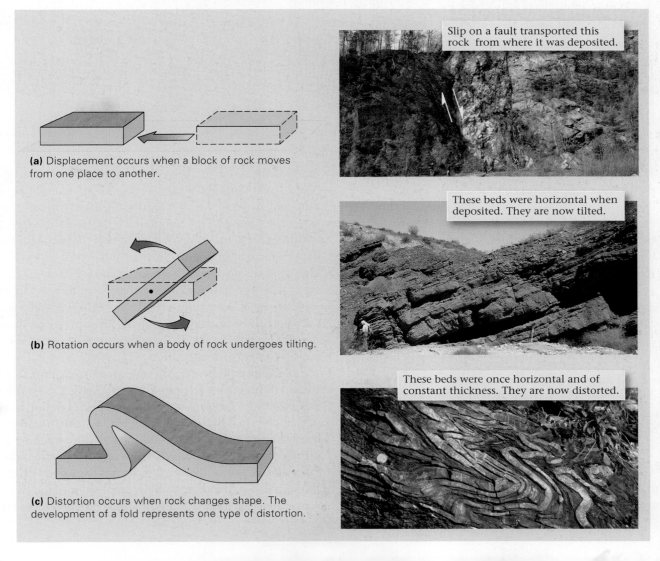

(a) Displacement occurs when a block of rock moves from one place to another.

(b) Rotation occurs when a body of rock undergoes tilting.

(c) Distortion occurs when rock changes shape. The development of a fold represents one type of distortion.

Slip on a fault transported this rock from where it was deposited.

These beds were horizontal when deposited. They are now tilted.

These beds were once horizontal and of constant thickness. They are now distorted.

FIGURE 9.4 Different kinds of strain in rock. Strain is a measure of the distortion, or change in shape, that takes place in rock during deformation.

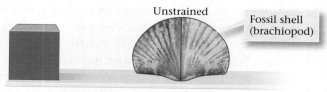

(a) An unstrained cube and an unstrained fossil shell (brachiopod).

(b) Horizontal stretching changes the cube into a horizontal brick and elongates the shell.

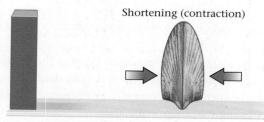

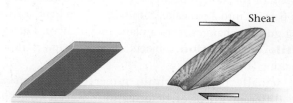

(c) Horizontal shortening changes the cube into a vertical brick and makes the shell narrower.

(d) Shear strain tilts the cube and transforms it into a parallelogram, and changes angular relationships in the shell.

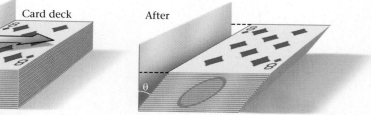

(e) You can produce shear strain by moving a deck of cards so that each card slides a little more with respect to the one below.

- *Deformation rate*: A sudden change in shape causes brittle deformation, whereas a slow change in shape causes ductile deformation. For example, if you hit a thin marble bench with a hammer, it shatters, but if you leave the bench alone for a century, it gradually sags without breaking.

- *Composition*: Some rock types are softer than others; for example, halite (rock salt) deforms ductilely under conditions in which granite deforms brittlely.

Considering that pressure and temperature both increase with depth in the Earth, geologists find that in typical continental crust, rocks generally behave brittlely above about 10 to 15 km, and ductilely below this depth; we call this zone the brittle-ductile transition. Earthquakes in continental crust only happen above this depth because these earthquakes involve brittle breaking.

Force, Stress, and the Causes of Deformation

Up to this point, we've focused on picturing the consequences of deformation. Describing the causes of deformation is a bit more challenging in the context of an introductory geology book. In captions for displays about mountain building, museums and national parks typically dispense with the issue by using the phrase "The mountains were caused by forces deep within the Earth." But what does this mean? Isaac Newton defined force by noting that if you apply a force to an object, the object speeds up, slows down, or changes direction. Applying this concept to geology, we see that phenomena such as plate interactions and continent-continent collisions apply forces to rock and thus cause rock to change location, orientation, or shape. In other words, the application of forces in the Earth indeed causes deformation.

Geologists, however, use the word *stress* instead of *force* when talking about the cause of deformation. We define the **stress** acting on a plane as the force applied *per unit area* of the plane. The need to distinguish between stress and force arises because the actual consequences of applying a force depend not just on the amount of force but also on the area over which the force acts. A pair of simple experiments shows why (Fig. 9.6a). Experiment 1: Stand on a single, empty aluminum can. All of your weight—a force—focuses entirely on the can, and the can

FIGURE 9.5 Brittle versus ductile deformation.

(a) Brittle deformation occurs when you drop a plate and it shatters.

Cracks separate the plate into pieces.

Crack (joint)

(b) Cracks in an Australian outcrop are due to brittle deformation.

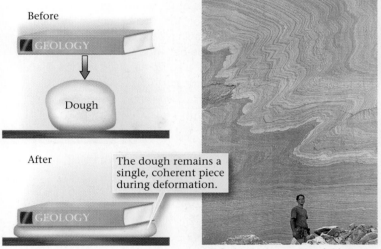

(c) Ductile deformation occurs when you squash a ball of dough.

The dough remains a single, coherent piece during deformation.

(d) Swirls in the marble of this Brazilian cliff formed ductilely.

crushes. Experiment 2: Place a board atop 100 cans and stand on the board. In this case, your weight is distributed across 100 cans, and the cans don't crush. In both experiments, the force caused by the weight of your body was the same, but in Experiment 1 the force was applied over a small area so a large stress developed, whereas in Experiment 2 the same force was applied over a large area so only a small stress developed. How does this concept apply to geology? During mountain building, the force of one plate interacting with another is distributed across the area of contact between the two plates, so the deformation resulting at any specific location actually depends on the stress developed at that location, not on the total force involved in the plate interaction.

Different kinds of stress occur in rock bodies (Fig. 9.6b–e). **Compression** takes place when a rock is squeezed, **tension** occurs when a rock is pulled apart, and **shear stress** develops when one side of a rock body moves sideways past the other side. **Pressure** refers to a special stress condition in which the same push acts on all sides of an object.

Note that "stress" and "strain" have different meanings to geologists, even though we tend to use them interchangeably in everyday English: stress refers to the amount of force applied per unit area of a rock, whereas strain refers to the change in shape of a rock. Thus, stress *causes* strain. Specifically, compression causes shortening, tension leads to stretching, and shear stress produces shear strain. Pressure can cause an object to become smaller, but will not cause it to change shape. With our knowledge of stress and strain, we can now look at the nature and origin of various classes of geologic structures.

TAKE-HOME MESSAGE

Mountain building produces stress (compression, tension, shear), which in turn causes deformation. Stress breaks rock under brittle conditions or bends and distorts rock, without breaking it, under ductile conditions. Strain is a measure of distortion.

9.3 JOINTS AND VEINS: NATURAL CRACKS IN ROCKS

If you look at the photographs of rock outcrops in this book, you'll notice thin black lines that cross the rock faces. These lines represent traces of natural cracks along which the rock broke and separated into two pieces during brittle deformation. Geologists refer to such natural cracks as **joints** (Fig. 9.7a, b). Rock bodies do *not* slide past each other on joints. (Since joints are roughly planar structures, we define their orientation by their strike and dip, as described in Box 9.1).

Joints develop in response to tensile stress in brittle rock: a rock splits open because it has been pulled slightly apart. Joints may form for a variety of geologic reasons. For example, some joints form when a rock cools and contracts, because contraction makes one part of a rock pull away from the adjacent part. Others develop when rock formerly at depth undergoes a decrease in pressure as overlying rock erodes away, and thus changes shape slightly. Still others form when rock layers bend.

If groundwater seeps through joints for a long period of time, minerals such as quartz or calcite may precipitate out of the groundwater and fill the joint. Such mineral-filled joints are called **veins** and look like white stripes cutting across a body of rock (Fig. 9.7c). Some veins contain small quantities of valuable metals, such as gold.

BOX 9.1

Describing the Orientation of Structures

When discussing geologic structures, it's important to be able to communicate information about their orientation. For example, does a fault exposed in an outcrop at the edge of town continue beneath the nuclear power plant 3 km to the north, or does it go beneath the hospital 2 km to the east? If we knew the fault's orientation, we might be able to answer this question. To describe the orientation of a geologic structure, geologists picture the structure as a simple geometric shape, then specify the angles that the shape makes with respect to a horizontal plane (a flat surface parallel to sea level), a vertical plane (a flat surface

perpendicular to sea level), and the north direction (a line of longitude).

Let's start by observing *planar* structures such as faults, beds, and joints. We call these structures planar because they resemble a geometric plane. A planar structure's orientation can be specified by its strike and dip. The **strike** is the angle between an imaginary horizontal line (the strike line) on the structure and the direction to true north (**Fig. 1a, b**). We measure the strike with a special type of compass (**Fig. 1c**). The **dip** is the angle of the structure's slope—more precisely, the angle between a horizon-

tal plane and the dip line, an imaginary line parallel to the steepest slope on the structure, as measured in a vertical plane perpendicular to the strike. We measure the dip angle with a clinometer, a type of protractor that measures slope angles. A horizontal plane has a dip of 0°, and a vertical plane has a dip of 90°. We represent strike and dip on a geologic map using the symbol shown in Figure 1b.

A *linear* structure resembles a line rather than a plane; examples of linear structures include scratches or grooves on a rock surface. Geologists specify the orientation of linear structures by giving their plunge and bearing (**Fig. 1d**). The plunge is the angle between a line and horizontal in the vertical plane that contains the line. A horizontal line has a plunge of 0°, and a vertical line has a plunge of 90°. The bearing is the compass heading of the line—more precisely, the angle between the projection of the line on the horizontal plane and the direction to true north.

FIGURE 1 Specifying the orientation of planar and linear structures.

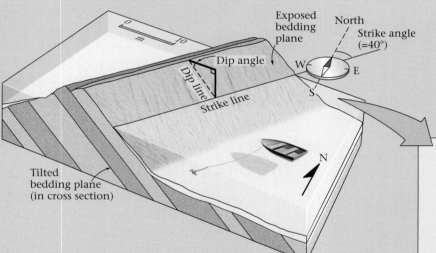

(a) We use strike and dip to measure the orientation of planar structures such as these tilted beds.

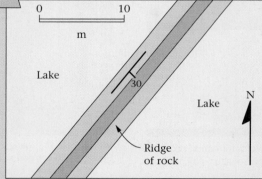

(b) On a map, the line segment represents the strike direction and the tick on the segment represents the dip direction. The number indicates the dip angle as measured in degrees.

(c) Geologists use a Brunton compass to measure strike and dip.

The edge of the compass is parallel to the strike.

The intersection of the water and the rock is horizontal.

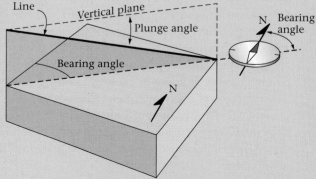

(d) To specify the orientation of a line, we use plunge and bearing.

Geotechnical engineers, people who study the geologic setting of construction sites, pay close attention to jointing when recommending where to put roads, dams, and buildings. Water flows much more easily through joints than it does through solid rock, so it would be a bad investment to situate a water reservoir over rock containing lots of joints—the water would leak down into the joints. Also, building a road on a steep cliff composed of jointed rock could be risky, for joint-bounded blocks separate easily from bedrock, and the cliff might collapse.

TAKE-HOME MESSAGE

Joints are cracks that form when rock undergoes slight stretching. Mineral-filled cracks are veins. Joints affect the strength of outcrops and the ability of water to flow through rock.

9.4 FAULTS: SURFACES OF SLIP

After the San Francisco earthquake of 1906, geologists found a rupture that had torn the landscape near the city. Where this rupture crossed orchards, it offset rows of trees, and where it crossed a fence, it broke the fence in two; the western side of the fence moved northward by about 2 m (see Fig. 8.4a). The rupture represents the trace of the San Andreas Fault. As we have seen, a **fault** is a fracture on which sliding occurs. Slip events, or faulting, generate earthquakes. Faults, like joints, are planar structures, so we represent their orientation by strike and dip.

Faults riddle the Earth's crust. Some are currently active in that sliding has been occurring on them in recent geologic time, but most are inactive, meaning that sliding on them ceased millions of years ago. Some faults, such as the San Andreas, intersect the ground surface and thus displace the ground when they move. Others accommodate the sliding of rocks in the crust at depth and remain invisible at the surface unless they are later exposed by erosion.

FIGURE 9.6 There are several kinds of stress.

A force applied to a small area (the top of a can) produces a large stress, so the can crushes.

The same force applied to a large area (many cans) produces a small stress, so the cans support your weight.

(a) The difference between stress and force. Stress is the force per unit area.

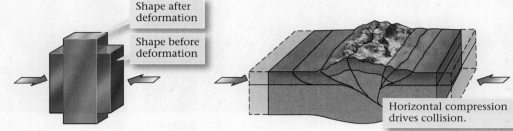

Shape after deformation

Shape before deformation

Horizontal compression drives collision.

(b) Compression takes place when an object is squeezed.

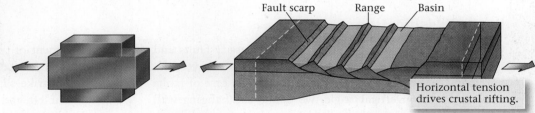

Fault scarp   Range   Basin

Horizontal tension drives crustal rifting.

(c) Tension occurs when the opposite ends of an object are pulled in opposite directions.

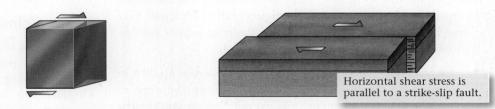

Horizontal shear stress is parallel to a strike-slip fault.

(d) Shear stress develops when one surface of an object slides relative to the other surface.

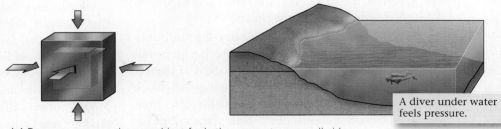

A diver under water feels pressure.

(e) Pressure occurs when an object feels the same stress on all sides.

FIGURE 9.7 Examples of joints and veins.

(a) Prominent vertical joints cut red sandstone beds in Arches National Park, Utah, as seen from the air.

(b) Vertical joints on a cliff face in shale near Ithaca, New York.

(c) Milky white quartz veins cut across gray limestone beds.

Fault Classification

Geologists have developed terminology to classify faults and describe movement on them. The fault plane can be vertical, horizontal, or at some angle in between. In the case of faults that dip at a nonvertical angle, we define the hanging-wall block as the rock above the fault plane, and the footwall block as the rock below the fault plane (Fig. 9.8a). If you stand in a tunnel along a fault plane, the hanging-wall block looms over your head, and the footwall block lies under your feet. We distinguish several types of faults (Fig. 9.8b–d).

- *Dip-slip versus strike-slip versus oblique-slip faults*: On dip-slip faults, sliding occurs up or down the slope of the fault (therefore, up or down the dip); on **strike-slip faults**, one block slides past another horizontally (therefore, parallel to the strike line); and on oblique-slip faults, sliding occurs diagonally on the fault plane.

- *Types of dip-slip faults*: We subdivide dip-slip faults into two kinds, depending on which way the hanging-wall block moves relative to the footwall block. On **thrust faults** and **reverse faults**, the hanging-wall block moves up the slope of the fault. Thrust faults differ from reverse faults only in terms of the fault plane's slope (or dip)—thrust faults have a dip of less than about 30°, whereas reverse faults have a dip of more than 30°. On **normal faults**, the hanging-wall block moves down the slope of the fault.

- *Types of strike-slip faults*: Geologists distinguish between two types of strike-slip faults, based on the relative

movement of one side of the fault with respect to the other. If you stand facing the fault, you can say it is a left-lateral strike-slip fault if the block on the far side slipped to your left, and that it is a right-lateral strike-slip fault if the block on the far side slipped to your right.

Recognizing Faults

How do you recognize a fault when you see one? The most obvious criterion is the occurrence of **displacement**, or offset, meaning the amount of movement across a fault plane. Displacement disrupts the layers in rocks, so that layers on one side of a fault are not continuous with layers on the other side (Fig. 9.9a, b).

Faults may also leave their mark on the landscape. Those that intersect the ground surface while they are active can displace natural landscape features (such as stream valleys or glacial moraines; Fig. 9.9c) and human-made features (such as highways, fences, or rows of trees in orchards). Displacement on a dip-slip or oblique-slip fault will make a small step on the ground surface; this step is called a **fault scarp** (Fig. 9.10a). And because faults tend to break up rock, the fault may be preferentially eroded. If this happens, the fault trace (the line of intersection between the fault and the ground surface) will be marked by a linear valley (see satellite photo, p. 231).

Fault surfaces and their borders typically look different from bedding planes. For example, faulting under brittle conditions may crush or break adjacent rock. If this shattered

FIGURE 9.8 The different categories of faults.

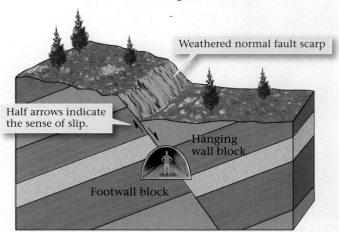

(a) The block of crust above a nonvertical fault is the hanging wall, whereas the block below is the footwall.

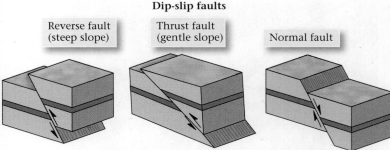

Dip-slip faults

Reverse fault (steep slope) Thrust fault (gentle slope) Normal fault

(b) Displacement on a dip-slip fault is parallel to the slope (dip) of the fault.

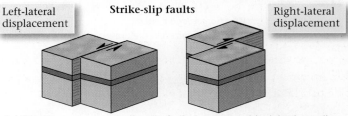

Left-lateral displacement **Strike-slip faults** Right-lateral displacement

(c) Displacement on a strike-slip fault moves one block horizontally, with respect to the other. There is no up-and-down motion.

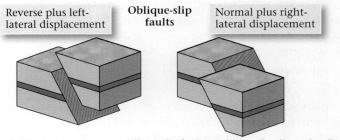

Reverse plus left-lateral displacement **Oblique-slip faults** Normal plus right-lateral displacement

(d) Displacement on an oblique-slip fault combines dip-slip and strike-slip displacement. One block moves diagonally relative to the other.

rock consists of visible angular fragments, then it is called fault breccia (Fig. 9.10b), but if it consists of a fine powder, then it is called fault gouge. Some fault surfaces are polished and grooved by the movement of the hanging wall past the foot-wall. Polished fault surfaces are called slickensides, and linear grooves on fault surfaces are slip lineations (Fig. 9.10c). We specify the orientation of a slip lineation by giving its plunge and bearing (see Box 9.1).

TAKE-HOME MESSAGE

Faults are brittle fractures on which sliding occurs. They can displace rock layers and, sometimes, the ground surface. Geologists classify faults based on the relative displacement across the fault. Faulting can break up, polish, and scratch rock.

9.5 FOLDS: CURVING ROCK LAYERS

Imagine a carpet lying flat on the floor. Push on one end of the carpet, and it will wrinkle or contort into a series of wave-like curves. Stresses developed during mountain building and other tectonic processes can similarly wrinkle or contort bedding and foliation (or other planar features) in rock. The result—a bend or curve in the shape of a rock layer—is called a **fold**.

Not all folds look the same—some look like arches, some like troughs, and some have other shapes. To describe these shapes, we first label the parts of a fold (Fig. 9.11a). The **hinge** refers to the portion of the fold where curvature is greatest, and the **limbs** are the sides of the fold that show less curvature. The **axial surface** is an imaginary surface that encompasses the hinges of successive layers. With these terms in hand, we can now describe types of folds.

- *Anticlines, synclines, and monoclines*: Folds that have an arch-like shape in which the limbs dip away from the hinge are called **anticlines** (Fig. 9.11a), whereas folds with a trough-like shape in which the limbs dip toward the hinge are called **synclines** (Fig. 9.11b). On the ground surface, the oldest beds crop out near the hinge, and the youngest away from the hinge, in an anticline. The opposite is true in a syncline. A **monocline** has the shape of a carpet draped over a stair step (Fig. 9.11c).

- *Nonplunging and plunging folds*: If the hinge is horizontal, the fold is called a nonplunging fold, but if the hinge is tilted, the fold is called a plunging fold (Fig. 9.11d, e). Layers of rock are parallel to each other on both sides of a nonplunging fold. They trace out a U-shape on the map of a plunging fold.

FIGURE 9.9   Recognizing fault displacement in the field.

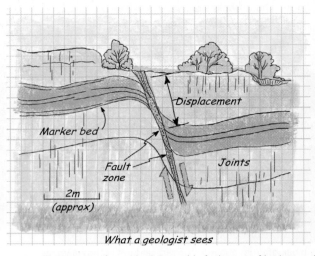

(a) A steep normal fault has displaced a distinctive red bed (a marker layer). Note that the displacement formed a 0.5-m-wide fault zone of broken rock.

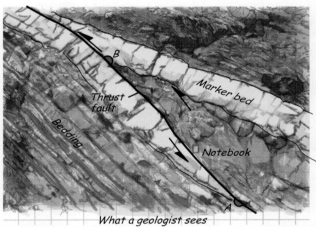

(b) Slip on a thrust fault caused one part of the light-colored marker bed to be shoved over another part, as emphasized in the drawing. Note that the beds are tilted to the right. The distance between the red dots is the displacement.

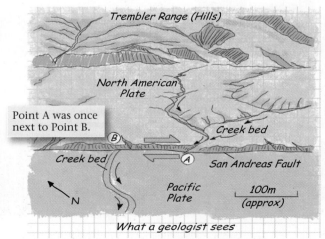

(c) An aerial photograph of a portion of the San Andreas Fault. As emphasized by the sketch, the fault offsets a stream channel in a right-lateral sense.

FIGURE 9.10 Features of exposed fault surfaces.

A scarp due to slip on a normal fault in Nevada.

(a) A fault scarp develops where faulting displaces the land surface.

(b) This fault breccia consists of irregular fragments of light-colored rock.

Striation orientation

(c) Striations on the surface of a strike-slip fault look like grooves.

FIGURE 9.11 Geometric characteristics of folds.

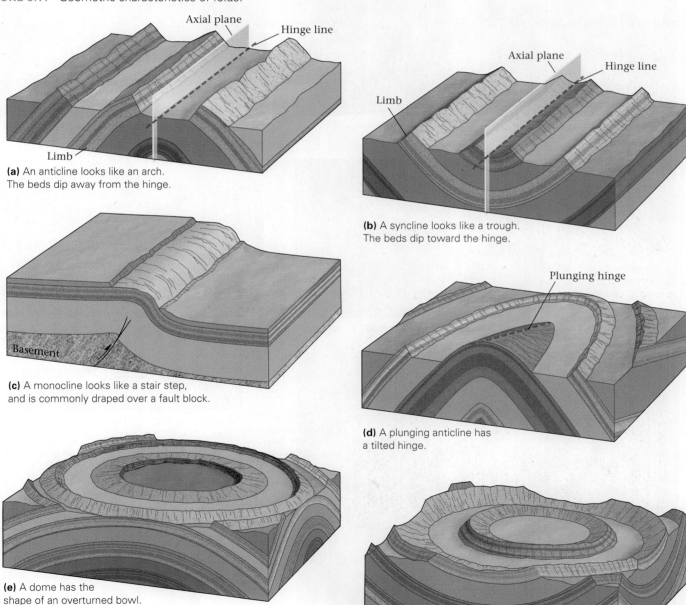

Axial plane

Hinge line

Limb

(a) An anticline looks like an arch. The beds dip away from the hinge.

Axial plane

Hinge line

Limb

(b) A syncline looks like a trough. The beds dip toward the hinge.

Basement

(c) A monocline looks like a stair step, and is commonly draped over a fault block.

Plunging hinge

(d) A plunging anticline has a tilted hinge.

(e) A dome has the shape of an overturned bowl.

(f) A basin has the shape of an upright bowl.

FIGURE 9.12 Characteristics of folds on outcrops and in the landscape.

(a) This anticline, exposed in a road cut near Kingston, New York, involves beds of Paleozoic limestone.

(b) This syncline, exposed in a road cut in Maryland, involves beds of Paleozoic sandstone and shale.

(c) This fold, exposed along the coast of Brazil, occurs in Precambrian gneiss.

(d) A train of folds exposed in sea cliffs in eastern Ireland includes anticlines and synclines. The folds affect beds of Paleozoic sandstone and shale.

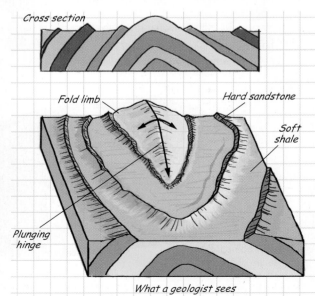

(e) The plunging anticline of Sheep Mountain, Wyoming, is easy to see because of the lack of vegetation. Resistant rock layers (sandstone) stand out as ridges, whereas weak rock layers (shale) erode away.

■ *Domes and basins*: A fold with the shape of an overturned bowl is called a **dome**, whereas a fold shaped like an upright bowl is called a **basin** (Fig. 9.11f, g). Domes and basins both show circular outcrop patterns that look like bull's-eyes—the oldest layer occurs in the center of a dome, whereas the youngest layer is located in the center of a basin. Some small domes and basins are found at outcrops, but others measure hundreds of kilometers across.

Now see if you can identify the various folds shown in **Figure 9.12a–e**.

Folds develop in two principal ways (**Fig. 9.13a, b**). During formation of flexural-slip folds, a stack of layers bends, and slip occurs between the layers. The same phenomenon happens when you bend a deck of cards—to accommodate the change in shape, the cards slide with respect to each other. Passive-flow folds form when the rock, overall, is so soft that it behaves like weak plastic and slowly flows; these folds develop simply because different parts of the rock body flow at different rates.

Folds form for a variety of reasons (**Fig. 9.14a–d**). Some layers wrinkle up, or buckle, in response to end-on compression. Others form where shear stress gradually moves one part of a layer up and over another part. Still others develop where rock layers move up and over step-like bends in a fault and must curve to conform with the fault's shape. Finally, some folds form when new slip on a fault causes a block of basement to move up so that the overlying sedimentary layers must warp.

9.6 TECTONIC FOLIATION IN ROCKS

In an undeformed sandstone, the grains of quartz are roughly spherical, and in an undeformed shale, clay flakes press together into the plane of bedding so that shales tend to split parallel to the bedding. During ductile deformation, however, internal changes take place in a rock that gradually modify the original shape and arrangement of grains. For example, quartz grains may transform into cigar shapes, elongate ribbons, or tiny pancakes, and clay flakes may recrystallize or reorient so that they lie at an angle to the bedding. Overall, deformation can produce inequant grains and can cause them to align parallel to each other. We refer to layering developed by the alignment of grains in response to deformation as **foliation** (Fig. 9.15a).

We introduced foliation, such as slaty cleavage, schistosity, and gneissic layering, in Chapter 6 while discussing the effects of metamorphism. Here we add to the story by noting that foliation forms in response to flattening and shearing in ductilely deforming rocks—in other words, foliation indicates that the rock has developed a strain (Fig. 9.15a, b). For example, in rocks with slaty cleavage, the cleavage planes lie perpendicular to the direction of the

FIGURE 9.13 Fold development in flexural-slip and passive-flow folding.

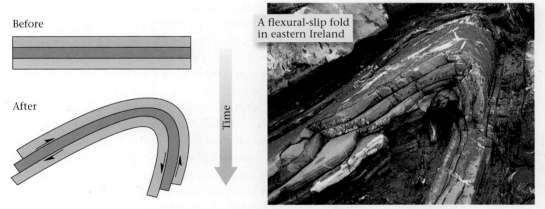

A flexural-slip fold in eastern Ireland

(a) During the formation of flexural-slip folds, layers maintain constant thickness, so for the fold to form, layers must bend. To accommodate the bending, each bed slips relative to its neighbor.

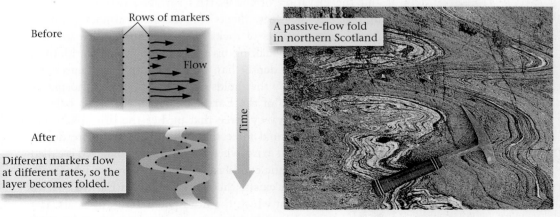

A passive-flow fold in northern Scotland

(b) During the formation of passive-flow folds, the rock slowly flows. Different points along a marker line flow at different rates, causing the layer to become folded. Note that the layer's thickness doesn't stay constant during folding.

FIGURE 9.14   Folding is caused by several different processes.

Before   **After**

(a) If a layer is shortened along its length, it buckles.

(b) If a layer is sheared, one part of the layer moves over another part to produce a fold.

Ramp

(c) When layers move up and over step-shaped faults, they must bend into folds.

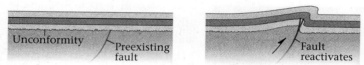

Unconformity   Preexisting fault   Fault reactivates

(d) Faulting at depth may push up a block of crust and cause overlying beds to bend into a monocline.

shortening strain, so the cleavage may be parallel to the axial plane of folds. In schists and gneisses, the foliation commonly lies parallel to or at a slight angle to the direction of shear, because shear smears grains out into the plane of shearing.

TAKE-HOME MESSAGE

During ductile deformation, grains within a rock may be reshaped and realigned. As a result, the rock develops a new foliation (layering formed as a result of deformation) in which grains have a preferred orientation. Slaty cleavage is an example.

9.7 UPLIFT AND THE FORMATION OF MOUNTAIN TOPOGRAPHY

Leonardo da Vinci, the Renaissance artist and scientist, enjoyed walking in the mountains, sketching ledges and examining the rocks he found there. In the process, he discovered marine shells (fossils) in limestone beds cropping out a kilometer above sea level, and suggested that the rock containing the fossils had risen from below sea level up to its present elevation. Modern geologists agree with da Vinci, and now refer to the process by which the surface of the Earth moves vertically from a lower to a higher elevation as **uplift**. Mountain building requires substantial uplift of the Earth's surface. In this section we look at why uplift occurs, how erosion carves rugged landscapes out of uplifted crust, and why Earth's mountains can't get much higher than Mt. Everest.

Crustal Roots and Mountain Heights

In the mid-1800s, Sir George Everest undertook a survey of India, which at the time formed part of the British Empire. Everest, aside from contributing his name to Earth's highest mountain, discovered that the mass of the Himalaya Mountains was great enough to pull the plumb bob (a lead weight at the end of a string) away from vertical. But he was surprised to note that the amount of deflection caused by the mountains was actually *less* than it should have been, given their size. Why? A British scientist, George Airy, keeping in mind that Earth's crustal rocks are less dense than its mantle rocks, came up with an explanation. Airy suggested that the crust of the Earth is thicker beneath the Himalayas than elsewhere, and thus that a low-density "crustal root" protrudes downward beneath the range (see **Geology at a Glance**, pp. 256–257). A mountain range with a low-density crustal root has less mass overall than a mountain range underlain by dense mantle, and so it exerts less pull on a plumb bob.

Work in the twentieth century has confirmed that thicker continental crust lies beneath collisional and convergent mountain ranges. Whereas typical continental crust has a thickness of about 35 to 40 km (measured from the surface to the Moho), the crust beneath some mountain belts may reach a thickness of 50 to 70 km. Mountain building in these belts shortened the crust horizontally and thickened it vertically.

The relationship between continental crustal thickness and the elevation of the Earth's surface exists because the lithosphere, in effect, "floats" on the asthenosphere. Continental crust consists of rocks (such as granite) that are less dense than those making up the asthenosphere (peridotite). Because it is less dense than the asthenosphere, continental crust is like a buoy that holds up the continental lithosphere. Thus, the surface of the Earth rises until there is a balance between the buoyancy force that pushes the lithosphere up, and the gravitational force that pulls the lithosphere down. The condition that exists when this balance is achieved is called **isostasy**, or isostatic equilibrium. Put another way, where isostatic equilibrium exists, the elevation of the Earth's surface reflects the level at which the lithosphere naturally floats. If a geologic event happens that increases the thickness of buoyant

FIGURE 9.15 The development of tectonic foliation in rock.

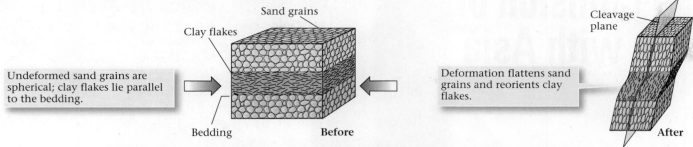

(a) Compression shortens beds, flattens sand grains, and reorients clay flakes. Clay flakes were originally parallel to bedding, but become parallel to slaty cleavage during deformation. Folding may accompany cleavage formation.

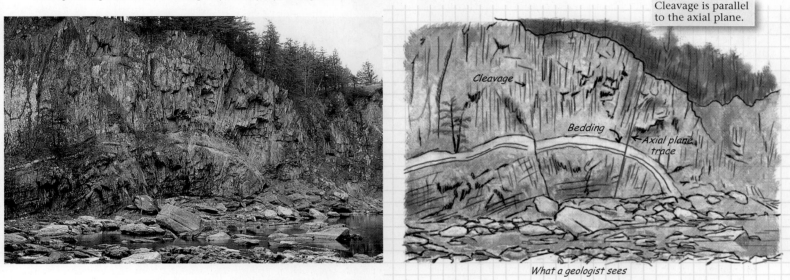

(b) An example of slaty cleavage developed in Paleozoic strata exposed in a stream cut in New York. Relict bedding is still visible, but note that the rock breaks more easily on the cleavage. This cleavage formed in association with folding.

crust, then the Earth's surface will rise in order to reestablish isostatic equilibrium.

To picture the concept of isostasy, let's consider an analogy. Imagine placing a set of wooden blocks in a bathtub. (Wooden blocks float because they are less dense than water—by Archimedes' principle, the total mass of a block is the same as the mass of the volume of water it displaces.) After bobbing about a bit, the blocks achieve isostatic equilibrium. When this happens, you will see that the surface of a thicker block is higher than that of a thinner block. In fact, if you arrange the blocks in order with the thickest at the center and the thinnest at the ends, you will see a cross section that resembles that of a collisional mountain range with the highest elevation occurring at the top of the thickest block (Fig. 9.16a–c).

What Goes Up Must Come Down

The image we most often associate with a mountain range is that of rugged topography with spire-like peaks, knife-edge ridges, precipitous cliffs, and deep valleys. This type of landscape develops because of landslides and because of erosion by ice and water that, over time, sculpts uplifted land.

The specific style of topography found within a mountain range depends on the climate. If conditions in the mountains become cool enough, glaciers form and, as they flow, carve pointed peaks and steep-sided valleys, as we will see in Chapter 18. In many ranges, we can see glacially carved landscapes even though there are no glaciers today. Such landscapes formed during the last ice age and stopped forming when the ice melted away, less than 14,000 years ago (Fig. 9.17a). At lower elevations, or in warmer climates, mountain landscapes reflect the consequences of river erosion (Fig. 9.17b). Soil formation and vegetation growth may blunt escarpments, so the land surface is smoother.

Mountains much higher than Mt. Everest cannot exist on the Earth, for two reasons. First, erosion attacks mountains as soon as they rise. Second, rock does not have infinite strength. As mountains rise, the weight of overlying rock presses down on rock buried at depth in the crust. While this is happening,

The Collision of India with Asia

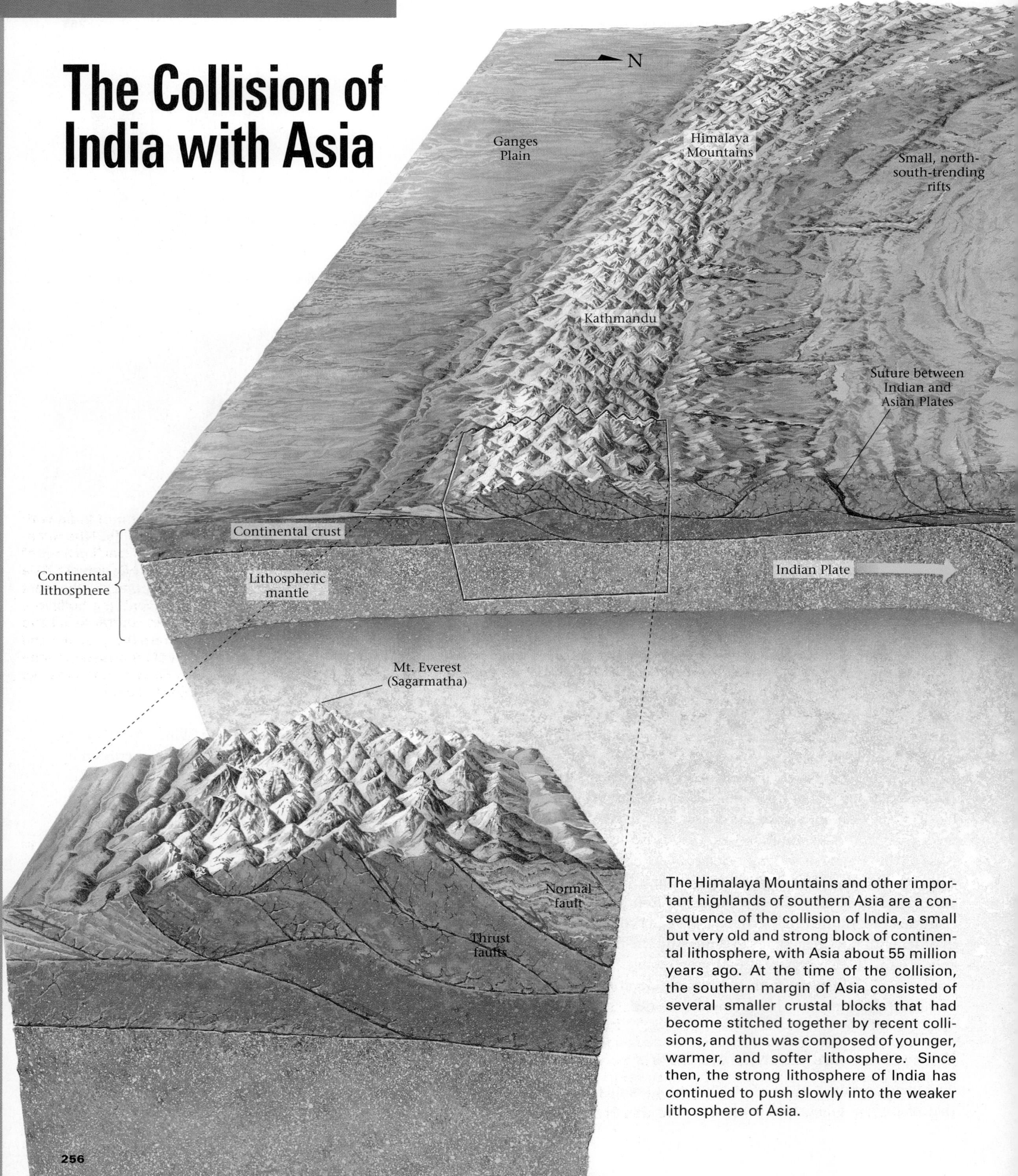

N

Ganges
Plain

Himalaya
Mountains

Small, north-
south-trending
rifts

Kathmandu

Suture between
Indian and
Asian Plates

Continental crust

Continental
lithosphere

Lithospheric
mantle

Indian Plate

Mt. Everest
(Sagarmatha)

Normal
fault

Thrust
faults

The Himalaya Mountains and other important highlands of southern Asia are a consequence of the collision of India, a small
but very old and strong block of continental lithosphere, with Asia about 55 million
years ago. At the time of the collision,
the southern margin of Asia consisted of
several smaller crustal blocks that had
become stitched together by recent collisions, and thus was composed of younger,
warmer, and softer lithosphere. Since
then, the strong lithosphere of India has
continued to push slowly into the weaker
lithosphere of Asia.

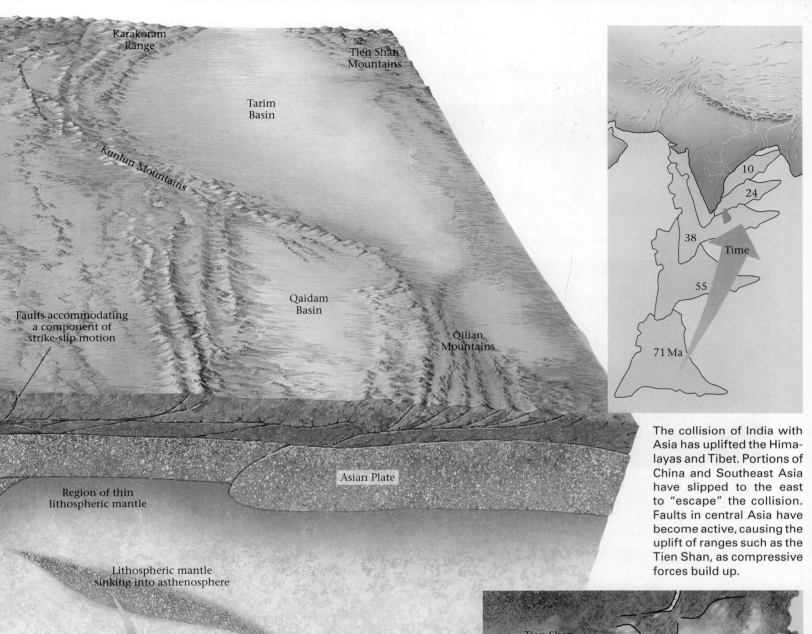

Karakoram Range

Tien Shan Mountains

Tarim Basin

Kunlun Mountains

Faults accommodating a component of strike-slip motion

Qaidam Basin

Qilian Mountains

Region of thin lithospheric mantle

Asian Plate

Lithospheric mantle sinking into asthenosphere

The collision of India with Asia has uplifted the Himalayas and Tibet. Portions of China and Southeast Asia have slipped to the east to "escape" the collision. Faults in central Asia have become active, causing the uplift of ranges such as the Tien Shan, as compressive forces build up.

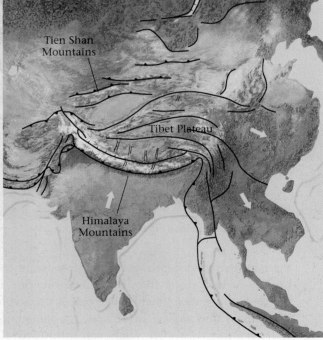

Tien Shan Mountains

Tibet Plateau

Himalaya Mountains

The development of large thrust faults has uplifted the curving Himalayan chain where Asia begins to thrust over India. Why the broad plateau of Tibet has risen remains something of a mystery. In part, the uplift may be a consequence of the thickening of the crust as it is squashed horizontally; continental crust is relatively weak, and so may spread laterally (like soft cheese in the sun), leading to the formation of normal faults and small rifts in the upper crust and a plastic-like flow in the deep crust. The uplift may also be due to the heating of the region when slabs of the underlying lithospheric mantle drop off and sink, to be replaced by hot asthenosphere.

As India has pushed into Asia, it may have squeezed blocks of China and Southeast Asia sideways, toward the east; this motion is accommodated by slip on strike-slip faults. The collision may also have caused reverse faults in the interior of Asia to become active, uplifting a succession of small mountain ranges, such as the Tien Shan.

FIGURE 9.16 The concept of isostasy as applied to collisional mountain ranges.

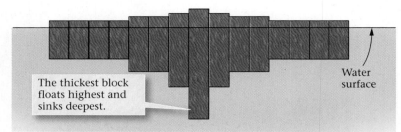

The thickest block floats highest and sinks deepest.

Water surface

(a) Because of isostasy, blocks of wood floating in water sink to a depth such that the mass of the water displaced is the same as the mass of the block.

Mt. Everest

(c) The Himalayas are the world's highest mountain range. The crust beneath them is almost twice the normal thickness.

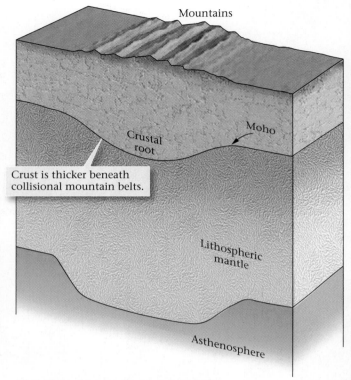

Mountains

Moho

Crustal root

Crust is thicker beneath collisional mountain belts.

Lithospheric mantle

Asthenosphere

(b) Collision thickens the crust. Mountains form where the low-density crust is thicker.

FIGURE 9.17 Manifestations of erosion in mountain ranges. When land rises, water and ice start cutting into it.

(a) Glaciers carved these rugged peaks in Switzerland.

(b) Streams cut valleys into weathered bedrock in Brazil.

the rock buried at depth gradually grows warmer and softer because of heat rising from the Earth's interior. Eventually, the weight of the mountain range causes the warm, soft rock at depth to flow slowly. Effectively, the mountains begin to collapse under their own weight and spread laterally like soft cheese that has been left out in the summer sun. Geologists call this process **orogenic collapse**. Together, uplift, erosion, and orogenic collapse bring metamorphic and plutonic rocks that were once deep in the crust up toward the Earth's surface, a process called exhumation (see Chapter 7).

FIGURE 9.18 Characteristics of convergent-margin orogens.

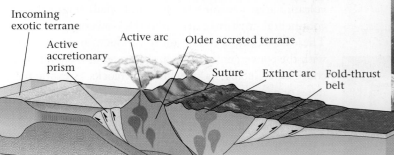

(a) At some convergent margins, compression between the downgoing and overriding plates uplifts a mountain range in which volcanism occurs. Subduction may bring in exotic crustal blocks that collide and become incorporated in the orogen.

(b) The Andes in Chile formed along a convergent margin.

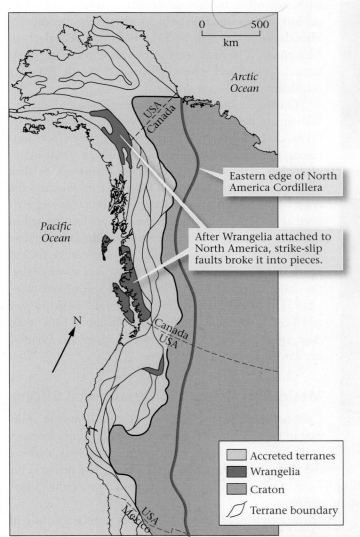

(c) The western portion of the North American Cordillera consists of accreted terranes that attached to the continent during the Mesozoic. During docking, a distinct terrane called Wrangelia (highlighted in red) was sliced into pieces that were displaced by strike-slip faults.

TAKE-HOME MESSAGE

Mountains exist where uplift of the crust occurred. Beneath some belts, crust thickened; the thickened crust is relatively buoyant, so the lithosphere of these belts "floats" higher. Once uplifted, erosion can sculpt rugged topography. Mountain belts may eventually collapse under their own weight.

9.8 CAUSES OF MOUNTAIN BUILDING

Before plate tectonics theory became established, geologists were just plain confused about how mountains formed. In the context of the new theory, however, the many processes driving mountain building became clear: mountains form in response to convergent-boundary deformation, continental collisions, and rifting. Below, we look at these different settings and the types of mountains and geologic structures that develop in each one.

Mountains Related to Convergence

Along the margins of continental convergent plate boundaries, where oceanic lithosphere subducts beneath a continent, a continental volcanic arc forms, and compression between the two plates causes a mountain range to rise (Fig. 9.18a, b). If plate movements push the continent tightly against the subduction

FIGURE 9.19   Characteristics of collisional orogens.

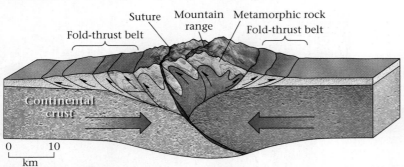

(a) During collision, continents squeeze together and deform. Thrusting brings metamorphic rock up to shallower levels.

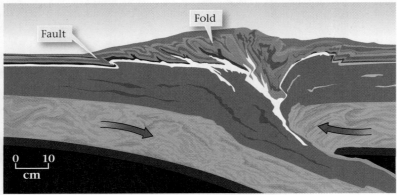

(b) Geologists simulate collision in the laboratory using layers of colored sand. Dragging the left side of the model under the right produces structures and uplift as shown in this sketch of a model.

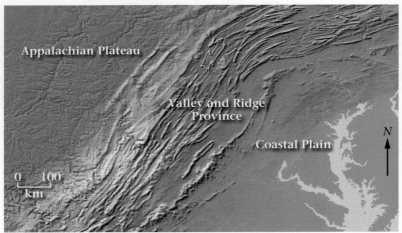

(c) The ridges and valleys of the Appalachians represent eroded folds of a Paleozoic collisional orogen. Ridges consist of durable sandstone beds.

zone, compression on the continent side of the volcanic arc generates a fold-thrust belt. In such belts, numerous thrust faults and associated folds develop, accommodating significant horizontal shortening of the crust. Convergence along the western coast of South America, for example, has generated a fold-thrust belt on the eastern side of the Andes.

During convergent-margin orogeny, offshore island volcanic arcs, oceanic plateaus, and small fragments of continental crust may drift into the convergent margin. These blocks are too buoyant to subduct, so they collide with the convergent margin and accrete, or attach, to the continent. Geologists refer to such blocks as exotic terranes or accreted terranes. The western half of the North American Cordillera consists largely of accreted terranes that attached during the Mesozoic when the region was a convergent orogen (Fig. 9.18c).

Mountains Related to Continental Collision

Once the oceanic lithosphere between two continents completely subducts, the continents themselves collide with each other. Continental collision results in the formation of large mountain ranges such as the present-day Himalayas or the Alps and the Paleozoic Appalachian Mountains. The final stage in the growth of the Appalachians happened when Africa and North America collided.

During collision, intense compression generates fold-thrust belts on the margins of the orogen (Fig. 9.19a–c). In the interior of the orogen, where one continent overrides the edge of the other, high-grade metamorphism occurs, accompanied by formation of flow folds and tectonic foliation. During this process, the crust below the orogen thickens to as much as twice its normal thickness. Gradually, rocks squeeze upward in the hanging walls of large thrust faults and later become exposed by exhumation. Finally, as noted earlier, rock at depth in the orogen heats up and becomes so weak that the mountain belt may collapse and spread out sideways. The broad Tibet Plateau may have formed in part when crust, thickened during the collision of India with Asia, spread to the northeast (see Geology at a Glance, pp. 256–257).

Mountains Related to Continental Rifting

Continental rifts are places where continents are splitting in two. When rifts first form, uplift occurs, because as the lithosphere thins, hot asthenosphere rises, making the remaining lithosphere warmer and less dense. This lithosphere is more buoyant and thus rises to reestablish isostatic equilibrium.

As rifting continues, stretching causes normal faulting in the brittle crust above (Fig. 9.20a). Movement on the normal faults drops down blocks of crust, producing deep, sediment-filled basins separated by narrow, elongate mountain ranges

FIGURE 9.20 Rift-related orogens.

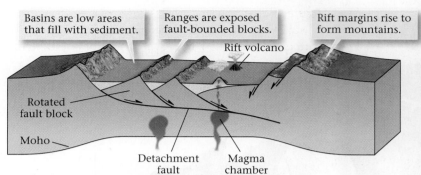

Basins are low areas that fill with sediment.

Ranges are exposed fault-bounded blocks.

Rift margins rise to form mountains.

Rift volcano

Rotated fault block

Moho

Detachment fault

Magma chamber

(a) Rifting leads to the development of numerous narrow mountain ranges. Rift-margin mountains also may form.

(b) The Basin and Range Province of the western United States is a Cenozoic rift.

that contain tilted rocks. These ranges are sometimes called "fault-block mountains." In addition, the rising asthenosphere beneath the rift partially melts, generating magmas that rise to form volcanoes within the rift. Today, the East African Rift clearly shows the configuration of rift-related mountains and volcanoes. And in North America, rifting yielded the broad Basin and Range Province of Utah, Nevada, and Arizona (Fig. 9.20b).

TAKE-HOME MESSAGE

Mountain belts form in association with subduction, collision, and rifting. At convergent margins, slivers of crust may attach to a continent over time. Collision tends to generate folds and cause regional metamorphism. Rifting produces tilt-block mountains.

9.9 THE BASINS AND DOMES OF CRATONS

A **craton** consists of crust that has not been affected by orogeny for at least about the last 1 billion years. Because orogeny happened so long ago in cratons, their crust has become quite cool, and therefore relatively strong and stable. We can divide cratons into two provinces: **shields**, in which Precambrian metamorphic and igneous rocks crop out at the ground surface, and cratonic platforms, where a relatively thin layer of Phanerozoic sediment covers the Precambrian rocks (Fig. 9.21).

In shield areas, we find intensively deformed metamorphic rocks with abundant examples of flow folds and tectonic foliation. That's because the crust making the cratons was deformed during a succession of orogenies in the Precambrian.

Recent studies of the Canadian Shield, which occupies much of the eastern two-thirds of Canada, for example, reveal the traces of Himalaya-like collision zones, Andean-like convergent boundaries, and East African-like rifts, all formed more than 1 billion years ago (some over 3 billion years ago). These orogens are so old that erosion has worn away the original topography, in the process exposing deep crustal rocks at the Earth's surface.

In the cratonic platform, the pattern of contacts between stratigraphic formations defines regional domes and basins. These are broad areas that gradually sank or rose, respectively (Fig. 9.22a, b). For example, in Missouri, strata arch across a broad dome, the Ozark Dome, whose diameter is 300 km. Individual sedimentary layers thin toward the top of the dome, because less sediment accumulated on the dome than in adjacent basins. Erosion during more recent geologic history has produced the characteristic bull's-eye pattern of a dome, with the oldest rocks (Precambrian granite) exposed near the center. In the Illinois Basin, strata appear to warp downward into a huge bowl that is also about 300 km across. Of note, strata get thicker toward the basin center, indicating that the floor of the basin sank so there was more room for sediment to accumulate. The basin also has a bull's-eye shape, but here the youngest strata are exposed in the center. Locally, faulting has displaced basement blocks, leading to the development of folds in overlying strata.

TAKE-HOME MESSAGE

Cratons are portions of continents that consist of old (Precambrian) and relatively stable crust. Parts of cratons may be covered by Paleozoic sedimentary rocks. Variations in the dip and thickness of these strata define regional basins, arches, and domes.

FIGURE 9.21 North America's craton consists of a shield, where Precambrian rock is exposed and a platform, where Paleozoic sedimentary rock covers the Precambrian.

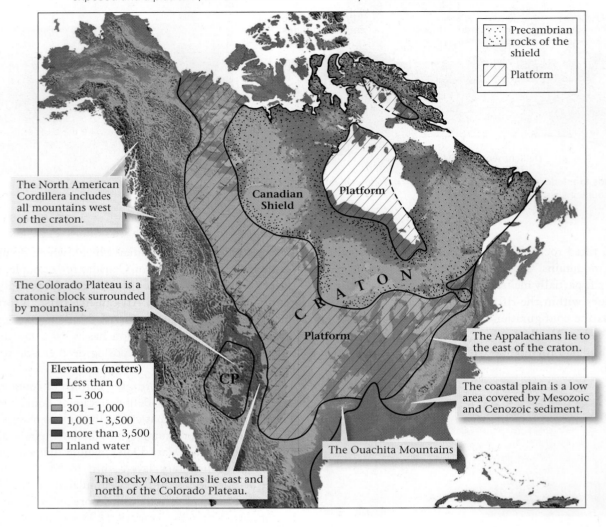

9.10 MEASURING MOUNTAIN BUILDING IN PROGRESS

Mountains are not just "old monuments," as John Muir mused. The rumblings of earthquakes and the eruptions of volcanoes attest to present-day movements in mountains. Geologists can measure the rates of these movements through field studies and satellite technology. For example, geologists can determine where coastal areas have been rising relative to the sea level by locating ancient beaches that now lie high above the water. And they can tell where the land surface has risen relative to a river by identifying places where a river has recently carved a new valley down into sediments that it had previously deposited. In addition, geologists now use the satellite **global positioning system (GPS)** to measure rates of uplift and horizontal shortening in orogens. With present technology, we can "see" the Andes shorten horizontally at a rate of a couple of centimeters per year (Fig. 9.23), and we can "watch" as mountains along this convergent boundary rise by a couple of millimeters per year.

FIGURE 9.22 Domes and basins of the North American cratonic platform.

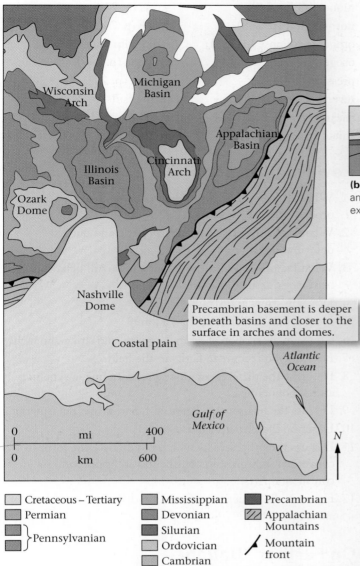

Precambrian basement is deeper beneath basins and closer to the surface in arches and domes.

Cretaceous – Tertiary	Mississippian	Precambrian
Permian	Devonian	Appalachian Mountains
}Pennsylvanian	Silurian	Mountain front
	Ordovician	
	Cambrian	

(a) A geologic map of the U.S. mid-continent platform region showing the locations of basins and domes.

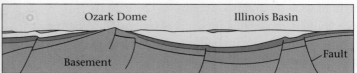

(b) A cross section showing how strata thin toward the crest of a dome and thicken toward the center of a basin. The cross section is vertically exaggerated.

FIGURE 9.23 GPS measurements of shortening in the Andes. The lines indicate the velocity of the red dots relative to the interior of South America. The line at the yellow dot indicates relative plate motion.

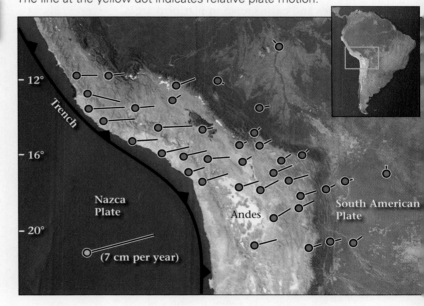

Chapter Summary

- Mountains occur in linear ranges called mountain belts, orogenic belts, or orogens. An orogen forms during an orogeny, or mountain-building event.

- Mountain building causes rocks to bend, break, shorten, stretch, and shear. Because of such deformation, rocks change their location, orientation, and shape.

- During brittle deformation, rocks crack and break into two or more pieces. During ductile deformation, rocks change shape without breaking.

- Rocks undergo three kinds of stress: compression, tension, and shear. Strain refers to the way rocks change shape when subjected to a stress.

- Deformation results in the development of geologic structures.

- Joints are natural cracks in rock, formed in response to tension under brittle conditions. Veins develop when minerals precipitate out of water passing through joints.

- Faults are fractures on which there has been shearing. Geologists distinguish among normal, reverse, thrust, strike-slip, and oblique faults.

- Folds are curved layers of rock. Anticlines are arch-like, synclines are trough-like, monoclines resemble the shape of a carpet draped over a stair step, basins are shaped like a bowl, and domes are shaped like an overturned bowl.
- Tectonic foliation forms when grains flatten, rotate, or grow so that they align parallel with one another.
- Large mountain ranges are underlain by buoyant roots. The height of such mountains is controlled by isostasy.
- Once uplifted, mountains are sculpted by erosion. When crust thickens during mountain building, the deep crust eventually becomes warm and weak, leading to orogenic collapse.
- Mountain belts formed by convergent-margin tectonism may incorporate accreted terranes. Continental collision generates metamorphic rocks and tectonic foliation. Fold-thrust belts form on the continental edge of collisional and convergent-margin orogens. Tilted blocks of crust in rifts become narrow, elongate mountain ranges.
- Cratons are the old, relatively stable parts of continental crust. They include shields and platforms. Broad regional domes and basins form in platform areas.
- With modern GPS technology, it is now possible to measure the slow uplift of mountains.

GEOPUZZLE REVISITED

Mountain ranges form in response to the interaction of lithosphere plates. Some ranges form along convergent plate boundaries, some in association with rifting, and some where two continents collide after the sea floor between them has been subducted. Stress (compression, tension, or shear) that develops during mountain building causes deformation, producing such geologic structures as folds, faults, and foliations. Mountain building involves uplift, and in many cases igneous activity and metamorphism. As soon as a range has risen above sea level, erosion attacks it. In fact, if the geologic processes causing uplift slow, erosion can eventually bevel a range nearly back to sea level. Thus, mountain ranges do not last forever.

Key Terms

anticline (p. 249)	ductile deformation (p. 243)
axial surface (p. 249)	fault (pp. 241, 247)
basin (p. 253)	fault scarp (p. 248)
brittle deformation (p. 243)	fold (pp. 241, 249)
compression (p. 245)	foliation (pp. 241, 253)
craton (p. 261)	global positioning system (GPS) (p. 262)
deformation (p. 241)	
dip (p. 246)	hinge (p. 249)
displacement (p. 248)	isostasy (p. 254)
dome (p. 253)	joints (pp. 241, 245)

limb (of fold) (p. 249)	strain (p. 242)
monocline (p. 249)	stress (p. 244)
normal fault (p. 248)	strike (p. 246)
orogenic collapse (p. 258)	strike-slip fault (p. 248)
orogen (p. 241)	syncline (p. 249)
pressure (p. 245)	tension (p. 245)
reverse fault (p. 248)	thrust fault (p. 248)
shear stress (p. 245)	uplift (pp. 241, 254)
shield (p. 261)	vein (p. 245)

Review Questions

1. What changes do rocks undergo during formation of an orogenic belt such as the Alps?
2. What is the difference between brittle and ductile deformation?
3. What factors determine whether a rock will behave in brittle or ductile fashion?
4. How are stress and strain different?
5. How is a fault different from a joint?
6. Compare the motion of normal, reverse, and strike-slip faults.
7. How do you recognize faults in the field?
8. Describe the differences among an anticline, a syncline, and a monocline.
9. Discuss the relationship between foliation and deformation.
10. Describe the principle of isostasy.
11. Discuss the processes by which mountain belts form in convergent margins, in continental collisions, and in continental rifts.
12. How are the structures of a craton different from those of an orogenic belt?

On Further Thought

1. Imagine that a geologist sees two outcrops of resistant sandstone, as depicted in the cross section sketch. The region between the outcrops is covered by soil. A distinctive bed of cross-bedded sandstone occurs in both outcrops, so the geologist correlated the western outcrop (on the left) with the eastern outcrop. The curving lines in the bed indicate the shape of the cross beds.

   (a) Keeping in mind how cross beds form (see Chapter 7), sketch how the cross bedded bed connected from one outcrop to the other, before erosion? What geologic structure have you drawn?

   (b) Is the bedrock directly beneath the geologist older than or younger than the sandstone bed of the outcrop?

West   East

ANOTHER VIEW Topography of North America.

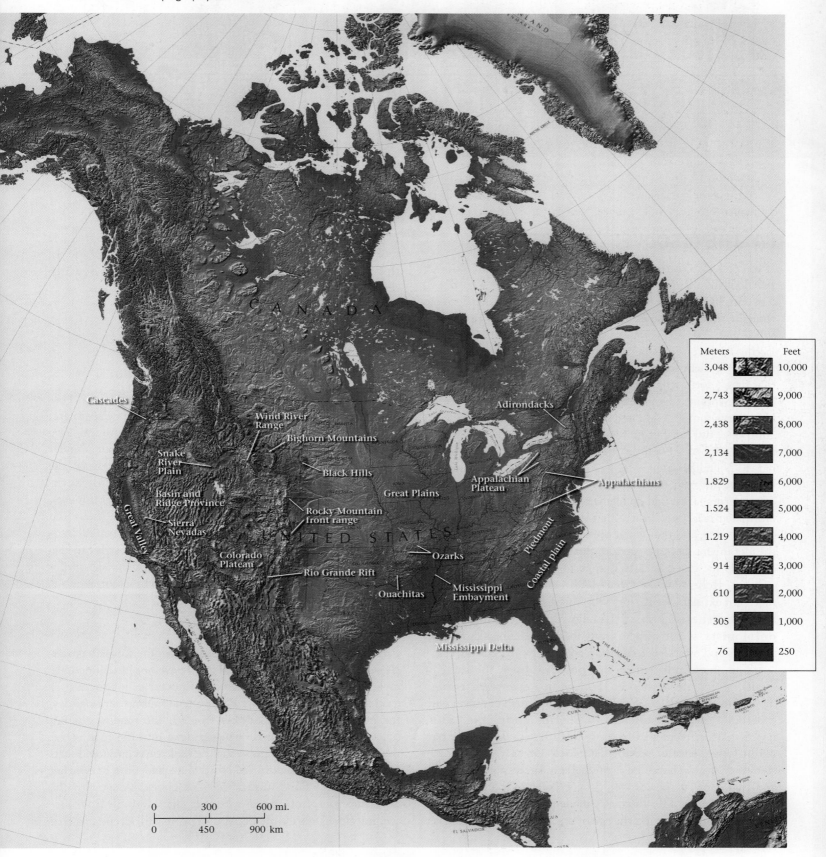

Meters		Feet
3,048		10,000
2,743		9,000
2,438		8,000
2,134		7,000
1,829		6,000
1,524		5,000
1,219		4,000
914		3,000
610		2,000
305		1,000
76		250

Memories of Past Life: Fossils and Evolution

E.1 THE DISCOVERY OF FOSSILS

If you pick up a piece of sedimentary rock, you may find shapes that look like shells, bones, leaves, or footprints (Fig. E.1a). The origin of these shapes mystified early thinkers. Some thought that the shapes had simply grown underground, in solid rock. Today, geologists consider such **fossils** (from the Latin word *fossilis*, which means dug up) to be remnants or traces of ancient living organisms now preserved in rock, and that fossils form when organisms become buried by sediment. This viewpoint, though first proposed by the Greek historian Herodotus in 450 B.C.E. and revived by Leonardo da Vinci in 1500 C.E., did not become widely accepted until the publication in 1669 of a book on the subject by a Danish physician named Nicholaus Steno (1638–1686). Steno noted the similarity between fossils known as "dragon's tongues" and the teeth of modern sharks, and concluded that fossil-containing rocks originated as loose sediment that had incorporated the remains of organisms, and that when the sediment hardened into rock, the organisms became part of the rock. The understanding of fossils increased greatly thanks to the efforts of a British scientist, Robert Hooke (1635–1703), who realized that most represent extinct species, meaning species that lived in the past but no longer exist. During the following two centuries, **paleontologists**, scientists who study fossils, described thousands of fossils and established museum collections (Fig. E.1b, c).

The nineteenth century saw **paleontology**, the study of fossils, ripen into a science. Work with fossils went beyond description alone when William Smith, a British engineer who supervised canal construction in England during the 1830s, noted that different fossils occur in different layers of strata within a sequence of sedimentary rocks. In fact, Smith realized that strata contained a predictable succession of fossils, and that a given species would be present only in a specific interval of strata. This discovery made it possible to use fossils as a basis for determining the age of one sedimentary rock layer

relative to another. Fossils, therefore, became an indispensable tool for studying geologic history and the evolution of life. In this interlude, we introduce fossils, an understanding of which serves as essential background for the next two chapters.

E.2 FOSSILIZATION

What Kinds of Rocks Contain Fossils?

Most fossils are found in sediments or sedimentary rocks. Fossils form when organisms die and become buried by sediment, or when organisms travel over or through sediment and leave their mark. Rocks formed from sediments deposited under anoxic (oxygen-free) conditions in quiet water (such as lake beds or lagoons) preserve particularly fine specimens. Rocks made from sediments deposited in high-energy environments, on the other hand—where strong currents tumble shells and bones and break them up—contain at best only small fragments of fossils mixed with other clastic grains. Fossils can also occur in deposits formed from air-fall ash, for the ash settles just like sediment and can bury an organism or a footprint.

Forming a Fossil

Paleontologists refer to the process of forming a fossil as **fossilization**. To see how a typical fossil develops in sedimentary rock, let's follow the fate of an old dinosaur as it searches for food along a riverbank (Fig. E.2). On a scalding summer day, the hungry dinosaur, plodding through the muddy ground, succumbs to the heat and collapses dead into the mud. Over the coming days, scavengers strip the skeleton of meat and scatter the bones. But before the bones have had time to weather away, the river floods and buries the bones, along with the dinosaur's footprints, under a layer of silt. More silt from succeeding floods buries the bones and prints still deeper, so that

FIGURE E.1   Paleontologists collect fossils for study.

(a) This bedding plane contains the imprint of species that lived hundreds of millions of years ago and are now extinct.

(b) Paleontologists painstakingly chip and dig away at rock layers in search of fossils.

(c) A drawer of labeled fossils in a museum. Paleontologists study such collections to help identify new specimens.

the bones cannot be reworked by currents or disrupted by burrowing organisms. Later, sea level rises and a thick sequence of marine sediment buries the fluvial sediment.

Eventually, the sediment containing the bones turns to rock (siltstone). The footprints remain outlined by the contact between the siltstone and mud, while the bones reside within the siltstone. Minerals precipitating from groundwater passing through the siltstone gradually replace some of the chemicals constituting the bones, until the bones themselves have become like rock. The buried bones and footprints are now fossils. One hundred million years later, uplift and erosion expose the dinosaur's grave. Part of a fossil bone protrudes from a rock outcrop. A lucky paleontologist observes the fragment and starts excavating, gradually uncovering enough of the bones to permit reconstruction of the beast's skeleton. Further digging uncovers footprints. The dinosaur rises again, but this time in a museum. In recent years, bidding wars have made some fossil finds extremely valuable. For example, a skeleton of a *Tyrannosaurus rex*, a 67-million-year-old dinosaur, sold at auction in 1997 for $7.6 million. The specimen, named Sue after its discoverer, now stands in the Field Museum of Chicago.

Similar tales can be told for fossil seashells buried by sediment settling in the sea, for insects trapped in hardened tree sap (amber), and for mammoths drowned in the muck of a tar pit. In all cases, fossilization involves the burial and preservation of an organism or the trace (a footprint or burrow) of an organism.

The Many Different Kinds of Fossils

Perhaps when you think of a fossil, you picture either a dinosaur bone or the imprint of a seashell in rock. In fact, paleontologists distinguish many different kinds of fossils, according to the specific way in which the organism was fossilized. Let's look at examples of these categories.

- *Frozen or dried body fossils*: In a few environments, whole bodies of organisms may be preserved. Most of these fossils are fairly young, by geologic standards—their ages can be measured in thousands, not millions, of years. Examples include woolly mammoths that became incorporated in the permafrost (permanently frozen ground) of Siberia (Fig. E.3a).
- *Body fossils preserved in amber or tar*: Insects landing on the bark of trees may become trapped in the sticky sap or resin the trees produce. This golden syrup envelops the insects and over time hardens into **amber**, the semiprecious "stone" used for jewelry. Amber can preserve insects, as well as other delicate organic material such as feathers, for 40 million years or more (Fig. E.3b).

FIGURE E.2 How a dinosaur eventually becomes a fossil. This takes many steps, over a long period of time.

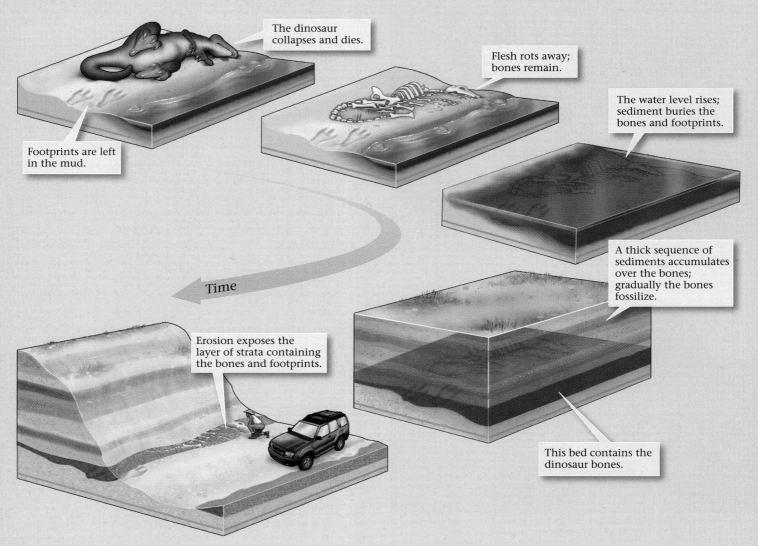

The dinosaur collapses and dies.

Footprints are left in the mud.

Flesh rots away; bones remain.

The water level rises; sediment buries the bones and footprints.

A thick sequence of sediments accumulates over the bones; gradually the bones fossilize.

Time

Erosion exposes the layer of strata containing the bones and footprints.

This bed contains the dinosaur bones.

Tar similarly acts as a preservative. In isolated regions where oil has seeped to the surface, the more volatile components of the oil evaporate away and bacteria degrade what remains, leaving behind a sticky residue. At one such locality, the La Brea Tar Pits in Los Angeles, tar accumulated in a swampy area. While grazing, drinking, or hunting at the swamp, animals became mired in the tar and sank into it. Their bones have been remarkably well preserved for over 40,000 years (Fig. E.3c).

- *Preserved or replaced bones, teeth, and shells*: Bones (the internal skeletons of vertebrate animals) and shells (the external skeletons of invertebrate animals) consist of durable minerals, which may survive in rock (Fig. E.3d).

Some bone or shell minerals are not stable, and they recrystallize. But even when this happens, the shape of the bone or shell remains in the rock.

- *Permineralized organisms*: **Permineralization** refers to the process by which minerals precipitate in porous material, such as wood or bone, from groundwater solutions that have seeped into the pores. **Petrified wood**, for example, forms by permineralization of wood, so that cell interiors are replaced with silica, causing the wood to become chert. In fact, the word *petrified* literally means turned to stone. The more resistant cellulose of the wood transforms into an organic film that remains after permineralization, so that the fine detail of the wood's cell structure can be seen in a petrified log (Fig. E.3e).

FIGURE E.3 Examples of different kinds of fossils.

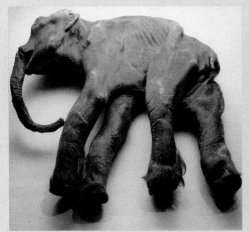

(a) A 100,000-year-old frozen mammoth found in Siberia still has flesh and fur.

(b) A piece of amber containing two fossil insects.

(c) A fossil skeleton of a 2-m-high giant ground sloth from the La Brea Tar Pits in California.

(d) Fossil dinosaur bones on a tilted bed of sandstone in Utah.

(e) Petrified wood from Arizona. It is so hard that it remains after the rock that surrounded it has eroded away.

(f) Molds and casts of a shell. Even fine detail is preserved.

(g) The carbonized impressions of fern fronds in a shale.

(h) Dinosaur footprints in a mudstone in Arizona.

(i) Worm burrows on siltstone in Ireland (lens cap for scale).

- *Molds and casts of bodies*: As sediment compacts around a shell, it conforms to the shape of the shell or body. If the shell or body later disappears, because of weathering and dissolution, a cavity called a **mold** remains (Fig. E.3f). (Sculptors use the same term to refer to the receptacle into which they pour bronze or plaster.) A mold preserves the delicate shape of the organism's surface; it looks like an indentation on a rock bed. The sediment that had filled the mold also preserves the organism's shape; this **cast** protrudes from the surface of the adjacent bed.

- *Carbonized impressions of bodies*: Impressions are simply flattened molds created when soft or semisoft organisms (leaves, insects, shell-less invertebrates, sponges, feathers, jellyfish) get pressed between layers of sediment. Chemical reactions eventually remove the organic material, leaving only a thin film of carbon on the surface of the impression (Fig. E.3g).

- *Trace fossils*: These include footprints, feeding traces, burrows, and dung (coprolites) that organisms leave behind in sediment (Fig. E.3h, i).

- *Chemical fossils*: Living things consist of complex organic chemicals. When buried with sediment and subjected to diagenesis, some of these chemicals are destroyed, but some either remain intact or break down to form different, but still distinctive, chemicals. A distinctive chemical derived from an organism and preserved in rock is called a chemical fossil. (Such chemicals may also be called molecular fossils or **biomarkers**.)

Paleontologists also find it useful to distinguish among different fossils on the basis of their size. Macrofossils are fossils large enough to be seen with the naked eye. But some rocks and sediments also contain abundant **microfossils**, which can be seen only with a microscope or even an electron microscope. Microfossils include remnants of plankton, algae, bacteria, and pollen.

Fossil Preservation

Not all living organisms become fossils when they die. In fact, only a small percentage do, for it takes special circumstances—namely, one or more of the following four—to produce a fossil and allow it to survive.

- *Death in an anoxic (oxygen-poor) environment*: A dead squirrel by the side of the road won't become a fossil. As time passes, birds, dogs, or other scavengers may come along and eat the carcass. And if that doesn't happen, maggots, bacteria, and fungi will infest the carcass and gradually digest it. Flesh that has not been eaten, or does not rot, can react with oxygen in the atmosphere and transform into carbon dioxide gas. The remaining skeleton weathers in air and turns to dust. Thus, before the dead squirrel can become incorporated in sediment, it has vanished. In order for fossilization to occur, a carcass must settle into an oxygen-poor environment, where oxidation reactions happen slowly, where scavenging organisms aren't as abundant, and where bacterial metabolism takes place very slowly. In such environments the organism won't rot away before it has a chance to be buried and preserved.

- *Rapid burial*: If an organism dies in a depositional environment where sediment accumulates rapidly, it may be buried before it has time to rot, oxidize, be eaten, be completely broken up, or be consumed by burrowing organisms. For example, if a storm suddenly buries an oyster bed with a thick layer of silt, the oysters die and become part of the sedimentary rock derived from the sediment.

- *The presence of hard parts*: Organisms without durable shells or skeletons, or other "hard parts," usually won't be fossilized, for soft flesh decays long before hard parts do under most depositional conditions. For this reason, paleontologists know much more about the fossil record of bivalves (a class of organisms, including clams and oysters, with strong shells) than they do about the fossil record of jellyfish (which have no shells) or spiders (which have very fragile shells).

- *Lack of metamorphism*: Metamorphism may destroy fossils, either by dissolving them away, or by causing recrystallization or neocrystallization sufficient to obscure the fossil's shape.

By carefully studying modern organisms, paleontologists have been able to provide rough estimates of the **preservation potential** of organisms, meaning the likelihood that an organism will be buried and eventually transformed into a fossil. For example, in a typical modern-day shallow-marine environment, such as the mud-and-sand sea floor close to a beach, about 30% of the organisms have sturdy shells and thus a high preservation potential, 40% have fragile shells and a low preservation potential, and the remaining 30% have no hard parts at all and are not likely to be fossilized except in special circumstances. Of the 30% with sturdy shells, though, few happen to die in a depositional setting where they actually *can* become fossilized.

Extraordinary Fossils: A Special Window to the Past

Though only hard parts survive in most fossilization environments, paleontologists have discovered a few special locations where rock contains relicts of soft parts as well; such fossils are known as **extraordinary fossils** (Fig. E.4a, b). We've already seen how extraordinary fossils such as insects and even feath-

FIGURE E.4   Extraordinary fossils.

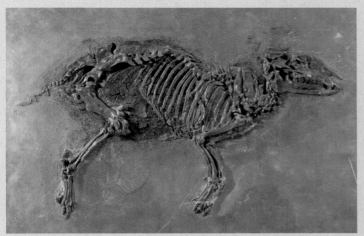

(a) A 50 million-year-old mammal fossil was chiseled from oil shale near Messel, Germany. It still contains the remains of skin.

(b) *Archaeopteryx* from the 150 million-year-old Solenhofen Limestone of Germany. The imprints of feathers are clearly visible.

ers can be preserved in amber, and how complete skeletons have been found in tar pits. In some cases, such finds have yielded examples of ancient DNA. Extraordinary fossils may also be preserved in sediment that accumulates on the anoxic floor of lakes or lagoons or the deep ocean. Here, oxidation cannot occur, and flesh does not rot before burial. Carcasses of animals that settle into the mud gradually become fossils, but because they were buried before the destruction of their soft parts, fossil impressions of their soft parts surround the fossils of their bones.

E.3 CLASSIFYING LIFE

The classification of fossils follows the same principles used for the classification of living organisms. These principles were first proposed in the eighteenth century by Carolus Linnaeus,

a Swedish biologist. The study of how to classify organisms is now referred to as **taxonomy**. Linnaeus's scheme has a hierarchy of divisions. First, all life is divided into kingdoms, and each kingdom consists of one or more phyla. A phylum, in turn, consists of several classes; a class, of several orders; an order, of several families; a family, of several genera; and a genus, of one or more species. Kingdoms, therefore, are the broadest category and species the narrowest. Biologists now recognize six kingdoms (Fig. E.5):

- *Archaea*: microorganisms typically found in extreme environments such as hot springs and in very acidic or very alkaline water;
- *Bacteria*: single-celled organisms which do not contain a nucleus; some of these may cause infection.
- *Protista*: various unicellular and simple multicellular organisms; these include algae (such as diatoms) and forams, two of the major plankton types in the oceans;
- *Fungi*: mushrooms and yeast, etc.;
- *Plantae*: trees, grasses, and ferns, etc.; and
- *Animalia*: sponges, corals, snails, dinosaurs, ants, and people, etc.

More recently, biologists have realized that Archaea are as different from Bacteria as both are from the other kingdoms. The last four kingdoms share enough common characteristics, at the

FIGURE E.5   The basic "domains" of life on Earth. Archaea and Bacteria are both prokaryotes, and don't have nuclei. All other life forms are eukaryotes, because they have cells with a nucleus.

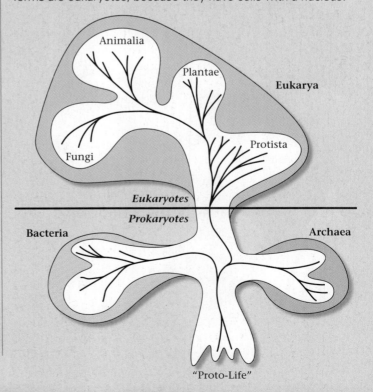

cellular level, to be lumped together as a group called Eukarya. Thus, biologists now divide life into three domains: Archaea, Bacteria, and Eukarya. The cells of eukaryotic organisms have distinct nuclei and internal membranes, whereas those of Bacteria and Archaea do not. The latter two domains contain organisms with prokaryotic cells (cells that have no nucleus).

E.4 CLASSIFYING FOSSILS

Paleontologists traditionally distinguish different fossil species from each other according to **morphology** (the form or shape) alone. A fossil clam is a fossil clam because it looks like one! In some cases, the characteristics that distinguish one fossil species from another may be quite subtle (such as the number of ridges on the surface of a shell, or the relative length of different leg bones), but even beginners can distinguish the major groups of fossils from one another on sight.

There's nothing magical about classifying fossils. Major characteristics at the level of class or higher are fairly easy to distinguish just from the way they look. For instance, skeletons

FIGURE E.6   Common types of invertebrate fossils.

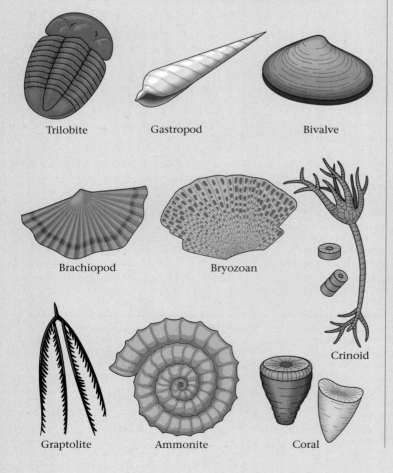

Trilobite   Gastropod   Bivalve

Brachiopod   Bryozoan

                                      Crinoid

Graptolite   Ammonite   Coral

of fish are hard to mistake for skeletons of mammals, and snail shells are hard to mistake for clam shells. But classification can also be difficult, especially if specimens are incomplete, and in many cases classification can be controversial.

Figure E.6 shows examples of some of the major types of invertebrate fossils. With this chart, you should be able to identify many of the fossils you'll find in a bed of limestone. Particularly common invertebrate fossils include the following:

- *Trilobites*: These have a segmented shell that is divided lengthwise into three parts.
- *Gastropods* (snails): Most fossil specimens of gastropods have a shell that does not contain internal chambers.
- *Bivalves* (clams and oysters): These have a shell that can be divided into two similar halves.
- *Brachiopods* (lamp shells): The top and bottom parts of these shells have different shapes. Shells typically have ridges radiating out from the hinge.
- *Bryozoans*: These are colonial animals. Their fossils resemble a screen-like grid of cells.
- *Crinoids* (sea lilies): These organisms look like a flower but actually are animals. Their shells have a stalk consisting of numerous circular plates stacked on top of each other.
- *Graptolites*: These look like tiny carbon saw blades in a rock. They are remnants of colonial animals that floated in the sea.
- *Cephalopods*: These include ammonites, with a spiral shell, and nautiloids, with a straight shell. Their shells contain internal chambers and have ridged surfaces.
- *Corals*: These include colonial organisms that form distinctive mounds or columns.

E.5 THE FOSSIL RECORD

A Brief History of Life

As we will see in Chapter 10, the fossil record defines the long-term evolution of life on planet Earth. Archaea and Bacteria fossils appear in rocks as old as about 3.7 billion years, indicating that these organisms are the earliest forms of life on Earth. Researchers once thought that Archaea somehow developed by chemical reactions in concentrated "soups" that formed when seawater trapped in shallow pools evaporated. Growing evidence suggests, however, that they may instead have developed in warm groundwater beneath the Earth's surface or at hydrothermal vents on the sea floor.

For the first billion years or so of life history, Archaea and Bacteria were the only types of life on Earth. Then, about 2.5 Ga (billion years ago), organisms of the Protista kingdom first appeared. Early multicellular organisms, shell-less inverte-brates of the animal kingdom, and fungi came into existence per-haps 1.0 to 1.5 Ga. Within each kingdom, life radiated (divided) into different phyla, and within each phylum, life radiated into different classes. The great variety of shelly invertebrate classes appeared a little over 540 million years ago, an event called the **Cambrian explosion**, after the geologic era in which it occurred. Many organisms that persist today appeared at dif-ferent times since then: first came invertebrates, then fish, land plants, amphibians, reptiles, and finally birds and mammals.

Is the Fossil Record Complete?

By some estimates, more than 250,000 species of fossils have been collected and identified to date, by thousands of geologists working on all continents during the past two centuries. These fossils define the framework of life evolution on planet Earth. But the record is not complete—known fossils cannot account for every intermediate step in the evolution of every organism. Considering that as many as 5 million species may be living on Earth today (not counting bacteria), over the billions of years that life has existed there may have been 5 billion to 50 billion species. Clearly, known fossils represent at most a tiny percent-age of these species. Why is the record so incomplete?

First, despite all the fossil-collecting efforts of the past two centuries, paleontologists have not even come close to sampling every cubic centimeter of sedimentary rock exposed on Earth. Just as biologists have not yet identified every living species of insect, paleontologists have not yet identified every species of fossil. New species and even genera of fossils continue to be discovered every year.

Second, not all organisms are represented in the rock record, because not all organisms have a high preservation potential. As noted earlier, fossilization occurs only under special conditions, and thus only a minuscule fraction of the organisms that have lived on Earth become fossilized. There may be few, if any, fos-sils of a vast number of extinct species, so we have no way of knowing that they ever existed.

Finally, as we will learn in Chapter 10, the sequence of sed-imentary strata that exists on Earth does not account for every minute of time since the formation of our planet. Sediments accumulate only in environments where conditions are appro-priate for deposition and not for erosion—sediments do not accumulate, for example, on the dry great plains or on moun-tain peaks, but do accumulate in the sea and in the floodplains and deltas of rivers. Because Earth's climate changes through time and because the sea level rises and falls, certain locations on continents are sometimes sites of deposition and sometimes aren't, and on occasion are sites of erosion. Therefore, strata accumulate only episodically.

E.6 EVOLUTION AND EXTINCTION

Darwin's Grand Idea

As a young man in England in the early nineteenth century, Charles Darwin had been unable to settle on a career but had developed a strong interest in natural history. Therefore, he jumped at the opportunity to serve as a naturalist aboard H.M.S. *Beagle* on an around-the-world surveying cruise. During the five years of the cruise, from 1831 to 1836, Darwin made detailed observations of plants, animals, and geology in the field and amassed an immense specimen collection from South America, Australia, and Africa. Just before departing on the voyage, a friend gave him a copy of Charles Lyell's 1830 textbook, *Prin-ciples of Geology*, which argued in favor of James Hutton's pro-posal that the Earth had a long history and that geologic time extended much further into the past than did human civiliza-tion. A visit to the Galápagos Islands, off the coast of Peru, led to a turning point in Darwin's thinking. The naturalist was most impressed with the variability of Galápagos finches. He marveled not only at the fact that different varieties of the bird occurred on different islands, but at how each variety had adapted to utilize a particular food supply. With Lyell's writings in mind, Darwin developed a hypothesis that the finches had begun as a single species but had branched into several different species when isolated on different islands. This proposal implied that a species could change, or undergo **evolution**, throughout long periods of time and that new species could appear.

The crux of Darwin's argument is simply this: populations of organisms cannot increase in numbers forever, because they are limited by competition for scarce resources in the environ-ment. In nature, only organisms capable of survival can pass on their characteristics to the next generation. In each new gen-eration, some individuals have characteristics that make them more fit, whereas some have characteristics that make them less fit. The fitter organisms are more likely to survive and produce offspring. Thus, beneficial characteristics that they possess get passed on to the next generation. Darwin called this process **natural selection**, because it occurs on its own in nature. According to Darwin, when natural selection takes place over long periods of time—geologic time—it eventually produces new organisms that differ so significantly from their distant ancestors that the new organisms can be considered to consti-tute a new species. If environmental conditions change, or if competitors enter the environment, species that do not evolve and become better adapted to survive eventually die off.

Darwin's view of evolution has been successfully supported by many observations, and so far has not been definitively disproved by any observation or experiment. Also, it can be used to make testable predictions. Thus, scientists now refer to Darwin's idea as the **theory of evolution** (see Box P.1 in the

Prelude for the definition of a theory). In the century and a half since Darwin published his work, the science of genetics has developed and has provided insight into *how* evolution works. With the discovery of DNA in 1953, biologists began to understand the molecular nature of mutations, and thus of evolution.

The theory of evolution provides a conceptual framework in which to understand paleontology. By studying fossils in sequences of strata, paleontologists are able to observe progressive changes in species through time, and can determine when some species die out and other species appear. But because of the incompleteness of the fossil record, questions remain as to the rates at which evolution takes place during the course of geologic time. Originally, it was assumed that evolution happened at a constant, slow rate—this concept is called **gradualism**. More recently, however, researchers have suggested that evolution takes place in fits and starts: evolution occurs very slowly for quite a while (the species are in equilibrium), and then during a relatively short period it takes place very rapidly. This concept is called **punctuated equilibrium**. Factors that could cause sudden pulses of evolution include (1) a sudden mass extinction event, during which many organisms disappear, leaving ecological niches open for new species to colonize; (2) a sudden change in the Earth's climate that puts stress on organisms—organisms that evolve to survive the new stress survive, whereas others become extinct; (3) the sudden formation of new environments, as may happen when rifting splits apart a continent and generates a new ocean with new coastlines; and (4) the isolation of a breeding population.

Extinction: When Species Vanish

Extinction occurs when the last members of a species die, so there are no parents to pass on their genetic traits to offspring. Some species become extinct as they evolve into new species, whereas others just vanish, leaving no hereditary offspring. These days, we take for granted that species become extinct, because a great number have, unfortunately, vanished from the Earth during human history. Before the 1770s, however, few geologists thought that extinction occurred; they thought that fossils that didn't resemble known species must have living relatives somewhere on Earth. Considering that large parts of the Earth remained unexplored, this idea wasn't so far-fetched. But by the end of the eighteenth century, it became clear that numerous fossil organisms did not have modern-day counterparts. The bones of mastodons and woolly mammoths, for example, were too different from those of elephants to be of the same species, but the animals were too big to hide.

Twentieth-century studies concluded that many different phenomena can contribute to extinction. Some extinctions may happen suddenly, when all members of a species die off in a short time, whereas others may occur over longer periods, when the replacement rate of a population simply becomes lower than the mortality rate. By examining the number of species on Earth through time (in other words, by studying variations in the diversity of life, or biodiversity), paleontologists have found that the rate of extinction varies through time. Generally, the rate is fairly slow, but on occasion a **mass extinction event** occurs, during which a large number of species worldwide disappear. At least five major mass extinction events have happened during the past half billion years (Fig. E.7). These events define the boundaries between some of the major intervals into which geologists divide time. For example, a major extinction event marks the end of the Cretaceous Period, 65 million years ago. During this event, all dinosaur species (with the exception of their modified descendants, the birds) vanished, along with many marine invertebrate species. A huge extinction event also brought the Permian to a close. Some researchers have suggested that extinction events are periodic, but this idea remains controversial.

Following are some of the geologic factors that may cause extinction.

- *Global climate change*: At times, the Earth's mean temperature has been significantly colder than today's, whereas at other times it has been much hotter. Because of a change in climate, an individual species may lose its habitat, and if it cannot adapt to the new habitat or migrate to stay with its old one, the species will disappear.

- *Tectonic activity*: Tectonic activity causes mountain building, the gradual vertical movement of the crust over broad regions, changes in sea-floor spreading rates, and changes

FIGURE E.7 This graph shows how the diversity of life has changed with time. Sudden drops indicate periods when mass extinctions occurred.

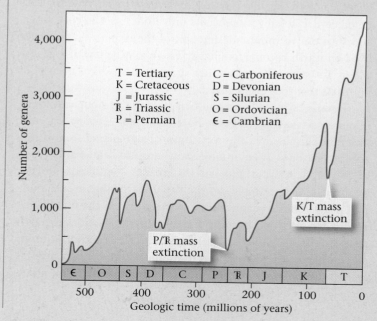

T = Tertiary   C = Carboniferous
K = Cretaceous   D = Devonian
J = Jurassic   S = Silurian
Ŧ = Triassic   O = Ordovician
P = Permian   € = Cambrian

P/Ŧ mass extinction

K/T mass extinction

Number of genera

€ | O | S | D | C | P | Ŧ | J | K | T

500   400   300   200   100   0

Geologic time (millions of years)

in the amount of volcanism. These phenomena can make sea level rise or fall, or can modify the distribution and area of habitats. Species that cannot adapt die off.

- *Asteroid or comet impact*: Many geologists have concluded that impacts of large meteorites or asteroids with the Earth have been catastrophic for life. An impact would send so much dust and debris into the atmosphere that it would blot out the Sun and plunge the Earth into darkness and cold. Such a change, though relatively short lived, could interrupt the food chain.

- *Voluminous volcanic eruption*: Several times during Earth history, incredible quantities of lava have spilled out on the surface. These eruptions, perhaps due to the rise of superplumes in the mantle, were accompanied by the release of enough greenhouse gas into the atmosphere to alter the climate.

- *The appearance of a predator or competitor*: Some extinctions may happen simply because a new predator appears on the scene. Some researchers suggest that this phenomenon explains the mass extinction that occurred during the past 20,000 years, when a vast number of large mammal species vanished from North America. The timing of these extinctions appears to coincide with the appearance of the first humans (fierce predators) on the continent. If a more efficient competitor appears, the competitor steals an ecological niche from the weaker species, whose members can't obtain enough food and thus die out.

Key Terms

amber (p. 267)

biomarker (p. 270)

Cambrian explosion (p. 273)

cast (p. 270)

evolution (p. 273)

extinction (p. 274)

extraordinary fossil (p. 270)

fossil (p. 266)

fossilization (p. 266)

gradualism (p. 274)

mass extinction event (p. 274)

microfossil (p. 270)

mold (p. 270)

morphology (p. 272)

natural selection (p. 273)

paleontologists (p. 266)

paleontology (p. 266)

permineralization (p. 268)

petrified wood (p. 268)

preservation potential (p. 270)

punctuated equilibrium (p. 274)

taxonomy (p. 271)

theory of evolution (p. 273)

ANOTHER LOOK   This Cretaceous fish fossil was found in a sandstone bed in Brazil.

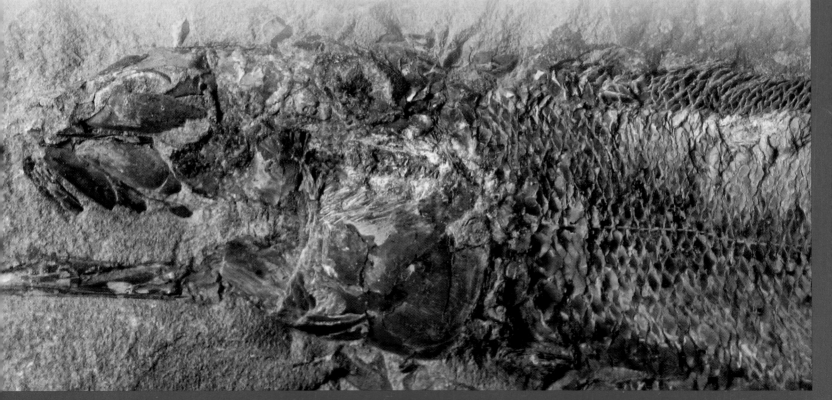

CHAPTER **10**

Deep Time: How Old Is Old?

How do geologists determine the age of rocks? This cliff face in Missouri shows two rock units. The sedimentary layers on the left were deposited on the rhyolite to the right. From this relationship, we can determine the relative age of the two units—the rhyolite is older. But to determine the age of the rhyolite in years—its numerical age—we must date it isotopically.

GEOPUZZLE

Geologists state that the Earth is very, very old—4.57 billion years old, to be exact. How was this age determined? Is there any place on Earth where we can see a succession of rocks representing the entirety of Earth history?

If the Eiffel Tower were now representing the world's age, the skin of paint on the pinnacle-knob at its summit would represent man's share of that age; and anybody would perceive that that skin was what the tower was built for. I reckon they would, I dunno.

—Mark Twain (1835–1910)

10.1 INTRODUCTION

In May of 1869, a one-armed Civil War veteran named John Wesley Powell set out with a team of nine geologists and scouts to explore the previously unmapped expanse of the Grand Canyon, the greatest gorge on Earth. Though Powell and his companions battled fearsome rapids and the pangs of starvation, most managed to emerge from the mouth of the canyon three months later. During their voyage, seemingly insurmountable walls of rock both imprisoned and amazed the explorers, and led them to pose important questions about the Earth and its history, questions that even casual tourists to the canyon ponder today: Did the Colorado River sculpt this marvel, and if so, how long did it take? When did the rocks making up the walls of the canyon form? Was there a time *before* the colorful layers accumulated? These questions pertain to **geologic time**, the span of time since Earth's formation.

In this chapter, we first learn the geologic principles that allowed geologists to develop the concept of geologic time and thus a frame of reference for describing the relative ages of rocks, fossils, structures, and landscapes. This information sets the stage for introducing the geologic column, the way that geologists divide time into intervals. Then we look at the tools geologists use to determine the numerical age of the Earth and its features in years; specifically, we introduce isotopic (radiometric) dating and the geologic time scale. With the concept of geologic time in hand, a hike down a trail into the Grand Canyon becomes a trip into the distant past, into what authors call *deep time*. The geological discovery that our planet's history extends billions of years into the past changed humanity's perception of the Universe as profoundly as did the astronomical discovery that the limit of space extends billions of light years beyond the edge of our Solar System.

TAKE-HOME MESSAGE

By the end of this chapter, you should understand the geologic principles used to determine relative ages of geologic features, how isotopic study provides the numerical age (age in years) of rocks, and how time is divided into intervals of the geologic column.

10.2 THE CONCEPT OF GEOLOGIC TIME

Setting the Stage for Studying the Past

Before the dawn of modern science, most cultures assumed that geologic time was not much longer than human history, and that the Earth essentially formed as we see it today. This view was challenged by James Hutton (1726–1797), who based his conclusions on relationships that he observed in the rocky crags of his native Scotland. Hutton lived during the Age of Enlightenment, when the discovery of physical laws by Sir Isaac Newton led people to look to natural, not supernatural, processes to explain the Universe. In this context, as Hutton wandered around Scotland, he came up with an idea that provides the foundation of geology and led to his title as the "father of geology." This idea, the principle of **uniformitarianism**, states simply that physical processes we observe today also operated in the past at roughly the same rates and were responsible for the formation of the geologic features we see in outcrops. Put more concisely, uniformitarianism means *the present is the key to the past*. If this principle is correct, Hutton reasoned, the Earth must be much older than human history, for observed geologic processes work very slowly.

Hutton was not a particularly clear writer, and it took the efforts of subsequent geologists to clarify the implications of the principle of uniformitarianism and to publicize them. Once this had been accomplished, geologists around the world began to apply their growing understanding of geologic processes to define and interpret geologic events of the Earth's past.

Relative versus Numerical Age

Like historians, geologists want to establish both the sequence of events that created an array of geologic features (such as rocks, structures, and landscapes) and the exact dates on which the events happened. We specify the age of one feature with respect to another as its **relative age** and the age of a feature given in years as the **numerical age** (or, in older literature, the "absolute age"). Geologists developed ways of defining relative age long before they did so for numerical age, so we will look at the principles leading to relative-age determination first.

TAKE-HOME MESSAGE

The principle of uniformitarianism (the present is the key to the past) implies that the Earth must be very old, for geologic processes happen slowly. Geologists distinguish between relative age (is one feature older or younger than another?) and numerical age (age in years).

FIGURE 10.1   The major geologic principles.

(a) Uniformitarianism: The processes that formed cracks in the dried-up mud puddle on the left also formed the mudcracks preserved in the solid Paleozoic rock on the right. We can see these ancient mudcracks because erosion removed the overlying bed.

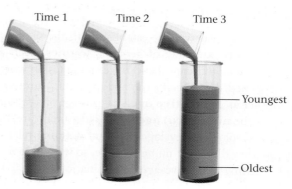

(b) Original horizontality: Gravity causes sediment to accumulate in fairly horizontal sheets.

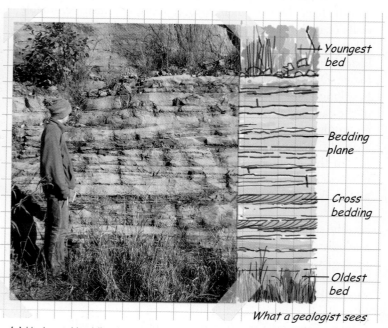

(c) Horizontal bedding in an outcrop of Paleozoic sandstone in Wisconsin.

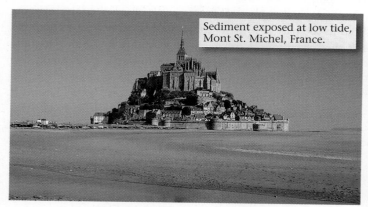

(d) Superposition: In a sequence of strata, the oldest bed is on the bottom, and the youngest on top. Pouring sand into a glass illustrates this point.

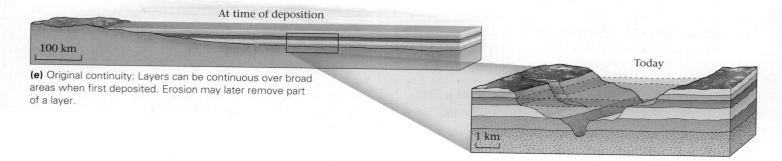

(e) Original continuity: Layers can be continuous over broad areas when first deposited. Erosion may later remove part of a layer.

FIGURE 10.1 (CONTINUED)

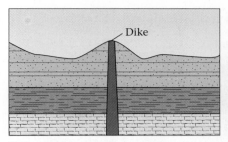

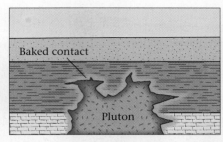

(f) Cross-cutting relations: The dike cuts across the sedimentary beds, so the dike is younger.

(g) Baked contact: The pluton baked the adjacent rock, so the adjacent rock is older.

The pebbles of basalt in a conglomerate must be older than the conglomerate.

Xenoliths of sandstone must be older than the basalt containing them.

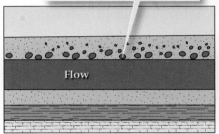

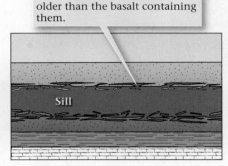

(h) Inclusions: Rock that occurs as an inclusion in another rock must be the older of the two.

10.3 PHYSICAL PRINCIPLES FOR DEFINING RELATIVE AGE

Nineteenth-century geologists, such as Charles Lyell, recast the ideas of earlier researchers into formal, usable geologic principles. These principles, defined below, continue to provide the basic framework within which geologists read the record of Earth history and determine relative ages.

- *The principle of uniformitarianism*: As noted earlier, physical processes we observe operating today also operated in the past, at roughly comparable rates (Fig. 10.1a). Again, the present is the key to the past.
- *The principle of **original horizontality***: Sediments on Earth settle out of a fluid in a gravitational field. Typically, the surfaces on which sediments accumulate (such as a floodplain or the bed of a lake or sea) are fairly horizontal. Therefore, layers of sediment when originally deposited are also fairly horizontal (Fig. 10.1b, c). If sediments were deposited on a steep slope, they would likely slide downslope before lithification, and so would not be preserved as sedimentary layers. With this principle in mind, we realize that when we see folds and tilted beds, we are seeing the consequences of deformation that *post-dates* deposition.

- *The principle of **superposition***: In a sequence of sedimentary rock layers, each layer must be younger than the one below, for a layer of sediment cannot accumulate unless there is already a substrate on which it can collect. Thus, the layer at the bottom of a sequence is the oldest, and the layer at the top is the youngest (Fig. 10.1d).
- *The principle of original continuity*: Sediments generally accumulate in continuous sheets. If today you find a sedimentary layer cut by a canyon, then you can assume that the layer once spanned the canyon but was later eroded by the river that formed the canyon (Fig. 10.1e).
- *The principle of **cross-cutting relations***: If one geologic feature cuts across another, the feature that has been cut is older. For example, if an igneous dike cuts across a sequence of sedimentary beds, the beds must be older than the dike (Fig. 10.1f). If a fault cuts across and displaces layers of sedimentary rock, then the fault must be younger than the layers.
- *The principle of baked contacts*: An igneous intrusion "bakes" (metamorphoses) surrounding rocks. The rock that has been baked must be older than the intrusion (Fig. 10.1g).
- *The principle of inclusions*: If a layer of sediment deposited on an igneous layer includes pebbles of the igneous rock, then the sedimentary layer must be younger (Fig. 10.1h). If an igneous intrusion contains fragments of another rock, the fragments must be older than the intrusion. The fragments (xenoliths) in an igneous body and the pebbles in the sedimentary layer are inclusions, or pieces of one material incorporated in another. The rock containing the inclusion must be younger than the inclusion.

We can use these principles to determine the relative ages of the features shown in Figure 10.2a. In so doing, we *develop a geologic history* of the region, defining the relative ages of events that took place there.

For this example, we propose the following geologic history for this region (Fig. 10.2b): deposition of the sedimentary sequence, in order from layers 1 to 7; intrusion of the sill; folding of the sedimentary layers and the sill; intrusion of the granite pluton; faulting; intrusion of the dike; formation of the land surface.

FIGURE 10.2 Interpreting the geologic history of a region, using geologic principles as a guide.

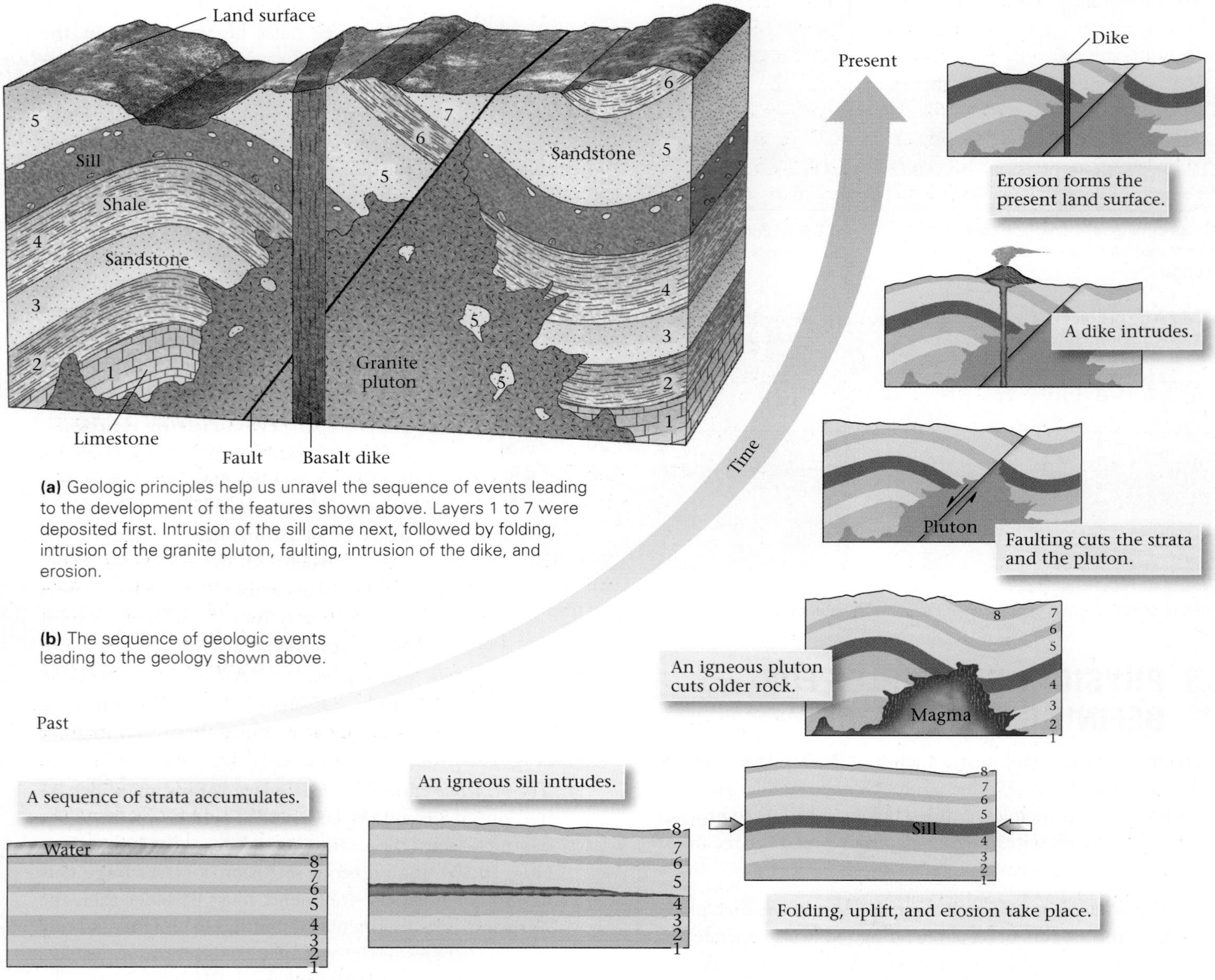

(a) Geologic principles help us unravel the sequence of events leading to the development of the features shown above. Layers 1 to 7 were deposited first. Intrusion of the sill came next, followed by folding, intrusion of the granite pluton, faulting, intrusion of the dike, and erosion.

(b) The sequence of geologic events leading to the geology shown above.

Adding Fossils to the Story: Fossil Succession

As Britain entered the Industrial Revolution in the late eighteenth and early nineteenth centuries, new factories demanded coal to fire their steam engines. The government decided to build a network of canals to transport coal and iron, and hired an engineer named William Smith (1769–1839) to survey the excavations. These excavations provided fresh exposures of bedrock, which previously had been covered by vegetation. Smith learned to recognize distinctive layers of sedimentary rock and to identify the fossil assemblage (the group of fossil species) that they contained. He also realized that a particular assemblage of

fossil species can be found only in a limited interval of strata, and not above or below this interval. Thus, once a fossil species disappears at a horizon in a sequence of strata, it never reappears higher in the sequence. In other words, extinction is forever. Smith's observation has been repeated at millions of locations around the world, and has been codified as the principle of **fossil succession**.

To see how this principle works, examine **Figure 10.3**, which depicts a sequence of strata. Bed 1 at the base contains fossil species A, Bed 2 contains fossil species A and B, Bed 3 contains B and C, Bed 4 contains C, and so on. From these data, we can define the range of specific fossils in the sequence, meaning the interval in the sequence in which the fossils occur. Note that the sequence contains a definable succession of fossils (A, B, C, D, E, F), that the range in which a particular species occurs may overlap with the range of other species, and that once a species vanishes, it does not reappear higher in the sequence.

Because of the principle of fossil succession, we can define the relative ages of strata by looking at fossils. For example, if we find a bed containing Fossil A, we can say that the bed is older than a bed containing, say, Fossil F. Geologists have now identified and determined the relative ages of hundreds of thousands of fossil species.

TAKE-HOME MESSAGE

Relative-age determination is based on geologic principles, including: uniformitarianism (the present is the key to the past), superposition (younger strata overlie older strata), cross-cutting relations (younger features cut older ones), and fossil succession (fossil species occur in a predictable order).

FIGURE 10.3 The principle of fossil succession.

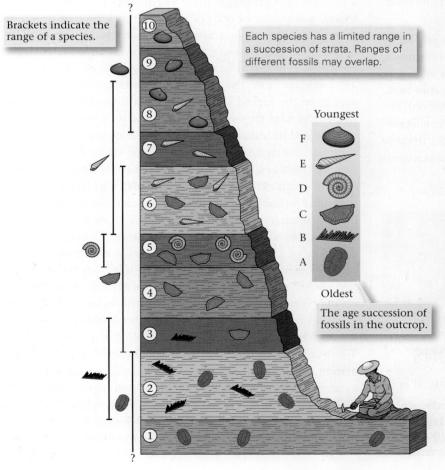

Brackets indicate the range of a species.

Each species has a limited range in a succession of strata. Ranges of different fossils may overlap.

Youngest

F
E
D
C
B
A

Oldest

The age succession of fossils in the outcrop.

10.4 UNCONFORMITIES: GAPS IN THE RECORD

James Hutton often strolled along the coast of Scotland because the shore cliffs provided good exposures of rock, stripped of soil and shrubbery. He was particularly puzzled by an outcrop along the shore at Siccar Point. One sequence of rock exposed there consisted of alternating beds of gray sandstone and shale, but another consisted of red sandstone and conglomerate (Fig. 10.4a, b). The beds of gray sandstone and shale were nearly vertical, whereas the beds of red sandstone and conglomerate were roughly horizontal. Further, the horizontal layers seemed to lie across the truncated ends of the vertical layers, like a handkerchief lying across a row of books. We can imagine that as Hutton examined this odd geometric relationship, the tide came in and deposited a new layer of sand on top of the rocky shore. With the principle of uniformitarianism in mind, Hutton suddenly realized the significance of what he saw. Clearly, the gray sandstone–shale sequence had been deposited, turned into rock, tilted, and truncated by erosion *before* the red sandstone–conglomerate beds were deposited.

Hutton deduced that the surface between the gray and red rock sequences represented a long interval of time during which new strata had not been deposited at Siccar Point and older strata may have been eroded away. We now call such a surface, representing a period of nondeposition and possibly erosion, an **unconformity**. Geologists recognize three kinds of unconformities:

- *Angular unconformity*: Rocks below an angular unconformity were tilted or folded before the unconformity developed (Fig. 10.5a). Thus, an angular unconformity cuts across the underlying layers; the layers below have a different orientation from the layers above. (The outcrop at Siccar Point reveals an angular unconformity.) Angular unconformities form where rocks were tilted by either folding or faulting, before being exposed at the Earth's surface.

- *Nonconformity*: A nonconformity is a type of unconformity at which sedimentary rocks overlie intrusive igneous rocks and/or metamorphic rocks (Fig. 10.5b). The igneous or metamorphic rocks must have cooled, been uplifted, and been exposed by erosion to form the substrate or basement on which new sedimentary rocks were deposited.

- *Disconformity*: Imagine that a sequence of sedimentary beds has been deposited beneath a shallow sea. Then sea level drops, exposing the recently deposited beds for some time. During this time, no new sediment accumulates, and some of the preexisting sediment gets eroded away. Later, sea level rises, and a new sequence of sediment

FIGURE 10.4 Discovering unconformities: James Hutton spotted this one at Siccar Point, Scotland.

(a) At Siccar Point, gently dipping Devonian redbeds lie on vertical beds of Ordovician sandstone.

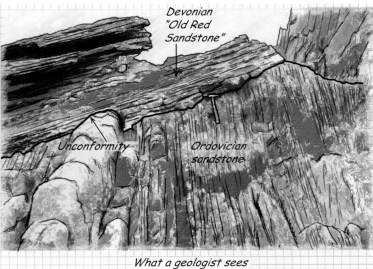

What a geologist sees

(b) Geologists conclude that the contact is an unconformity, and now estimate that, in this example, the rocks below are 50 million years older than the rocks above.

■ accumulates over the old. The boundary between the two sequences is a disconformity (**Fig. 10.5c**). Even though the beds above and below the disconformity are parallel, the contact between them represents an interruption in deposition.

The succession of strata at a particular location provides a record of Earth history there. But because of unconformities, *the record preserved in the rock layers is incomplete*. It's as if geologic history is being chronicled by a tape recorder that turns on only intermittently—when it's on (times of deposition), the rock record accumulates, but when it's off (times of nondeposition and possibly erosion), an unconformity develops. Because of unconformities, no single location on Earth contains a complete record of Earth history.

TAKE-HOME MESSAGE

At a given location, sediments do not accumulate continuously. Surfaces representing intervals of nondeposition and possible erosion are unconformities. Because of unconformities, the geologic record at any given location is incomplete.

10.5 STRATIGRAPHIC FORMATIONS AND THEIR CORRELATION

Geologists summarize information about the sequence of sedimentary strata at a location by drawing a **stratigraphic column**. Typically, we draw columns to scale, so that the rela-

tive thicknesses of layers portrayed on the column reflect the thicknesses of layers in the outcrop. Then, for ease of reference, geologists divide the sequence of strata represented on a column into **stratigraphic formations** (formations, for short), a sequence of beds of a specific rock type or group of rock types that can be traced over a fairly broad region. The boundary surface between two formations is a type of geologic **contact**. Typically, a formation has a specific geologic age.

Let's see how the concept of a stratigraphic formation applies to the Grand Canyon. The walls of the canyon look striped, because they expose a variety of rock types that differ in color and in resistance to erosion. Geologists identify major contrasts distinguishing one interval of strata from another, and use them as a basis for dividing the strata into formations, each of which may consist of many beds (**Fig. 10.6a–c**). Note that some formations include a single rock type, whereas others include interlayered beds of two or more rock types. Also, note that not all formations have the same thickness, and the thickness of a single formation can vary with location. Typically, geologists name a formation after a locality where it was first identified. If the formation consists of only one rock type, we may incorporate that rock type in the name (for example, Kaibab Limestone), but if the formation contains more than one rock type, we use the word formation in the name (such as Toroweap Formation). Note that in the formal name of a formation, all words are capitalized. Several adjacent formations in a succession may be lumped together as a stratigraphic group.

While he was excavating canals in England, William Smith discovered that formations cropping out at one locality resembled formations cropping out at another, in that their

FIGURE 10.5 The three kinds of unconformities and their formation.

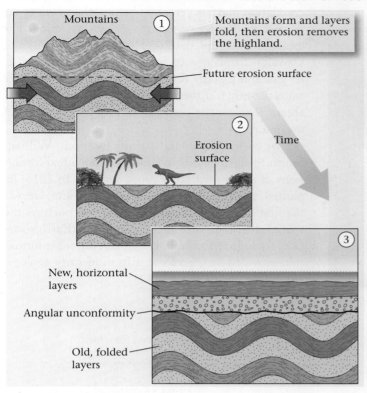

(a) An angular unconformity: (1) layers undergo folding; (2) erosion produces a flat surface; (3) sea level rises and new layers of sediment accumulate.

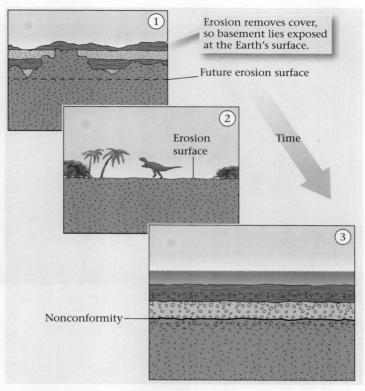

(b) A nonconformity: (1) a pluton intrudes; (2) erosion cuts down into the crystalline rock; (3) new sedimentary layers accumulate above the erosion surface.

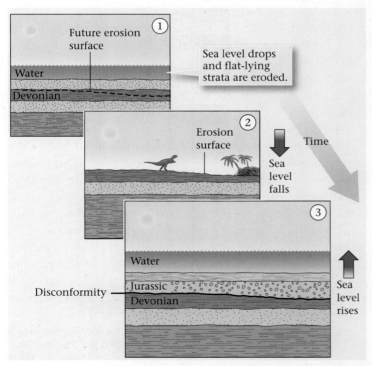

(c) A disconformity: (1) layers of sediment accumulate; (2) sea level drops and an erosion surface forms; (3) sea level rises and new sedimentary layers accumulate.

beds looked similar and contained similar fossil assemblages. In other words, Smith was able to define the age relationship between the strata at one locality and the strata at another, a process called **correlation**.

How does correlation work? Typically, geologists correlate formations between *nearby* regions based on similarities in rock type. We call this method lithologic correlation (Fig. 10.7a, b). For example, the sequence of strata on the southern rim of the Grand Canyon clearly correlates with the sequence on the northern rim, because they contain the same rock types in the same order. Before deformation and/or erosion, the layers that correlate may be continuous. In some cases, a sequence contains a key bed, or marker bed, a particularly unique layer that provides a definitive basis for correlation.

To correlate rock units over *broad* areas, we must rely on fossils to define the relative ages of sedimentary units. We call this method fossil correlation. Geologists use fossil correlation for studies of broad areas because sources of sediments and depositional environments may change from one location to another. The beds deposited at one location during a given time interval may look quite different from the beds deposited at another location during the same time interval. But if fossils of the same relative age occur at both locations, we can say

FIGURE 10.6   The stratigraphic column and stratigraphic formation: examples from the Grand Canyon in Arizona.

The light layer at the top is the Kaibab Limestone.

(a) The walls of the Grand Canyon expose several formations.

that the strata at the two locations correlate. Fossil correlation may also come in handy when rock types are not distinctive enough to allow correlation. For example, imagine that the Santuit Sandstone and Oswaldo Sandstone of Figure 10.7a look the same. In Column C, only fossils may distinguish one layer from the other, if the intervening Milo Limestone is absent. The contact between the two formations is an unconformity.

By correlating strata at many locations, William Smith realized that he could trace individual formations of strata over fairly broad regions. In 1815, he plotted the distribution of formations and created the first modern **geologic map**, which portrays the spatial distribution of rock units at the Earth's surface. As **Figure 10.8b** shows, we can use the information provided by a geologic map to identify geologic structures.

Permian	Kaibab Limestone
	Toroweap Formation
	Coconino Sandstone
	Hermit Shale
Pennsylvanian	Supai Group
Mississippian	Surprise Canyon Fm.
Devonian	Redwall Limestone
	Temple Butte Fm.
Cambrian	Muav Limestone
	Bright Angel Shale
	Tapeats Sandstone
Precambrian	Grand Canyon Supergroup
	Zoroaster Granite Vishnu Schist

Irregularity of this edge represents resistance to erosion. Stronger units protrude as ledges.

m — 300
— 150
— 0

〜 Unconformity
Sandstone
Limestone
Shale
Siltstone

(b) This stratigraphic column is a chart that represents the sequence of strata in the Grand Canyon. The vertical scale indicates thicknesses.

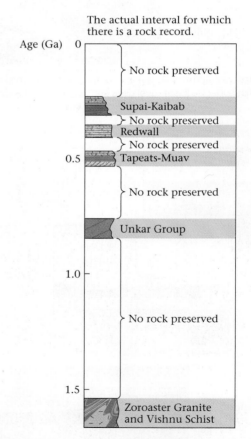

The actual interval for which there is a rock record.

Age (Ga)   0

No rock preserved

Supai-Kaibab
No rock preserved
Redwall
No rock preserved

0.5   Tapeats-Muav

No rock preserved

Unkar Group

1.0

No rock preserved

1.5

Zoroaster Granite and Vishnu Schist

(c) Because of unconformities, the strata at the Grand Canyon represent only a partial record of geologic time. By plotting the interval of time represented by each formation on a numerical time scale, we see large gaps in the record.

FIGURE 10.7 The principles of correlation.

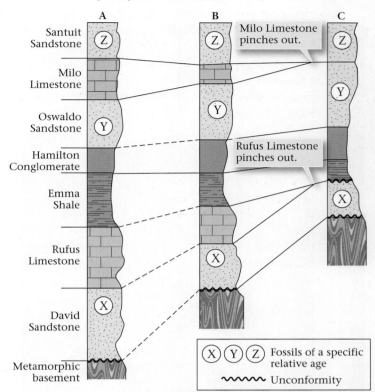

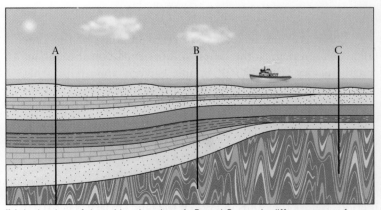

(a) Stratigraphic columns can be correlated by matching rock types (lithologic correlation). The Hamilton Conglomerate is a marker horizon. Because some strata pinch out, Column C contains unconformities. Fossil correlation indicates that the youngest beds in C are Santuit Sandstone.

(b) At the time of deposition, locations A, B, and C were in different parts of a basin. The basin floor was subsiding fastest at A.

TAKE-HOME MESSAGE

A recognizable sequence of beds that can be mapped across a broad region is called a stratigraphic formation. The boundary surface between two formations is a type of contact. Geologists correlate formations regionally on the basis of rock type and fossil content.

FIGURE 10.8 A geologic map depicts the distribution of rock units and structures.

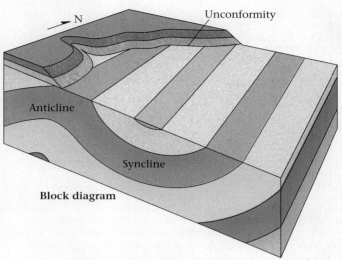

(a) A block diagram provides a three-dimensional representation. Here, we see an angular unconformity over folds.

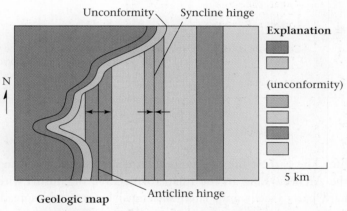

(b) A geologic map shows what the distribution of units would look like if viewed from above. Contacts occur between units.

10.6 THE GEOLOGIC COLUMN

As stated earlier, no one locality on Earth provides a complete record of our planet's history, because stratigraphic columns can contain unconformities. But by correlating rocks from locality to locality at millions of places around the world, geologists have pieced together a *composite* stratigraphic column, called the **geologic column**, that *represents* the entirety of Earth history (Fig. 10.9a, b). The column is divided into segments, each of which represents a specific interval of time. The largest subdivisions break Earth history into the Hadean, Archean, Proterozoic, and Phanerozoic **Eons**. (The first three together constitute the **Precambrian**.) The suffix "-zoic" means life, so Phanerozoic means visible life, and Proterozoic means earlier

FIGURE 10.9   Global correlation of strata led to the development of the geologic column.

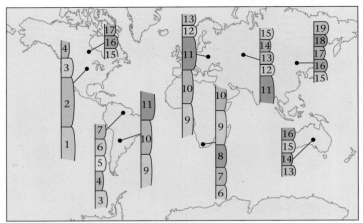

	Eon	Era	Period	Epoch
19			Quaternary	Holocene
18				Pleistocene
17		Cenozoic		Pliocene
16			Tertiary	Miocene
15				Oligocene
14				Eocene
13				Paleocene
12				
11	Phanerozoic	Mesozoic	Cretaceous	
10			Jurassic	
9			Triassic	
8			Permian	Pennsylvanian
7			Carboniferous	Mississippian
6		Paleozoic	Devonian	
5			Silurian	
4			Ordovician	
3			Cambrian	
2	Precambrian	Proterozoic		
1		Archean		

(a) Each of these small columns represents the stratigraphy at a given location. By correlating these columns, geologists determined their relative ages, filled in the gaps in the record, and produced the geologic column.

(b) By correlation, the strata from localities around the world were stacked in a chart representing geologic time to create the geologic column. Geologists assigned names to time intervals, but since the column was built without knowledge of numerical ages, it does not depict the duration of these intervals.

life. The earliest life, bacteria and archaea, appeared during the Archean Eon. In the Phanerozoic Eon, organisms with hard parts (shells and, later, skeletons) became widespread, so there are abundant fossils from this eon, whereas in Precambrian time, only small organisms with no shells existed, so Precambrian fossils are hard to find.

The Phanerozoic Eon is subdivided into **eras**. In order from oldest to youngest, they are the Paleozoic (ancient life), Mesozoic (middle life), and Cenozoic (recent life) Eras. We can further divide each era into **periods** and each period into **epochs**.

Where do the names of the periods come from? They refer either to localities where a fairly complete stratigraphic column representing that time interval was first identified (for example, rocks representing the Devonian Period crop out near Devon, England) or to a characteristic of the time (rocks from the Carboniferous Period contain a lot of coal). The terminology was not set up in a planned fashion that would make it easy to learn. Instead, it grew haphazardly in the years between 1760 and 1845, as geologists began to refine their understanding of geologic history and fossil succession.

The succession of fossils preserved in strata of the geologic column defines the course of life's evolution throughout Earth history (Fig. 10.10). Simple bacteria appeared during the Archean Eon, but complex shell-less invertebrates did not evolve until the late Proterozoic. The appearance of invertebrates with shells defines the Precambrian-Cambrian boundary. At this time there was a sudden diversification in life, with many new genera appearing over a relatively short interval—this event is called the **Cambrian explosion**.

The first vertebrates, fish, appeared during the Ordovician Period. Before the Silurian Period, the land surface was barren of multicellular life, but in the Silurian Period, land plants spread over the continents. Amphibians appeared during the Devonian. Though reptiles evolved during the Pennsylvanian Period, the first dinosaurs did not pound across the land until the Triassic. Dinosaurs continued to inhabit the Earth until their sudden extinction at the end of the Cretaceous Period. For this reason, geologists refer to the Mesozoic Era as the Age of Dinosaurs. Small mammals also appeared during the Triassic Period, but the diversification (development of many different

FIGURE 10.10 Life evolution in the context of the geologic column. The Earth formed at the beginning of the Hadean Eon.

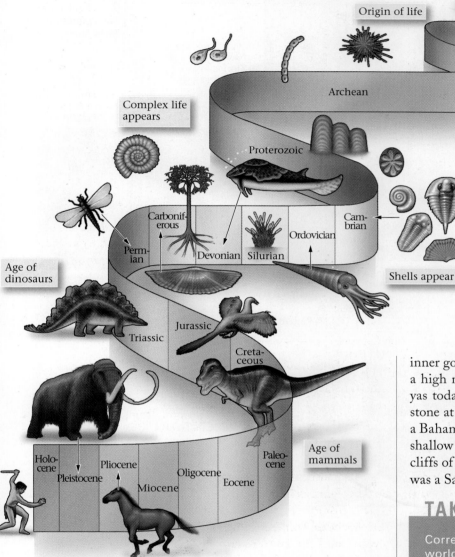

Big bang

Origin of life

Hadean

Archean

Complex life appears

Proterozoic

Carbonif- erous

Perm- ian

Devonian Silurian

Ordovician

Cam- brian

Shells appear

Age of dinosaurs

Jurassic

Triassic

Creta- ceous

Holo- cene

Pliocene

Pleistocene

Miocene

Oligocene

Eocene

Paleo- cene

Age of mammals

of vegetation in this region, you can easily see bedrock exposures on the walls of cliffs and canyons, and some of these exposures are so beautiful that they have become national parks. The oldest sedimentary rock of the region crops out at the base of the Grand Canyon, whereas the youngest form the cliffs of Cedar Breaks and Bryce Canyon (see **Geotour 10** on p. GT-22).

Walking through these parks is thus like walking through time—each rock layer gives an indication of the climate and topography of the region in the past (see **Geology at a Glance**, pp. 290–291). For example, when the Precambrian metamorphic and igneous rocks exposed in the inner gorge of the Grand Canyon first formed, the region was a high mountain range, perhaps as dramatic as the Himalayas today. When the fossiliferous beds of the Kaibab Limestone at the rim of the canyon first developed, the region was a Bahama-like carbonate reef and platform, bathed in a warm, shallow sea. And when the rocks making up the towering red cliffs of sandstone in Zion Canyon were deposited, the region was a Sahara-like desert, blanketed with huge sand dunes.

TAKE-HOME MESSAGE

Correlation of stratigraphic sequences from around the world allowed production of a chart, the geologic column, that represents the entirety of Earth history. The column, developed using only relative age relations, is subdivided into eons, periods, and epochs.

species) of mammals to fill a wide range of ecological niches did not happen until the beginning of the Cenozoic Era, so geologists call the Cenozoic the Age of Mammals. Birds also appeared during the Mesozoic (specifically, at the beginning of the Cretaceous Period), but underwent great diversification in the Cenozoic Era. Note that some species existed only for a short interval of the geologic column, and thus are diagnostic of a particular period or epoch. The fossils of such species are called **index fossils**. The best index fossils, for correlation purposes, are from species that were widespread.

To conclude our discussion, let's see how the geologic column comes into play when correlating strata across a region. We return to the Colorado Plateau of Arizona and Utah, in the southwestern United States (**Fig. 10.11a, b**). Because of the lack

10.7 HOW DO WE DETERMINE NUMERICAL AGE?

Geologists since the days of Hutton could determine the relative ages of geologic events, but they had no way to specify numerical ages. Thus, they could not define a time line for Earth history or determine the duration of events. This situation changed with the discovery of radioactivity. Simply put, radioactive elements decay at a constant rate that can be measured in the lab

FIGURE 10.11 Correlation of strata among the national parks of Arizona and Utah.

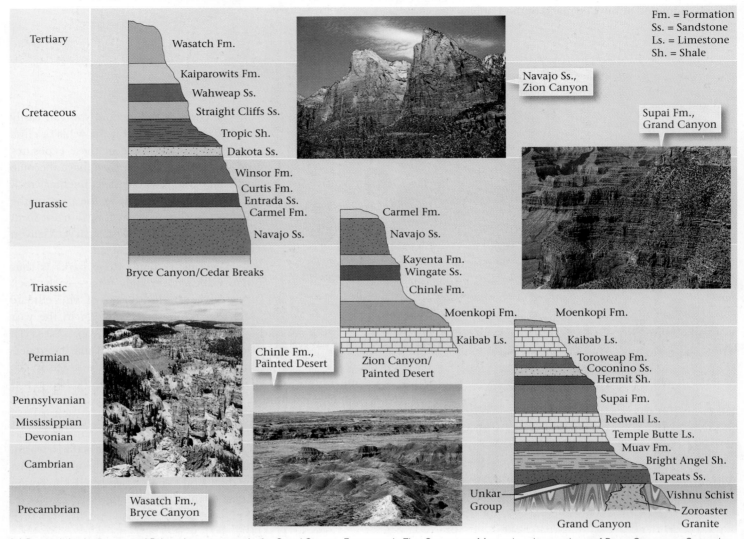

(a) Precambrian basement and Paleozoic strata occur in the Grand Canyon. Exposures in Zion Canyon are Mesozoic, whereas those of Bryce Canyon are Cenozoic.

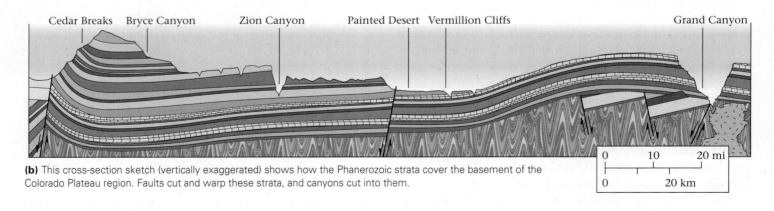

(b) This cross-section sketch (vertically exaggerated) shows how the Phanerozoic strata cover the basement of the Colorado Plateau region. Faults cut and warp these strata, and canyons cut into them.

and can be specified in years. In the 1950s, geologists first developed techniques for using measurements of radioactive elements to calculate the ages of rocks. This process of determining the numerical age of rocks was called radiometric dating, and more recently, **isotopic dating**. Geologists refer to the study of numerical ages as geochronology. Techniques of isotopic dating have been vastly improved over the years. We begin our discussion of this technique by first learning about radioactive decay.

Isotopes, Radioactive Decay, and the Concept of a Half-Life

All atoms of a given element have the same number of protons in their nucleus—we call this number the atomic number (see Appendix). However, not all atoms have the same number of neutrons in their nucleus. Therefore, not all atoms of a given element have the same atomic weight—roughly the number of protons plus neutrons. Different versions of an element, called **isotopes** of the element, have the same atomic number but a different atomic weight (see Appendix). For example, all uranium atoms have 92 protons, but the uranium-238 isotope (abbreviated ^{238}U) has an atomic weight of 238 and thus has 146 neutrons, whereas the ^{235}U isotope has an atomic weight of 235 and thus has 143 neutrons.

Some isotopes of an element are stable, meaning that they last essentially forever. Radioactive isotopes are unstable: after a given time, they undergo a change called **radioactive decay**, which converts them into a different element. Radioactive decay can take place by a variety of reactions that change the atomic number of the nucleus and thus form a different element. In these reactions, the isotope that undergoes decay is the **parent isotope**, while the decay product is the **daughter isotope**. For example, rubidium-87 (^{87}Rb) decays to strontium-87 (^{87}Sr), potassium-40 (^{40}K) decays to argon-40 (^{40}Ar), and uranium-238 (^{238}U) decays to lead-206 (^{206}Pb).

Physicists cannot specify how long an individual radioactive isotope will survive before it decays, but they can measure how long it takes for half of a group of parent isotopes to decay. This time is called the **half-life** of the isotope. Figure 10.12a–c can help you visualize the concept of a half-life. Imagine a crystal containing 16 radioactive parent isotopes. (In real crystals, the number of atoms would be much larger.) After one half-life, 8 isotopes have decayed, so the crystal now contains 8 parent and 8 daughter isotopes. After a second half-life, 4 of the remaining parent isotopes have decayed, so the crystal contains 4 parent and 12 daughter isotopes. And after a third half-life, 2 more parent isotopes have decayed, so the crystal contains 2 parent and 14 daughter isotopes. For a given decay reaction, the half-life is a constant.

FIGURE 10.12 The concept of a half-life, in the context of radiaoctive decay.

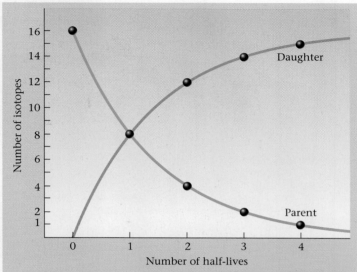

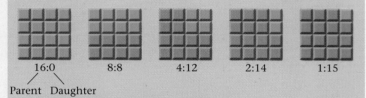

(a) This graph shows how the number of parent isotopes decreases and the number of daughter isotopes increases, as time passes. The rate of change decreases with time.

(b) The ratio of parent-to-daughter isotopes changes with the passage of each successive half-life.

(c) In a cluster of isotopes undergoing decay, there is no way to predict which parent will decay next.

Radiometric Dating Techniques

Like the ticktock of a clock, radioactive decay proceeds at a known rate and thus provides a basis for telling time. In other words, because an element's half-life is a constant, we can calculate the age of a mineral by measuring the ratio of parent to daughter isotopes in the mineral.

How do geologists actually obtain an isotopic date? First, we must find the right kind of elements to work with. Although there are many different pairs of parent and daughter isotopes among the known radioactive elements, only a few have long enough half-lives, and occur in sufficient abundance in minerals,

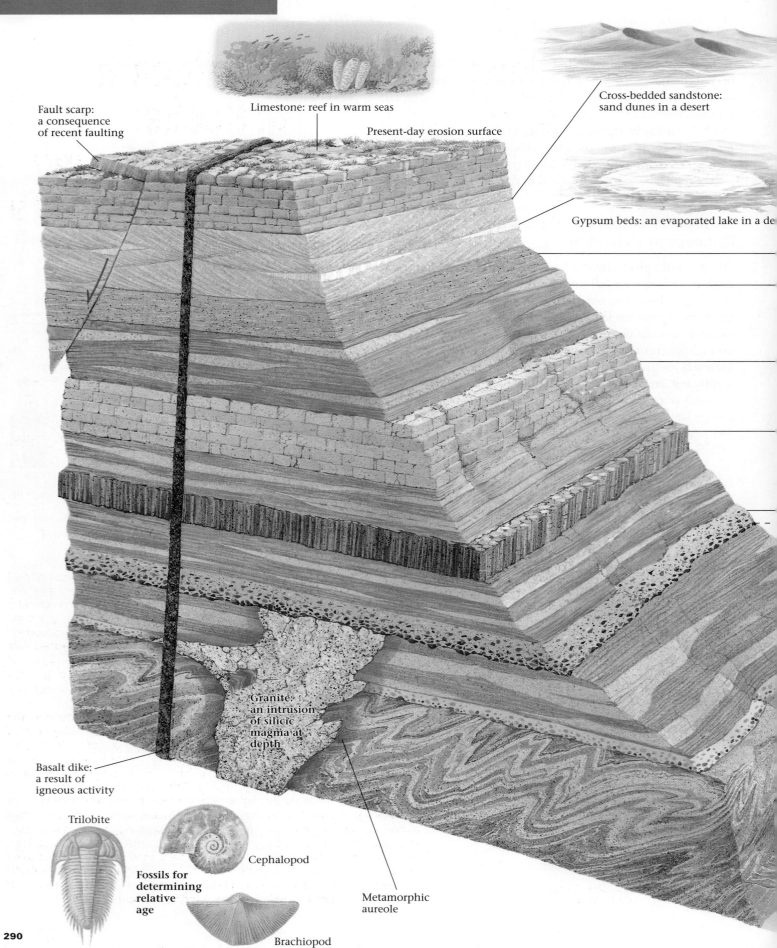

Limestone: reef in warm seas

Cross-bedded sandstone: sand dunes in a desert

Fault scarp: a consequence of recent faulting

Present-day erosion surface

Gypsum beds: an evaporated lake in a de

Granite: an intrusion of silicic magma at depth

Basalt dike: a result of igneous activity

Trilobite

Cephalopod

Fossils for determining relative age

Metamorphic aureole

Brachiopod

The Record in Rocks: Reconstructing Geologic History

Ignimbrite (welded tuff): an explosive volcanic eruption

Limestone: reef in warm seas

Redbeds: sand and mud deposited in a river channel and bordering floodplain

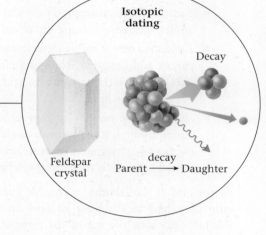

Isotopic dating

Decay

Feldspar crystal

Parent → Daughter

decay

Basalt lava: flows from a volcano

Conglomerate: debris eroded from a cliff

— Unconformity

Redbeds: sand and mud deposited by distributaries of a delta plain

Conglomerate: deposits of a pebble beach

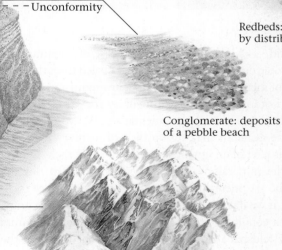

Gneiss: metamorphism at depth beneath a mountain belt

When geologists examine a sequence of rocks exposed on a cliff, they see a record of Earth history that can be interpreted by applying the basic principles of geology, searching for fossils, and using radiometric dating. In this canyon, we see evidence for many geologic events. The layers of sediment (and the sedimentary structures they contain), the igneous intrusions, and the geologic structures tell us about past climates and past tectonic activity.

The insets show the way the region looked in the past, based on the record in the rocks. For example, the presence of gneiss at the base of the canyon indicates that at one time the region was a mountain belt, for the protoliths of the gneiss were buried deeply. Unconformities indicate that the region underwent uplift and erosion. Sedimentary successions record transgressions and regressions of the sea, igneous rocks are evidence of volcanic and intrusive activity, and faults indicate deformation.

Clearly, the land surface portrayed in this painting was sometimes a river floodplain or a delta (indicated by redbeds), sometimes a shallow sea (limestone), and sometimes a desert dune field (cross-bedded sandstones). And at several times in the past, volcanic activity occurred in the region. We can gain insight into the age of the sedimentary rocks by studying the fossils they contain, and the age of the igneous and metamorphic rocks by using isotopic dating methods.

to be useful for isotopic dating. Particularly useful elements are listed in **Table 10.1**. Each radioactive element has its own half-life. Note that carbon dating is *not* used for dating rocks because carbon isotopes used in dating occur only in organisms and have a very short half life.

Second, we must identify the right kind of minerals to work with. Not all minerals contain radioactive elements, but fortunately some common minerals do. Now we can set to work using the following steps.

- *Collecting the rocks:* Geologists collect unweathered rocks for dating, for the chemical reactions that happen during weathering may allow isotopes to leak.

- *Separating the minerals:* The rocks are crushed, and the appropriate minerals are separated from the debris.

- *Extracting parent and daughter isotopes:* To separate out the parent and daughter isotopes from minerals, geologists either dissolve the minerals in acid or evaporate portions of them with a laser.

- *Analyzing the parent-daughter ratio:* Geologists pass the atoms through a mass spectrometer, an instrument that uses a strong magnet to separate isotopes from one another according to their respective weights (Fig. 10.13).

At the end of the laboratory process, geologists can define the ratio of parent to daughter isotopes in a mineral, and from this ratio calculate the age of the mineral. Needless to say, the description of the procedure here has been simplified—in reality, obtaining an isotopic date is time-consuming and expensive and requires complex calculations.

What Does an Isotopic Date Mean?

At high temperatures, atoms in a crystal lattice vibrate so rapidly that chemical bonds can break and reattach relatively easily. As a consequence, parent and daughter isotopes escape from or move into crystals, so parent-daughter ratios are meaningless. Because isotopic dating is based on the parent-daughter ratio, the "isotopic clock" starts only when crystals become cool enough for both parent and daughter isotopes to be locked into the lattice. The temperature below which isotopes are no longer free to move is called the **closure temperature** of a mineral. When we specify an isotopic date for a rock, we are defining the time at which a specific mineral in the rock cooled below its closure temperature.

With the concept of closure temperature in mind, we can interpret the meaning of isotopic dates. In the case of igneous rocks, isotopic dating tells you when a magma or lava cooled

TABLE 10.1 **Isotopes Used in the Radiometric Dating of Rocks**

Parent → Daughter	Half-Life (years)	Minerals Containing the Isotopes
$^{147}Sm \rightarrow {}^{143}Nd$	106 billion	Garnets, micas
$^{87}Rb \rightarrow {}^{87}Sr$	48.8 billion	Potassium-bearing minerals (mica, feldspar, hornblende)
$^{238}U \rightarrow {}^{206}Pb$	4.5 billion	Uranium-bearing minerals (zircon, uraninite)
$^{40}K \rightarrow {}^{40}Ar$	1.3 billion	Potassium-bearing minerals (mica, feldspar, hornblende)
$^{235}U \rightarrow {}^{207}Pb$	713 million	Uranium-bearing minerals (zircon, uraninite)

Sm = samarium, Nd = neodymium, Rb = rubidium, Sr = strontium, U = uranium, Pb = lead, K = potassium, Ar = argon

FIGURE 10.13 In an isotopic dating laboratory, samples are analyzed using a mass spectrometer. This instrument measures the ratio of parent to daughter isotopes.

to form a solid, cool igneous rock. In the case of metamorphic rocks, an isotopic date tells you when a rock cooled from the high temperature of metamorphism down to a low temperature.

Can we isotopically date a sedimentary rock directly? No. If we date the minerals in a sedimentary rock, we determine only when the minerals making up the sedimentary rock first crystallized as part of an igneous or metamorphic rock, not the time when the minerals were deposited as sediment nor the time when the sediment lithified to form a sedimentary rock. For example, if we date the feldspar grains contained in a granite pebble in a conglomerate, we're dating the time the granite cooled below feldspar's closure temperature, not the time the pebble was deposited by a stream. The age of mineral grains in sediment, however, can be useful. In recent years, geologists have used the ages of detrital (clastic) grains to learn the age of the rocks in the sediment's source region.

Other Methods of Determining Numerical Age

The rate of tree growth depends on the season. During the spring, trees grow rapidly and produce lighter, less dense wood, but during the winter, trees grow slowly or not at all, and produce darker, denser wood. Thus, wood contains recognizable annual growth rings. Such tree rings provide a basis for determining age. If you've ever wondered how old a tree that's just been cut down might be, just look at the stump and count the rings. Notably, by correlating clusters of distinctive rings in the older parts of living trees with comparable clusters of rings in dead logs, scientists can extend the tree-ring record back for many thousands of years, allowing geologists to track climate changes back into prehistory.

Seasonal changes also affect rates of such phenomena as shell growth, snow accumulation, clastic sediment deposition, chemical sediment precipitation, and production of organic material. Geologists have learned to use growth rings in shells, as well as rhythmic layering in sediments and in glacial ice (Fig. 10.14), to date events numerically back through recent Earth history.

TAKE-HOME MESSAGE

Radiometric dating provides numerical ages (in years). To obtain a radiometric date, geologists measure the ratio of parent radioactive isotopes to stable daughter products. The date gives the time at which a mineral cooled below its closure temperature.

10.8 HOW DO WE ADD NUMERICAL AGES TO THE GEOLOGIC COLUMN?

We have seen that isotopic (radiometric) dating can be used to date the time when igneous rocks formed and when metamorphic rocks metamorphosed, but not when sedimentary rocks were deposited. So how do we determine the numerical age of a sedimentary rock? We must answer this question if we want to add numerical ages to the geologic column—remember, the column was originally constructed by studying only the *relative* ages of fossil-bearing sedimentary rocks.

Geologists obtain dates for sedimentary rocks by studying cross-cutting relationships between sedimentary rocks and datable igneous or metamorphic rocks. For example, if we find a sedimentary rock layer deposited unconformably on a datable igneous rock, we know that the sedimentary rock must be younger than the igneous rock. If we find a datable igneous dike that cuts across beds of sedimentary rock, then the datable rock must be younger. And if a datable ash layer spread out over a layer of sediment and then was buried by another layer of sediment, the datable rock or ash must be younger than the underlying sediment and older than the overlying sediment (Fig. 10.15).

FIGURE 10.14 Seasonal layers in the ice of a glacier.

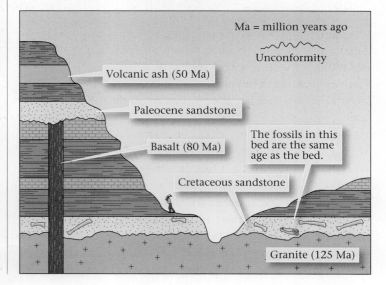

FIGURE 10.15 The Cretaceous sandstone bed was deposited on the granite, so it must be younger than 125 Ma. The dike cuts the bed, so the bed must be older than 80 Ma. Thus, the Cretaceous bed was deposited between 125 and 80 Ma. The Paleocene sandstone was unconformably deposited over the dike and lies beneath a 50-million-year-old layer of ash. Therefore, it must have been deposited between 80 and 50 Ma.

Ma = million years ago

Unconformity

Volcanic ash (50 Ma)

Paleocene sandstone

The fossils in this bed are the same age as the bed.

Basalt (80 Ma)

Cretaceous sandstone

Granite (125 Ma)

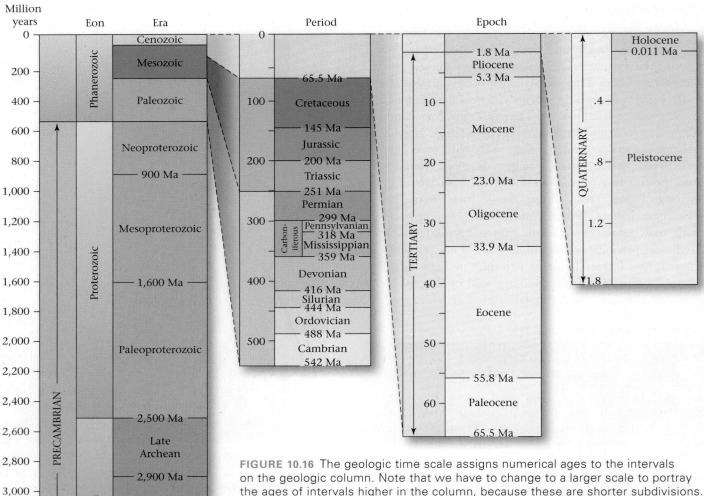

FIGURE 10.16 The geologic time scale assigns numerical ages to the intervals on the geologic column. Note that we have to change to a larger scale to portray the ages of intervals higher in the column, because these are shorter subdivisions. This time scale utilizes numbers favored by the International Commission on Stratigraphy. For further information, see the **GeoWhen Database** on the web.

from around the world shows that the Cretaceous Period began about 145 million years ago and ended 65 million years ago. So the bed in Figure 10.15 was deposited during the middle part of the Cretaceous, not at the beginning or end.

The discovery of new data may cause the numbers defining the boundaries of periods to change, which is why the term *numerical age* is preferred to *absolute age*. In fact, around 1995, new research showed that the Cambrian-Precambrian boundary occurred at about 542 million years ago, in contrast to previous, less definitive studies that had placed the boundary at 570 million years ago. **Figure 10.16** shows the currently favored numerical ages of periods and eras in the geologic column. This dated column is commonly called the **geologic time scale**. Because of the numerical constraints provided by the geologic time scale, when geologists say that the first dinosaurs appeared during the Triassic Period, they mean that dinosaurs appeared later than 251 million years ago.

Geologists have searched the world for localities where they can recognize cross-cutting relations between datable igneous rocks and sedimentary rocks. By isotopically dating the igneous rocks, they have been able to provide numerical ages for the boundaries between all geologic periods. For example, work

TAKE-HOME MESSAGE

Sedimentary rocks cannot be dated directly. So attaching numerical ages to the periods of the geologic column required geologists to radiometrically date igneous or metamorphic rocks in known cross-cutting relations with fossiliferous sedimentary rocks. Such work yielded the geologic time scale.

10.9 WHAT IS THE AGE OF THE EARTH?

The mind grows giddy gazing so far back into the abyss of time.

—John Playfair (1747–1819),
British geologist who popularized the works of Hutton

Isotopic Dating of the Earth

During the eighteenth and nineteenth centuries, before the discovery of isotopic dating, scientists came up with a great variety of clever solutions to the question "How old is the Earth?"—all of which have since been proven wrong. Lord William Kelvin, a nineteenth-century physicist renowned for his discoveries in thermodynamics, made the most influential scientific estimate of the Earth's age of his time. Kelvin calculated how long it would take for a planet to cool down from a temperature as hot as the Sun's, and concluded that the Earth is about 20 million years old.

Kelvin's view contrasted with the idea being promoted by followers of Hutton, Lyell, and Darwin, who argued that if the concepts of uniformitarianism and evolution were correct, the Earth must be much older, because physical processes that shape the Earth and form its rocks and the process of natural selection that yields the diversity of species take a long time. Geologists and physicists debated the age issue for many years. The route to a solution came in 1896, when Henri Becquerel announced the discovery of radioactivity. Geologists immediately realized that the Earth's interior was producing some heat from the decay of radioactive material. This realization uncovered one of the flaws in Kelvin's argument: Kelvin had assumed that no new heat was produced after the planet first formed. Because radioactivity constantly generates new heat in the Earth, the planet has cooled down much more slowly than Kelvin had calculated. The discovery of radioactivity not only invalidated Kelvin's estimate of the Earth's age, it also led to the development of isotopic dating.

Since the 1950s, geologists have scoured the planet to identify its oldest rocks. Samples from several localities (Wyoming, Canada, Greenland, and China) have yielded dates as old as 3.96 billion years. And sandstones found in Australia contain clastic grains that yielded dates of between 4.1 and 4.2 billion years, indicating that rock as old as 4.2 billion years did once exist. Models of the Earth's formation suggest that all objects in the Solar System developed at roughly the same time from the same nebula. Radiometric dating of meteors and Moon rocks have yielded ages as old as 4.57 billion years; geologists take this number to be the approximate age of the Earth, leaving more than enough time for the rocks and life forms of the Earth to have formed and evolved.

We don't find 4.57-billion-year-old rocks in the crust because during the first half-billion years of Earth history, rocks in the crust remained too hot for the isotopic clock to start (their temperature stayed above the closure temperature), and/or most rock that formed before about 4.0 Ga was destroyed by an intense meteorite bombardment at about 4.0 Ga. Geologists have named the time interval between the birth of the Earth and the formation of the oldest dated rock the Hadean Eon.

Picturing Geologic Time

The number 4.57 billion is so staggeringly large that we can't begin to comprehend it. If you lined up this many pennies in a row, they would make an 87,400-km-long line that would wrap around the Earth's equator more than twice. Notably, at the scale of our penny chain, human history is only about 100 city blocks long.

Another way to grasp the immensity of geologic time is to make a scale model, which we do by equating the entire 4.57 billion years to a single calendar year. On this scale, the oldest rocks preserved on Earth date from early February, and the first bacteria appear in the ocean on February 21. The first shelly invertebrates appear on October 25, and the first amphibians crawl out onto land on November 20. On December 7, the continents coalesce into the supercontinent of Pangaea. The first mammals and birds appear about December 15, along with the dinosaurs, and the Age of Dinosaurs ends on December 25. The last week of December represents the last 65 million years of Earth history, including the entire Age of Mammals. The first human-like ancestor appears on December 31 at 3 P.M., and our species, *Homo sapiens,* shows up an hour before midnight. The last ice age ends a minute before midnight, and all of recorded human history takes place in the last 30 seconds. To put it another way, human history occupies the last 0.000001% of Earth history.

TAKE-HOME MESSAGE

Because of heat-generating processes during early Earth history, the oldest rocks on Earth cooled below their closure temperature a half billion years after the Earth formed. Thus, the age of the Earth, 4.57 billion years, comes from dating meteorites.

Chapter Summary

- The concept of geologic time, the time span since the Earth's formation, developed when geologists suggested that geologic features were formed by the same processes we see today.

- Relative age specifies whether one geologic feature is older or younger than another; numerical age provides the age of a geologic feature in years.

- Using such principles as uniformitarianism, original horizontality, superposition, and cross-cutting relations, we can construct the geologic history of a region.

- The principle of fossil succession states that the assemblage of fossils in a sequence of strata changes from base to top. Once a species becomes extinct, it never reappears.

- Strata are not deposited continuously at a location. An interval of nondeposition and/or erosion is called an unconformity. Geologists recognize three kinds: angular unconformity, nonconformity, and disconformity.

- A stratigraphic column shows the succession of strata in a region. A given succession of strata that can be traced over a fairly broad region is called a stratigraphic formation. The process of determining the relationship between strata at one location and strata at another is called correlation. A geologic map shows the distribution of formations.

- A composite chart that represents the entirety of geologic time is called the geologic column. The column's largest subdivisions, each of which represents a specific interval of time, are eons. Eons are subdivided into eras, eras into periods, and periods into epochs.

- The numerical age of rocks can be determined by isotopic (radiometric) dating. This is because radioactive elements decay at a constant rate known as a half-life.

- The isotopic date of a mineral specifies the time at which the mineral cooled below a certain temperature. We can use isotopic dating to determine when an igneous rock solidified and when a metamorphic rock cooled from high temperatures. To date sedimentary strata, we must examine cross-cutting relations with dated igneous or metamorphic rock.

- Other methods for dating materials include counting growth rings in trees and seasonal layers in glaciers.

- From the radiometric dating of meteors and Moon rocks, geologists conclude that the Earth formed about 4.57 billion years ago. Our species, *Homo sapiens*, has been around for only 0.000001% of geologic time.

GEOPUZZLE REVISITED

The age of the Earth comes from isotopic dating of meteorites thought to have formed at the same time as the Earth. This technique, based on measuring the ratio of parent radioactive elements to stable daughter products, tells us when minerals drop below a certain temperature. The Earth was too hot for the first half billion years of its existence for the isotopic clocks of crustal rocks to start ticking, so the oldest known rocks on Earth are only about 4 billion years old. Layers of strata record Earth's history. But as geologic environments change, a given locality that accumulates sediment during part of this history may be exposed and subjected to erosion later on. Thus, there is no single place on Earth where a sequence of strata records all of Earth history. But correlation of strata from around the world has let geologists produce a composite chart, the geologic column, that represents all of geologic time.

Key Terms

Cambrian explosion (p. 286)
closure temperature (p. 292)
contact (p. 282)
correlation (p. 283)
cross-cutting relations (p. 279)
daughter isotope (p. 289)
eon (p. 285)
epoch (p. 286)
era (p. 286)
fossil succession (p. 280)
geologic column (p. 285)
geologic map (p. 284)
geologic time (p. 277)
geologic time scale (p. 294)
half-life (p. 289)

index fossil (p. 287)
isotope (p. 289)
isotopic dating (p. 289)
numerical age (p. 277)
original horizontality (p. 279)
parent isotope (p. 289)
period (p. 286)
Precambrian (p. 285)
radioactive decay (p. 289)
relative age (p. 277)
stratigraphic column (p. 282)
stratigraphic formation (p. 282)
superposition (p. 279)
unconformity (p. 281)
uniformitarianism (p. 277)

Review Questions

1. Compare numerical age and relative age.

2. Describe the principles that allow us to determine the relative ages of geologic events.

3. How does the principle of fossil succession allow us to determine the relative ages of strata?

4. How does an unconformity develop? Describe the differences among the three kinds of unconformities.

5. Describe two different methods of correlating rock units. How was correlation used to develop the geologic column? What is a stratigraphic formation?

6. What does the process of radioactive decay entail?

7. How do geologists obtain an isotopic date? What are some of the pitfalls in obtaining a reliable one?

8. Why can't we date sedimentary rocks directly?

9. How are growth rings and ice cores useful in determining the ages of geologic events?

10. What is the age of the oldest rocks on Earth? What is the age of the oldest rocks known? Why is there a difference?

On Further Thought

1. Imagine an outcrop exposing a succession of alternating sandstone and conglomerate beds. A geologist studying the outcrop notes the following:

   - The sandstone beds contain fragments of land plants, but the fragments are too small to permit identification of species.
   - A layer of volcanic ash overlies the sandstone bed. Radiometric dating indicates that this ash is 300 Ma.
   - A paleosol occurs at the base of the ash layer.
   - A basalt dike, dated at 100 Ma, cuts both the ash and the sandstone-conglomerate sequence.
   - Pebbles of granite in the conglomerate yield radiometric dates of 400 Ma.

   On the basis of these observations, how old is the sandstone conglomerate? (Specify both the numerical age range and the period or periods of the geologic column during which it formed.) If the igneous rocks were not present, could you still specify the maximum or minimum ages of the sedimentary beds? Explain.

THE VIEW FROM SPACE   The pages of Earth history stand on end in Namibia, southwestern Africa. Here, in a false-color image, the Ugab River cuts across layer upon layer of strata that were tilted to near-vertical by a Precambrian mountain-building event. Subsequent erosion exposed the strata, and the desert climate keeps it clear of vegetation. The field of view is 45 km.

CHAPTER **11**

A Biography of Earth

Geology students examining an outcrop of Archean rock near Marquette, Michigan. Field observations allow geologists to piece together a history of the Earth.

GEOPUZZLE

Is life on Earth as old as the Earth? Has land always been land? When did the mountain belts that we see today first form?

I weigh my words well when I assert that the man who should know the true history of the bit of chalk which every carpenter carries about in his breeches pocket, though ignorant of all other history, is likely, if he will think his knowledge out to its ultimate results, to have a truer and therefore a better conception of this wonderful universe and of man's relation to it than the most learned student who [has] deep-read the records of humanity [but is] ignorant of those of nature.

—Thomas Henry Huxley, from *On a Piece of Chalk* (1868)

11.1 INTRODUCTION

In 1868, a well-known British scientist, Thomas Henry Huxley, presented a public lecture on geology to an audience in Norwich, England. Seeking a way to convey his fascination with Earth history to people with no previous geologic knowledge, he focused his audience's attention on the piece of chalk he had been writing with (see epigraph above). And what a tale the chalk has to tell! Chalk, a type of limestone, consists of microscopic marine algae shells and shrimp feces. The specific chalk that Huxley held came from beds deposited in Cretaceous time (the name *Cretaceous*, in fact, derives from the Latin word for chalk). These beds now form the White Cliffs of Dover, England. Geologists in Huxley's day knew of similar chalk beds in outcrops throughout much of Europe, and had discovered that the chalk contains not only plankton shells but also fossils of bizarre swimming reptiles, fish, and invertebrates—species absent in the seas of today. Clearly, when the chalk was deposited, warm seas holding unfamiliar creatures covered some of what is dry land today.

Clues in his humble piece of chalk allowed Huxley to demonstrate to his audience that the geography and inhabitants of the Earth in the past differed markedly from those today, and thus that *the Earth has a history*. We now see a complex, evolving Earth System in which physical and biological components interact pervasively in ways that transformed a formerly barren, crater-pocked surface into countless environments supporting a diversity of life. In this chapter, we offer a concise geologic biography of our planet, from its birth 4.57 billion years ago to the present. We show how continents came into existence and have waltzed across the globe ever since. We also describe mountain-building events, changes in Earth's climate and sea level through time, and the evolution of life. To simplify the discussion, we use the following abbreviations: Ga (for billion years ago), Ma (million years ago), and Ka (thousand years ago).

TAKE-HOME MESSAGE

By the end of this chapter you should have a sense of how the crust grew, continents assembled, mountains formed, atmosphere became breathable, and life evolved, all in the context of a planet that geologists conclude has existed for 4.57 billion years.

11.2 THE HADEAN EON: HELL ON EARTH?

James Hutton, the eighteenth-century Scottish geologist who was the first to provide convincing evidence that the Earth was very old, could not measure Earth's age directly, and indeed speculated that there may be "no vestige of a beginning." But isotopic dating studies conducted in recent decades have shown that it is possible, in fact, to assign a numerical age to our planet's formation. Specifically, dates obtained for a class of meteorites thought to be remnants of the planetesimal cloud out of which the Earth formed consistently yield an age of 4.57 billion years. Geologists currently take this age to be the Earth's birth date. But a clear record of Earth history, as recorded in continental crustal rocks, does not begin until about 3.80 Ga, for crustal rocks older than 3.80 billion years are exceedingly rare. Geologists refer to the mysterious time interval between the birth of Earth and 3.80 Ga as the **Hadean Eon** (from Hades, the Greek god of the underworld) because during this interval the Earth's surface was, at times, an inferno.

The Hadean Eon began with the formation of the Earth by the accretion of planetesimals (see Chapter 1). As the planet grew, collection and compression of matter into a dense ball generated substantial heat. Further, each time another meteorite collided with the Earth, its kinetic energy transformed into more heat. Radioactive decay produced still more heat within the newborn planet. Eventually, the Earth became hot enough to partially melt, and when this happened, by about 4.5 Ga, it underwent internal **differentiation**—gravity pulled molten iron down to the center of the Earth, where it accumulated to form the core. A mantle, composed of ultramafic rock, remained as a thick shell surrounding the core (see Chapter 2). Researchers suggest that soon after—or perhaps during—differentiation, a Mars-sized protoplanet collided with the Earth. The energy of this collision blasted away a significant fraction of Earth's mantle; the resulting debris formed a ring of silicate-rock debris orbiting the Earth. This ring coalesced rapidly to form the Moon.

In the wake of differentiation and Moon formation, the Earth was so hot that its surface was mostly an ocean of seething magma, supplied by intense eruption of melts rising from the mantle (Fig. 11.1). Here and there, rafts of solid rock formed temporarily on the surface of the magma ocean, but these eventually sank and remelted. This stage lasted at least until about 4.4 Ga. After that time, because of the decrease in radioactive heat generation, Earth *might* have become cool enough for solid rocks to form at its surface. The evidence for this statement comes from western Australia, where geologists have found 4.4 Ga grains of a durable mineral called zircon mixed with the quartz of sandstone beds. The zircons presumably formed originally in an igneous rock.

FIGURE 11.1 A speculative painting depicting the Hadean Earth's surface as a magma ocean pummeled by meteorites. The Moon was much closer then, but might not have been visible through the dense atmosphere.

TAKE-HOME MESSAGE

Little is known about the Hadean, the first 700 million years of Earth history. For part of this time, the surface may have been molten. Crust and oceans may have formed at about 4 Ga, but the planet's surface was destroyed around 3.9 Ga by meteorite bombardment.

11.3 THE ARCHEAN EON: THE BIRTH OF THE CONTINENTS AND THE APPEARANCE OF LIFE

During the Hadean Eon, outgassing of the Earth's mantle began to take place. This means that volatile (gassy) elements or compounds originally incorporated in mantle minerals were released and erupted at the Earth's surface, along with lava. The gases accumulated to constitute an unbreathable atmosphere of water (H_2O), methane (CH_4), ammonia (NH_3), hydrogen (H_2), nitrogen (N_2), carbon dioxide (CO_2), sulfur dioxide (SO_2), and other gases. Some researchers have speculated that gases from comets colliding with Earth may have contributed additional gases to the early atmosphere. If the Hadean Earth's surface was sufficiently cool for an extensive solid crustal rock to form beginning at 4.4 Ga, then the first oceans may have accumulated soon thereafter, when water in the atmosphere condensed and fell as rain.

Though mineral grains as old as 4.4 Ga exist, the oldest *whole rock* yet found on Earth has an age of only 4.03 Ga. What destroyed the pre–4.03 Ga crust (and oceans, if they existed) of the Earth? The answer comes from studies of cratering on the Moon. These studies suggest that the Moon—and, therefore, all inner planets of the Solar System—underwent intense meteor bombardment between 4.0 and 3.9 Ga. Researchers speculate that this bombardment would have pulverized and/or melted almost all crust that had existed on Earth at the time, and would have destroyed the existing atmosphere and ocean. Only after the bombardment ceased could long-lasting crust, atmosphere, and oceans begin to form. The discovery of 3.85 Ga marine sedimentary rocks in Greenland suggests that the appearance of land and sea happened quite soon after bombardment ceased. What did the Earth's surface look like at this time? An observer probably would have found small, barren land masses; spotted with volcanoes, poking up above an acidic sea. But both land and sea might have been partially covered by ice, and both would have been obscured by murky, dense (CO_2- and SO_2-rich) air.

The date that marks the boundary between the **Archean** (from the Greek word for ancient) **Eon** and the Hadean Eon has been placed at about 3.8 Ga. Effectively, it marks the time at which substantial quantities of crustal rocks, including rocks that originated as marine sediments, formed. With the advent of the Archean, crust was locally cool and stable enough for isotopic clocks to start ticking. The oldest abundant rocks to have survived intact are from this time.

Geologists still argue about whether plate tectonics operated in the early part of the Archean Eon in the same form as today. Some researchers picture an early Archean Earth with rapidly moving small plates, numerous volcanic island arcs, and abundant hot-spot volcanoes. Others propose that early Archean lithosphere was too warm and buoyant to subduct, and that plate tectonics could not have operated until the later part of the Archean; these authors argue that plume-related volcanism was the main source of new crust until the late Archean. Regardless of which model ultimately proves more accurate, it is clear that the Archean was a time of significant change in the map of the Earth. During that time, the volume of continental crust increased significantly.

How did the continental crust come to be? According to one model, relatively buoyant (felsic and intermediate) crustal rocks formed both at subduction zones and at hot-spot volcanoes. Frequent collisions sutured volcanic arcs and hot-spot volcanoes together, creating progressively larger blocks called protocontinents (Fig. 11.2a, b). Some of these protocontinents underwent stretching, forming rifts that filled with basalt. As time passed, protocontinents became cooler and stronger, and between 3.2 and 2.7 Ga, the first cratons, long-lived blocks of durable continental crust, had developed. By the end of the Archean Eon, about 80% of continental crust had formed (Fig. 11.2c). Volcanic activity during this time continued to supply gases to the atmosphere.

Clearly, the Archean Eon saw many firsts in Earth history. Not only did the first continents appear during the Archean,

FIGURE 11.2 Crust formation during the Archean Eon.

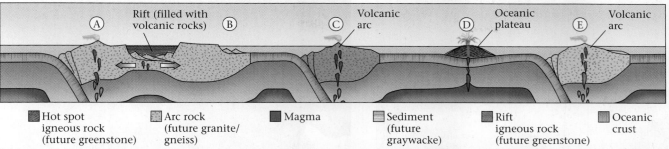

| ■ Hot spot igneous rock (future greenstone) | ▨ Arc rock (future granite/ gneiss) | ■ Magma | ▨ Sediment (future graywacke) | ■ Rift igneous rock (future greenstone) | ▨ Oceanic crust |

(a) In the Archean, island arcs and hot-spot volcanoes built small blocks of buoyant crust. Rifting of these blocks may have produced flood basalts, and erosion of them produced sediment. Not to scale.

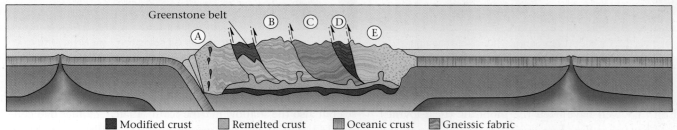

| ■ Modified crust | ▨ Remelted crust | ▨ Oceanic crust | ▨ Gneissic fabric |

(b) Buoyant blocks collided and sutured together, forming protocontinents. Melting at depth produced granite. Eventually, regions of crust cooled, stabilized, and became cratons. Not to scale.

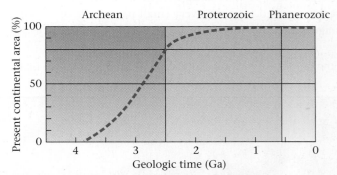

(c) As time progressed, the area of the Earth covered by continental crust increased. Most crust had formed by the beginning of the Proterozoic.

but probably also the first life. Geologists use three sources of evidence to identify early life:

- *Chemical (molecular) fossils, or biomarkers*: These are durable chemicals that are produced only by the metabolism of living organisms.

- *Isotopic signatures*: By analyzing the ratio of ^{12}C to ^{13}C in carbon-rich sediment, geologists can determine if the sediment once contained the bodies of organisms, because organisms preferentially incorporate ^{12}C.

- *Fossil forms*: Given appropriate depositional conditions, fossils of bacteria or archaea cells can be preserved in rock.

The search for the earliest evidence of life continues to make headlines in the popular media. Most geologists currently conclude that life has existed on Earth since at least 3.5 Ga, and perhaps since 3.8 Ga, for rocks of this age contain clear isotopic signatures of organisms. The oldest undisputed fossil forms of bacteria and archaea occur in 3.2 Ga rocks (Fig. 11.3a). Some rocks of this age contain **stromatolites,** distinctive mounds of sediment produced by mats of cyanobacteria. Stromatolites form because cyanobacteria secrete a mucous-like substance to which sediment settling from water sticks. As the mat gets buried, new cyanobacteria colonize the top of the sediment, building a mound upward (Fig. 11.3b). Biomarkers in Archean sediments indicate that photosynthetic organisms appeared by 2.7 Ga.

What specific environment on the Archean Earth served as the cradle of life? Laboratory experiments conducted in the 1950s led many researchers to think that life began in warm pools of surface water, beneath a methane- and ammonia-rich atmosphere streaked by bolts of lightning. The only problem with this hypothesis is that more recent evidence suggests that the early atmosphere consisted mostly of CO_2 and N_2, with relatively little methane and ammonia. Thus, some researchers suggest instead that submarine hot-water vents, so-called black smokers, served as the hosts of the first organisms. These vents emit clouds of ion-charged solutions from which sulfide minerals precipitate and build chimneys. The earliest life in the

FIGURE 11.3 Archean life forms.

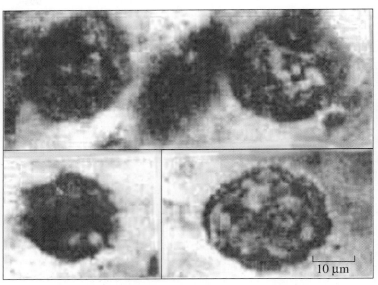

(a) These shapes in 3.2 Ga chert from South America are thought to be fossil bacteria or archaea.

(b) This weathered outcrop of 1.85 Ga dolostone near Marquette, Michigan, reveals the layer-like structure of stromatolites. The delicate ridges represent the fossilized remnants of bacterial mats. Similar stromatolites also occur in exposures of Archean rocks.

Archean Eon may well have been thermophilic (heat-loving) bacteria or archaea that dined on pyrite at dark depths in the ocean alongside these vents. Wet clays underground may have provided the substrate for the earliest life.

As the Archean Eon came to a close, the first continents had formed, and life had colonized not only the depths of the sea but also the shallow marine realm. Plate tectonics had commenced, continental drift was taking place, collisional mountain belts were forming, and erosion was occurring. The atmosphere was gradually accumulating oxygen, although probably this gas still accounted for only a very small percentage of the air. The stage was set for another major change in the Earth System.

FIGURE 11.4 Major crustal provinces of Earth. The black lines indicate the borders of regions underlain by Precambrian crust. Shields are regions where broad areas of Precambrian rocks are exposed.

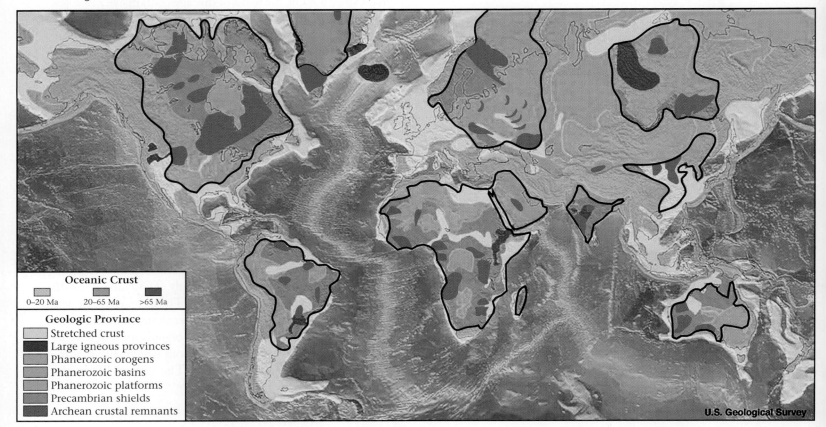

Oceanic Crust

| 0–20 Ma | 20–65 Ma | >65 Ma |

Geologic Province
- Stretched crust
- Large igneous provinces
- Phanerozoic orogens
- Phanerozoic basins
- Phanerozoic platforms
- Precambrian shields
- Archean crustal remnants

U.S. Geological Survey

11.4 THE PROTEROZOIC EON: TRANSITION TO THE MODERN WORLD

The **Proterozoic** (from the Greek words meaning first life) **Eon** spans roughly 2 billion years, from about 2.5 Ga to the beginning of the Cambrian Period 542 Ma—thus, it encompasses almost half of Earth's history. The eon received its now misleading name before fossil organisms were discovered in Archean rock. During Proterozoic time, Earth's surface environment changed from the unfamiliar world of possibly small, fast-moving plates, small continents, and an oxygen-poor atmosphere to the more familiar world of mostly large plates, large continents, and an oxygen-rich atmosphere.

First, let's look at changes to the continents. New continental crust continued to form during the Proterozoic Eon, but at progressively slower rates—in fact, by the middle of the eon, over 90% of the Earth's continental crust had formed. Collisions between the Archean continents, as well as with volcanic

island arcs and hot-spot volcanoes, resulted in the assembly of large continents whose interiors were located far from orogenic belts. These continents cooled and strengthened to become cratons. The large cratons that exist today had formed by the end of the Proterozoic (Fig. 11.4).

Let's consider the geology of a large craton a little more closely, by examining the interior of North America. The Precambrian rocks constituting this craton crop out extensively in Canada. Geologists refer to this region as the Canadian Shield (see Fig. 9.21), for a **shield** consists of a broad, low-lying region of exposed Precambrian rocks. In the United States, Phanerozoic strata bury most of the Precambrian rocks, forming a province called a **cratonic platform**, or continental platform—here we see Precambrian rocks only where they have been exposed by erosion of the sedimentary cover. The North American craton includes several Archean blocks that were sutured together along huge collisional belts. Each of these blocks and belts has been given a name (Fig. 11.5). The southern portion of the craton grew when a series of volcanic island arcs and continental slivers accreted (attached) to the margin of the Canadian Shield between 1.8 and 1.6 Ga. In the Midwest, huge granite plutons intruded much of this accreted region, and rhyolite covered it, between 1.5 and 1.3 Ga.

Successive collisions ultimately brought together most continental crust on Earth into a single supercontinent, named **Rodinia**, by around 1 Ga. The last major collision during the formation of Rodinia produced a large orogen called the Grenville orogen. If you look at a popular (though not totally

FIGURE 11.5 The North American craton consists of a collage of different belts and blocks stitched together during collisional and accretionary orogenies of Precambrian time.

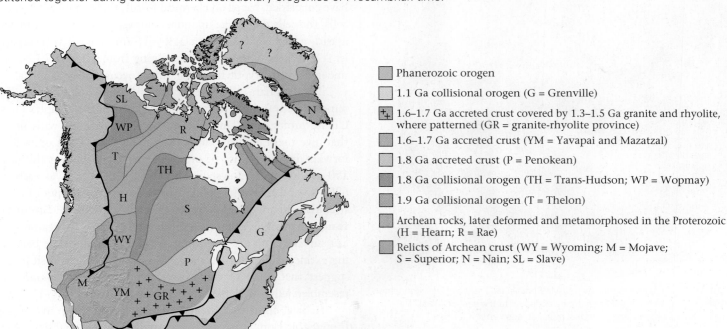

- Phanerozoic orogen
- 1.1 Ga collisional orogen (G = Grenville)
- 1.6–1.7 Ga accreted crust covered by 1.3–1.5 Ga granite and rhyolite, where patterned (GR = granite-rhyolite province)
- 1.6–1.7 Ga accreted crust (YM = Yavapai and Mazatzal)
- 1.8 Ga accreted crust (P = Penokean)
- 1.8 Ga collisional orogen (TH = Trans-Hudson; WP = Wopmay)
- 1.9 Ga collisional orogen (T = Thelon)
- Archean rocks, later deformed and metamorphosed in the Proterozoic (H = Hearn; R = Rae)
- Relics of Archean crust (WY = Wyoming; M = Mojave; S = Superior; N = Nain; SL = Slave)

1,000 km

FIGURE 11.6 Supercontinents in the late Precambrian.

(a) Rodinia formed around 1 Ga and lasted until about 700 Ma. North America and Greenland together comprise Laurentia.

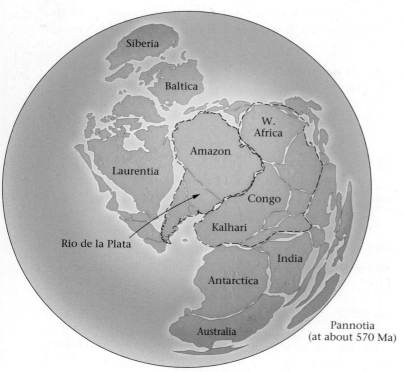

(b) According to one model, by 570 Ma Rodinia had broken apart; continents that once lay to the west of Laurentia ended up to the east of Africa. The resulting supercontinent, Pannotia, broke up soon after it formed.

proven) reconstruction of Rodinia, you can identify the crustal provinces that would eventually become the familiar continents of today (Fig. 11.6a). In this reconstruction, the future Antarctica and Australia lay somewhere along the western coast of North America.

Several studies suggest that sometime between 800 and 600 Ma, Rodinia "turned inside out," in that Antarctica, India, and Australia broke away from western North America and swung around and collided with the future South America, possibly forming a short-lived supercontinent that some geologists refer to as Pannotia (Fig. 11.6b).

Much of the transformation of the Earth's atmosphere from the oxygen-poor volcanic gas mix of the Archean Eon to an oxygen-rich mix had occurred by about 1.8 Ga (Fig. 11.7a). This transition led to the global deposition of banded iron formation (BIF), now used as iron ore, because iron that had been dissolved in the sea combined with oxygen and precipitated on the sea floor. This new oxygen-rich atmosphere had a profound effect on the Earth, for it permitted a great diversification of life and the eventual conquest of the land by living organisms. Life could become more complex because oxygen-dependent (aerobic) metabolism produces energy much more efficiently than does oxygen-free (anaerobic) metabolism—eating sulfide minerals may sustain bacteria, but it can't keep multicellular organisms alive! The addition of substantial oxygen to the atmosphere also set the stage for the eventual move of life onto the land, hundreds of millions of years after the Proterozoic. Oxygen provides the raw material from which ozone (O_3) forms—ozone protects the land surface from harmful ultraviolet rays.

So far, we've talked only of *prokaryotic* organisms, those which consist of single cells that do not contain a nucleus. Geologists have concluded, based on chemical fossils, that the first *eukaryotic* cells, the cells that make up more complex organisms, originated as early as 2.7 Ga. Controversial fossil forms of eukaryotic organisms have been found in rocks ranging from 2.1 to 1.2 Ga, but undeniable eukaryotic fossil organisms occur in 1.0 Ga rocks.

The last half-billion years of the Proterozoic Eon saw the remarkable transition from simple organisms into complex ones. Ciliate protozoans (single-celled organisms coated with fibers that give them mobility) appear at about 750 Ma. A great leap forward in complexity of organisms occurred during the next 150 million years, for sediments deposited perhaps as early as 620 Ma and certainly by 565 Ma contain several types of multicellular organisms that together constitute the **Ediacaran fauna**, named for a region in southern Australia. Ediacaran species survived into the beginning of the Cambrian before becoming extinct. Though their nature is enigmatic, their fossil forms suggest that some of these invertebrate (shell-less) organisms resembled jellyfish, while others resembled worms (Fig. 11.7b).

Radical climate shifts occurred on Earth at the end of the Proterozoic Eon. Specifically, accumulations of glacial sediments occur worldwide in the latest Proterozoic stratigraphic sequences.

FIGURE 11.7 Major changes in the Earth System during the Proterozoic Eon.

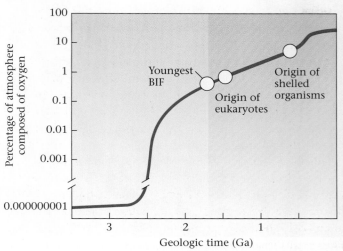

(a) The change in oxygen content of the Earth's atmosphere through time. Today, oxygen accounts for 21% of the atmosphere; in the Archean Eon, it constituted just 0.000000001%.

(b) *Dicksonia*, a fossil of the Ediacaran fauna. These complex, soft-bodied marine organisms appeared in the late Proterozoic.

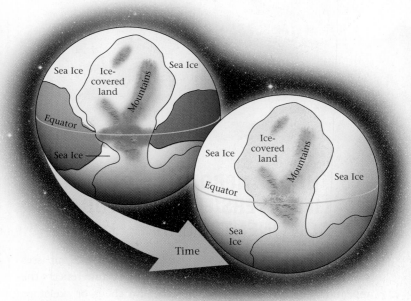

(c) Layers of Proterozoic glacial till crop out in Africa, indicating that low-latitude land masses were glaciated during the Proterozoic.

(d) This planet may have frozen over completely to form "snowball Earth." Glaciers first grew on land, and eventually the sea surface froze

What's strange about the occurrence of these sediments is that they can be found even in regions that were located at the equator during these times (**Fig. 11.7c**). This observation implies that the *entire* planet was cold enough for glaciers to form at the end of the Proterozoic. Geologists still are debating the history of these global ice ages (see Chapter 18), but in one model, glaciers covered the land, and perhaps the entire ocean surface froze, resulting in **snowball Earth** (**Fig. 11.7d**). The shell of ice cut off the oceans from the atmosphere, causing oxygen levels in the sea to drop drastically, so many life forms died off.

What brought an end to snowball Earth conditions? According to one model, the icy sheath also prevented atmos-

pheric CO_2 from dissolving in seawater, but it did not prevent volcanic activity from continuing to add CO_2 to the atmosphere. Earth would have remained a snowball forever, were it not for volcanic CO_2, for CO_2 is a greenhouse gas, meaning that it traps heat in the atmosphere much as glass panes trap heat in a greenhouse (see Chapter 19). Thus, as the CO_2 concentration increased, Earth warmed up and eventually the glaciers melted. Life may have survived snowball Earth conditions only near submarine black smokers and near hot springs. When the ice vanished, life rapidly expanded into new environments, and new species, such as the Ediacaran fauna, could evolve.

FIGURE 11.8 Land and sea in the early Paleozoic Era.

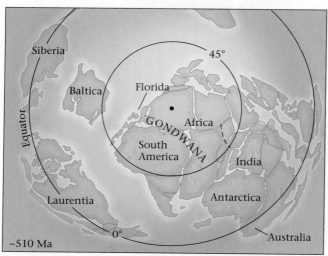

(a) The distribution of continents in the Cambrian Period (510 Ma), as viewed looking down on the South Pole.

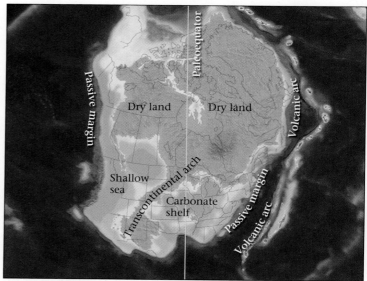

(b) A paleogeographic map of North America shows the regions of dry land and shallow sea in the Early Cambrian Period.

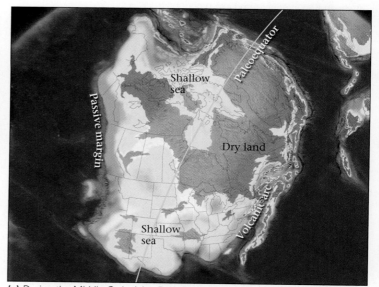

(c) During the Middle Ordovician Period, shallow seas covered much of North America. A volcanic arc formed off the east coast.

> ### TAKE-HOME MESSAGE
>
> During the Proterozoic, stable crustal blocks formed, and then sutured together to form large continents and, toward the end of the Proterozoic, supercontinents. Multicellular organisms appeared, and the atmosphere began to be rich in oxygen.

11.5 THE PHANEROZOIC EON: LIFE DIVERSIFIES, AND TODAY'S CONTINENTS FORM

The **Phanerozoic** (Greek for visible life) **Eon** encompasses the last 542 million years of Earth history. Its name reflects the appearance of diverse organisms with hard shells or skeletons that became the well-preserved fossils you can find easily in rock outcrops. Geologists divide the Phanerozoic into three eras—the Paleozoic (Greek for ancient life), the Mesozoic (middle life), and the Cenozoic (recent life)—to emphasize the changes in Earth's living population throughout the eon. The solid Earth also changed during this time. Continental blocks rearranged, and new supercontinents formed, only to break apart again. Numerous orogenies built mountain ranges, the most recent of which persist today (see Geotour 11 on p. GT-24). Shell-secreting organisms, burrowing organisms, and plants invaded the sedimentary environment and changed the style of deposition. Next, we look at the changes in the planet's map (its paleogeography) and in its life forms during the three eras of the Phanerozoic Eon.

11.6 THE PALEOZOIC ERA: FROM RODINIA TO PANGAEA

The Early Paleozoic Era (Cambrian–Ordovician Periods, 542–444 Ma)

Paleogeography. At the beginning of the Paleozoic Era, Pannotia broke up, yielding smaller continents including Laurentia (composed of North America and Greenland), Gondwana (South America, Africa, Antarctica, India, and Australia), Baltica (Europe), and Siberia (Fig. 11.8a). Passive-margin basins formed along the edges of these new continents. In addition, sea level rose, so that vast areas of continents were flooded with shallow seas called **epicontinental seas** (Fig. 11.8b). In many places, water depths in epicontinental seas reached only a few meters, creating a well-lit marine environment in which life abounded. Deposition in these seas, therefore, yielded layers of fossiliferous sediment. Sea level, however, did not stay high for the entire early Paleozoic Era; regressions and transgressions took place, the former marked by unconformities and the latter by accumulations of sediment.

The geologically peaceful world of the early Paleozoic Era in Laurentia abruptly came to a close in the Middle Ordovician Period, for at this time its eastern margin rammed into a volcanic island arc and other crustal fragments. The resulting collision, called the Taconic orogeny, deformed and metamorphosed strata of the continent's margin and produced a mountain range (Fig. 11.8c).

Life evolution.   Beginning with the Cambrian Period, life on Earth left a clear record of evolution, because so many organisms developed shells of durable minerals. The fossil record indicates that soon after the Cambrian began, life underwent remarkable diversification. This event, which paleontologists refer to as the **Cambrian explosion** of life, took about 20 million years. What caused this explosion? No one can say for sure, but considering that it occurred roughly at the time a supercontinent broke up, it may have had something to do with the production of new ecological niches and the isolation of populations that resulted when small continents formed and drifted apart.

The first animals to appear in the Cambrian Period had simple tube- or cone-shaped shells, but soon thereafter, the shells became more complex. Small fossils called conodonts, which

resemble tiny jaws, are found in Cambrian strata. Their presence suggests that creatures with jaws appeared at this time; shells on other organisms may have evolved as a means of protection against such predators. By the end of the Cambrian, trilobites were grazing the sea floor. Trilobites shared the environment with mollusks, brachiopods, nautiloids, gastropods, graptolites, and echinoderms (Fig. 11.9; see Interlude E). Thus, a complex food chain arose, which included plankton, bottom feeders, and at the top, giant predators. Many of the organisms crawled over or swam around reefs composed of mounds of sponges with mineral skeletons. The Ordovician Period saw the first crinoids and the first vertebrate animals, jawless fish. At the end of the Ordovician, mass extinction took place, perhaps because of the brief glaciation and associated sea-level lowering of the time.

The Middle Paleozoic Era (Silurian–Devonian Periods, 444–359 Ma)

Paleogeography.   As the world entered the Silurian Period, the global climate warmed (leading to so-called greenhouse conditions), sea level rose, and the continents flooded once again. In some places, where water in the epicontinental seas was clear and could exchange with water from the oceans, huge reef complexes grew, forming a layer of fossiliferous limestone on the continents.

More orogeny took place during the middle Paleozoic Era. A succession of collisions on the eastern side of Laurentia during Silurian and Devonian time produced the Caledonian orogen (affecting eastern Greenland, western Scandinavia, and Scotland) and the Acadian orogen in the region that became the Appalachians (Fig. 11.10a).

Throughout the middle Paleozoic, the western margin of North America continued to be a passive-margin basin. Finally, in the Late Devonian, the quiet environment of the west-coast passive margin ceased, possibly because of a collision with an island arc. During this event, known as the Antler orogeny, deep-marine strata were shoved eastward on thrust faults, moving up and over shallow-water strata. It was the first of many orogenies to affect the western margin of the continent. The Caledonian, Acadian, and Antler orogens all shed deltas of sediment onto the continents (Fig. 11.10b).

Life evolution.   Life on Earth underwent radical changes in the middle Paleozoic Era. In the sea, new species of trilobites, eurypterids, gastropods, crinoids, and bivalves replaced species that had disappeared during the mass extinction at the end of the Ordovician Period. On land, vascular plants, with woody tissues, seeds, and veins (for transporting water and food), rooted for the first time. With the evolution of veins and wood, plants could grow much larger, and by the Late Devonian Period the land surface hosted swampy forests with tree-sized relatives of club mosses and ferns. Also at this time, spiders, scorpions, insects,

FIGURE 11.9   A museum diorama illustrates what early Paleozoic marine organisms may have looked like.

Nautiloid

Brachiopod   Trilobite   Coral

FIGURE 11.10 Paleogeography and fossils of Silurian and Devonian time.

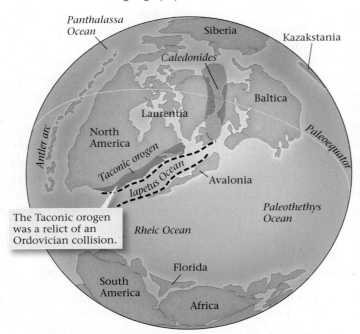

(a) During the Silurian and Devonian Periods, Laurentia collided with Baltica, Avalonia, and South America in succession, as oceans in between were consumed. The Antler arc formed off the west coast.

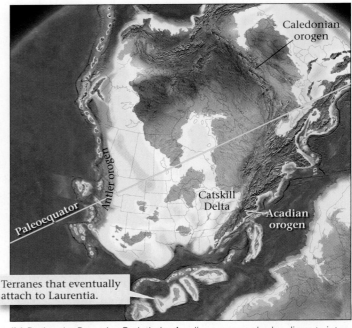

(b) During the Devonian Period, the Acadian orogeny shed sediments into a shallow sea to form the Catskill Delta on the east coast of Laurentia. The Antler orogen shed sediments in the west.

and crustaceans began to exploit both dry-land and freshwater habitats, and jawed fish such as sharks and bony fish began to cruise the oceans. Finally, at the very end of the Devonian Period, the first amphibians crawled out onto land and inhaled air with lungs (Fig. 11.10c).

The Late Paleozoic Era (Carboniferous–Permian Periods, 359–251 Ma)

Paleogeography. After the peak of the greenhouse conditions and high sea levels in the middle Paleozoic Era, the climate cooled significantly (leading to so-called icehouse conditions) in the late Paleozoic. Seas gradually retreated from the continents, so that during the Carboniferous Period, regions that had hosted the limestone-forming reefs of epicontinental seas now became coastal areas and river deltas in which sand, shale, and organic debris accumulated. In fact, during the Carboniferous Period, Laurentia lay near the equator, so it enjoyed tropical and semitropical conditions that favored lush growth in swamps. This growth left thick piles of woody debris that transformed into coal after burial. Much of Gondwana and Siberia, in contrast, lay at high latitudes, and by the Permian Period they were covered by ice sheets.

The late Paleozoic Era also saw a succession of continental collisions, culminating in the formation of a single supercontinent, **Pangaea** (Fig. 11.11a). The largest collision occurred

(c) A Late Devonian fossil skeleton of *Tiktaalik*, this lobe-finned fish was one of the first animals to walk on land.

during Carboniferous and Permian time, when Gondwana rammed into Laurentia and Baltica, causing the Alleghanian orogeny of North America (Fig. 11.11b). During this event, the final stage in the development of the Appalachians, eastern North America squashed against northwestern Africa.

Along the continental side of the Alleghanian orogen, a wide band of deformation called the Appalachian fold-thrust belt formed. In this province, you can see a distinctive style of deformation, called thin-skinned deformation, characterized by displacement on thrust faults in sedimentary strata *above* the Precambrian basement (Fig. 11.12). At depth, the thrust system

FIGURE 11.11   Paleogeography at the end of the Paleozoic Era.

(a) At the end of the Paleozoic, almost all land had combined into a single supercontinent called Pangaea.

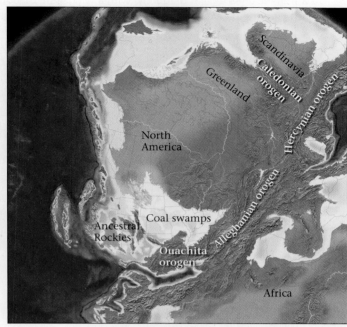

(b) During the Alleghanian and Hercynian orgenies, a huge mountain belt formed. The Caledonian orogen had formed earlier. Coal swamps bordered interior seas, and the Ancestral Rockies rose.

merges with a near-horizontal sliding surface, or detachment. Movement on the faults produces folds in overlying beds.

Forces generated during the Alleghanian orogeny were so strong that faults in the continental crust clear across North America moved, producing uplifts (local high areas) and sediment-filled basins in the Midwest and even in the region of the present-day Rocky Mountains (Colorado, New Mexico, and Wyoming). Geologists refer to the late Paleozoic uplifts of the Rocky Mountain region as the **Ancestral Rockies**.

The assembly of Pangaea involved a number of other collisions around the world as well. Notably, Africa collided with southern Europe to form the Hercynian orogen. Also, a rift or

small ocean in Russia closed to create the Ural Mountains, and parts of China along with other fragments of Asia attached to southern Siberia.

Life evolution.   The fossil record indicates that during the late Paleozoic Era, plants and animals continued to evolve toward more familiar forms. In coal swamps, fixed-wing insects including huge dragonflies flew through a tangle of ferns, club mosses, and scouring rushes, and by the end of the Carboniferous Period insects such as the cockroach, with foldable wings, appeared (Fig. 11.13). Forests containing gymnosperms ("naked seed" plants such as conifers) and cycads (trees with a palm-like stalk peaked by a fan of fern-like fronds) became widespread in the Permian Period. Amphibians and, later, reptiles populated the land. The appearance of reptiles marked the evolution of a

FIGURE 11.12   Features of the Appalachian Mountains in the eastern United States.

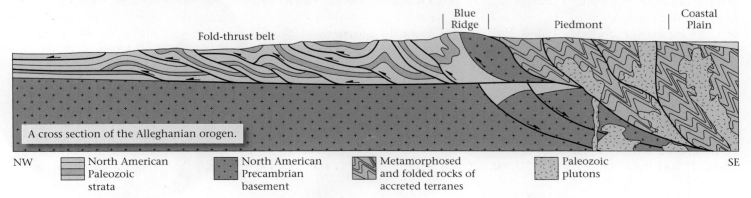

A cross section of the Alleghanian orogen.

NW				SE
	North American Paleozoic strata	North American Precambrian basement	Metamorphosed and folded rocks of accreted terranes	Paleozoic plutons

FIGURE 11.13 A museum diorama of a Carboniferous coal swamp includes a giant dragonfly, with a wingspan of about 1 m. The inset photo gives a sense of its size relative to a human.

radically new component in animal reproduction: eggs with a protective covering. Such eggs permitted reptiles to reproduce without returning to the water, and thus allowed the group to populate previously uninhabitable environments. The late Paleozoic Era came to a close with two major mass-extinction events,

during which over 90% of marine species disappeared. Why these particular events occurred remains a subject of debate, but there is interesting evidence that the terminal Permian mass extinction occurred either as a result of a huge meteorite impact or as a result of an episode of an extraordinary amount of volcanic activity. Either event could have clouded the atmosphere and disrupted the food chain.

TAKE-HOME MESSAGE

At the beginning of the Paleozoic, the late Precambrian supercontinent broke apart. North America was then bordered by passive margins, until collisions brought together all land to form Pangaea. The interiors of continents flooded during sea-level rise. The planet also saw life diversify into complex forms, first in the sea and then later on land.

11.7 THE MESOZOIC ERA: WHEN DINOSAURS RULED

The Early and Middle Mesozoic Era (Triassic–Jurassic Periods, 251–145 Ma)

Paleogeography. Pangaea, the supercontinent formed at the end of the Paleozoic Era, lasted for about 100 million years, until rifts formed and the supercontinent began to break up

FIGURE 11.14 The breakup of Pangaea began in the Late Triassic Period. The North Atlantic Ocean had started to form, while the South Atlantic remained closed.

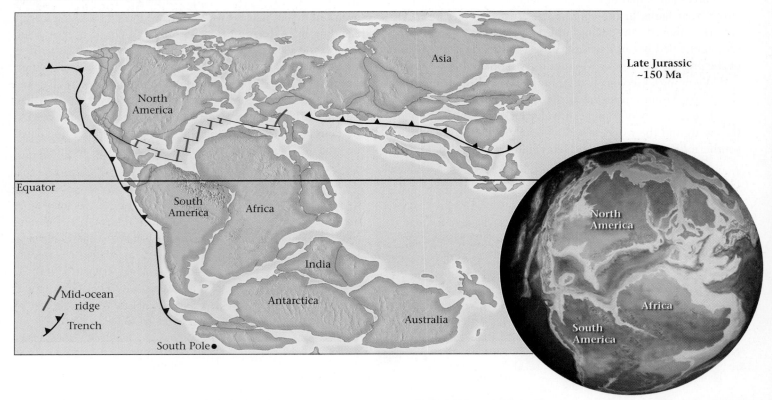

during the Late Triassic and Early Jurassic Periods. By the end of the Jurassic Period, rifting had succeeded in forming the Mid-Atlantic Ridge, and the North Atlantic Ocean started to grow (Fig. 11.14). At first, the Atlantic was narrow and shallow, and evaporation made its water so salty that thick evaporite deposits, which now underlie much of the Gulf Coast region, accumulated.

According to the record of sedimentary rocks, during the Triassic and Early Jurassic, Earth had a warm climate, but during the Late Jurassic and Early Cretaceous, it had a cool climate. For the early Mesozoic, the interior of Pangaea remained a nonmarine environment in which red sandstones and shales, now exposed in the spectacular cliffs of Zion National Park, were deposited. By the Middle Jurassic Period, sea level began to rise. During the resulting transgression, a shallow sea submerged much of the Rocky Mountain region.

On the western margin of North America, convergent-margin tectonics became the order of the day. Beginning with Late Permian and continuing through Mesozoic time, subduction generated volcanic island arcs and caused them, along with microcontinents and hot-spot volcanoes, to collide with North America. Thus, North America grew in land area by the addition (accretion) of crustal fragments and island arcs—exotic terranes—on its western margin. From the end of the Jurassic through the Cretaceous Period, a major continental volcanic arc, the Sierran arc, formed along the western margin of North America; we'll learn more about this arc later.

Life evolution. During the early Mesozoic Era, a variety of new species of established plant and animal groups appeared, filling the ecological niches left vacant by the Late Permian mass extinction. Swimming reptiles plied the oceans, and corals became the predominant reef builders. On land, gymnosperms and reptiles diversified, and the Earth saw its first turtles and flying reptiles. More dramatic, at the end of the Triassic Period, the first true dinosaurs evolved. Dinosaurs differed from other reptiles in that their legs were positioned under their bodies rather than off to the sides, and they were possibly warm-blooded. By the end of the Jurassic Period, gigantic sauropod dinosaurs (which weighed up to 100 tons), along with other familiar species such as stegosaurus, thundered across the landscape, and the first feathered birds took to the skies (Fig. 11.15). The earliest ancestors of mammals appeared at the end of the Triassic Period, in the form of small, rat-like creatures.

FIGURE 11.15 During the Jurassic Period, giant dinosaurs inhabited the land.

The Late Mesozoic Era (Cretaceous Period, 145–65 Ma)

Paleogeography. During the Cretaceous Period, the Earth's climate continued to shift to warmer, greenhouse conditions, and sea level rose significantly, reaching levels that had not been attained for the previous 200 million years. Great seaways flooded most of the continents (Fig. 11.16). In the latter part of the Cretaceous Period, a shark could have swum from the Gulf of Mexico to the Arctic Ocean, or across much of western Europe.

The breakup of Pangaea continued through the Cretaceous Period, with the opening of the South Atlantic Ocean and the separation of South America and Africa from Antarctica and Australia. India broke away from Gondwana and headed rapidly northward toward Asia (Fig. 11.17a, b). Along the continental margins of the newly formed Mesozoic oceans, new passive-margin basins developed that filled with great thicknesses of sediments. For example, along the Gulf Coast, a wedge of sediment over 15 km thick has accumulated.

In western North America, the Sierran arc, a large continental volcanic arc that had initiated at the end of the Jurassic Period, continued to be active. This arc resembled the one that currently lies along the Andes on the western edge of South America. Though the volcanoes of the arc have long since eroded away, we can see their roots in the form of the plutons that now constitute the granitic batholith of the Sierra Nevada. Compressional forces along the western North American convergent boundary activated large thrust faults east of the arc, an event geologists refer to as the Sevier orogeny. This orogeny produced a fold-thrust belt whose remnants you can see today in the Canadian Rockies and in western Wyoming.

FIGURE 11.16 During the Late Cretaceous Period, the Sierran volcanic arc developed along the west coast of North America. The Sevier fold-thrust belt formed to the east. Part of the interior, along with coastal areas, flooded.

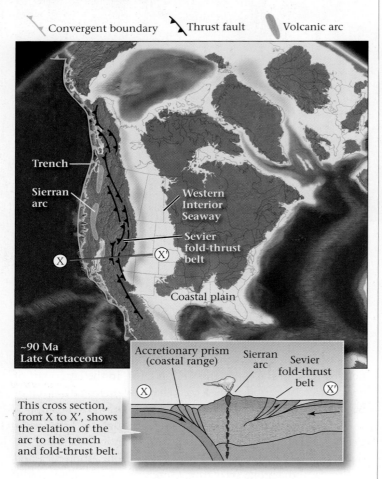

At the end of the Cretaceous Period, continued compression along the convergent boundary of western North America caused large faults in the region of Wyoming, Colorado, eastern Utah, and northern Arizona to slip (Fig. 11.17c). In contrast to the faults of fold-thrust belts, these faults penetrated deep into the Precambrian rocks of the continent, and thus movement on them generated basement uplifts. During the formation of these uplifts, overlying layers of earlier strata warped into large monoclines, folds whose shape resembles the drape of a carpet over a step. This event, which geologists call the **Laramide orogeny**, formed the structure of the present Rocky Mountains in the United States (Fig. 11.17d).

By mapping the sea floor, geologists have discovered that sea-floor spreading may have been particularly fast, and that huge submarine plateaus composed of hot-spot volcanic rocks formed during the Cretaceous Period. The presence of hot spots implies that the Cretaceous may have been a time when par-

ticularly large mantle plumes, called **superplumes**, reached the base of the lithosphere. These produced large igneous provinces, areas in which vast quantities of basalt erupted. Hot-spot and mid-ocean-ridge activity may have influenced the climate and sea level. Large mid-ocean ridges and large plateaus displaced seawater, and volcanic eruptions released huge quantities of CO_2 into the atmosphere. An increase in the concentration of CO_2 in the atmosphere leads to an increase in atmospheric temperature, for CO_2 in the air traps infrared radiation rising from the Earth and causes the air to warm up. This increase would melt ice sheets, and thus contribute to the rise of sea level.

Life evolution. In the seas of the late Mesozoic world, modern fish appeared and became dominant. What set them apart from earlier fish were their short jaws, rounded scales, symmetrical tails, and specialized fins. Huge swimming reptiles and gigantic turtles (with shells up to 4 m across) preyed on the fish. On land, cycads largely vanished, and angiosperms (flowering plants), including hardwood trees, began to compete successfully with conifers for dominance of the forest. Dinosaurs reached their peak of success at this time, inhabiting almost all environments on Earth. Social herds of grazing dinosaurs roamed the plains, preyed on by the fearsome *Tyrannosaurus rex* (a Cretaceous, not a Jurassic, dinosaur, despite what Hollywood says!). Pterosaurs, with wingspans of up to 11 m, soared overhead, and birds began to diversify. Mammals also diversified and developed larger brains and more specialized teeth, but they remained small and rat-like.

The "K-T boundary event." Geologists first recognized the K-T boundary (K stands for Cretaceous and T for Tertiary) from eighteenth-century studies that identified an abrupt global change in fossil assemblages. Until the 1980s, most geologists assumed the faunal turnover took millions of years. But modern dating techniques indicate that this change happened almost instantaneously and that it signaled a sudden mass extinction of most species on Earth. The dinosaurs, which had ruled the planet for over 150 million years, simply vanished, along with 90% of some plankton species in the ocean and up to 75% of plant species. What kind of catastrophe could cause such a sudden and extensive mass extinction? From data collected in the 1970s and 1980s, most geologists have concluded that the Cretaceous Period came to a close, at least in part, as a result of the impact of a 10-km-wide meteorite at the site of the present-day Yucatán Peninsula in Mexico, where a 100-km-wide by 16-km-deep scar called the Chicxulub crater lies buried by younger sediment (Fig. 11.18).

The impact caused so much destruction because it not only formed a crater, blasting huge quantities of debris into the sky, but probably also generated 2-km-high tsunamis that inundated the shores of continents and generated a blast of hot air that

FIGURE 11.17 Paleogeography in Late Cretaceous through Eocene time.

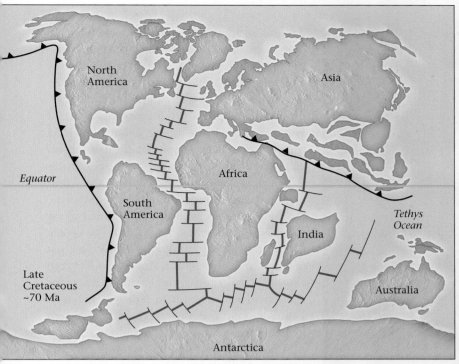

(a) By the Late Cretaceous Period, the Atlantic Ocean had formed, and India was moving rapidly northward to eventually collide with Asia.

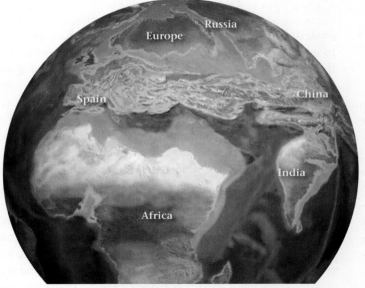

(b) In this Eocene paleogeographic reconstruction, southern Europe and Asia are beginning to form from a collage of many crustal blocks.

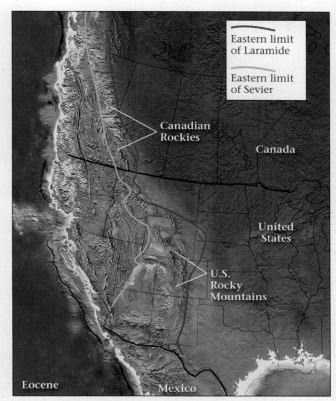

(c) During the Laramide orogeny, deformation shifted eastward in the United States, moving from the Sevier belt to the Rocky Mountains, and the style of deformation changed.

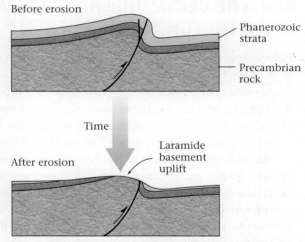

(d) The Laramide orogeny produced "basement-cored uplifts." In these, faults lifted up blocks of basement, causing the overlying strata to bend into a stair step–like fold.

set forests on fire. The blast and the blaze together could have ejected so much debris into the atmosphere that for months there would have been perpetual night and winter-like cold. In addition, chemicals ejected into the air could have combined with water to produce acid rain. These conditions would cause photosynthesis to all but cease, and thus would break the food chain and trigger extinctions.

TAKE-HOME MESSAGE

The Mesozoic was the age of dinosaurs, for these beasts roamed all continents during this eon. In North America, a major convergent plate boundary formed along the west coast, while rifting of Pangaea led to the formation of the Atlantic Ocean on the east coast.

FIGURE 11.18 A painting illustrates the collision of a huge bolide with the Earth at the end of the Cretaceous Period.

11.8 THE CENOZOIC ERA: THE FINAL STRETCH TO THE PRESENT

Paleogeography. During the last 65 million years, the map of the Earth has continued to change, gradually producing the configuration of continents we see today. The final stages of the Pangaea breakup separated Australia from Antarctica and Greenland from North America, and formed the North Sea between Britain and continental Europe. The Atlantic Ocean continued to grow because of sea-floor spreading on the Mid-Atlantic Ridge, and thus the Americas moved westward, away from Europe and Africa. Meanwhile, the continents that once constituted Gondwana drifted northward as the intervening Tethys Ocean was consumed by subduction (see Fig. 11.17). Collisions of the former Gondwana continents with the southern margins of Europe and Asia resulted in the formation of the largest orogenic belt on Earth today, the **Alpine-Himalayan chain** (Fig. 11.19). India and a series of intervening volcanic island arcs and microcontinents collided with Asia to form the Himalayas and the Tibetan Plateau to the north, while Africa along with some volcanic island arcs and microcontinents collided with Europe to produce the Alps in the west and the Zagros Mountains of Iran in the east.

As the Americas moved westward, convergent plate boundaries evolved along their western margins. In South America, convergent-boundary activity built the Andes, which

remains an active orogen to the present day. In North America, convergent-boundary activity continued without interruption until the Eocene Epoch, yielding, as we have seen, the Laramide orogen. Then, because of the rearrangement of plates off the western shore of North America, beginning about 40 Ma, a transform boundary replaced the convergent boundary in the western part of the continent. When this happened, volcanism and compression ceased in western North America, the San Andreas fault system formed along the coast of the United States, and the Queen Charlotte fault system developed off the coast of Canada. Along the San Andreas and Queen Charlotte faults today, the Pacific Plate moves northward with respect to North America at a rate of about 6 cm per year. In the western United States, convergent-boundary tectonics continues only in Washington, Oregon, and northern California, where subduction of the Juan de Fuca Plate generates the volcanism of the Cascade volcanic chain.

As convergent tectonics ceased in the western United States south of the Cascades, the region began to undergo rifting (stretching) in roughly an east-west direction. The result was the formation of the **Basin and Range Province** (see **Geotour 11** on p. GT-24), a broad continental rift that has caused the region to stretch to twice its original width (Fig. 11.20). The Basin and Range gained its name from its topography—the province contains long, narrow mountain ranges separated from each other by flat, sediment-filled basins. This geometry reflects the normal faulting resulting from stretching: crust of the region was broken up by faults, and movement on the faults created elongate depressions.

Recall that in the Cretaceous Period, the world experienced greenhouse conditions and sea level rose so that extensive areas of continents were submerged. During the Cenozoic Era, however, the global climate rapidly shifted to icehouse conditions, and by the early Oligocene Epoch, Antarctic glaciers reappeared for the first time since the Triassic. The climate continued to grow colder through the Late Miocene Epoch, leading to the formation of grasslands in temperate climates. About 2.5 Ma, the Isthmus of Panama formed, separating the Atlantic completely from the Pacific, changing the configuration of oceanic currents, perhaps leading the Arctic Ocean to freeze over.

During the overall cold climate of the last 2 million years, the Quaternary Period, continental glaciers have expanded and retreated across northern continents at least twenty times, resulting in the **Pleistocene ice age** (Fig. 11.21). Each time the glacier grew, sea level fell so much that the continental shelf became exposed to air, and a land bridge formed across the Bering Strait, west of Alaska, providing migration routes for animals and people from Asia into North America; a partial land bridge also formed from southeast Asia to Australia, making human migration to Australia easier. Erosion and deposition by the glaciers created much of the landscape we see today in northern temperate regions. About 11,000 years ago, the climate warmed, and we entered the interglacial time interval we are still experiencing today (see Chapter 18). This most recent interval of time is the Holocene.

Life Evolution. When the skies finally cleared in the wake of the K-T boundary catastrophe, plant life recovered, and soon forests of both angiosperms and gymnosperms reappeared. A new group of plants, the grasses, sprang up and began to dominate the plains in temperate and subtropical climates by the middle of the Cenozoic Era. The dinosaurs, however, were gone for good. Mammals rapidly diversified into a variety of forms to take their place. In fact, most of the modern groups of mammals that exist today originated at the beginning of the Cenozoic Era, giving this time the nickname "Age of Mammals." During the latter part of the era, remarkably huge mammals appeared (such as mammoths, giant beavers, giant bears,

and giant sloths), but these became extinct during the past 10,000 years, perhaps because of hunting by humans.

It was during the Cenozoic that our own ancestors first appeared. Ape-like primates diversified in the Miocene Epoch (about 20 Ma), and the first human-like primate appeared at about 4 Ma, followed by the first members of the human genus, *Homo*, at about 2.4 Ma. Fossil evidence, primarily from Africa, indicates that *Homo erectus*, capable of making stone axes, appeared about 1.6 Ma, and the line leading to *Homo sapiens* (our species) diverged from *Homo neanderthalensis* (Neanderthal man) about 500,000 years ago. According to the fossil record, modern people appeared about 150,000 years ago. Thus, much of human evolution took place during the radically shifting climatic conditions of the Pleistocene Epoch.

We end our brief biography of Earth for now with the appearance of *Homo sapiens*. As summarized in **Geology at a Glance**, pp. 316–317, this history has been shaped by complex plate interactions (including continental drift and collisions), sea-level changes, atmospheric changes, and life evolution. Clearly, the Earth System has changed significantly over time. We'll pick up the thread of this story in Chapter 19, where we discuss ideas of how the Earth System may change in the future.

FIGURE 11.19 The two main active continental orogenic systems on the Earth today. The Alpine-Himalayan system formed when Africa, India, and Australia collided with Asia (inset). The Cordilleran and Andean systems reflect the consequences of convergent-boundary tectonism along the eastern Pacific Ocean.

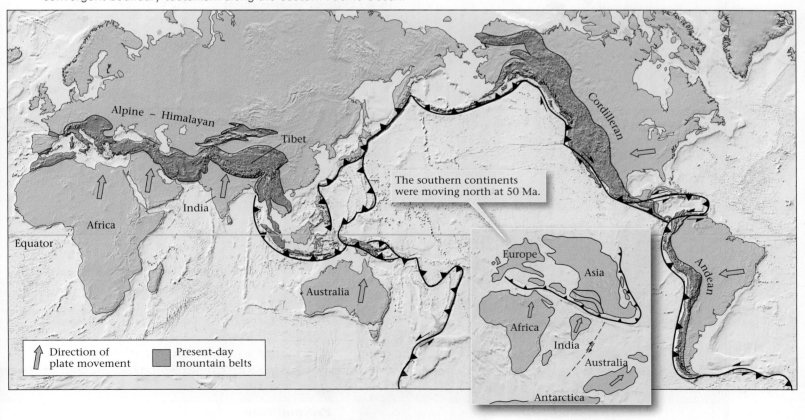

The Evolution of Earth

Earth has not had a static history; because of plate tectonics and its consequences (continental drift, sea-floor spreading, volcanism, and so on), the map of the Earth constantly changes. Distinct mountain-building events, or orogenies, have taken place during this process. The fossil record suggests that life first appeared within the first few hundred million years of our planet's existence, soon after a liquid-water ocean had accumulated, and like the planet itself has constantly changed ever since. The progressive change of the assemblage of species of life on Earth is called evolution.

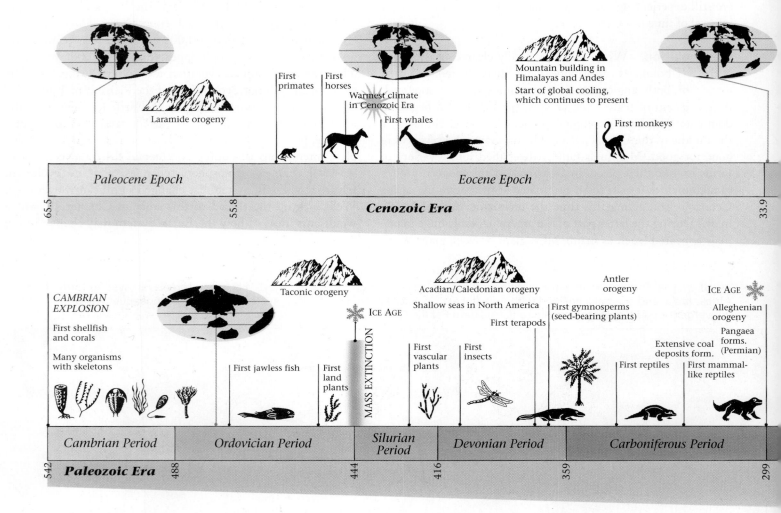

Laramide orogeny

First primates

First horses

Warmest climate in Cenozoic Era

First whales

Mountain building in Himalayas and Andes

Start of global cooling, which continues to present

First monkeys

Paleocene Epoch

Eocene Epoch

Cenozoic Era

65.5 55.8 33.9

CAMBRIAN EXPLOSION

First shellfish and corals

Many organisms with skeletons

Taconic orogeny

ICE AGE

First jawless fish

First land plants

MASS EXTINCTION

Acadian/Caledonian orogeny

Shallow seas in North America

First terapods

First vascular plants

First insects

Antler orogeny

First gymnosperms (seed-bearing plants)

ICE AGE

Alleghenian orogeny

Pangaea forms. (Permian)

Extensive coal deposits form.

First reptiles

First mammal-like reptiles

Cambrian Period **Ordovician Period** **Silurian Period** **Devonian Period** **Carboniferous Period**

542 488 444 416 359 299

Paleozoic Era

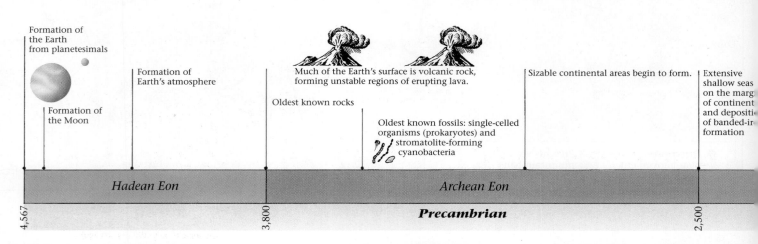

Formation of the Earth from planetesimals

Formation of the Moon

Formation of Earth's atmosphere

Much of the Earth's surface is volcanic rock, forming unstable regions of erupting lava.

Oldest known rocks

Oldest known fossils: single-celled organisms (prokaryotes) and stromatolite-forming cyanobacteria

Sizable continental areas begin to form.

Extensive shallow seas on the margins of continents and deposition of banded-iron formation

Hadean Eon **Archean Eon**

4,567 3,800 2,500

Precambrian

The earliest life forms were microscopic. By the end of the Precambrian, complex multicellular organisms had formed, and a burst of evolution at the Precambrian-Cambrian boundary yielded a diversity of invertebrates with shells. During the past half-billion years, several new major groups of organisms have appeared, and countless species have become extinct. Evidence suggests that evolution is not a continuous, gradual process, but occurs in pulses, separated by intervals of time during which the assemblage of species is fairly stable.

The last few hundred million years of Earth history have seen life leave the ocean and spread across the land. During the Mesozoic Era, dinosaurs roamed the Earth—then vanished abruptly 65 million years ago, perhaps as a result of a meteorite's colliding with the Earth. Since then, mammals have diversified into a great variety of species. And the last 100,000 years or so have witnessed the evolution of our own species, *Homo sapiens*. Considering the changes that people have brought about to the Earth System, the appearance of our species is clearly a major event in the history of the planet. The time scale is the 2004 time scale (International Committee on Stratigraphy).

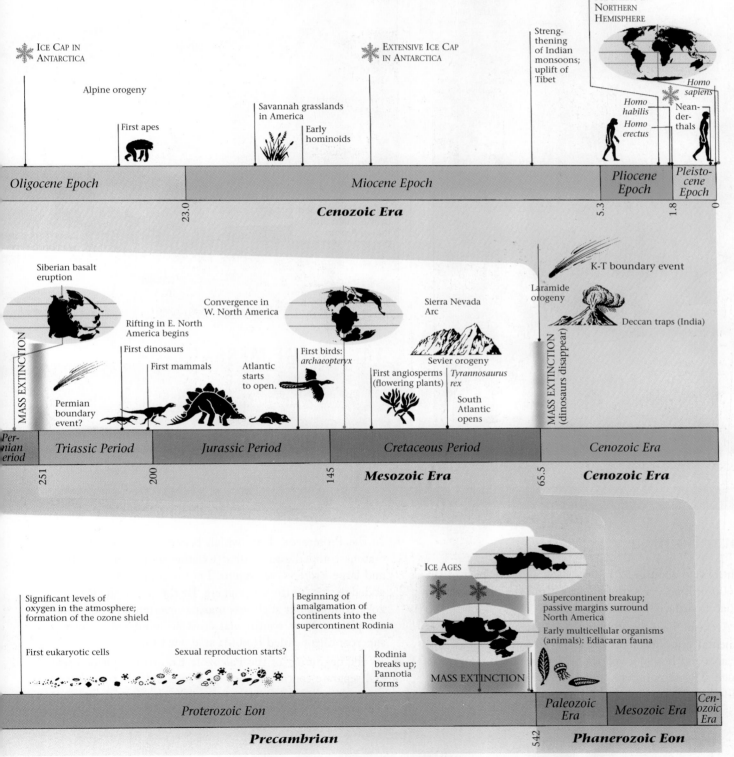

Geologic time (million years ago)

317

FIGURE 11.20 The Basin and Range Province is a rift. Its opening caused rotation of the Sierra Nevada. The opening of the Rio Grande Rift caused rotation of the Colorado Plateau, which is an unrifted block of cratonic crust. The inset shows a cross section along the red line.

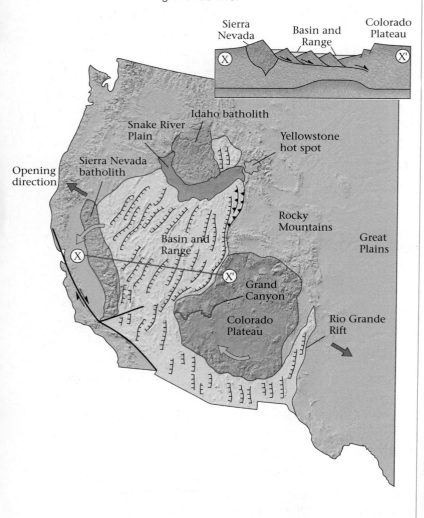

FIGURE 11.21 The maximum advance of the Pleistocene ice sheet in North America.

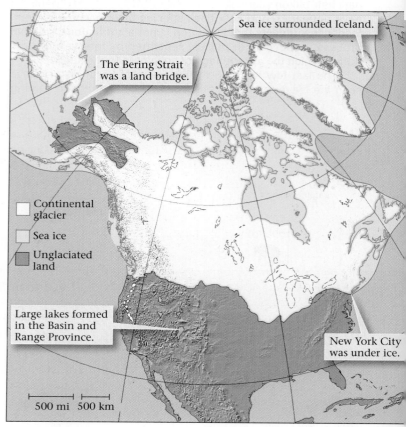

TAKE-HOME MESSAGE

The Cenozoic began with a bang—literally—as the age of dinosaurs came to a close, probably in the wake of a huge meteorite impact. During the Cenozoic, the modern mountain belts and plate boundaries of the planet became established, and mammals diversified.

Chapter Summary

- Earth formed about 4.57 billion years ago. For part of the first 600 million years, the Hadean Eon, the planet was so hot that its surface was a magma ocean.

- The Archean Eon began about 3.8 Ga, when permanent continental crust formed. The crust assembled out of volcanic arcs and hot-spot volcanoes that were too buoyant to subduct. The atmosphere contained little oxygen, but the first life forms—bacteria and archaea—appeared.

- In the Proterozoic Eon, which began at 2.5 Ga, Archean cratons collided and sutured together along orogenic belts and large Proterozoic cratons. Photosynthesis by organisms added oxygen to the atmosphere. By the end of the Proterozoic, complex but shell-less marine invertebrates populated the planet. Most continental crust accumulated to form a supercontinent called Rodinia at about 1 Ga.

- At the beginning of the Paleozoic Era, rifting yielded several separate continents. Sea level rose and fell a number of times, creating sequences of strata in continental interiors.

Continents began to collide and coalesce again, leading to orogenies and, by the end of the era, another supercontinent, Pangaea. Early Paleozoic evolution produced many invertebrates with shells, and jawless fish. Land plants and insects appeared in the middle Paleozoic. And by the end of the eon, there were land reptiles and gymnosperm trees.

- In the Mesozoic Era, Pangaea broke apart and the Atlantic Ocean formed. Convergent-boundary tectonics dominated along the western margin of North America. Dinosaurs appeared in Late Triassic time and became the dominant land animal through the Mesozoic Era. During the Cretaceous Period, sea level was very high, and the continents flooded. Angiosperms appeared at this time, along with modern fish. A huge mass-extinction event, which wiped out the dinosaurs, occurred at the end of the Cretaceous Period, probably because of the impact of a large meteorite.

- In the Cenozoic Era, continental fragments of Pangaea began to collide again. The collision of Africa and India with Asia and Europe formed the Alpine-Himalayan orogen. Convergent tectonics has persisted along the margin of South America, creating the Andes, but ceased in North America when the San Andreas Fault formed. Rifting in the western United States during the Cenozoic Era produced the Basin and Range Province. Various kinds of mammals filled niches left vacant by the dinosaurs, and the human genus, *Homo*, appeared and evolved throughout the radically shifting climate of the Pleistocene Epoch.

GEOPUZZLE REVISITED

The Earth has a long and complex history. According to geologic studies, our planet formed at about 4.57 Ga. The first land and water may have appeared as early as 4.04 Ga, but the record of the earliest crust was destroyed by meteorite bombardment about 3.9 Ga. The first long-lived continental crust appeared about 3.8 Ga. Oceans also appeared about 3.8 Ga and have lasted ever since. Subtle evidence suggests that life appeared by around 3.5 Ga. Thus, life did not become established until the Earth was a billion years old. Most continental crust formed by about 2.5 Ga. Some has survived intact, but most has been recycled by tectonic processes. The amount of dry land, however, changes over time due to the rise and fall of sea level. As continents drift, collide, and break apart, mountain belts form and then later erode away. The present ranges on Earth have been around only since the beginning of the Cenozoic.

Key Terms

Alpine-Himalayan chain (p. 314)
Ancestral Rockies (p. 309)
Archean Eon (p. 300)
Basin and Range Province (p. 314)
Cambrian explosion (p. 307)
cratonic platform (p. 303)
differentiation (p. 299)
Ediacaran fauna (p. 304)
epicontinental sea (p. 307)
Hadean Eon (p. 299)

Laramide orogeny (p. 312)
Pangaea (p. 308)
Phanerozoic Eon (p. 306)
Pleistocene ice age (p. 314)
Proterozoic Eon (p. 303)
Rodinia (p. 303)
shield (p. 303)
snowball Earth (p. 305)
stromatolite (p. 301)
superplume (p. 312)

Review Questions

1. Why are there no rocks on Earth that yield isotopic dates older than 4 billion years?

2. Describe the condition of the crust, atmosphere, and oceans during the Hadean Eon.

3. How did the atmosphere and tectonic conditions change during the Proterozoic Eon?

4. What evidence do we have that the Earth nearly froze over twice during the Proterozoic Eon?

5. How did the Cambrian explosion of life change the nature of the living world?

6. How did the Alleghanian and Ancestral Rockies orogenies affect North America?

7. What are the major classes of organisms that appeared in the Paleozoic?

8. Describe the plate-tectonic conditions that led to the formation of the Sierran arc and the Sevier thrust belt. What happened during the Laramide orogeny?

9. What life forms appeared during the Mesozoic?

10. What may have caused the flooding of the continents during the Cretaceous Period?

11. What could have caused the K-T extinctions?

12. What continents formed as a result of the breakup of Pangaea?

13. What are the causes of the uplift of the Himalayas and the Alps?

On Further Thought

1. During intervals of the Paleozoic, large areas of continents were submerged by shallow seas. Using *Google Earth*™ or a comparable program, tour North America from space. Do any present-day regions *within* North America consist of continental crust that was submerged by seawater? What about regions offshore? (*Hint*: Look at the region just east of Florida.)

2. Geologists have concluded that 80% to 90% of Earth's continental crust had formed by 2.5 Ga. But if you look at a geological map of the world, you find that only about 10% of the Earth's continental crustal surface is labeled "Precambrian." Why?

THE VIEW FROM SPACE The present-day Bahamas serve as an example of what the interior of the United States might have looked like during intervals of the Paleozoic. Shallow land areas were submerged and became the site of shallow-marine sedimentation.

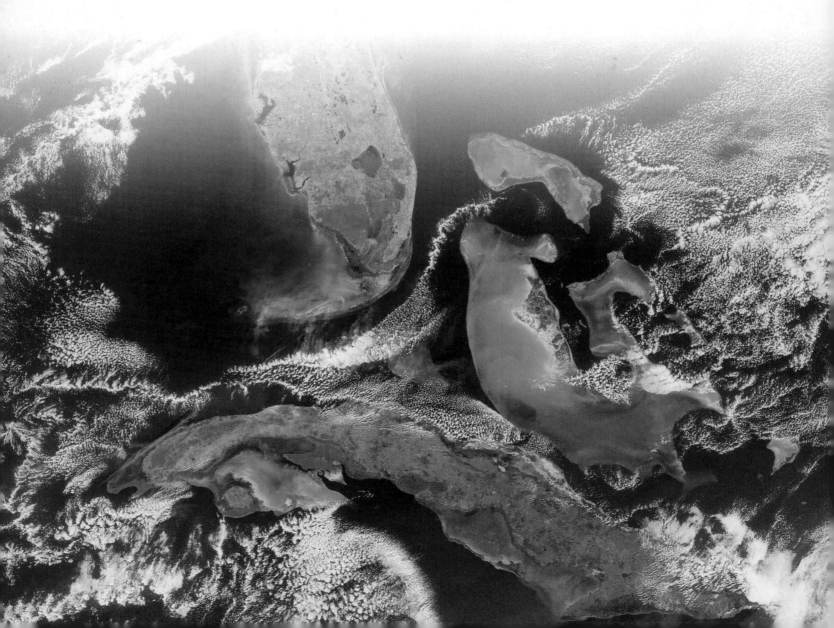

Riches in Rock: Energy and Mineral Resources

Workers struggle to extinguish the inferno erupting from an oil well fire in Kuwait. Hundreds of wells were set ablaze at the end of the Gulf War in 1991. The flames are a dramatic display of the energy held in oil.

GEOPUZZLE

The average citizen of a developed nation uses about 15,000 kg of energy and mineral resources (oil, stone, salt, iron, copper, etc.) per year. Where does all this material come from and will supplies of it always exist?

12.1 INTRODUCTION

In the extreme chill of a midwinter night, a pan of water freezes almost instantly. But the low temperature doesn't stop a wolf from stalking its prey—the wolf's legs carry it through the ice and snow, and its body radiates heat. To keep living in this way, the wolf must find materials from which to produce the energy that keeps its body warm. In this context, **energy** is simply the capacity to do work, to cause something to happen, or to cause change in a system. We refer to any material that can be of use, either to make something or to provide energy, as a **resource**. The resources that a wolf seeks include mice, insects, berries, and water. The wolf's energy comes from the breaking of chemical bonds in molecules of sugar, protein, and carbohydrates in food.

The earliest humans were hunter-gatherers and could survive on pretty much the same resources that a wolf needs. But when people discovered how fire could be used for cooking and heating, how weapons could facilitate hunting, how shelters made the winter nights more comfortable, and how harvests could be improved by using plows and fertilizers, their need for resources started to increase dramatically. In the twenty-first century, an average citizen of an industrialized country uses more than 100 times the amount of resources used by a hunter-gatherer.

Where do people obtain resources? Most come from the Earth, and are either geologic materials or the products of geologic processes. That is why we discuss resources in a geology book. In this chapter, we distinguish between two kinds of resources—**energy resources**, which can be used to produce heat and electricity or move machines, and **mineral resources**, which provide natural inorganic chemicals from which products can be manufactured. To understand where to find energy and mineral resources, why we run the risk of running out of some resources, and how the production and use of resources affects the environment, we must understand the geologic origin of resources and the settings where they can be found or disposed of. Searching for resources, and dealing with the consequences of their production and use, provides jobs for tens of thousands of geologists worldwide.

TAKE-HOME MESSAGE

By the end of this chapter you should understand where energy and mineral resources come from and why it takes special geologic circumstances for reserves to accumulate. You will also see that some resources are running out, and how resource use may threaten the environment.

12.2 SOURCES OF ENERGY IN THE EARTH SYSTEM

Humans obtain energy from a variety of sources. Some are physical (such as solar radiation and gravity), whereas others are biological (involving plant growth):

- *Energy directly from the Sun*: Solar energy, resulting from nuclear fusion reactions in the Sun, bathes the Earth's surface. It may be converted directly into electricity, using solar-energy panels, or it may be used to heat water.

- *Energy directly from gravity*: The gravitational attraction of the Moon, and to a lesser extent the Sun, causes ocean tides, the daily up-and-down movement of the sea surface. The flow of water in and out of channels during tidal changes can drive turbines.

- *Energy involving both solar energy and gravity*: Solar radiation heats the air, which becomes less dense. In a gravitational field, this warm air rises. Cool, denser air then sinks. The resulting air movement, wind, powers sails and windmills. Solar energy also evaporates water, which enters the atmosphere. When the water condenses, it rains on the land, where it accumulates in streams that flow downhill in response to gravity. This moving water can power waterwheels and turbines.

- *Energy via photosynthesis*: Green plants absorb some of the solar energy that reaches the Earth's surface. Their green color comes from a pigment called chlorophyll. With the aid of chlorophyll, plants produce sugar through a chemical reaction called **photosynthesis**. Plants use the sugar to manufacture more complex chemicals, or they metabolize it to provide themselves with energy.

  Burning plant matter in a fire releases potential energy stored in the chemical bonds of organic chemicals. During burning, the molecules react with oxygen and break apart to produce carbon dioxide, water, and carbon (soot). People have burned wood to produce energy for centuries. More recently, plant material (**biomass**) has been used to produce ethanol, a flammable alcohol.

- *Energy from chemical reactions*: A number of inorganic chemicals can burn to produce light and energy. A dynamite explosion is an extreme example of such energy production. Recently, researchers have been studying electrochemical devices, such as hydrogen fuel cells, that produce electricity directly from chemical reactions.

- *Energy from fossil fuels*: Oil, gas, and coal come from organisms that lived millions of years ago, and thus store solar energy that reached the Earth long ago. We refer to these substances as **fossil fuels**, to emphasize that they were derived from ancient organisms and have been preserved in rocks for geologic time. (In a general sense, a fuel is any material that can be used to produce energy.)

- *Energy from nuclear fission*: Atoms of radioactive elements can split into smaller pieces, a process called nuclear fission (see Appendix). During fission, a tiny amount of mass transforms into a large amount of energy, called nuclear energy. This type of energy runs nuclear power plants and nuclear submarines.

- *Energy from Earth's internal heat*: Some of Earth's internal energy dates from the birth of the planet, while some comes from radioactive decay in minerals. This internal energy heats underground water. The resulting hot water, when transformed to steam, provides geothermal energy, which can drive turbines.

During the course of civilization, the sources of energy that people use have changed (**Fig. 12.1**). Prior to the industrial age, direct burning of wood and other biomass provided most of humanity's energy needs. But by the second half of the nineteenth century, deforestation had nearly destroyed this resource, and energy needs had increased so dramatically that other fuels

FIGURE 12.1 Energy needs have increased in the past 150 years. Different energy resources have been used to fill those needs.

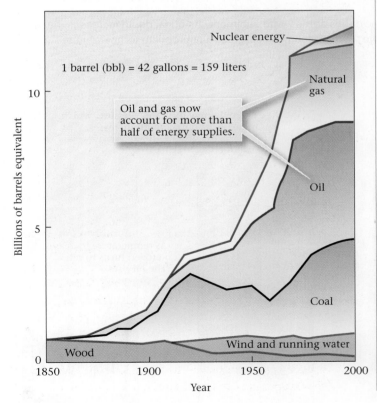

came into use. In the succeeding sections we discuss the geology of various energy sources, beginning with oil and gas, the dominant fuels of the present day.

TAKE-HOME MESSAGE

Energy comes from several sources—solar radiation, gravity, chemical reactions, radioactive decay, and internal heat. Solar radiation and gravity together drive wind and water movement. Products of photosynthesis can be preserved in rocks as fossil fuels.

12.3 OIL AND GAS

What Are Oil and Gas?

For reasons of economics and convenience, industrialized societies today rely primarily on oil (petroleum) and natural gas for their energy needs. Oil and natural gas consist of **hydrocarbons**, chain-like or ring-like molecules made of carbon and hydrogen atoms. Chemists consider hydrocarbons to be a type of organic chemical.

Some hydrocarbons are gaseous and invisible, some resemble a watery liquid, some appear syrupy, and some are solid. The viscosity (ability to flow) and the volatility (ability to evaporate) of a hydrocarbon product depend on the size of its molecules. Hydrocarbon products composed of short chains of molecules tend to be less viscous (meaning they can flow more easily) and more volatile (meaning they evaporate more easily) than products composed of long chains, because the long chains tend to tangle up with each other. Thus, short-chain molecules occur in gaseous form (natural gas) at room temperature, moderate-length-chain molecules occur in liquid form (gasoline and oil), and long-chain molecules occur in solid form (tar).

Where Do Oil and Gas Form?

The carbon and hydrogen atoms that comprise oil and natural gas come from the cells of algae and plankton—*not* from trees or dinosaurs. Transformation of tiny floating organisms into flamable liquids or gases takes several steps and can only happen under special conditions. The process begins when the organisms die, and their bodies accumulate, along with clay, on the floor of a quiet lake or sea. The bodies are so tiny that they cannot settle in places where the water is moving. If water at the site of deposition contains dissolved oxygen, the organic chemicals making up the bodies either oxidize or are eaten by microbes and decompose. But in oxygen-poor water, the organic chemicals survive and mix with clay to form an ooze which, when deeply buried, become beds of *black* organic shale. The organic chemicals in such beds are the raw materials from which hydrocarbons form, so black organic shale is called a **source rock**.

If a source rock is buried deeply enough (2 to 4 km), it becomes warmer, since temperature increases with depth in the Earth. Chemical reactions slowly transform the organic material in black shale into a waxy substance called **kerogen** (Fig. 12.2). Shale containing kerogen is called **oil shale**. If oil shale gets buried still deeper and warms to temperatures of greater than about 90°C, the kerogen molecules break down to form oil and natural gas molecules. At temperatures over about 160°C, any remaining oil breaks down to form natural gas, and at temperatures over 250°C, organic matter transforms into graphite. Thus, oil itself forms only in a relatively narrow range of temperatures, called the **oil window**, which generally can exist only in the topmost 6 to 9 km of the crust.

TAKE-HOME MESSAGE

Oil and gas are hydrocarbons whose atoms come from algae and plankton. When the organisms die, their bodies mix with clay to produce organic ooze on the floor of quiet lakes and seas. Burial transforms the ooze into black shale, a source rock. At temperatures of between 90° and 160°C, the organic material in black shale turns into oil.

12.4 MAKING AN OIL RESERVE

Oil and gas do not occur in all rocks at all locations. That's why the desire to control oil fields, regions that contain significant amounts of accessible oil underground, has played a role in triggering wars. A known supply of oil and gas held underground is a **hydrocarbon reserve**; if the reserve consists dominantly of oil, it is an oil reserve. Reserves are not randomly distributed around the Earth (Fig. 12.3). In fact, countries bordering the Persian Gulf have almost 60% of the world's reserves, whereas the United States, by far the largest consumer of oil, has only a few percent of the total. Reserves are specified in "barrels" (bbl); 1 bbl = 42 gallons = 159 liters.

We now show how the development of a reserve requires a source rock, a reservoir rock, a migratory pathway, and a trap. An association of all of these components in an appropriate thermal environment is called a hydrocarbon system.

Source Rocks and Hydrocarbon Generation

Previously we saw that the chemicals that become oil and gas start out in algae and plankton cells. These accumulate along with clay to form an organic ooze, which when lithified becomes black organic shale. Geologists refer to organic-rich shale as a source rock because it provides the organic chemicals that

FIGURE 12.2 The formation of oil. The process begins when organic debris settles with sediment. As burial depth increases, heat and pressure transform the sediment into black shale in which organic matter becomes kerogen. At appropriate temperatures, kerogen becomes oil, which then seeps upward.

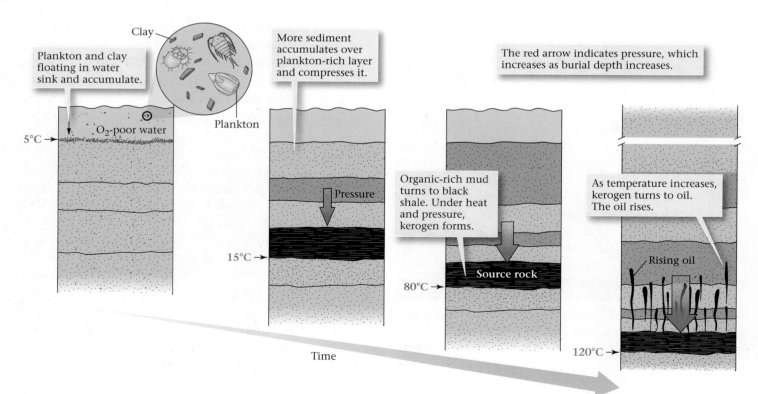

FIGURE 12.3 The distribution of oil reserves around the world.

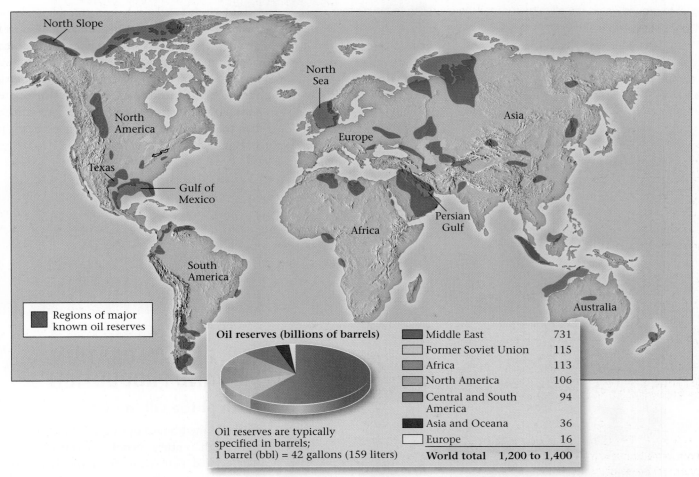

Oil reserves (billions of barrels)	
Middle East	731
Former Soviet Union	115
Africa	113
North America	106
Central and South America	94
Asia and Oceana	36
Europe	16
World total	**1,200 to 1,400**

Oil reserves are typically specified in barrels; 1 barrel (bbl) = 42 gallons (159 liters)

ultimately become oil and gas. If black shale resides in the oil window, the kerogen it contains transforms into oil and gas. This process is called hydrocarbon generation.

Reservoir Rocks and Hydrocarbon Migration

Wells drilled into source rocks do not yield much oil because waxy kerogen can't flow easily from the rock into the well. So oil companies instead drill into a **reservoir rock**, rock that contains (or could contain) an abundant amount of *easily extractable* oil and gas.

To be a reservoir rock, a rock must have high porosity and permeability. **Porosity** refers to the amount of open space (pore space) in a rock. Pore space can hold oil or gas, much as the holes in a sponge can hold water. Shale has low porosity (<10%), and thus is not a reservoir rock. Sandstone may have high porosity (up to 40%) and thus could be a reservoir rock. **Permeability** refers to the degree to which pore spaces connect to each other. Even if a rock has high porosity, it is not necessarily permeable. In a permeable rock, the holes and cracks are linked, so a fluid can flow slowly through the rock, following a tortuous pathway. Keeping the concepts of porosity and per-

meability in mind, we can see that a poorly cemented sandstone makes a good reservoir rock, because it is both porous and permeable.

To fill the pores of a reservoir rock, oil and gas must first migrate (move) from the source rock into a reservoir rock (Fig. 12.4). Why do hydrocarbons migrate? Oil and gas are less dense than water, so they try to rise toward the Earth's surface to get above groundwater, just as salad oil rises above the vinegar in a bottle of salad dressing. Natural gas, being less dense, ends up floating above oil. In other words, buoyancy drives oil and gas upward. Typically, a hydrocarbon system must have a good migration pathway, such as a permeable set of fractures, in order for large volumes of hydrocarbons to move. The process of migration generally takes millions of years.

Traps and Seals

If oil or gas escapes from the reservoir rock and ultimately reaches the Earth's surface, where it leaks away at an oil seep, none will be left underground to extract. Thus, for an oil reserve to exist, oil and gas must be held underground in the reservoir rock, by means of a geologic configuration called a **trap**.

FIGURE 12.4   Initially, oil resides in the source rock. Because it is buoyant relative to groundwater, the oil migrates into the overlying reservoir rock. The oil accumulates beneath a seal rock in a trap.

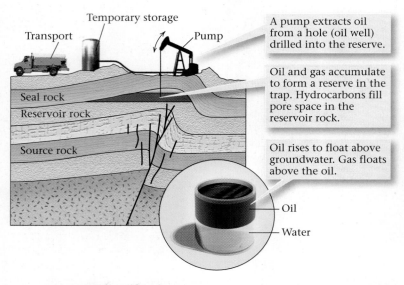

Present

A pump extracts oil from a hole (oil well) drilled into the reserve.

Oil and gas accumulate to form a reserve in the trap. Hydrocarbons fill pore space in the reservoir rock.

Oil rises to float above groundwater. Gas floats above the oil.

Oil

Water

A trap forms and oil accumulates beneath the seal rock.

Fault causes fracturing; a migration pathway develops. Oil migrates up from the source.

Tectonic stress causes fault to slip and overlying beds to fold.

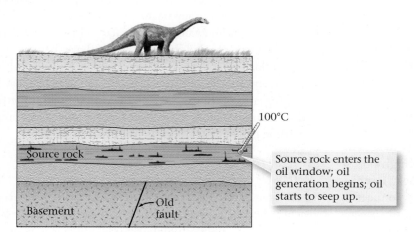

Source rock enters the oil window; oil generation begins; oil starts to seep up.

Past

An oil or gas trap has two components. First, a **seal rock** (a relatively impermeable rock such as shale, salt, or unfractured limestone) must lie above the reservoir rock and stop the hydrocarbons from rising farther. Second, the seal and reservoir rock bodies must be arranged in a geometry that collects the hydrocarbons in a restricted area. Geologists recognize several types of hydrocarbon trap geometries, four of which are described in **Box 12.1**.

TAKE-HOME MESSAGE

Oil and gas reserves develop only where there is an appropriate source rock (black shale), a migration pathway, a reservoir rock (such as porous and permeable sandstone), and a trap that keeps oil and gas underground. Hydrocarbons rise because they are buoyant.

12.5 OIL EXPLORATION AND PRODUCTION

Birth of the Oil Industry

In the United States during the first half of the nineteenth century, people collected "rock oil" (later called petroleum, from the Latin words *petra*, meaning rock, and *oleum*, meaning oil) at seeps and used it to grease wagon axles and to make patent medicines. But such oil was rare and expensive. In 1854, George Bissel, a New York lawyer, came to the realization that oil might have broader uses, particularly as fuel for lamps (to replace whale oil). Bissel and a group of investors asked Edwin Drake, a colorful character who had drifted among many professions, to find a way to drill for oil in rocks beneath a hill near Titusville, Pennsylvania, where an oily film floated on the water of springs. Using the phony title "Colonel" to add respectability, Drake hired drillers, built a wooden drilling rig, and obtained a steam-powered drill. Work was slow and the investors became discouraged, but the very day that a letter arrived ordering Drake to stop drilling, his drillers found that the hole, which had reached a depth of 21.2 m, had filled with oil. They set up a pump and—on August 27, 1859, for the first time in history—pumped oil out of the ground. No one had given much thought to

BOX 12.1

THE REST OF THE STORY

Types of Oil and Gas Traps

Geologists who work for oil companies spend much of their time trying to identify underground traps. No two traps are exactly alike, but we can classify most into the following four categories.

- *Anticline trap*: In some places, sedimentary beds are not horizontal, as they are when originally deposited, but have been bent by the forces involved in mountain building. These bends, as we have seen, are called folds. An anticline is a type of fold with an arch-like shape (**Fig. 1a**; see Chapter 9). If the layers in the anticline include a source rock overlain by a reservoir rock that is overlain by a seal rock, then we have the recipe for an oil reserve. The oil and gas rise from the source rock, enter the reservoir rock, and rise to the crest of the anticline, where they are trapped by a seal.

- *Fault trap*: A fault is a fracture on which there has been sliding. If the slip on the fault crushes and grinds the adjacent rock to make an impermeable layer along the fault, then oil and gas may migrate upward along bedding in the reservoir rock until they stop at the fault surface (**Fig. 1b**). Alternatively, a fault trap develops if the slip on the fault juxtaposes an impermeable rock layer against a reservoir rock.

- *Salt-dome trap*: In some sedimentary basins, the sequence of strata contains a thick layer of salt, deposited when the basin was first formed and seawater covering the basin was shallow and very salty. Sandstone, shale, and limestone overlie the salt.

The salt layer is not as dense as sandstone or shale, so it is buoyant and tends to rise up slowly through the overlying strata. Once the salt starts to rise, the weight of surrounding strata squeezes the salt out of the layer and up into a growing, bulbous **salt dome**. As the dome rises, it bends up the adjacent layers of sedimentary rock. Oil and gas in reservoir rock layers migrate upward until they are trapped against the boundary of the salt dome, for salt is not permeable (**Fig 1c**).

- *Stratigraphic trap*: In a stratigraphic trap, a tilted reservoir rock bed "pinches out" (thins and disappears along its length) between two impermeable layers. Oil and gas migrating upward along the bed accumulate at the pinch-out (**Fig. 1d**).

FIGURE 1 Examples of oil traps. A trap is a configuration of a seal rock over a reservoir rock, in a geometry that keeps the oil underground.

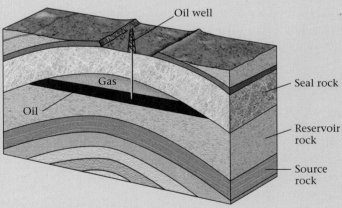

(a) Anticline trap. Oil and gas rise to the crest of the fold.

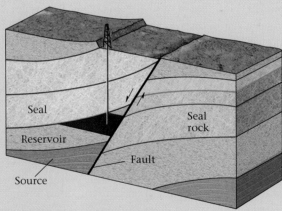

(b) Fault trap. Oil and gas collect in tilted strata adjacent to the fault.

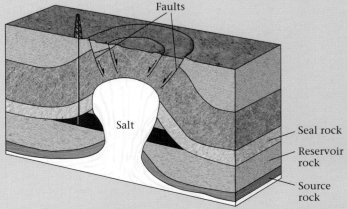

(c) Salt-dome trap. Oil and gas collect in strata on the flanks of the dome, beneath salt.

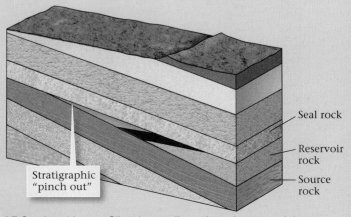

(d) Stratigraphic trap. Oil and gas collect where the reservoir layer pinches it out.

the issue of storing the oil, so workers dumped it into empty whisky barrels. This first oil well yielded 10 to 35 barrels a day, which sold for about $20 a barrel.

Within a few years, thousands of oil wells had been drilled in many states, and by the turn of the twentieth century civilization had begun its addiction to oil. Initially, most oil went into the production of kerosene for lamps. Later, as electricity took over from kerosene as the primary source for illumination, gasoline derived from oil became the fuel of choice for the newly invented automobile. Oil was also used to fuel electric power plants and heat homes. In its early years, the oil industry was in perpetual chaos. When wildcatters (people who search for new oil fields) discovered one, there would be a short-lived boom during which the price of oil could drop to pennies a barrel. Now oil production and distribution have become a global industry governed by the complex interplay of politics, profits, supply, and demand.

The Modern Search for Oil

Wildcatters discovered the earliest oil fields either by blind luck or by searching for surface seeps. But when all known seeps had been drilled and blind luck became too risky, oil companies realized that finding new oil reserves would require systematic exploration. The modern-day search for oil is a complex, sometimes dangerous, and often exciting procedure with many steps. Most modern-day exploration depends on the study of seismic-reflection images. These are produced by using either small explosions, a special vibrating truck, or pulses of pressurized air to send seismic waves into the crust (see Interlude D). The waves reflect off the contacts between rock layers underground and return to the surface, where sensitive instruments record their arrival. A computer measures the time between the generation of the seismic waves and their return to the surface, and from this information defines the depth to the contacts that reflected the waves. The computer then constructs an image showing the arrangement of rock layers underground.

If geological studies identify a trap and the presence of good source rocks and reservoir rocks, geologists make a recommendation to drill. Once the decision has been made, drillers go to work. These days, they use rotary drills to grind a hole down through rock. A rotary drill consists of a rotating pipe tipped by a bit, a bulb of metal studded with industrial diamonds or hard metal prongs (**Fig. 12.5a**). As the bit rotates, it scratches and gouges the rock, turning it into powder and chips. Drillers pump drilling mud, a slurry of water mixed with clay, down the center of the pipe. The mud squirts out of holes in the drill bit, cooling the bit and flushing rock cuttings up and out of the hole. The weight of the drilling mud also keeps oil down in the hole and prevents gushers, fountains of oil formed when underground pressure causes the oil to rise

out of the hole on its own. On completion of a hole, workers remove the drilling rig and set up a pump (**Fig. 12.5b**).

Once extracted directly from the ground, crude oil flows first into storage tanks and then into a pipeline (see **Geotour 12** on p. GT-26) or tanker (**Fig. 12.5c**), which transports it to a refinery. At a refinery, workers distill crude oil into several separate components by heating it gently in a vertical pipe called a distillation column (**Fig. 12.5d**). Lighter molecules rise to the top of the column, while heavier molecules stay at the bottom. The heat may also "crack" larger molecules to make smaller ones.

TAKE-HOME MESSAGE

Searching for oil is a complex and expensive process. Geologists use seismic-reflection profiles to locate possible traps. Drilling taps reserves and pumping brings crude oil to the surface, where it is processed at refineries that crack hydrocarbon molecules.

12.6 ALTERNATIVE RESERVES OF HYDROCARBONS

So far, we've focused our discussion of hydrocarbons on reserves that can be pumped from the subsurface in the form of a liquid (oil). Hydrocarbons, however, also occur in other forms that are increasingly being used and that may provide energy sources when liquid oil starts to run out.

Natural Gas

Natural gas consists of volatile short-chain hydrocarbons, including methane, ethane, propane, and butane. Gas may occur in association with oil; it occurs in the pores of reservoir rock above oil, because it "floats" over the oil. But where temperatures in the subsurface are so high that oil molecules break apart to form gas, gas-only fields develop.

Gas burns more cleanly than oil. Burning gas produces primarily carbon dioxide and water, whereas burning oil also produces complex organic pollutants. Thus, gas has become the preferred fuel for home cooking and heating. But transporting gas requires expensive, pressurized pipelines or container ships, so even though gas is much more abundant than oil, the world's population still consumes more oil than gas. However, in recent years many industries and electrical generation plants have switched to natural gas, so demand for gas has increased.

Tar Sands (Oil Sands) and Oil Shale

In several locations around the world, most notably Alberta (in western Canada) and Venezuela, vast reserves of very viscous, tar-like "heavy oil" exist. This heavy oil, also known as bitumen,

FIGURE 12.5 The steps involved in searching for, recovering, transporting, and refining oil.

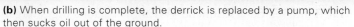

The head of the pump goes up and down.

Drill derrick

Rising drilling mud

Drill pipe

Pulley

Device to rotate pipe

Drill hole

Drill bit

Grinding head

(a) A rotating pipe tipped by a drill bit grinds a hole into the ground. Drilling mud, pumped down through the pipe, comes out through holes in the bit, flushing cuttings out of the hole.

(b) When drilling is complete, the derrick is replaced by a pump, which then sucks oil out of the ground.

(c) The Trans-Alaska pipeline transports oil from fields on the Arctic coast to a tanker port on the southern coast of Alaska.

(d) Distilling columns of an oil refinery transform crude oil into gasoline and other hydrocarbon products.

has the consistency of molasses, and thus cannot be pumped directly from the ground. It fills the pore spaces of sand or of poorly cemented sandstone, constituting up to 12% of the sediment or rock volume. Sand or sandstone containing high concentrations of bitumen is known as **tar sand** or oil sand.

The hydrocarbon system that led to the generation of tar sands began with the production and burial of a source rock in a large sedimentary basin. When subjected to temperatures of the oil window, the source rock yielded oil and gas, which migrated into sandstone layers and then up the dip of tilted layers to the edge of the basin, where it became caught in stratigraphic traps (Box 12.1). Over time, microbes (bacteria) attacked the oil reserve underground, digesting the lighter,

smaller hydrocarbon molecules and leaving behind only the larger tar molecules.

Production of usable oil from tar sand is difficult and expensive, but not impossible. It takes about 2 tons of tar sand to produce one barrel of oil. Oil companies mine near-surface deposits in open-pit mines, then heat the tar sand in a furnace to extract the oil. To extract oil from deeper deposits of tar sand, oil companies drill wells and pump steam or solvents down into the sand to liquefy the oil enough so that it can be pumped out.

Vast reserves of organic-rich shale that have not been subjected to temperatures of the oil window, or if they were, did not stay within the oil window long enough to complete the transformation to oil, still contain a high proportion of kerogen.

Shale that contains at least 15% to 30% kerogen is called oil shale. During the 1850s, researchers developed techniques to produce liquid oil from oil shale. The process involves heating the oil shale to a temperature of 500°C; at this temperature, the shale decomposes and the kerogen transforms into liquid hydrocarbon and gas. As is the case with tar sand, production of oil from oil shale is possible, but very expensive.

Gas Hydrate

Gas hydrate is a chemical compound consisting of a methane (CH_4) molecule surrounded by a cage-like arrangement of water molecules. An accumulation of gas hydrate occurs as a whitish solid that resembles regular water ice. Gas hydrate forms when anaerobic bacteria (bacteria that live in the absence of oxygen) eat organic matter, such as dead plankton, that has been incorporated into sea-floor sediment. When the bacteria digest organic matter, they produce methane as a byproduct, and the methane bubbles into the cold seawater that fills pore spaces in sediments. Under pressures found at water depths greater than 300 m, the methane dissolves in water and produces gas hydrate molecules. Gas hydrate occurs as layers interbedded with sediment, and/or as a cement holding together the sediment, at depths of between 90 and 900 m beneath the sea floor. Geologists estimate that an immense amount of methane lies trapped in gas hydrate layers. So far, however, techniques for safely recovering gas hydrate from the sea floor have not been devised.

TAKE-HOME MESSAGE

Large reserves of hydrocarbons occur in tar sands and oil shale as natural gas and as gas hydrate. But these alternative sources are difficult to process and/ or transport, and thus cannot be exploited until new technologies evolve and the price of oil increases.

12.7 COAL: ENERGY FROM THE SWAMPS OF THE PAST

Coal, a black, brittle, sedimentary rock that burns, consists of elemental carbon mixed with minor amounts of organic chemicals, quartz, and clay. Note that coal and oil do not have the same composition or origin. In contrast to oil, coal forms from plant material (wood, stems, leaves) that once grew in coal swamps, regions that resembled the wetlands of modern tropical to semitropical coastal areas (Fig. 12.6a). As sea level rose, the position of the swamp moved inland, and older portions of the swamp were buried (Fig. 12.6b). Like oil and gas, coal is a fossil fuel because it stores solar energy that reached Earth long ago.

Significant coal deposits could not form until vascular land plants appeared in the Late Silurian Period, about 420 million years ago. The most extensive deposits of coal in the world occur in Carboniferous-age strata (deposited between 354 and 286 million years ago). In fact, geologists coined the name *Carboniferous* because strata representing this interval of the geologic column contain so much coal. The abundance of Carboniferous coal reflects the equatorial position of the continents in the past and the height of sea level (shallow seas bordered by coal swamps covered vast parts of continental interiors). Extensive coal deposits can also be found in strata of the Cretaceous Period, the time interval between 144 and 65 million years ago.

The Formation of Coal

How do the remains of plants transform into coal? First, the vegetation of an ancient swamp must fall and be buried in an oxygen-poor environment, such as stagnant water, so that it can be incorporated in a sedimentary sequence without first reacting with oxygen or being eaten. Compaction and partial decay of the vegetation transforms it into **peat** (Fig. 12.6c). Peat, which contains about 50% carbon, itself serves as a fuel in many parts of the world.

For peat to become coal, the peat must be buried deeply (4 to 10 km) by sediment. Such deep burial can happen where the surface of the continent gradually sinks, and a sedimentary basin forms that can collect sediment. Eventually, many kilometers of sediment containing numerous peat layers accumulate. At depth in the pile, the weight of overlying sediment compacts the peat and squeezes out any remaining water. Then, because temperature increases with depth in the Earth, deeply buried peat gradually heats up. Heat accelerates chemical reactions that gradually destroy plant fiber and release elements such as hydrogen, nitrogen, and sulfur in the form of gas. These gases seep out of the peat layer, leaving behind a residue concentrated with carbon. Once the proportion of carbon in the residue exceeds about 70%, we have coal. With further burial and higher temperatures, chemical reactions remove additional hydrogen, nitrogen, and sulfur, yielding progressively higher concentrations of carbon.

The Classification of Coal

Geologists classify coal according to the concentration of carbon. With increasing burial, peat transforms into a soft, dark-brown coal called lignite. At higher temperatures (about 100°C to 200°C), lignite becomes dull, black bituminous coal, which contains 85% carbon. At still higher temperatures (about 200°C to 300°C), bituminous coal is transformed into shiny, black anthracite coal, which contains 95% carbon. The progressive

FIGURE 12.6 The formation of coal. Coal forms when plant debris becomes deeply buried.

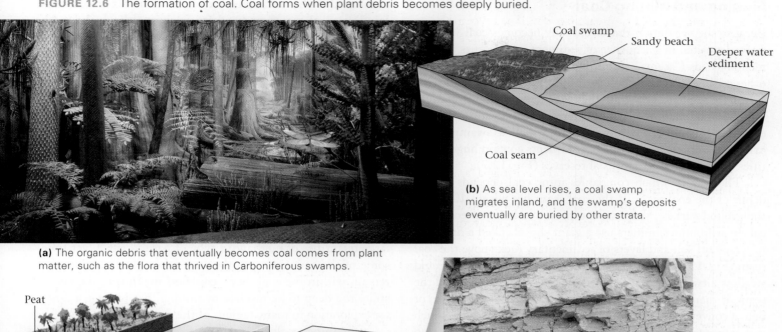

(a) The organic debris that eventually becomes coal comes from plant matter, such as the flora that thrived in Carboniferous swamps.

(b) As sea level rises, a coal swamp migrates inland, and the swamp's deposits eventually are buried by other strata.

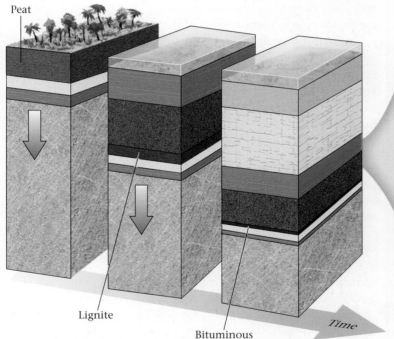

(c) A layer of peat accumulates beneath a swamp. Upon burial, the peat becomes lignite. After further burial, the lignite alters to form bituminous coal.

(d) Coal is found in sedimentary beds interlayered with other strata (sandstone, shale, and limestone).

transformation of peat to anthracite coal, which occurs as the coal layer is buried more deeply and becomes warmer, reflects the completeness of chemical reactions that remove water, hydrogen, nitrogen, and sulfur from the organic chemicals of the peat and leave behind carbon (Fig. 12.6c). As the carbon content of coal increases, we say the **coal rank** increases.

Of note, the formation of anthracite coal requires high temperatures that develop only in basins bordering mountain belts.

Here, mountain-building processes push thick sheets of rock along thrust faults up and over the coal-bearing sediment, so the sediment ends up at depths of 8 to 10 km. Temperatures at such depths reach 300°C. In the interiors of mountain belts, temperatures become even higher, and rock begins to undergo metamorphism. During metamorphism, almost all elements except carbon leave the coal. The remaining carbon atoms rearrange to form graphite.

Finding and Mining Coal

Because the vegetation that eventually becomes coal was initially deposited in a sequence of sediment, coal occurs as sedimentary beds ("seams," in mining parlance) interlayered with other sedimentary rocks (Fig. 12.6d). To find coal, geologists search for sequences of strata that were deposited in tropical to semitropical, shallow-marine to terrestrial (fluvial or deltaic) environments—the environments in which a swamp could exist. The sedimentary strata of continents contain huge quantities of discovered coal, or coal reserves (Fig. 12.7).

The way in which companies mine coal depends on the depth of the coal seam. If the coal seam lies within 100 m of the ground surface, strip mining proves to be most economical. In strip mines, miners use a giant shovel called a drag line to scrape off soil and layers of sedimentary rock above the coal seam (Fig. 12.8a). Drag lines are so big that their shovel could swallow a two-car garage. Once the drag line has exposed the seam, it then scrapes out the coal and dumps it into trucks or onto a conveyor belt that carries it to a storage area from which it is transported by rail (Fig. 12.8b). When the coal has been scraped out, the operator fills the remaining pit or trench with the rock that had been stripped to expose the coal and covers the rock back up with topsoil. Grass or trees may eventually grow over the "reclaimed" mine (see Geotour 12 on p. GT-26).

Coal at depths greater than 100 m can be obtained only by underground mining. Miners dig a shaft down to the depth of the coal seam and then create a maze of tunnels, using huge grinding machines that chew their way into the coal. Underground coal mining can be very dangerous, not only because the sedimentary rocks forming the roof of the mine are weak and can collapse, but also because methane gas released by chemical reactions in coal can accumulate in the mine, leading to the danger of a small spark triggering a deadly mine explosion. Unless they breathe through filters, underground miners also risk contracting black-lung disease from the inhalation of coal dust. The dust particles wedge into tiny cavities of the lungs and gradually cut off the oxygen supply or cause pneumonia.

Coalbed Methane and Coal Gasification

The natural process by which coal forms underground yields large quantities of methane, a type of natural gas. Over time, some of the gas escapes to the atmosphere, but vast amounts remain within the coal. Such **coalbed methane**, trapped in strata too deep to be reached by mining, has become a target for exploration in many regions of the world.

Obtaining coalbed methane from deep layers of strata involves drilling, rather than mining. Drillers penetrate a coal bed with a hole, and then start pumping out groundwater. As a result of pumping out water, the pressure in the vicinity of the drillhole decreases relative to the surrounding bed. Methane bubbles into the hole and then up to the ground surface, where condensers force it into tanks for storage.

FIGURE 12.7 The distribution of coal reserves. Vast quantities lie buried in continental sedimentary basins.

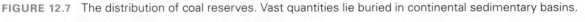

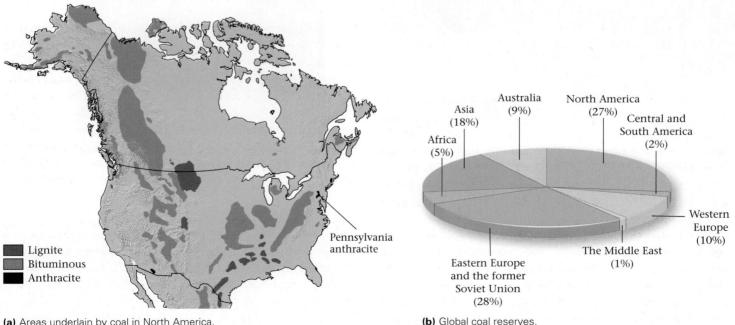

(a) Areas underlain by coal in North America.

(b) Global coal reserves.

FIGURE 12.8 The process of mining coal.

A drag line stripping coal in an Indiana mine.

Undistributed land • Reclaimed land • High wall • Undisturbed land • Spoil bank • Bedding plane • Coal seam

(a) The configuration of a coal strip mine. After removing the overburden, the drag line strips the coal. Then the trench is refilled and the ground surface replanted.

(b) Special machinery grinds into an underground coal seam and dumps coal on a conveyor. This seam is about 2.5 m thick.

Traditional coal burning produces clouds of smoke, containing fly ash (solid residue left after the carbon in coal has been burned) and noxious gases. This problem can be partially solved if coal is first transformed into gases, as well as solid byproducts, *before* burning. The gases burn quite cleanly and the byproducts can be disposed of relatively easily. The process of producing clean-burning gases from solid coal is called **coal gasification**.

Coal gasification involves the following steps. First, pulverized coal is placed in a large container. Then, a mixture of steam and oxygen passes through the coal at high pressure. As a result, the coal heats up to a high temperature but does not ignite; under these conditions, chemical reactions break down the carbon molecules in coal and produce hydrogen gas and

other gases such as carbon monoxide. Solid ash, as well as sulfur and mercury, concentrate at the bottom of the container and can be removed before the gases are burned.

Underground Coalbed Fires

Coal burns not only in furnaces but also in surface and subsurface mines, as long as the fire has access to oxygen. Coal mining during the past two centuries has exposed underground coal to the air and has provided many more opportunities for fires to begin. Once started, a coalbed fire that has spread underground (sucking in oxygen from joints in overlying rock) may be very difficult to extinguish.

Major coalbed fires can be truly disastrous. The most notorious of these fires in North America began as a result of trash burning in a mine near the town of Centralia, Pennsylvania. For more than forty years the fire has progressed underground, eventually burning coal beneath the town itself. The fire produces toxic fumes that rise through the ground and make the overlying landscape uninhabitable, and it also causes the land surface to collapse. In recent years, many coalbed fires have started in China, where industrialization has led to increased mining. Over one hundred major coalbed fires are burning in northern China.

TAKE-HOME MESSAGE

Coal forms from the remains of plant material. When buried deeply, this organic material undergoes reactions that concentrate carbon; higher-rank coal contains more carbon. Coal occurs in sedimentary successions and must be mined underground or in open pits.

12.8 NUCLEAR POWER

How Does a Nuclear Power Plant Work?

So far, we have looked at fuels (oil, gas, and coal) that release energy when chemically burned. During such burning, a chemical reaction between the fuel and oxygen releases the potential energy stored in the chemical bonds of the fuel. Nuclear power, however, comes from a different process: the fission, or breaking, of the nuclear bonds that hold protons and neutrons together in the nucleus provides the energy. Fission splits an atom into smaller pieces.

Nuclear power plants were first built to produce electricity during the 1950s. A **nuclear reactor**, the heart of the plant, commonly lies within a dome-shaped shell (containment building) made of reinforced concrete (Fig. 12.9a, b). The reactor contains nuclear fuel, pellets of concentrated uranium oxide or a comparable radioactive material, packed into metal tubes called fuel rods. Fission occurs when a neutron strikes a radioactive atom, causing it to split. For example, radioactive uranium-235 splits into barium-141 and krypton-92, plus three neutrons and energy. During fission, a tiny fraction of the matter composing the original atom transforms into thermal and electromagnetic energy. The neutrons released during the fission of one atom strike other atoms, thereby triggering more fission in a self-perpetuating process called a chain reaction. Pipes carry water close to the heat-generating fuel. The heat transforms the water into high-pressure steam. The pipes then carry this steam to a turbine, where it rotates fan blades; the rotation drives a dynamo that generates electricity. Eventually the steam goes into cooling towers, where it condenses back into water that can be reused in the plant or returned to the environment.

The Geology of Uranium

^{235}U, an isotope of uranium containing 143 neutrons, is the most common fuel for conventional nuclear power plants. But ^{235}U accounts for only about 0.7% of naturally occurring uranium; most uranium consists of ^{238}U, an isotope with 145 neutrons. Thus, to make a fuel suitable for use in a power plant, the ^{235}U concentration in a mass of natural uranium must be increased by a factor of 2 or 3, an expensive process called enrichment.

Where does uranium come from? The Earth's radioactive elements, including uranium, probably developed during the explosion of a supernova before the existence of the Solar System. Uranium atoms from this explosion became part of the nebula out of which the Earth formed and thus were incorporated into the planet. They gradually rose into the upper crust, carried in granitic magma.

Even though granite contains uranium, it does not contain very much. But nature has a way of concentrating uranium. Hot water circulating through a pluton after intrusion dissolves the uranium and precipitates it as the mineral pitchblende (UO_2) in veins. Uranium may be further concentrated once plutons, and the associated uranium-rich veins, weather and erode at the ground surface. Sand derived from a weathered pluton washes down a stream, and as it does so, uranium-rich grains stay behind because they are so heavy, relative to quartz and feldspar grains. The world's richest uranium deposits, in fact, occur in ancient stream-bed gravels. Uranium deposits may also form when groundwater percolates through uranium-rich sedimentary rocks; the uranium dissolves in the water and moves with the water to another location, where it precipitates out of solution and fills the pores of the host sedimentary rock.

Challenges of Using Nuclear Energy

Maintaining safety at nuclear power plants requires work. Operators must constantly cool the nuclear fuel with circulating water, and the rate of nuclear fission must be regulated by the insertion of boron-steel control rods, which absorb neutrons and thus decrease the number of collisions between neutrons and radioactive atoms. Without control rods, the number of neutrons dashing around in the nuclear fuel would progressively increase, causing the rate of fission and accompanying heat production to increase. Eventually, the fuel would become so hot that it would melt. Such a **meltdown** might cause a steam explosion that could breach the containment building and scatter radioactive debris into the air. (Note that a meltdown is *not* the same as an atomic bomb explosion.)

One near-meltdown has occurred since the beginning of the nuclear age—at the power plant in Chernobyl, Ukraine, in April 1986. The fuel pile became too hot, triggering a steam explosion that raised the roof of the containment building and spread fragments of the reactor and its fuel around the plant grounds. Within six weeks, twenty people had died from radiation sickness. Radioactive material entered the atmosphere and dispersed over Eastern Europe and Scandinavia. No one yet knows the full effects of this fallout on the health of exposed populations. These days, there is concern that even a safely operated plant could pose a hazard if it became the target of a terrorist attack.

Nuclear waste is the radioactive material produced in a nuclear plant. It includes spent fuel, which contains radioactive elements, as well as water and equipment that have come in contact with radioactive materials. Radioactive elements emit gamma rays and X-rays that can damage living organisms and cause cancer. Some radioactive material decays quickly (in decades to centuries), but some remains dangerous for thousands of years or more.

Nuclear waste cannot just be stashed in a warehouse or buried in a town landfill. If the waste were simply buried, groundwater passing through the dump site might transport radioactive

FIGURE 12.9 Producing electricity at a nuclear power plant.

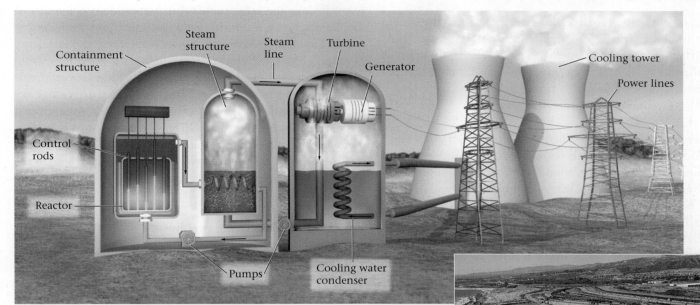

(a) In a nuclear power plant, a reactor heats water, which produces high-pressure steam. The steam drives a turbine that, in turn, drives a generator to produce electricity. A condenser transforms the steam back into water.

(b) This nuclear power plant, in California, has two reactors, each in its own containment building.

elements into municipal water supplies or nearby lakes or streams. Ideally, waste should be sealed in containers that will last for thousands of years (the time needed for the short-lived radioactive atoms to undergo decay) and stored in a place where it will not come in contact with the environment. Finding an appropriate place is not easy. So far, experts disagree about which is the best way to dispose of nuclear waste. The U.S. government favors storing waste at Yucca Mountain in the Nevada desert. The mountain consists of fairly dry rhyolite and the area is far from a population center.

TAKE-HOME MESSAGE

Controlled fission in reactors produces nuclear power. The fuel consists of uranium or other elements obtained by mining. Reactors run the risk of meltdown, though this has not yet happened. Reactors also produce radioactive waste that is difficult to store.

12.9 OTHER ENERGY SOURCES

Biofuels

In recent years, farmers have begun to produce rapidly growing crops, such as corn and sugar cane, specifically for the purpose of producing biomass for fuel production. The resulting liquids are called **biofuels**.

The most commonly used biofuel, ethanol (a type of alcohol), is currently produced by the fermentation of sugar from corn or sugar cane. Ethanol can substitute for gasoline in car engines. The process of producing ethanol includes the following steps. First, producers grind grain into a fine powder, mix it with water, and cook it to produce a mash of starch. Then, they add an enzyme to the mash, which converts the mash into sugar. The sugar, when mixed with yeast, ferments. Fermentation produces ethanol and CO_2. Finally, the fermented mash is distilled to concentrate the ethanol.

In recent years, researchers have begun to develop processes that yield ethanol from cellulose. This approach has the great advantage of permitting perennial grasses to become a source of liquid fuel, thus making ethanol a renewable energy source. Recent technologies have also led to the commercial production of biodiesel, a fuel produced by chemical modification of fats and vegetable oils. Biodiesel can be mixed with petroleum-derived diesel to run trucks and buses.

Geothermal Energy

As the name suggests, **geothermal energy** refers to the internal heat of the Earth. Geothermal energy exists because the crust becomes progressively hotter with increasing depth. Where the geothermal gradient is steep (meaning temperature increases rapidly with depth), we find high temperatures at relatively shallow depths. Groundwater absorbs heat from hot rock and itself becomes hot.

Power companies use geothermal energy to produce heat and electricity in two ways. In some places, they simply pump hot groundwater out of the ground and run it through pipes to heat houses and spas. Elsewhere, the groundwater is so hot that when it rises to the Earth's surface and decompresses, it turns to steam. This steam can drive turbines and generate electricity (Fig. 12.10). In volcanic areas such as Iceland or New Zealand, geothermal energy provides a major portion of energy needs. But on a global basis, it doesn't, because few cities lie near geothermal resources.

Hydroelectric and Wind Power

For millennia, people have used flowing water and air to produce energy. In fact, many towns were founded next to rivers, where streams could rotate the waterwheels that powered mills and factories. And in agricultural areas, farmers used windmills to pump water for irrigation. In the past century, engineers have begun to employ the same basic technology to drive generators that produce electricity. Energy derived from flowing fluids is clean ("green"), in the sense that its produc-

tion does not release chemical or radioactive pollutants, and is renewable, in the sense that its production does not consume limited resources. But its use does impact the environment.

In a modern hydroelectric power plant, the potential energy of water is converted into kinetic energy as the water flows from a higher elevation to a lower elevation. The flowing water turns turbine blades placed in a pipe, and the turbine drives an electrical generator. Most hydroelectric plants rely on water from a reservoir held back by a dam (Fig. 12.11a). Dam construction increases the available potential energy of the water, because the water level in a filled reservoir is higher than the level of the valley floor that the dam spans.

Hydroelectric energy is clean, and reservoirs may have the added benefit of providing flood control, irrigation water, and recreational opportunities. But the construction of dams and reservoirs may bring unwanted changes to a region, so the benefits of their construction must be weighed against potential harm. Damming a river may flood a spectacular canyon, eliminate exciting rapids, and may destroy an ecosystem. Further, reservoirs trap sediment and nutrients, thus disrupting the supply of these materials to downstream floodplains or deltas, a process that may adversely affect agriculture.

Not all hydroelectric power generation utilizes flowing river water—engineers have been developing new means to tap the energy in tides. One approach involves building floodgates across tidal inlets. Water flows into the inlet when the tide rises, and becomes trapped when the floodgate closes. Later, when the tide goes out and the water surface outside the floodgate drops to a lower level, engineers open valves, allowing water in the inlet to flow back to the sea via a pipe that carries it through a power-generating turbine. More recently, engineers have developed technologies to place huge fan blades underwater in near-shore regions where tidal currents naturally flow; the blades slowly turn in the current and run generators.

When you think of wind energy, you may picture a classic Dutch windmill driving a water pump. Modern efforts to harness the wind are on a much larger scale. To produce wind-

FIGURE 12.10 Geothermal power plants utilize hot groundwater. In some cases, the water is so hot that it becomes steam when it rises and undergoes decompression. Most geothermal plants are in volcanic areas.

FIGURE 12.11 Alternative energy sources.

(a) Gravity causes water held back by the Grand Coulee Dam to flow through turbines and generate electricity.

(b) A wind farm in southwestern England. The towers are about 50 m high.

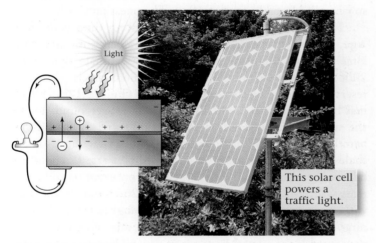

This solar cell powers a traffic light.

(c) A photovoltaic cell produces electricity directly from solar radiation.

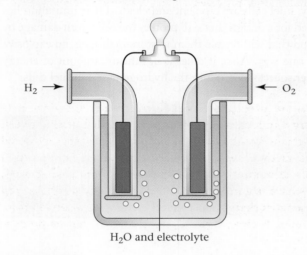

H_2O and electrolyte

(d) A hydrogen fuel cell.

generated electricity, engineers identify regions, either onshore or just offshore, with steady breezes. In these regions, they build wind farms that consist of numerous towers, each of which holds a wind turbine, a giant fan blade that turns even in a gentle breeze (**Fig. 12.11b**). Some towers are on the order of 100 m (300 ft) tall, with fan blades that are over 40 m (120 ft) long. The largest wind farm currently has close to two hundred towers. As with any type of energy source, wind power has drawbacks. Cluttering the horizon with towers may spoil a beautiful view, and the constant loud hum of the towers may be bothersome. The towers may also be a hazard to migrating birds, and offshore towers may interfere with marine life.

Solar Energy

The Sun drenches the Earth with energy in quantities that dwarf the amounts stored in fossil fuels. Were it possible to harness this energy directly, humanity would have a reliable and totally clean solution for powering modern technology. But using solar energy is not quite so simple. At present, energy consumers have two options for the direct use of solar energy.

A solar collector is a device that collects energy to produce heat. One class of solar collectors includes mirrors or lenses that focus light striking a broad area into a smaller area. On a small scale, such devices can be used for cooking. Another class of solar collectors consists of a black surface placed beneath a glass plate. The black surface absorbs light that has passed through the glass plate and heats up. The glass does not let the heat escape. When a consumer runs water between the glass and the black surface, the water heats up.

Photovoltaic cells (solar cells), the second option, convert light energy directly into electricity (**Fig. 12.11c**). Most photovoltaic cells consist of two wafers of silicon pressed together. One wafer contains atoms of arsenic, and the other wafer contains atoms of boron. When light strikes the cell, arsenic atoms release electrons that flow over to the boron atoms. If a wire loop connects the back side of one wafer to the back side of the other, this phenomenon produces an electrical current.

Fuel Cells

In a fuel cell, chemical reactions produce electricity directly. Let's consider a hydrogen fuel cell. Hydrogen gas flows through a tube across an anode (a strip of platinum) that has been placed in a water solution containing an electrolyte (a substance that enables the solution to conduct electricity). At the same time, a stream of oxygen gas flows onto a separate platinum cathode that also has been placed into the solution. A wire connects the anode and the cathode to provide an electrical circuit (Fig. 12.11d). In this configuration, hydrogen reacts with oxygen to produce water and electricity. Fuel cells are efficient and clean. Their limitation lies in the need for a design that will protect the cells from damage by impact and that will enable them to store hydrogen, an explosive gas, in a safe way. Also, it takes a significant amount of energy from other sources to produce the hydrogen used in fuel cells.

TAKE-HOME MESSAGE

Geothermal energy utilizes hot groundwater that has been warmed by heat from Earth's interior. Flowing water (in rivers or tides), flowing air, solar panels, and fuel cells can also produce energy. Biomass can be transformed into burnable alcohol and biodiesel.

12.10 ENERGY CHOICES, ENERGY PROBLEMS

The Oil Crunch

Energy usage in industrialized countries grew with dizzying speed through the mid-twentieth century, and during this time people came to rely increasingly on oil (Fig. 12.12a). Eventually, oil supplies within the borders of industrialized countries could no longer match the demand, and these countries began to import more oil than they produced themselves. Through the 1960s, oil prices remained low (about $1.80 a barrel), so this was not a problem. In 1973, however, a complex tangle of politics and war led the Organization of Petroleum-Exporting Countries (OPEC) to limit its oil exports. In the United States, fear of an oil shortage turned to panic, and motorists began lining up at gas stations, in many cases waiting for hours to fill their tanks. The price of oil rose to $18 a barrel, and newspaper headlines proclaimed an "Energy Crisis!" Governments in industrialized countries instituted new rules to encourage oil

conservation. During the last two decades of the twentieth century, the oil market stabilized, though political events occasionally led to price jumps and short-term shortages. Since 2004, oil prices rose overall, passing the $147/bbl mark in 2008, but the price collapsed in late 2008. Will a day come when shortages arise not because of an embargo or limitations on refining capacity but because there is no more oil to produce? To answer these questions, we need to examine the world oil supply.

To understand the issues involved in predicting the future of energy supplies, we must first classify energy resources. As noted earlier, we call a particular resource *renewable* if nature can replace it within a short time relative to a human life span (in months or, at most, decades). A resource is *nonrenewable* if nature takes a very long time (hundreds to perhaps millions of years) to replenish it. Oil, gas, and coal are nonrenewable resources, in that the rate at which humans consume these materials far exceeds the rate at which nature replenishes them, so we will inevitably run out of oil. The question is, when?

Historians in the future may refer to our time as the **Oil Age**, because so much of our economy depends on oil. How long will this Oil Age last? Geologists estimate that as of 2000, there were about 850 to 1,000 billion barrels of proven reserves, meaning oil that had been found (Fig. 12.12b). There may be an additional 2,000 billion barrels not yet found. Thus, the world probably holds between 1,000 and 3,000 billion barrels of obtainable oil (not counting oil in tar sands or oil shales). Presently, humanity guzzles oil at a rate of about 31 billion barrels per year. At this rate of consumption, oil supplies will last until sometime between 2050 and 2150, even with more offshore drilling. Already, geologists see the beginning of the end of the Oil Age, for the rate of consumption now exceeds the rate of discovery of new oil by a factor of 3. In the end, the Oil Age will probably have lasted less than three centuries (Fig. 12.12c). On a time line representing the four thousand years since the construction of the Egyptian pyramids, this looks like a very short blip. We may indeed be living during a unique interval of human history.

Can Other Energy Sources Meet the Need?

Are there alternatives to oil? Perhaps. Certainly the world contains vast fossil-fuel supplies in other forms, enough to last for centuries if they can be exploited in an economical and environmentally sound way. These supplies include tar sand, oil shale, natural gas, and coal. All told, perhaps 1.5 trillion barrels of hydrocarbons are trapped in tar sands and 3 trillion barrels trapped in oil shale. In addition, coal deposits could meet our energy needs for the next few centuries. But extracting tar sand, oil shale, and coal requires the construction of huge mines, which might scar the landscape. Natural gas is probably our best alternative in the near future.

FIGURE 12.12 World energy reserves. Oil is the dominant source of energy, at present, but other sources have more abundant reserves.

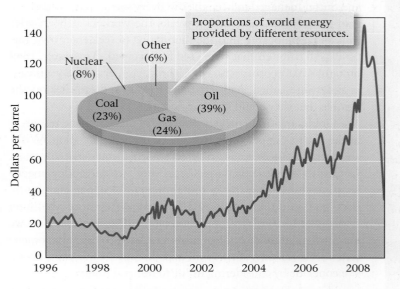

(a) Oil is currently the dominant source of energy worldwide. Recently, prices have been fluctuating dramatically.

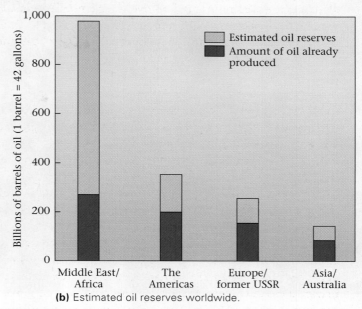

(b) Estimated oil reserves worldwide.

Can nuclear power or hydroelectric power fill the need? Vast supplies of uranium, the fuel of traditional nuclear plants, remain untapped, and in addition nuclear engineers have designed alternative plants, powered by breeder reactors, that essentially produce new fuel. But many people view nuclear plants with suspicion because of concerns about radiation, accidents, terrorism, and waste storage, and these concerns have slowed the industry. A substantial increase in hydroelectric power production is not likely, as most appropriate rivers have already been dammed, and industrialized countries have little appetite for taming any more. Similarly, the growth of geothermal-energy output seems limited.

Because of the problems that would result from relying more on coal, hydroelectric, and nuclear energy, researchers have been increasingly exploring clean energy options. One possibility is solar power. Unfortunately, affordable technologies for large-scale solar energy do not yet exist, and the idea of covering the landscape with solar cells is not attractive. Similarly, we can turn to wind power for relatively small-scale energy production, but covering the landscape with windmills is no more appealing. Fusion power may be possible, but physicists and engineers have not yet figured out a way to harness it. Clearly society will be facing difficult choices in the not-so-distant future about where to obtain energy, and we will need to invest in the research required to discover new alternatives.

Environmental Issues

Environmental concerns about energy resources begin right at the source. Oil drilling requires substantial equipment, the use

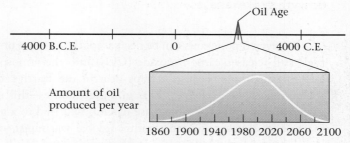

(c) The predicted history of the Oil Age, which may be a blip along the time line of human history.

of which can damage the land. Oil spills from pipelines or trucks sink into the subsurface and contaminate groundwater, and oil spills from ships create slicks that spread over the sea surface and foul the shoreline (Fig. 12.13). Coal and uranium mining also scar the land and can lead to the production of acid mine runoff, a dilute solution of sulfuric acid that forms when sulfur-bearing minerals such as pyrite (FeS_2) in mines react with rainwater. The runoff enters streams and kills fish and plants. Collapse of underground coal mines may cause the ground surface to sink.

Numerous air-pollution issues also arise from the burning of fossil fuels, which sends soot, carbon monoxide, sulfur dioxide, nitrous oxide, and unburned hydrocarbons into the air. Coal, for example, commonly contains sulfur, primarily in the form of pyrite, which enters the air as sulfur dioxide (SO_2) when coal is burned. This gas combines with rainwater to form dilute sulfuric acid (H_2SO_4), or acid rain. For this reason, many countries now regulate the amount of sulfur that coal can contain when it is burned.

FIGURE 12.13 A marine oil spill. Volatile components evaporate, leaving a tar-like residue.

But even if pollutants can be decreased, the burning of fossil fuels still releases carbon dioxide (CO_2) into the atmosphere. CO_2 is important because it traps heat in the Earth's atmosphere much like glass traps heat in a greenhouse—this is the **greenhouse effect**. An increase of CO_2 will lead to a global increase in atmospheric temperature (global warming), which in turn may alter the distribution of climatic belts. We'll learn more about this issue in Chapter 19.

TAKE-HOME MESSAGE

Oil is a nonrenewable resource that may run out in less than a century. Use of energy resources has environmental consequences.

12.11 INTRODUCING MINERAL RESOURCES

In June 1845, James Marshall arrived by horse at Sutter's Fort, in central California, to make a new life. Marshall convinced Captain John Sutter, a former army officer, to finance the construction of a sawmill in the foothills of the Sierra Nevada Mountains. Marshall's scruffy crew finished the mill by the beginning of 1848. As Marshall stood admiring the new building, he noticed a glimmer of metal in the gravel that littered the bed of the adjacent stream. He picked it up, banged it between two rocks to test its hardness, and shouted, "Boys, by God, I

believe I have found a gold mine!" For a short while, Marshall and Sutter managed to keep the discovery secret. But word of the gold soon spread, and within weeks the workers at Marshall's mill had disappeared into the mountains to seek their own fortunes. All through that year, gold fever spread. As a result, 1849 brought 40,000 prospectors to California. These "forty-niners," as they were called, had left their friends and families on the gamble that they could strike it rich.

Gold-containing rocks are but one of many mineral resources in the Earth's upper crust that are of use to civilization. Without such resources, industrialized societies could not function. Geologists divide mineral resources into two categories: metallic mineral resources (rocks containing gold, copper, aluminum, iron, and so on) and nonmetallic mineral resources (building stone, gravel, sand, gypsum, phosphate, salt, and so on). Now we look at the nature of mineral resources, the geologic phenomena responsible for their formation, and the way people mine them. We conclude by considering limits to mineral reserves.

12.12 METALS, ORES, AND ORE DEPOSITS

Metal and Its Discovery

We use metals for many purposes—nails, wires, car bodies, girders—because of their properties. **Metals** are opaque, shiny, smooth solids that can conduct electricity and can be bent, drawn into wire, or hammered into thin sheets. They look and behave quite differently from wood, plastic, meat, or rock. This is because, unlike other substances, the atoms that make up metals are held together by metallic bonds, which allow electrons to flow from atom to atom fairly easily. Despite the mobility of their electrons, metals are solids, so in metals, atoms lie fixed in a regular lattice defining a crystal structure.

The first metals that people used—copper, silver, and gold—can occur in rock as native metals (Fig. 12.14 inset). Native metals consist only of metal atoms, and thus look and behave like metal. Gold nuggets, for example, are chunks of native metal that have eroded free of bedrock. Over the ages, people have collected nuggets of native metal from stream beds and pounded them together with stone hammers to make arrowheads, scrapers, and later, coins and jewelry (Fig. 12.14). But if we had to rely solely on native metals as our source of metal, we would have access to only a tiny fraction of our current metal supply. Most of the metal atoms we use today originated as ions bonded to nonmetallic elements in a great variety of minerals that themselves look nothing like metal. Only because of the chance discovery by some prehistoric genius that certain rocks, when heated to high temperatures in fire (a process called smelting), decompose to yield metal plus a nonmetallic residue called slag, do we now have the ability to produce sufficient metal for the needs of industrialized society. The earliest smelters plied

Gold nuggets in white quartz veins.

Gold bracelets on display in a Kuwaiti marketplace.

FIGURE 12.14 From gold nuggets to gold jewelry. For each ounce of gold, miners process 30 tons of rock. Riches do come from the Earth—but at a cost.

their trade in 4000 B.C.E. and produced copper. Iron smelting began about 1300 B.C.E.; aluminum smelting did not become widespread until about 1880 C.E.

What Is an Ore?

The minerals from which metals can be extracted are called ore minerals, or economic minerals. These minerals contain metal in high concentrations and in a form that can be easily extracted. Galena (PbS), for example, is about 50% lead, so we consider it to be an ore mineral of lead (Fig. 12.15a). We obtain most of

our iron from the hematite and magnetite. And copper comes from a variety of minerals, none of which look like copper (Fig. 12.15b). Geologists have identified different kinds of ore minerals. Many ore minerals are sulfides, in which the metal occurs in combination with sulfur (S), or oxides, in which the metal occurs in combination with oxygen (O).

To obtain the metals needed for industrialized society, we mine **ore**, rocks containing native metals or a concentrated accumulation of ore minerals. To be an ore, a rock must not only contain ore minerals, it must also have a sufficient amount to make the rock worth mining. Iron constitutes only about 6.2% of the

FIGURE 12.15 Examples of ore minerals.

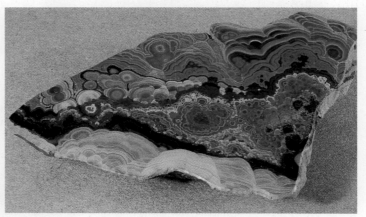

(a) This lead ore, from Missouri, consists of galena (PbS) crystals that grew in dolostone.

(b) Most copper comes from ore minerals that look nothing like metallic copper. This ore consists of azurite (blue) and malachite (green).

FIGURE 12.16   Various processes that form ore deposits

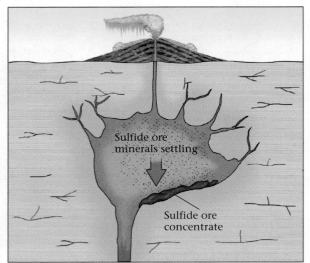

(a) Massive sulfide deposits can form when sulfide ore minerals sink to the bottom of a magma chamber.

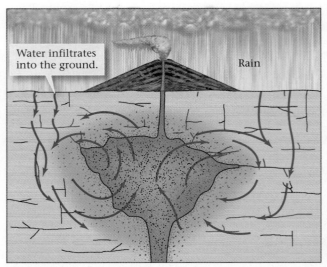

(b) Hydrothermal deposits form when water circulating around and through magma dissolves and redistributes metals (arrows indicate flowing water).

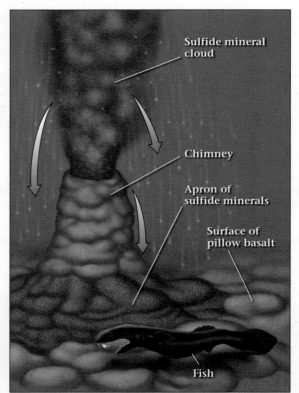

(c) Massive-sulfide deposits also form when ore minerals precipitate around hydrothermal vents (black smokers) along a mid-ocean ridge.

continental crust's weight but makes up 30 to 60% of iron ore. The concentration of a useful metal in an ore determines the grade of the ore—the higher the concentration, the higher the grade. Whether or not an ore of a given grade is worth mining depends on the price of metal in the market.

How Do Ore Deposits Form?

Simply put, an **ore deposit** is an economically significant occurrence of ore. The various kinds of ore deposits differ from each other in terms of how they formed and which ore minerals they contain. Below, we introduce a few examples.

Magmatic deposits.   When a magma cools, sulfide ore minerals crystallize and then, because of their high density, sink to the bottom of the magma chamber, where they accumulate. An accumulation formed in this way is a magmatic deposit. When the magma freezes solid, a concentration of sulfide ore minerals occurs at the base of the igneous body (Fig. 12.16a). Geologists call such a concentration a massive-sulfide deposit.

Hydrothermal deposits.   Hydrothermal activity involves the circulation of hot-water solutions through a magma or through the rocks surrounding an igneous intrusion. These fluids dissolve metal ions and carry them elsewhere. When the solution enters a region of lower pressure, lower temperature, different acidity, and/or different availability of oxygen, the metals come out of solution and form ore minerals that precipitate in fractures and pores, creating a hydrothermal deposit (Fig. 12.16b). Such deposits may form within an igneous intrusion or in surrounding country rock. If the resulting ore minerals disperse through the intrusion, we call the deposit a disseminated deposit, but if they precipitate to fill cracks in preexisting rock, we call the deposit a vein deposit; veins are simply mineral-filled cracks.

In recent years, geologists have discovered that hydrothermal activity at the submarine volcanoes along mid-ocean ridges leads to the eruption of hot water, containing high concentrations of dissolved metal and sulfur, from a vent. When this hot water comes in contact with cold seawater, the dissolved components instantly precipitate as tiny crystals of metal sulfide minerals (Fig. 12.16c), creating black clouds called black smokers. The minerals in the cloud eventually sink and form a pile of massive sulfide ore around the vent.

Secondary-enrichment deposits. Sometimes groundwater passes through ore-bearing rock long after the rock first formed. This groundwater dissolves some of the ore minerals and carries the dissolved ions away. When the water eventually flows into a different chemical environment (such as one with a different amount of oxygen or acid), it precipitates new ore minerals, commonly in concentrations that exceed that of the original deposit. A new ore deposit formed from metals that were dissolved and carried away from a preexisting ore deposit is called a secondary-enrichment deposit.

MVT ores. In recent years, geologists have recognized that some groundwater beneath mountain belts sinks several kilometers down into the crust and follows a curving flow path that eventually brings it back to the surface, perhaps hundreds of kilometers away from the mountain range. At the depths reached by the groundwater, temperatures become high enough for the water to dissolve metals. When the water returns to the surface and enters cooler rock, these metals precipitate in ore minerals. Ore deposits formed in this way typically contain lead- and zinc-bearing minerals. They appear in dolomite beds of the Mississippi Valley region, and thus have come to be known as Mississippi Valley–type, or MVT, ore.

Sedimentary deposits of metals. Some ore minerals accumulate in sedimentary environments under special circumstances. For example, between 2.0 and 2.5 billion years ago, the atmosphere, which previously had contained little oxygen, evolved into the oxygen-rich atmosphere we breathe today. This change affected the chemistry of seawater and led to the precipitation of iron-oxide minerals that settled as sediment on the sea floor. The resulting iron-rich sedimentary layers are known as banded-iron formation (BIF) (Fig. 12.17), because after lithification they consist of alternating beds of gray iron oxide (magnetite or hematite) and red beds of jasper (iron-rich chert).

The chemistry of seawater in some parts of the ocean today leads to the deposition of manganese-oxide minerals on the sea floor. These minerals grow into lumpy accumulations called manganese nodules. Mining companies have begun to explore technologies for vacuuming up these nodules.

Residual mineral deposits. Recall from Interlude B that as rainwater sinks into the Earth, it leaches (dissolves) certain elements and leaves behind others, as part of the process of forming soil. In rainy tropical environments, the residuum left

FIGURE 12.17 A Precambrian banded-iron formation in northern Michigan.

Hematite layer

Jasper layer

behind in soils after leaching includes concentrations of iron or aluminum. Locally, these metals become so concentrated that the soil itself becomes an ore deposit. We refer to such deposits as residual mineral deposits. Most of the aluminum ore mined today comes from bauxite, a residual mineral deposit created by the extreme leaching of granite.

Placer deposits. Ore deposits may develop when rocks containing native metals, such as gold, erode and create a mixture of sand grains and metal flakes or nuggets (pebble-sized fragments). The heavy metal grains accumulate in sand or gravel

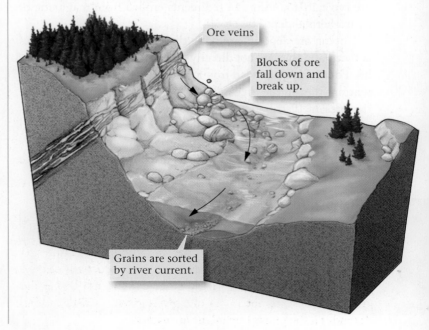

FIGURE 12.18 Placer deposits form where erosion produces clasts of native metals. Sorting by the stream concentrates the metals.

Ore veins

Blocks of ore fall down and break up.

Grains are sorted by river current.

bars along the course of rivers, for the moving water carries away lighter mineral grains but can't move the metal grains as easily. Concentrations of metal grains in stream sediments are a type of placer deposit (Fig. 12.18). Panning further concentrates gold flakes or nuggets—swirling water in a pan causes the lighter sand grains to wash away, leaving the gold behind. Placer deposits may eventually be buried and lithify to become part of a new sedimentary rock.

Where Are Ore Deposits Found?

The Inca empire of fifteenth-century Peru boasted elaborate cities and temples, decorated with fantastic masks, jewelry, and sculpture made of gold. Then, around 1532, Spanish conquistadors arrived, led by commanders who quipped, "We Spaniards suffer from a disease that only gold can cure." The Incas, already weakened by civil war, were no match for the armor-clad Spaniards and their guns. Within six years the Inca empire had vanished and Spanish ships were transporting Inca treasure back to Spain. Why did the Incas possess so much gold? Or to ask the broader question, what geologic factors control the distribution of ore? We can find the answer once again by considering the consequences of plate tectonics.

Several of the ore-deposit types mentioned in this chapter occur in association with igneous rocks. As we learned in Chapter 4, igneous activity does not happen randomly around the Earth, but rather concentrates at convergent plate boundaries (in the overriding plate of a subduction zone), at divergent plate boundaries (along mid-ocean ridges), at continental rifts, or at hot spots. Thus, magmatic and hydrothermal deposits, and secondary enrichment deposits derived from these, occur along plate boundaries, along rifts, or at hot spots. Placer deposits are typically found in the sediments eroded from magmatic or hydrothermal deposits.

Some ore deposits are not a direct result of plate tectonics activity, and thus are not directly associated with plate boundaries. For example, banded iron formed along passive continental margins during the Precambrian—its occurrence today reflects the present distribution of Precambrian rocks. Bauxite forms wherever aluminum-rich bedrock occurs, thick soils form, and extreme leaching takes place. Thus, many bauxite deposits form on granite bedrock in stable continental areas that now lie in tropical regions.

TAKE-HOME MESSAGE

Ores can be processed economically to produce metals. Ores form by: settling from melt, precipitating from hot water, deposition from currents, interaction with groundwater, and extreme weathering. The distribution of ores can be explained by plate tectonics.

12.13 ORE-MINERAL EXPLORATION AND PRODUCTION

Imagine an old prospector clanking through the desert with a broken-down donkey, eyeing the hillsides for "shows" of ore, exposures of ore minerals at the ground surface. If he finds a show, he pries out chunks of the rock with a pick, and the poor donkey hauls the rock back to town for an assay, a test to determine how much extractable metal the rock contains. Mining laws permitted a prospector to, literally, "stake a claim" by marking off an area of ground with stakes. The prospector would then have the exclusive right to dig up ore at that spot and sell it. What does a show look like? Typically, prospectors looked for milky-white quartz veins and/or exposures in which rocks were stained by the oxidation of metal-containing minerals.

These days, large mining companies employ geologists to survey potential ore-bearing regions systematically. The geologists focus their studies on rocks that developed in settings appropriate for ore formation. Once such a region has been identified, they measure the local strength of Earth's gravity and magnetic field. These measurements help narrow the search, because ore minerals tend to be denser and more magnetic than average rocks. Geologists also sample rocks and soils to test for metal content, and may even analyze plants in the area to detect traces of metals, for plants absorb metals through their roots. Once geologists have identified a possible ore deposit, they drill holes to sample subsurface rock and to determine the ore deposit's shape and extent. Ore-mineral exploration takes geologists into jungles, deserts, and tundras worldwide.

If calculations show that the mining of an ore deposit will yield a profit, and if environmental concerns can be accommodated, a company digs a mine. Mines can be below or above ground (see Geotour 12 on p. GT-26), depending on how close the ore deposit is to the surface. To make an open-pit mine, workers first drill a series of holes into the solid bedrock and then fill the holes with high explosives. They must space the holes carefully and must set off the charges in a precise sequence, so that the bedrock shatters into appropriate-sized blocks for handling. When the dust settles, large front-end loaders dump the ore into giant trucks (Fig. 12.19a). The trucks transport waste rock (rock that doesn't contain ore) to a tailings pile and the ore to a crusher, a giant set of moving steel jaws that smash rock into small fragments. Workers then separate ore minerals from other minerals and send the ore-mineral concentrate to a processing plant, where the ore undergoes smelting or treatment with acidic solutions to separate metal atoms from other atoms.

If the ore deposit lies more than about a hundred meters below the Earth's surface, miners must make an underground mine. To do so, miners sink a vertical shaft in which they install an elevator. At the level in the crust where the ore body appears, they build a maze of tunnels into the ore by drilling holes into

FIGURE 12.19 The mining of ore deposits.

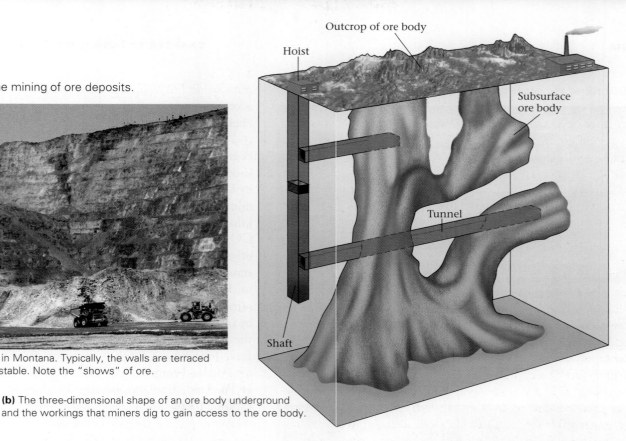

(a) An open-pit mine in Montana. Typically, the walls are terraced to make them more stable. Note the "shows" of ore.

(b) The three-dimensional shape of an ore body underground and the workings that miners dig to gain access to the ore body.

the rock and then blasting (Fig. 12.19b). The rock removed must be carried back to the surface. Rock columns between the tunnels hold up the ceiling of the mine.

TAKE-HOME MESSAGE

Geologists study outcrops for ore shows and drill holes, and make maps of rock magnetism in order to discover ore deposits. Mining takes place either in open-pit mines or in shafts and tunnels underground.

12.14 NONMETALLIC MINERAL RESOURCES

So far this chapter has focused on resources that contain metal. But society uses many other geological materials, sometimes called industrial minerals, as well. From the ground, we get the stone used to make roadbeds and buildings, the chemicals for fertilizers, the gypsum in drywall, the salt filling salt shakers, and the sand used to make glass—the list is endless. This section looks at a few of these geologic materials and explains where they come from.

Dimension Stone

The Parthenon, a colossal stone temple supported by forty-six carved columns, has stood atop a hill overlooking the city of Athens for almost 2,500 years. No wonder—stone, rock used

for building and other practical purposes, outlasts nearly all other materials. We use stone to make facades, roofs, curbs, steps, countertops, and floors. The names that architects give to various types of stone may differ from the formal names used by geologists. For example, architects refer to any polished carbonate rock as marble, whether or not it has been metamorphosed. Likewise, they refer to any rock containing feldspar and quartz as granite, regardless of whether the rock has an igneous or a metamorphic texture.

To obtain intact slabs and blocks of rock—known as dimension stone in the trade—for architectural purposes, workers carefully cut it out of the walls of quarries (Fig. 12.20a). (Note that a quarry provides stone, whereas a mine supplies ore.) To cut stone slabs, quarry operators split blocks from bedrock by hammering a series of wedges into the rock, or cut it off bedrock by using a wireline saw, a thermal lance, or a water jet. A wireline saw consists of a loop of braided wire moving between two pulleys. The movement of the wire drags the abrasive along the rock and grinds a slice into it. A thermal lance looks like a long blowtorch—a flame of burning diesel fuel stoked by high-pressure air pulverizes rock, and thereby cuts a slot. More recently, quarry operators have begun to use an abrasive water jet, which squirts out water and abrasives at very high pressure, to cut rock.

Crushed Stone and Concrete

Crushed stone forms the substrate of highways and railroads and is the raw material for manufacturing cement, concrete, and asphalt. In crushed-stone quarries (Fig. 12.20b), operators use

FIGURE 12.20   Stone production in quarries.

(a) An active quarrying operation that produces large blocks of cut dimension stone.

(b) A crushed limestone quarry in Illinois. The stone is sorted by fragment size.

high explosives to break up bedrock into rubble that they then transport by truck to a jaw crusher, which reduces the rubble into usable-sized fractions.

Much modern construction utilizes mortar and concrete, human-made rock-like materials formed when a slurry composed of sand and/or gravel mixed with cement and water is allowed to harden. The hardening takes place when a complex assemblage of minerals grows by chemical reactions in the slurry; these minerals bind together the grains of sand or gravel in mortar or concrete. (Note that the word *mortar* refers to the substance that holds bricks or stone blocks together, whereas *concrete* refers to the substance that workers shape into roads or walls by spreading it out into a layer or by pouring it into a form.) The **cement** in mortar or concrete starts out as a powder composed of lime (CaO), quartz (SiO$_2$), alumi-

num oxide (Al$_2$O$_3$), and iron oxide (Fe$_2$O$_3$). Typically, lime accounts for 66% of cement, silica for 25%, and the remaining chemicals for about 9%.

It appears that the ancient Romans were the first to use cement—they made it from a mixture of volcanic ash and limestone. During recent centuries, cement has been produced by heating specific types of limestone (which happened to contain calcite, clay, and quartz in the correct proportions) in a kiln up to a temperature of about 1,450°C; the heating releases CO$_2$ gas and produces "clinker," chunks consisting of lime and other oxide compounds. Manufacturers crush the clinker into cement powder and pack it in bags for transport. But natural limestone with the exact composition of cement is fairly rare, so most cement used today is "Portland cement," made by intentionally mixing limestone, sandstone, and shale in just the right proportions to provide the proper chemical makeup. Isaac Johnson, an English engineer, came up with the recipe for Portland cement in 1844, and he named it after the town of Portland, England, because he thought it resembled rock exposed there.

Nonmetallic Minerals in Your Home

We use an astounding variety of nonmetallic geologic resources without ever realizing where they come from. Consider the materials in a house or apartment. The concrete foundation consists of cement, made from limestone mixed with sand or gravel. The bricks in the exterior walls originated as clay, formed from the chemical weathering of silicate rocks and perhaps dug from the floodplain of a stream. To make bricks, workers mold wet clay into blocks, which they then bake. Baking drives out water and causes metamorphic reactions that recrystallize the clay.

The glass used to glaze windows consists largely of silica, formed by first melting and then freezing pure quartz sand from a beach deposit or a sandstone formation. Gypsum board (drywall), used to construct interior walls, is a sandwich of gypsum powder between sheets of paper. Gypsum (CaSO$_4$ · 2H$_2$O) occurs in evaporite strata precipitated from seawater or saline lake water. Evaporites provide other useful minerals as well, such as halite and borax. Truly, without the geologic resources of the Earth, modern society would grind to a halt.

TAKE-HOME MESSAGE

Society uses a great variety of nonmetallic geologic materials. These include dimension stone, crushed stone, concrete (made from roasted limestone), evaporites (including gypsum), and clay (to make bricks).

12.15 GLOBAL MINERAL NEEDS

How Long Will Resources Last?

The average citizen of an industrialized country uses 25 kilograms (kg) of aluminum, 10 kg of copper, and 550 kg of iron and steel in a year's time (Table 12.1). If you combine these figures with the quantities of energy resources and nonmetallic geologic resources a person uses, you get a total of about 15,000 kg (15 metric tons) of resources used per capita each year. Thus, the population of the United States consumes about 4 billion metric tons of geologic material per year. To supply this demand, workers must mine, quarry, or pump 18 billion metric tons of rock per year. For each ounce of gold, miners process 30 tons of rock. By comparison, the Mississippi River transports 190 million metric tons of sediment per year into the Gulf of Mexico.

Mineral resources, like oil and coal, are nonrenewable resources. Once mined, an ore deposit or a limestone hill disappears forever. Natural geologic processes do not happen fast enough to replace the deposits as quickly as we use them. Geologists have calculated reserves for various mineral deposits just as they have for oil. Judging from current definitions of reserves (which depend on today's prices) and rates of consumption, supplies of some metals may run out in only decades to centuries (Table 12.2). But these estimates may change as supplies become depleted and prices rise, making previously uneconomical deposits worth mining. And supplies could increase if geologists discover new reserves or if new mining techniques are developed.

Further, increased efforts at conservation and recycling can cause a dramatic decrease in rates of consumption, and thereby stretch the lifetime of existing reserves.

Ore deposits do not occur everywhere, because their formation requires special geologic conditions. As a result, some countries possess vast supplies, whereas others have none. In fact, no single country owns all the mineral resources it needs, so nations must trade with one another to maintain supplies, and global politics inevitably affects prices. Many wars have their roots in competition for mineral reserves, and it is no surprise that the outcomes of some wars have hinged on who controls these reserves.

TABLE 12.1 Yearly Per Capita Usage of Geologic Materials in the United States

4,100 kg	Stone
3,860 kg	Sand and gravel
3,050 kg	Petroleum
2,650 kg	Coal
1,900 kg	Natural gas
550 kg	Iron and steel
360 kg	Cement
220 kg	Clay
200 kg	Salt
140 kg	Phosphate
25 kg	Aluminum
10 kg	Copper
6 kg	Lead
5 kg	Zinc

1 kg = 2.205 pounds

TABLE 12.2 Expected Lifetimes of Currently Known Ore Resources (in years)

Metal	World Resources	U.S. Resources
Iron	120	40
Aluminum	330	2
Copper	65	40
Lead	20	40
Zinc	30	25
Gold	30	20
Platinum	45	1
Nickel	75	less than 1
Cobalt	50	less than 1
Manganese	70	0
Chromium	75	0

Mining and the Environment

Mining leaves a big footprint in the Earth System. Some of the gaping holes that open-pit mining creates in the landscape are so big that astronauts can see them from space. Both open-pit and underground mining yield immense quantities of waste rock, which miners dump in tailings piles, some of which grow into artificial hills 200 meters high and many kilometers long. Lacking soil, tailings piles tend to remain unvegetated for a long time. Mining also exposes ore-bearing rock to the atmosphere, and since many ore minerals are sulfides, they react with rainwater to produce acid mine runoff, which can severely damage vegetation downstream (Fig. 12.21a). Ore processing tends to release noxious chemicals that can mix with rain and spread over the countryside, damaging ecosystems. Before the installation of modern environmental controls, smoke from ore-processing plants caused severe air pollution and acid rain. Plumes of smoke from the old smelters in Sudbury, Ontario, for example, created a wasteland for many kilometers downwind (Fig. 12.21b). In recent years, people have made efforts to reclaim mining spoils, and new technologies have been developed to extract metals in ways that are less deleterious to the environment.

TAKE-HOME MESSAGE

Mineral resources are nonrenewable, so minerals have limited reserves. Also, reserves are not distributed uniformly around the planet, so supplies are not necessarily accessible to consumers. Mineral utilization has significant environmental consequences.

(a) The orange color in this mine runoff comes from dissolved iron in the water.

(b) Acidic smelter smoke killed off vegetation near Sudbury, Ontario, in the 1970s. A large tailings pile can be seen in the distance.

FIGURE 12.21 Environmental consequences of producing metallic mineral resources.

Chapter Summary

- Energy comes from a variety of sources: directly from the Sun; from tides, flowing water, or wind; from chemical reactions; from nuclear fission; and from Earth's internal heat. Plants use photosynthesis to store energy. Buried organic matter becomes fossil fuel.

- Oil and gas are hydrocarbons. The viscosity and volatility of a hydrocarbon depend on the length of its molecules.

- Oil and gas form from the bodies of plankton and algae, which settle out in a quiet-water, oxygen-poor depositional environment and form black organic shale. Later, chemical reactions at elevated temperatures convert the organic matter into kerogen, then oil.

- In order to create a usable oil reserve, oil must migrate from a source rock into a reservoir rock. Unless the reservoir rock is overlain by an impermeable seal rock, the oil will escape to the ground surface. The subsurface configuration of strata that leads to the entrapment of oil is called an oil trap.

- Substantial volumes of hydrocarbons exist in tar sand, oil shale, and gas hydrate.

- For coal to form, plant material must be deposited in an oxygen-poor environment so that it does not decompose. Compaction creates peat, which, when buried deeply and heated, transforms into coal.

- Coal is classified into ranks, based on the amount of carbon it contains. Coal occurs in beds, interlayered with other sedimentary rocks, and can be mined by either strip mining or underground mining.

- Coalbed methane and coal gasification provide additional sources of energy.

- Nuclear power plants generate energy by using the heat released by the fission of uranium. The heat turns water into steam, and the steam drives turbines. Some economic uranium deposits occur as veins in igneous rock bodies, while others are found in sedimentary beds.

- Nuclear reactors must be carefully controlled to avoid overheating or meltdown. The disposal of radioactive nuclear waste can create environmental problems.

- Geothermal energy uses Earth's internal heat to transform groundwater into steam that drives turbines; hydroelectric power plants use the potential energy of water; and solar cells convert sunlight to electricity.

- We now live in the Oil Age, but oil supplies may last only for another century.

- Use of energy resources has many negative environmental consequences.

- Industrial societies use many types of minerals, all of which must be extracted from the upper crust. We distinguish two general categories: metallic resources and nonmetallic resources.

- Metals come from ore. An ore is a rock containing native metals or ore minerals (minerals with a high proportion of metal) in sufficient quantities to be worth mining. An ore deposit is an accumulation of ore.

- Magmatic deposits form when sulfide ore minerals settle to the floor of a magma chamber. In hydrothermal deposits, ore minerals precipitate from hot-water solutions. Secondary-enrichment deposits form when groundwater carries metals away from a preexisting deposit. MVT deposits precipitate from groundwater that has passed long distances through the crust. Sedimentary deposits precipitate out of the ocean. Residual mineral deposits in soil are the result of severe leaching in tropical climates. Placer deposits develop when heavy metal grains accumulate in sediment along a stream.

- Many ore deposits are associated with igneous activity in subduction zones, along mid-ocean ridges, along continental rifts, or at hot spots.

- Nonmetallic resources include dimension stone for decorative purposes, crushed stone for cement and asphalt production, clay for brick making, sand for glass production, and many others. A large proportion of materials in your home have a geologic ancestry.

- Mineral resources are nonrenewable. Many are now or may soon be in short supply.

GEOPUZZLE REVISITED

The natural resources that sustain modern civilization come from the Earth and reflect the consequences of geologic processes. For example, hydrocarbons (oil and gas) come from the buried remains of algae and plankton, coal comes from the remnants of ancient plants, and metals come from special types of rocks called ores. Since natural resources can require long intervals of geologic time to form, many are being consumed faster than they can be replenished, and thus may eventually run out.

Key Terms

biofuel (p. 335)
biomass (p. 322)
cement (p. 346)
coal (p. 330)
coalbed methane (p. 332)
coal gasification (p. 333)
coal rank (p. 331)

energy (p. 322)
energy resource (p. 322)
fossil fuel (p. 322)
gas hydrate (p. 330)
geothermal energy (p. 336)
greenhouse effect (p. 340)
hydrocarbon reserve (p. 324)

hydrocarbon (p. 323)
kerogen (p. 324)
meltdown (of nuclear reactor) (p. 334)
metal (p. 340)
mineral resources (p. 322)
nuclear reactor (p. 334)
nuclear waste (p. 334)
Oil Age (p. 338)
oil shale (p. 324)
oil window (p. 324)
ore (p. 341)

ore deposit (p. 341)
peat (p. 330)
permeability (p. 325)
photosynthesis (p. 322)
porosity (p. 325)
reservoir rock (p. 325)
resource (p. 322)
salt dome (p. 327)
seal rock (p. 326)
source rock (p. 324)
tar sand (p. 329)
trap (p. 325)

Review Questions

1. What are the fundamental sources of energy?
2. What is the source of the organic material in oil?
3. What is the oil window, and why does oil form only there?
4. How is organic matter trapped and transformed to create an oil reserve?
5. What are tar sand and oil shale, and how can oil be extracted from them?
6. What are gas hydrates, and where do they occur?
7. Where is most of the world's oil found? At present rates of consumption, how long will it last?
8. How is coal formed?
9. Explain how coal is transformed in rank from peat to anthracite.
10. Describe the operation of a nuclear reactor.
11. Where does uranium form in the Earth's crust? Where does it usually accumulate in minable quantities?
12. What are some of the drawbacks of nuclear energy?
13. What is geothermal energy? Why is it not more widely used?
14. What is the difference between a renewable and a nonrenewable resource?
15. What is the likely future of hydrocarbon production and use in the twenty-first century?
16. Why don't we use an average granite as a source for useful metals?
17. Describe various kinds of economic mineral deposits.
18. What procedures are used to locate and mine mineral resources today?
19. How is stone cut from a quarry?
20. What are the ingredients of cement? How is Portland cement made?

21. Will the supply of mineral resources run out? Can a country survive without importing minerals?

22. What are some environmental hazards of large-scale mining?

On Further Thought

1. Much of the oil production in the United States takes place at offshore platforms along the coast of the Gulf of Mexico. Many of the traps within this province are associated with salt domes. Consider the geologic setting of the Gulf Coast, in the context of the theory of plate tectonics, and explain why an immensely thick sequence of sediment accumulated in this region, and why so many salt-dome traps formed.

2. Imagine that an ore deposit of a certain metal contains 0.6% grade ore. This means that 0.6 percent by weight of a block of ore consists of the metal. The pure metal, on the open market, sells for $8,000/ton. It costs $15/ton to mine the ore, $15/ton to transport the ore to the processing plant, and $15/ton to process the ore and produce pure metal. Start-up costs (building the mine and building the processing factory) are about $100 million. How much profit does the company make when it sells a ton of metal? How much ore does the operation have to mine to pay back the start-up costs? Considering that a giant dump truck in a mine can carry 200 tons of ore at a time, how many dump-truck loads will have been transported at the break-even point? If the mine has 8 trucks that can each make 6 loads a day, about how many years will it take to break even?

THE VIEW FROM SPACE   The Bingham open-pit mine near Salt Lake City, Utah is 1.2 km deep and 4 km wide, making it the largest excavation in the world. In this photo, the pit itself is largely hidden by vast tailing piles, the debris left behind once the ore has been extracted. During its century of operation, the mine has yielded about 17 million tons of copper, 700 tons of gold, 6,000 tons of silver, and 400 tons of molybdenum. In 2006 alone, the value of metal produced was $1.8 billion. To obtain this metal, miners have to move up to 450,000 tons of rock a year, for typical ore contains less than 0.3% copper. The ore formed due to hydrothermal fluid circulation that accompanied granitic magmatism about 36 million years ago.

An Introduction to Landscapes and the Hydrologic Cycle

A **digital elevation map (DEM)** of Oahu, Hawaii. The map was produced using radar data from satellites. This landscape formed when a volcano emerged from the sea and was eroded. Erosion produced sediment, which accumulated to form the flat land of Honolulu.

FIGURE F.1 Examples of the great variety of landscapes on Earth.

(a) A rock and sand seascape along the coast of Brazil.

(b) The peaks of the Grand Tetons in Wyoming.

(c) Buttes of sandstone in Monument Valley, Arizona.

(d) Steep cliffs in Australia's Blue Mountains.

F.1   INTRODUCTION

It's no wonder that artists and writers across the ages have sought inspiration from the **landscape**—the character and shape of the land surface in a region—for landscapes display a diversity that rivals that of human emotion (Fig. F.1a–d). Geologists, like artists and writers, savor the impression of a dramatic landscape, but on seeing one, they can't help but ask, "How did it come to be, and how will it change in the future?"

The subject of landscape development and evolution, and the **landforms** (individual shapes such as mesas, valleys, cliffs, and dunes) that constitute it, dominates many of the subsequent chapters of this book, each of which focuses on an aspect of the surface and near-surface realm. This interlude, a general introduction to the topic, explains the driving forces behind landscape development, and identifies factors that control which landscape develops in a given locality. We also use this opportunity

to describe the hydrologic cycle, the pathway water molecules follow as they move from ocean to air to land and back to ocean. We discuss the hydrologic cycle here because so many processes on or near Earth's surface involve water in its various forms.

F.2   SHAPING THE EARTH'S SURFACE

If the Earth's surface were totally flat, the great diversity of landscapes that embellish our vistas would not exist. But the surface isn't flat, because a variety of geologic processes cause portions of the surface to move up or down relative to adjacent regions. We refer to the upward movement of the land surface as **uplift**, and the sinking or downward movement of the land surface as **subsidence**.

When uplift or subsidence takes place, the elevation difference does not remain the same forever, because other components of the Earth System kick into action—material at higher elevations becomes unstable and susceptible to **downslope movement** (the tumbling or sliding of rock and sediment from higher elevations to lower ones); moving water, ice, and air cause **erosion** (the grinding away and removal of the Earth's surface); and where moving fluids slow down, **deposition** of sediment takes place. Downslope movement, erosion, and deposition redistribute rock and sediment, ultimately stripping it from higher areas and collecting it in low areas. As a consequence, a great variety of both **erosional landforms** (those carved by erosion) and **depositional landforms** (those built from an accumulation of sediment) can form.

The energy that drives landscape evolution comes from three sources: **internal energy**, the heat within the Earth, drives plate motions and mantle plumes that, in turn, cause displacement of the crust's surface; **external energy**, the energy coming to the Earth from the Sun, causes materials at the surface of the Earth to warm up; and **gravitational energy** pulls rock and water from higher to lower elevations. External and gravitational energy, working together, drive convective flow in the atmosphere and oceans and drive the hydrologic cycle. Landscape evolution, in fact, reflects a "battle" between (1) tectonic processes such as collision, convergence, and rifting, which create **relief**, an elevation difference between two locations, in an area, which we can represent on a topographic map (Box F.1); and (2) processes such as downslope movement, erosion, and deposition, which destroy relief by removing material from high areas and depositing it in low ones. If, in a particular region, the rate of uplift exceeds the rate of erosion, the land surface rises; if the rate of subsidence exceeds the rate of deposition, the land surface sinks. Without uplift and subsidence, Earth's surface would long ago have been beveled to a flat plain. Without erosion and deposition, high and low areas would have lasted for the entirety of Earth history.

How rapidly do uplift, subsidence, erosion, and deposition take place? The Earth's surface can rise or sink by as much as 3 m during a single major earthquake. But averaged over time, the rates of uplift and subsidence range between 0.01 and 10 mm per year (Fig. F.2a). Similarly, erosion can carve out several meters of **substrate**, the material just below the ground surface, during a single flood, storm, or landslide (Fig. F.2b). And deposition during a single event can create a layer of debris tens of meters thick in a matter of minutes to days. But averaged over time, erosional and depositional rates also vary between 0.10 and 10 mm per year. Although these rates seem small, a change in surface elevation of just 0.5 mm (the thickness of a fingernail) per year can yield a net change of 5 km in 10 million years. Uplift can build a mountain range, and erosion can whittle one down to near sea level—it just takes time!

F.3 FACTORS CONTROLLING LANDSCAPE DEVELOPMENT

Imagine traveling across a continent. On your journey, you pass plains, swamps, hills, valleys, mesas, and mountains. Some of these features are erosional landforms, in that they result from the breakdown and removal of rock or sediment and develop where **agents of erosion** such as water, wind, or ice carve into the substrate. Of these three agents, water has the greatest effect on a global basis. Other features are depositional landforms, in that they result from the deposition of sediment where the fluid carrying the sediment evaporates, slows down, or melts. The specific landforms that develop at

FIGURE F.2 The processes of uplift, subsidence, erosion, and deposition can be slow or rapid.

(a) Uplifted beach terraces form where the coast is rising relative to sea level. Present-day wave erosion is forming a new terrace and cutting a cliff on the edge of the old one.

(b) So much erosion can take place during a single hurricane that houses built along the beach become undermined.

Topographic Maps and Profiles

We can distinguish one landform from another by its shape—for example, as you will see in succeeding chapters, a river-carved valley simply does not look like a glacially carved valley. Landform shapes are manifested by variations in elevation within a region. Geologists use the term **topography** to refer to such variations.

How can we convey information about topography—a three-dimensional feature—on a two-dimensional sheet of paper? Geologists do this by means of a **topographic map**, which uses contour lines to represent variations in elevation (**Fig. 1a, b**). A **contour line** is an imaginary line along which all points have the same elevation. For example, if you walk along the 200-m contour line on a hill slope, you stay at exactly the same elevation. As another example, the shoreline on a flat, calm body of water is a contour line. In other words, you can picture the contour line as the intersection between the land surface and an imaginary horizontal plane.

The elevation difference between two adjacent contour lines on a topographical map is called the **contour interval**. For a given topographic map, the contour interval is constant, so the spacing between contour lines represents the steepness of a slope. Specifically, closely spaced contour lines represent a steep slope, whereas widely spaced contour lines represent a gentle slope.

We can also represent variations in elevation by means of a topographic profile (**Fig. 1c**). A profile is the trace of the ground surface as it would appear on a vertical plane that sliced into the ground—put another way, it's the shape of the ground surface as viewed from the side.

If we add a representation of geologic features under the ground surface, then we have a geologic cross section. In some cases, geologists gain insight into subsurface geology simply by looking at the shape of a landform. For example, a steep cliff in a region of dipping sedimentary strata may indicate the presence of a resistant (difficult to erode) layer; low areas may be underlain with nonresistant (easy to erode) layers (**Fig. 2**).

FIGURE 1 Topographic maps and profiles.

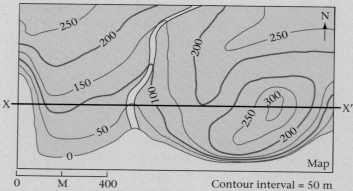

Contour interval = 50 m

(a) A topographic map depicts the shape of the land surface through the use of contour lines. The difference in elevation between two adjacent lines is the contour interval.

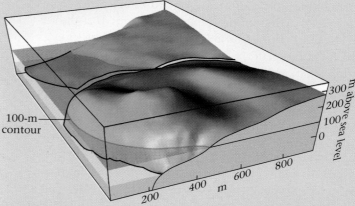

(b) A contour line is the intersection of a horizontal plane with the land surface. This block diagram shows the map area.

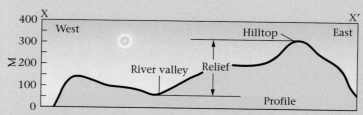

(c) A topographic profile (along section line X-X') shows the shape of the land surface as seen in a vertical slice.

FIGURE 2 A geologic cross section can show the topographic profile and the material underground.

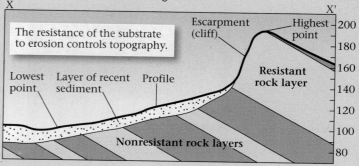

The resistance of the substrate to erosion controls topography.

a given locality, which together make up the landscape, reflect several factors.

- *Eroding or transporting agent*: Water, ice, and wind all cause erosion and transport sediment. But the shapes of landforms formed by each are different, because of differences in the abilities of these agents to carve into the substrate and to carry debris.
- *Relief*: The elevation difference, or relief, between adjacent places in a landscape determines the height and steepness of slopes. Steepness, in turn, controls the velocity of ice or water flow and determines whether rock or soil stays in place or tumbles downslope.
- *Climate*: The average mean temperature and the volume of precipitation in a region determine whether running water, flowing ice, or wind serves as the main agent of erosion or deposition.
- *Substrate composition*: The material comprising the substrate determines how the substrate responds to erosion.

For example, strong rocks can stand up to form steep cliffs, while soft sediment collapses to generate gentle slopes.

- *Life activity*: Some life activity weakens the substrate (by burrowing, wedging, or digesting), while some holds it together (by binding it with roots).
- *Time*: Landscapes evolve through time, in response to continued erosion and/or deposition. For instance, a gully that has just started to form in response to the flow of a stream does not look the same as a deep canyon that develops after the same stream has existed for a long time.

Although water, wind, and ice sculpt most landscapes, human activities have had an increasingly important impact on the Earth's surface. We have dug pits (mines) where once there were mountains, built hills (tailings piles and landfills) where once there were valleys, and made steep slopes gentle and gentle slopes steep (Fig. F.3a–c). By constructing concrete walls, we modify the shapes of coastlines, change the courses of rivers, and fill new lakes (reservoirs). In cities, buildings and

FIGURE F.3 Human influence on a geologic scale.

(a) The pyramids of Egypt are human-made hills that rise above the desert sands. They have lasted for thousands of years.

(b) In the process of making highway cuts, deep valleys are cut through high ridges. This example is near Denver.

(c) This stone dam holds back a reservoir in Colorado. Think about how long it would take a glacier to pile up so much sediment.

pavements completely seal the ground and cause water that might once have seeped down into the ground to spill into streams instead. And in the country, agriculture, grazing, water usage, and deforestation substantially alter the rates at which natural erosion and deposition take place. For example, agriculture greatly increases the rate of erosion, because for much of the year farm fields have no vegetation cover.

F.4 THE HYDROLOGIC CYCLE

Nothing that is can pause or stay—
  The moon will wax, the moon will wane
  The mist and cloud will turn to rain,
  The rain to mist and cloud again,
  Tomorrow be today.

—Henry Wadsworth Longfellow (1807–1882)

As is evident from the previous section, water in its various forms (liquid, gas, and solid) plays a major role in erosion and deposition on the Earth's surface. As we will see in succeeding chapters, it also modifies the upper crust underground, drives storms in the atmosphere, transports heat in the oceans, and hosts life everywhere. Our planet's water occupies several distinct reservoirs. Atmospheric water occurs as vapor, mist, or ice crystals. Surface water collects in oceans, lakes, streams, puddles, and swamps. Frozen water forms snowfields and glaciers. Subsurface water dampens soil and rock near the surface, or sinks deeper to a realm where it fills pores and cracks as **groundwater**. A significant amount of water also resides in living organisms—about 50 to 65% of your body consists of water. The oceans contain 95% of Earth's free water (water not incorporated in mineral crystals), and thus make up the biggest reservoir. Water constantly flows from one reservoir to another, a never-ending process called the **hydrologic cycle** (see Geology at a Glance, pp. 358–359). Perhaps without realizing it, Longfellow, an American poet fascinated with reincarnation, provided an accurate if somewhat romantic image of the hydrologic cycle.

What drives the hydrologic cycle? Simply put, the whole process takes place because energy from solar radiation bathes the Earth, and because gravity always pulls mass from higher elevations to lower elevations. To get a clearer sense of how the hydrologic cycle operates, let's follow the fate of seawater that has just reached the surface of the ocean. Solar radiation heats the water, and the increased thermal energy of the vibrating water molecules allows them to evaporate and drift upward in a gaseous state to become part of the atmosphere. About 30% of the total ocean volume evaporates every year. Atmospheric water vapor moves with the wind to higher elevations, where it cools, undergoes condensation, and rains or snows. About 76% of this water precipitates (falls out of the air) directly back into the ocean. The remainder precipitates onto land. Most of this water becomes trapped temporarily in the soil, or in plants and animals, and soon returns directly to the atmosphere by evapotranspiration, which is the sum of evaporation from bodies of water, evaporation from the ground surface, and release of water from plants and animals. Rainwater that did not become trapped in the soil or in living organisms either enters lakes or rivers and ultimately flows back to the sea as surface water, becomes trapped in glaciers, or sinks deeper into the ground to become groundwater. Groundwater also flows and ultimately returns to the Earth's surface reservoirs. In sum, during the hydrologic cycle, water moves among the ocean, the atmosphere, reservoirs on or below the land, and living organisms.

The *average* length of time that water stays in a particular reservoir during the hydrologic cycle is called the residence time. Water in different reservoirs has different residence times. For example, a typical molecule of water remains in the oceans for 4,000 years or less, in lakes and ponds for 10 years or less, in rivers for 2 weeks or less, and in the atmosphere for 10 days or less. Groundwater residence times are highly variable and depend on how deep the groundwater flows. Water can stay underground for anywhere from 2 weeks to 10,000 years before it inevitably moves on to another reservoir.

F.5 LANDSCAPES OF OTHER PLANETS

The dynamic, ever-changing landscape on Earth contrasts markedly with those of other terrestrial planets. Each of the terrestrial planets and moons has its own unique surface landscape features, reflecting the interplay between the object's particular tectonic and erosional processes. Let's look at a few examples: the Moon, Mars, and Venus.

Our Moon has a static, pockmarked landscape generated exclusively by meteorite impacts and volcanic activity billions of years ago. Because no plate tectonics occurs on the Moon, no new mountains form, and because no atmosphere or ocean exists, the Moon has no hydrologic cycle and undergoes no erosion from rivers, glaciers, or winds. Therefore, the lunar surface has remained largely unchanged for most of its history (Fig. F.4a).

Landscapes on Mars differ from those on the Moon because Mars *does* have an atmosphere (though much less dense than that of Earth), whose winds generate huge dust storms, some of which obscure nearly the entire surface of the planet for months at a time. The landscapes of Mars also differ from the Moon's because Mars probably once had surface water (Box F.2). Four kinds of materials cover the Martian surface: volcanic flows and deposits (primarily of basalt), debris from impacts, wind-blown sediment, and possibly water-laid sediment. Martian winds not only deposit sediment, they slowly erode impact craters and polish surface rocks.

Landscapes on Mars also differ from those on Earth, because Mars does not have plate tectonics. So, unlike Earth, Mars has no mountain belts or volcanic arcs. In fact, most landscape features on Mars, with the exception of wind-related ones, are over 3 billion years old. Thermal activity also led to the eruption of gargantuan hot-spot volcanoes, such as the 22-km-high Olympus Mons (Fig. F.4b). Mars also boasts the largest known canyon, the Valles Marineris, a gash over 3,000 km long and 8 km deep—no comparable feature exists on Earth. Because Mars has no vegetation and no longer has rain, its surface does not weather and erode like that of Earth, so it still bears the scars of impact by meteors earlier in the history of the Solar System, scars that have long since disappeared on Earth.

Venus is closer in size to the Earth and may still have operating mantle plumes. Virtually the entire surface of Venus was covered by volcanic deposits about 300 to 1,600 million years ago, making the planet's surface much younger than that of the Moon or of Mars. Further, Venus has a dense atmosphere that protects it from impacts by smaller objects. Because relatively little cratering has taken place since the resurfacing event, volcanic and tectonic features dominate the landscape of Venus (Fig. F.4c). Satellites have used radar to reveal a variety of volcanic constructions (such as shield volcanoes, lava flows, calderas). Rifting on Venus produced faults. Liquid water cannot survive the scalding temperatures of Venus's surface, so no hydrologic cycle operates there and no life exists. Because of the density of the atmosphere, winds are too slow to cause much erosion or deposition.

After this brief side trip to other planets, let's now return to Earth. Our planet has the greatest diversity of landscapes in the Solar System.

FIGURE F.4 Landscapes on other planets.

(a) A closeup view of the lunar landscape, with the lunar rover and an astronaut for scale.

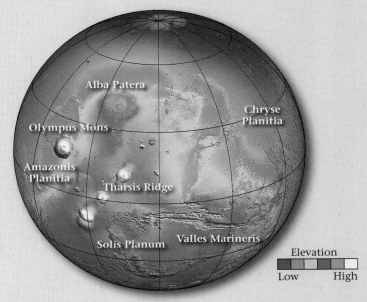

(b) A digital elevation model depicting the surface of Mars. Note the huge bulge of Tharsis Ridge, the giant Olympus Mons volcano, and the deep Valles Marineris canyon.

(c) A radar image of Venus. A thick blanket of clouds obscure this planet's surface, so it can't be seen through a telescope.

The Hydrologic Cycle

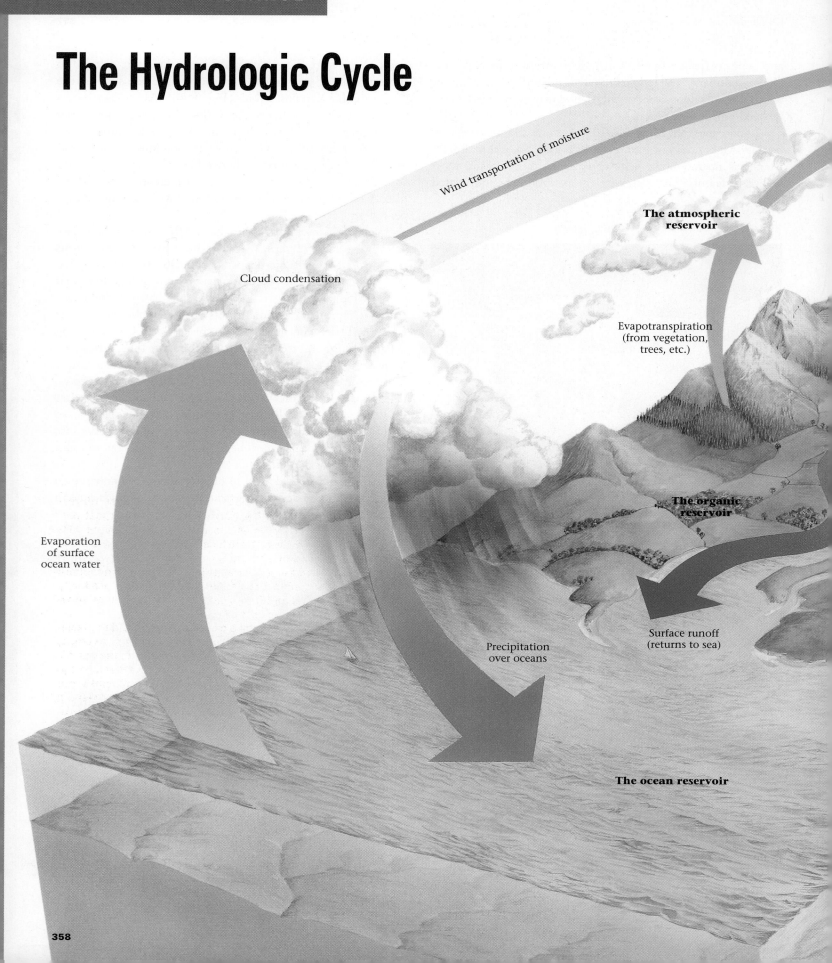

Wind transportation of moisture

Cloud condensation

The atmospheric reservoir

Evapotranspiration (from vegetation, trees, etc.)

Evaporation of surface ocean water

The organic reservoir

Surface runoff (returns to sea)

Precipitation over oceans

The ocean reservoir

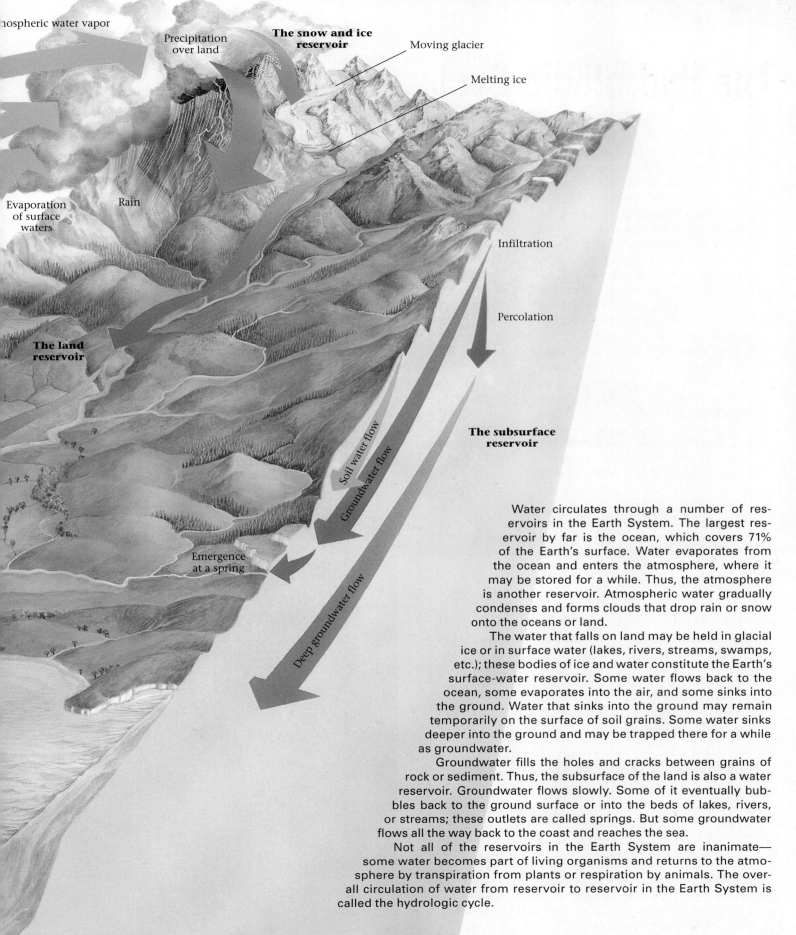

hospheric water vapor

Precipitation over land

The snow and ice reservoir

Moving glacier

Melting ice

Evaporation of surface waters

Rain

Infiltration

Percolation

The land reservoir

Soil water flow

Groundwater flow

The subsurface reservoir

Emergence at a spring

Deep groundwater flow

Water circulates through a number of reservoirs in the Earth System. The largest reservoir by far is the ocean, which covers 71% of the Earth's surface. Water evaporates from the ocean and enters the atmosphere, where it may be stored for a while. Thus, the atmosphere is another reservoir. Atmospheric water gradually condenses and forms clouds that drop rain or snow onto the oceans or land.

The water that falls on land may be held in glacial ice or in surface water (lakes, rivers, streams, swamps, etc.); these bodies of ice and water constitute the Earth's surface-water reservoir. Some water flows back to the ocean, some evaporates into the air, and some sinks into the ground. Water that sinks into the ground may remain temporarily on the surface of soil grains. Some water sinks deeper into the ground and may be trapped there for a while as groundwater.

Groundwater fills the holes and cracks between grains of rock or sediment. Thus, the subsurface of the land is also a water reservoir. Groundwater flows slowly. Some of it eventually bubbles back to the ground surface or into the beds of lakes, rivers, or streams; these outlets are called springs. But some groundwater flows all the way back to the coast and reaches the sea.

Not all of the reservoirs in the Earth System are inanimate—some water becomes part of living organisms and returns to the atmosphere by transpiration from plants or respiration by animals. The overall circulation of water from reservoir to reservoir in the Earth System is called the hydrologic cycle.

Water on Mars?

In 1877, an Italian astronomer named Giovanni Schiaparelli studied the surface of Mars with a telescope and announced that long, straight *canali* crisscrossed the planet's surface. *Canali* should have been translated into the English word "channel," but perhaps because of the recent construction of the Suez Canal, newspapers of the day translated the word into the English "canal," with the implication that the features had been constructed by intelligent beings. An eminent American astronomer began to study the "canals" and suggested that they had been built to carry water from polar ice caps to Martian deserts.

Late-twentieth-century satellite mapping of Mars showed that the "canals" do not exist—they were simply optical illusions. There are no lakes, oceans, rainstorms, or flowing rivers on the surface of Mars today. The atmosphere of Mars has such low density, and thus exerts so little pressure on the planet's surface, that any liquid water released at the surface would quickly evaporate.

Thus, Mars has no hydrologic cycle the way the Earth does. But three crucial questions remain: Does liquid water ever form, even for short periods, on the Martian surface today? Did Mars ever have a significant amount of running water or standing water in the past? If the planet once had significant water, where is the water now? The question of the presence of water lies at the heart of an even more basic question: Given that the simplest life as we know it requires water, is there, or was there, life on Mars?

Many planetary geologists believe that the case for liquid water on Mars is quite strong. Much of the evidence comes from comparing landforms on the planet's surface with landforms of known origin on Earth. High-resolution images of Mars reveal a number of landforms that look as if they formed in response to flowing water. Examples include networks of channels resembling river networks on Earth (**Fig. 1**), scour features, deep gullies, and streamlined deposits of sediment.

Studies by the *Odyssey* satellite in 2003, and by Mars rovers (*Spirit* and *Opportunity*) that landed on the planet in 2004, added intriguing new data to the debate.

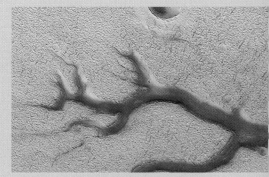

FIGURE 1 A satellite image reveals the presence of a river-like network of channels on the surface of Mars.

Odyssey detected hints that hydrogen, an element in water, exists beneath the surface of the planet over broad regions, and the Mars rovers have documented the existence of hematite and gypsum, minerals that form in the presence of water. The rovers also have found sedimentary deposits that appear to have been deposited in water. The Phoenix lander, in 2008, confirmed the existence of water ice by digging into the surface to expose some. Researchers speculate that Mars was much wetter in its past, perhaps billions of years ago. But since the atmosphere became less dense, the water evaporated and now lies hidden underground or trapped in polar ice caps.

Key Terms

agents of erosion (p. 353)

contour interval (p. 354)

contour line (p. 354)

deposition (p. 353)

depositional landform (p. 353)

downslope movement (p. 353)

erosion (p. 353)

erosional landform (p. 353)

external energy (p. 353)

gravitational energy (p. 353)

groundwater (p. 356)

hydrologic cycle (p. 356)

internal energy (p. 353)

landform (p. 352)

landscape (p. 352)

relief (p. 353)

subsidence (p. 352)

substrate (p. 353)

topographic map (p. 354)

topography (p. 354)

uplift (p. 352)

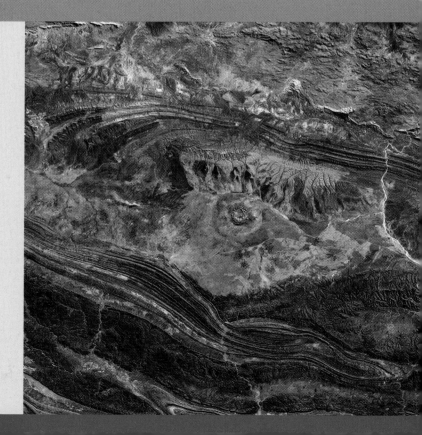

THE VIEW FROM SPACE Textures of the land, as viewed from a satellite, illustrate how substrate controls erosion. Elongate ridges visible here in central Australia are layers of resistant strata on the limbs of folds. The circular structure at the center of the image is Gosses Bluff, the relict of a 142 Ma meteorite impact site.

CHAPTER 13

Unsafe Ground: Landslides and Other Mass Movements

What goes up must come down. As a consequence, rock and regolith forming the substrate of hill slopes occasionally give way and slide downslope. The results can be disastrous, as happened when a mudslide buried homes in La Conchita, California.

GEOPUZZLE

Picture a mountain slope covered by an ancient forest. The next day, the forest along with tens of meters of its substrate (soil and bedrock) are gone, having tumbled downhill to form a massive pile of debris in the river valley below. What triggers such catastrophic landslides? Can they be prevented?

13.1 INTRODUCTION

It was Sunday, May 31, 1970, a market day, and thousands of people had crammed into the Andean town of Yungay, Peru, to shop. Suddenly they felt the jolt of an earthquake, strong enough to topple some masonry houses. But worse was to come. This earthquake broke an 800-m-wide ice slab off the end of a glacier at the top of Nevado Huascarán, a nearby 6.6-km-high mountain peak. Gravity pulled the ice slab 3.7 km down the mountain's steep slopes. As it tumbled, the ice disintegrated into a chaotic avalanche of chunks traveling at speeds of over 300 km per hour. Near the base of the mountain, most of the avalanche fed into a valley and thickened into a dense cloud as high as a ten-story building. Friction ripped up rocks and soil along the way and transformed the ice into water, which when mixed with rock and dust created 50 million cubic meters of mud, a slurry viscous enough to carry boulders larger than houses. This mass, sometimes floating on a compressed air cushion, traveled over 14.5 km in less than four minutes.

Most of the debris came to rest on top of the village of Ranrahica, at the mouth of the valley. But part of it shot up the sides of the valley and literally flew over the ridge bordering Yungay. As the town's inhabitants and visitors stumbled out of earthquake-damaged buildings, they heard a deafening roar and looked up to see the churning mud cloud bursting above the nearby ridge. In a moment, the town was completely buried under several meters of mud and rock. When the dust had settled, only the top of the church and a few palm trees remained visible to show where Yungay once lay (Fig. 13.1a, b). Eighteen thousand people are forever entombed beneath the debris mass. Today, the site is a grassy meadow with a hummocky (irregular and lumpy) surface, spotted with crosses left by mourning relatives.

People often assume that the earth beneath them is *terra firma*, a solid foundation on which they can build their lives. But the catastrophe at Yungay says otherwise—much of the Earth's surface is dangerous, unstable ground, land capable of moving downslope in a matter of seconds to weeks. Geologists refer to the gravitationally caused downslope transport of rock, regolith (soil, sediment, and debris), snow, and ice as **mass movement**, or mass wasting. Like earthquakes, volcanic eruptions, storms, and floods, mass movements are a type of **natural hazard**, meaning a natural feature of the environment that can cause damage to living organisms and to buildings. Unfortunately, mass movement becomes more of a threat to people every year, because as the world's population grows, cities expand into areas of unsafe ground. But mass movement also plays a critical role in the rock cycle, for it's the first step in the transportation of sediment. And it plays a critical role in the evolution of landscapes by modifying the shapes of slopes.

In this chapter, we look at the types, causes, and consequences of mass movement, and at the precautions society can take to protect people and property from its dangers. You might want to think about this information when selecting a site for your home or when voting on land-use propositions for your community.

TAKE-HOME MESSAGE

By the end of this chapter, you should be aware of the different kinds of mass movements that contribute to erosion and pose hazards for humans. And, you should have ideas about how to recognize and avoid these hazards.

FIGURE 13.1 The May 1970 Yungay landslide disaster in Peru.

(a) Before the landslide, the town of Yungay perched on a hill near the ice-covered mountain Nevado Huascarán.

(b) The landslide completely buried the town beneath debris. A landslide scar is visible on the mountain in the distance.

13.2 TYPES OF MASS MOVEMENT

Though in everyday language people commonly refer to any mass-movement event as a "landslide," geologists and civil engineers tend to distinguish different types of mass movement based on four factors: the type of material involved (rock, regolith, or snow and ice), the velocity of the movement (fast, intermediate, or slow), the character of the moving mass (chaotic cloud, slurry, or coherent body), and the environment in which the movement takes place (subaerial or submarine). Below, we look at mass movements that occur, roughly in order from slow to very fast (see also Geotour 13 on p. GT-28).

Creep and Solifluction

Creep refers to the slow, gradual downslope movement of regolith on a slope. (Commonly, creep affects soil, so geologists often refer to the process as "soil creep.") Creep happens when regolith alternately expands and contracts in response to freezing and thawing, wetting and drying, or warming and cooling. Freezing causes expansion because water increases in volume by about 9% as it turns to ice; wetting causes expansion because clay mineral grains absorb water and become thicker; and warming causes expansion because heat makes atoms in materials move farther apart. To see how the process of creep works, let's focus on the consequences of seasonal freezing and thawing.

In the winter, when water freezes, the regolith expands, and particles move outward, perpendicular to the slope. During the spring thaw, water becomes liquid again, and gravity makes the particles sink vertically and thus migrate downslope slightly (Fig. 13.2a, b). You can't see creep by staring at a hill slope because it occurs too slowly, but over a period of years, creep causes trees, fences, gravestones, walls, and foundations built

FIGURE 13.2 The process and consequences of slow mass movements (creep and solifluction).

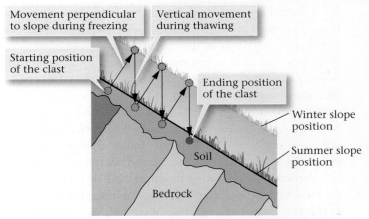

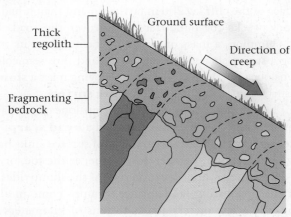

(a) Creep due to freezing and thawing: The clast rises perpendicular to the ground during freezing, and sinks vertically during thawing. After three years, it migrates to the position shown.

(b) As rock layers weather and break up, the resulting debris creeps downslope.

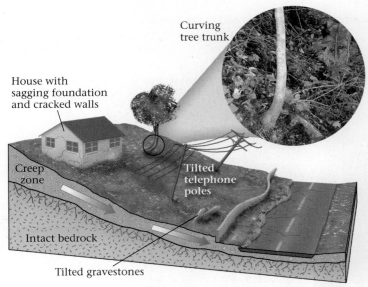

(c) Soil creep causes walls to bend and crack, building foundations to sink, trees to bend, and power poles and gravestones to tilt.

(d) Solifluction on a hill slope in the tundra.

on a hillside to tilt downslope. Notably, trees that continue to grow after they have been tilted display a pronounced curvature at their base (Fig. 13.2c).

In Arctic or high-elevation regions, regolith freezes solid to great depth during the winter. In the brief summer thaw, only the uppermost 1 to 3 m of the ground thaws. Since meltwater cannot sink into the permanently frozen ground, or permafrost, the melted layer becomes soggy and weak and flows slowly downslope in overlapping sheets. Geologists refer to this kind of creep, characteristic of cold, treeless tundra regions, as **solifluction** (Fig. 13.2d).

Slumping

Near Pacific Palisades, along the coast of southern California, Highway 1 runs between the beach and a 120-m-high cliff. Between March 31 and April 3, 1958, a 1-km-long section of the highway disappeared beneath a mass of regolith that had moved down the adjacent cliff. When the movement stopped, the face of the cliff lay 200 m farther inland than it had before. It took weeks for bulldozers to uncover the road. During such slumping, a mass of regolith detaches from its substrate along a spoon-shaped sliding surface and slips semicoherently downslope (Fig. 13.3). We call the moving mass a **slump**, and the surface on which it slips a **failure surface**.

The distinct, curving step at the upslope edge of a slump, where the regolith detaches, is called a **head scarp.** Immediately below the head scarp, the land surface sinks below its previous elevation. Farther downslope, at the toe, or end, of the slump, the ground elevation rises as the slump rides up and over the preexisting land surface. Slumps come in all sizes, from only a few meters across to tens of kilometers across, and they move at speeds of millimeters per day to meters per minute.

Mudflows, Debris Flows, and Lahars

Rio de Janeiro, Brazil, originally occupied only the flatlands bordering beautiful crescent beaches between towering mountains. But in recent decades, its population has grown so much that the city has expanded up the steep sides of the mountains, and in many places densely populated communities of makeshift shacks cover the slopes. These communities, which have no storm drains, were built on the thick regolith that resulted from long-term weathering of bedrock in Brazil's tropical climate. In 1988, particularly heavy rains saturated the regolith, which turned into a viscous slurry of mud that flowed downslope. Whole communities disappeared overnight, replaced by a hummocky muddle of mud and debris. And at the base of the cliffs, the flowing mud caused high-rise buildings and homes to collapse (Fig. 13.4a).

In areas such as the hill-slope communities of Rio, where neither vegetation nor drainage systems protect the ground from rainfall, water mixes with regolith to create a slurry that moves downslope. If the slurry consists of just mud, it's a **mudflow**, but if the mud is mixed with larger rock fragments, it's a **debris flow**. The speed at which mud or debris flows can move depends on the slope angle and on the water content. Flows move faster if they are wetter (that is, less viscous) and if they move on steeper slopes. On a gentle slope, viscous mud flows like molasses, but on a steep slope, low-viscosity mud may move at tens of kilometers per hour. Because mud and debris flows have greater viscosity than clear water, they can carry large rock chunks, as well as houses and cars. They typically follow channels downslope, and at the base of the slope they spread out into a broad lobe.

Particularly devastating mudflows spill down volcanoes and the river valleys bordering volcanoes. These mudflows, known as **lahars**, consist of a mixture of volcanic ash and water from the snow and ice that melts in a volcano's heat or from heavy rains (Fig. 13.4b; see Chapter 7).

FIGURE 13.3   The process of slumping on a hill slope. Note the scarps that form at the head of the slump.

FIGURE 13.4 Examples of mudflows and lahars.

(a) The aftermath of a mudflow in Rio de Janeiro. As the population grows, more people build on dangerous slopes.

(b) The aftermath of a lahar that flowed down the side of Mt. St. Helens, following an eruption in 1982.

Landslides (Rock and Debris Slides)

In the early 1960s, engineers built a huge new dam across a river on the northern side of Monte Toc, in the Italian Alps, to create a reservoir for generating electricity. This dam, the Vaiont Dam, was an engineering marvel, a concrete wall rising 260 m (as high as an 85-story skyscraper) above the valley floor (Fig. 13.5a). Unfortunately, the dam's builders did not recognize the hazard posed by nearby Monte Toc. The side of Monte Toc facing the reservoir is underlain by dipping limestone beds interlayered with weak shale beds. These beds dip parallel to the surface of the mountain and curve under the reservoir (Fig. 13.5b). As the reservoir filled, the flank of the mountain cracked, shook, and rumbled, and local residents began to call Monte Toc *la montagna che cammina* (the mountain that walks).

After several days of rain, Monte Toc began to rumble so much that on October 9, 1963, engineers lowered the water level in the reservoir. They thought the wet ground might slump a little into the reservoir, but no more than that, so no one ordered the evacuation of the town of Longarone, a few kilometers down the valley. Unfortunately, the engineers underestimated the

FIGURE 13.5 The Vaiont Dam disaster—a catastrophic landslide that displaced the water in a reservoir with rock debris.

(a) Vaiont Dam and the debris now behind it. The exposed failure surface on Monte Toc is visible in the distance.

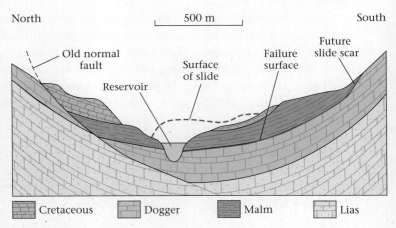

(b) A cross section parallel to the face of Vaiont Dam before the landslide. The failure surface lies at the base of the weak Malm Shale. The dashed line shows the top of the slide debris pile, after the slide had moved.

FIGURE 13.6 Examples of avalanches.

(a) Aftermath of a 1999 avalanche in the Austrian Alps. Masses of snow buried several homes.

(b) Trees that were flattened by an avalanche are exposed now that the snow has melted.

problem. At 10:30 that evening, a huge chunk of Monte Toc—600 million tons of rock—detached from the mountain and slid downslope into the reservoir. Some debris rocketed up the opposite wall of the valley to a height of 260 m above the original reservoir level. The displaced water of the reservoir spilled over the top of the dam and rushed down into the valley below. When the flood had passed, nothing remained of Longarone or its 1,500 inhabitants. Though the dam itself still stands, it holds back only debris.

Geologists refer to such a sudden movement of rock and debris down a nonvertical slope as a **landslide**. If the mass consists only of rock, it may also be called a **rock slide** (the case in the Vaiont Dam disaster), and if it consists mostly of regolith, it may also be called a **debris slide**. Once a landslide has taken place, it leaves a landslide scar on the slope and forms a debris pile at the base of the slope.

Slides happen when bedrock and/or regolith detaches from a slope and shoots downhill on a failure surface (slide surface) roughly parallel to the slope. Slide surfaces typically form along weak bedding or foliation planes, or along weak fractures. At the Vaiont Dam, for example, the plane of weakness that would become the failure surface was a weak shale bed. Slides may move at speeds of up to 300 km per hour; they are particularly fast when a cushion of air gets trapped beneath, so there is virtually no friction between the slide and its substrate, and the mass moves like a hovercraft. Rock and debris slides sometimes have enough momentum to climb the opposite side of the valley into which they fell.

Avalanches

In the winter of 1999, an unusual weather system passed over the Austrian Alps. First it snowed; then the temperature warmed and the snow began to melt. But then the weather turned cold again, and the melted snow froze into a hard, icy crust. This cold snap ushered in a blizzard that blanketed the ice crust with tens of centimeters (1 to 2 ft) of snow. On the mountaintops the wind built the snow into huge overhanging drifts called cornices. Skiers delighted in the bounty of white, but not for long. No one knows how it started—perhaps a gust of wind or a falling branch was enough—but the world witnessed the aftermath. With the frozen snow layer underneath acting as a failure surface, the heavy layer of new snow began to slide down the mountain, accelerating as it moved and then disintegrating and mixing with air. It became a roaring cloud—an avalanche—traveling at hurricane speeds and flattening everything in its path. Trees and ski lodges toppled like toothpicks before the mass finally reached the valley floor and came to a halt (Fig. 13.6a).

Avalanches, in a general sense, are turbulent clouds of debris mixed with air that rush down steep hill slopes at high velocity. If the debris consists of snow, as in the Austrian avalanche, it's a snow avalanche. If it consists of fragments of rock and dust, it's a debris avalanche. The moving air-debris mass is denser than clear air and thus hugs the ground and acts like an extremely strong and viscous wind that can knock down and

FIGURE 13.7   Examples of rockfalls.

(a) Successive rockfalls have littered the base of this sandstone cliff with boulders. Note the talus at the base of the cliff.

(b) A large rock slide buried a portion of the forest bordering a lake in the Uinta Mountains, Utah.

blow away anything in its path. As illustrated by the Austrian example, snow avalanches pose a particular threat when frozen snow layers get buried and thus can become a failure surface for the overlying snow. Typically, avalanches happen again and again in the same area, creating pathways, called avalanche chutes, in which no mature trees grow (Fig. 13.6b). Avalanches of sediment, mixed with water, can occur underwater. These are more commonly referred to as turbidity currents (see Chapter 5).

Rockfalls and Debris Falls

Rockfalls and **debris falls**, as their names suggest, occur when a mass free-falls from a steep (vertical) cliff (Fig. 13.7a, b). Friction and collision with other rocks may bring some blocks to a halt before they reach the bottom of the slope; these blocks pile up to form a **talus**, a sloping apron of rocks along the base of the cliff. Rock or debris that has fallen a long way can reach speeds of 300 km per hour, and may have so much momentum that it keeps going when it touches bottom and triggers a debris avalanche that can cross a valley floor and rise up the other side. Like avalanches, large, fast rockfalls push the air in front of them, creating a short blast of hurricane-like wind. The wind alone from a 1996 rockfall in Yosemite National Park flattened over 2,000 trees. Commonly, rockfalls happen when a rock separates from a cliff face along a joint.

Submarine Mass Movements

Not all mass movements are subaerial and visible to observers. Huge volumes of debris slide down slopes along the coasts of islands and continents, and these movements, hidden by a blanket of water, may go undetected. In some cases, however, the movements trigger huge tsunamis.

As is the case with mass movements on land, submarine mass movements occur at a variety of scales and rates, and the material within the moving mass may remain nearly intact as a single large slump block, may break up into several slump blocks that move separately, may mix with some water to become a slurry-like debris, or may mix with lots of water to become a submarine avalanche. We introduced submarine avalanches in Chapter 6, noting that they are also called turbidity currents. Geologists have recognized in recent years that some submarine slumps are truly monumental in size, and their movement significantly modifies the shape of the sea floor. For example, the broad apron of hummocky sea floor that surrounds Hawaii is underlain by the debris of countless slumps that has accumulated over millions of years (Fig. 13.8).

FIGURE 13.8   A map of the larger submarine landslides around the Hawaiian Islands. Tan areas delineate the slides.

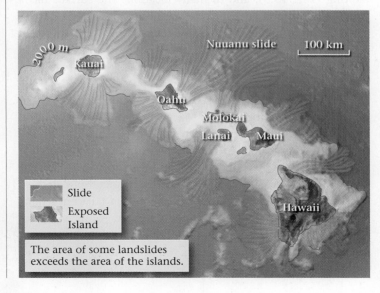

Some of the slumps are more than 100 km across, and some took big bites out of the edges of islands, leaving head scarps that now form dramatic coastal cliffs.

TAKE-HOME MESSAGE

Geologists distinguish among different kinds of mass movement according to the speed and character of the flow. Slow movements include creep, solifluction, and slumping. Mudflows and debris flows move faster, and avalanches and rockfalls move the fastest. All mass movements are hazards. Submarine occurrences may generate tsunamis.

13.3 WHY DO MASS MOVEMENTS OCCUR?

In order for the mass movements described above (see **Geology at a Glance,** pp. 370–371) to take place, the following conditions must develop. First, materials at the Earth's surface must be weakened by fracturing and weathering. Second, relief (the difference in elevation between two adjacent locations) must develop, for where relief exists, gravity can pull material from a higher elevation down to a lower elevation. Let's look at these conditions in more detail.

Weakening the Surface: Fragmentation and Weathering

If the Earth's surface were covered with intact (unbroken) rock, mass movements would be of little concern, for intact rock has great strength and could form stalwart mountain faces that would never tumble, even if they were vertical. But the rock of Earth's upper crust has been broken by jointing and faulting (Fig. 13.9), and in many locations the surface has a cover of regolith created by the weathering of rock in Earth's corrosive atmosphere. Fragmented rock and regolith are much weaker than intact rock and can indeed collapse fairly easily in response to Earth's gravitational pull. Thus, jointing, faulting, and weathering make mass movements possible.

Why are regolith and fractured rock so much weaker than intact rock? Intact rock is held together by the strong chemical bonds within mineral crystals, by mineral cement, or by the interlocking of grains. In contrast, a joint-bounded or fault-bounded block is held in place only by friction between the block and its surroundings. Dry regolith is unconsolidated, so it holds together because of friction between adjacent grains and/or because weak electrical charges cause grains to attract each other. Slightly wet regolith holds together because of water's surface tension, caused by the attraction of water molecules to each other.

FIGURE 13.9   The limestone in this outcrop on the coast of western Ireland has broken into blocks bounded by vertical joints and horizontal bedding planes. Storm waves moved some of the blocks.

Slope Stability: The Battle between Downslope Force and Resistance Force

Mass movements do not take place on all slopes, and even on slopes where such movements are possible, they occur only occasionally. Geologists distinguish between stable slopes, on which sliding is unlikely, and unstable slopes, on which sliding will likely happen. When material starts moving on an unstable slope, we say that slope failure has occurred. Whether a slope fails or not depends on the balance between two forces—the downslope force, caused by gravity, and the resistance force, which inhibits sliding. If the downslope force exceeds the resistance force, the slope fails and mass movement results.

To understand slope stability, imagine a block sitting on a slope. We can represent the gravitational attraction between this block and the Earth by an arrow (a vector) that points straight down, toward the Earth's center of gravity. This arrow can be resolved into two components, the downslope force parallel to the slope and the normal force perpendicular to the slope. The resistance force can be represented by an arrow pointing uphill. If the downslope force exceeds the resistance force, then the block moves; otherwise, it stays in place (**Fig. 13.10a, b**). Note that for a given mass, the magnitude of the downslope force increases as the slope angle increases, so downslope forces are greater on steeper slopes.

What causes the resistance force? As we saw above, chemical bonds in mineral crystals, cement, and the jigsaw-puzzle-like interlocking of crystals hold intact rock in place, friction holds an unattached block in place, electrical charges and friction hold dry regolith in place, and surface tension holds slightly wet regolith in place.

Because of resistance force, granular debris tends to pile up and create the steepest slope it can without collapsing. The angle of this slope is called the **angle of repose**, and for most dry, unconsolidated materials (such as dry sand) it typically has a value of around 30°. The angle depends partly on the shape and size distribution of grains, which determine how the grains fit together, and partly on the amount of friction and cohesion across grain boundaries. For example, larger angles of repose tend to form on slopes composed of large, irregularly shaped grains, for these grains interlock with each other (Fig. 13.11). Wet sand has a much steeper angle of repose than does dry sand, as you know if you've had the opportunity to build a sand castle.

In many locations, the resistance force is less than might be expected because a weak surface exists at some depth below ground level. This weak surface separates unstable rock and debris above from the stronger substrate below. Geologists recognize several different kinds of weak surfaces that are likely to become failure surfaces (Fig. 13.12a–c). These include: wet clay layers; wet, unconsolidated sand layers; surface-parallel joints (also known as exfoliation joints); weak bedding planes; and metamorphic foliation planes.

Failure surfaces that dip parallel to the slope are particularly likely to fail because the downslope force is parallel to the surface. As an example, consider the 1959 landslide that occurred in Madison Canyon, in southwestern Montana. On August 17 of that year, shock waves from a strong earthquake jarred the region. The southern wall of the canyon is underlain by metamorphic rock with a strong foliation that provided a plane of weakness. When the ground vibrated, rock detached along a foliation plane and tumbled downslope. Unfortunately, twenty-eight campers lay sleeping on the valley floor. They were awakened by the hurricane-like winds blasting in front of the moving mass, but seconds later were buried under a 45-m-thick layer of rubble.

FIGURE 13.10 Forces that trigger downslope movement.

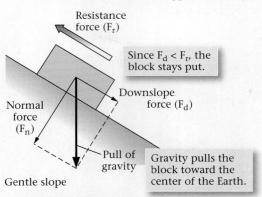

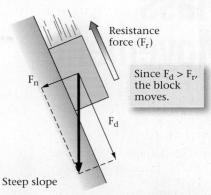

(a) Gravity can be divided into a normal force and a downslope force. If the resistance force, caused by friction, is greater than the downslope force, the block does not move.

(b) If the slope angle increases, the downslope force due to gravity increases. If the downslope force becomes greater than the resistance force, the block starts to move.

Fingers on the Trigger: Factors Causing Slope Failure

What triggers an individual mass-wasting event? In other words, what causes the balance of forces to change so that the downslope force exceeds the resistance force, and a slope suddenly fails? Here, we look at various phenomena—natural and human-made—that trigger slope failure.

Shocks, vibrations, and liquefaction. Earthquake tremors, the passing of large trucks, or blasting in construction sites may cause a mass that was on the verge of moving to actually start moving. For example, an earthquake-triggered slide dumped debris into Lituya Bay, in southeastern Alaska, in 1958. The debris displaced the water in the bay, creating a 300-m-high (1,200-ft-high) splash that washed the slopes on the opposite side of the bay clean of their forests and carried fishing boats many kilometers out to sea. The vibrations of an earthquake break bonds that hold a mass in place and/or cause the mass and the slope to separate slightly, thereby decreasing friction. As a consequence, the resistance force decreases, and the downslope force sets the mass in motion. Shaking can also cause **liquefaction** of wet sediment by either increasing water pressure in spaces between grains so that the grains are pushed apart or by breaking the cohesion between the grains.

Changing slope angles, slope loads, and slope support. Factors that make a slope steeper or heavier may cause the slope to fail. For example, when a river eats away at the base of the slope, or a contractor excavates at the base of a slope, the slope becomes steeper and the downslope force increases. But while this happens, the resistance force stays the same. Therefore, if the excavation continues, the downslope force will eventually exceed the resistance force and the slope will fail.

FIGURE 13.11 The angle of repose is the steepest slope that a pile of unconsolidated sediment can have and remain stable. The angle depends on the shape and size of grains.

30°   45°

Well-rounded sand has a small angle.

Irregularly shaped gravel has a large angle.

Mass Movement

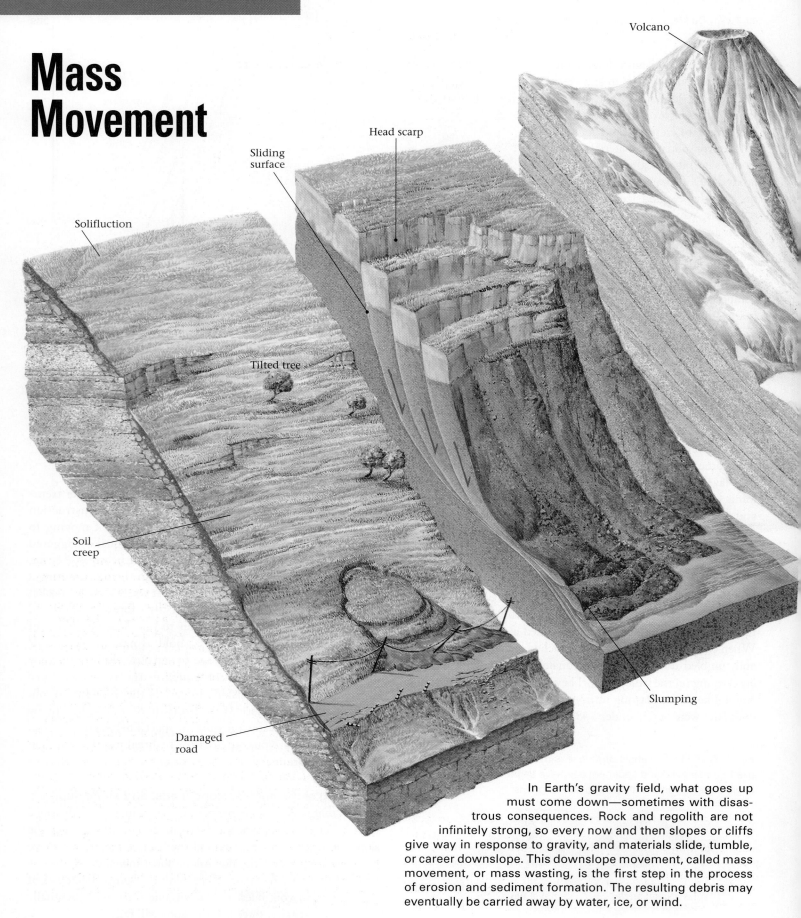

Solifluction

Sliding surface

Head scarp

Volcano

Tilted tree

Soil creep

Damaged road

Slumping

In Earth's gravity field, what goes up must come down—sometimes with disastrous consequences. Rock and regolith are not infinitely strong, so every now and then slopes or cliffs give way in response to gravity, and materials slide, tumble, or career downslope. This downslope movement, called mass movement, or mass wasting, is the first step in the process of erosion and sediment formation. The resulting debris may eventually be carried away by water, ice, or wind.

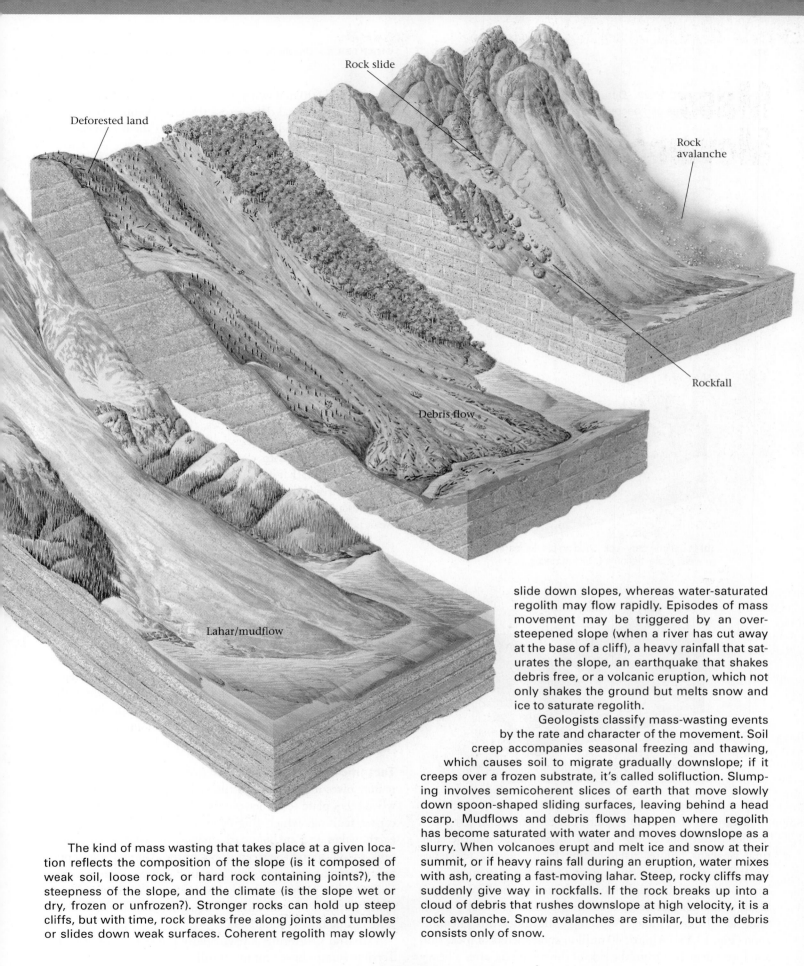

Deforested land

Rock slide

Rock avalanche

Rockfall

Debris flow

Lahar/mudflow

slide down slopes, whereas water-saturated regolith may flow rapidly. Episodes of mass movement may be triggered by an over-steepened slope (when a river has cut away at the base of a cliff), a heavy rainfall that saturates the slope, an earthquake that shakes debris free, or a volcanic eruption, which not only shakes the ground but melts snow and ice to saturate regolith.

Geologists classify mass-wasting events by the rate and character of the movement. Soil creep accompanies seasonal freezing and thawing, which causes soil to migrate gradually downslope; if it creeps over a frozen substrate, it's called solifluction. Slumping involves semicoherent slices of earth that move slowly down spoon-shaped sliding surfaces, leaving behind a head scarp. Mudflows and debris flows happen where regolith has become saturated with water and moves downslope as a slurry. When volcanoes erupt and melt ice and snow at their summit, or if heavy rains fall during an eruption, water mixes with ash, creating a fast-moving lahar. Steep, rocky cliffs may suddenly give way in rockfalls. If the rock breaks up into a cloud of debris that rushes downslope at high velocity, it is a rock avalanche. Snow avalanches are similar, but the debris consists only of snow.

The kind of mass wasting that takes place at a given location reflects the composition of the slope (is it composed of weak soil, loose rock, or hard rock containing joints?), the steepness of the slope, and the climate (is the slope wet or dry, frozen or unfrozen?). Stronger rocks can hold up steep cliffs, but with time, rock breaks free along joints and tumbles or slides down weak surfaces. Coherent regolith may slowly

FIGURE 13.12 Different kinds of weak surfaces can become failure surfaces.

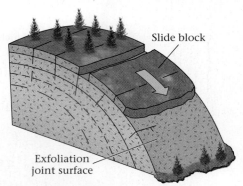

(a) Exfoliation joints form parallel to slope surfaces in granite and become failure surfaces.

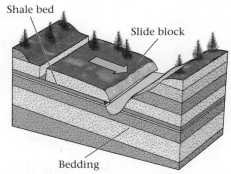

(b) In sedimentary rock, bedding planes (particularly in weak shale) become failure surfaces.

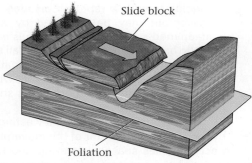

(c) In metamorphic rock, foliation planes (particularly in mica-rich schist) become failure surfaces.

Addition of water to regolith not only makes the regolith heavier, thereby increasing the downslope force, but at the same time weakens failure surfaces, thereby decreasing resistance force. Therefore, heavy rain can trigger mass movement. The largest observed landslide in U.S. history—the Gros Ventre Slide, which took place in 1925 on the flank of Sheep Mountain, near Jackson Hole, Wyoming—illustrates this phenomenon (Fig. 13.13). Almost 40 million cubic meters of rock, soil, and forest detached from the side of the mountain after a heavy rain and slid 600 m down a slope, filling the valley and creating a 75-m-high natural dam across the Gros Ventre River. River erosion had removed support at the base of the hill so the rain-soaked mass became unstable.

In some cases, excavation results in the formation of an over-hang. When such **undercutting** has occurred, rock making up the overhang eventually breaks away from the slope and falls. Overhangs commonly develop along seacoasts and rivers, where the water cuts into a slope (Fig. 13.14a, b).

Changing the slope strength. The stability of a slope depends on the strength of the material composing it. If the material weakens with time, the slope becomes weaker and eventually collapses. Three factors influence the strength of slopes: weathering, vegetation cover, and water.

With time, chemical weathering produces weaker minerals, and physical weathering breaks rocks apart. Thus, a formerly intact rock composed of strong minerals is transformed into a weaker rock or into regolith.

We've seen that thin films of water cause cohesion between grains. Water in larger quantities, though, decreases cohesion, because it fills pore spaces entirely and keeps grains apart. Saturation of regolith with water during a rainstorm weakens the regolith so much that it may begin to move downslope as a slurry. If the water weakens a specific subsurface layer, then the layer becomes a failure surface. Similarly, if groundwater rises above a weak failure surface, overlying rock or regolith may start to slide.

In the case of slopes underlain with regolith, vegetation tends to strengthen the slope, because the roots hold other-wise unconsolidated grains together. Also, plants absorb water from the ground, thus keeping it from turning into slippery mud. The removal of vegetation therefore has the net result of making slopes more susceptible to downslope mass movement. In 2003, terrifying wildfires, stoked by strong winds, destroyed ground-covering vegetation in many areas of California. When heavy rains followed, the barren ground of this hilly region became saturated with water and turned into mud, which then flowed downslope, damaging and destroying many homes and roads. Deforestation in tropical rain forests, similarly, leads to catastrophic mass wasting of the forest's substrate.

Tectonic setting. Most unstable ground on Earth ultimately owes its existence to the activity of plate tectonics. As we've seen, plate tectonics causes uplift, generating relief, and causes faulting, which fragments the crust. And, of course, earthquakes on plate boundaries trigger devastating landslides. Spend a day along the steep slopes of the Alpine Fault, a plate boundary that transects New Zealand, and you can hear mass movement in progress: during heavy rains, rockfalls and land-slides clatter with astounding frequency, as if the mountains were falling down around you.

Coastal California is particularly prone to mass wasting. Terrible slumps have sent multimillion-dollar homes tumbling

FIGURE 13.13 Stages leading to the 1925 Gros Ventre Slide in Wyoming.

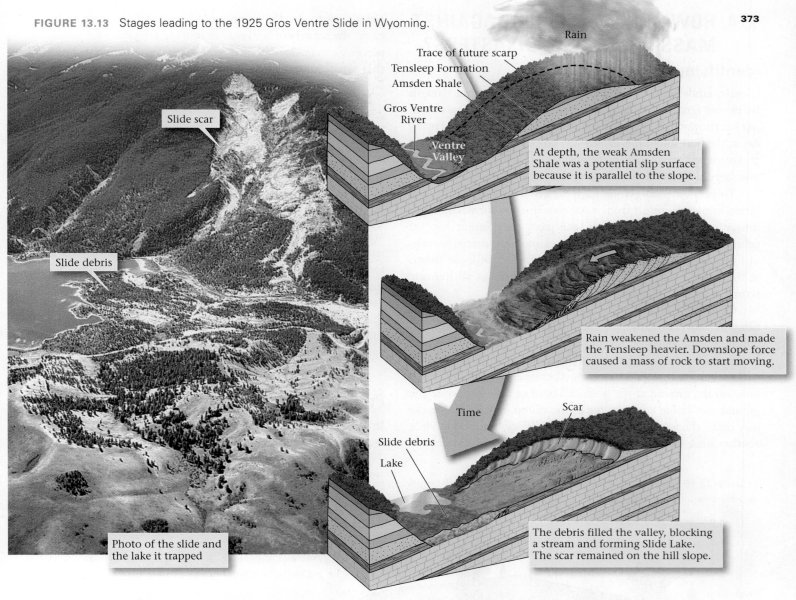

Rain

Trace of future scarp
Tensleep Formation
Amsden Shale

Gros Ventre River

Ventre Valley

At depth, the weak Amsden Shale was a potential slip surface because it is parallel to the slope.

Rain weakened the Amsden and made the Tensleep heavier. Downslope force caused a mass of rock to start moving.

Time

Scar

Slide debris

Lake

The debris filled the valley, blocking a stream and forming Slide Lake. The scar remained on the hill slope.

Slide scar

Slide debris

Photo of the slide and the lake it trapped

and have closed highways and railroads for weeks. A number of phenomena combine to cause California's landslides and mudflows: (1) California sits astride an active plate boundary, so faulting not only breaks and weakens crustal materials, but also jars them with seismic energy. (2) Wildfires and human-caused deforestation have removed vegetation that could stabilize slopes. (3) Torrential winter rains saturate the regolith with water. (4) Development of cities and suburbs has oversteepened slopes and has affected water content of the substrate.

TAKE-HOME MESSAGE

Weathering and fragmentation weaken slope materials and make them more susceptible to mass movement. Failure occurs when downslope pull exceeds the resistance force. This may happen due to shocks, changing slope angles and strength, and changing slope support.

FIGURE 13.14 Undercutting and collapse of a sea cliff.

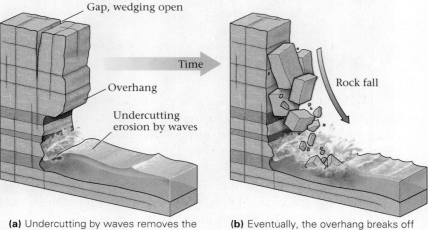

Gap, wedging open

Time

Overhang

Rock fall

Undercutting erosion by waves

(a) Undercutting by waves removes the support beneath an overhang.

(b) Eventually, the overhang breaks off along joints, and a rockfall takes place.

13.4 HOW CAN WE PROTECT AGAINST MASS-MOVEMENT DISASTERS?

Identifying Regions at Risk

Clearly, landslides, mudflows, and slumps are natural hazards we cannot ignore. Too many of us live in areas where mass wasting has the potential to kill people and destroy property. In many cases, the best solution is avoidance: don't build, live, or work in an area where mass movement will take place. But avoidance is possible only if we know where the hazards are.

To pinpoint dangerous regions, geologists look for landforms known to result from mass movements, for where these movements have happened in the past, they might happen again in the future. Features such as slump head scarps, swaths of forest in which trees have been tilted, piles of loose debris at the base of hills, and hummocky land surfaces all indicate recent mass wasting.

Geologists may also be able to detect regions that are beginning to move downslope (Fig. 13.15). For example, roads, buildings, and pipes on top of unstable ground begin to crack. Power lines may be too tight or too loose because the poles to which they are attached move together or apart. Visible cracks form on the ground at the potential head of a slump, while the ground may bulge up at the toe of the slump. Subsurface cracks may drain the water from an area and kill off vegetation, while another area may sink and form a swamp. Slow movements

cause trees to develop pronounced curves at their bases. In some cases, the activity of land masses moving too slowly for people to perceive can be documented with sensitive surveying techniques that can detect a subtle tilt of the ground or changes in distance between nearby points.

Even without evidence of recent movement, a danger may still exist, for just because a steep slope hasn't collapsed in the recent past doesn't mean it won't in the future. In recent years, geologists have begun to identify such potential hazards by using computer programs that evaluate factors that trigger mass wasting. They then produce maps that portray the degree of risk for a certain location. These factors include the following: slope steepness; strength of substrate; degree of water saturation; orientation of bedding, joints, or foliation relative to the slope; nature of vegetation cover; potential for heavy rains; potential for undercutting to occur; and likelihood of earthquakes. From such hazard-assessment studies, geologists compile landslide-potential maps, which rank regions according to the likelihood that a mass movement will occur. In any case, common sense suggests that you should avoid building on or below particularly dangerous slide-prone slopes.

Preventing Mass Movements

In areas where a hazard exists, people can take certain steps to remediate the problem and stabilize the slope (Fig. 13.16a–h).

FIGURE 13.15 Surface features warn that a large slump is beginning to develop. Cracks that appear at the head scarp may drain water and kill trees. Power-line poles tilt and the lines become tight. Fences, roads, and houses on the slump begin to crack.

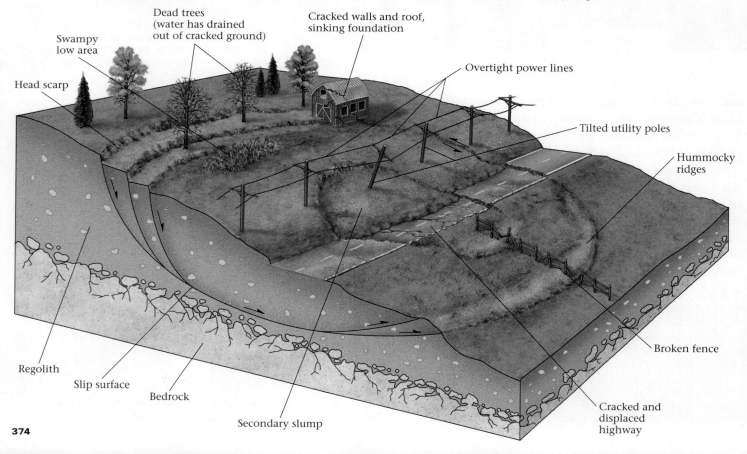

FIGURE 13.16 A variety of remedial steps can stabilize unstable ground.

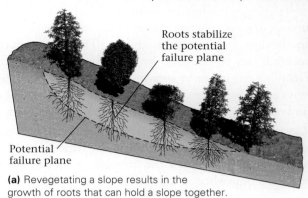

Roots stabilize the potential failure plane

Potential failure plane

(a) Revegetating a slope results in the growth of roots that can hold a slope together.

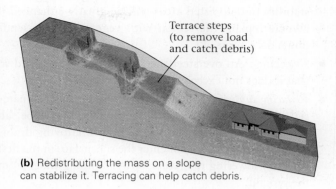

Terrace steps (to remove load and catch debris)

(b) Redistributing the mass on a slope can stabilize it. Terracing can help catch debris.

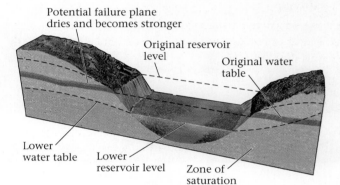

Potential failure plane dries and becomes stronger

Original reservoir level

Original water table

Lower water table   Lower reservoir level   Zone of saturation

(c) Lowering the level of the water table can strengthen a potential failure surface.

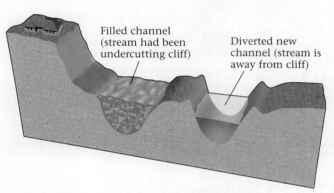

Filled channel (stream had been undercutting cliff)

Diverted new channel (stream is away from cliff)

(d) Relocating a river channel can prevent undercutting.

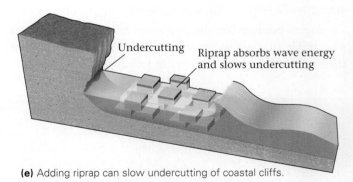

Undercutting   Riprap absorbs wave energy and slows undercutting

(e) Adding riprap can slow undercutting of coastal cliffs.

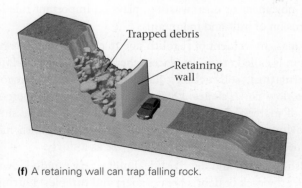

Trapped debris

Retaining wall

(f) A retaining wall can trap falling rock.

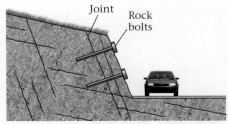

Joint   Rock bolts

(g) Bolting or screening a cliff face can hold loose rocks in place.

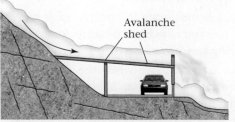

Avalanche shed

(h) An avalanche shed diverts debris or snow over a roadway.

- *Revegetation*: Since bare ground is much more susceptible to downslope movement than vegetated ground, stability in deforested areas will be greatly enhanced if owners replant the region with vegetation that sends down deep roots.

- *Regrading*: An oversteepened slope can be regraded so that it does not exceed the angle of repose.

- *Reducing subsurface water*: Because water weakens material beneath a slope and adds weight to the slope, an unstable situation may be remedied by improving drainage so that water does not enter the subsurface in the first place, by lowering reservoirs, or by pumping water from the ground.

- *Preventing undercutting*: In places where a river undercuts a cliff face, engineers can divert the river. Similarly, along coastal regions they may build an offshore breakwater or pile riprap (loose boulders or concrete) along the beach to absorb wave energy before it strikes the cliff face.

- *Constructing safety structures*: In some cases, the best way to prevent mass wasting is to build a structure that stabilizes a potentially unstable slope or protects a region downslope

from debris if a mass movement does occur. For example, civil engineers can build retaining walls or bolt loose slabs of rock to more coherent masses in the substrate in order to stabilize highway embankments. The danger from rockfalls can be decreased by covering a road cut with chainlink fencing or by spraying it with concrete. Highways at the base of an avalanche chute can be covered by an avalanche shed, whose roof keeps debris off the road.

- *Controlled blasting of unstable slopes*: When it is clear that unstable ground threatens a particular region, the best solution may be to blast the unstable ground or snow loose at a time when its movement can do no harm.

TAKE-HOME MESSAGE

Various features of the landscape, as well as detailed measurements by satellites, may help geologists to identify unstable slopes and send out warnings. Systematic study, in fact, allows production of landslide-potential maps. Engineers may use a variety of techniques to stabilize slopes physically.

Chapter Summary

- Rock or regolith on unstable slopes has the potential to move downslope under the influence of gravity. This process, called mass movement or mass wasting, plays an important role in the erosion of hills and mountains.

- Slow mass movement of regolith due to expansion and contraction is called creep. In places where slopes are underlain by permafrost, solifluction causes a melted layer of regolith to flow down slopes. During slumping, a semicoherent mass of material moves down a spoon-shaped failure surface. Mudflows and debris flows occur where regolith has become saturated with water and moves as a slurry.

- Landslides (rock and debris slides) move very rapidly down a slope; the rock or debris breaks apart and tumbles. During avalanches, debris mixes with air and moves downslope as a turbulent cloud. And in a debris fall or rockfall, the material free-falls down a vertical cliff.

- Intact, fresh rock is too strong to undergo mass movement. Thus, for mass movement to be possible, rock must be weakened by fracturing and/or weathering.

- Unstable slopes start to move when the downslope force exceeds the resistance force that holds material in place. The steepest angle at which a slope of unconsolidated material can remain without collapsing is the angle of repose.

- Downslope movement can be triggered by shocks and vibrations, a change in the steepness of a slope, a change in the strength of a slope, deforestation, weathering, or heavy rain.

- Geologists produce landslide-potential maps to identify areas susceptible to mass movement. Engineers can help prevent mass movements using a variety of techniques.

GEOPUZZLE REVISITED

Gravity constantly applies a downslope force. For a time, the strength of material making up the substrate of a slope may be strong enough to resist this relentless pull. But heavy rains, ground shaking, undercutting, deforestation, and/or a change in the water table depth can destabilize a slope until it finally gives way and a landslide takes place. Roots, retaining walls, and other natural or human-built features may delay a landslide, but in the context of geologic time, gravity always wins. Mass wasting events, such as landslides, contribute to the erosion of uplifted land on islands and continents. They can also take place under the sea.

Key Terms

angle of repose (p. 369)
avalanche (p. 366)
creep (p. 363)
debris fall (p. 367)
debris flow (p. 364)
debris slide (p. 366)
failure surface (p. 364)
head scarp (p. 364)
lahar (p. 364)
landslide (p. 366)

liquefaction (p. 369)
mass movement (p. 362)
mudflow (p. 364)
natural hazard (p. 362)
rockfall (p. 367)
rock slide (p. 366)
slump (p. 364)
solifluction (p. 364)
talus (p. 367)
undercutting (p. 372)

Review Questions

1. What factors distinguish the various types of mass movement?

2. How does a slump differ from creep? How does it differ from a mudflow or debris flow?

3. How does a rock or debris slide differ from a slump? What conditions trigger a snow avalanche?

4. How does a small amount of water between grains help hold material together?

5. What force is responsible for downslope movement? What force helps resist that movement?

6. What evidence suggests that large, submarine slumps have developed?

7. How does the angle of repose change with grain shape? How does it change with water content?

8. What factors trigger downslope movement?

9. How do geologists predict whether an area is susceptible to mass wasting?

10. What steps can people take to reduce the risk of mass wasting?

On Further Thought

1. Imagine that you have been asked by the World Bank to determine whether or not it makes sense to build a dam in a steep-sided, east-west-trending valley in a small central Asian nation. The local government has lobbied for the dam, because the climate of the country has gradually been getting drier, and the farms of the area are running out of water. The World Bank is considering making a loan to finance construction of the dam, a process that would employ thousands of now-jobless people. Initial investigation shows that the rock of the valley floor consists of schist containing a strong foliation that dips south. Outcrop studies reveal that abundant fractures occur in the schist along the valley floor; the surface of most fractures are coated with slickensides. Moderate earthquakes have rattled the region. What would you advise the bank? Explain the hazards and what might happen if the reservoir were filled.

2. Marine geologists have been studying the nature and distribution of submarine slumps around the world in order to understand the tsunami threat that submarine slumping poses. Initial results indicate that submarine slumps occur both along active margins (continental margins that coincide with convergent plate boundaries) and along passive margins. Slumps occur much more frequently, but are smaller, along convergent plate boundaries than along passive margins. Suggest an explanation for this observation. Is there a significant tsunami threat in the Atlantic Ocean?

THE VIEW FROM SPACE   The huge Ganges Chasma landslide, which formed along the Valles Marineris, a canyon wall on Mars, has taken part of a meteorite crater with it. The field of view is about 60 km across.

CHAPTER **14**

Running Water: The Geology of Streams and Floods

Water that falls on land drains back to the sea via rivers. Here, the Niagara River drops dramatically over Niagara Falls, giving a visible display of the power of running water.

GEOPUZZLE

Why do stream networks develop, and why do streams sometimes overflow their channels and flood the surrounding landscape?

As many fresh streams meet in one salt sea, as many lines close in the dial's center, so may a thousand actions, once afoot, end in one purpose.

—William Shakespeare (1564–1616)

14.1 INTRODUCTION

By the 1880s, Johnstown, built along the Conemaugh River in scenic western Pennsylvania, had become a significant industrial town. Recognizing the attraction of the Conemaugh region as a summer retreat from the heat and pollution of nearby Pittsburgh, speculators built a mud and gravel dam across the river, upstream of Johnstown, to fill a pleasant reservoir of cool water. Industrialists and bankers bought the reservoir and established the exclusive South Fork Hunting and Fishing Club. Unfortunately, the dam had been poorly designed, and debris tended to block its spillway (a passageway for surplus water), setting the stage for a monumental tragedy. On May 31, 1889, torrential rain drenched Pennsylvania, and the reservoir surface rose until water flowed over the dam and down its face. Despite frantic attempts to strengthen the dam, the soggy structure abruptly collapsed, and the reservoir emptied into the Conemaugh River Valley. A 20-m-high wall of water roared downstream and slammed into Johnstown, instantly transforming bridges and buildings into twisted wreckage (Fig. 14.1). When the water subsided, 2,300 people were dead, and Johnstown became the focus of national sympathy. It took years for the town to recover, and many residents simply picked up and left.

The unlucky inhabitants of Johnstown experienced one of the more destructive consequences of "running water," meaning water that flows down sloping land surfaces in response to gravity. Running water generally flows in **streams**, ribbons of water confined to **channels**, or troughs, cut into the land. Note that geologists may call any channelized body of flowing water a stream, regardless of its size; in common usage, though, a large stream is referred to as a river. Streams remove, or drain, excess water (runoff) from the landscape and carry it eventually to the sea, just as culverts drain water from parking lots. In the process, streams erode the landscape, transport sediment and nutrients, and leave behind deposits of sediment. They also provide a home for living organisms, and for human society, they supply avenues for commerce, water for agriculture, and sources for power. When the volume of water flowing in a stream exceeds the volume of the channel, a **flood** occurs and water covers an area that is normally dry land—that's what happened in Johnstown.

Earth is the only planet in the Solar System that currently has streams of running water. (Mars may have had them in the past; see Interlude F.) In this chapter, we see how streams operate in the Earth System. First we learn about the origin of running water and the architecture of streams. Then we look at the processes of stream erosion and deposition, and at the landscapes that form in response to these processes. Finally, we consider the nature and consequences of flooding.

TAKE-HOME MESSAGE

By the end of this chapter, you should understand how drainage networks form and evolve, how streams erode the land in some places and deposit sediment elsewhere, how floods originate, and how flood hazards can be avoided.

14.2 DRAINING THE LAND

Forming Streams and Drainage Networks

Where does the water in a stream come from? During the hydrologic cycle (Interlude F), atmospheric water eventually condenses and falls back to the Earth's surface as rain or snow that accumulates in various reservoirs. Some rain or snow remains on the land as surface water (in puddles, swamps, lakes, snowfields, and glaciers), some flows downslope as a thin film called **sheetwash**, and some sinks into the ground where it either becomes trapped in soil (as soil moisture) or descends below the water table to become groundwater. Streams can receive input of water from all of these reservoirs (Fig. 14.2). Specifically, gravity pulls surface water (including meltwater) downhill into stream channels, and the pressure exerted by the weight of new rainfall squeezes existing soil moisture back out of the ground, and groundwater seeps out of the channel walls.

Running water collects in stream channels, because a channel is lower than the surrounding area and gravity always moves material from higher to lower elevation. But how does a stream channel form in the first place? The process of channel formation begins when sheetwash starts flowing downslope. Like any flowing fluid, sheetwash erodes its substrate (the material it flows over). The efficiency of such erosion depends on the velocity of the flow—faster flows erode more effectively. In nature, the ground is not perfectly planar, not all substrate

FIGURE 14.1 During the disastrous 1889 flood in Johnstown, Pennsylvania, floodwaters tumbled large houses.

FIGURE 14.2 Excess surface water comes from rain, melting ice or snow, and groundwater springs. On flat ground, water accumulates in puddles or swamps, but on slopes, it flows downslope in streams.

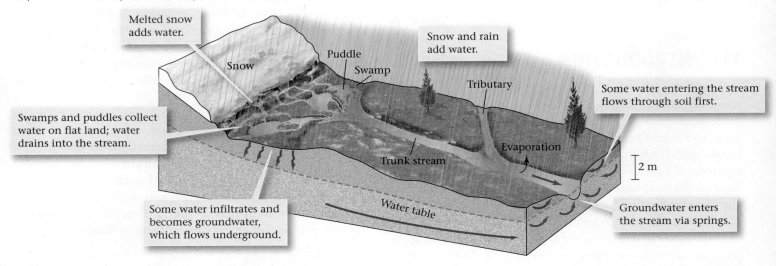

has the same resistance to erosion, and the amount of vegetation that covers and protects the ground varies with location. Thus, the velocity of sheetwash also varies with location. Where the flow happens to be a bit faster, or the substrate a little weaker, erosion scours (digs) a channel. Since this channel is lower than the surrounding ground, sheetwash in adjacent areas starts to head toward it. With time, the extra flow deepens the channel relative to its surroundings, a process called **downcutting**, and a stream forms.

As its flow increases, a stream channel begins to lengthen at its origin, a process called **headward erosion** (Fig. 14.3). Headward erosion occurs because the flow is more intense at the entry to the channel (upslope) than in the surrounding sheetwashed areas. As the main channel lengthens, new side channels form nearby. These side channels, or **tributaries**, merge with the main channel, because once a channel forms, the surrounding land slopes into it. An array of linked streams evolves, with the smaller tributaries flowing into a single or trunk stream. The array of interconnecting streams together constitute the **drainage network**.

Like transportation networks, drainage networks reach into all corners of a region to provide conduits for the removal of water. The configuration of tributaries and trunk streams defines the map pattern of a drainage network. This pattern depends on the shape of the landscape and the composition of the substrate. Geologists recognize several types of networks on the basis of their map pattern (Fig. 14.4; and see **Geotour 14** on p. GT-30):

- *Dendritic*: When rivers flow over a fairly uniform substrate with a fairly uniform initial slope, they develop a dendritic network, which looks like the pattern of branches connecting to the trunk of a deciduous tree.

- *Radial*: Drainage networks forming on the surface of a cone-shaped mountain flow outward from the mountain peak, like spokes on a wheel. Such a pattern defines a radial network.

- *Rectangular*: In places where a rectangular grid of fractures (vertical joints) breaks up the ground, channels form along the preexisting fractures, and streams join each other at right angles, creating a rectangular network.

- *Trellis*: In places where a drainage network develops across a landscape of parallel valleys and ridges, major tributaries flow down a valley and join a trunk stream that cuts across the ridges. The resulting map pattern resembles a garden trellis, so the arrangement of streams constitutes a trellis network.

- *Parallel*: On a uniform slope, several streams with parallel courses develop simultaneously.

FIGURE 14.3 Headward erosion in Canyonlands National Park, Utah.

FIGURE 14.4 Different types of drainage networks.

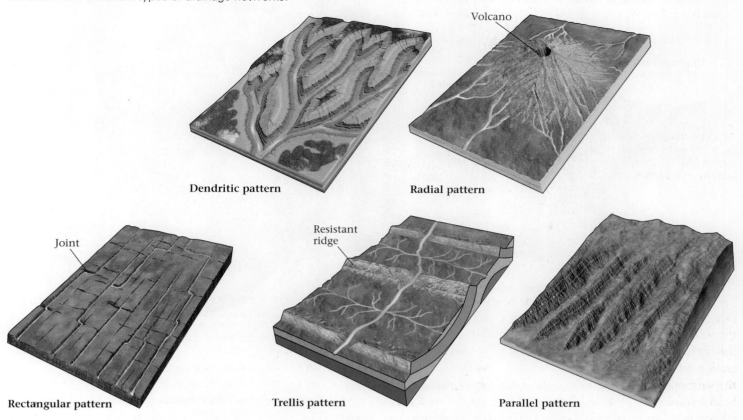

Dendritic pattern

Radial pattern

Volcano

Joint

Rectangular pattern

Resistant ridge

Trellis pattern

Parallel pattern

Drainage Basins and Divides

A drainage network collects water from a broad region, variously called a **drainage basin**, catchment, or watershed, and feeds it into the trunk stream, which carries the water away. The highland, or ridge, that separates one watershed from another is a **drainage divide** (Fig. 14.5a). A continental divide separates drainage that flows into one ocean from drainage that flows into another. For example, if you straddle the North American continental divide and pour a cup of water out of each hand, the water in one cup flows to the Atlantic, and the water in the other flows to the Pacific. Three divides bound the Mississippi drainage basin, which drains the interior of the United States (Fig. 14.5b).

Streams That Last, Streams That Don't: Permanent and Ephemeral Streams

Permanent streams flow all year long, whereas **ephemeral streams** flow only for part of the year—in fact, some ephemeral streams flow only for a brief time after a heavy rain. Most permanent streams exist where the floor (or bed) of the stream channel lies *below* the water table, the top surface of groundwater (Fig. 14.6a; see Chapter 16). In these streams, which occur in humid or temperate climates, water comes not only from

upstream or from surface runoff, but also from springs through which groundwater seeps. If the bed of a stream lies *above* the water table, then the stream can be permanent only when the rate at which water arrives from upstream exceeds the rate at which water infiltrates into the ground below. For example, the downstream portion of the Colorado River in the dry Sonoran Desert of Arizona flows all year, because enough water enters it from the river's wet headwaters upstream in Colorado. Streams that do not have a sufficient upstream source, and whose beds lie above the water table, are ephemeral, because the water that fills a channel due to a heavy rain or a spring thaw eventually sinks into the ground and/or evaporates, and the stream dries up (Fig. 14.6b). Streams whose watersheds lie entirely within an arid region tend to be ephemeral. The dry bed of an ephemeral stream is variously called a dry wash, an arroyo, or a wadi.

TAKE-HOME MESSAGE

Water in streams comes from standing bodies spilling through outlets, sheetwash on the surface, and groundwater. Stream channels form by downcutting and lengthen by headward erosion. Eventually, networks of tributaries flow into a trunk stream and drain the land.

FIGURE 14.5 Drainage divides and basins.

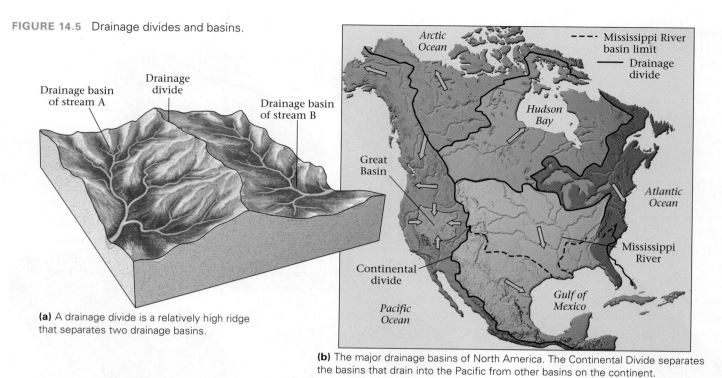

(a) A drainage divide is a relatively high ridge that separates two drainage basins.

(b) The major drainage basins of North America. The Continental Divide separates the basins that drain into the Pacific from other basins on the continent.

FIGURE 14.6 The contrast between permanent and ephemeral streams.

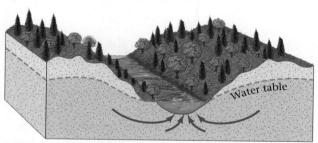

(a) The floor of a permanent stream in a temperate climate lies below the water table. Springs add water from below, so the stream contains water even between rains.

An empty stream channel is called a dry wash.

(b) The channel of an ephemeral stream lies above the water table, so the stream flows only when water enters the stream faster than it can infiltrate into the ground.

14.3 DESCRIBING FLOW IN STREAMS: DISCHARGE AND TURBULENCE

Imagine two streams—a larger one in which water flows slowly, and a smaller one in which water flows rapidly. Which stream carries more water? The answer is not obvious. To answer this question completely, geologists must calculate the streams' discharge. Technically speaking, we define **discharge** as the volume of water passing through an imaginary cross section of the stream in a unit of time. We can calculate stream discharge by using a simple formula: $D = A_c \times v_a$. In this formula, A_c is the area of the stream, as measured in an imaginary plane perpendicular to the stream flow, and v_a is the *average* velocity

with which water moves in the downstream direction. Stream discharge can be determined at a stream-gauging station, where instruments measure the velocity and depth of the water at several points across the stream (Fig. 14.7a).

A stream's average discharge reflects the size of its drainage basin and the climate. The Amazon River drains a huge rainforest and has the largest average discharge in the world—about 200,000 m³/s, or 15% of the total amount of runoff on Earth. The "mighty" Mississippi's is only 17,000 m³/s. The discharge of a given stream varies along its length. For example, the discharge in a temperate region *increases* in the downstream direction, because each tributary that enters the stream adds more water, whereas the discharge in an arid region may *decrease* downstream, as progressively more water seeps into the ground or evaporates. Human activity can affect discharge. If people divert the river's water for irrigation, the river's discharge decreases downstream. Finally, the discharge at a given location can vary with time: in a temperate climate, a stream's discharge during the spring may be double or triple the amount during a hot summer, and a flood may increase the discharge to more than a hundred times normal.

The average velocity of stream water (v_a) can be difficult to calculate because the water doesn't all travel at the same velocity for two reasons. First, friction along the sides and floor of the stream slows the flow. Thus, water near the channel walls or the stream bed (the floor of the stream) moves more slowly than water in the middle of the flow (Fig. 14.7b). In fact, the fastest-moving part of the stream flow lies near the surface in the center of the channel. Second, turbulence—the twisting, swirling motion of a fluid—can create eddies (whirlpools) in which water curves and flows upstream or circles in place (Fig. 14.7c). Turbulence develops because the shearing motion of one water volume against its neighbor causes the neighbor to spin, and because obstacles, such as boulders, deflect water volumes.

TAKE-HOME MESSAGE

Stream discharge indicates the amount of water passing through a cross section of the stream in a given time. Discharge depends on factors such as drainage area and climate. Water in streams tends to be turbulent, complicating calculation of average velocity.

14.4 THE WORK OF RUNNING WATER

How Do Streams Erode?

The energy that makes running water move comes from gravity. As water flows downslope from a higher to a lower elevation, the gravitational potential energy stored in water transforms into kinetic energy. About 3% of this energy goes into the work of eroding the walls and beds of stream channels. Running water causes erosion in four ways:

- *Scouring*: Running water removes loose fragments of sediment, a process called scouring.
- *Breaking and lifting*: The push of flowing water can break chunks of solid rock off the channel floor or walls. In addition, the flow of a current over a clast can cause the clast to rise, or lift off the substrate.
- *Abrasion*: Clean water has little erosive effect, but sand- or gravel-laden water acts like sandpaper and grinds or rasps away at the channel floor and walls, a process called abrasion. In places where turbulence creates long-lived

FIGURE 14.7 Flow velocity and turbulence in streams.

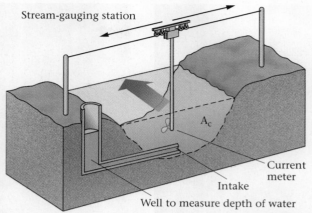

(a) At a stream-gauging station, geologists measure the area (A_c) and depth of the stream. They measure velocity using a current meter at various points in the stream.

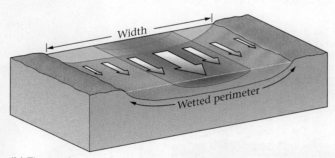

(b) The maximum velocity occurs in the center of a straight channel. Friction with the wetted perimeter (the area where water touches the channel walls) slows the water.

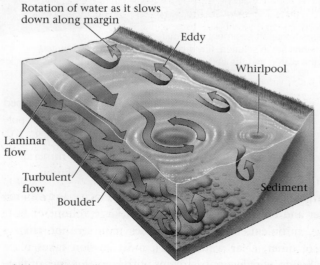

(c) In a turbulent flow, the water swirls in curving paths and becomes caught in eddies.

FIGURE 14.8 Erosion and transportation in streams.

(a) Two potholes in the bed of a stream near Ithaca, New York.

During saltation, a clast in the flow falls and collides with one on the bed, which bounces up into the water like a billiard ball.

(b) This slot canyon in Arizona formed when many potholes linked together.

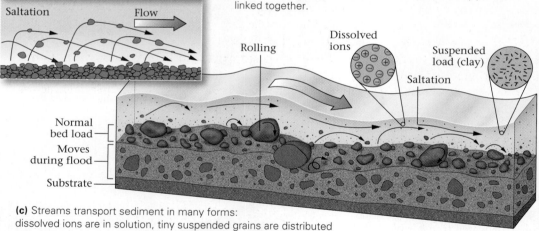

(c) Streams transport sediment in many forms: dissolved ions are in solution, tiny suspended grains are distributed through the water, and the bed load slides, rolls, and/or undergoes saltation.

How Do Streams Transport Sediment?

The Mississippi River received the nickname "Big Muddy" for a reason—its water can become chocolate brown, because of all the clay and silt it carries. All streams carry sediment, though not in the same quantities. Geologists refer to the total volume of sediment carried by a stream as its sediment load. The sediment load consists of three components (Fig. 14.8c):

- *Dissolved load*: Running water dissolves soluble minerals from the sediment or rock of its substrate, and groundwater seeping into a stream through the channel walls brings dissolved minerals with it. These ions constitute a stream's dissolved load.

- *Suspended load*: The suspended load of a stream consists of tiny solid grains (silt or clay size) that swirl along with the water without settling to the floor of the channel.

- *Bed load*: The bed load of a stream consists of larger particles (such as sand, pebbles, or cobbles) that bounce or roll along the stream floor. Typically, bed-load movement involves saltation, a process during which grains on the channel floor get knocked into the water column momentarily, follow a curved trajectory downstream, and gradually sink to the bed again, where they strike other grains and knock them like billiard balls into the water column.

whirlpools, abrasion by sand or gravel carves a bowl-shaped depression, called a pothole, into the floor of the stream; potholes may link together to form a channel (Fig. 14.8a, b).

- *Dissolution*: Running water dissolves soluble minerals as it passes, and carries the minerals away in solution.

The efficiency of erosion depends on the velocity and volume of water and on its sediment content. A large volume of fast-moving, turbulent, sandy water causes more erosion than a trickle of quiet, clear water. Thus, most erosion takes place during floods, which supply streams with large volumes of fast-moving, sediment-laden water.

When describing a stream's ability to carry sediment, geologists specify its competence and capacity. The **competence** of a stream refers to the maximum particle size it carries; a stream with high competence can carry large particles, whereas one with low competence can carry only small particles. Therefore, a fast-moving, turbulent stream has greater competence (it can carry bigger particles) than a slow-moving stream, and a stream in flood has greater competence than a stream with normal flow. In fact, the huge boulders that litter the bed of a mountain creek move only during floods. The **capacity** of a stream refers to the total quantity of sediment it can carry. A stream's capacity depends on both its competence and its discharge.

Depositional Processes: When Streams Lose Their Loads

A raging torrent of water can carry coarse and fine sediment—the finer clasts rush along with the water as suspended load, whereas the coarser clasts bounce, roll, and tumble as bed load. If the flow velocity decreases, either because the gradient (downstream slope) of the stream bed becomes shallower or because the channel broadens out and friction between the bed and the water increases, then the competence of the stream decreases and sediment settles out. The size of the clasts that settle at a particular locality depends on the decrease in flow velocity at the locality. For example, if the stream slows by a small amount, only large clasts settle; if the stream slows by a greater amount, medium-sized clasts settle; and if the stream slows almost to a standstill, the fine grains settle. Thus, coarser sediment tends to settle out farther upstream, where the gradient of the stream is steeper and water flows faster, whereas finer grains settle out farther downstream, where the water flows more slowly. Because of this process of sediment sorting, stream deposits tend to be segregated by size—gravel accumulates in one location, and mud in another. Geologists refer to sediments transported by a stream as fluvial deposits (from the Latin *fluvius*, meaning river) or **alluvium**. Fluvial deposits may accumulate along the stream bed in elongate mounds called **bars** (Fig. 14.9a, b). Some stream channels make broad

curves—water slows along the inner edge of a curve, so crescent-shaped point bars bordering the shoreline develop. During floods, a stream may overtop the banks of its channel and spread out over its **floodplain**, a broad flat area bordering the stream. Friction slows the water on the floodplain, so a sheet of silt and mud settles out. Where a stream empties, at its mouth, into a standing body of water, the water slows and a wedge of sediment called a **delta**, accumulates.

TAKE-HOME MESSAGE

Streams erode into the substrate by scouring, breaking and lifting, abrasion, and dissolution. They carry sediment as dissolved, suspended, or bed loads. Competence, the ability to carry sediment, depends on velocity. Where the velocity of flow decreases, sediment settles out.

14.5 HOW DO STREAMS CHANGE ALONG THEIR LENGTH?

Longitudinal Profiles

In 1803, the United States under President Thomas Jefferson's leadership bought the Louisiana Territory, a vast tract of land encompassing the western half of the Mississippi drainage basin. At the time, the geography of the territory was a mystery. To fill the blank on the map, Jefferson asked Meriwether Lewis and William Clark to lead a voyage of exploration across the Louisiana Territory to the Pacific.

FIGURE 14.9 Sediment deposits in and along streams. The velocity of flow affects the size of clasts carried.

(a) Gravel in a stream bed of a mountain stream in Denali National Park, Alaska. The large clasts were carried during floods.

(b) Point bars of mud deposited along a gentle, slowly moving stream in Brazil.

Lewis and Clark, along with about forty men, began their expedition at the mouth of the Missouri River where it joins the Mississippi. At this juncture, the Missouri is a wide, languid stream of muddy water. The group found the lower reaches (segments) of the Missouri, where the channel is deep and the water smooth, to be easy going. But the farther upstream they went, the more difficult their voyage became, for the stream gradient became progressively steeper, and its discharge became less. When Lewis and Clark reached the site of what is now Bismarck, North Dakota, they had to abandon their original boats and haul smaller vessels up rough, rapidly flowing water and around waterfalls, where water drops over an escarpment. When they reached what is now southwestern Montana, they deserted these boats as well and followed stream valleys on foot or by horseback, struggling up steep gradients until they reached the continental divide.

If Lewis and Clark had been able to plot a graph showing their elevation above sea level relative to their distance along the Missouri, they would have found that the **longitudinal profile** of the Missouri, a cross-sectional image showing the variation in the river's elevation along its length, is roughly a concave-up curve (Fig. 14.10a, b). This curve illustrates that

a stream's gradient is steeper near its headwaters (source) than near its mouth. Near its headwaters, an idealized stream flows down deep valleys or canyons; near its mouth, it flows over nearly horizontal plains. Real longitudinal profiles are not perfectly smooth curves, but rather include little plateaus and steps, representing interruptions by lakes or waterfalls.

The Base Level

Streams progressively deepen their channels by downcutting, but there is a depth below which a stream cannot downcut. The lowest elevation a stream channel's floor can reach at a given locality is the **base level** of the stream. A local base level occurs upstream of a drainage network's mouth, whereas the ultimate base level (that is, the lowest elevation along the drainage network's longitudinal profile) is determined by sea level. A trunk stream cannot downcut so its surface is deeper than sea level, for if it did, it would have to flow upslope to enter the sea.

Lakes or reservoirs can act as local base levels along a stream, for where the stream enters such standing bodies of water, it almost slows to a halt and cannot downcut further. A ledge of resistant rock can also act as a local base level, for

FIGURE 14.10 Drainage basins, and the change in character of a stream along its longitudinal profile.

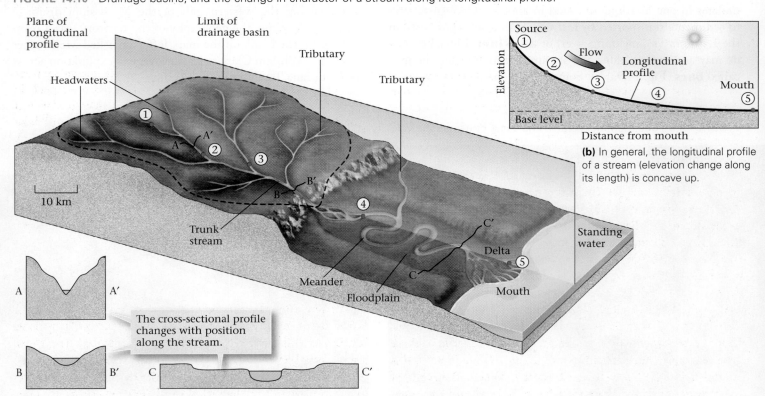

(b) In general, the longitudinal profile of a stream (elevation change along its length) is concave up.

The cross-sectional profile changes with position along the stream.

(a) A drainage network collects water from a broad drainage basin, or watershed, via numerous tributaries. These carry water to a trunk stream and eventually to a standing body of water. Points 1 to 5 refer to locations along the longitudinal profile (inset).

the stream level cannot drop below the ledge until the ledge erodes away. Finally, where a tributary joins a larger stream, the channel of the larger stream acts as the base level for the tributary. Local base levels do not last forever, because eventually erosion removes the obstruction.

TAKE-HOME MESSAGE

Streams have steeper regional gradients toward their sources, and gentler gradients near their mouths, so longitudinal profiles tend to be concave up. A stream's mouth cannot be lower than its base level. Sea level is the ultimate base level for drainage networks.

14.6 STREAMS AND THEIR DEPOSITS IN THE LANDSCAPE

Valleys and Canyons

About 17 million years ago, a large block of crust, the region now known as the Colorado Plateau (located in what is now Arizona, Utah, Colorado, and New Mexico), began to rise. Before the rise, rivers in the area had been flowing over a plain not far above sea level, causing little erosion. But as the land uplifted, the rivers began to downcut steadily. Over time, river erosion produced deep canyons. Today, the deepest of these is known as the Grand Canyon. In places, the Colorado River at the floor of the canyon lies 1.6 km below the surface of the plateau. The formation of the Grand Canyon illustrates a general phenomenon. In regions where the land surface lies well above the base level, a stream can create a deep trough, much deeper than the channel itself, in the land. If the walls of the trough slope gently, the landform is a valley, but if they slope steeply, the landform is a canyon.

Whether stream erosion produces a valley or a canyon depends on the rate at which downcutting occurs relative to the rate at which mass wasting causes the walls on either side of the stream to collapse. In places where a stream erodes down through its substrate faster than the walls of the stream collapse, a steep-walled slot canyon develops. Such canyons typically form in hard rock, which can hold up steep cliffs for a long time (Fig. 14.11a). In places where the walls collapse as fast as the stream downcuts, landslides and slumps gradually cause the slope of the walls to approach the angle of repose. When this happens, the stream channel lies at the floor of a valley whose cross-sectional shape resembles the letter V (Fig. 14.11b); this landform is called a **V-shaped valley**. Where the walls of the stream consist of alternating layers of hard and soft rock, the walls develop a stair-step shape such as that of the Grand Canyon (Fig. 14.11c).

FIGURE 14.11 The shape of a canyon or valley depends on the resistance of its walls to erosion slumping.

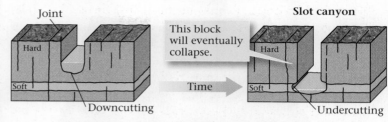

(a) If downcutting by the stream happens faster than mass wasting on the walls, a slot canyon forms. The canyon widens as the stream undercuts the walls.

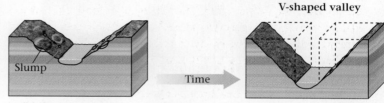

(b) If mass wasting takes place as fast as downcutting occurs, a V-shaped valley develops.

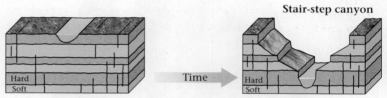

(c) Downcutting through alternating hard and soft layers produces a stair-step canyon.

In places where active downcutting occurs, the valley floor remains relatively clear of sediment, for the stream—especially when it floods—carries away sediment that has fallen or slumped into the channel from the stream walls. But if the stream's base level rises or its discharge decreases, the valley floor fills with alluvium, producing an alluvium-filled valley (Fig. 14.12a). The surface of the deposit becomes a broad floodplain. If the stream's base level then drops again and/or the discharge increases, the stream will start to cut down into its own alluvium, a process that generates **stream terraces** bordering the present floodplain (Fig. 14.12b).

Rapids and Waterfalls

Recall that when Lewis and Clark struggled up the Missouri River, they came to reaches that could not be navigated by boat. Their path was blocked by **rapids**, particularly turbulent water with a rough surface (Fig. 14.13a). Rapids form where water flows over steps or large clasts in the channel floor. Rapids also form where the channel abruptly narrows or its gradient changes, suddenly accelerating the water. The turbulence in rapids can create eddies, waves, and whirlpools that roil and

FIGURE 14.12 The evolution of alluvium-filled stream valleys, and the development of terraces.

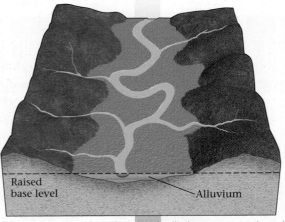

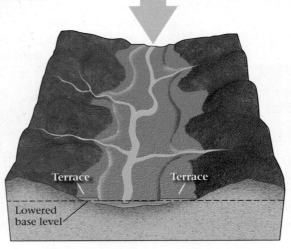

(a) A rise in the base level or a decrease in discharge causes the valley to fill with alluvium.

(b) Later, if the base level falls or the discharge increases, the stream downcuts through the alluvium and a new, lower floodplain develops. The remnants of the original alluvial plain remain as a pair of terraces.

FIGURE 14.13 Examples of rapids and waterfalls.

(a) These rapids in the Grand Canyon formed when a flood from a side canyon dumped debris into the channel of the Colorado River.

(b) Iguaçu Falls, at the Brazil-Argentina border, spills across a ledge of basalt.

churn the water into whitewater, a mixture of bubbles and water. Modern-day whitewater boaters and rafters thrill to the unpredictable movement of a boat caught in rapids.

A **waterfall** forms where the gradient of a stream becomes so steep that the water free-falls down the stream bed (Fig. 14.13b). The energy of falling water may scour a depression, called a plunge pool, at the base of the waterfall. Though a waterfall may appear to be a permanent entity of the landscape, all waterfalls eventually disappear as headward erosion slowly eats back the resistant ledge. We can see a classic example of headward erosion at Niagara Falls. As water flows from Lake Erie to Lake Ontario, it drops over a 55-m-high ledge of hard Silurian dolostone, which overlies a weak shale (Fig. 14.14a, b). Erosion of the shale undercuts the dolostone. Gradually, the overhang of dolostone becomes unstable and collapses, with the result that the waterfall migrates upstream. Before the industrial age, the edge of Niagara Falls cut upstream at an average rate of 1 m per year, but since then, the diversion of water from the Niagara River into a hydroelectric power station has decreased the rate of headward erosion to half that.

Alluvial Fans and Braided Streams

Where a fast-moving stream abruptly emerges from a mountain canyon into an open plain at the range front, the water that was once confined to a narrow channel can spread out over a broad surface. As a consequence, the water slows and abruptly drops its sedimentary load, and a gently sloping apron of sediment (sand, gravel, and cobbles), called an **alluvial fan**, accumulates (Fig. 14.15a). The stream then divides into an array of small channels that spread out over the fan.

In some localities, streams carry abundant coarse sediment during floods but cannot carry this sediment during normal flow. Thus, during normal flow, the sediment settles out and the channel becomes choked with sediment. As a consequence, the stream divides into numerous strands weaving back and forth between elongate mounds or bars of gravel and sand. The result is a **braided stream**—the name emphasizes that the streams intertwine like strands of hair in a braid (Fig. 14.15b).

Meandering Streams and Their Floodplains

A riverboat cruising along the lower reach of the Mississippi River cannot sail in a straight line, for the river channel winds back and forth in a series of snake-like curves called **meanders** (Fig. 14.16a; and see **Geotour 14**). In fact, the boat has to go

FIGURE 14.14 The formation of Niagara Falls, at the border between Ontario, Canada, and New York State. The falls tumble over the Lockport Dolomite, a relatively strong rock layer.

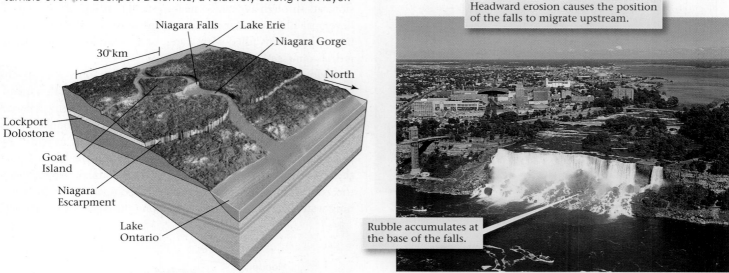

(a) Niagara Falls formed where the outlet of Lake Erie flowed over the Niagara escarpment.

(b) The American Falls, a part of Niagara Falls.

FIGURE 14.15 Examples of depositional landforms produced from stream sediment.

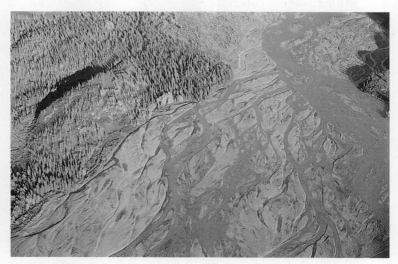

(a) An alluvial fan in Death Valley, California consists of sand, gravel, and debris flows. The curving black line is a road.

(b) A braided stream, carrying meltwater from a glacier, near Denali, Alaska, deposits elongate bars of gravel.

500 km along the river channel to travel 100 km as the crow flies. Meandering streams such as the lower Mississippi form where running water travels over a broad floodplain underlain with a soft substrate, in a region where the river has a very gentle gradient. The development of meanders increases the volume of the stream by increasing its length.

How do meanders form and evolve? Even if a stream starts out with a straight channel, natural variations in the water depth

FIGURE 14.16 The character and evolution of meandering streams.

(a) A meandering stream wanders across a floodplain.

cause the fastest-moving current to swing back and forth. The water erodes the side of the stream more effectively where it flows faster, so it begins to cut away faster on the outer arc of the curve. Thus, each curve begins to migrate sideways and grow more pronounced until it becomes a meander (**Fig. 14.16b**). On the outside edge of a meander, erosion continues to eat away at the channel wall, creating a cut bank. On the inside edge, water slows down so that its competence decreases and sediment accumulates in a wedge-shaped deposit called a **point bar**. With continued erosion, a meander may curve through more than 180°, so that the cut bank at the meander's entrance approaches the cut bank at its end, leaving a meander neck, a narrow isthmus of land separating the portions of the meander. When erosion eats through a meander neck, a straight reach called a cutoff develops. The meander that has been cut off is called an oxbow lake if it remains filled with water, or an abandoned meander if it dries out (see Fig. 14.16a).

Most meandering stream channels cover only a relatively small portion of a broad **floodplain**. In many cases, a floodplain terminates at its sides along a bluff, or escarpment. During a flood, water spills out from the stream channel onto the floodplain, and large floods may cover the entire region from bluff to bluff. As the water leaves the channel, friction between the ground and the thin sheet of water moving over the floodplain slows down the flow. This slowdown decreases the competence of the running water, so sediment settles out along the edge of the channel.

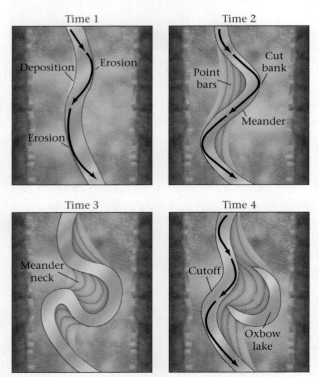

(b) Meanders evolve because erosion occurs faster on the outer bank of a curve, and deposition takes place on the inner curve. Eventually, a cutoff isolates an oxbow lake.

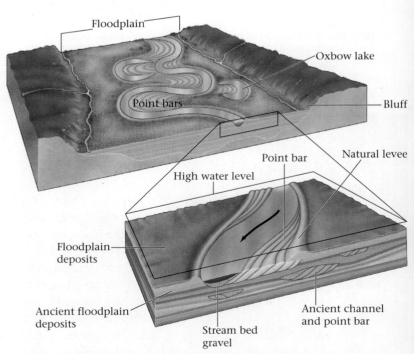

(c) Landforms along meandering streams include natural levees, point bars, and floodplains. Older deposits record the position of ancient channels and floodplains.

FIGURE 14.17 Examples of different delta shapes.

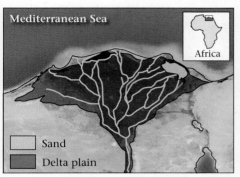

(a) The Nile is a Δ-shaped delta.

(b) The Niger is an arc-like delta.

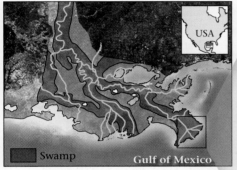

(c) The Mississippi is a bird's-foot delta.

Sediment accumulates at the outlet of the Mississippi River.

(Fig. 14.17a–c). The existence of several toes indicates that the main course of the river in the delta has shifted on several occasions. These shifts occur when a delta builds so far out into the sea that the slope of the stream becomes too gentle to allow the river to flow. At this point, the river overflows a natural levee upstream and begins to flow in a new direction, an event called an avulsion. The distinct lobes of the Mississippi Delta, a bird's-foot delta, suggest that avulsions have happened several times during the past 9,000 years (Fig. 14.18). New Orleans, built along one of the Mississippi's distributaries, may eventually lose its riverfront, for a break in a levee upstream of the city could divert the Mississippi into the Atchafalaya River channel.

The shape of a delta depends on many factors. Deltas that form where the strength of the river current exceeds that of ocean currents have a bird's-foot shape, since the sediment can be carried far offshore. In contrast, deltas that form where the ocean currents are strong have a Δ shape, for the ocean currents redistribute sediment in bars running parallel to the shore. And in places where waves and currents are strong enough to remove sediment as fast as it arrives, a river has no delta at all.

Over time, the accumulation of this sediment creates a pair of low ridges, called **natural levees**, on either side of the stream. Natural levees may grow so large that the floor of the channel may become higher than the surface of the floodplain (Fig. 14.16c).

Deltas: Deposition at the Mouth of a Stream

At the mouth of the Nile in North Africa, the river divides into a fan of small streams called **distributaries**, and the stripe of fertile alluvium broadens into a triangular patch. The Greek historian Herodotus noted that this triangular patch resembles the shape of the Greek letter delta (Δ), and so the region became known as the Nile Delta.

Deltas develop where the running water of a stream enters standing water, the current slows, the stream loses competence, and sediment settles out. Geologists refer to any wedge of sediment formed at a river mouth as a delta, even though relatively few have the triangular shape of the Nile Delta. Some deltas define arc-like lobes, whereas others consist of many elongate lobes that protrude into the sea; the latter are called bird's-foot deltas, because they resemble the scrawny toes of a bird

With time, the sediment of a delta compacts, and the lithosphere beneath the delta subsides. As a consequence, the surface of a delta slowly sinks. In a natural delta, distributaries provide sediment that fills the resulting space so that the delta's surface remains at or just above sea level. But if people build artificial levees to constrain the river to its channel, sediment gets carried directly to the seaward edge of the delta and the delta's interior "starves" (does not receive sediment). When this happens, the land surface drops below sea level. Because of this process, much of New Orleans lies below sea level, so high waters from Hurricane Katrina in 2005 caused the city to flood extensively (see Chapter 15).

TAKE-HOME MESSAGE

Erosion carves valleys and canyons, with shapes that depend on the balance between slope-collapse and downcutting rates. Streams choked with sediment become braided; those following snake-like paths are meandering; those emptying into standing water build deltas.

FIGURE 14.18   A map showing ancient lobes of the Mississippi Delta. A major flood could divert water from the Mississippi into the channel of the Atchafalaya.

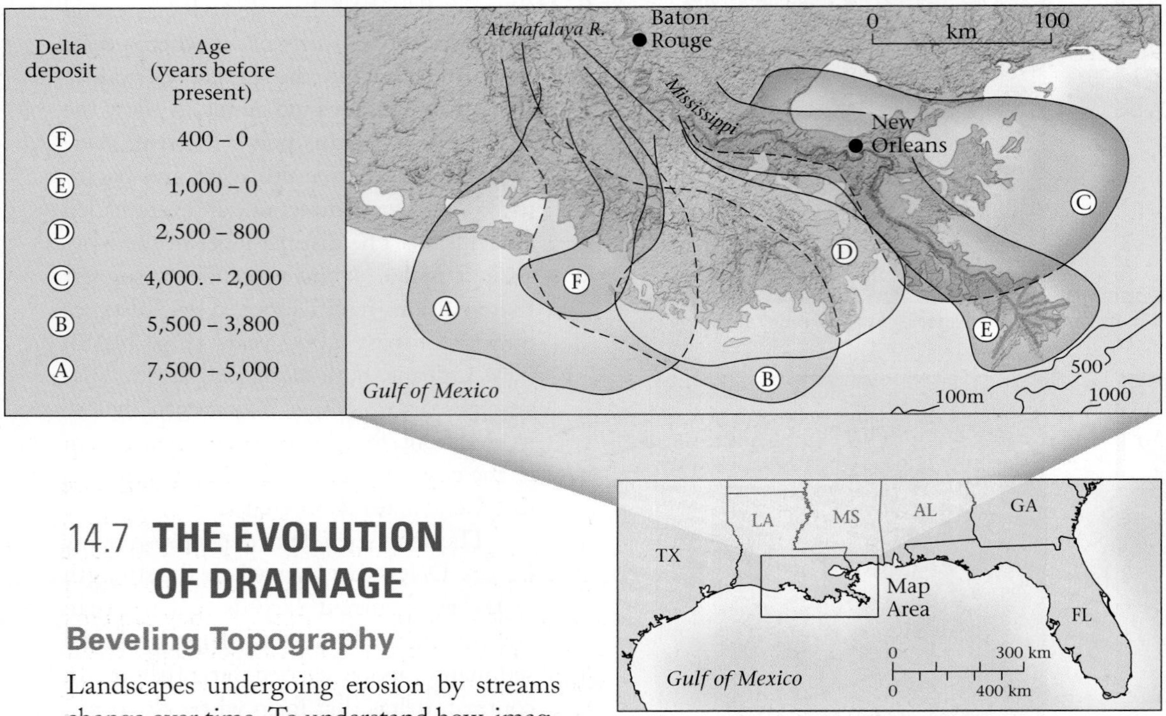

Delta deposit	Age (years before present)
F	400 – 0
E	1,000 – 0
D	2,500 – 800
C	4,000. – 2,000
B	5,500 – 3,800
A	7,500 – 5,000

14.7 THE EVOLUTION OF DRAINAGE

Beveling Topography

Landscapes undergoing erosion by streams change over time. To understand how, imagine a region where uplift of the land surface has just formed a mountain range. Running water soon cuts stream channels, and a drainage network that transports water from the mountain range to the sea develops. At first, the streams in the mountain range have steep gradients, drop over numerous rapids and waterfalls, and flow in deep valleys. As time passes, however, slumps and landslides move debris down mountainsides and into stream channels, and floods carry the debris away. All the while, downcutting continues to lower the channel floor. Overall, erosion and transport transforms the rugged mountains into low, rounded hills. As this happens, once-narrow valleys broaden into wide floodplains with gentle gradients. As more time passes, even the hills erode away, and the landscape evolves almost into a new plain at an elevation close to the streams' base level (Fig. 14.19). Geologists sometimes refer to such plains as peneplains (*pene* is Latin for almost).

At any stage during the evolution of an uplifted landscape into a low-lying plain, the base level of the streams may become relatively lower. This can happen if renewed mountain building uplifts the land over which the streams flow, or if global sea level drops. When the base level drops, the streams start to downcut again to form new canyons or valleys. Geologists refer to the process of renewed downcutting by a stream as **stream rejuvenation**. In cases where a stream had a meandering course before rejuvenation, renewed downcutting may preserve the shapes of meanders, so that the new canyon formed by rejuvenation has a meandering course (Fig. 14.20). The goose-

necks of the San Juan River in southern Utah illustrate this phenomenon.

Stream Piracy

Stream piracy sounds like pretty violent stuff. In reality, **stream piracy**, or stream capture, simply refers to a situation in which headward erosion causes one stream to intersect the course of another stream. When this happens, the pirate stream "captures" the water in the stream it intersects, and the water of the captured stream starts flowing into the pirate stream (Fig. 14.21a, b).

Superposed and Antecedent Streams

In some locations, the structure and topography of the landscape does *not* appear to control the path or course of a stream. As an example, imagine a stream that carves a deep canyon straight across a strong mountain ridge—why didn't the stream find a way around the ridge? We distinguish two types of streams that cut across resistant topographic highs:

- *Superposed streams*: Imagine a region in which drainage initially forms on a layer of flat strata that unconformably overlies folded strata. Streams carve channels into the flat strata; when they eventually erode down through the unconformity and start to downcut into the folded strata, they maintain their earlier courses, ignoring the structure of the folded strata and cutting across resistant ridges. Geologists call such streams **superposed streams**, because their preexisting geometry has been laid down on the rock structure below (Fig. 14.22a, b).

- *Antecedent streams*: In some cases, tectonic activity (such as subduction or collision) causes a mountain range to rise up beneath an already established stream. If the stream downcuts as fast as the range rises, it can maintain its course and will cut right across the range. Geologists call such streams **antecedent streams** (from the Greek *ante*, meaning before), to emphasize that they existed before the range uplifted. Note that if the range rises faster than the stream downcuts, the new highlands divert (change) the stream's course so that it flows along the edge of the range (Fig. 14.23a–c).

FIGURE 14.19 Evolution of a fluvial landscape when base level drops.

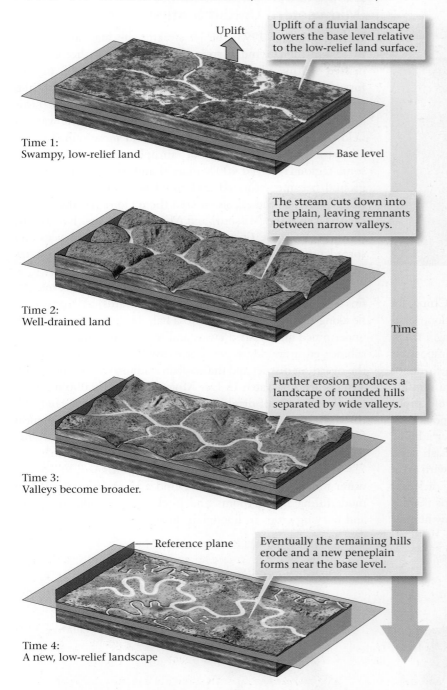

Uplift

Uplift of a fluvial landscape lowers the base level relative to the low-relief land surface.

Time 1:
Swampy, low-relief land

Base level

The stream cuts down into the plain, leaving remnants between narrow valleys.

Time 2:
Well-drained land

Time

Further erosion produces a landscape of rounded hills separated by wide valleys.

Time 3:
Valleys become broader.

Reference plane

Eventually the remaining hills erode and a new peneplain forms near the base level.

Time 4:
A new, low-relief landscape

TAKE-HOME MESSAGE

Stream-carved landscapes evolve over time as gradients diminish and the ridges and hills between valleys erode away. Superposed streams attain their shape before cutting down into rock structure, whereas antecedent streams cut while the land beneath them uplifts.

14.8 RAGING WATERS

[And Enhil, the ruler of the gods, said,] "The earth bellows like a herd of wild oxen. The clamor of human beings disturbs my sleep. Therefore, I want Adad [god of the skies] to cause heavy rains to pour down upon the Earth, both day and night. I want a great flood to come like a thief upon the Earth, steal the food of these people and destroy their lives."

—from the *Epic of Gilgamesh* (written in Sumeria, c. 2100 B.C.E.)

The Inevitable Catastrophe

Up to now, this chapter has focused on the variety of features and processes of a river system (see Geology at a Glance, pp. 400–401). Now we turn our attention to the havoc that a stream can cause when it floods. Floods can be catastrophic—they can strip land of crops and buildings, they can bury land in mud and silt, and they can submerge cities. A flood occurs when the volume of water flowing down a stream exceeds the volume of the stream channel, so water either rises out of the normal channel and spreads out over the floodplain or delta plain, or it fills a canyon to a greater depth than normal. Because of its increased discharge, a stream in flood can flow faster, be more turbulent, have greater competence, and exert more pressure on structures in its path than it can when its flow is normal.

Floods happen (1) during abrupt, heavy rains, when water falls on the ground faster than it can infiltrate and, therefore, becomes surface runoff; (2) after a long period of continuous rain, when the ground has become saturated with water and can hold no more; (3) when heavy snows from the previous winter melt rapidly in response to a sudden warm spell; or (4) when an artificial or natural dam holding back a lake suddenly collapses and the contents of the lake suddenly flow downstream. In all these cases, the discharge increases dramatically.

Geologists find it convenient to divide floods into two general categories. A **seasonal flood** gradually submerges a floodplain and covers a broad region (Fig. 14.24a–c). Such floods typically take hours to days to develop, allowing people living in the floodplain to evacuate before the water gets too deep. These floods, which sometimes are known also as floodplain floods, generally occur during the rainy season or when spring melting takes place. Because seasonal floods affect a broad area and may last for days or weeks, they can cause staggering losses of life

FIGURE 14.20 The development of incised meanders due to a relative drop of the base level.

The "goosenecks" of the San Juan River, Utah, are incised meanders.

and property—a 1931 flood of the Yangtze River in China, for example, destroyed fields and disrupted transportation, leading to a famine that killed 3.7 million people. During a **flash flood**, water in a stream rises so fast that it may be impossible for people to escape from the path of the flood. Flash floods typically happen as a result of unusually intense local rainfall, or as a result of a dam collapse. They tend to be more of a problem in narrow valleys or canyons (and thus may be known also as canyon floods), where water cannot spread out over a broad area and thus rapidly becomes very deep. In fact, a flash flood may arrive as a wall of water, slamming downstream with great force. Because flash floods contain water from only a single storm or single reservoir, the supply of water that feeds them is limited, so they tend to last only a few minutes to a few hours.

Case Study: A Seasonal Flood (Midwestern United States)

In the spring of 1993, the jet stream, the high-altitude (10 to 15 km high) wind current that controls weather systems, drifted southward. For weeks, the jet stream's cool, dry air formed an invisible wall that trapped warm, moist air from the Gulf of Mexico over the central United States. When this moist air rose to higher elevations, it cooled, and the water it held condensed and fell as rain, rain, and more rain. In fact, almost a whole year's supply of rain fell in just that spring— some regions received 400% more than usual. Eventually, the ground became saturated and could no longer absorb additional water, so the excess entered the region's streams, which carried it into the Missouri and Mississippi rivers. Eventually, the water in these rivers rose above the height of levees and spread out over the floodplain. By July, parts of nine states were under water (Fig. 14.24a).

The roiling, muddy flood uprooted trees, cars, and even coffins (which floated up from inundated graveyards). All barge traffic along the Mississippi came to a halt, bridges and roads were undermined and washed away, and towns along the river were submerged in muddy water. For example, in Davenport, Iowa, the riverfront district and the baseball stadium were covered with 4 m (14 feet) of water. In Des Moines, Iowa, 250,000 residents lost their supply of drinking water when floodwaters contaminated the municipal water supply with raw sewage and chemical fertilizers. Rowboats replaced cars as the favored mode of transportation in towns where only the rooftops remained visible.

For seventy-nine days, the flooding continued. When the water finally subsided, it left behind a thick layer of silt and mud, filling living rooms and kitchens in floodplain towns and burying crops in floodplain fields. In the end, more than

FIGURE 14.21 The concept of stream capture or "piracy."

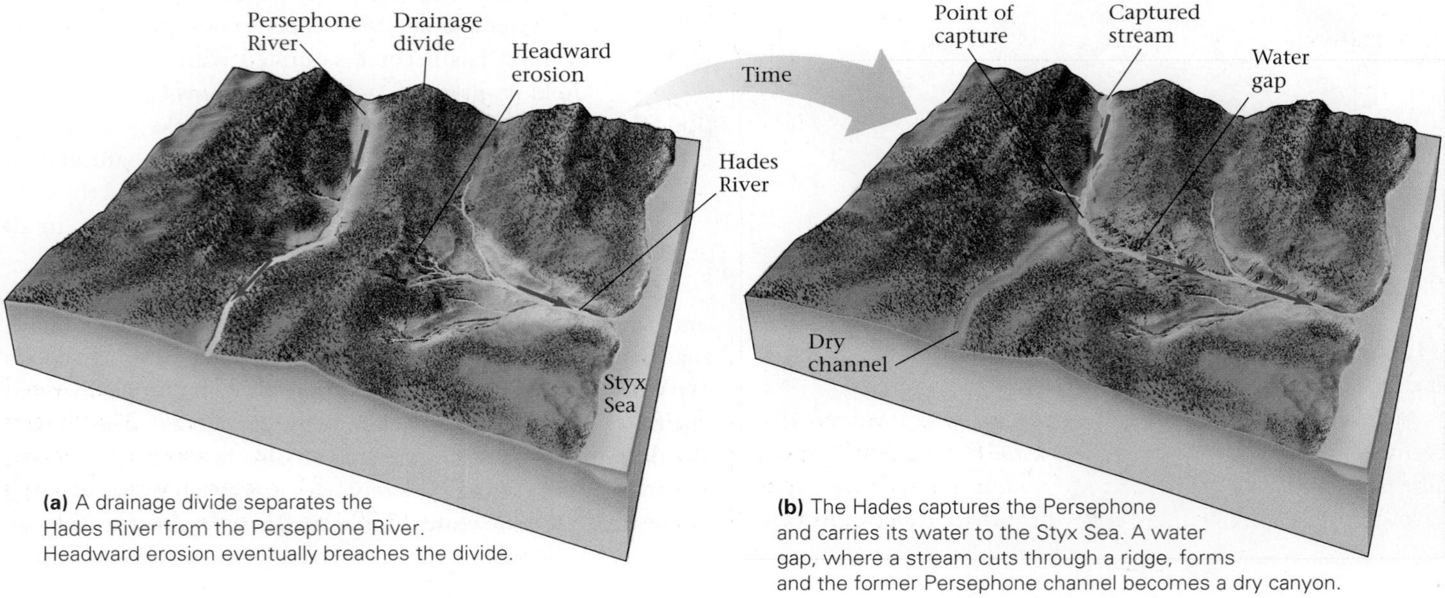

Persephone River   Drainage divide   Headward erosion

Time

Point of capture   Captured stream

Water gap

Hades River

Dry channel

Styx Sea

(a) A drainage divide separates the Hades River from the Persephone River. Headward erosion eventually breaches the divide.

(b) The Hades captures the Persephone and carries its water to the Styx Sea. A water gap, where a stream cuts through a ridge, forms and the former Persephone channel becomes a dry canyon.

FIGURE 14.22 Formation of superposed drainage.

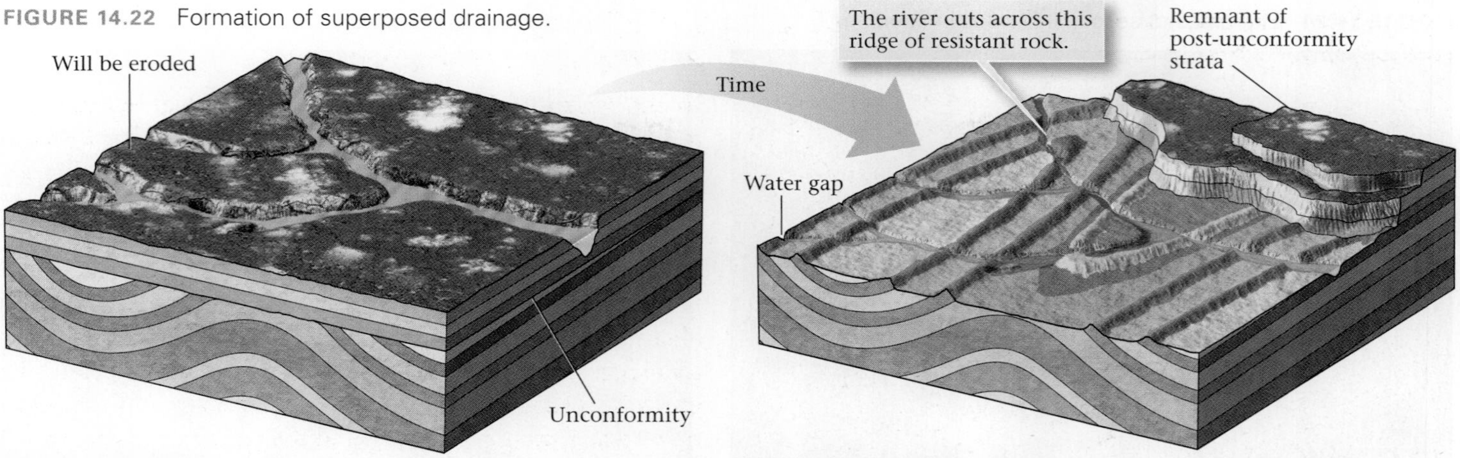

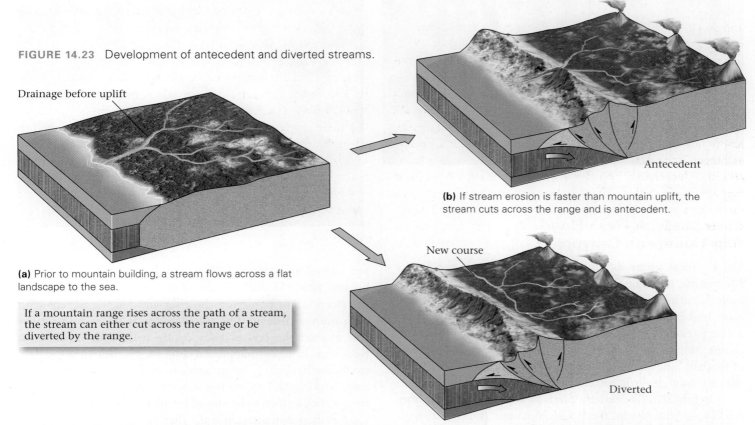

Will be eroded

The river cuts across this ridge of resistant rock.

Remnant of post-unconformity strata

Time

Water gap

Unconformity

(a) A superposed stream establishes its geometry while flowing over a uniform substrate above an unconformity.

(b) When erosion exposes underlying rock with a different structure, the river is superposed on the structure. As a result, it cuts across resistant ridges, instead of flowing around them.

FIGURE 14.23 Development of antecedent and diverted streams.

Drainage before uplift

Antecedent

(b) If stream erosion is faster than mountain uplift, the stream cuts across the range and is antecedent.

New course

(a) Prior to mountain building, a stream flows across a flat landscape to the sea.

If a mountain range rises across the path of a stream, the stream can either cut across the range or be diverted by the range.

Diverted

(c) If uplift happens faster than erosion, the stream is diverted and flows along the edge of the range.

40,000 square km of the floodplain had been submerged, fifty people died, at least 55,000 homes were destroyed, and countless acres of crops were buried. Officials estimated that the flood caused over $12 billion in damage.

While the Mississippi River flooding of 1993 was remarkable in terms of the extent of the area affected, 2008 proved to be a time of even more intense flooding for specific localities of the watershed. For example, the Cedar River in Iowa, a tributary of the Mississippi, overtopped its banks and covered 1,300 blocks in the city of Cedar Rapids, Iowa. Muddy and contaminant-laden water rose 10 m (32 ft) above flood stage, exceeding 1993 levels by 1 m (3 ft), contaminating the city's water supply, and driving 24,000 people from their homes.

FIGURE 14.24 Examples of seasonal floodplain flooding.

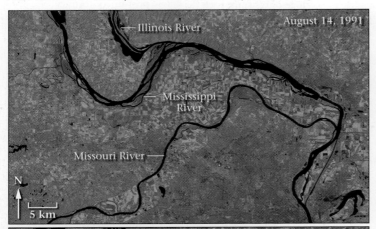

(b) Great Falls, Montana, was submerged by floodwaters in 1975.

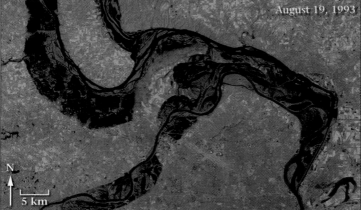

(a) Satellite photos (before and during) show the extent of flooding during the 1993 Mississippi River flood.

(c) In 2007, contaminated floodwaters spread disease in Indonesia.

Case Study: A Flash Flood (Big Thompson Canyon)

On a typical sunny day in the Front Ranges of the Rocky Mountains, north of Denver, the Big Thompson River seems quite harmless. Clear water, dripping from melting ice and snow higher in the mountains, flows down its course through a narrow canyon, frothing around boulders. In places, vacation cabins, campgrounds, and motels line the river, for the pleasure of tourists. The landscape seems immutable, but, as is the case with so many geologic features, permanence is an illusion.

On July 31, 1976, easterly winds blew warm, moist air from the Great Plains toward the Rocky Mountain front. As this air rose over the mountains, towering thunderheads built up, and at 7:00 P.M. rain began to fall. It poured, in quantities that even old-timers couldn't recall. In a little over an hour, 7.5 inches (19 cm) drenched the watershed of the Big Thompson River. The river's discharge grew to more than four times the maximum recorded at any time during the previous century. The river rose quickly, in places reaching depths several meters above normal. Turbulent water swirled down the canyon at up to 8 m

per second and churned up so much sand and mud that it became a viscous slurry. Slides of rock and soil tumbled down the steep slopes bordering the river and fed the torrent with even more sediment. The water undercut house foundations and washed the houses away, along with their inhabitants. Roads and bridges disappeared (Fig. 14.25). Boulders that had stood like landmarks for generations bounced along in the torrent, striking and shattering other rocks along the way—the largest rock known to be moved by the flood weighed 275 tons. Cars drifted downstream until they finally wrapped like foil around obstacles. When the flood subsided, the canyon had changed forever, and 144 people had lost their lives.

Living with Floods

Mark Twain once wrote of the Mississippi that we "cannot tame that lawless stream, cannot curb it or confine it, cannot say to it, 'go here or go there,' and make it obey." Was Twain right?

Since ancient times, people have attempted to confine rivers to a set course so as to prevent undesired flooding. In the twentieth century, flood-control efforts intensified as the population living along rivers increased. For example, since the passage of the 1927 Mississippi River Flood Control Act (drafted after a disastrous flood took place that year), the U.S. Army Corps of Engineers has labored to control the Mississippi. First, engineers built about 300 dams along the river's tributaries so that excess runoff could be stored in reservoirs and later be released slowly. Second, they built artificial levees of sand and mud, and built concrete floodwalls. Artificial levees and floodwalls increase the channel's volume and protect portions of the floodplain (Fig. 14.26a, b).

But although the Corps' strategy worked for floods up to a certain size, it was insufficient to handle the 1993 flood, for the volume of water drenching the Midwest earlier that year was so great that the reservoirs were filled to capacity, and additional runoff headed downstream. The river rose until it spilled over the top of some artificial levees and undermined others. Undermining occurs when high floodwaters increase the water pressure on the river side of the levee, forcing water through sand under the levee. In susceptible areas, water begins to spurt out of the ground on the dry side of the levee, thereby washing away the levee's support. The levee finally becomes so weak that it collapses.

When it comes to flooding, defensive efforts merely delay the inevitable, for it is unfeasible and too expensive to build levees high enough to handle all conceivable floods. And in some cases, building artificial levees may be counterproductive, since they constrain water to a smaller area and thus make floodwaters rise to a higher level than they would if they were free to spread over a wide floodplain. Thus, researchers are beginning to consider ways to prevent floods besides building levees and reservoirs along the channel. For example, moving levees farther away from the channel edge, thereby creating "floodways" that revert into wetlands, could help. Wetlands help prevent floods by absorbing water like a sponge.

When making decisions about investing in flood-control measures, mortgages, or insurance, planners need a basis for defining the hazard or risk posed by flooding. If floodwaters submerge a locality every year, a bank officer would be ill advised to approve a loan that would promote building there. But if floodwaters submerge the locality very rarely, then the loan may be worth the risk. Geologists characterize the risk of flooding in two ways. The **annual probability** of flooding indicates the likelihood that a flood of a given size or larger will happen at a specified locality during any given year. For example, if we say that a flood of a given size has an annual probability of 1%, then we mean that there is a 1 in 100 chance that a flood of at least this size will happen in any given year. The **recurrence interval** of a flood of a given size is defined as the *average* number of years between successive floods of at least this size. For example,

FIGURE 14.25 During the 1976 Big Thompson River flash flood, this house was carried off its foundation and dropped on a bridge.

if a flood of a given size happens once in 100 years, *on average*, then it is assigned a recurrence interval of 100 years and is called a 100-year-flood. Note that annual probability and recurrence interval are related:

$$\text{annual probability} = \frac{1}{\text{recurrence interval}}$$

For example, the annual probability of a 50-year-flood is 1/50, which can also be written as 0.02 or 2%.

Note again that the recurrence interval is simply an *average* determined by studying the past history of flooding in a stream. Some floodplain residents have the impression that since a 100-year flood just happened, another won't happen for 100 years. This is a false impression! Two 100-year floods can happen in the same year, can happen 80 years apart, or can happen 210 years apart.

The recurrence interval for a flood along a particular river reflects the size of a flood. For example, the discharge of a 100-year flood is larger than that of a 2-year flood, because the 100-year event happens less frequently (Fig. 14.27a). To define this relationship, geologists construct graphs that plot flood discharge on the vertical axis against recurrence interval on the horizontal axis (Fig. 14.27b).

Knowing the discharge during a flood of a specified annual probability, and knowing the shape of the river channel and the elevation of the land bordering the river, hydrologists can predict the extent of land that will be submerged by such a flood. Such data, in turn, permit hydrologists to produce flood-hazard maps. In the United States, the Federal Emergency Management Agency (FEMA) produces Flood Insurance Rate Maps that show the 1% annual probability (100-year) flood area and the 0.2% annual probability (500-year) flood risk zones (Fig. 14.27c).

FIGURE 14.26   Holding back rivers to prevent floods.

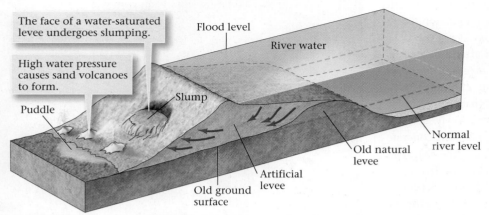

The face of a water-saturated levee undergoes slumping.

High water pressure causes sand volcanoes to form.

Puddle

Slump

Flood level

River water

Normal river level

Old natural levee

Artificial levee

Old ground surface

(a) Engineers build artificial levees to keep floodwater in the channel. Levees weaken and may fail when infiltrated by water.

High-water marks of past floods.

(b) A concrete floodwall at Cape Girardeau, Missouri. When floods threaten, a crane drops a gate into the slot to hold out the river.

TAKE-HOME MESSAGE

Seasonal floods submerge floodplains and delta plains at certain times of the year. Flash floods are sudden and short lived. We can specify the probability that a certain-size flood will happen in a given year, but flood-control efforts meet with mixed success.

14.9 RIVERS: A VANISHING RESOURCE?

As *Homo sapiens* evolved from hunter-gatherers into farmers, areas along rivers became attractive places to settle. Considering the multitudinous resources that rivers provide, it's no coincidence that early civilizations gathered in river valleys and on floodplains. Unfortunately, over time, humans have increasingly tended to abuse or overuse the Earth's rivers. Here we note four pressing environmental issues.

Pollution

The capacity of some rivers to carry pollutants has long been exceeded, transforming them into deadly cesspools. Pollutants include raw sewage and storm drainage from urban areas, spilled oil, toxic chemicals from industrial sites, and excess fertilizer and animal waste from agricultural fields.

Dam Construction

In 1950, there were about 5,000 large (over 15 m high) dams worldwide, but today there are over 38,000. Damming rivers has both positive and negative results. Reservoirs provide irrigation water and hydroelectric power, and they trap some floodwaters and create popular recreation areas. But sometimes their construction destroys wild rivers (the whitewater streams of hilly and mountainous areas) and alters the ecosystem of a drainage network by providing barriers to migrating fish, by decreasing the nutrient supply to organisms downstream, by removing the source of sediment for the delta, and by eliminating seasonal floods that replenish nutrients in the landscape.

Overuse of Water

Because of growing populations, our thirst for river water continues to increase, but the supply of water does not. The use of water has grown especially in response to the "Green Revolution" of the 1960s, during which huge new tracts of land came under irrigation. Today, 65% of the water taken out of rivers is used for agriculture, 25% for industry, and 9% for drinking and sewage transport. Civilization needed three times as much river water in 1995 as it did in 1950.

As a result, in some places human activity consumes the entire volume of a river's water, so that the channel contains little more than a saline trickle, if that, at its mouth. For example, except during unusually wet years, the Colorado River

FIGURE 14.27 The relationships among recurrence interval, discharge, and the area affected by flooding.

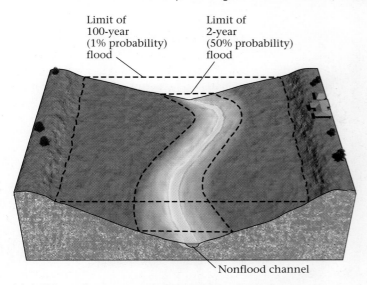

Limit of
100-year
(1% probability)
flood

Limit of
2-year
(50% probability)
flood

Nonflood channel

(a) A 100-year flood covers a larger area than a 2-year flood, and occurs less frequently.

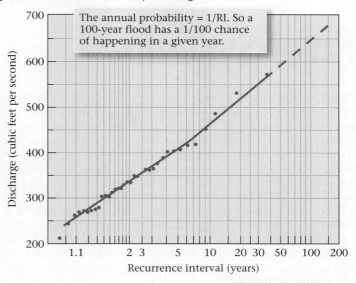

The annual probability = 1/RI. So a 100-year flood has a 1/100 chance of happening in a given year.

Discharge (cubic feet per second)

Recurrence interval (years)

(b) A flood-frequency graph shows the relationship between the recurrence interval and the discharge for an idealized river.

contains almost no water where it crosses the Mexican border, for huge pipes and canals carry the water instead to Phoenix and Los Angeles (Fig. 14.28). In the case of the Colorado River, states along its banks have established legal agreements that divide up the river's water. Unfortunately, the agreements were written during wet years when the river had unusually large discharge. Thus, the amount of water specified in the agreements actually exceeds the amount of water the river carries in most years.

Effects of Urbanization and Agriculture

Although the consumption of water for agricultural and industrial purposes decreases the overall supply of river water, urbanization may actually increase the short-term supply, causing local flooding. This is because cities cover the ground with impermeable concrete or blacktop, so rainfall does not soak into the ground but rather runs into storm sewers and then into streams. Stream discharge during a rainfall thus increases much more rapidly than it would without urbanization. Similarly, although damming rivers decreases the amount of silt a river carries downstream, agriculture may increase the sediment supply, because agriculture decreases the vegetative cover on the land, so that when it rains, soil washes into streams.

TAKE-HOME MESSAGE

People have greatly modified streams by constructing dams and levees, by adding pollutants, and by modifying the amount of water that enters streams. These changes have created significant problems in drainage networks.

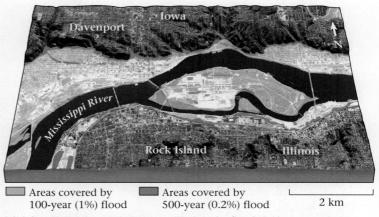

Iowa

Davenport

Mississippi River

Rock Island

Illinois

Areas covered by
100-year (1%) flood

Areas covered by
500-year (0.2%) flood

2 km

(c) A flood hazard map shows areas likely to be flooded. Here, even large floods are confined to the floodplain.

FIGURE 14.28 The Central Arizona Project canal shunts water from the Colorado River to Phoenix.

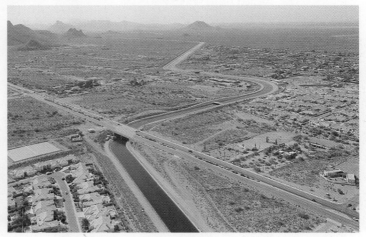

River Systems

Rivers, or streams, drain the landscape of surface runoff. Typically, an array of connected streams called a drainage network develops, consisting of a trunk stream into which numerous tributaries flow. The land drained is the network's watershed. A stream starts from a source, or headwaters, in the mountains, perhaps collecting water from rainfall or from melting ice and snow. In the mountains, streams carve deep, V-shaped valleys and tend to have steep gradients. For part of its course, a river may flow over a steep, bouldery bed, forming rapids. It may drop off an escarpment, creating a waterfall. Rivers gradually erode landscapes and carry away debris, so after a while, if there is no renewed uplift, mountains evolve into gentle hills. Through time, rivers can bevel once-rugged mountain ranges into nearly flat plains.

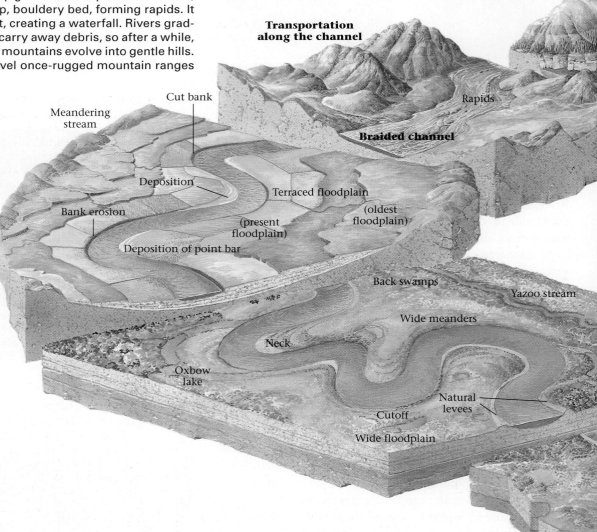

Transportation along the channel

Rapids

Braided channel

Meandering stream

Cut bank

Deposition

Terraced floodplain

(oldest floodplain)

Bank erosion

(present floodplain)

Deposition of point bar

Back swamps

Yazoo stream

Wide meanders

Neck

Oxbow lake

Natural levees

Cutoff

Wide floodplain

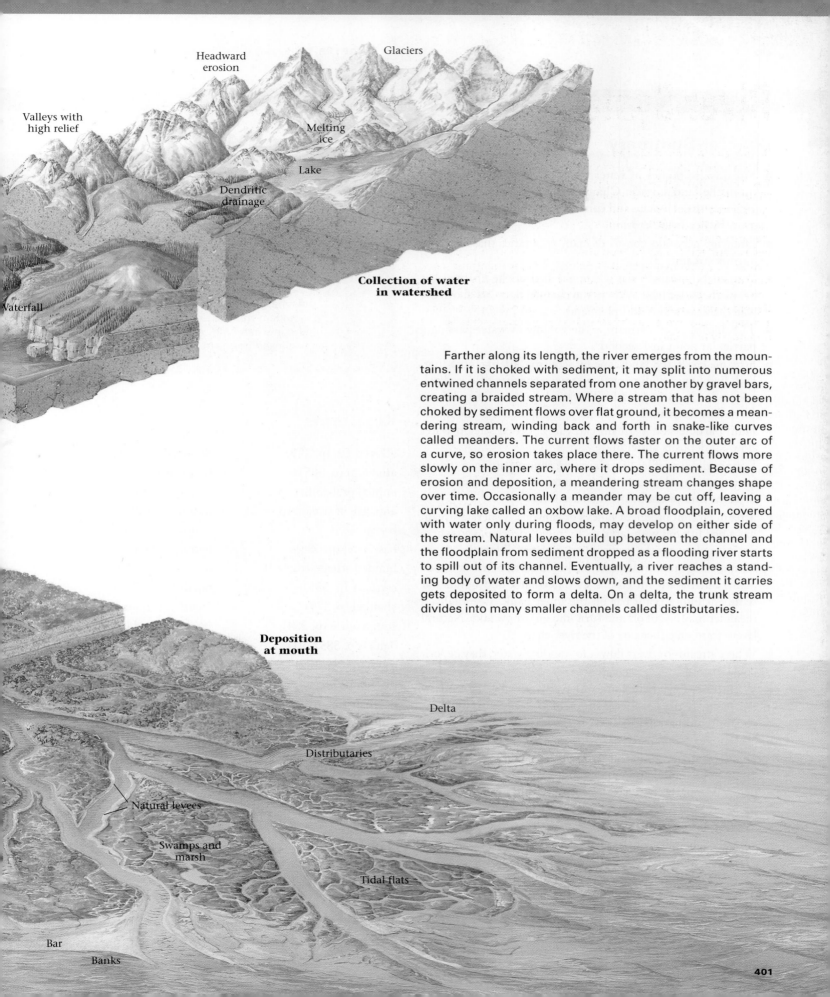

Headward
erosion

Glaciers

Valleys with
high relief

Melting
ice

Lake

Dendritic
drainage

Waterfall

**Collection of water
in watershed**

Farther along its length, the river emerges from the mountains. If it is choked with sediment, it may split into numerous entwined channels separated from one another by gravel bars, creating a braided stream. Where a stream that has not been choked by sediment flows over flat ground, it becomes a meandering stream, winding back and forth in snake-like curves called meanders. The current flows faster on the outer arc of a curve, so erosion takes place there. The current flows more slowly on the inner arc, where it drops sediment. Because of erosion and deposition, a meandering stream changes shape over time. Occasionally a meander may be cut off, leaving a curving lake called an oxbow lake. A broad floodplain, covered with water only during floods, may develop on either side of the stream. Natural levees build up between the channel and the floodplain from sediment dropped as a flooding river starts to spill out of its channel. Eventually, a river reaches a standing body of water and slows down, and the sediment it carries gets deposited to form a delta. On a delta, the trunk stream divides into many smaller channels called distributaries.

**Deposition
at mouth**

Delta

Distributaries

Natural levees

Swamps and
marsh

Tidal flats

Bar

Banks

Chapter Summary

- Streams are bodies of water that flow down channels and drain the land surface. Channels develop when sheetwash cuts into the substrate and concentrates the water flow; they grow by headward erosion.
- Drainage networks consist of many tributaries that flow into a trunk stream.
- Permanent streams exist where the bed of the channel lies below the water table. Where the channel floor lies above the water table, streams are ephemeral.
- The discharge of a stream is the volume of water passing a particular point in 1 second.
- Streams erode the landscape by scouring, lifting, abrading, and dissolving. Sediment in a stream consists of dissolved, suspended, and bed loads. The total quantity of sediment carried by a stream is its capacity. Competence is the maximum particle size a stream can carry. When stream water slows, it deposits alluvium.
- Longitudinal profiles of streams are concave up, for a stream has steeper gradients at its headwaters than near its mouth. Streams cannot cut below the base level.
- Stream erosion produces valleys or canyons. Where a stream has a bed littered with large rocks, rapids develop, and where a stream plunges off a vertical face, a waterfall forms.
- Meandering streams wander back and forth across a floodplain. Such a stream erodes its outer bank and builds out sediment into a point bar on the inner bank. Eventually, a meander may be cut off and turn into an oxbow lake. Natural levees form on either side of the river channel.
- Where streams or rivers flow into standing water, they deposit deltas.
- With time, fluvial erosion can bevel landscapes to a nearly flat plain. If the base level drops or the land surface rises, stream rejuvenation takes place.
- If an increase in rainfall or spring melting causes more water to enter a stream than the channel can hold, a flood results. Some floods submerge broad floodplains. Flash floods happen very rapidly. Officials try to prevent floods by building reservoirs and levees.
- We can specify the likelihood of flooding by giving the annual probability, or the recurrence interval.
- Human activities are significantly modifying fluvial systems.

GEOPUZZLE REVISITED

As soon as the land surface of a region rises above the ultimate base level (sea level), water starts flowing toward lower elevations. Eventually, the faster flow carves channels, with tributary channels flowing into a trunk channel. Streams eventually cut down to the base level. The channel volume reflects the usual discharge of the stream. If heavy rain, melting, or a dam rupture provides more water than the channel can hold, water spills over the channel walls and floods the surrounding landscape.

Key Terms

alluvial fan (p. 389)
alluvium (p. 385)
annual probability (p. 397)
antecedent stream (p. 392)
bar (p. 385)
base level (p. 386)
braided stream (p. 389)
capacity (p. 385)
channel (p. 379)
competence (p. 384)
delta (pp. 385, 391)
discharge (p. 382)
distributaries (p. 391)
downcutting (p. 380)
drainage basin (p. 381)
drainage divide (p. 381)
drainage network (p. 380)
ephemeral stream (p. 381)
flash flood (p. 394)
flood (p. 379)

floodplain (p. 390)
headward erosion (p. 380)
longitudinal profile (p. 386)
meander (p. 389)
natural levees (p. 391)
permanent stream (p. 381)
point bar (p. 390)
rapids (p. 387)
recurrence interval (p. 397)
seasonal flood (p. 393)
sheetwash (p. 379)
stream (p. 379)
stream piracy (p. 392)
stream rejuvenation (p. 392)
stream terrace (p. 387)
superposed stream (p. 392)
tributary (p. 380)
V-shaped valley (p. 387)
waterfall (p. 388)

Review Questions

1. What role do streams have in the hydrologic cycle? Indicate various sources of water in streams.
2. Describe the different types of drainage networks.
3. What factors determine whether a stream is permanent or ephemeral?
4. How does discharge vary according to the stream's length, climate, and position along the stream course?
5. Describe how streams and running water erode the Earth's surface.
6. What are the components of sediment load in a stream?
7. Distinguish between a stream's competence and its capacity.
8. What factors determine the position of the base level? What is the difference between a local base level and the ultimate base level?
9. Why do canyons form in some places and valleys in others?
10. How does a braided stream differ from a meandering stream?
11. Describe how meanders form and are cut off and the landforms that result from the process.
12. Describe how deltas grow and develop. How do they differ from alluvial fans?
13. How does a stream-eroded landscape evolve as time passes?
14. How are superposed and antecedent drainages similar? How are they different?

15. What is the difference between a seasonal flood and a flash flood?
16. How do people try to prevent flood damage? Are the methods always successful? Why?
17. What is the recurrence interval of a flood? Why can't someone say that "the hundred-year flood happened last year, so I'm safe for another hundred years"?

On Further Thought

1. The northeastern two-thirds of Illinois, in the midwestern United States, was last covered by glaciers only 14,000 years ago. The rest of the state was last covered by glaciers over 100,000 years ago. Until the advent of modern agriculture, the recently glaciated area was a broad, grassy swamp, cut by very few stream channels. In contrast, the area that was glaciated overt 100,000 years ago is not swampy and has been cut by numerous stream valley. Why?
2. Records indicate that flood crests for a given amount of discharge along the Mississippi River have been getting higher since 1927, when a system of levees began to block off portions of the floodplain. Why?
3. The Ganges River carries an immense amount of sediment load, which has been building a huge delta in the Bay of Bengal. Look at the region using an atlas or *Google Earth*™, think about the nature of the watershed supplying water to the drainage network that feeds the Ganges, and explain why this river carries so much sediment.

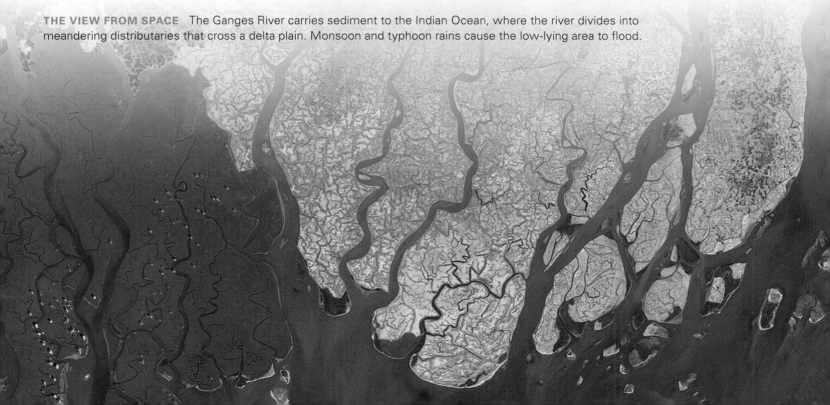

THE VIEW FROM SPACE   The Ganges River carries sediment to the Indian Ocean, where the river divides into meandering distributaries that cross a delta plain. Monsoon and typhoon rains cause the low-lying area to flood.

Surfers off Maui ride the power of
waves where the sea meets the shore.

CHAPTER **15**

Restless Realm:
Oceans and Coasts

GEOPUZZLE

In 1992, 29,000 plastic bath toys
fell off a cargo ship and spilled
into the middle of the Pacific
Ocean. Some of the toys are about
to be washed onto the Atlantic
Ocean beaches of the UK. Why?

15.1 INTRODUCTION

When seen from space, Earth glows blue, for the oceans cover 70.8% of its surface (Fig. 15.1). The sea provides the basis for life, tempers Earth's climate, and spawns its storms. It is a vast reservoir for water and chemicals that cycle into the atmosphere and crust, and for sediment washed off the continents. In this chapter, we first examine the fundamental characteristics of ocean basins and seawater. Then we focus on the landforms that develop along the **coast**, the region where the land meets the sea and where over 60% of the global population lives today. Finally, we consider how to cope with the hazards of living on the coast.

TAKE-HOME MESSAGE

By the end of this chapter, you should understand why ocean basins exist, how water circulates in these basins, and why so many different kinds of coastal landscapes develop. You should also be aware of the many natural hazards that threaten coastal areas.

15.2 LANDSCAPES BENEATH THE SEA

The oceans exist because oceanic lithosphere and continental lithosphere differ markedly from each other in terms of composition and thickness (Fig. 15.2; see Chapter 1). The surface of the denser and thinner oceanic lithosphere lies deeper than the surface of the relatively buoyant, thicker continental litho-sphere, resulting in oceanic basins (low areas) that collect water. On the present-day map of the world, the ocean encircles the globe. For purposes of reference, however, cartographers divide the ocean into several major parts, with somewhat arbitrary boundaries and significantly different volumes (see Fig. 15.1).

Have you ever wondered what the ocean floor would look like if all the water evaporated? Marine geologists can now provide a clear image of the ocean's **bathymetry**, or variation in depth, based originally on sonar measurements and more recently on measurements made by satellites. Such studies indicate that the ocean contains broad bathymetric provinces, distinguished from one another by their water depth. We look at these now.

Continental Shelves, Slopes, and Rises

Imagine you're in a submersible cruising just above the floor of the western half of the North Atlantic. If you start at the shoreline of North America and head east, you will find that a **continental shelf**, a relatively shallow portion of the ocean in which water depth does not exceed 500 m, fringes the continent. Across the width of the shelf, about 100 to 500 km, the ocean floor slopes seaward at only about 0.3°, an almost imperceptible amount. At its eastern edge, the continental shelf merges with the continental slope, which descends to depths of nearly 4 km with a slope of about 2°. From about 4 km down to about 4.5 km, a province called the continental rise, the angle decreases until at 4.5 km deep, you find yourself above a vast, nearly horizontal plain called the **abyssal plain**.

Broad continental shelves, like that of eastern North America, form along **passive continental margins**, margins that

FIGURE 15.1 The oceans of the world. The Pacific is the largest, covering almost half the planet. The Arctic region is an ocean covered by a thin coating of ice, whereas the Antarctic region is a continent.

FIGURE 15.2 Contrasts between continental lithosphere and oceanic lithosphere. The crustal portion of continental lithosphere differs markedly from that of oceanic lithosphere.

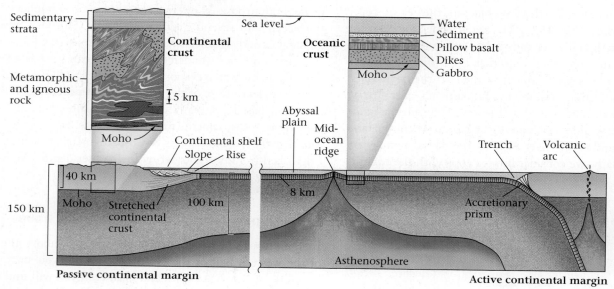

are not plate boundaries and thus host few earthquakes (Fig. 15.3a; see Chapter 2). Passive margins originate after rifting breaks a continent in two. When rifting stops and sea-floor spreading begins, the stretched lithosphere at the boundary between the ocean and continent gradually cools and sinks. Sand and mud that washed off the continent, along with the shells of marine creatures, buries the sinking crust, slowly producing a pile of sediment up to 20 km thick. The flat surface of this sedimentary pile constitutes the continental shelf.

If you were to take your submersible to the western coast of South America and cruise out into the Pacific, you would find a very different continental margin. After crossing a narrow continental shelf, the sea floor falls off at the relatively steep angle of 3.5° down into a trench. South America does not have a broad continental shelf because it is an **active continental margin**, a margin that coincides with a plate boundary and thus hosts many earthquakes. In the case of South America, the edge of the Pacific Ocean is a convergent plate boundary. The narrow shelf along a convergent plate boundary forms where an apron of sediment spreads out over the top of an accretionary prism, the pile of material scraped off the downgoing subducting plate.

At many locations, relatively narrow and deep valleys called **submarine canyons** cut into continental shelves and slopes (Fig. 15.3b). Some submarine canyons start offshore of major rivers, and for good reason: the rivers cut into the continental shelf at times when sea level was low and the shelf was exposed. But river erosion cannot explain the total depth of these canyons—some slice almost 1,000 m down into the continental margin, far greater than the maximum sea-level change. Much

of the erosion of submarine canyons results from the flow of turbidity currents, avalanches of sediment mixed with water (see Chapter 6).

The Bathymetry of Oceanic Plate Boundaries

You can see all three types of plate boundaries by studying the bathymetry of the ocean floor (see Fig. 15.3 and **Geotour 15** on p. GT-32). Sea-floor spreading at a divergent boundary yields a mid-ocean ridge, a 2-km-high submarine mountain belt. Because crust stretches and breaks as sea-floor spreading continues, the axis of a ridge may be bordered by escarpments, a result of normal faulting. Oceanic transform faults, strike-slip faults along which one plate shears sideways past another, typically link segments of mid-ocean ridges (see Chapter 2). Transforms are delineated by narrow belts of steep escarpments and broken-up rock. These fracture zones can be traced into the oceanic plate away from the ridge axis. Subduction at convergent boundaries yields a trench, a deep, elongate trough bordering a volcanic arc. Some trenches reach depths of over 8 km—the deepest point in the ocean, –11,035 m, lies in the Mariana Trench of the western Pacific. Some trenches border continents; others border island arcs, which are curving chains of active volcanic islands.

Abyssal Plains and Seamounts

As oceanic crust ages and moves away from the axis of the mid-ocean ridge, two changes take place. First, the lithosphere cools, and as it does so, its surface sinks. Second, a blanket of sedi-

FIGURE 15.3 Bathymetric features of the sea floor. The maps are produced by computer using measurements from satellites or submersibles.

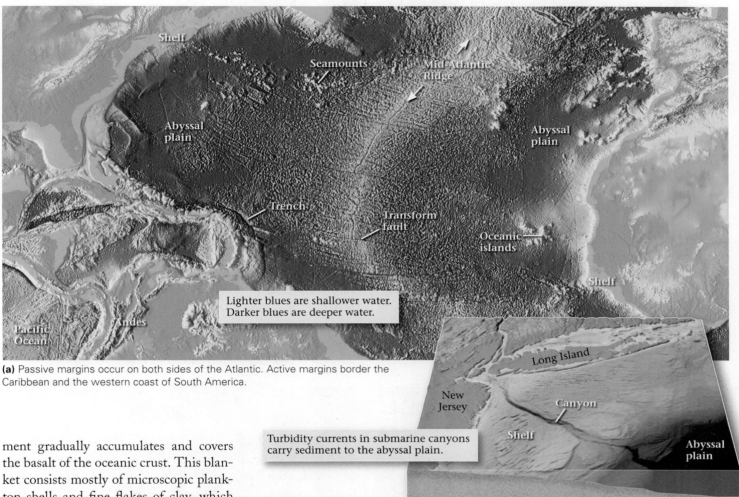

Lighter blues are shallower water.
Darker blues are deeper water.

(a) Passive margins occur on both sides of the Atlantic. Active margins border the Caribbean and the western coast of South America.

Turbidity currents in submarine canyons carry sediment to the abyssal plain.

(b) A submarine canyon along the coast of New Jersey.

ment gradually accumulates and covers the basalt of the oceanic crust. This blanket consists mostly of microscopic plankton shells and fine flakes of clay, which slowly fall like snow from the ocean water and settle on the sea floor. Because the ocean crust gets progressively older away from the ridge axis, sediment thickness increases away from the ridge axis. In dozens of locations around the world, hot-spot eruptions on oceanic lithosphere produced volcanoes. Presently active oceanic hot-spot volcanoes, as well as the remnants of extinct ones, rise above sea level as oceanic islands; those that lie below sea level are called **seamounts**.

TAKE-HOME MESSAGE

Ocean basins exist because oceanic lithosphere differs from continental lithosphere and "floats" lower in the asthenosphere. Sea-floor bathymetric features (mid-ocean ridges, trenches, islands, seamounts) reflect plate-tectonic processes. Abyssal plains are underlain by sediment-covered older lithosphere. Continental shelves form over passive-margin basins.

15.3 OCEAN WATER AND CURRENTS

Composition and Temperature

If you've ever had a chance to swim in the ocean, you may have noticed that you float much more easily in ocean water than you do in freshwater. That's because ocean water contains an average of 3.5% dissolved salt; in contrast, typical freshwater contains only 0.02% salt. Dissolved ions fit between water molecules without changing the volume of the water, so adding salt to water increases the water's density, and you float higher in a denser liquid. There's so much salt in the ocean that if all the water suddenly evaporated, a 60-m-thick layer of salt would coat the ocean floor. This layer would consist of about 75% halite (NaCl), with lesser amounts of gypsum ($CaSO_4 \bullet H_2O$), anhydrite ($CaSO_4$), and other salts. Oceanographers refer to the concentration of salt in water as salinity. Although ocean salinity averages 3.5%, salinity varies with location, ranging from about 1.0% to about 4.1%. Salinity reflects the balance between the addition of freshwater by rivers or rain and the removal of freshwater by evaporation, for when seawater evaporates, salt stays behind. Salinity also depends on water temperature, for warmer water can hold more salt in solution than can cold water.

When the *Titanic* sank after striking an iceberg in the North Atlantic, most of the unlucky passengers and crew who jumped or fell into the sea died within minutes, because the seawater temperature at the site of the tragedy approached freezing, and cold water removes heat from a body very rapidly. Yet swimmers can play for hours in the Caribbean, where sea-surface temperatures reach 28°C (83°F). Though the *average* global sea-surface temperature hovers around 17°C, it ranges between freezing near the poles to almost 35°C in restricted tropical seas. The correlation of average temperature with latitude exists because the intensity of solar radiation varies with latitude.

Water temperature in the ocean also varies markedly with depth. Waters warmed by the Sun are less dense and tend to remain at the surface because warmer water is less dense. An abrupt boundary, below which water temperatures decrease sharply, reaching near freezing at the sea floor, exists at a depth of about 300 m in the tropics. There is no pronounced boundary in polar seas, since surface waters there are already so cold.

Currents: Rivers in the Sea

Since first setting sail on the open ocean, people have known that the water of the ocean does not stand still, but rather flows or circulates at velocities of up to several kilometers per hour in fairly well-defined streams called **currents**. Oceanographic studies demonstrate that circulation in the sea occurs at two levels: surface currents affect the upper hundred meters of water, and deep currents keep even water at the bottom of the sea in motion.

Surface currents occur in all the world's oceans (Fig. 15.4). They result from interaction between the sea surface and the wind—as moving air molecules shear across the surface of the water, the friction between air and water drags the water along. Because of the Earth's rotation, which generates

FIGURE 15.4 The major surface currents of the world's oceans.

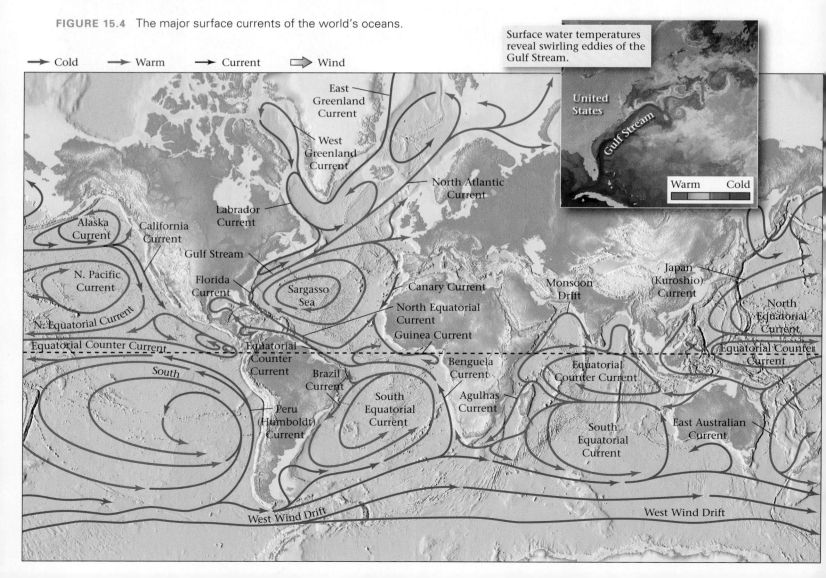

the **Coriolis effect** (Fig. 15.5a, b), surface currents in the northern hemisphere veer toward the right and surface currents in the southern hemisphere veer toward the left of the average wind direction.

Deep currents exist because at various locations, water must sink or rise in the vertical direction. Oceanographers have now identified downwelling zones, places where near-surface water sinks, and upwelling zones, places where subsurface water rises. What causes upwelling and downwelling? First, along coastal regions, these phenomena exist because as the wind blows, it drags surface water along. If surface water moves toward the coast, then an oversupply of water develops along the shore and excess water must sink—that is, downwelling takes place. Alternatively, if surface water moves away from the coast, then a deficit of water develops near the coast and water rises to fill in the gap—upwelling takes place. Upwelling of subsurface water also occurs along the equator because the winds blow steadily from east to west. Wind causes surface water to move. Due to the Coriolis effect, the surface water deflects to the north in the northern hemisphere and to the south in the southern hemisphere. Upwelling replaces this deficit. As a result of upwelling, nutrients come to the surface and foster an abundance of life in equatorial water.

Upwelling and downwelling can also be driven by contrasts in water density, caused by differences in temperature and salinity; we refer to the rising and sinking of water driven by density contrasts as **thermohaline circulation**. During thermohaline circulation, denser water (colder and/or saltier) sinks, whereas water that is less dense (warmer and/or less salty) rises. As a result, the water in polar regions sinks and flows back along the bottom of the ocean toward the equator. This process divides the ocean vertically into a number of distinct water masses, which mix only very slowly with one another (Fig. 15.6a).

The combination of surface currents and thermohaline circulation, like a conveyor belt, moves water and heat among

FIGURE 15.5 The Coriolis effect occurs because the velocity of a point at the equator, in the direction of the Earth's spin, is greater than that of a point near the pole.

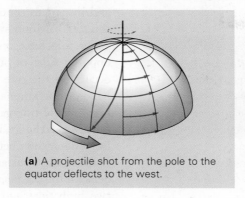

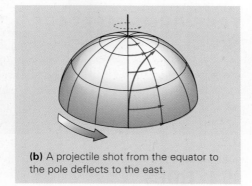

(a) A projectile shot from the pole to the equator deflects to the west.

(b) A projectile shot from the equator to the pole deflects to the east.

FIGURE 15.6 Global-scale upwelling and downwelling of ocean water.

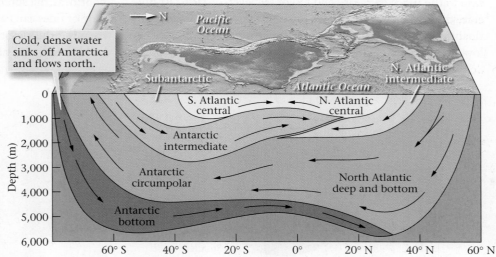

(a) Due to variations in density, the oceans are stratified into distinct water masses.

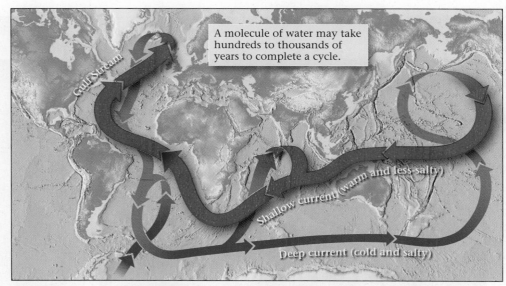

(b) Thermohaline circulation results in a global-scale "conveyor belt" that circulates water throughout the entire ocean system. Because of this circulation, the ocean mixes entirely in a 1,500-year period.

the various ocean basins (Fig. 15.6b). This transport of heat moderates global climate.

TAKE-HOME MESSAGE

Ocean water contains about 3.5% dissolved salt. Salinity and temperature vary with depth and location. The wind drives surface currents, whereas deep water circulates due to the density variations resulting from salinity and temperature variation.

15.4 TIDES

A ship captain hoping to set sail from a port must pay close attention to the twice daily rise and fall of sea level, or **tide**. At high tide, the harbor's water is deep enough that the ship can easily float over obstacles, but at low tide, the water may be so shallow that the ship could run aground. Tides develop in response to the **tide-generating force**, which in simple terms is due in part to the gravitational attraction of the Sun and Moon, and in part to centrifugal force produced by the revolution of the Earth-Moon system around its center of mass. The tide-generating force results in two tidal bulges in the oceans on opposite sides of our planet (Fig. 15.7a). Of these, the larger occurs on the side of the Earth closer to the Moon, for the Moon's gravitational pull is stronger than that of other contributors to the tide-generating force. When a shore location lies under a tidal bulge, it experiences a high tide, and when it underlies a depression between two bulges, it experiences a low tide. If the Earth's surface were smooth and completely submerged beneath the oceans, each point on the surface would experience two high tides and two low tides per day, as the Earth spins relative to the tidal bulges. But the complex shape of the sea floor near the shore means that the specific timing and magnitude of tides can vary along a coast. Specifically, the tidal reach, meaning the elevation difference between sea level at high tide and low tide, can range from less than a meter to several meters.

FIGURE 15.7 Ocean tides and their manifestation.

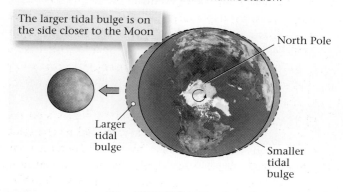

(a) Tides develop as the Earth spins relative to the two tidal bulges.

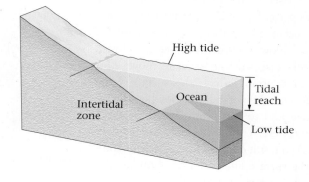

(b) Tidal reach is the elevation difference between sea level at high tide and low tide.

(c) Mont-Saint-Michel, off the western coast of France, is a castle-adorned island during high tide.

(d) At low tide, muddy tidal flats surround Mont-Saint-Michel.

The largest tidal reach on Earth, 16.8 m (54.6 feet), occurs in the Bay of Fundy on the Atlantic coast of Canada.

The manifestation of tides along a shore depends not only on the tidal reach, but on the slope of the shore. Along a vertical cliff, the intertidal zone (the region of shore submerged at high tide and exposed at low tide) appears as a ring of stained, barnacle- and seaweed-encrusted rock. Where the coast has a gentle slope, the shoreline (meaning the boundary between land and sea) moves substantially inland during a rising tide, and seaward during a falling tide (Fig. 15.7b). Vast muddy tidal flats may be exposed during low tide (Fig. 15.7c, d). Tidal flats can be hazardous for people who head out across the mud flats in search of shellfish, for the rising tide can trap them offshore.

FIGURE 15.8 Ocean waves build in response to the shear of wind blowing over the water surface.

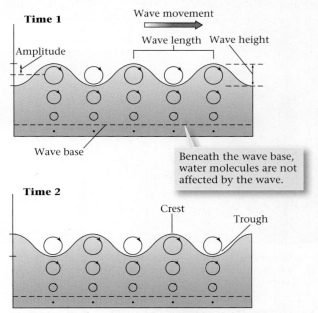

(a) Within a deep-ocean wave, water molecules follow a circular path. The radius of the circle decreases with depth.

15.5 WAVE ACTION

Wind-driven waves make the ocean surface a restless, ever-changing vista. They develop because of the shear between the molecules of air in the wind and the molecules of water at the surface of the sea. When you watch a wave travel across the open ocean, you may get the impression that the whole mass of water constituting the wave moves with the wave. But drop a cork overboard and watch it bob up and down and back and forth; it does not move along with a wave. Within a wave, away from shore, a particle of water moves in a circular motion, as viewed in cross section. The radius of the circle is greatest at the ocean's surface, where it equals the amplitude of the wave. With increasing depth, though, the radius of the circle decreases until, at a depth equal to about half the wavelength (the horizontal distance between two wave troughs), there is no wave movement at all (Fig. 15.8a). Submarines traveling below this **wave base** cruise through smooth water, even when ships toss about on the surface above.

The character of waves in the open ocean depends on the strength of the wind (how fast the air moves) and on the fetch of the wind (over how long a distance it blows). When the wind first begins to blow, it generates ripples in the water surface, pointed waves whose amplitude and wavelength are small. With continued blowing over a long fetch, swells—larger waves with amplitudes of 2 to 10 m and wavelengths of 40 to 500 m—begin to build. Such waves can travel beyond the region where they formed and become swells that can move long distances. Overlapping of waves and interaction between waves and currents

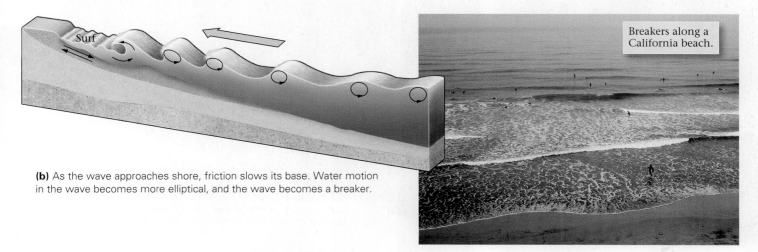

Breakers along a California beach.

(b) As the wave approaches shore, friction slows its base. Water motion in the wave becomes more elliptical, and the wave becomes a breaker.

may build localized rogue waves over 35 m (100 feet) that can wash over the deck of an ocean liner. Hurricane wave amplitudes may grow to over 25 m.

Waves have no effect on the ocean floor, as long as the floor lies below the wave base. However, near the shore, where the wave base just touches the floor, it causes a slight back-and-forth motion of sediment. Closer to shore, as the water gets shallower, friction between the wave and the sea floor slows the deeper part of the wave, and the motion of water in the wave becomes more elliptical (Fig. 15.8b). Eventually, water at the top of the wave curves over the base, and the wave becomes a breaker, ready for surfers to ride. Breakers crash onto the shore in the surf zone, sending a surge of water up the beach. This upward surge, or swash, continues until friction brings motion to a halt. Then gravity draws the water back down the beach as backwash.

Waves may make a large angle with the shoreline as they're coming in, but they bend as they approach the shore, a phenomenon called **wave refraction**; right at the shore, their crests make no more than about a 5° angle with the shoreline (Fig. 15.9a). To understand why this happens, imagine a wave approaching the shore so that its crest initially makes an angle of 45° with the shoreline. The end of the wave closest to the shore touches bottom first and slows down because of friction, while the end farther offshore continues to move at its original velocity, swinging the whole wave around so that it's more parallel with the shoreline. Because of wave refraction, wave energy focuses on headlands (places where higher land protrudes into the sea), and is weaker in embayments (places set back from the sea). Thus, erosion happens faster at headlands, forming cliffs, whereas deposition takes place in embayments, forming a beach (Fig. 15.9b).

FIGURE 15.9 Wave refraction and its consequences along the shore.

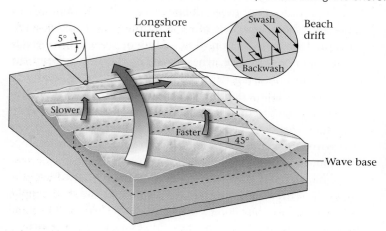

(a) Wave refraction occurs when waves approach the shore at an angle. If the wave reaches the beach at an angle, it causes a longshore current and beach drift of sand.

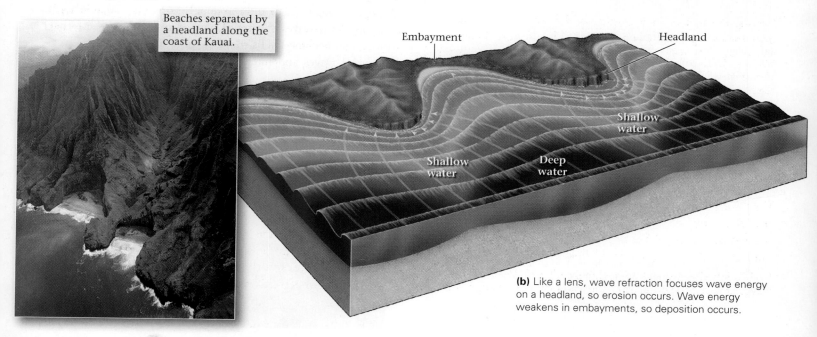

(b) Like a lens, wave refraction focuses wave energy on a headland, so erosion occurs. Wave energy weakens in embayments, so deposition occurs.

Although wave refraction decreases the angle at which waves move close to shore, it does not necessarily eliminate the angle. Where waves do arrive at the shore obliquely, water in the nearshore region has a component of motion that trends parallel to the shore. This **longshore current** causes swimmers floating in the water just offshore to drift gradually in a direction parallel to the beach. And where waves roll onto the shore at an angle, sediment in the surf follows a sawtooth pattern of movement that results in a gradual *net* transport beach sediment parallel to the beach—such movement is called **longshore drift** (or beach drift). This sawtooth pattern happens because the swash of a wave moves perpendicular to the wave crest, so an oblique wave carries sediment diagonally up the beach, but the backwash must flow straight down the slope of the beach due to gravity.

Waves pile water up on the shore incessantly. As the excess water moves back to the sea, it creates a strong, localized seaward flow perpendicular to the beach called a rip current. Rip currents are the cause of many drownings every year along beaches, because they suddenly carry unsuspecting swimmers away from the beach.

TAKE-HOME MESSAGE

The friction of the wind against the sea surface causes waves to form. Within a wave, water moves in a circle; the amount of motion decreases with depth. Near shore, water piles up into breakers that refract when they approach the shore causing longshore drift.

15.6 WHERE LAND MEETS SEA: COASTAL LANDFORMS

Coasts, the belt of land bordering the sea, vary dramatically in terms of topography and associated landforms. Some coasts boast dramatic cliffs, whereas others host broad sandy beaches or a dense cover of mangrove trees. The type of coastal landform that develops depends on many factors, including the composition and elevation of the shore, the climate, and the sediment supply. Let's look at the characteristics of several coastal types (see also Geotour 15).

Beaches and Tidal Flats

For millions of vacationers, the ideal holiday includes a trip to a **beach**, a gently sloping fringe of sediment along the shore. Some beaches consist of pebbles or boulders, whereas others consist of sand grains (Fig. 15.10a, b). Cobble beaches exist only where nearby cliffs or streams supply large rock fragments, for storm waves smash cobbles against one another with enough force to

shatter them—with time, cobbles break into smaller fragments. Sediment on a beach tends to be no smaller than sand sized, however, because the swash and backwash of waves winnows out finer sediment and carries it into deeper, quieter water, where it settles.

Beaches are complex, evolving systems containing many components. A beach profile, a cross section drawn perpendicular to the shore, illustrates these components (Fig. 15.10c). Starting from the sea and moving landward, a beach consists of a foreshore zone, or intertidal zone, across which the tide rises and falls. The beach face, a steeper, concave part of the foreshore zone, forms where the swash of the waves actively scours the sand. The backshore zone extends from a small step, or escarpment, cut by high-tide swash, to the front of the dunes or cliffs that lie farther inshore. The backshore zone includes one or more berms, horizontal to landward-sloping terraces that receive sediment only during storms.

Geologists commonly refer to beaches as "rivers of sand," to emphasize that beach sand moves along the coast over time—it is not a permanent substrate. Wave action at the shore moves an active sand layer on the sea floor on a daily basis. Inactive sand, buried below this layer, moves only during severe storms or not at all. Longshore drift, described earlier, can transport sand hundreds of kilometers along a coast in a matter of centuries. Where the coastline indents landward, longshore drift stretches beaches out into open water to create a **sand spit**. Some sand spits grow across the opening of a bay to form a baymouth bar (Fig. 15.10d).

The scouring action of waves piles sand up in a narrow ridge away from the shore called an **offshore bar**, which parallels the shoreline. In regions with an abundant sand supply, offshore bars rise above the mean high-water level and become **barrier islands** (Fig. 15.10e). The lower area between a barrier island and the mainland forms a quiet-water **lagoon**, a body of shallow seawater separated from the open ocean. Though developers have covered some barrier islands with expensive resorts, barrier islands are temporary features in the time frame of centuries to millennia.

Tidal flats, regions of mud and silt exposed or nearly exposed at low tide but totally submerged at high tide, develop in regions protected from strong wave action (Fig. 15.10f). They are typically found along the margins of lagoons or on shores protected by barrier islands. Here, mud and silt accumulate to form thick, sticky layers.

Rocky Coasts

More than one ship has met its end smashed and splintered in the spray and thunderous surf of a rocky coast, where bedrock cliffs rise directly from the sea. Lacking the protection of a beach, rocky coasts feel the full impact of ocean breakers. The water pressure generated during the impact of a breaker can pick

FIGURE 15.10 Characteristics of beaches, barrier islands, and tidal flats.

(a) A gravel beach along the Olympic Peninsula, Washington. The clasts were derived from erosion of adjacent cliffs.

(b) A sand beach on the western coast of Puerto Rico. Wave action has carried away finer sediment.

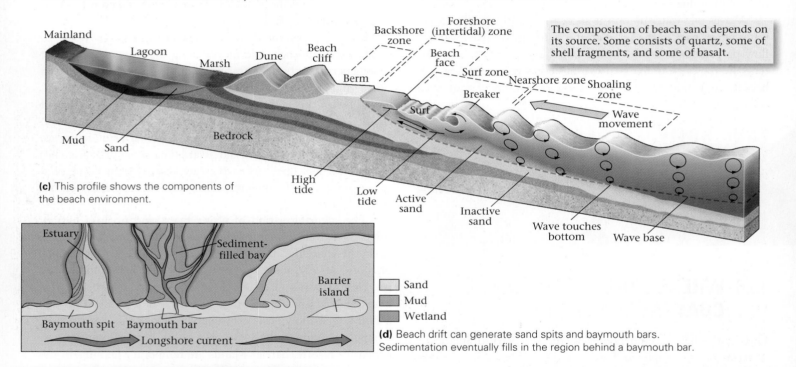

The composition of beach sand depends on its source. Some consists of quartz, some of shell fragments, and some of basalt.

(c) This profile shows the components of the beach environment.

| Sand |
| Mud |
| Wetland |

(d) Beach drift can generate sand spits and baymouth bars. Sedimentation eventually fills in the region behind a baymouth bar.

(e) Barrier islands off the coast of North Carolina.

(f) At low tide, boats at anchor rest in the mud of a tidal flat along the coast of Wales.

up boulders and smash them together until they shatter, and it can squeeze air into cracks, creating enough pressure to wedge rocks free. Further, because of its turbulence, the water hitting a cliff face carries suspended sand, and thus can abrade the cliff. The combined effects of shattering, wedging, and abrading, together called wave erosion, gradually undercut a cliff face and make a **wave-cut notch** (Fig. 15.11a). Undercutting continues until the overhang becomes unstable and breaks away at a joint,

building a pile of rubble at the base of the cliff that waves immediately attack and break up. By this process, wave erosion cuts away at a rocky coast, so that the cliff gradually migrates inland. Such cliff retreat eventually leaves behind a **wave-cut bench**, or platform, which becomes visible at low tide (Fig. 15.11b).

Many rocky coasts have an irregular shoreline, with headlands protruding into the sea and embayments set back from the sea. Such irregular coastlines tend to change over time. Because

FIGURE 15.11 Erosion landforms of rocky shorelines.

(a) A wave-cut notch.

(b) A wave-cut bench at the foot of the cliffs at Etretat, France.

(c) Landforms of a rocky shore. Beaches collect in embayments, whereas erosion concentrates at headlands.

(d) Coastal erosion along Australia's southern coast produced a sea arch (left). Eventually, the bridge will collapse, and only sea stacks will remain (right). These two are among several which together are known locally as the Twelve Apostles.

of refraction, waves curve and attack the sides of a headland, slowly eating through it to create a sea arch connected to the mainland by a narrow bridge (Fig. 15.11c). Eventually the arch collapses, leaving isolated sea stacks just offshore (Fig. 15.11d). Once formed, a sea stack protects the adjacent shore from waves. Therefore, sand collects in the lee of the stack, slowly building a tombolo, a narrow ridge of sand that links the sea stack to the mainland.

Organic Coasts

In temperate and tropical climates, the landforms of some coasts are produced by the growth of living organisms. These include coastal wetlands and coral reefs.

A **coastal wetland** is a vegetated flat-lying stretch of coast that floods with shallow water but does not feel the impact of strong waves and thus hosts numerous salt-resistant plants. In temperate climates, coastal wetlands include marshes (dominated by grasses; Fig. 15.12a), whereas in tropical or semitropical climates (between 30° north and 30° south of the equator), wetlands include mangrove swamps (Fig. 15.12b). Mangrove tree roots can filter salt out of water, so the trees have evolved the ability to survive in freshwater or saltwater. Some mangrove species form a broad network of roots above the water surface, making the plant look like an octopus standing on its tentacles. Dense stands of mangroves counter the effects of stormy weather and thus prevent coastal erosion.

In tropical climates, where the coastal water is warm (18°C to 30°C), clear, and drenched by sunlight, a variety of calcium carbonate ($CaCO_3$)-secreting animals (such as corals, clams, and sponges) grow in abundance. Corals come in many forms—some look like brains, others like elk antlers, and still others like delicate fans. When you examine a coral close up, you'll see that it consists of a colony of tiny invertebrate animals that are related to jellyfish. An individual coral animal, or polyp, has a tube-like body with a head of tentacles. The polyp feeds on nutrients that it gleans from seawater, and on nutrients and oxygen provided by algae that live in the coral's tissue. Polyps secrete $CaCO_3$, which forms a shell that protects the polyp. The shells of the multitude of polyps in a colony grow together into a solid mass. When the polyps die, the shells remain, and a new layer of polyps grows on top of the old. Through time, corals can build large mounds. Debris of broken corals and other shell fragments fills in the space between mounds. With time, the mounds and debris becomes a **coral reef** (Fig. 15.13a). Because of the environmental conditions necessary for large reefs to grow, most occur in water depths of less than 60 m and at latitudes of less than 30° (Fig 15.13b).

Geologists distinguish different kinds of coral reef, based on the reef's position relative to shore. A fringing reef forms directly along the coast, a barrier reef develops offshore, and an atoll makes a circular ring surrounding a lagoon. As Darwin first recognized back in 1859, coral reefs associated with islands in the Pacific start out as fringing reefs and then later become barrier reefs and finally atolls (Fig. 15.13c). This progression reflects the continued growth of the reef as the island around which it formed gradually sinks. Eventually, the reef itself sinks too far below sea level to remain alive and the island becomes a guyot, or flat-topped seamount.

Estuaries

Along some coastlines, a relative rise in sea level causes the sea to flood river valleys that merge with the coast, resulting in **estuaries**, where seawater and river water mix. You can recognize some estuaries on a map by the dendritic pattern of their

FIGURE 15.12 Examples of coastal wetlands.

(a) A salt marsh along the coast of Cape Cod.

(b) A mangrove swamp in northeastern Brazil.

FIGURE 15.13 The character and evolution of coral reefs.

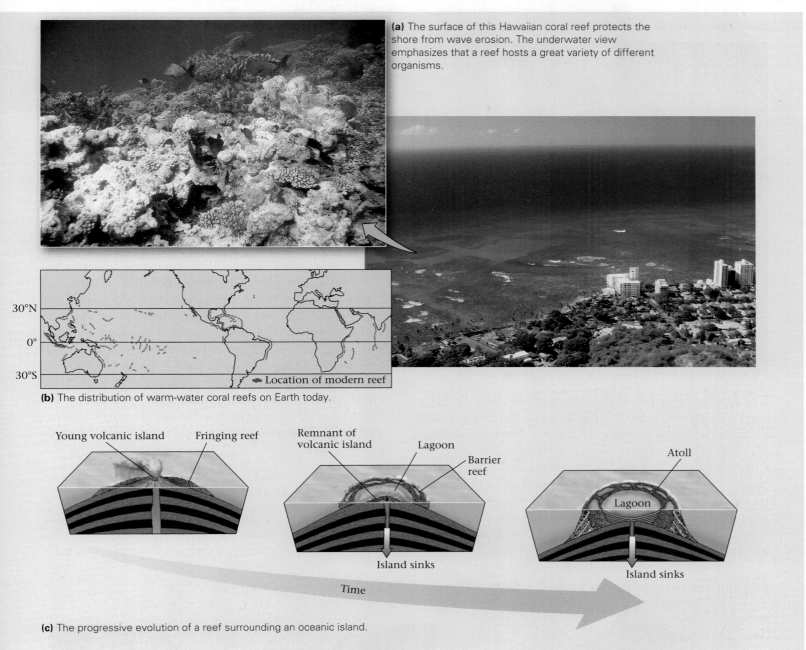

(a) The surface of this Hawaiian coral reef protects the shore from wave erosion. The underwater view emphasizes that a reef hosts a great variety of different organisms.

(b) The distribution of warm-water coral reefs on Earth today.

(c) The progressive evolution of a reef surrounding an oceanic island.

coastline (Fig. 15.14). Oceanic and fluvial water interact in two ways within an estuary. In quiet estuaries, protected from wave action or river turbulence, the water becomes stratified, with denser oceanic saltwater flowing upstream as a wedge beneath less dense fluvial freshwater. In turbulent estuaries, such as Chesapeake Bay, oceanic and fluvial water combine to create nutrient-rich brackish water with a salinity between that of oceans and rivers. Estuaries are complex ecosystems inhab-

ited by unique species of shrimp, clams, oysters, worms, and fish that can tolerate large changes in salinity.

Fjords

During the last ice age, glaciers carved deep valleys in coastal mountain ranges. When the ice age came to a close, the glaciers melted away, leaving deep, U-shaped valleys (see Chapter 18).

FIGURE 15.14 The Chesapeake Bay estuary formed when the sea flooded river valleys. The region is sinking relative to other coast areas, because it overlies a buried meteor crater.

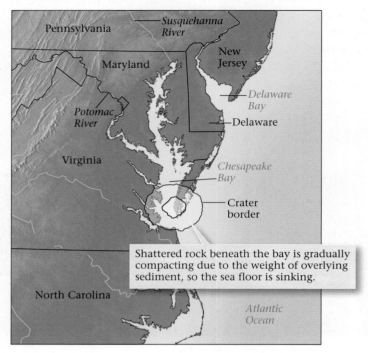

> Shattered rock beneath the bay is gradually compacting due to the weight of overlying sediment, so the sea floor is sinking.

The water stored in the glaciers, along with the water within the vast ice sheets that covered continents during the ice age, flowed back into the sea and caused sea level to rise. The rising sea filled the deep valleys, creating **fjords**, or flooded glacial valleys. Coastal fjords are fingers of the sea surrounded by mountains; because of their deep water and steep walls of polished rock, they are distinctively beautiful (Fig. 15.15). Some of the world's most spectacular fjords decorate the western coasts of Norway, British Columbia, and New Zealand. Smaller examples appear along the coast of Maine and southeastern Canada.

TAKE-HOME MESSAGE

Beaches form where there is abundant sediment; over time, sediment moves and builds spits and bars. Rocky coasts evolve due to wave erosion, fjords are submerged glacial valleys and estuaries are submerged river valleys. In warm climates, reefs grow offshore.

15.7 CAUSES OF COASTAL VARIABILITY

Plate Tectonic Setting

The tectonic setting of a coast plays a role in determining whether the coastline is marked by steep cliffs or broad plains (see Geology at a Glance, pp. 420–421). Along an active margin, compression squeezes the crust and pushes it up, producing mountains such as the Andes along the western coast of South America. Along a passive margin, the cooling and sinking of the lithosphere may create a broad **coastal plain**, a region of flat land that merges with the continental shelf, as exists along the Gulf Coast and southeastern Atlantic coast of the United States. But not all passive margins have coastal plains. At some, the margin of the rift that gave birth to the passive margin remains at a high elevation, even tens of millions of years after rifting ceased. For example, highlands initiated during Cretaceous rifting persist along portions of the Brazilian coast.

Relative Sea-Level Changes

Sea level, relative to the land surface, changes over geologic time. For example, during the last ice age, because so much water was trapped in glaciers on land, sea level was lower than it is today.

FIGURE 15.15 Fjord landscapes form where relative sea-level rise drowns glacially carved valleys.

> Glacial valleys have a U-shape, so fjords have steep sides.

Fjord

A fjord in Norway

FIGURE 15.16 Because of sea-level drop during the ice age, there was more dry land.

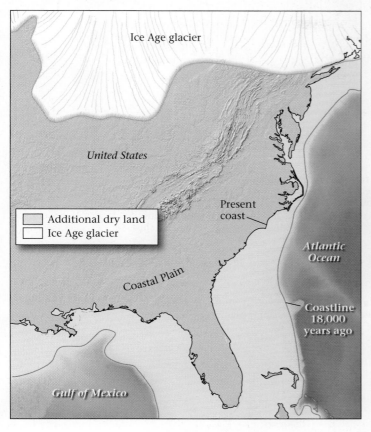

As a consequence, large areas of continental shelves were dry land (Fig. 15.16).

Geologists refer to coasts where the land is rising or rose, relative to sea level, as **emergent coasts**. At emergent coasts, steep slopes typically border the shore. A series of step-like terraces form along some emergent coasts (Fig. 15.17a, b). These terraces reflect episodic changes in relative sea level. Those coasts at which the land sinks relative to sea level become **submergent coasts** (Fig. 15.17c, d). At submergent coasts, landforms include estuaries and fjords that developed when the sea flooded coastal valleys. Chesapeake Bay in eastern North America demonstrates the consequences of submergence (see Fig. 15.14).

Sediment Supply and Climate

The quantity and character of sediment supplied to a shore affects its character. At erosional coasts, the sea washes sediment away faster than it can be supplied. Such coasts recede landward and may become rocky. Accretionary coasts, ones that receive more sediment than erodes away, grow seaward and develop broad beaches.

Climate also affects the character of a coast. Shores that enjoy generally calm weather erode less rapidly than those constantly savaged by storms. A sediment supply large enough to generate an accretionary coast in a calm environment may be insufficient to prevent the development of an erosional coast in a stormy environment. The climate also affects biological activity along coasts. For example, in the warm water of tropical climates, mangrove swamps flourish along the shore, and coral reefs form offshore. In temperate climates, salt marshes develop, and in arctic regions, the coast may be a stark environment of lichen-covered rock and barren sediment.

TAKE-HOME MESSAGE

Not all coasts look the same. Some are rimmed by beaches of moving sand. Some are dominated by wave-cut rocky cliffs, and some host living wetlands. Estuaries are submerged river valleys, and fjords are submerged glacial valleys.

15.8 COASTAL PROBLEMS AND SOLUTIONS

Contemporary Sea-Level Changes

People tend to view a shoreline as a permanent entity. But in fact, shorelines are ephemeral geologic features. On a time scale of hundreds to thousands of years, a shoreline moves inland or seaward depending on whether relative sea level rises or falls. In places where sea level is rising today, shoreline towns will eventually be submerged. For example, if present rates of sea-level rise along the east coast of the United States continue, major coastal cities such as Washington, New York, Miami, and Philadelphia may be inundated within the next millennium (Fig. 15.18).

Beach Destruction—Beach Protection?

In a matter of hours, a storm can radically alter a landscape that took centuries or millennia to form. The backwash of storm waves sweeps vast quantities of sand seaward, leaving the beach a skeleton of its former self. The surf submerges barrier islands and shifts them toward the lagoon. Waves and wind together rip out mangrove swamps and salt marshes and break up coral reefs, thereby destroying the organic buffer that normally protects the coast, leaving it vulnerable to erosion for years to come. Of course, major storms also destroy human constructions: erosion undermines shoreside buildings, causing them to

Oceans and Coasts

The oceans of the world provide a diverse array of environments illustrating the full complexity of the Earth System. Tectonic processes and surface processes constantly battle with each other to produce submarine and subaerial landscapes.

Water in the ocean circulates in currents that transport heat from equator to pole. Interactions between the atmosphere and the ocean build waves that ripple the surface. Waves erode shorelines and transport sediment. Sea-floor features define the location of plate boundaries and hot spots. Coastal landforms depend on tectonic setting, climate, and sediment supply. Specifically, passive margins differ markedly from active convergent margins; equatorial coasts differ from polar coasts. A great variety of organisms inhabit all these realms.

Wave erosion cuts notches at the base of cliffs and bevels wave-cut benches.

Along sandy shores, sand builds beaches, sand spits, and bars.

Turbidities flowing down submarine canyons produce submarine fans.

In tropical environments, mangroves live along the shore and coral reefs grow offshore.

Along rocky coasts, sea cliffs, sea arches, and sea stacks evolve.

At a passive margin, a broad continental shelf develops. Submarine slumping may occur along the shelf.

The ocean teems with life.

At divergent plate boundaries, a mid-ocean ridge rises. Transform faults, marked by fracture zones, link segments of the ridge.

The Global Conveyor

Surface winds drive surface currents in large gyres.

- 0
- 1
- 2
- 3 Km
- 4
- 5
- 6

Cold water sinks at polar regions.

Coastal Landforms

Tidewater glaciers produce icebergs.

At high latitudes, fjords form when the rising sea floods glacially carved valleys.

A river transports sediment to a delta.

Hot spots build chains of oceanic islands. Only the youngest island of the chain is active.

Bathymetry of the Sea Floor

Volcanic arcs form along convergent-margin coasts.

Seamounts and guyots are relicts of hot spots.

Trench

At a convergent boundary, a trench bordered by an accretionary prism develops.

Waves and Beaches

The wind forms ocean waves. As a wave passes, water moves in a circular motion.

Near the shore, the top of the wave breaks over the base of the wave. Swash carries sand up the beach, and backwash carries sand back.

Sand may pile into dunes that build out over a lagoon, in which mud had accumulated.

FIGURE 15.17 Features of emergent coastlines (relative sea level is falling) and submergent coastlines (relative sea level is rising).

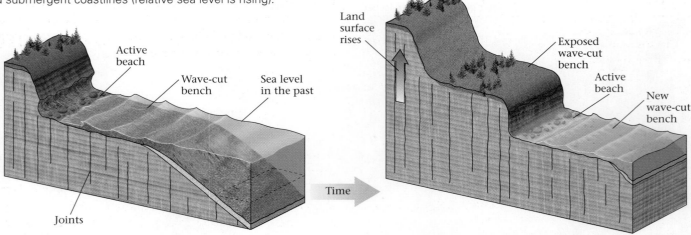

(a) Wave erosion produces a wave-cut bench along an emergent coast.

(b) As the land rises, the bench becomes a terrace, and a new wave-cut bench forms.

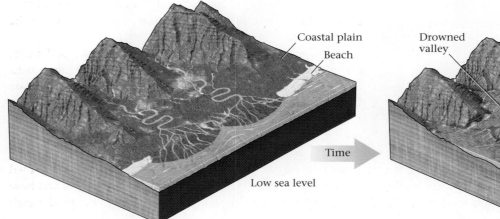

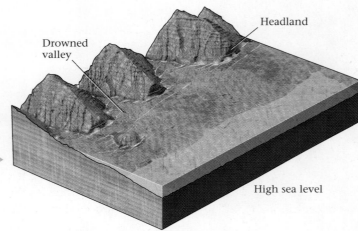

(c) A coast before sea level rises. Rivers drain valleys and deposit sediment on a coastal plain.

(d) As a submergent coast forms, sea level rises and floods the valleys, and waves erode the headlands.

collapse into the sea; wave impacts smash buildings to bits; and the **storm surge**—very high water levels created when storm winds push water toward the shore—floats buildings off their foundations (Fig. 15.19a, b).

But even less dramatic events, such as the loss of river sediment, a gradual rise in sea level, a change in the shape of a shoreline, or the destruction of coastal vegetation, can alter the balance between sediment accumulation and sediment removal from a beach, leading to beach erosion. Locally, beaches retreat landward at rates of 1 to 2 m per year.

In many parts of the world, beachfront property has great value. But if a hotel loses its beach sand, it probably won't stay in business. Thus, property owners construct artificial barriers to protect their stretch of coastline or to shelter the mouth of a harbor from waves. These barriers alter the natural movement of sand in the beach system and thus change the shape

of the beach, sometimes with undesirable results. For example, people may build groins, concrete or stone walls protruding perpendicular to the shore, to prevent beach drift from removing sand (Fig. 15.20a). Sand accumulates on the updrift side of the groin, forming a long triangular wedge, but sand erodes away on the downdrift side. Needless to say, the property owner on the downdrift side doesn't appreciate this process.

A pair of walls called jetties may protect the entrance to a harbor (Fig. 15.20b). But jetties erected at the mouth of a river channel effectively extend the river into deeper water, and thus may lead to the deposition of an offshore sandbar. A breakwater, built parallel or at an angle to the beach, prevents the full force of waves from reaching a harbor. With time, however, sand builds up in the lee of the breakwater and the beach grows seaward, clogging the harbor (Fig. 15.20c). And to protect expensive shoreside homes, people build seawalls out of riprap (large

FIGURE 15.18 Future sea-level rise, due to melting of polar ice, would flood many coastal cities.

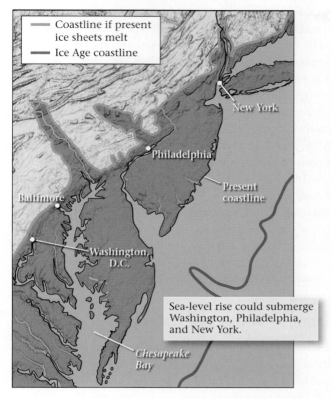

Sea-level rise could submerge Washington, Philadelphia, and New York.

quantities of sand to replenish a beach. This procedure, called beach nourishment, can be hugely expensive and at best provides only a temporary fix, for the backwash and beach drift that removed the sand in the first place continue unabated as long as the wind blows and the waves break.

Pollution and the Destruction of Organic Coasts

Bad cases of beach pollution create headlines. Because of longshore current, garbage dumped in the sea in an urban area may drift along the shore and be deposited on a tourist beach far from its point of introduction. Oil spills—most commonly from ships that flush their bilges but also from tankers that have run aground or foundered in stormy seas—have severely contaminated shorelines at several places around the world. Organic coasts are particularly susceptible to changes in the environment. The loss of such landforms can increase a coast's vulnerability to erosion and, considering that they provide spawning grounds for marine organisms, can upset the food chain of the global ocean.

In wetlands and estuaries, sewage, chemical pollutants, and agricultural runoff cause havoc. Toxins settle along with clay and concentrate in the sediments, where they contaminate burrowing marine life and then move up the food chain. Fertilizers and sewage that enter the sea with runoff increase the nutrient content of water, triggering algae blooms that absorb oxygen and therefore kill animal and plant life. Coastal wetlands face destruction by development—they have been filled or drained to be converted into farmland or suburbia and have been used as garbage dumps. In most parts of the world, such

stone or concrete blocks) or reinforced concrete on the landward side of the backshore zone (Fig. 15.20d).

In some places, people have given up trying to decrease the rate of beach erosion, and instead have worked to increase the rate of sediment supply. To do this, they truck or ship in vast

FIGURE 15.19 Examples of beach erosion.

September 9, 2008

September 15, 2008

(b) Wave erosion has completely removed the beach, and has started to erode a beach cliff, along the coast of Cape Cod.

(a) When Hurricane Ike hit Galveston, Texas, in September 2008, many beachfront homes were washed away.

FIGURE 15.20 Techniques used to preserve beaches.

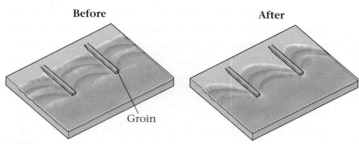

(a) The construction of groins may produce a sawtooth beach.

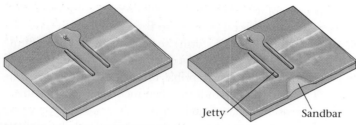

(b) Jetties extend a river farther into the sea, but may cause a sandbar to form at the end of the channel.

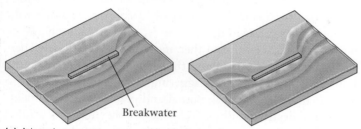

(c) A beach may grow seaward behind a breakwater.

(d) Riprap will slow erosion of a parking lot along a California beach.

activities have destroyed between 20% and 70% of coastal wetlands during the last century.

Reefs, which depend on the health of delicate coral polyps, may be devastated by even slight changes in their environment. For example, pollutants, hydrocarbons, sewage, and agricultural runoff either poison the corals or cause algal blooms that rob

seawater of oxygen and thus cause corals to suffocate. Factors that cause rivers to carry more sediment to the sea, and even beach-nourishment projects, can increase the amount of suspended sediment carried by nearshore water, and this sediment kills corals by clogging the pores in polyps and by decreasing the amount of light that can pass through seawater. Changes in water temperature or salinity caused by dumping waste water from power plants into the sea or by global warming of the atmosphere also destroy reefs, for reef-building organisms are very sensitive to temperature changes. People can destroy reefs directly by dragging anchors across reef surfaces, by touching reef organisms, or by quarrying reefs to obtain construction materials. In the last decade, marine biologists have noticed that reefs around the world have lost their color and died. This process, called reef bleaching, may be due to the removal or death of symbiotic algae in response to the warming of seawater, or may be a result of the dust carried by winds from desert or agricultural areas. Reef bleaching, along with various bacterial diseases, has destroyed substantial areas of reefs worldwide since 1980.

Hurricanes—A Coastal Calamity

Global-scale convection of the atmosphere, influenced by the Coriolis effect, causes currents of warm air to flow steadily from east to west in tropical latitudes. As the air flows over the ocean, it absorbs moisture. Because air becomes less dense as it gets warmer, tropical air eventually begins to rise like a balloon. As the air rises, it cools, and the water vapor it contains condenses to form clouds (mists of very tiny water droplets). If the air contains sufficient moisture, the clouds grow into a cluster of large thunderstorms, which consolidate to form a single, very large storm. Because of the Coriolis effect, this large storm evolves into a rotating swirl called a tropical disturbance. If the disturbance remains over warm ocean water, as can happen in late summer and early fall, rising warm moist air continues to feed the storm, fostering more growth. Eventually a spiral of rapidly circulating clouds forms, and the tropical disturbance becomes a tropical depression. Additional nourishment causes the tropical depression to spin even faster and grow broader, until it becomes a tropical storm and receives a name. If a tropical storm becomes powerful enough, it becomes a hurricane.

Formally defined, a **hurricane** (named for the Carib god of evil) is a huge rotating storm formed in the tropical latitudes (south of 20°N) of the Atlantic Ocean, where strong prevailing winds blow over warm water. It resembles a giant counterclockwise spiral in map view (Fig. 15.21a) and hosts sustained winds that exceed 119 km per hour (74 mph). These storms, which range from several hundred kilometers to 1,500 km (930 miles) wide, first drift westward toward the Caribbean or North America at speeds of up to 60 km per hour (37 mph). They may eventually turn north and head into the North Atlantic or into the interior of North America, where they die when they run out of a supply of warm water (Fig. 15.21b). Occasionally, hurricanes cross Central America and hook north over the Pacific. Similar storms form in tropical latitudes of other oceans—those of the western Pacific are called typhoons, and those that form in the Indian Ocean are called cyclones.

A typical hurricane (or typhoon or cyclone) consists of several spiral arms extending inward to a central zone of relative calm known as the hurricane's **eye** (Fig. 15.21c). A rotating vertical cylinder of clouds, the eye wall, surrounds the eye. Winds spiral toward the eye, so like an ice skater who spins faster when she brings her arms inward, the winds accelerate toward the interior of the storm and are fastest along the eye wall. Thus, hurricane-force winds affect a belt that is only 15% to 35% as wide as the whole storm (Fig. 15.21d). On the side of the eye where winds blow in the same direction as the whole storm is moving, the ground speed of winds is greatest, because the storm's overall speed adds to the rotational motion.

Meteorologists classify hurricanes using the Saffir-Simpson scale. Category 1 storms are the weakest, with winds between 119 and 153 km per hour (74 to 95 mph), and Category 5 storms are the strongest, with winds of over 250 km per hour (155 mph). The fastest winds yet to be measured in an Atlantic hurricane topped 300 km per hour (190 mph). In the last decade, there have been an average of eight or nine Atlantic hurricanes per year. New evidence suggests that in the past 35 years, the proportion of Category 4 and 5 hurricanes, typhoons, and cyclones has increased and now averages eighteen per year, globally. This increase may be due to the warming of ocean waters.

Hurricanes pose extreme danger in the open ocean, because their winds cause huge waves to build, and thus have led to the foundering of countless ships. They also wreak havoc in coastal regions, and even inland, though they die out rapidly after moving onshore. Hurricanes cause coastal damage in several ways:

FIGURE 15.21 Characteristics and paths of hurricanes in the western North Atlantic.

(a) A 2005 satellite photograph of Hurricane Katrina entering the Gulf of Mexico, where very warm waters fed the storm.

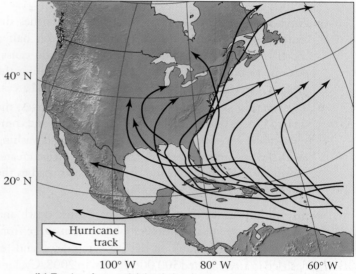

(b) Tracks of several Atlantic hurricanes show how most drift westward, then northward.

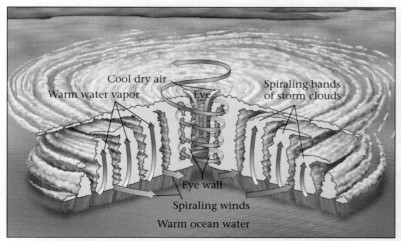

(c) In this cutaway drawing, we can see the spirals of clouds, the eye, and the eye wall of a hurricane. Dry air descends in the eye.

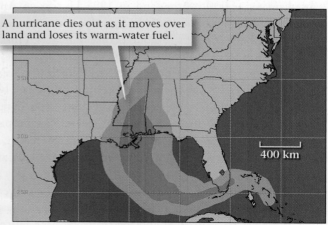

(d) A wind-swath map of Hurricane Katrina. Red areas represent hurricane winds; orange areas represent tropical-storm winds.

- *Wind*: Winds of weaker hurricanes tear off branches and smash windows. Stronger hurricanes uproot trees, rip off roofs, and collapse walls.

- *Waves*: Winds shearing across the sea surface during a hurricane generate huge waves. In the open ocean, these waves, which may be up to 30 m high, can capsize ships. Near shore, waves batter and erode beaches, rip boats from moorings, and destroy coastal property.

- *Storm surge*: Rising air in a hurricane causes a region of extremely low air pressure beneath. This decrease in pressure causes the surface of the sea to bulge upward over an area with a diameter of 60 to 80 km. Sustained winds blowing in an onshore direction build this bulge even higher. When the hurricane reaches the coast, this bulge of water, or storm surge, swamps the land. If the bulge hits the land at high tide, the sea surface will be especially high and will affect a broader area. Wind-driven waves on top of the storm surge can batter buildings hundreds of meters or farther inland.

- *Rain, stream flooding, and landslides*: Rain drenches the Earth's surface beneath a hurricane. In places, half a meter or more of rain falls in a single day. Rain causes streams to flood, even far inland, and by saturating the ground, can trigger landslides, especially in coastal areas.

- *Disruption of social structure*: When the storm passes, the hazard is not over. By disrupting transportation and communication networks, breaking water mains, and washing away sewage-treatment plants, hurricane damage creates severe obstacles to search and rescue, and can lead to the spread of disease, fire, and looting.

Nearly all hurricanes that reach the coast cause death and destruction, but some are truly catastrophic. Storm surge from a 1970 cyclone making landfall on the low-lying delta lands of Bangladesh led to an estimated 500,000 deaths. In 2008, Cyclone Nargis killed over 100,000 people in Myanmar (Burma). Storm surge from a hurricane that struck Galveston, Texas, in 1900 caused up to 12,000 deaths, and in 1992 Hurricane Andrew leveled extensive areas of southern Florida. It caused over $30 billion in damage, and left 250,000 people homeless. Hurricane Ike left a swath of destruction in the Caribbean before destroying Galveston in 2008; in little more than a century, this city has been completely overwhelmed by storm waves twice. Hurricane Katrina, in 2005, stands as the most destructive hurricane to strike the United States. Because of where Katrina passed, worst-case damage-prediction scenarios became reality, particularly in the historic city of New Orleans. Let's look at this storm's history.

Tropical Storm Katrina came into existence over the Bahamas and headed west. Just before landfall in southeastern Florida, winds strengthened and the storm became Hurricane Katrina. This hurricane sliced across the southern tip of Flor-ida, causing several deaths and millions of dollars in damage. It then entered the Gulf of Mexico and passed directly over the Loop Current, an eddy (circular flow) of summer-heated water from the Caribbean that had entered the Gulf of Mexico. Water in the Loop Current reaches temperatures of 32°C (90°F), and thus stoked the storm, injecting it with a burst of energy sufficient for the storm to morph into a Category 5 monster whose swath of hurricane-force winds reached a width of 325 km (200 miles). When it entered the central Gulf of Mexico, Katrina turned north and began to bear down on the Louisiana-Mississippi coast. The eye of the storm passed just east of New Orleans, and then across the coast of Mississippi. Storm surges broke records, in places rising 7.5 m (25 feet) above sea level, and they washed coastal communities off the map along a broad swath of the Gulf Coast (Fig. 15.22a, b). In addition to the devastating wind and surge damage, Katrina led to the drowning of New Orleans.

To understand what happened to New Orleans, we must consider the city's geologic history. New Orleans grew on the Mississippi Delta, between the banks of the Mississippi River on the south and Lake Pontchartrain (actually a bay of the Gulf of Mexico) on the north. The older parts of the town grew up on the relatively high land of the Mississippi's natural levee. Younger parts of the city, however, spread out over the topographically lower delta plain. As decades passed, people modified the surrounding delta landscape by draining wetlands, by constructing artificial levees that confined the Mississippi River, and by extracting groundwater. Sediment beneath the delta compacted, and the delta's surface has been starved of new sediment, so large areas of the delta sank below sea level. Today, most of New Orleans lies in a bowl-shaped depression as much as 2 m (7 feet) *below* sea level—the hazard implicit in this situation had been recognized for years (Fig. 15.22c).

The winds of Hurricane Katrina ripped off roofs, toppled trees, smashed windows, and triggered the collapse of weaker buildings, but their direct consequences were not catastrophic. However, when the winds blew storm surge into Lake Pontchartrain, its water level rose beyond most expectations and pressed against the system of artificial levees and flood walls that had been built to protect New Orleans. Hours after the hurricane eye had passed, the high water of Lake Pontchartrain found a weakness along the floodwall bordering a drainage canal and pushed out a section. Preliminary studies suggest that the foundation of the floodwall was anchored in a very weak bed of swamp deposits, rather than in the firmer sand below—under pressure, the base of the wall pushed outward and gave way. Breaks eventually formed in a few other locations as well. So, a day after the hurricane was over, New Orleans began to flood. As the water line climbed the walls of houses, brick by brick, residents fled first upstairs, then to their attics, and finally to their roofs. Water spread across the city until the bowl of New Orleans filled to the same level as Lake Pontchartrain, submerging 80% of the city (Fig. 15.22d).

FIGURE 15.22 The devastation of coastal areas by Hurricane Katrina.

(a) Surge from the storm destroyed homes along the Alabama coast.

(b) Officials survey the storm damage.

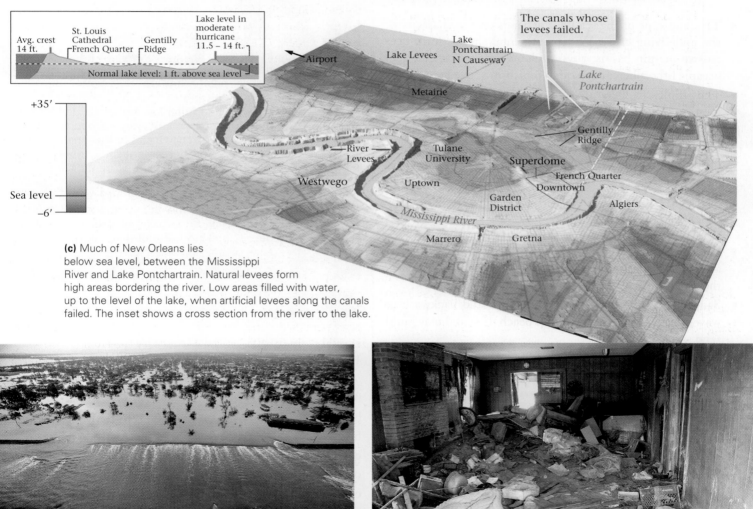

(c) Much of New Orleans lies below sea level, between the Mississippi River and Lake Pontchartrain. Natural levees form high areas bordering the river. Low areas filled with water, up to the level of the lake, when artificial levees along the canals failed. The inset shows a cross section from the river to the lake.

(d) Water flowing across the levees bordering the 17th Street Canal after the hurricane had passed.

(e) The flooded interior of a New Orleans home.

Floodwaters washed some houses away and filled others with debris (Fig. 15.22e). The disaster grew to be of national significance, as the trapped population sweltered without food, drinking water, or adequate shelter. With no communications, no hospitals, and few police, the city almost descended into anarchy. It took days for outside relief to reach the city, and by then, many had died and parts of New Orleans, a cultural landmark and major port, had become uninhabitable.

TAKE-HOME MESSAGE

Coastal landscapes are an ever-changing problem for society. Beaches and bars erode, pollution destroys wetlands, and hurricanes ravage the landscape. People try to protect beaches (for example, with breakwaters), but such construction leads to other problems. Adding sand may only be a temporary fix.

Chapter Summary

- The landscape of the sea floor depends on the character of the underlying crust. Wide continental shelves form over passive-margin basins; abyssal plains develop on old, cool oceanic lithosphere; and seamounts form above hot spots.
- The salinity, temperature, and density of seawater vary with location and depth.
- Ocean water circulates in currents. Surface currents are driven by the wind. The vertical upwelling and downwelling of water produces deep currents. Some of this movement is a consequence of variations in temperature and salinity.
- Tides—the daily rise and fall of sea level—are caused by a tide-generating force dominated by the gravitational pull of the Moon.
- Wind shear produces waves. Water particles in a wave follow a circular motion as a wave passes. Waves refract (bend) when they approach the shore because of drag on the sea floor.
- Sand on beaches moves with the swash and backwash of waves. Longshore current causes beach drift and may build sand spits.
- At rocky coasts, wave erosion yields such features as wave-cut beaches and sea stacks.
- Some shores are wetlands where marshes or mangrove swamps grow. Coral reefs grow along coasts in warm, clear water.
- The differences in coasts reflect their tectonic setting, whether sea level is rising or falling, sediment supply, and climate.
- To protect beach property, people build groins, jetties, breakwaters, and seawalls.
- Human activities have led to the pollution of coasts and reef destruction.
- Hurricanes, fed by the warm, moist water of tropical latitudes, produce winds of between 119 and 300 km per hour. The force of the winds, along with storm surge and heavy rains, can destroy coastal areas.

GEOPUZZLE REVISITED

In the past 15 years, ocean currents have carried the armada of yellow duckies and blue turtles nearly around the world. Some of the toys have arrived on the shores of South America and Australia. Others floated through the Bering Strait and were carried by sea ice around the Arctic Ocean to the Atlantic. Their amazing voyage emphasizes how mobile ocean water is. Ocean currents serve as a major conveyor of heat in the Earth System. Also, as the toys have found out, wave action where the sea meets the land not only transports water, but also transports sediment and debris, making it a tool that can sculpt coastlines of amazing beauty.

Key Terms

abyssal plain (p. 405)	estuary (p. 416)
active continental margin (p. 406)	eye (of hurricane) (p. 425)
	fjord (p. 418)
barrier island (p. 413)	hurricane (p. 424)
bathymetry (p. 405)	lagoon (p. 413)
beach (p. 413)	longshore current (p. 413)
coast (p. 405)	longshore drift (p. 413)
coastal plain (p. 418)	offshore bar (p. 413)
coastal wetland (p. 416)	passive continental margin (p. 405)
continental shelf (p. 405)	
coral reef (p. 416)	sand spit (p. 413)
Coriolis effect (p. 409)	seamount (p. 407)
current (p. 408)	storm surge (p. 422)
emergent coasts (p. 419)	submarine canyon (p. 406)

Review Questions

1. How much of the Earth's surface is covered by oceans?

2. Describe the typical topography of a passive continental margin, from the shoreline to the abyssal plain.

3. Where does the salt in the ocean come from? How does the salinity in the ocean vary?

4. What factors control the direction of surface currents in the ocean? What is the Coriolis effect, and how does it affect oceanic circulation? Explain thermohaline circulation.

5. What causes the tides? What is the tidal reach?

6. Describe the motion of water molecules in a wave. How does wave refraction cause longshore currents?

7. Describe the components of a beach profile.

8. How does beach sand migrate as a result of longshore currents? Explain the sediment budget of the coast.

9. Describe how waves affect a rocky coast, and how such coasts evolve.

10. What is an estuary? What is the difference between an estuary and a fjord?

11. Describe the different kinds of reefs, and how a reef surrounding an oceanic island changes with time.

12. Explain the difference between emergent and submergent coasts.

13. In what ways do people try to modify or "stabilize" coasts? How do the actions of people threaten the natural systems of coastal areas?

14. Why do hurricanes form, and in what way can they affect coastal regions?

On Further Thought

1. In 1789, the crew of the H.M.S. *Bounty* mutinied. Near Tonga, in the Friendly Islands (approximately 20°S and 175°W), the crew, led by Fletcher Christian, forced the ship's commanding officer, Lieutenant Bligh, along with those crewmen who remained loyal to Bligh, into a rowboat and set them adrift in the Pacific Ocean. The castaways, amazingly survived, and 47 days later landed at Timor (near Sumatra), 6,700 km to the west. Why did they end up where they did? Considering how long the journey took them, how fast were they moving?

2. A hotel chain would like to build a new beach-front hotel along a north-south-trending stretch of beach where a strong long-shore current flows from south to north. The neighbor to the south has constructed an east-west-trending groin on the property line. Will this groin pose a problem? If so, what solutions could the hotel try?

3. Observations made during the last decade suggest that sea level is rising in response to global warming. This rise happens partly because water expands when it heats up, and partly because glacial ice sheets in Greenland and elsewhere are melting. Much of southern Florida lies at elevations of less than 6 m above sea level. What changes will take place to the region as sea level rises? To answer, keep in mind the concepts of emergent and submergent coasts, and assume that southern Florida will continue to lie in the subtropical realm.

THE VIEW FROM SPACE The Great Barrier Reef forms shoals off the coast of northeastern Australia. It is the largest reef on Earth.

Reef

Mainland

50 km

CHAPTER **16**

A Hidden Reserve: Groundwater

Groundwater can turn a desert green, as shown by these circular irrigated fields sprouting in the sands of Jordan. A water well lies at the center of each circle.

GEOPUZZLE

In centuries past, the social center of a town would be the town's well, consisting of a hole dug down several meters into the ground. Usually, townspeople would line the well with stone, and to obtain water they would lower a bucket down the well, then lift up the water. Where did the water come from?

FIGURE 16.1 Development of sinkholes in central Florida.

(a) The Winter Park sinkhole, as seen from a helicopter.

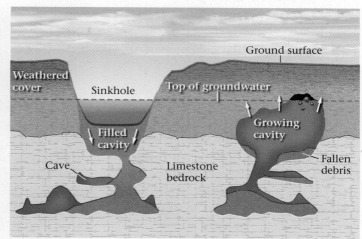

(b) As overburden slowly washes into underlying caves, a cavity forms. When the roof of this cavity collapses, a sinkhole forms (not to scale).

(c) A satellite view of central Florida's sinkholes.

We will never know the worth of water till the well is dry.

—Thomas Fuller (British collector of proverbs, 1654–1734)

16.1 INTRODUCTION

Imagine Rosa May Owen's surprise when, on May 8, 1981, she looked out her window and discovered that a large sycamore tree in the backyard of her Winter Park, Florida, home had suddenly disappeared. It wasn't a particularly windy day, so the tree hadn't blown over—it had just vanished! When Owen went outside to investigate, she found that more than the tree had disappeared. Her whole backyard had become a deep, gaping hole. The hole continued to grow for a few days until finally it swallowed Owen's house and six other buildings, as well as the deep end of the municipal swimming pool and several Porsches in a car dealer's lot (Fig. 16.1a).

What had happened in Winter Park? The bedrock beneath the town consists of limestone, a fairly soluble rock. **Groundwater**, the water that resides under the surface of the Earth, had gradually dissolved the limestone, producing open rooms, or caverns, underground. On May 8, the roof of an open cavity underneath Owen's backyard began to collapse, forming a circular depression called a **sinkhole** (Fig. 16.1b). The sycamore tree and the rest of the neighborhood simply dropped down into the sinkhole. It would have taken too much effort to fill in the hole with soil, so the community allowed it to fill with water, and now it's a circular lake, the centerpiece of a municipal park. Similar lakes appear throughout central Florida (Fig. 16.1c).

The Winter Park sinkhole is one of the more dramatic reminders that significant quantities of water reside underground. Whereas we can easily see Earth's surface water (in lakes, rivers, streams, marshes, and oceans) and atmospheric water (in clouds and rain), groundwater lies hidden beneath the surface in the pores and cracks found within sediment or rock. Nevertheless, groundwater accounts for about two-thirds of the world's freshwater supply and has increasingly become a major supply of water for homes, agriculture, and industry. This chapter provides a basic picture of where groundwater comes from, how it flows, and how it interacts with the rock and sediment through which it flows.

TAKE-HOME MESSAGE

By the end of this chapter, you should understand the nature of groundwater and the water table, the difference between porosity and permeability, why some wells flow and some go dry, and how groundwater becomes contaminated. Also, you should understand how caves and karst landscapes evolve.

16.2 WHERE DOES GROUNDWATER RESIDE?

The Underground Reservoir of the Hydrologic Cycle

As we saw in Interlude F, water moves among various reservoirs (the ocean, the atmosphere, rivers and lakes, groundwater, living organisms, soil, and glaciers) during the hydrologic cycle. Of the water that falls on land, some evaporates directly back into the atmosphere, some gets trapped in glaciers, and some becomes runoff that enters a network of streams and lakes that drains to the sea. The remainder sinks, via the process of infiltration, into the ground; in effect, the upper part of the crust behaves like a giant sponge that can soak up water.

Of the water that does infiltrate, some descends only into the soil and wets the surfaces of grains and organic material making up the soil. This water, called **soil moisture**, later evaporates back into the atmosphere or gets sucked up by the roots of plants and transpires back into the atmosphere. But some water sinks deeper and fills available spaces in cracks and between grains of sediment or rock. This water, as well as water trapped in rock at the time the rock formed, makes up groundwater. Groundwater slowly flows underground for anywhere from a few months to tens of thousands of years before returning to the surface to pass once again into other reservoirs of the hydrologic cycle.

Porosity: The Home of Groundwater

Contrary to popular belief, only a small proportion of groundwater flows freely in the underground lakes and streams of cavern networks. Most groundwater resides within the pore space of what might at first glance look like solid rock or sediment. Generally speaking, a **pore** is any open space (as opposed to solid material) within a body of sediment or rock, and **porosity** refers to the total volume of empty space in a material, usually expressed as a percentage. For example, if we say that a block of sandstone has 30% porosity, then 30% of a block of what looks like solid rock actually consists of open space.

Geologists further distinguish between primary and secondary porosity. Primary porosity consists of space that remains between solid grains or crystals after sediment accumulates or rocks form (Fig. 16.2a, b). For example, in sand or gravel, primary porosity exists for the simple reason that rounded clasts can't fit together tightly. Secondary porosity refers to new open space in rocks, produced some time after the rock first formed. For example, when rocks fracture, the opposing walls of the fracture do not fit together tightly, so narrow spaces remain in between (Fig. 16.2c). Secondary porosity also forms when groundwater flowing through rock dissolves and removes minerals, thereby producing openings called solution cavities.

Permeability: The Ease of Flow

If solid rock completely surrounds a pore, the water in the pore cannot flow to another location. For groundwater to flow, pores must be linked by conduits (openings). **Permeability,** a measure of the ease with which fluids can flow through a porous material, depends on the degree to which pores are interconnected (Fig. 16.2d). Water flows easily through a permeable material, whereas water flows slowly or not at all through an impermeable material. Porosity and permeability are not the same. A material whose pores are isolated from one another can have high porosity but low permeability. The permeability of a material depends on several factors.

- *Number of available conduits*: As the number of conduits increases, permeability increases.
- *Size of the conduits*: More fluid can travel through wider conduits than through narrower ones.
- *Straightness of the conduits*: Water flows more rapidly through straight conduits than it does through crooked ones.

Note that the factors that control permeability in rock or sediment resemble those that control the ease with which traffic moves through a city. Traffic can flow quickly through cities with many straight, multilane boulevards, whereas it flows slowly through cities with a few narrow, crooked streets.

Aquifers and Aquitards

With the concept of permeability in mind, hydrogeologists (geologists who study groundwater) distinguish between **aquifers**, sediment or rocks that transmit water easily, and **aquitards**, sediment or rocks that do not transmit water easily and thereafter retard the motion of water. Aquifers that intersect the surface of the Earth are called unconfined aquifers, because water can percolate directly from the surface down into the aquifer, and water in the aquifer can rise to the surface. Aquifers that are separated from the surface by an aquitard are called confined aquifers, as the water they contain is isolated from the surface (Fig. 16.3a).

The Water Table

Are porous rocks and sediments filled with water right up to the ground surface everywhere? The answer is no. Geologists define a boundary, called the **water table**, above which pore spaces contain mostly air and below which they contain only water (Fig. 16.3b; and see **Geotour 16** on p. GT-34). In technical jargon, the region of the subsurface above the water table is the unsaturated zone, and the region below the water table is the saturated zone; the unsaturated zone can be damp, but its pores are not full of water. Typically, surface tension, the electrostatic attraction of water molecules to mineral surfaces, causes water to

FIGURE 16.2 Where does water reside and flow inside rock? Porosity defines the volume of open space between grains in rock. Permeability refers to the degree to which pores are connected.

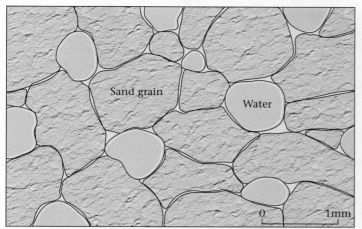

(a) Isolated pores in a sandstone occur in the spaces between grains. Water or air can fill pores.

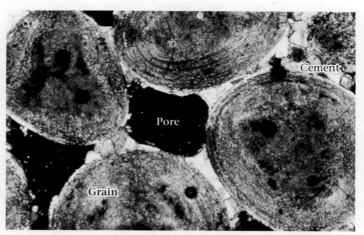

(b) A thin section photograph of a rock composed of calcite spheres only partially cemented by blocky calcite crystals.

These fractures have been enlarged by dissolution.

(c) An outcrop in Ireland shows fractures in limestone that, underground, provide secondary porosity.

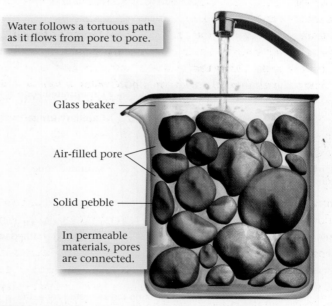

Water follows a tortuous path as it flows from pore to pore.

Glass beaker

Air-filled pore

Solid pebble

In permeable materials, pores are connected.

(d) Gravel contains pore space, because clasts don't fit together tightly. The connection of pores produces permeability, so water can flow through gravel.

seep up from the water table (just as water rises in a thin straw), filling pores in the capillary fringe, a thin layer between the saturated and unsaturated zones.

The depth of the water table in the subsurface varies greatly with location. The surface of a permanent stream, lake, or marsh, for example, defines the water table at that location, for water saturates the soil or rock below (**Fig. 16.3c, d**). Elsewhere,

the water table lies hidden below the ground surface: in humid regions, it typically lies within a few meters of the ground surface, whereas in arid regions, it may lie tens of meters to over 200 m below the surface.

Rainfall affects the water-table depth—the level sinks during a dry season. If the water table drops below the floor of a river or lake, the river or lake dries up, because the water it contains infiltrates into the ground. Arid regions have no permanent streams, for the water table lies below the stream bed. Streams in such regions flow only during storms, when surface water fills the channel at a faster rate than the water can infiltrate.

FIGURE 16.3   Water in the ground—aquifers, aquitards, and the water table.

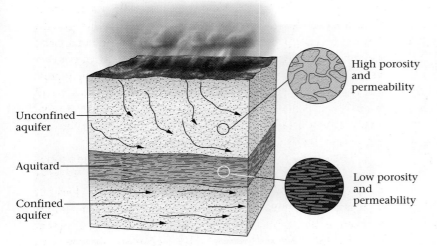

High porosity and permeability

Unconfined aquifer

Aquitard

Confined aquifer

Low porosity and permeability

(a) An aquifer is a high-porosity, high-permeability rock. Some aquifers are confined, some are unconconfined.

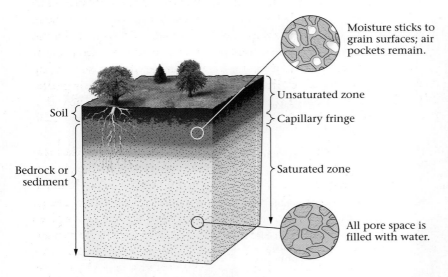

Moisture sticks to grain surfaces; air pockets remain.

Soil

Unsaturated zone

Capillary fringe

Bedrock or sediment

Saturated zone

All pore space is filled with water.

(b) The character of the water table, indicating the saturated zone, the unsaturated zone, and the capilary fringe between them.

Topography of the Water Table

In hilly regions, the water table is not a planar surface. Rather its shape mimics, in a subdued way, the shape of the overlying topography (Fig. 16.4a). This means that the water table lies at a higher elevation beneath hills than it does beneath valleys. But the relief (the vertical distance between the highest and lowest elevations) of the water table is not so great as that of the overlying land, so the surface of the water table tends to be smoother than that of the landscape.

At first thought, it may seem surprising that the elevation of the water table varies as a consequence of ground-surface topography. After all, when you pour a bucket of water into a pond, the surface of the pond immediately adjusts to remain horizontal. The elevation of the water table varies because groundwater moves so slowly through rock and sediment that it cannot *quickly* assume a horizontal surface. When it rains on a hill and water infiltrates down to the water table, the water table rises a little. When it doesn't rain, the water table sinks slowly, but so slowly that rain will fall again and make the water table rise before the water table has had time to sink very far.

In some locations, an isolated lens or layer of an impermeable material lies within the unsaturated zone between the ground surface and the regional water table. Where this happens, downward infiltrating water gets trapped to form an isolated mound or layer that lies above the regional water table. Such groundwater is, in effect, "perched" on an aquitard, like a bird that is perched on a branch above the ground. We refer to the top surface of groundwater that has been trapped above the regional water table as a **perched water table** (Fig. 16.4b).

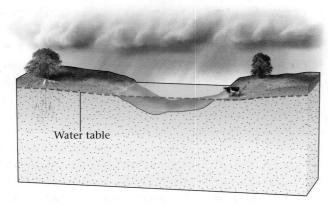

Water table

(c) The surface of a permanent pond or stream represents the water table.

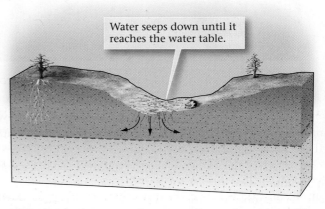

Water seeps down until it reaches the water table.

(d) During the dry season, the water table can drop substantially, causing ponds and streams to dry up.

FIGURE 16.4 Possible configurations of the water table. The position of the water table can be affected by local topography and by local stratigraphy.

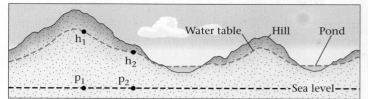

(a) The shape of a water table beneath hilly topography. Point h_1 on the water table is higher than Point h_2, relative to a reference elevation (sea level). The pressure at p_1 is, therefore, less than the pressure at p_2.

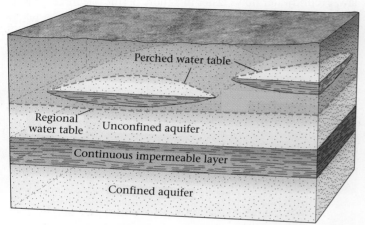

(b) A perched water table occurs where a lens of groundwater becomes trapped above a localized aquitard that lies above the regional water table.

TAKE-HOME MESSAGE

Most underground water fills pores and cracks in rock or sediment. Porosity refers to the total amount of open space within a material, whereas permeability indicates the degree to which pores connect. Aquifers have high porosity and permeability; aquitards don't.

16.3 GROUNDWATER FLOW

Groundwater Flow Paths

What happens to water that has infiltrated down into the ground and now lies below the water table? Does it just sit, unmoving, like the water in a stagnant puddle, or does it flow and eventually find its way back to the surface? Countless measurements confirm that groundwater enjoys the latter fate: groundwater indeed flows, and in some cases it moves great distances underground. Let's examine the factors that drive this flow.

In the unsaturated zone, the region between the ground surface and the water table, water percolates straight down, like the water passing through a drip coffee maker, for this water moves only in response to the downward pull of gravity. But in the saturated zone, the region below the water table, water flow is more complex, for in addition to the downward pull of gravity, flow directions reflect differences in pressure. Pressure may cause groundwater to flow sideways, or even upward. (If you've ever watched water spray up from a fountain, you've seen how pressure can push water upward.) Thus, to understand the nature of groundwater flow, we must first understand the origin of pressure in groundwater. For simplicity, we'll consider only the case of groundwater in an unconfined aquifer.

Pressure in groundwater at a specific point underground is caused by the weight of all the overlying water from that point up to the water table. (The weight of overlying rock does not contribute to the pressure in groundwater, for the contact points between mineral grains bear the rock's weight.) Thus, a point at a greater depth below the water table feels more pressure than does a point at lesser depth. If the water table is horizontal, the pressure acting on an imaginary horizontal reference plane at a specified depth below the water table is the same everywhere. But if the water table is not horizontal, as shown in Figure 16.4a, the pressure at points on a horizontal reference plane at depth changes with location. For example, the pressure acting at point p_1, which lies below the hill in Figure 16.4a, is greater than the pressure acting at point p_2, which lies below the valley, even though both p_1 and p_2 are at the same elevation (sea level, in this case).

Both the elevation of a volume of groundwater and the pressure within the water provide energy that, if given the chance, will cause the water to flow. Physicists refer to such stored energy as potential energy (see Appendix). The potential energy available to drive the flow of a volume of groundwater at a given location is called the **hydraulic head**. To measure the hydraulic head at a point in an aquifer, hydrogeologists drill a vertical hole down to the point and then insert a pipe in the hole. The height above a reference elevation (for example, sea level) to which water rises in the pipe represents the hydraulic head—water rises higher in the pipe where the head is higher. As a rule, *groundwater flows from regions where it has greater hydraulic head to regions where it has lesser hydraulic head*. In simple terms, this statement generally implies that groundwater flows from locations where the water table is higher to locations where the water table is lower.

When hydrogeologists took into account both the effect of gravity and the effect of pressure, they discovered that groundwater flows along concave-up curved paths (Fig. 16.5a). These curved paths eventually take groundwater from regions where the water table is high (below a hill) to regions where the water table is low (below a valley). Because of flow-path shape, some groundwater may flow deep down into the crust—in some cases, to depths of 10 km—along the first part of its path, and then may flow back up, toward the ground surface, along the final part of its path. The location where water enters the

FIGURE 16.5 The flow of groundwater.

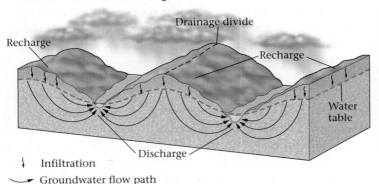

(a) Groundwater flows from recharge areas to discharge areas. Typically, the flow follows curving paths.

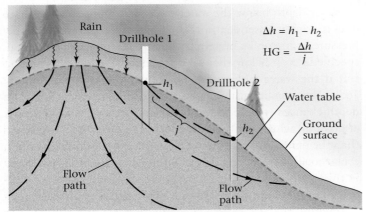

$$\Delta h = h_1 - h_2$$

$$HG = \frac{\Delta h}{j}$$

(b) The level to which water rises in a drillhole is the hydraulic head (h). Groundwater flows in response to a hydraulic gradient (HG).

ground and flows down is called the **recharge area**; the location where groundwater flows back up to the surface is called the **discharge area** (Fig. 16.5a).

Rates of Groundwater Flow

Flowing water in an ocean current moves at up to 3 km per hour; water in a steep river channel can reach speeds of up to 30 km per hour. In contrast, groundwater moves at a snail's pace—typical rates range between 0.01 and 1.4 m *per day* (about 4 to 500 m per year). Groundwater moves much more slowly than surface water for two reasons. First, groundwater moves by navigating through a complex, crooked network of tiny conduits, so it must travel a much greater distance than it would if it could follow a straight path. Second, friction and/or electrostatic attraction between groundwater and conduit walls slows down the water flow.

Why is the rate of groundwater flow so variable? In the mid-nineteenth century, a French engineer named Henry Darcy studied this problem. He discovered that the rate at which groundwater flows between two points in the subsurface depends

on two factors: the permeability of the material between the two points and a quantity called the hydraulic gradient. **Hydraulic gradient** is defined as the difference in hydraulic head between two points, divided by the distance between the two points as measured along the flow path (Fig. 16.5b). For the region close to the water table, we can simplistically picture hydraulic gradient as the "slope" of the water table (meaning the difference in elevation between two points on the water table, divided by the horizontal distance between the two points). Where the water table has a steep slope, the hydraulic gradient is greater than where the water table has a gentle slope. Hydrogeologists now refer to Darcy's discovery as **Darcy's law**. According to Darcy's law, groundwater flows faster in a more permeable material than in a less permeable material, and groundwater flows faster where the water table has a steep slope than where the water table has a gentle slope.

TAKE-HOME MESSAGE

Gravity and pressure cause groundwater to flow slowly from recharge to discharge areas. In essence, the rate of flow depends on the water table's slope and on permeability. Groundwater can follow curving flow paths that take it deep into the crust.

16.4 TAPPING THE GROUNDWATER SUPPLY

Groundwater can rise to the ground surface in wells or at springs. **Wells** are holes that people dig or drill to obtain water. **Springs** are natural outlets from which groundwater flows. Wells and springs provide a welcome source of water but must be treated with care if they are to last.

Wells

In an **ordinary well**, the base of the well penetrates an aquifer below the water table. Water from the pore space in the aquifer seeps into the well and fills it; the water surface in the well is the water table. Drilling into an aquitard, or into material that lies above the water table, will not supply water, and thus yields a dry well. Some ordinary wells are seasonal, in that they only function during the rainy season, when the water table rises above the base of the well.

To obtain water from an ordinary well, you either pull water up in a bucket or pump the water out. As long as the rate at which groundwater fills the well exceeds the rate at which water is removed, the level of the water table near the well remains about the same. However, if users pump water out of the well too fast, then the water table sinks down around the well, a pro-

FIGURE 16.6 Pumping groundwater at a normal well affects the water table.

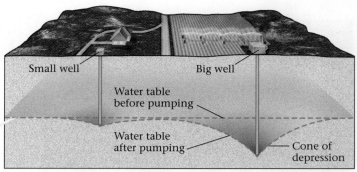

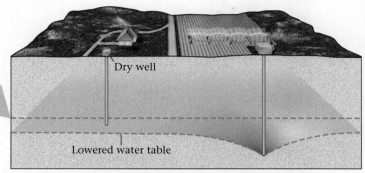

(a) If groundwater is extracted faster than it can be replaced, a cone of depression forms around the well.

(b) Pumping by the big well may lower the water table enough to cause the nearby small well to go dry.

FIGURE 16.7 Artesian wells, where water rises from the aquifer without pumping.

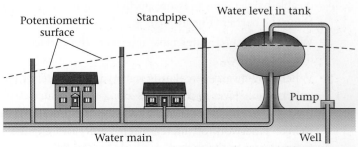

(a) A flowing artesian well in a Missouri field.

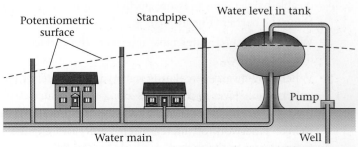

(b) The configuration of a city water supply. Water rises in vertical pipes up to the level of the potentiometric surface.

cess called drawdown, so that the water table becomes a downward-pointing, cone-shaped surface called a **cone of depression** (Fig. 16.6a). Drawdown may cause shallower wells that have been drilled nearby to run dry (Fig. 16.6b).

An **artesian well**, named for the province of Artois in France, penetrates confined aquifers, in which water is under enough pressure to cause the water to rise on its own to a level above the surface of the aquifer. If this level lies below the ground surface, the well is a nonflowing artesian well. But if the level lies above the ground surface, the well is a flowing artesian well, and water actively fountains out of the ground (Fig. 16.7a). Artesian wells occur in special situations where a confined aquifer lies beneath a sloping aquitard.

We can understand why artesian wells exist if we look first at the configuration of a city water supply (Fig. 16.7b). Water companies pump water into a high tank that has a significant hydraulic head relative to the surrounding areas. If the water were connected by a water main to a series of vertical pipes, pressure caused by the elevation of the water in the high tank would make the water rise in the pipes until it reached an imaginary surface, called a potentiometric surface, that lies above the ground. This pressure drives water through water mains to household water systems without requiring pumps. In an artesian system, water enters a tilted, confined aquifer that intersects the ground in the hills of a high-elevation recharge area (Fig. 16.7c). The confined groundwater flows down to the adjacent plains, which lie at a lower elevation. The potentiometric surface to which the water would rise were it not confined lies above this aquifer; in fact, it may lie above the ground surface of the plains. Pressure in the confined aquifer can push water up a well. Where the potentiometric surface lies below ground, the well is a nonflowing artesian well, but where the surface lies above the ground, the well is a flowing artesian well.

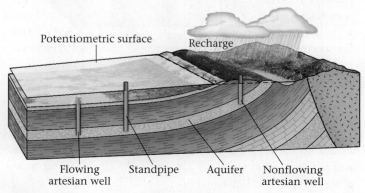

(c) The configuration of a regional artesian system.

FIGURE 16.8 Examples of geological settings in which springs form.

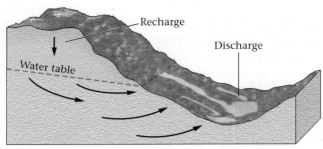

(a) Groundwater reaches the ground surface in a discharge zone.

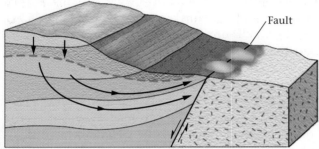

(b) Where groundwater reaches an impermeable barrier, it rises.

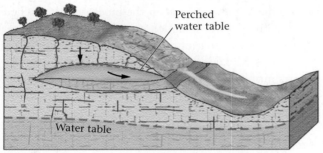

(c) Groundwater seeps where a perched water table intersects a slope.

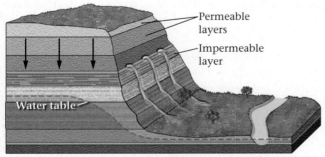

(d) Groundwater seeps out of a cliff face at the top of a relatively impermeable bed.

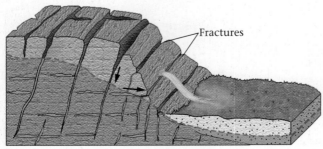

(e) A network of interconnected fractures channels water to the surface of a hill.

A spring on the wall of the Grand Canyon.

Springs

More than one town has grown up around a spring, a place where groundwater naturally flows or seeps onto the Earth's surface, for springs provide fresh, clear groundwater for drinking or irrigation without the expense of drilling or digging. Springs form under a variety of conditions:

- Where the ground surface intersects the water table in a discharge area (Fig. 16.8a): such springs typically occur in valley floors where they may add water to lakes or streams.

- Where flowing groundwater collides with a steep, impermeable barrier, and pressure pushes it up to the ground along the barrier (Fig. 16.8b).

- Where a perched water table intersects the surface of a hill (Fig. 16.8c).

- Where downward-percolating water runs into a relatively impermeable layer and migrates along the top surface of the layer to a hillslope (Fig. 16.8d).

- Where a network of interconnected fractures channels groundwater to the surface of a hill (Fig. 16.8e).

- **Artesian springs** form if the ground surface intersects a natural fracture (joint) that taps a confined aquifer in which the pressure is sufficient to drive the water to the surface.

Springs can provide water in regions that would otherwise be uninhabitable. For example, oases in deserts may develop around a spring. An **oasis** is a wet area, where plants can grow, in an otherwise bone-dry region.

TAKE-HOME MESSAGE

Groundwater can be obtained at wells (built by people) and springs (natural outlets). In ordinary wells, water must be lifted to the surface, but in artesian wells and springs, it rises due to its hydraulic head. Pumping of groundwater lowers the water table.

16.5 HOT SPRINGS AND GEYSERS

Hot springs, springs that emit water ranging in temperature from about 30°C to 104°C, are found in two geologic settings. First, they occur where very deep groundwater, heated in warm bedrock at depth, flows up to the ground surface. This water brings heat with it as it rises. Such hot springs form in places where faults or fractures provide a high-permeability conduit for deep water, or where the water emitted in a discharge region followed a trajectory that first carried it deep into the crust. Second, hot springs develop in **geothermal regions**, places where magma and/or very hot rock resides close to the Earth's surface (Fig. 16.9a). In hot springs, groundwater is a steaming tea of water and dissolved minerals. Hot groundwater contains more dissolved minerals because water becomes a more effective solvent when hot.

FIGURE 16.9 Geothermal waters and examples of their manifestation in the landscape.

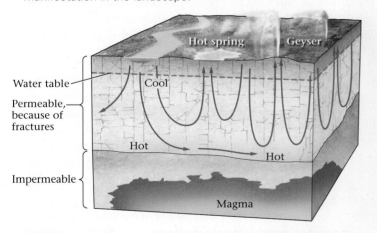

(a) Geysers and hot springs from where groundwater, heated at depth, rises to the surface.

(b) Terraces of minerals precipitated at Mammoth Hot Springs, Yellowstone.

(c) Colorful bacteria- and archaea-laden pools, Yellowstone.

(d) The Old Faithful geyser, in Yellowstone, which erupts predictably.

Numerous distinctive geologic features form in geothermal regions as a result of the eruption of hot water. In places where the hot water rises into soils rich in volcanic ash and clay, a viscous slurry forms and fills bubbling mud pots. Bubbles of steam rising through the slurry cause it to splatter about in goopy drops. Where geothermal waters spill out of natural springs and then cool, dissolved minerals in the water precipitate, forming mounds or terraces of travertine and other minerals (Fig. 16.9b). Geothermal waters may accumulate in brightly colored pools—the gaudy greens, blues, and oranges of these pools come from thermophilic (heat-loving) bacteria and archaea that eat the minerals dissolved in the groundwater (Fig. 16.9c).

The most spectacular consequence of geothermal waters is a **geyser** (from the Icelandic word for gush), a fountain of steam and hot water that erupts episodically from a vent in the ground (Fig. 16.9d). To understand why a geyser erupts, we first need a picture of its underground plumbing. Beneath a geyser lies a network of irregular fractures in very hot rock; groundwater sinks and fills these fractures. Adjacent hot rock then superheats the water: it raises the temperature above the temperature at which water at the ground surface will boil. Eventually, this superhot water rises through a conduit to the surface. When some of this water transforms into steam, the resulting expansion causes water higher up to spill out of the conduit at the ground surface. When this spill happens, pressure in the conduit, from the weight of overlying water, suddenly decreases. A sudden drop in pressure causes the superhot water at depth instantly to turn to steam, and this steam quickly rises, ejecting all the water and steam above it out of the conduit in a geyser eruption. Once the conduit empties, the eruption ceases, and the conduit fills once again with water that gradually heats up, starting the eruptive cycle all over again.

TAKE-HOME MESSAGE

In geothermal regions, or in localities where discharged groundwater followed a flow path deep into the crust, hot springs appear. Geysers develop under special circumstances where pressure builds up sufficiently to eject water and steam forcefully.

16.6 GROUNDWATER USAGE PROBLEMS

Since prehistoric times, groundwater has been an important resource that people have relied on for drinking, irrigation, and industry. Groundwater feeds the lushness of desert oases in the Sahara, the amber grain in the North American high plains, and the growing cities of arid regions. Though groundwater accounts for about 95% of the liquid freshwater on the planet, accessible groundwater cannot be replenished quickly in important locations, and this leads to shortages. In modern times, the problem has been exacerbated by the contamination of existing groundwater. Such pollution, caused when toxic wastes and other impurities infiltrate down to the water table, may be invisible to us but may ruin a water supply for generations to come. In this section, we'll take a look at problems associated with the use of groundwater supplies.

Depletion of Groundwater Supplies

Is groundwater a renewable resource? In a time frame of 10,000 years, the answer is yes, for the hydrologic cycle will eventually resupply depleted reserves. But in a time frame of 100 to 1,000 years—the span of a human lifetime or a civilization—groundwater in many regions may be viewed as a nonrenewable resource. By pumping water out of the ground at a rate faster than nature replaces it, people effectively mine the groundwater supply. In fact, in portions of the desert Sunbelt region of the United States, supplies of young groundwater have already been exhausted, and deep wells now extract 10,000-year-old groundwater. Such ancient water has been in rock so long that some of it has become an unusable soup of dissolved minerals. A number of other problems accompany the depletion of groundwater.

- *Lowering the water table*: When we extract groundwater from wells at a rate faster than it can be resupplied, the water table drops. First, a cone of depression forms locally around the well; then the water table gradually becomes lower in a broad region. As a consequence, existing wells, springs, and rivers dry up (Fig. 16.10a, b). To continue tapping into the water supply, we must drill progressively deeper.

  The water table can also drop when people divert surface water from the recharge area. Such a problem has developed in the Everglades of southern Florida, a huge swamp where, before the expansion of Miami and the development of agriculture, the water table lay at the ground surface. Diversion of water from the Everglades' recharge area into canals has significantly lowered the water table, causing parts of the Everglades to dry up (Fig. 16.10c, d).

- *Reversing the flow direction of groundwater*: The cone of depression that develops around a well creates a local slope to the water table. The resulting hydraulic gradient may be large enough to reverse the flow direction of nearby groundwater (Fig. 16.11a, b). Such reversals can allow pollutants, seeping out of a septic tank, to contaminate a well.

- *Saline intrusion*: In coastal areas, fresh groundwater lies in a layer above saltwater that entered the aquifer from the adjacent ocean (Fig. 16.11c, d). (Saltwater is denser than freshwater, so the fresh groundwater floats above it.) If

FIGURE 16.10 Effects of human modification of the water table.

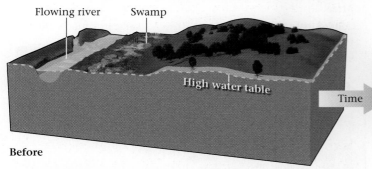

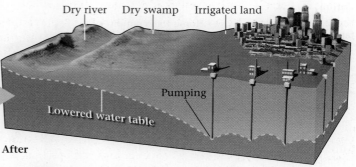

Before

(a) Before humans start pumping groundwater, the water table is high. A swamp and permanent stream exist.

After

(b) Pumping for consumers in a nearby city causes the water table to sink in, so the swamp dries up.

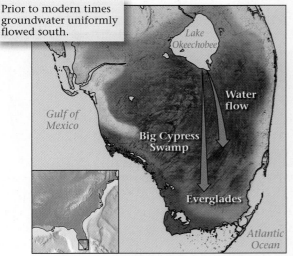

(c) The Florida Everglades before the advent of urban growth and intensive agriculture.

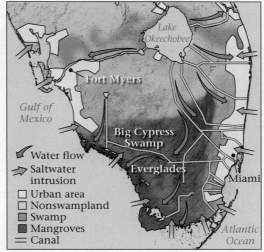

(d) Channelization and urbanization have removed water from recharge areas, disrupting flow paths.

people pump water out of a well too quickly, the boundary between the saline water and the fresh groundwater rises. And if this boundary rises above the base of the well, then the well will start to yield useless saline water.

■ *Pore collapse and land subsidence*: When groundwater fills the pore space of a rock, it holds the grains of the rock or regolith apart, for water cannot be compressed. The extraction of water from a pore eliminates the support holding the grains apart, because the air that replaces the water *can* be compressed. As a result, the grains pack more closely together. Such pore collapse permanently decreases the porosity and permeability of a rock, and thus lessens its value as an aquifer (Fig. 16.11e, f).

Pore collapse also decreases the volume of the aquifer, with the result that the ground above the aquifer sinks. Such land subsidence may cause fissures at the surface to develop and the ground to tilt. Buildings constructed over regions undergoing land subsidence may themselves tilt, or their foundations may crack. The Leaning Tower of Pisa, in Italy, tilts because the removal of groundwa-

ter caused its foundation to subside. In the San Joaquin Valley of California, the land surface subsided by 9 m between 1925 and 1975, because water was removed to irrigate farm fields. In coastal areas, land subsidence may even make the land surface sink below sea level. The flooding of Venice, Italy, is due in part to land subsidence accompanying the withdrawal of groundwater.

To avoid such problems, communities have sought to prevent groundwater depletion, either by directing surface water into recharge areas or by pumping surface water back into the ground. For example, some communities have excavated to lower the land surface of a park, so that the park acts as a catchment for storm water. This water can then infiltrate down into the soil.

Natural Groundwater Quality

Much of the world's groundwater is crystal clear, pure enough to drink right out of the ground, for rocks and sediment are natural filters capable of efficiently removing suspended solids (mud and solid waste) from groundwater. But in some places,

FIGURE 16.11   Some causes of groundwater problems.

Before

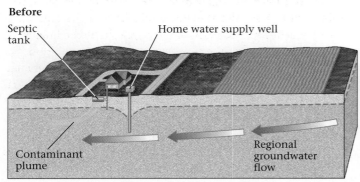

Septic tank

Home water supply well

Contaminant plume

Regional groundwater flow

(a) Before pumping, effluent from a septic tank drifts with the regional groundwater flow, and the home well pumps clean water.

After

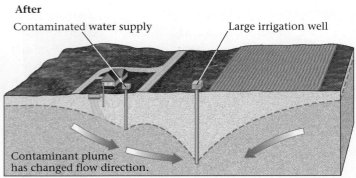

Contaminated water supply

Large irrigation well

Contaminant plume has changed flow direction.

(b) After pumping by a nearby irrigation well, effluent flows into the home well in response to the new local slope of the water table.

Before

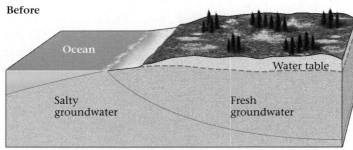

Ocean

Water table

Salty groundwater

Fresh groundwater

(c) Before pumping, fresh groundwater forms a lens below the ground.

After

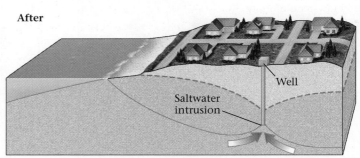

Well

Saltwater intrusion

(d) If the fresh water is pumped too fast, saltwater from below is sucked up into the well. This is saltwater intrusion.

Before

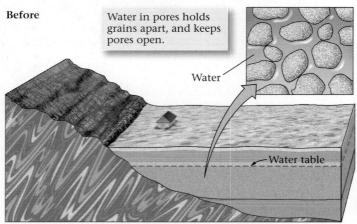

Water in pores holds grains apart, and keeps pores open.

Water

Water table

(e) When intensive irrigation removes groundwater, pore space in an aquifer collapses.

After

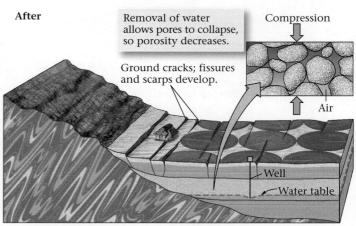

Removal of water allows pores to collapse, so porosity decreases.

Ground cracks; fissures and scarps develop.

Compression

Air

Well

Water table

(f) As a result, the land surface sinks, leading to the formation of ground fissures and cracked houses.

groundwater either is unusable or must be treated before being used, even though it has not been contaminated by human-created materials. Specifically, groundwater quality can be affected by the presence of natural dissolved ions that enter solution as groundwater flows through rock or sediment. For example, groundwater that has passed through salt-containing strata may become salty and unsuitable for irrigation or drinking. Groundwater that has passed through limestone or dolomite contains dissolved calcium (Ca^{+2}) and magnesium (Mg^{+2}) ions; this water,

called hard water, can be a problem because carbonate minerals precipitate from it to form "scale" that clogs pipes. Groundwater that has passed through iron-bearing rocks may contain dissolved iron oxide that precipitates to form rusty stains. Some groundwater contains dissolved hydrogen sulfide, which comes out of solution when the groundwater rises to the surface; hydrogen sulfide is a poisonous gas that has a rotten-egg smell. In recent years, concern has grown about arsenic, a highly toxic chemical that enters groundwater when arsenic-bearing minerals dissolve.

Human-Caused Groundwater Contamination

In addition to natural chemicals added to groundwater by interaction with the rock through which it flows, human activities have increasingly added a great variety of unhealthy materials (pesticides, gasoline and oil, acids, radioactive wastes, exotic organic chemicals, bacteria, and viruses) to groundwater. Such **groundwater contamination** may be due to leaking septic systems and landfills, leaking fuel and chemical storage tanks, the intentional pumping of waste down wells, the leaching of chemicals from minerals exposed by mining, and the infiltration of chemicals and wastes washed from city streets, agricultural fields, and livestock facilities (Fig. 16.12a). Typically, contamination disperses from its source to form a diffuse cloud of contaminants mixed with, or dissolved in, groundwater. Geologists refer to such a cloud as a contaminant plume (Fig. 16.12b). The concentration of contamination tends to decrease toward the margins of the plume.

The best way to avoid contamination is to prevent contaminants from entering groundwater in the first place. This can be done by locating potential sources of contamination on impermeable bedrock so that they are isolated from the aquifer. If such a site is not available, the storage area should be lined with a thick layer of clay, for the clay not only acts as an aquitard, but it can hold on to contaminants. For this reason, environmental engineers commonly place landfills on top of a liner of packed clay, or place waste in durable, sealed containers. Government agencies have studied various options to store the containers safely. One option involves stockpiling them in tunnels cut into salt domes, for salt is impermeable. Another option involves storing waste containers in tunnels above the water table. For example, American nuclear waste may be stored in a network of tunnels 300 m beneath Yucca Mountain, Nevada. This mountain consists of dry, fairly impermeable tuff, below which the water table lies at great depth.

Fortunately, in some cases, natural processes can clean up groundwater contamination. Chemicals may be absorbed by clay, oxygen in the water may oxidize the chemicals, and bacteria in the water may metabolize the chemicals, thereby turning them

FIGURE 16.12 Contamination plumes in groundwater.

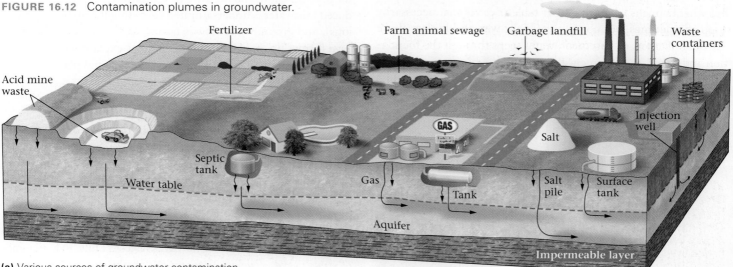

(a) Various sources of groundwater contamination.

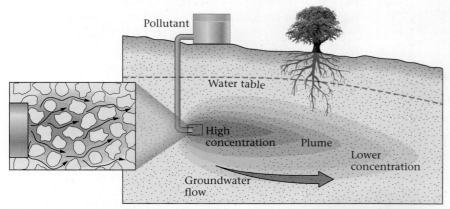

(b) A contaminant plume as seen in cross section. The darker the color, the greater the concentration of contaminant.

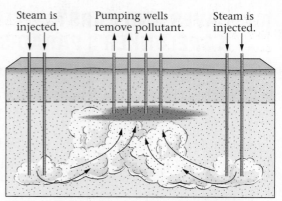

(c) Steam flushing can clean groundwater. Injected steam drives contaminated water into pumping wells.

into harmless substances. Where contaminants do make it into an aquifer, environmental engineers drill test wells to determine which way and how fast the contaminant plume is flowing; once they know the flow path, they can close wells in the path to prevent consumption of contaminated water. Alternatively, engineers attempt to clean the groundwater by drilling a series of extraction wells to pump it out of the ground. If the contaminated water does not rise fast enough, engineers drill injection wells to force clean water or steam into the ground beneath the contaminant plume (Fig. 16.12c). The injected fluids then push the contaminated water up into the extraction wells. More recently, environmental engineers have begun exploring techniques of chemical- and bio-remediation. For example, injecting oxygen and nutrients into a contaminated aquifer can foster growth of bacteria, which break down molecules of contaminants. Iron filings have been buried in the path of flowing groundwater to react with and remove some contaminants. Needless to say, cleaning techniques are expensive and generally only partially effective.

Unwanted Effects of Rising Water Tables

We've seen the negative consequences of sinking water tables, but what happens when the water table rises? Is that necessarily good? Sometimes, but not always. If the water table rises above the floor of a house's basement, water seeps through the foundation and floods the basement floor. Catastrophic damage occurs when a rising water table weakens the base of a hill slope. The addition of water into pore space buoys up the overlying rock or regolith, and therefore makes the slope unstable and likely to slip. Thus, rising water tables can trigger landslides and slumps.

TAKE-HOME MESSAGE

Groundwater usage can cause problems. Overpumping lowers the water table, causes land subsidence, and causes saltwater intrusion. Contamination can ruin a groundwater supply for years to come. Remediation of groundwater is extremely expensive.

16.7 CAVES AND KARST: A SPELUNKER'S PARADISE

Dissolution and the Development of Caves

In 1799, as legend has it, a hunter by the name of Hutchins was tracking a bear through the woods of Kentucky when the bear suddenly disappeared on a hill slope. Baffled, Hutchins plunged through the brambles trying to sight his prey. Suddenly he felt a draft of surprisingly cool air flowing down the slope from uphill. Now curious, Hutchins climbed up the hill and found a dark portal into the hill slope beneath a ledge of rocks. Bear tracks were all around—was the creature inside? Hutchins returned later with a lantern and stepped into the passageway. It led into a large, open room. Hutchins had discovered Mammoth Cave, an immense network—over 540 km in total length—of tunnel-like passages connecting underground rooms.

Large cave networks such as Mammoth Cave develop primarily in limestone bedrock, as a consequence of the dissolution of limestone by groundwater. Slightly acidic groundwater reacts with calcite to produce HCO_3^{--} and Ca^{--} ions, which readily dissolve. Of note, groundwater develops acidity because the water that ultimately becomes groundwater absorbed carbon dioxide (CO_2) partly from the atmosphere when it fell as rain, and more from organic-rich soil as it percolated down to the water table. When it absorbs CO_2, water transforms into carbonic acid.

Geologists have debated about the depth at which limestone caves form in the subsurface. Acidic rainwater clearly dissolves limestone near the ground surface, as indicated by the presence of extensive pitting on limestone bedrock surfaces and along joints in limestone bedrock. And some caves may also form at depth below the water table. It appears, however, that most dissolution takes place in limestone that lies just below the water table, for in this interval the acidity of the groundwater remains high, the mixture of groundwater and newly introduced rainwater is undersaturated (meaning it can dissolve more ions), and groundwater flow is fastest. The association between cave formation and the water table helps explain why openings in a cave system align along the same horizontal plane.

The Character of Cave Networks

Caves in limestone usually occur as part of a network. Networks include rooms, or chambers, which are large, open spaces sometimes with cathedral-like ceilings, and tunnel-shaped or slot-shaped passages (see Geology at a Glance, pp. 446–447). The shape of the cave network reflects variations in permeability and in the composition of the rock from which the caves formed. Larger open spaces develop where the limestone was most soluble and where groundwater flow was fastest. Thus, in a sequence of strata, caves develop preferentially in the more soluble limestone beds. Passages in cave networks typically follow preexisting joints, for the joints provide secondary porosity along which groundwater can flow faster (Fig. 16.13a). Because joints commonly occur in orthogonal systems (consisting of two sets of joints oriented at right angles to each other; see Chapter 9), passages form a grid.

Precipitation and the Formation of Speleothems

When the water table drops below the level of a cave, the cave becomes an open space filled with air. A dilute solution containing dissolved calcite then emerges from the rock above the cave and drips from the ceiling or trickles down the walls. As soon as this solution reenters the air, it evaporates a little and releases carbon dioxide. As a result, calcite precipitates out of the water. Rock formed by such precipitation is a type of travertine (chemical limestone) called dripstone, and the various intricately shaped formations that grow in caves by the accumulation of dripstone are called **speleothems**.

Cave explorers, or spelunkers, and geologists have developed a detailed nomenclature for different kinds of speleothems. Where water drips from the ceiling of the cave, the precipitated travertine adds to the tip of an icicle-like cone called a stalactite. Initially, calcite precipitates around the outside of the drip, forming a delicate, hollow stalactite called a soda straw. But eventually, the soda straw fills up, and water migrates down the margin of the cone to form a more massive, solid stalactite. Where the drips hit the floor, the resulting travertine builds an upward-pointing cone called a stalagmite. If the process of dripstone for-

FIGURE 16.13 Development of karst and dripstone.

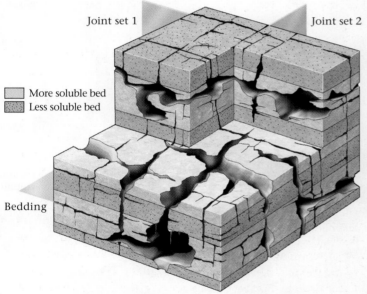

(a) Joints act as conduits for water in cave networks. Caves and passageways follow joints, and preferentially form in more soluble beds.

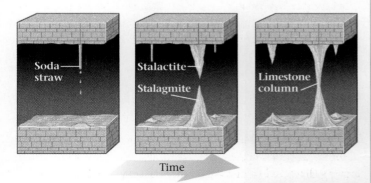

(b) The evolution of a soda straw stalactite into a limestone column.

mation in a cave continues long enough, stalagmites merge with overlying stalactites to create travertine columns (Fig. 16.13b). If the groundwater flows along the surface of a wall, it drapes the wall with cloth-like sheets of travertine called flowstone. Thin sheets of dripstone and flowstone tend to be translucent and, when lit from behind, glow with an eerie amber light.

The Formation of Karst Landscapes

Limestone bedrock underlies most of the Kras Plateau in Slovenia, along the eastern coast of the Adriatic Sea. The name *kras*, which means bare, rocky ground, is apt because of the rough, unvegetated land surface in the region. Throughout the area, the bedrock contains abundant caves; the landscape is pockmarked by deep circular-to-elongate depressions, or sinkholes, which, as we have seen, can form where the roof of a cave collapses. Such sinkholes are called collapse sinkholes. (Sinkholes can also form where acidic water accumulates in a pool at the surface, seeps down, and dissolves bedrock below.) The sinkholes of the Kras Plateau are separated from one another by hills or walls of bedrock (Fig. 16.14). Locally, where most of a cave collapsed, a natural bridge spans the cave remnant. Where the water table rises above the floor of a sinkhole, the sinkhole fills to become a lake. And where surface streams intersect cracks or holes that link to the caves below, the water disappears into the subsurface and becomes an underground stream (Fig. 16.14b). Such disappearing streams reemerge from a cave entrance downstream.

Geologists refer to landscapes such as the Kras Plateau in which surface features reflect the dissolution of bedrock below as **karst landscapes**, from the Germanized version of kras (see Geotour 16).Karst landscapes form in a series of stages (Fig. 16.15a–c).

- *The establishment of a water table in limestone*: The story of a karst landscape begins after the formation of a thick interval of limestone. If relative sea level drops, a water table can develop in the limestone below the ground surface. If, however, the limestone was buried deeply, it must first be uplifted before it can contain the water table.

- *The formation of a cave network*: Once the water table has been established, dissolution begins and a cave network develops.

- *A drop in the water table*: If the water table later becomes lower, either because of a decrease in rainfall or because nearby rivers cut down through the landscape and drain the region, newly formed caves dry out. Downward-percolating groundwater emerges from the roofs of the caves; dripstone and flowstone precipitate.

- *Roof collapse*: If rocks fall off the roof of a cave for a long time, the roof eventually collapses. Such collapse creates sinkholes and troughs, leaving behind hills, ridges, and natural bridges of limestone.

Caves and Karst Landscapes

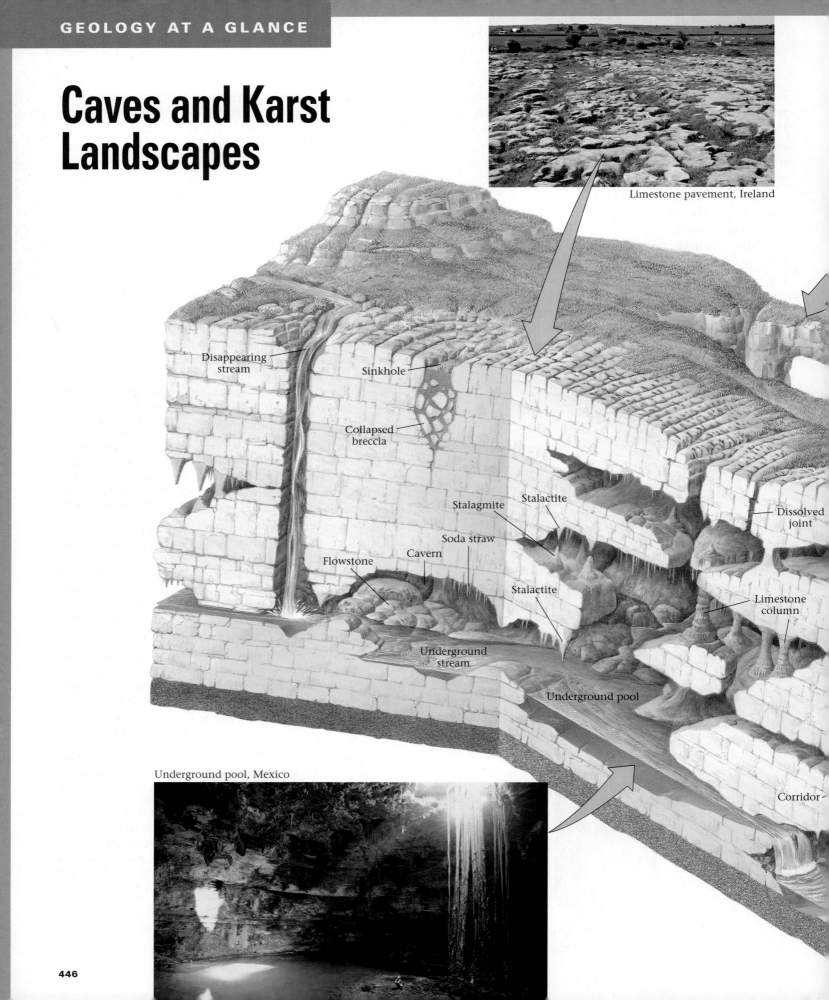

Limestone pavement, Ireland

Disappearing stream

Sinkhole

Collapsed breccia

Stalagmite

Stalactite

Soda straw

Dissolved joint

Flowstone

Cavern

Stalactite

Limestone column

Underground stream

Underground pool

Corridor

Underground pool, Mexico

Natural Bridge, Virginia

Spelunker crawling in a cave

Limestone, a sedimentary rock made of the mineral calcite, is soluble in acidic water. Much of the water that falls to the ground as rain, or seeps through the ground as groundwater, tends to be acidic, so in regions of the Earth where bedrock consists of limestone, we find signs of dissolution. Underground openings that develop by dissolution are called caves or caverns. Some of these may be large, open rooms, whereas others are long, narrow passages. Underground lakes and streams may form on the floor. A cave's location depends on the orientation of bedding and joints, for these features localize the flow of groundwater.

Caves originally form at or near the water table (the subsurface boundary between rock or sediment in which pores contain air, and rock or sediment in which pores contain water). As the water table drops, caves empty of most water and become filled with air. In many locations, groundwater drips from the ceiling of a cave or flows along its walls. As the water evaporates and thus loses its acidity (because of the evaporation of dissolved carbon dioxide), new calcite precipitates. Over time, this calcite builds into cave formations, or speleothems, such as stalactites, stalagmites, columns, and flowstone.

Distinctive landscapes, called karst landscapes, develop at the Earth's surface over limestone bedrock. In such regions, the ground may be rough where rock has dissolved along joints; where the roofs of caves collapse, sinkholes develop. If a surface stream flows through an open joint into a cave network, we say that the stream is "disappearing." The water from such streams may flow underground for a distance and then reemerge elsewhere as a spring. In some places, the collapse of subsurface openings leaves behind natural bridges.

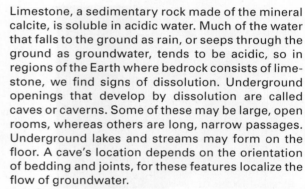

merging spring

Some karst landscapes contain many round sinkholes separated by hills. In regions where vertical joints control roof collapse, steep-sided residual bedrock towers remain between sinkholes. A karst landscape with such spires is called tower karst. The surreal collection of pinnacles constituting the tower karst landscape in the Guilin region of China has inspired generations of artists to portray them on scroll paintings (Fig. 16.16).

Life in Caves

Despite their lack of light, caves are not sterile, lifeless environments. Caves that are open to the air provide a refuge for bats as well as for various insects and spiders. Similarly, fish and crustaceans enter caves where streams flow in or out. Species living in caves have evolved some unusual characteristics. For example, cave fish lose their pigment and in some cases their eyes. Recently, explorers discovered caves in Mexico in which warm, mineral-rich groundwater currently flows. Colonies of bacteria metabolize sulfur-containing minerals in this water, and create thick mats of living ooze in the complete darkness of the cave. Long gobs of this bacteria slowly drip from the ceiling. Because of the mucus-like texture of these drips, they have come to be known as snotites.

TAKE-HOME MESSAGE

Reaction with natural acids dissolves limestone underground to form caverns. Most dissolution takes place near the water table. If the water table sinks, dripping of water in caves can produce speleothems. Collapse of a cavern network produces karst terrain.

FIGURE 16.14 Karst landscapes.

Sinkholes of the Kras Plateau.

A disappearing stream flows into a cave.

FIGURE 16.15 The progressive formation of caves and a karst landscape.

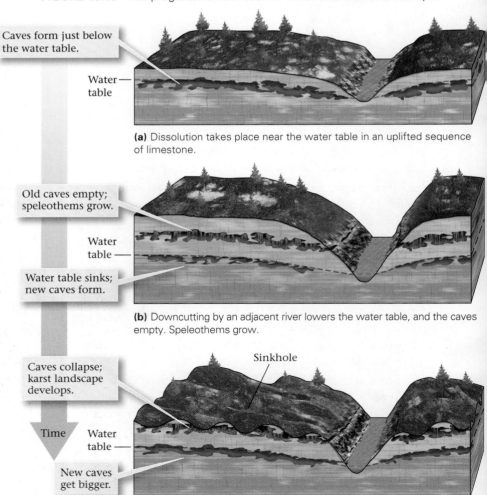

Caves form just below the water table.

Water table

(a) Dissolution takes place near the water table in an uplifted sequence of limestone.

Old caves empty; speleothems grow.

Water table

Water table sinks; new caves form.

(b) Downcutting by an adjacent river lowers the water table, and the caves empty. Speleothems grow.

Caves collapse; karst landscape develops.

Sinkhole

Time

Water table

New caves get bigger.

(c) After roof collapse, the landscape becomes pockmarked with sinkholes.

FIGURE 16.16 Tower karst forms a spectacular landscape in southern China.

Over the centuries, Chinese artists have painted countless scrolls depicting forested towers of karst.

The landscape is treeless today, a consequence of industrialization policies in the 1950s.

Chapter Summary

- During the hydrologic cycle, water infiltrates the ground and fills the pores and cracks in rock and sediment. This subsurface water is called groundwater. The amount of open space in rock or sediment is its porosity; the degree to which pores are interconnected, so that water can flow through, defines its permeability.

- Geologists classify rock and sediment according to their permeability. Aquifers are relatively permeable, and aquitards are relatively impermeable.

- The water table is the surface in the ground above which pores contain mostly air, and below which pores are filled with water. The shape of a water table is a subdued imitation of the shape of the overlying land surface.

- Groundwater flows wherever the water table has a hydraulic gradient, and moves slowly from recharge areas to discharge areas. Darcy's law shows that the rate of movement depends on permeability and on the hydraulic gradient.

- Groundwater can be extracted in wells. An ordinary well penetrates below the water table, but in an artesian well, water rises on its own. Pumping water out of a well too fast causes drawdown, yielding a cone of depression. At a spring, groundwater exits the ground on its own.

- Hot springs and geysers release hot water to the Earth's surface. This water may have been heated by residing very deep in the crust, or by the proximity of a magma chamber.

- Groundwater is a precious resource, used for municipal water supplies, industry, and agriculture. In recent years, some regions have lost their groundwater supply because of overuse or contamination.

- When limestone dissolves just below the water table, underground caves are the result. Soluble beds and joints determine the location and orientation of caves. If the water table drops, caves empty out. Limestone precipitates out of water dripping from cave roofs, and creates speleothems (such as stalagmites and stalactites). Regions that contain abundant caves, some of which have collapsed to form sinkholes, are called karst landscapes.

GEOPUZZLE REVISITED

Water resides underground in the pores and cracks of rock and regolith. Below the water table, this water completely fills open space. When it rains, some water infiltrates the ground and percolates downward until it reaches the water table—it is this water that becomes groundwater. The level of water in a well defines the water table.

Key Terms

Review Questions

1. How do porosity and permeability differ?
2. What factors affect the level of the water table? What factors affect the flow direction of the water below the water table?
3. How does the rate of groundwater flow compare with that of moving ocean water or river currents?
4. What factors control the rate of groundwater flow?
5. How does excessive pumping affect the local water table?
6. How is an artesian well different from an ordinary well?
7. Why do hot springs form and geysers erupt?
8. Is groundwater a renewable or nonrenewable resource?
9. Describe some of the ways in which human activities can adversely affect the water table.
10. What are some sources of groundwater contamination? How can it be prevented?
11. Describe the process leading to the formation of caves and the speleotherms within caves.

On Further Thought

1. You are part of a cave-exploration team that is trying to map a cave network in a temperate region of flat-lying limestone beds that underlie a plateau. A set of NW-SE-trending systematic joints cuts the limestone beds. A river cuts through the region, and the entrance to the cave is along the valley wall. The surface of the river lies about 400 m below the surface of the plateau. What do you predict will be the trend of tunnels in the cave network, and how far below the surface of the plateau do you think the cave network extends?

THE VIEW FROM SPACE   A satellite view of the Arecibo radio telescope in Puerto Rico. The dish was built by smoothing the surface of a 300-m-wide sinkhole in a karst terrain.

CHAPTER 17

Dry Regions:
The Geology of Deserts

In a desert, very little vegetation grows because it rarely rains. But not all deserts are seas of sand. Here in the Sonoran Desert of northwestern Mexico, a saguaro cactus stands guard over a landscape of stony plains and barren rock cliffs.

GEOPUZZLE

Hollywood movies typically portray deserts as endless seas of hot sand, in which nothing at all grows. Is this portrayal accurate?

The bare hills are cut out with sharp gorges, and over their stone skeletons scanty earth clings . . . A white light beat down, dispelling the last tract of shadow, and above hung the burnished shield of hard, pitiless sky.

—Clarence King (1842–1901, 1st director of the U.S. Geological Society, describing a desert)

TAKE-HOME MESSAGE

By the end of this chapter, you should understand what deserts are and why they form. You should also recognize erosional and depositional processes that shape desert landscapes, and the challenge of preventing more lands from becoming deserts.

17.1 INTRODUCTION

For generations, nomadic traders have used camels to traverse the Sahara Desert in northern Africa (Fig. 17.1a). The Sahara, the world's largest desert, receives so little rainfall that it has hardly any surface water or vegetation. So camels must be able to walk for up to three weeks without drinking or eating. They can survive these journeys because they sweat relatively little, thereby conserving their internal water supply; they have the ability to metabolize their own body fat to produce new water; and they can withstand severe dehydration. The survival challenges faced by a camel emphasize that deserts are lands of extremes—extreme dryness, heat, cold, and, in some places, beauty. Desert vistas include everything from sand seas to sagebrush plains, cactus-covered hills to endless stony pavements. Although less populated than other regions on Earth, deserts cover a significant percentage (about 25%) of the land surface, and thus constitute an important component of the Earth System (Fig. 17.1b). In this chapter, we take a look at the desert landscape. We learn why deserts occur where they do, and how erosion and deposition shape their surface. We conclude by exploring life in the desert and by examining the problem of desertification, the gradual transformation of temperate lands into desert.

17.2 WHAT IS A DESERT?

Formally defined, a **desert** is a region that is so arid (dry) that it contains no permanent streams, except for those that bring water in from temperate regions elsewhere. No more than 15% of a desert's surface has a cover of vegetation, for in general, desert conditions exist where less than 25 cm of rain falls per year.

Note that the definition of a desert depends on a region's aridity, not on its temperature. Geologists distinguish between cold deserts, where temperatures generally stay below about 20°C, and hot deserts, where summer daytime temperatures exceed 35°C. Cold deserts exist at high latitudes where the Sun's rays strike the Earth obliquely and thus don't provide much energy (Fig. 17.2), at high elevations where the air is too thin to hold much heat, or in lands adjacent to cold oceans where the cold water absorbs heat from the air above. Hot deserts are found at low latitudes where the Sun's rays strike the desert at a high angle, at low elevations where dense air can hold a lot of heat, and in regions distant from the cooling effect of cold ocean currents. The hottest recorded temperatures on Earth occur in low-latitude, low-elevation deserts—58°C (136°F) in Libya and 56°C (133°F) in Death Valley, California.

FIGURE 17.1 Deserts and their inhabitants.

(a) Camels can survive the harsh desert conditions of the Arabian Desert.

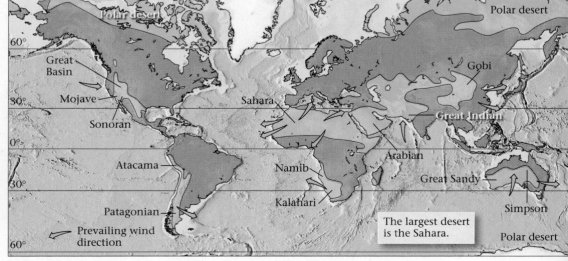

(b) The global distribution of deserts. Arid regions cover 25% of the land surface.

FIGURE 17.2 Subtropical deserts form because the air that convectively flows downward in the subtropics warms and absorbs water as it sinks.

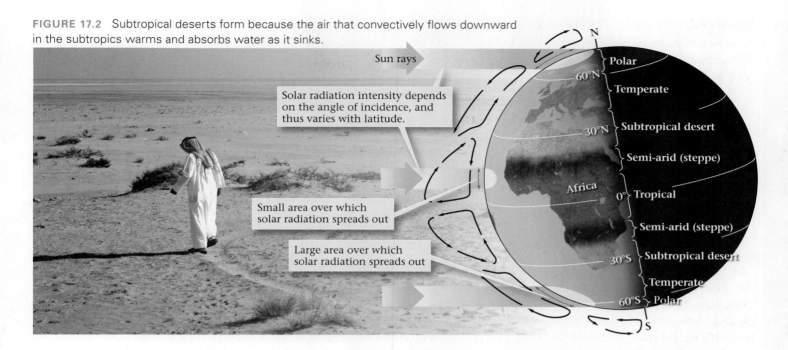

Each desert on Earth has unique characteristics of landscape and vegetation that distinguish it from others. Geologists group deserts into five different classes, based on the environment in which the deserts form.

- *Subtropical deserts*: Subtropical deserts (e.g., the Sahara, Arabian, Kalahari, and Australian) form because of the pattern of convection cells in the atmosphere. At the equator, the air becomes warm and humid, for sunlight is intense and water rapidly evaporates from the ocean. The hot, moisture-laden air rises to great heights above the equator. As this air rises, it expands and cools, and can no longer hold as much moisture. Water condenses and falls in downpours that feed the lushness of the equatorial rain forest. The now-dry air spreads laterally north or south at high altitude. When this air reaches latitudes of 20° to 30°, a region called the subtropics, it has become cold and dense enough to sink (Fig. 17.2; Box 17.1 provides further explanation). Because the air is dry, no clouds form, and intense solar radiation strikes the Earth's surface. The sinking dry air condenses and heats up, soaking up any moisture present. In the regions swept by this hot air on its journey back to the equator, evaporation rates greatly exceed rainfall rates, and thus the regions become deserts.

- *Deserts formed in rain shadows*: As winds blow at low elevation over the ocean toward the shore, the air in the wind absorbs moisture and becomes quite humid. Upon reaching a coastal mountain range, this humid air must rise (Fig. 17.3). As the air rises, it expands and cools. The water it contains condenses and falls as rain on the seaward flank of the mountains, nourishing a coastal rain forest. When the air finally reaches the inland side of the mountains, it has lost all its moisture and can no longer provide rain. As a consequence, a **rain shadow** forms, and the land beneath the rain shadow becomes a desert. A rain-shadow desert developed on the eastern side of the Cascade Mountains in Washington State.

- *Coastal deserts formed along cold ocean currents*: Cold ocean water cools the overlying air by absorbing heat, and decreases the capacity of the air to hold moisture. For example, the cold Humboldt Current, which carries water northward from Antarctica to the western coast of South America, cools and dries the air that blows east over the coast. Thus, rain rarely falls on the coastal areas of Chile and Peru. As a result, this region hosts a desert landscape, including one of the driest regions on Earth, the Atacama Desert (Fig. 17.4a, b). Portions of this narrow (less than 200 km wide) desert, which lies between the Pacific coast on the west and the Andes on the east, received no rain at all between 1570 and 1971.

- *Deserts formed in the interiors of continents*: As air masses move across a continent, they lose moisture by dropping rain, even in the absence of a coastal mountain range. Thus, when an air mass reaches the interior of a large continent such as Asia, it has grown quite dry, so the land beneath becomes arid. The largest example of such a continental-interior desert, the Gobi, lies in central Asia, over 2,000 km away from the nearest ocean.

THE REST OF THE STORY

BOX 17.1

Convection in the Atmosphere and Generation of Prevailing Wind

Light energy from the Sun passes straight down through the atmosphere until it reaches the ground, for air is transparent. The ground, however, consists of opaque material that absorbs some light energy and heats up. Warm ground radiates heat back up into the air, in the form of infrared waves. This infrared radiation heats air at the base of the atmosphere, causing the air to expand and become less dense. Warm air, because of its lower density, is buoyant and rises. As this happens, denser, cold air sinks to take its place. A pattern of circular flow, known as convection, develops (see Appendix).

Because the Earth is a sphere, not all areas receive the same amount of incoming solar energy, or insolation: portions of the Earth's surface hit by direct rays of the Sun receive more energy per square meter than portions hit by oblique rays (see Fig. 17.2). Higher latitudes thus receive less

energy than lower latitudes. (Because of the tilt of Earth's axis, the amount of solar radiation that any point on the surface receives changes during the year, and for that reason, we have seasons.)

The contrast in the amount of solar radiation received by different latitudes means that polar regions are cooler at the surface than are equatorial regions. In 1735, George Hadley, a British mathematician, realized that this contrast could cause air to circulate on a global scale by convection. Specifically, he suggested that warm air at the equator would rise and flow toward the poles, to be replaced by cool polar air, which would flow to the equator at lower elevations (**Fig. 1a**). Hadley's proposal, however, did not take into account an important factor, namely the Earth's rotation and the resulting Coriolis effect. The Coriolis effect, as we learned in Chapter 15, refers to the deflection

that an object is subject to as it moves from the equator to the pole of a rotating sphere (and vice versa).

Because of the Coriolis effect, circulating air in the troposphere splits into three globe-encircling convection cells in each hemisphere (**Fig. 1b**). The low-latitude cells, extending from the equator to a latitude of about 30°, are called Hadley cells, in honor of George Hadley. The mid-latitude cells are called Ferrel cells, in honor of the American meteorologist William Ferrel, who proposed them. The high-latitude cells are simply called polar cells. The global-scale flow in the six major convective cells on Earth creates belts in which surface air generally flows in a consistent direction. Such air flows are called prevailing surface winds—these are labeled on Figure 1b.

FIGURE 1 Global convection cells in the atmosphere.

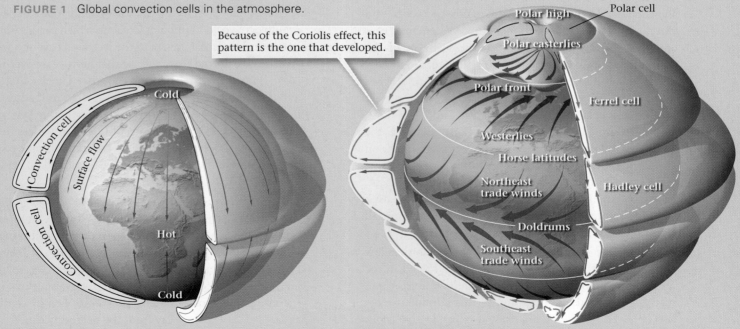

(a) If the Earth did not rotate, two simple convection cells would be established in the atmosphere.

(b) Because of the Coriolis effect, atmospheric circulation breaks into three convection cells within each hemisphere.

FIGURE 17.3 The formation of a rain-shadow desert. Moist air rises and drops rain on the coastal side of the range. By the time the air has crossed the mountains, it is dry.

FIGURE 17.4 The formation of a coastal desert.

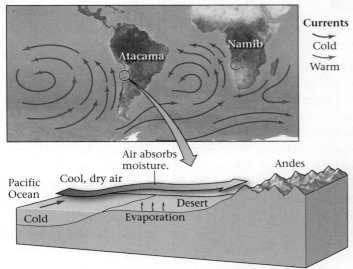

(a) Cold ocean currents cool and dry the air along the coast. This air absorbs moisture from adjacent coastal land.

(b) The Atacama Desert is the driest place on Earth.

- *Deserts of the polar regions*: So little precipitation falls in Earth's polar regions (north of the Arctic Circle, at 66°30'N, and south of the Antarctic Circle, at 66°30'S) that these areas are, in fact, arid. Polar regions are dry, in part, for the same reason that the subtropics are dry (the global pattern of air circulation means that the air flowing over these regions is dry) and, in part, for the same reason that coastal areas along cold currents are dry (cold air holds little moisture).

The distribution of deserts around the world changes through geologic time because of plate tectonics, for plate movements determine the latitude of land masses, the position of land masses relative to the coast, the proximity of land masses to mountain ranges, and the configuration of ocean currents. Because of continental drift, some regions that were deserts in the past are temperate or tropical regions now, and vice versa.

TAKE-HOME MESSAGE

Deserts are so arid that they host only very sparse vegetation. They occur in several settings: subtropical dry climates, rain shadows, coasts bordered by cold currents, continental interiors, and polar regions.

17.3 WEATHERING AND EROSION IN DESERTS

Without the protection of foliage to catch rainfall and slow the wind, and without roots to hold regolith in place, rain and wind can attack and erode the land surface of deserts. As a result, hillslopes are typically bare, and plains can be covered with stony debris or drifting sand.

Arid Weathering and Soil Formation

In the desert, as in temperate climates, physical weathering happens primarily when joints (natural fractures) split rock into pieces. Joint-bounded blocks eventually break free of bedrock and tumble down slopes, fragmenting into smaller pieces as they fall. In temperate climates, thick soil covers bedrock and rubble. In deserts, however, bedrock and rubble remain exposed, creating rugged, rocky escarpments and block-strewn slopes.

FIGURE 17.5 Evidence of chemical weathering in deserts.

(b) In the Painted Desert of Arizona, the different colors of the rock layers are due, in part, to the oxidation state of iron.

(a) Desert varnish is a dark coating on rock surfaces. Native Americans created petroglyphs by chipping through the varnish to reveal the lighter rock beneath.

Chemical weathering happens more slowly in deserts than in temperate or tropical climates, because less water is available to react with rock. Still, rain or dew provides enough moisture for *some* weathering to occur. This water seeps into rock and leaches (dissolves and carries away) calcite, quartz, and various salts. Leaching effectively rots the rock by transforming it into a poorly cemented aggregate. Over time, the rock crumbles and transforms into unconsolidated sediment.

Shiny **desert varnish**, a dark, rusty brown coating of iron oxide, manganese oxide, and clay, covers the surface of rock in deserts. Desert varnish was once thought to form when water from rain or dew seeped into a rock, dissolved iron and magne-

sium ions, and carried the ions back to the surface of the rock by capillary action. More recent studies, however, have shown that desert varnish forms when windborne clay settles on the surface of the rock; in the presence of moisture, microorganisms (bacteria) extract elements from the clay and transform it into iron or manganese oxide. The oxides bind together clay flakes. Such varnish can't form in humid climates, because rain washes the ions and dust away too fast. Desert varnish takes a long time to form. In fact, the thickness of a desert varnish layer provides an approximate estimate of how long a rock has been exposed at the ground surface. Native Americans used desert-varnished rock as a medium for art. By chipping away the varnish to reveal

FIGURE 17.6 Evidence of erosion by running water in deserts.

(a) These hills in the desert near Las Vegas, Nevada, are bone dry, but their shape indicates erosion by water. Note the numerous stream channels.

(b) Gravel and sand are left behind on the floor of a dry wash after a flash flood. The wash has steep walls because downcutting happens so fast.

the underlying lighter-colored rock, they were able to create figures or symbols on a dark background. The resulting drawings are called petroglyphs (Fig. 17.5a).

Because of the lack of plant cover in deserts, variations in bedrock color stand out. Slight variations in the concentration of iron, or in the degree of iron oxidation, result in spectacular color bands in rock layers. The Painted Desert of northern Arizona earned its name from the brilliant and varied hues of its shale (Fig. 17.5b). Locally derived soils typically retain the color of the bedrock from which they were derived.

Water Erosion

Although rain rarely falls in deserts, when it does come, it can radically alter a landscape in a matter of minutes. Since deserts lack plant cover, rainfall, sheetwash, and stream flow all are extremely effective agents of erosion. It may seem surprising, but water generally causes more erosion than does wind in most deserts.

Water erosion begins with the impact of raindrops, which eject sediment into the air. On a hill, the ejected sediment lands downslope. The ground quickly becomes saturated with water during a heavy rain, so water starts flowing across the surface, carrying the loose sediment with it. On hill slopes, networks of stream channels develop (Fig. 17.6a). Within minutes after a heavy downpour begins, dry stream channels may fill with a turbulent mixture of water and sediment, which rushes downstream as a flash flood. When the rain stops, the water sinks into the stream bed's gravel and disappears—such streams are called ephemeral streams (see Chapter 14). Because of the relatively high viscosity of the water, owing to its load of suspended sediment, and the velocity and turbulence of the flow, flash floods in deserts cause intense erosion. Dry stream channels in desert regions of the western United States are called **dry washes**, or arroyos (Fig. 17.6b). In the Middle East and North Africa they are called wadis.

Wind Erosion

In temperate and humid regions, plant cover protects the ground surface from the wind, but in deserts, the wind has direct access to the ground. Wind, just like flowing water, can carry sediment either as suspended load or as surface load (Fig. 17.7a). The wind's **suspended load** consists of fine-grained sediment, such as dust and silt, that floats in the air and moves with it. The suspended sediment can be carried so high into the atmosphere and so far downwind that it may move completely out of its source region.

Moderate to strong winds can roll and bounce sand grains along the ground, a process called **saltation**. Saltation begins when turbulence caused by wind shearing along the ground

surface lifts sand grains. The grains move downwind, following an asymmetric, arch-like trajectory, but eventually return to the ground, where they strike other sand grains, causing the new grains to bounce up and drift or roll downwind. The collisions between sand grains make the grains rounded and frosted. Saltating grains generally rise no more than 0.5 m. But where sand bounces on bedrock during a desert sandstorm, they may rise 2 m and can strip the paint off a car. Saltating grains comprise the wind's **surface load**.

The size of clasts that wind can carry depends on the wind velocity. Wind, therefore, does an effective job of sorting sediment, sending dust-sized particles skyward and sand-sized particles bouncing along the ground, while pebbles and larger grains remain behind. In some cases, wind carries away so much fine sediment that pebbles and cobbles become concentrated at the ground surface. An accumulation of coarser sediment left behind when fine-grained sediment blows away is called a **lag deposit** (Fig. 17.7b).

In many locations, the desert surface resembles a tile mosaic in that it consists of separate stones that fit together tightly, forming a fairly smooth surface layer above a soil composed of silt and clay. Such natural mosaics constitute **desert pavement** (Fig. 17.8a). Typically, desert varnish coats the top surfaces of the stones forming desert pavement. Geologists have come up

FIGURE 17.7 Sediment transport by wind in deserts.

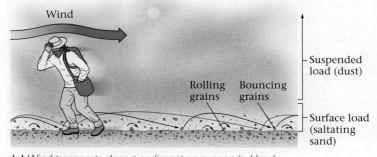

(a) Wind transports desert sediment as suspended load and surface load.

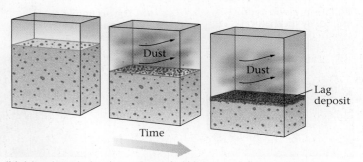

(b) A lag deposit develops when wind blows away finer sediment, leaving behind a layer of coarser grains.

with several explanations for the origin of desert pavements. Recently, researchers have suggested that pavements form as wind-blown dust slowly sifts down onto the stones, and then washes down between the stones. In this model, the stones forming the pavement were never buried, but have been progressively lifted up as soil collects and builds up beneath (Fig. 17.8b). Over time, the rocks at the surface crack into thin slices. Sheetwash, during downpours, washes away fine sediment between fragments, and when soils dry and shrink between rains, the clasts settle together, locking into an armor-like, jigsaw arrangement.

Just as sand blasting cleans the grime off the surface of a building, windblown sand and dust grind away at surfaces in the desert. Over long periods, such wind abrasion creates smooth faces, or facets, on pebbles, cobbles, and boulders. If a rock rolls or tips relative to the prevailing wind direction after it has been faceted on one side, or if the wind shifts direction, a new facet with a different orientation forms, and the two facets join at a

FIGURE 17.8 Desert pavement, and a hypothesis for how it forms by building up a soil from below.

(a) A well-developed desert pavement in the Sonoran Desert, Arizona.

sharp edge. Rocks whose surface has been faceted by the wind are faceted rocks, or **ventifacts** (Fig. 17.9). Recent exploration has revealed landscapes of ventifacts on Mars, where immense dust storms occasionally envelop the planet's entire surface.

Over time, in regions where the substrate consists of soft sediment, wind picks up and removes so much sediment that the land surface sinks. The process of lowering the land surface by wind erosion is called **deflation**. Shrubs can stabilize a small patch of sediment with their roots, so after deflation a forlorn shrub with its residual pedestal of soil stands isolated above a lowered ground surface (Fig. 17.10). In some places, deflation scours a deep, bowl-like depression called a blowout.

TAKE-HOME MESSAGE

Desert soils are thin to absent, so bare rock, debris, and desert pavements form at the surface. Water causes most erosion and sediment transport but rarely flows, so stream channels are dry washes. Wind also transports sediment and can carve ventifacts.

17.4 DEPOSITION IN DESERTS

We've seen that erosion relentlessly eats away at bedrock and sediment in deserts. Where does the debris go? Below, we examine the various desert settings in which sediment accumulates.

Talus Aprons and Alluvial Fans

With time, joint-bounded blocks of rock break off rock ledges and cliffs on the sides of hills. Under the influence of gravity, the resulting debris tumbles downslope and accumulates as a **talus apron** at the base of a hill. Talus aprons can survive for a long time in desert climates, so we typically see them fringing the base of cliffs in deserts (Fig. 17.11a).

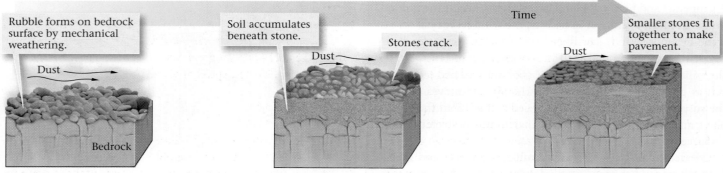

(b) Desert pavement forms in stages. First, loose pebbles and cobbles collect at the surface. Dust settles among the stones and builds up a soil layer below. The stones eventually crack into smaller pieces and settle to form a mosaic-like pavement.

FIGURE 17.9 The progressive development of a ventifact.

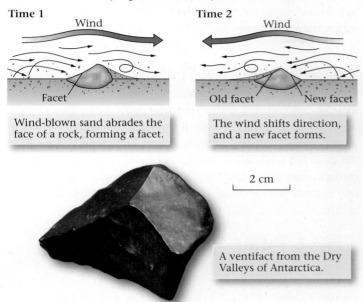

Time 1
Wind

Time 2
Wind

Facet

Old facet   New facet

Wind-blown sand abrades the face of a rock, forming a facet.

The wind shifts direction, and a new facet forms.

2 cm

A ventifact from the Dry Valleys of Antarctica.

FIGURE 17.10 Wind erosion in Death Valley has left bushes perched on little mounds, where the roots keep soil from blowing away.

FIGURE 17.11 Production and transportation of debris and sediment in deserts.

(a) This talus apron along the base of a desert cliff formed from rocks that broke off and tumbled down the cliff.

(b) Sediment in this alluvial fan in Death Valley was carried by flash floods. The fan forms at the mountain front where water slows.

Flash floods can carry sediment downstream and away from eroding cliffs in an ephemeral stream channel. When the turbulent water flows out into a plain at the mouth of a canyon, it spreads out over a broader surface, and friction causes the water flow to slow down. As a consequence, sediment in the water settles out. The resulting lenses of sediment cause the channel that has emerged from the mountains to subdivide into a number of subchannels (distributaries) that diverge outward in a broad fan. The fan of distributaries spreads the sediment, or alluvium, out into a broad **alluvial fan**, a wedge- or apron-shaped pile of sediment (Fig. 17.11b). Alluvial fans emerging from adjacent valleys may merge and overlap along the front of a mountain range, creating an elongate wedge of sediment called a bajada. Over long periods, the sediment of bajadas fills in adjacent valleys to depths of several kilometers.

Playas and Salt Lakes

Water from a flash flood may make it out to the center of an alluvium-filled basin, but if the supply of water is relatively small, it quickly sinks into the permeable alluvium without accumulating as a standing body of water. During a particularly large storm or an unusually wet spring, however, a temporary lake (a playa lake) may develop over the low part of a basin. During drier times, such desert lakes evaporate entirely, leaving behind a dry flat lake bed known as a **playa** (Fig. 17.12a, b). Over time, a smooth crust consisting of clay and various salts (halite, gypsum, and borax) accumulates on the surface of playas.

Where sufficient water flows into a desert basin, it creates a permanent lake. If the basin is an interior basin, with no outlet to the sea, the lake becomes very salty, because although its water escapes by evaporation in the desert sun, its salt cannot. The Great Salt Lake, in Utah, exemplifies the consequences of this process. Even though the streams feeding the lake are fresh enough to drink, their water contains trace amounts of

FIGURE 17.12 Playas form where a shallow, salty lake dries up.

(a) This playa in California formed at the base of a bajada.

(b) White salt crystals encrust the floor of a playa in Death Valley.

dissolved salt ions. Because the lake has no outlet, these ions have become concentrated in the lake over time, making it even saltier than the ocean.

Deposition from the Wind

As mentioned earlier, wind carries two kinds of sediment loads—a suspended load of dust-sized particles and a surface load of sand. Much of the dust is carried out of the desert and accumulates elsewhere. Sand, however, cannot travel far, and accumulates within the desert in piles called dunes, ranging in size from less than a meter to over 300 m high. In favorable locations, dunes accumulate to form vast sand seas hundreds of meters thick. We'll look at dunes in more detail later in this chapter.

TAKE-HOME MESSAGE

Sediment carried by water and wind in deserts accumulates in a variety of landforms. Alluvial fans form at the outlets of canyons, playas form where water temporarily collects in basins, and dunes form where large amounts of sand are available.

17.5 DESERT LANDSCAPES

The popular media commonly portray deserts as endless seas of sand, piled into dunes that hide the occasional palm-studded oasis. In reality, vast sand seas are merely one type of desert landscape. Some deserts are broad rocky plains, others sport a stubble of cacti and other hardy desert plants, and still others contain intricate rock formations that look like medieval castles. In this section, we'll see how the erosional and depositional processes described above lead to the formation of such contrasting landscapes (see **Geotour 17** on p. GT-36).

Rocky Cliffs and Mesas

In hilly desert regions, the lack of soil exposes rocky ridges and cliffs. As noted earlier, cliffs erode when rocks split away along vertical joints. When this happens, the cliff face retreats but retains roughly the same form. The process, commonly referred to as **cliff retreat**, or scarp retreat, occurs in fits and starts—a cliff may remain unchanged for decades or centuries, and then suddenly a block of rock falls off and crumbles into rubble at the foot of the cliff. Cliffs formed from stratified rocks commonly develop a step-like shape; strong layers (sandstone or limestone) become vertical cliffs, and weak layers (shale) become rubble-covered slopes.

With continued erosion and cliff retreat, a plateau of rock slowly evolves into a cluster of isolated hills, ridges, or columns (Fig. 17.13a). Flat-lying strata or flat-lying layers of volcanic

FIGURE 17.13 The formation of cliffs, buttes, and mesas in deserts.

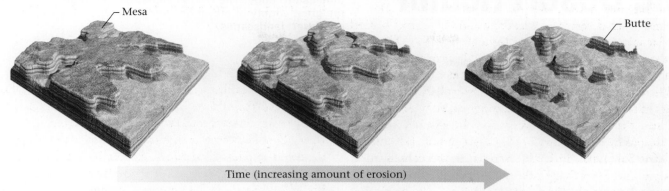

Time (increasing amount of erosion)

(a) Because of cliff retreat, a once-continuous layer of rock evolves into a series of isolated remnants. If the bedding is horizontal, the resulting landforms have flat tops.

(b) Cliffs on the buttes in Monument Valley, Arizona, formed when massive sandstone slabs broke free along large joints and tumbled downslope. The gentler slopes are shale.

(c) Hoodoos in Bryce Canyon, Utah, formed by the erosion of sandstone, siltstone, and shale.

rocks erode to make flat-topped hills. These go by different names, depending on their size. Large examples (with a top surface area of several square kilometers) are **mesas**, from the Spanish word for table. Medium-sized examples are **buttes** (Fig. 17.13b). Small examples, whose height greatly exceeds their top surface area, are **chimneys**. Erosion of strata has resulted in the skyscraper-like buttes of Monument Valley, Arizona, and the stark cliffs of Canyonlands National Park. Bryce Canyon National Park in Utah contains countless chimneys of brightly colored shale and sandstone—locally, these chimneys are known as hoodoos (Fig. 17.13c). Natural arches, such as those at Arches National Monument, form when erosion along joints leaves narrow walls of rock. When the lower part of the wall erodes while the upper part remains, an arch results (see **Geology at a Glance**, pp. 462–463).

In places where bedding dips at an angle to horizontal, flat-topped mesas and buttes don't form; rather, asymmetric ridges called cuestas develop. A joint-controlled cliff forms the steep front side of a cuesta, while the tilted top surface of a resistant bed forms the gradual slope on the backside (Fig. 17.14a). If the bedding dip is steep to near vertical, a narrow symmetrical ridge, called a hogback, forms.

With progressive cliff retreat on all sides of a hill, finally all that remains of the hill is a relatively small island of rock, surrounded by alluvium-filled basins. Geologists refer to such islands of rock by the German word **inselberg** (island mountain; Fig. 17.14b). Depending on the rock type or the orientation of stratification in the rock, and on rates of erosion, inselbergs may be sharp-crested, plateau-like, or loaf-shaped (steep sides and a rounded crest) (Fig. 17.14c).

The Desert Realm

The desert of the Basin and Range Province in Utah, Nevada, and Arizona consists of alternating basins (grabens or half-grabens) separated by narrow ranges (tilted fault blocks). The Sierra Nevada, underlain largely by granite, border the western edge of the province, while the Colorado Plateau, underlain by flat-lying sedimentary strata, borders the eastern edge. The overall climate of the region is dry. Because of the great variety of elevations and rock types, the region hosts different desert landscapes.

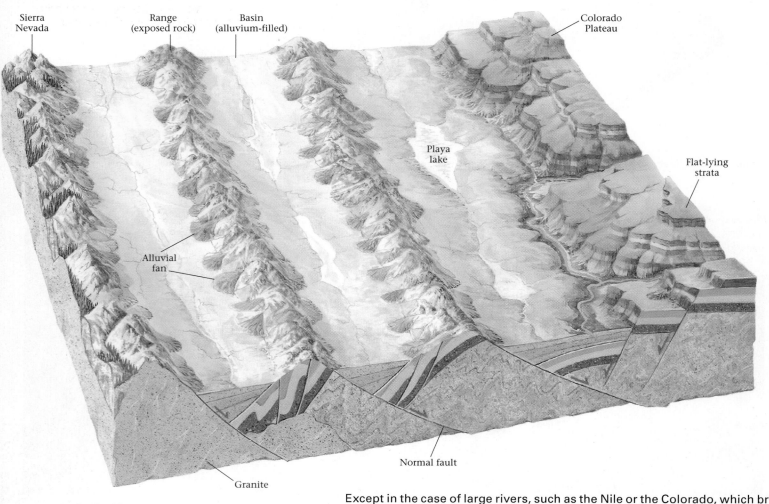

Sierra Nevada

Range (exposed rock)

Basin (alluvium-filled)

Colorado Plateau

Playa lake

Flat-lying strata

Alluvial fan

Granite

Normal fault

Where there is a large supply of sand, a variety of sand dunes develop. The geometry of a particular sand dune (such as barchan, longitudinal, or star) depends on the sand supply and the wind. Inside sand dunes, we find cross beds.

Barchan dune

Cross beds

Except in the case of large rivers, such as the Nile or the Colorado, which bring water into a desert region from a more temperate region, streams in deserts fill with water only after heavy rains. At other times, the stream channels are dry. These channels are called dry washes, arroyos, or wadis. When there is a heavy rain, water cannot be absorbed into the ground fast enough, so runoff enters dry washes and fills them very quickly, creating a flash flood. The turbulent, muddy water of a flash flood can transport even large boulders. This flash flood is rushing down a stream in the Sonoran Desert of Arizona.

Flash flood

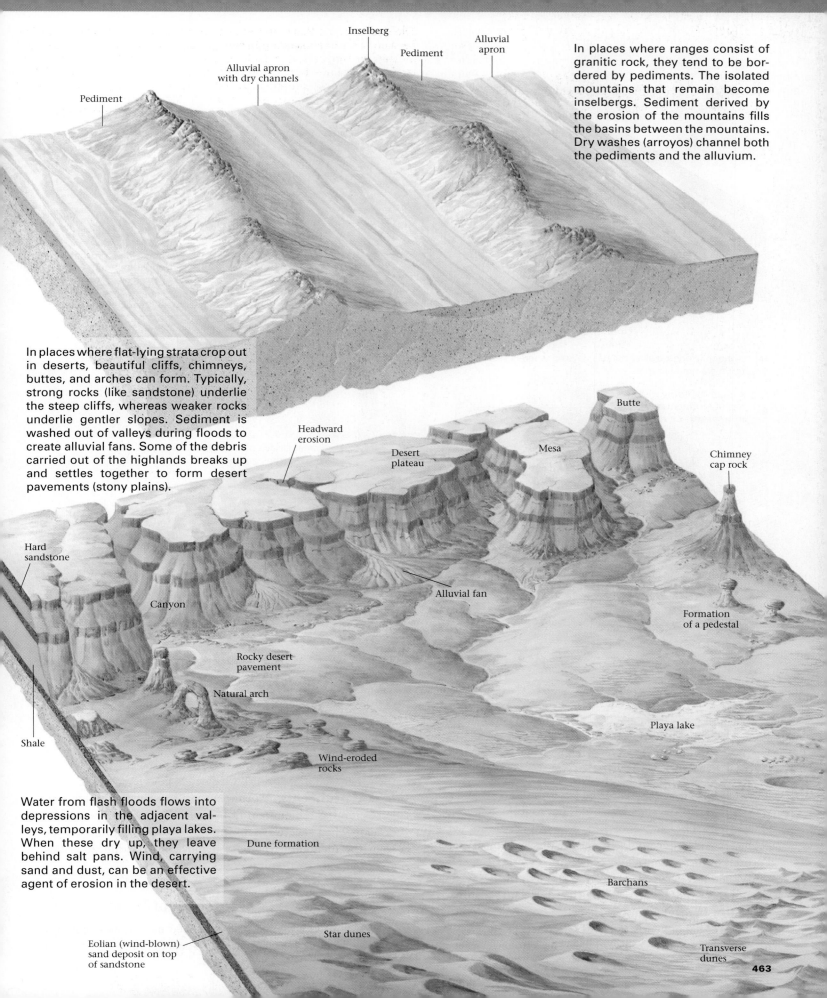

Pediment

Alluvial apron with dry channels

Inselberg

Pediment

Alluvial apron

In places where ranges consist of granitic rock, they tend to be bordered by pediments. The isolated mountains that remain become inselbergs. Sediment derived by the erosion of the mountains fills the basins between the mountains. Dry washes (arroyos) channel both the pediments and the alluvium.

In places where flat-lying strata crop out in deserts, beautiful cliffs, chimneys, buttes, and arches can form. Typically, strong rocks (like sandstone) underlie the steep cliffs, whereas weaker rocks underlie gentler slopes. Sediment is washed out of valleys during floods to create alluvial fans. Some of the debris carried out of the highlands breaks up and settles together to form desert pavements (stony plains).

Headward erosion

Desert plateau

Mesa

Butte

Chimney cap rock

Hard sandstone

Canyon

Alluvial fan

Formation of a pedestal

Rocky desert pavement

Natural arch

Playa lake

Shale

Wind-eroded rocks

Water from flash floods flows into depressions in the adjacent valleys, temporarily filling playa lakes. When these dry up, they leave behind salt pans. Wind, carrying sand and dust, can be an effective agent of erosion in the desert.

Dune formation

Barchans

Star dunes

Eolian (wind-blown) sand deposit on top of sandstone

Transverse dunes

463

FIGURE 17.14 The formation of cuestas and inselbergs, due to erosion and deposition in deserts.

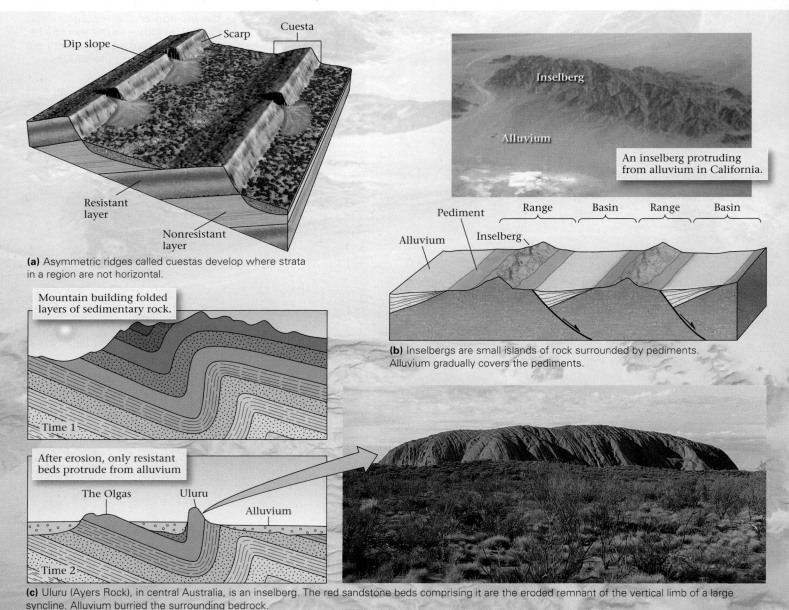

(a) Asymmetric ridges called cuestas develop where strata in a region are not horizontal.

Mountain building folded layers of sedimentary rock.

Time 1

After erosion, only resistant beds protrude from alluvium

Time 2

(b) Inselbergs are small islands of rock surrounded by pediments. Alluvium gradually covers the pediments.

An inselberg protruding from alluvium in California.

(c) Uluru (Ayers Rock), in central Australia, is an inselberg. The red sandstone beds comprising it are the eroded remnant of the vertical limb of a large syncline. Alluvium burried the surrounding bedrock.

Stony Plains and Pediments

The coarse sediment eroded from desert mountains and ridges washes into the lowlands and builds out to form gently sloping alluvial fans. The surfaces of these gravelly piles are strewn with pebbles, cobbles, and boulders, and are dissected by dry washes. Portions of these stony plains evolve into desert pavements.

When travelers began trudging through the desert Southwest of the United States during the nineteenth century, they found that in many locations the wheels of their wagons were rolling over flat or gently sloping bedrock surfaces. These bedrock surfaces extended outward like ramps from the steep cliffs of a mountain range on one side to alluvium-filled valleys on the other (see Fig. 17.14b). Geologists now refer to such surfaces as **pediments**. A pediment is a consequence of erosion, left behind as a mountain front gradually retreats. Pediments develop when sheetwash during floods carries sediment away from the mountain front. As it moves, the sediment grinds away the bedrock that it tumbles over.

Seas of Sand: The Geometry of Dunes

A sand dune is a pile of sand deposited by a moving current. Dunes in deserts form because of the wind. Some start to form where sand becomes trapped on the windward side of an obstacle, such as a rock or a shrub. Gradually the sand builds downwind into the lee of the obstacle. Once initiated, the dune itself affects the wind flow, and sand accumulates on the lee (downwind) side of the dune. Here, sand slides down the lee surface of the dune, so this surface is aptly named the slip face.

In places where abundant sand accumulates, sand seas bury the landscape. The wind builds the sand into dunes that display a variety of shapes and sizes, depending on the character of the wind and the sand supply (Fig. 17.15a). Where the sand is relatively scarce and the wind blows steadily in one direction, beautiful crescents called barchan dunes develop, with the tips of the crescents pointing downwind. If the wind shifts direction frequently, a group of crescents pointing in different directions overlap one another, creating a constantly changing star dune. Where enough sand accumulates to bury the ground surface completely and only moderate winds blow, sand piles into simple, wave-like shapes called transverse dunes. The crests of transverse dunes lie perpendicular to the wind direction. Strong winds may break through transverse dunes and change them into parabolic dunes, whose ends point in the upwind direction. Finally, if there is abundant sand and a strong, steady wind, the sand streams into longitudinal dunes, whose axis lies parallel to the wind direction.

In a sand dune, sand saltates up the windward side of the dune, blows over the crest of the dune, and then settles on the steeper, lee face of the dune. The slope of this face attains the angle of repose, the slope angle of a freestanding pile of sand. As sand collects on this surface, it may become unstable and slide down the slope—as we've noted, geologists refer to the lee side of a dune as the slip face (Fig. 17.15b). As progressively more sand accumulates on the slip face, the crest of the dune migrates downwind, and former slip faces become preserved inside the dune. In cross section, these slip faces appear as cross beds (Fig. 17.15c). The surfaces of dunes are not, in general, smooth surfaces, but rather are covered with small ripples.

TAKE-HOME MESSAGE

Erosion of thick layers of horizontal sedimentary rock in deserts yields buttes and mesas; erosion of tilted layers forms cuestas. Stony plains develop if finer sediment blows or washes away. Seas of sand contain many types of dunes.

FIGURE 17.15 The types of sand dunes and the cross beds within them.

(a) The various kinds of sand dunes.

(b) A sand dune with surface ripples.

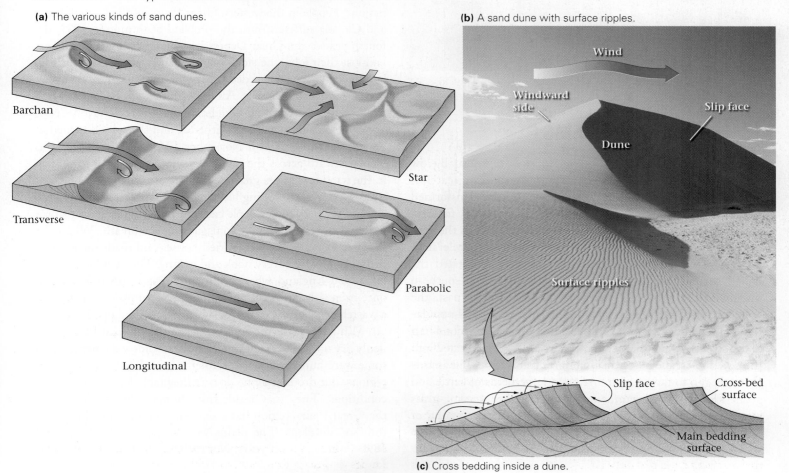

Barchan

Transverse

Longitudinal

Star

Parabolic

Wind

Windward side

Slip face

Dune

Surface ripples

Slip face

Cross-bed surface

Main bedding surface

(c) Cross bedding inside a dune.

FIGURE 17.16 Locations where desertification has taken place in recent times.

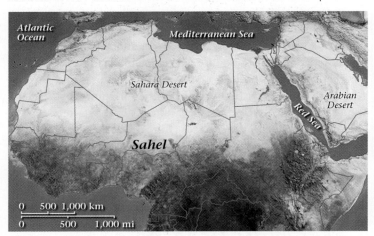

(a) The Sahel is the semi-arid land along the southern edge of Sahara. Large parts have undergone desertification.

(b) Drought in the Sahel has brought deadly consequences. Here, residents seek water from a dwindling pond.

(c) During the Dust Bowl era of the 1930s, storms blew away the topsoil.

17.6 DESERT PROBLEMS

In the modern era, natural droughts (periods of unusually low rainfall), aggravated by overpopulation, overgrazing, careless agriculture, and diversion of water supplies, have transformed semi-arid grasslands into true deserts, leading to tragic famines that have killed millions of people. **Desertification**, the process of transforming nondesert areas to desert, has accelerated.

The consequences of desertification have devastated the Sahel, the belt of semi-arid land that fringes the southern margin of the Sahara Desert (Fig. 17.16a). In the past, the Sahel provided sufficient vegetation to support a small population of nomadic people and animals. But during the second half of the twentieth century, large numbers of people migrated into the Sahel to escape overcrowding in central Africa. The immigrants began to farm and to maintain large herds of cattle and goats. Plowing and overgrazing removed soil-preserving grass and caused the soil to dry out. In addition, the trampling of animal hooves compacted the soil and collapsed its porosity, so the soil could no longer soak up water. In the 1960s and again

in the 1980s, a series of natural droughts hit the region, bringing catastrophe (Fig. 17.16b). Wind erosion stripped off the topsoil so hardly anything could grow. Without vegetation, the air grew drier, and the semiarid grassland of the Sahel became desert, with mass starvation as the result.

Desertification does not only happen in less industrialized nations. People in the western Great Plains of the United States and Canada suffered from the problem beginning in 1933, the fourth year of the Great Depression. Banks had failed, workers had lost their jobs, the stock market had crashed, and hardship burdened all. No one needed yet another disaster—but that year, even nature turned hostile. All through the fall, so little rain fell in the plains of Texas and Oklahoma that the region's grasslands and croplands browned and withered, and the topsoil turned to powdery dust. Then, on November 12 and 13, strong storms blew eastward across the plains. Without vegetation to protect it, the wind tore at the ground, stripped off the topsoil, and sent it skyward to form rolling black clouds that literally blotted out the sun (Fig. 17.16c). People caught in the resulting dust storm found themselves choking and gasping for breath. When the dust finally settled, it had buried houses and roads under huge drifts, and dirtied every nook and cranny. The dust blew east as far as New England, where it turned the snow brown. What had once been rich farmland in the southwestern plains turned into a wasteland that soon acquired a nickname, the Dust Bowl.

Why did the fertile soils of the southern Great Plains suddenly dry up? The causes were complex; some were natural and some were human-induced. Drought episodically visits certain regions, but drought alone doesn't inevitably lead to dust-bowl conditions. They may result from human activity. Typically, the Great Plains region has a semi-arid climate in which only thin soil develops. The plains were settled in the 1880s and 1890s, unusually wet years. Not realizing its true character, far more people moved into the region than it could sustain, and the land was farmed too intensively. When farmers used steel plows, they destroyed the fragile grassland root systems that held the thin soil in place. And when the drought of the 1930s came, it was catastrophic.

Desertification has dramatically increased the amount of blowing dust in the atmosphere. In recent years, satellite images have revealed that wind-blown dust from deserts can travel across oceans and affect regions on the other side. For example, dust blown off the Sahara can traverse the Atlantic and settle over the Caribbean (Fig. 17.17). This dust may carry pathogens that could kill coral.

The Dust Bowl of the 1930s and the fate of the Sahel in Africa remind us of how fragile the Earth's green blanket of vegetation really is. Global climate changes can shift climatic belts sufficiently to transform agricultural regions into deserts. Unfortunately, if the current trend in climate change continues, our present agricultural belts could someday become new Sahels.

TAKE-HOME MESSAGE

Desert populations have been burgeoning, so water supplies are problematic, and groundwater is being depleted. In lands bordering deserts, droughts and population pressures have led to desertification, the transformation of vegetated land into desert.

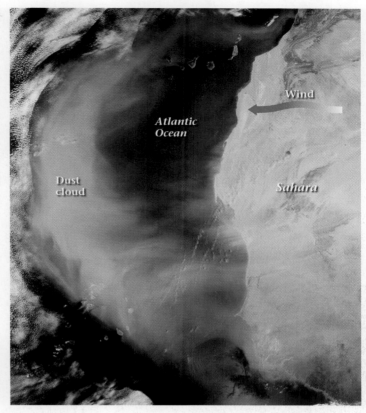

FIGURE 17.17 In this satellite image, a huge dust cloud that originated in the Sahara blows across the Atlantic.

Chapter Summary

- Deserts generally receive less than 25 cm of rain per year. Vegetation covers no more than 15% of their surface.

- Subtropical deserts form between 20° and 30° latitude, rain-shadow deserts are found on the inland side of mountain ranges, coastal deserts are located on the land adjacent to cold ocean currents, continental-interior deserts exist in land-locked regions far from the ocean, and polar deserts form at high latitude.

- In deserts, chemical weathering happens slowly, so rock bodies tend to erode primarily by physical weathering. Desert varnish forms on rock surfaces, and soils tend to accumulate soluble minerals.

- Water causes significant erosion in deserts, mostly during heavy downpours. Flash floods carry large quantities of sediments down ephemeral streams.

- Wind causes significant erosion in deserts, for it picks up dust and silt and carries them as suspended load, and causes sand to saltate. Where wind blows away finer sediment, a lag deposit remains. Wind-blown sediment abrades the ground.

- Desert pavements are mosaics of varnished stones armoring the surface of the ground.

- Talus aprons form when rock fragments accumulate at the base of a slope. Alluvial fans form at a mountain front where water in ephemeral streams deposits sediment. When temporary desert lakes dry up, they leave playas.

- In some desert landscapes, erosion causes cliff retreat, eventually resulting in the formation of mesas and inselbergs. Pediments of nearly flat or gently sloping bedrock surround some inselbergs.

- Where there is abundant sand, the wind builds it into dunes. Common types include barchan, star, transverse, parabolic, and longitudinal dunes.

- Changing climates and land abuse may cause desertification, the transformation of semi-arid land into deserts. Wind-blown dust from deserts may waft across oceans.

GEOPUZZLE REVISITED

Deserts are very arid regions that can sustain very little vegetation. They host only hardy plants and animals adapted to living with little water. Vast areas of sand dunes do cover some portions of some deserts, but desert landscapes can also host stony plains and rocky ridges. And, although many deserts are very hot, some occur at high elevations or at polar latitudes and can be very cold.

Key Terms

alluvial fan (p. 459)

butte (p. 461)

chimney (p. 461)

cliff retreat (p. 460)

deflation (p. 458)

desert (p. 452)

desertification (p. 466)

desert pavement (p. 457)

desert varnish (p. 456)

dry wash (p. 457)

inselberg (p. 461)

lag deposit (p. 457)

mesa (p. 461)

pediment (p. 464)

playa (p. 460)

rain shadow (p. 453)

saltation (p. 457)

surface load (p. 457)

suspended load (p. 457)

talus apron (p. 458)

ventifact (p. 458)

THE VIEW FROM SPACE   Under appropriate conditions, sand can build into large dunes. The dunes in this satellite image, some of which are 300 m high, developed in the Namib Desert of Namibia, in southwestern Africa.

Review Questions

1. What factors determine whether a region can be classified as a desert?
2. Explain why deserts form.
3. Have today's deserts always been deserts? Keep in mind the consequences of plate tectonics.
4. How do weathering processes in deserts differ from those in temperate or humid climates?
5. Describe how water modifies the landscape of a desert. Be sure to discuss both erosional and depositional landforms.
6. Explain the ways in which desert winds transport sediment.
7. Explain how the following features form: (a) desert varnish; (b) desert pavement; (c) ventifacts.
8. Describe the process of formation of the different types of depositional landforms that develop in deserts.
9. Describe the process of cliff (scarp) retreat and the landforms that result from it.
10. What are the various types of sand dunes? What factors determine which type of dune develops?
11. What is the process of desertification, and what causes it? How can desertification in Africa affect the Caribbean?

On Further Thought

1. You are working for an international non-governmental agency (NGO) and have been charged with the task of providing recommendations to an African nation that wishes to slow or halt the process of desertification within its borders. What are your recommendations?
2. The Namib Desert lies to the north and west of the Kalahari Desert, in southern Africa. The reason that the former region is a desert is not the same as the reason that the latter is a desert. Explain this statement.

CHAPTER **18**

Amazing Ice:
Glaciers and Ice Ages

Glaciers are rivers or sheets of ice. They carve beautiful landscapes and deposit hills of sediment. During the ice ages, glaciers covered vast areas of continents. Here, we see the jagged, melting surface of a glacier in Alaska.

GEOPUZZLE

If modern society had existed about 12,000 years ago, would it have been possible to build New York City (USA) or Edinburgh (Scotland) at their present locations? Why?

After witnessing the unveiling of the majestic peaks and glaciers and their baptism in the down-pouring sunbeams, it seemed inconceivable that nature could have anything finer to show us.

—John Muir (American naturalist, 1838–1914)
on seeing Glacier Bay, Alaska

18.1 INTRODUCTION

There's nothing like a good mystery, and one of the most puzzling in the annals of geology came to light in northern Europe early in the nineteenth century. When farmers of the region prepared their land for spring planting, they occasionally broke their plows on large boulders buried randomly through otherwise fine-grained sediment. Many of these boulders did not consist of local bedrock, but rather came from outcrops hundreds of kilometers away. Because the boulders had apparently traveled so far, they came to be known as **erratics** (from the Latin *errare*, to wander).

The mystery of the wandering boulders became a subject of great interest to early nineteenth-century geologists, who realized that deposits of *un*sorted sediment (containing a variety of different clast sizes) could not be stream alluvium, for running water sorts sediment by size. Most attributed the deposits to a vast flood that they imagined had somehow spread a slurry of boulders, sand, and mud across the continent. In 1837, however, a young Swiss geologist named Louis Agassiz proposed a radically different interpretation. Agassiz often hiked among **glaciers** (slowly flowing masses of ice that survive the summer melt) in the Alps near his home. He observed that glacial ice could carry enormous boulders as well as sand and mud, because ice is solid and has the strength to support the weight of rock. Agassiz realized that because flowing solid ice does not sort sediment, glaciers leave behind unsorted sediment when they melt. Working from these observations, he proposed that mysterious erratic-bearing sediments were deposits left by ice sheets, vast glaciers that had once covered much of the continent but had since disappeared. In Agassiz's view, Europe had at one time been in the grip of an **ice age,** a time when the climate was significantly colder and glaciers grew (Fig. 18.1).

Agassiz's radical proposal faced intense criticism for the next two decades. But by the late 1850s, most doubters had changed their minds, and the geologic community concluded that the notion that Europe had once endured Arctic-like climates was correct. Later in life, Agassiz moved to the United States and documented many glacier-related features in North America's landscape, proving that an ice age had affected vast areas of the planet. In fact, though glaciers cover only about 10% of the land on Earth today, during the most recent ice age, as much as 30% of the continents had a coating of ice. New York City, Montréal, and many of the great cities of Europe occupy land that once lay beneath hundreds of meters to a few kilometers of ice.

The work of Louis Agassiz brought the subject of glaciers and ice ages into the realm of geologic study and led people to recognize that major climate changes happen in Earth history. In this chapter, after introducing the nature of ice, we see how glaciers form and why they move. Next, we consider how glaciers modify landscapes by erosion and deposition. A substantial portion of this chapter concerns the Pleistocene ice ages, which happened during the last 2 million years, for their impact on the landscape can still be seen today. We conclude by considering hypotheses to explain why ice ages happen.

TAKE-HOME MESSAGE

When you finish this chapter, you should understand how glaciers form, move, and modify the landscape. You should also be familiar with the evidence for ice ages, and with the hypotheses that explain why ice ages happen.

18.2 THE NATURE OF GLACIERS

What Is Ice?

Ice consists of solid water, formed when liquid water cools below its freezing point. We can consider a single ice crystal to be a mineral, for it is a naturally occurring, inorganic solid, with a definite chemical composition (H_2O) and a regular crystal structure. Ice crystals have a hexagonal form, so snowflakes grow into six-pointed stars (Fig. 18.2a). We can think of a layer of fresh snow as a layer of sediment, and a layer of snow that has been compacted so that the grains stick together as a layer of sedimentary rock (Fig. 18.2b). We can also think of the ice that appears on the surface of a pond as an igneous rock, for it forms when molten ice (liquid water) solidifies. Glacier ice, in effect, is a

FIGURE 18.1 Agassiz envisoned that, in the past, vast areas of northern hemisphere continents looked like Antarctica does today.

FIGURE 18.2 The nature of ice and the formation of glaciers. Snow falls like sediment and metamorphoses to ice when buried.

(a) The hexagonal shape of snowflakes. No two are alike.

A boundary between layers

The layers in the photo at left are part of this glacier in the Alps.

(b) Layers of snow accumulate. They recrystallize to become ice.

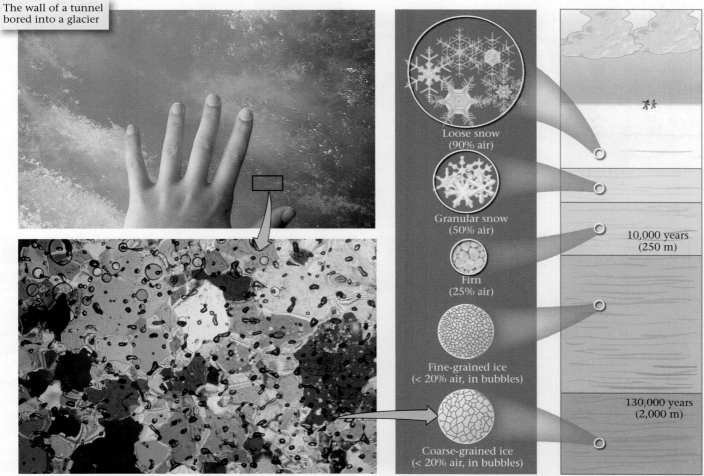

The wall of a tunnel bored into a glacier

Loose snow (90% air)

Granular snow (50% air)

Firn (25% air)

Fine-grained ice (< 20% air, in bubbles)

Coarse-grained ice (< 20% air, in bubbles)

10,000 years (250 m)

130,000 years (2,000 m)

(c) Glacial ice is blue. As revealed by a microscope, the ice has coarse grains and contains air bubbles.

(d) Snow compacts and melts to form firn, which recrystalizes into ice. Crystal size increases with depth.

metamorphic rock. It develops when preexisting ice recrystallizes in the solid state, meaning that the molecules in solid water rearrange to form new crystals. Typically, glacial ice has a bluish color and contains air bubbles (Fig. 18.2c).

How a Glacier Forms

In order for a glacier to form, three conditions must be met. First, the local climate must be sufficiently cold that winter snow does not melt entirely away during the summer. Second, there must be or must have been sufficient snowfall for a large amount to accumulate. And third, the slope of the surface on which the snow accumulates must be gentle enough that the snow does not slide away in avalanches, and must be protected enough that the snow doesn't blow away.

Glaciers develop in polar regions because, even though relatively little snow falls today, temperatures remain so low that most ice and snow survive all year. Glaciers develop in mountains, even at low latitudes, because temperature decreases with elevation, so that at high elevations, the mean temperature stays low enough for ice and snow to survive all year. Since the temperature of a region depends on latitude, the specific elevation at which glaciers form in mountains also depends on latitude. In Earth's present-day climate, glaciers form only at elevations above 5 km between 0° and 30° latitude and down to sea level at 60° to 90° latitude.

The transformation of snow to glacier ice takes place slowly, as younger snow progressively buries older snow (Fig. 18.2d). Freshly fallen snow consists of delicate hexagonal crystals with sharp points. The crystals do not fit together tightly, so this snow contains about 90% air. With time, the points of the snowflakes become blunt, because they either sublimate (evaporate directly into vapor) or melt, and the snow packs more tightly. As snow becomes buried, the weight of the overlying snow increases pressure, which causes remaining points of contact between snowflakes to melt. Gradually, the snow

FIGURE 18.3 A great variety of different types of glaciers form in mountainous areas.

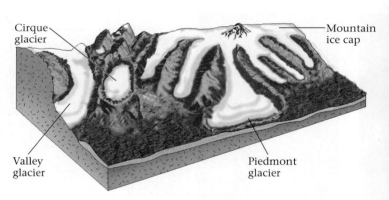

(a) Mountain glaciers are classified based on shape and position.

(b) A valley glacier and cirque glaciers in Switzerland.

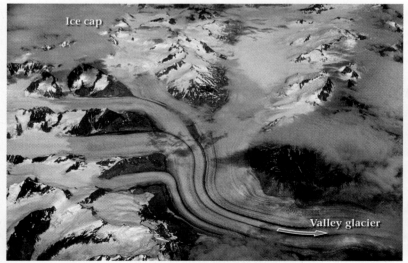

(c) Valley glaciers draining a mountain ice cap in Alaska.

(d) A piedmont glacier near the coast of Greenland.

transforms into a packed granular material called firn, which contains only about 25% air. Melting of firn grains at contact points produces water that crystallizes in the spaces between grains until eventually the firn transforms into a solid mass of glacial ice composed of interlocking ice crystals. As noted earlier, such glacial ice, which may still contain up to 20% air trapped in bubbles, has a bluish color. The transformation of fresh snow to glacier ice can take as little as tens of years in regions with abundant snowfall, or as long as thousands of years in regions with little snowfall.

Categories of Glaciers

Glaciers, formally defined, are streams or sheets of recrystallized ice that last all year long and flow under the influence of gravity. Today, they highlight coastal and mountain scenery in Alaska, western North America, the Alps of Europe, the Southern Alps of New Zealand, the Himalayas of Asia, and the Andes of South America, and they cover most of Greenland and Antarctica (see Geotour 18 on p. GT-38). Geologists distinguish between two main categories, mountain glaciers and continental glaciers.

Mountain glaciers (also called alpine glaciers), as the name suggests, exist in mountainous regions (Fig. 18.3a). Topographical features of the mountains control their shape; overall, mountain glaciers flow from higher elevations to lower elevations. Mountain glaciers include cirque glaciers, which fill bowl-shaped depressions, or **cirques**, on the flank of a mountain; valley glaciers, rivers of ice that flow down valleys (Fig. 18.3b); mountain ice caps, mounds of ice that submerge peaks and ridges at the crest of a mountain range (Fig. 18.3c); and piedmont glaciers (Fig. 18.3d), fans or lobes of ice that form where a valley glacier emerges from a valley and spreads out into the adjacent plain. Mountain glaciers range in size from a few hundred meters to a few hundred kilometers long and hundreds of meters to several kilometers wide.

Continental glaciers are vast ice sheets that spread over thousands of square kilometers of continental crust. Continental glaciers now exist only on Antarctica and Greenland (Fig. 18.4a, b). (Remember that Antarctica is a continent, so the ice at the South Pole rests on solid ground; in contrast, the ice at the North Pole is part of a thin sheet of sea ice floating on the Arctic Ocean.) Continental glaciers flow outward from their thickest point, where ice attains a thickness of up to 4 km, and become thinner toward their margins, where they may be only a few hundred meters thick. (Of note, similar ice sheets form on the poles of Mars; Box 18.1.)

Geologists also find it valuable to distinguish between types of glaciers on the basis of the thermal conditions in which the glaciers exist. **Temperate glaciers** occur in regions where atmospheric temperatures become high enough for the glacial ice to be at or near its melting temperature for part of the year.

FIGURE 18.4 Antarctica is an ice-covered continent.

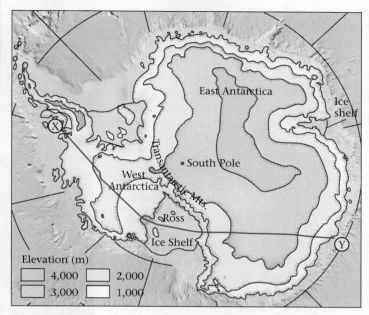

(a) A contour map of the Antarctic ice sheet. Valley glaciers carry ice from the ice sheet of East Antarctica down to the Ross Ice Shelf.

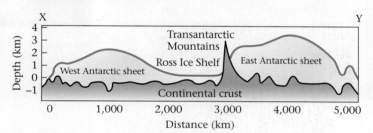

(b) A cross section X to Y of the Antarctic ice sheet. The Transantarctic Mountains separate East Antarctica from West Antarctica.

Polar glaciers occur in regions where atmospheric temperatures stay so low that the glacial ice remains well below melting temperature throughout the entire year.

The Flow of Glacial Ice

How do glaciers flow? Some move when meltwater accumulates at their bases, so that the mass of the glacier slides on a layer of water or on a slurry of water and sediment. During this kind of movement, known as **basal sliding**, the water or wet slurry holds the glacier above the underlying rock and reduces friction between the glacier and its substrate. Glaciers also move by means of **plastic deformation**, during which the mass of ice slowly changes shape internally, without breaking apart or completely melting. By studying ice deformation with a microscope, geologists have determined that plastic flow involves two processes—the rearrangement of water molecules within the crystal lattice and the slipping of ice crystals past one another.

BOX 18.1

THE REST OF THE STORY

Ice on Mars

The discovery that Mars has polar ice caps dates back to 1666, when the first telescopes allowed astronomers to resolve details of the red planet's surface. By 1719, astronomers had detected that Martian polar caps change in area with the season, suggesting that they partially melt and then refreeze. You can see these changes on modern images (**Fig. 1a, b**). The question of what the ice caps consist of remained a puzzle until fairly recently. Early studies revealed that the atmosphere of Mars consists mostly of carbon dioxide, so researchers first assumed that the ice caps consisted of frozen carbon dioxide. But data from modern spacecraft led to the conclusion that this initial assumption is false. It now appears that the Martian ice caps consist mostly of water ice (mixed with dust) in a layer from 1 to 3 km thick. During the winter, atmospheric carbon dioxide freezes and covers the northern polar cap with a 1-m-thick layer of dry ice. During the summer

this layer melts away. The southern polar cap is different, for its dry ice blanket is 8 m thick and doesn't melt away entirely in the summer. The difference between the north and south poles may reflect

elevation. The south pole is 6 km higher and therefore remains colder. Studies by the *Phoenix* spacecraft, which landed in 2008, indicate that ice may underlie the soil over broad areas of Mars.

FIGURE 1 The ice caps of Mars.

(a) During the winter, the ice caps expand to lower latitudes.

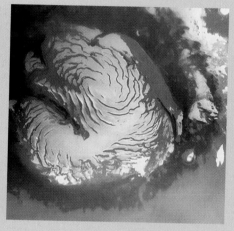

(b) A close-up of the northern polar cap in summer.

Note that the plastic deformation of ice in glaciers resembles the plastic deformation of metamorphic rock in mountain belts (see Chapter 9). However, metamorphic rock deforms plastically only at depths greater than 10 to 15 km in the crust, whereas ice deforms plastically at depths below about 60 m (Fig. 18.5). In other words, the brittle-plastic transition in ice lies at a depth of about 60 m. Above this depth, large cracks can form in ice, but below this depth, cracks cannot form because ice is too plastic. A large crack that develops by brittle deformation in a glacier is called a **crevasse**. In large glaciers, crevasses can be hundreds of meters long and tens of meters deep.

Why do glaciers flow? Ultimately, because ice is so weak, the pull of gravity alone can drive movement (Fig. 18.6a, b). Specifically, *glaciers flow in the direction in which their top surfaces slope*. Thus, valley glaciers flow down their valleys, and continental ice sheets spread outward from their thickest point. To picture the flow of mountain glaciers, imagine pouring honey on a sloping but bumpy board. The whole mass of honey flows down the board as long as the top surface of the honey has a downward tilt. To picture the movement of a continental ice sheet, imagine pouring honey on a table top. The honey spreads out until the puddle reaches an even thickness. In the case of a glacier, a thick pile of

ice builds up, and gravity causes the top of the pile to push down on the ice at the base. Eventually, the basal ice can no longer support the weight of the overlying ice and begins to deform plastically. When this happens, the basal ice starts squeezing out to the side, carrying the overlying ice with it. The greater the volume of ice that builds up, the wider the sheet of ice can become.

Glaciers generally flow at rates of between 10 and 300 m per year. Not all parts of a glacier move at the same rate. For example, friction between rock and ice slows a glacier, so the center of a valley glacier moves faster than its margins, and the top of a glacier moves faster than its base (Fig. 18.7a). If water builds up beneath a valley glacier to the point where it lifts the glacier off its substrate, the glacier undergoes a surge and flows much faster for a limited time (rarely more than a few months) until the water escapes. During surges, glaciers have been clocked at speeds of 20 to 110 m per day!

Glacial Advance and Retreat

Glaciers resemble bank accounts: snowfall adds to the account, whereas ablation—the removal of ice by sublimation (the evaporation of ice into water vapor), melting (the transformation of ice

FIGURE 18.5 Crevasses form in the brittle ice above the brittle-plastic transition boundary as a glacier moves.

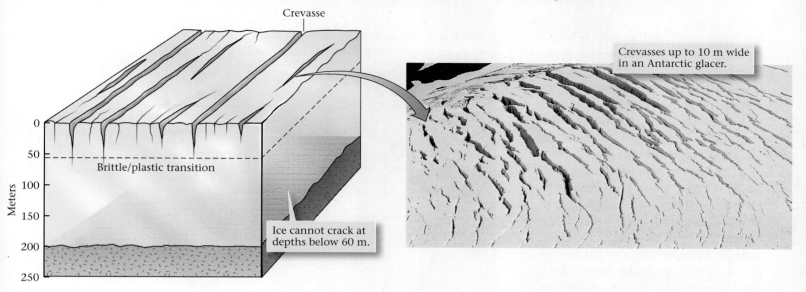

Crevasse

Crevasses up to 10 m wide in an Antarctic glacer.

0
50
Brittle/plastic transition
100
Meters
150
200
250

Ice cannot crack at depths below 60 m.

FIGURE 18.6 Forces that drive the movement of glaciers.

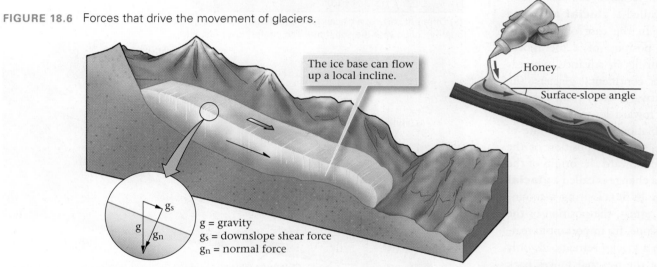

The ice base can flow up a local incline.

Honey

Surface-slope angle

g = gravity
g_s = downslope shear force
g_n = normal force

g_s

g g_n

(a) Movement of valley glaciers occurs if the top surface slopes down the valley, so that gravity produces a downslope shear force.

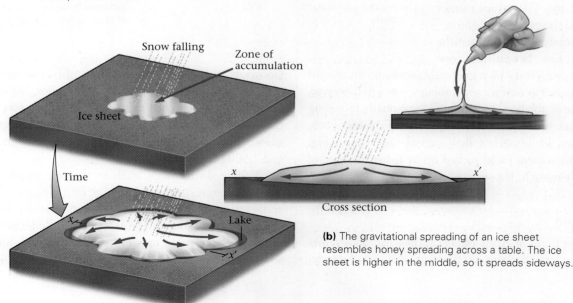

Snow falling

Zone of accumulation

Ice sheet

Time

x

x'

Lake

x

x'

Cross section

(b) The gravitational spreading of an ice sheet resembles honey spreading across a table. The ice sheet is higher in the middle, so it spreads sideways.

into liquid water), and calving (the breaking off of chunks of ice at the edge of the glacier)—subtracts from the account (Fig. 18.7b). Snowfall adds to the glacier in the **zone of accumulation**, whereas ablation subtracts in the **zone of ablation**; the boundary between these two zones is the **equilibrium line**.

The leading edge or margin of a glacier is called its **toe**, or terminus. If the rate of ablation equals the rate of accumulation, then the position of the toe remains fixed (Fig. 18.8a). If the rate at which ice accumulates in the zone of accumulation exceeds the rate at which ablation occurs below the equilibrium line, then the toe moves forward into previously unglaciated regions. Such a change is called a **glacial advance** (Fig. 18.8b). In the case of mountain glaciers, the position of a toe moves downslope during an advance, whereas in the case of continental glaciers, the toe moves outward, away from the glacier's origin, toward lower latitudes. If the rate of ablation exceeds the rate of accumulation, then the position of the toe moves back toward the origin of the glacier. Such a change is called a **glacial retreat** (Fig. 18.8c). During a mountain glacier's retreat, the position of the toe moves upslope. It's important to realize that when a glacier retreats, it's only the *position* of the toe that moves back toward the origin, for ice continues to flow toward the toe. Glacial ice cannot flow back toward the glacier's origin.

One final point before we leave the subject of glacial flow: beneath the zone of accumulation, a crystal of ice gradually moves down toward the base of the glacier as new ice accumulates above it. However, beneath the zone of ablation, a crystal of ice gradually moves up toward the surface of the glacier, as overlying ice ablates. Thus, as a glacier flows, ice crystals follow curved trajectories (Fig. 18.8a–c). For this reason, rocks picked up by ice at the base of the glacier slowly move to the surface.

Ice in the Sea

On the moonless night of April 14, 1912, the great ocean liner *Titanic* struck a large iceberg in the frigid North Atlantic.

FIGURE 18.7 Flow velocities vary with location in a glacier. Overall, ice flows from the zone of accumulation to the toe.

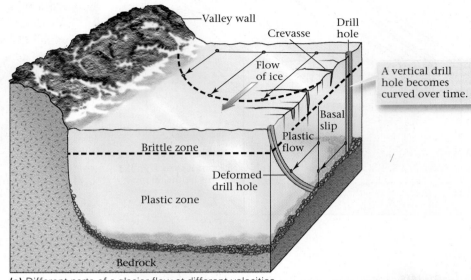

A vertical drill hole becomes curved over time.

(a) Different parts of a glacier flow at different velocities, due to friction with the substrate. The center flows fastest.

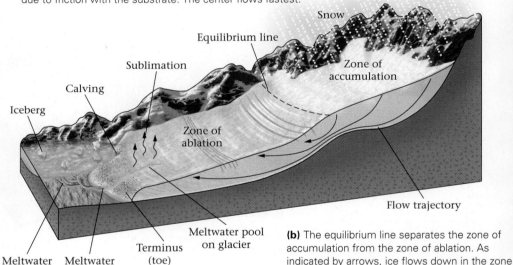

(b) The equilibrium line separates the zone of accumulation from the zone of ablation. As indicated by arrows, ice flows down in the zone of accumulation and up in the zone of ablation.

Lookouts had seen the ghostly mass of frozen water only minutes earlier and had alerted the ship's pilot, but the ship had been unable to turn fast enough to avoid disaster. The force of the blow split the steel hull, allowing water to gush in. Less than 3 hours later, the ship disappeared beneath the surface, and 1,500 people perished.

Where do icebergs, such as the one responsible for the *Titanic*'s demise, originate? In high latitudes, mountain glaciers and continental ice sheets flow down to the sea. Valley glaciers whose terminus lies in the water are called **tidewater glaciers**. Valley glaciers may protrude far enough into the ocean to become ice tongues, whereas continental glaciers entering

FIGURE 18.8 Glacial advance and retreat.

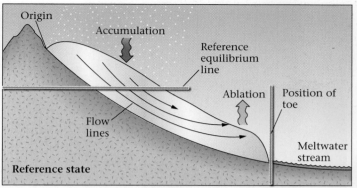

(a) The position of the toe represents a balance between addition by accumulation and loss by ablation.

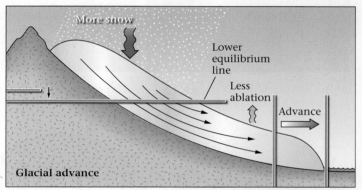

(b) If accumulation exceeds ablation, the glacier advances, the toe moves farther from the origin, and the ice thickens.

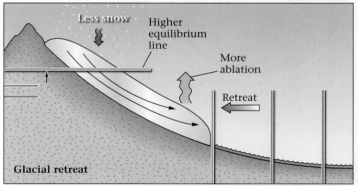

(c) If ablation exceeds accumulation, the glacier retreats and thins. The toe moves back, even though ice continues to flow toward the toe.

the sea become broad, flat sheets called ice shelves. Four-fifths of the ice in a tidewater glacier can lie below the sea surface, so in shallow water, an ice shelf remains grounded (Fig. 18.9a). But where the water becomes deep enough, the ice floats. At the boundary between glacier and ocean, blocks of ice from tidewater glaciers or ice shelves calve off and tumble into the water with an impressive splash. If a free-floating chunk rises 6 m above the water and is at least 15 m long, it is formally called

an **iceberg**. Since four-fifths of the ice lies below the surface of the sea, the base of a large iceberg may actually extend a few hundred meters below the surface (Fig. 18.9b).

Not all ice floating in the sea originates as glaciers on land. In polar climates, the surface of the sea itself freezes, forming **sea ice**. Some sea ice floats freely, but some protrudes outward from the shore. Sea ice may break up during summer months (Fig. 18.9c). Sea ice covers the Arctic Ocean and fringes Antarctica (Fig. 18.9d).

TAKE-HOME MESSAGE

Glaciers form when deeply buried snow turns to ice and recrystallizes. Mountain glaciers form at high elevation and can flow down valleys; ice sheets form in polar latitudes and spread over continents. Depending on the balance of accumulation to ablation, glaciers can advance or retreat.

18.3 CARVING AND CARRYING BY ICE

Glacial Erosion and Its Products

During the last ice age, valley glaciers cut deep, steep-sided valleys into the Sierra Nevada of California. In the process, domes of granite were cut in half, leaving a rounded surface on one side and a steep cliff on the other. Half Dome, in Yosemite National Park, formed in this way (Fig. 18.10a); its steep cliff has challenged countless rock climbers. Such glacial erosion also produces the knife-edge ridges and pointed spires of high mountains (Fig. 18.10b) and the broad expanses where rock outcrops have been stripped of overlying sediment and polished smooth (Fig. 18.10c).

Glaciers erode their substrate in several ways. During glacial incorporation, ice surrounds loose debris. During glacial plucking (or glacial quarrying), a glacier breaks off and then carries away fragments of bedrock. Plucking occurs when ice freezes onto rock that has already started to crack and separate from its substrate, so that movement of the ice lifts off pieces of the rock. At the toe of a glacier, ice may actually bulldoze sediment and trees slightly before flowing over them.

As glaciers flow, clasts embedded in ice act like the teeth of a giant rasp and grind away the substrate. This process, called glacial abrasion, produces very fine sediment (or rock flour), just as sanding wood produces sawdust. Rasping by embedded sand yields shiny, glacially polished surfaces, and rasping by large clasts produces long gouges, grooves, or scratches (1 cm to 1 m across) called **glacial striations** (Fig. 18.10d). As you might expect, striations run parallel to the flow direction of the ice.

Let's now look more closely at the erosional features associated with a mountain glacier. Over time, glacial erosion can radically change a preexisting stream-carved landscape (Fig. 18.11a).

FIGURE 18.9   Ice shelves, tidewater glaciers, and sea ice—the nature of coastal areas in glacial regions.

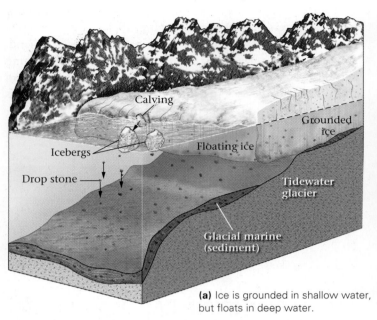

(a) Ice is grounded in shallow water, but floats in deep water.

(b) This artist's rendition of an iceberg emphasizes that most of the ice is underwater.

(c) In summer, some of the sea ice of Antarctica breaks up to form tabular icebergs.

(d) Sea ice covers most of the Arctic ocean (left) and surrounds Antarctica (right).

The process begins when freezing and thawing during the fall and spring help fracture the rock bordering the origin of the glacier. This rock falls on the ice or gets picked up at the base of the ice, and moves downslope with the glaciers. As a consequence, a bowl-shaped depression, or cirque, develops on the side of the mountain. If the ice later melts, a lake called a tarn may be trapped by the glacial sediment that collects at the base of the cirque. An **arête** (French for ridge), a residual knife-edge ridge of rock, separates two adjacent cirques. A pointed mountain peak surrounded by at least three cirques is called a **horn**. The Matterhorn, a peak in Switzerland, is a particularly beautiful example of a horn (**Fig. 18.11b**).

Glacial erosion severely modifies the shape of a valley. To see how, compare a river-eroded valley with a glacially eroded valley. If you look along the length of a river in unglaciated mountains, you'll see that it flows down a V-shaped valley, with the river channel forming the point of the V. The V develops because river erosion occurs only in the channel, and mass wasting causes the valley slopes to approach the angle of repose. But if you look down the length of a glacially eroded valley, you'll see that it resembles a U, with steep walls (**Fig. 18.11c**). A **U-shaped valley** forms because the combined processes of glacial abrasion and plucking not only lower the floor of the valley, they also bevel its sides.

Glacial erosion in mountains also modifies the intersections between tributaries and the trunk valley. In a river system, tributaries cut side valleys that merge with the trunk valley, such that

the mouths of the tributary valleys lie at the same elevation as the trunk valley. The ridges (spurs) between valleys taper to a point where they join the trunk valley floor. During glaciation, tributary glaciers flow down side valleys into a trunk glacier. But the trunk glacier cuts the floor of its valley down to a depth that far exceeds the depth cut by the tributary glaciers. Thus, when the glaciers melt away, the mouths of the tributary valleys perch at a higher elevation than the floor of the trunk valley. Such side valleys are called **hanging valleys** (**Fig. 18.11d**). The water in post-glacial streams that flow down a hanging valley must cascade over a spectacular waterfall to reach the post-glacial trunk stream.

FIGURE 18.10 Products of glacial erosion. Ice is a very aggressive agent of erosion.

(a) Half Dome, in Yosemite National Park, California.

(b) The spectacular sharp peaks of the Alps were carved by the action of ice.

(c) Glacially polished outcrop in Central Park, New York City.

Striation

(d) Glacial striations in Victoria, British Columbia.

Now let's look at the erosional features produced by continental ice sheets. To a large extent, these depend on the nature of the preglacial landscape. Where an ice sheet spreads over a region of low relief such as the Canadian Shield, glacial erosion creates a vast region of polished, flat, striated surfaces. Where an ice sheet spreads over a hilly area, it deepens valleys and smooths hills. Glacially eroded hills end up being elongate in the direction of flow and are asymmetric; glacial rasping smooths and bevels the upstream part of the hill, creating a gentle slope, whereas glacial plucking eats away at the downstream part, making a steep slope. Ultimately, the hill's profile resembles that of a sheep lying in a meadow—such a hill is called a **roche moutonnée**, from the French for sheep rock (Fig. 18.12a, b).

Fjords: Submerged Glacial Valleys

As noted earlier, where a valley glacier meets the sea, the glacier's base remains in contact with the ground until the water depth exceeds about four-fifths of the glacier's thickness. Thus, glaciers can continue carving a U-shaped valley even below sea level. In addition, during an ice age, water extracted from the sea becomes locked in the ice sheets on land, so sea level drops significantly. Therefore, the floors of valleys cut by coastal glaciers during the last ice age were cut much deeper than present sea level. Today, the sea has flooded these deep valleys, producing **fjords** (see Chapter 15). In the spectacular fjord-land regions along the coasts of Norway, New Zealand, Chile, and Alaska, the walls of submerged U-shaped valleys rise straight from the sea as vertical cliffs up to 1,000 m high (Fig. 18.13). A lake filling a glacially carved valley is also called a fjord.

TAKE-HOME MESSAGE

A glacier scrapes up and plucks rock from its substrate, and carries debris that falls on its surface. Glacial erosion polishes and scratches rock and carves distinctive landforms, such as U-shaped valleys. Since ice is solid, moving ice does not sort sediment.

FIGURE 18.11 Landscape features formed by the glacial erosion of a mountainous landscape.

Before glaciation, valleys are V-shaped, and tributary mouths are the same elevation as the trunk stream.

During glaciation, the valleys fill with ice.

Mt. Snowdon, Wales

After glaciation, the region contains U-shaped valleys, hanging valleys, truncated spurs, and horns.

V-shaped valley
Tributary valley
Trunk valley
Trunk valley
Tributary valley
Cirque
Tarn
Time
Cirque
Arête
Horn
U-shaped valley
Hanging valley
Truncated spur

(a) Stages in the development of a glacially carved mountainous landscape.

(b) The Matterhorn in Switzerland. The first ascent was in 1865.

(c) A U-shaped glacial valley in Switzerland. The town of Zermatt now occupies the valley floor.

(d) A waterfall spilling out of a U-shaped hanging valley.

FIGURE 18.12 A roche moutonnée is an asymmetric bedrock hill shaped by the flow of glacial ice.

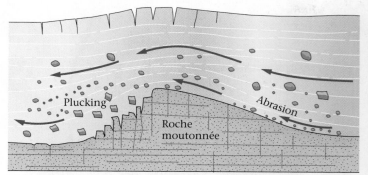

(a) Abrasion rasps the upstream side, and plucking carries away fracture-bounded blocks on the downstream side.

(b) An example of a roche moutonnée in the Sierra Nevada. The glacier flowed from right to left.

18.4 DEPOSITION ASSOCIATED WITH GLACIATION

The Glacial Conveyor: The Transport of Sediment by Ice

Glaciers can carry sediment of any size and, like a conveyor belt, transport it in the direction of flow (that is, toward the toe). The sediment load either falls onto the surface of the glacier from bordering cliffs or gets plucked and lifted from the substrate and incorporated into the moving ice (Fig. 18.14a). Because of the curving flow lines of glacial ice, rocks plucked off the floor may eventually reach the surface.

Sediment dropped on the glacier's surface moves with the ice and becomes a stripe of debris. Stripes formed along the side edges of the glacier are **lateral moraines** (Fig. 18.14b). (The word **moraine** was a local term used by Alpine farmers and shepherds for piles of rock and dirt.) When the glacier finally melts, lateral moraines lie stranded along the side of the glacially carved valley, like bathtub rings. Where two valley glaciers merge, the debris constituting two lateral moraines merges to become a **medial moraine**, a stripe running down the interior of a glacier (see Fig. 18.14b). Trunk glaciers into

FIGURE 18.13 One of the many spectacular fjords of Norway. The water is an arm of the sea that fills a glacially carved valley. Tourists are standing on Pulpit Rock (Prekestolen).

which numerous tributary glaciers may flow can contain several parallel medial moraines. Sediment that reaches the toe forms a pile called an **end moraine** (see Fig. 18.14a).

Types of Glacial Sedimentary Deposits

Several different types of sediment can be deposited in glacial environments. All of these types together constitute "glacial drift." The term dates from pre-Agassiz studies of glacial deposits, when geologists thought that the sediment had "drifted" into place during an immense flood. Specifically, glacial drift includes the following:

- *Till*: Sediment transported by ice and deposited on, beneath, or at the toe of a glacier is **glacial till**. Glacial till is unsorted, because the solid ice of glaciers can carry clasts of all sizes (Fig. 18.15a).

- *Erratics*: Glacial erratics are cobbles and boulders that have been dropped by a glacier. Some protrude from till piles, and others rest on glacially polished surfaces (Fig. 18.15b).

- *Glacial marine*: Where a sediment-laden glacier flows into the sea, icebergs calve off the toe and raft clasts out to sea. As the icebergs melt, they drop the clasts. Sediment consisting of ice-rafted clasts mixed with marine sediment makes up glacial marine.

FIGURE 18.14 The glacial conveyor and the formation of lateral and medial moraines on glaciers.

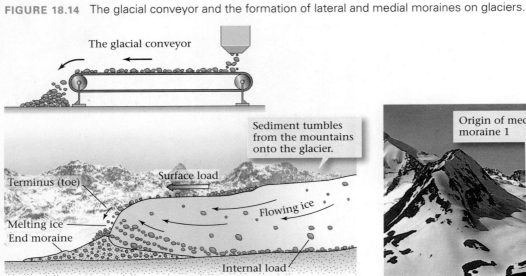

(a) Sediment falls on a glacier from bordering mountains and gets plucked up from below. Glaciers are like conveyor belts, moving sediment toward the toe of the glacier.

- *Glacial outwash*: Till deposited by a glacier at its toe may be picked up and transported by meltwater streams that sort the sediment. The clasts are deposited by a braided stream network in a broad area of gravel and sand bars called an outwash plain. This sediment is known as glacial outwash (**Fig. 18.15c**).

- *Loess*: When the warmer air above ice-free land beyond the toe of a glacier rises, the cold, denser air from above the glacier rushes in to take its place. A strong wind therefore blows at the margin of a glacier. This wind picks up fine clay and silt and transports it away from the glacier's toe. Where the winds die down, the sediment settles and forms a thick layer. This sediment, called **loess**, sticks together because of the electrical charges on clay flakes. Thus, steep escarpments develop by erosion of loess deposits (**Fig. 18.15d**).

- *Glacial lake-bed sediment*: Streams transport fine rock flour away from the glacial front. This sediment eventually settles in meltwater lakes, forming glacial lake-bed sediment that commonly contains varves. A varve is a pair of thin sediment layers deposited during a single year. One layer consists of silt brought in during spring floods, and the other of clay deposited in winter when the lake's surface freezes over and the water stands still (**Fig. 18.15e**).

Depositional Landforms of Glacial Environments

Picture a group of hunters, dressed in deerskin, standing at the toe of a continental glacier in what is now southern Canada, waiting for an unwary woolly mammoth to wander by. It's a sunny summer day 12,000 years ago, and milky, sediment-laden

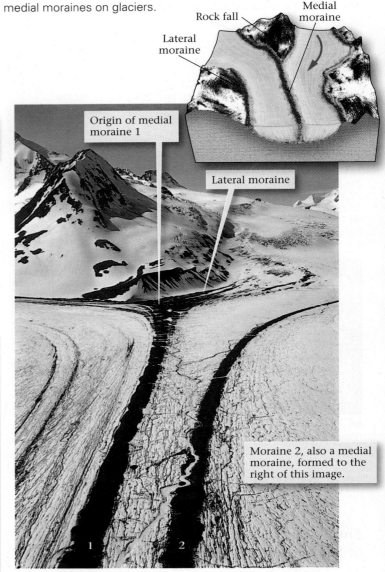

(b) A medial moraine forms where lateral moraines of two valley glaciers merge.

streams gush from tunnels and channels at the base of the glacier, and pour off the top of the glacier, as the ice melts. No mammoths venture by today, so the bored hunters climb to the top of the glacier for a view. The climb isn't easy, partly because of the incessant wind, and partly because deep crevasses interrupt their path. Reaching the top of the ice sheet, the hunters look northward, and the glare almost blinds them. Squinting, they see the white of snow, and where the snow has blown away, they see the rippled, glassy surface of bluish ice (see Fig. 18.1). Here and there, a rock protrudes from the ice. Now looking southward, they survey a stark landscape of low, sinuous ridges separated by hummocky (bumpy) plains (see **Geology at a Glance**, pp. 484–485). Braided streams, which carry meltwater out across this landscape, flow through the hummocky plains and supply a number of lakes. Dust fills the air because of the wind.

FIGURE 18.15 Sedimentation processes and products associated with glaciation. Glacial sediment is distinctive.

(a) This glacial till in Ireland is unsorted, because ice can carry sediment of all sizes.

(b) A glacial erratic in Central Park, New York City.

(c) Braided streams choked with glacial outwash in Alaska. The streams carry away finer sediment and leave the gravel behind.

(d) Loess deposits in Illinois, deposited in the Pleistocene.

(e) In the quiet water of an Alaskan glacial lake, fine-grained sediments accumulate. Alternating layers in the sediment (varves), now exposed in an outcrop near Puget Sound, Washington, reflect seasonal changes.

Glaciers and Glacial Landforms

Continental ice sheet

Crevasses

Ice shelf

Higher sea level

Lower sea level

Drop stones

Iceberg

Mountain (alpine) glaciation

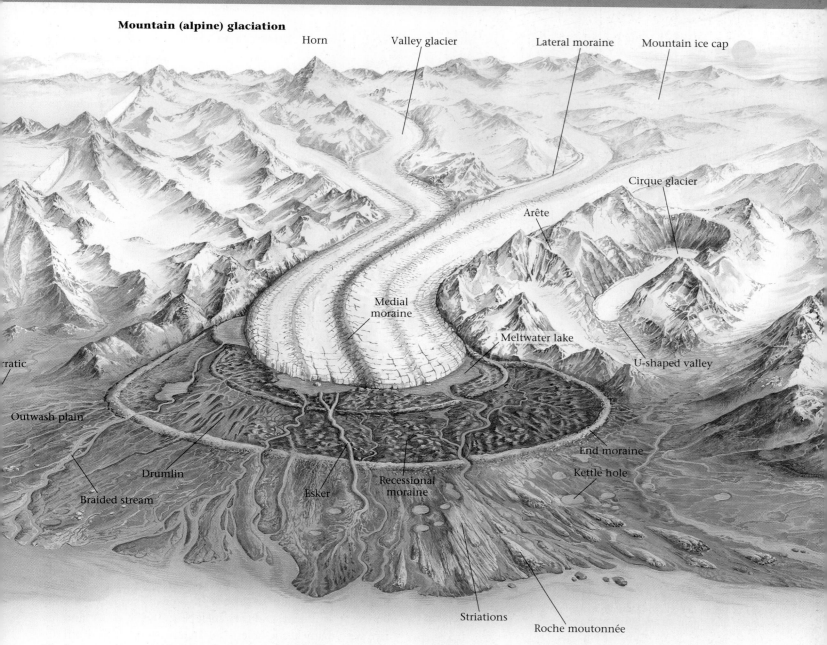

Horn · Valley glacier · Lateral moraine · Mountain ice cap · Cirque glacier · Arête · Medial moraine · Meltwater lake · U-shaped valley · End moraine · Kettle hole · Recessional moraine · Esker · Drumlin · Braided stream · Outwash plain · ...ratic · Striations · Roche moutonnée

Glaciers are rivers or sheets of ice that last all year and slowly flow. Continental glaciers, vast sheets of ice up to a few kilometers thick, covered extensive areas of land during times when Earth had a colder climate. Continental glaciers form when snow accumulates at high latitudes, then, when buried deeply enough, packs together and recrystallizes to make glacial ice. Ice, though solid, is weak, and thus ice sheets spread over the landscape, like syrup over a pancake. At the peak of the last ice age, ice sheets covered almost all of Canada, much of the United States, northern Europe, and parts of Russia.

The upper part of a sheet is brittle and may crack to form crevasses. Because ice sheets store so much of the Earth's water, sea level becomes lower during an ice age. When a glacier reaches the sea, it becomes an ice shelf. Rock that the glacier has plucked up along the way is carried out to sea with the ice; when the ice melts, the rocks fall to the sea floor as drop stones. At the edge of the shelf, icebergs calve off and float away.

A second class of glaciers, called mountain, or alpine, glaciers, exist in mountainous areas because snow can last all year at high elevations. During the ice age, mountain glaciers grew and flowed out onto the land surface beyond the

mountain front. The glacier at the right has started to recede after formerly advancing and covering more of the land. Glacial recession may happen when the climate warms, so ice melts away faster at the toe (terminus) of the glacier than can be added at the source. In front of the glacier, you can find consequences of glacial erosion such as striations on bedrock and roches moutonnées.

When the glacier pauses, till (unsorted glacial sediment) accumulates to form an end moraine. Meltwater lakes gather at the toe, and streams carry sediment and deposit it as glacial outwash. Sediment that accumulates in ice tunnels, exposed when the glacier melts, make up sinuous ridges called eskers. Even when the toe remains fixed in position for a while, the ice continues to flow, and thus molds underlying sediment into drumlins. Ice blocks buried in till melt to form kettle holes. (Though the examples shown here were left after the melting away of a piedmont glacier, most actually form during continental glaciation.)

In the mountains, the glacier is confined to a valley. Sediment falling on it from the mountains creates lateral and medial moraines. Glaciers carve distinct landforms in the mountains, such as cirques, arêtes, horns, and U-shaped valleys.

FIGURE 18.16 The formation of depositional landforms associated with continental glaciation.

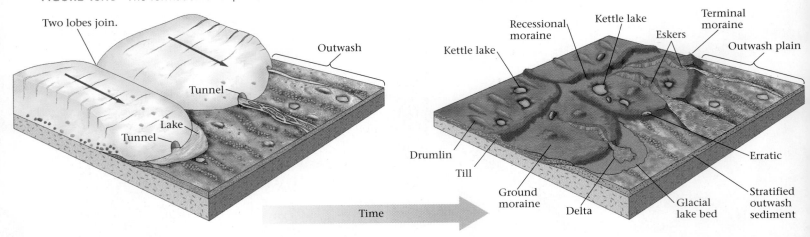

(a) The ice in continental glaciers flows toward the toe; sediment accumulates at the base and at the toe of the ice sheet.

(b) Several distinct depositional landforms form during glaciation; some developed under the ice, and some at the toe.

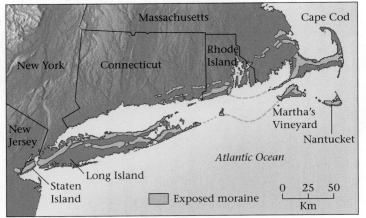

(c) Cape Cod, Long Island, and other landforms in the northeastern United States formed at the end of the continental ice sheet.

All the landscape features that the hunters observe as they look southward form by deposition in glacial environments, both mountain and continental (Fig. 18.16a, b). The low, sinuous ridges, called end moraines, develop when the toe of a glacier stalls in one position for a while; the ice keeps flowing to the toe, and, like a giant conveyor belt, transports sediment with it; the sediment accumulates in a pile at the toe (see Fig. 18.14a). The end moraine at the farthest limit of glaciation is called the **terminal moraine**. (The ridge of sediment that makes up Long Island, New York, and continues east-northeast into Cape Cod, Massachusetts, is part of the terminal moraine of the ice sheet that covered New England and eastern Canada during the last ice age; Fig. 18.16c.) End moraines that form when a glacier stalls for a while as it recedes are recessional moraines.

Till that has been released at the base of a flowing glacier and remains after the glacier has melted away is called lodgment till. Lodgement till left behind during rapid recession forms a thin, hummocky layer on the land surface. Clasts in lodgment till may be aligned and scratched during their movement or as ice flows over them. The flow of the glacier may mold till and other subglacial sediment into streamlined, elongate hills called **drumlins** (from the Gaelic word for hills). Drumlins tend to be asymmetric along their length, with a gentle downstream slope, tapered in the direction of flow, and a steeper upstream slope (Fig. 18.17a, b).

The hummocky surface of moraines reflects partly the variations in the amount of sediment supplied by the glacier, and partly the occurrence of **kettle holes**, circular depressions made when blocks of ice calve off the toe of the glacier, become buried by sediment, and then melt (Fig. 18.17c, d). A land surface with many kettle holes separated by round hills of till displays knob-and-kettle topography (Fig. 18.17e).

Sediment-choked water pours out of tunnels in the ice. This water feeds large braided streams that sort till and redeposit it as glacial outwash, stratified layers and lenses of gravel and sand. Thus, outwash plains form between recessional moraines and beyond the terminal moraine. Some of the meltwater pools to form lakes in lowlands between recessional moraines, or in ice-margin lakes between the glacier's toe and the nearest recessional moraine. Even long after the glacier has melted away, the lowlands between recessional moraines persist as lakes or swamps. When drained, glacial lake beds provide particularly fertile soil for agriculture. When a glacier eventually melts away, ridges of sorted sand and gravel that were deposited in subglacial meltwater tunnels snake across the ground moraine. These ridges are called **eskers** (Fig. 18.18a, b). Sediments in glacial outwash plains and in eskers serve as important sources of sand and gravel for road building and construction.

FIGURE 18.17 Drumlins and knob-and-kettle topography characterize some areas that were once glaciated.

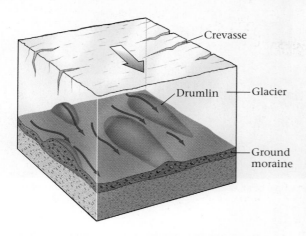

(a) The formation of a drumlin beneath a glacier.

View looking southeast

(b) Drumlins dominate this landscape near Rochester, New York.

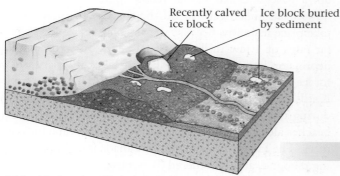

(c) Ice blocks calve off glaciers and become buried by sediment. When the ice melts, a kettle forms.

Time

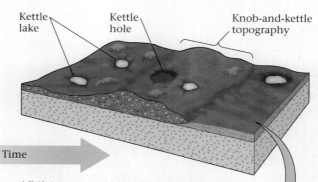

(d) If the water table is high, kettles fill with water and turn into roughly circular lakes.

TAKE-HOME MESSAGE

When ice melts, it drops unsorted sediment to form glacial till. Meltwater streams and wind can transport and sort the sediment to form outwash plains and loess deposits, respectively. Deposition by glaciers produces distinctive landforms, such as moraines.

(e) Knob-and-kettle topography make the surface of this moraine in Yellowstone Park, Wyoming, very hummocky.

18.5 OTHER CONSEQUENCES OF CONTINENTAL GLACIATION

Ice Loading and Glacial Rebound

When a large ice sheet (more than 50 km in diameter) grows on a continent, its weight causes the surface of the lithosphere to sink (Fig. 18.19). In other words, ice loading causes glacial subsidence. Lithosphere, the rela-tively rigid outer shell of the Earth, is able to sink because the underlying asthenosphere is soft enough to flow slowly out of the way. An ice-margin lake fills the low areas at the edge of the glacier.

FIGURE 18.18 Eskers are snake-like ridges of sand and gravel that form when sediment fills meltwater tunnels at the base of a glacier.

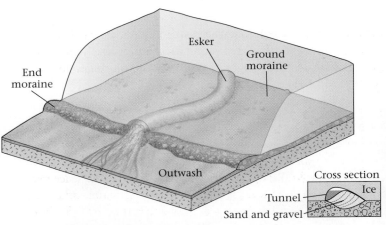

(a) At the time of formation, an esker develops beneath an ice sheet. In cross section (inset), cross-bedded sand accumulates in the tunnel. The stream exists at the toe.

(b) An example of an esker in an area once glaciated, but now farmed.

FIGURE 18.19 The concept of subsidence due to the weight of ice sheets. (Not to scale)

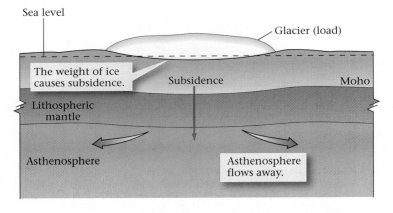

What happens when continental ice sheets melt away? Gradually, the surface of the underlying continent rises back up, a process called glacial rebound, and the asthenosphere flows back underneath to fill the space. This process doesn't take place instantly, because the asthenosphere flows so slowly (at rates of a few millimeters per year). In some places, the depression becomes a lake or bay. It takes thousands of years for ice-depressed continents to rebound. Thus, glacial rebound is still taking place in some regions that were burdened by ice during the last ice age. Where rebound affects coastal areas, beaches along the shoreline rise several meters and become terraces. The Hudson Bay region of Canada is currently undergoing rebound, and is bordered by terraces.

FIGURE 18.20 The link between sea level and global glaciation: glaciers store water on land, so when glaciers grow, sea level falls, and when glaciers melt, sea level rises.

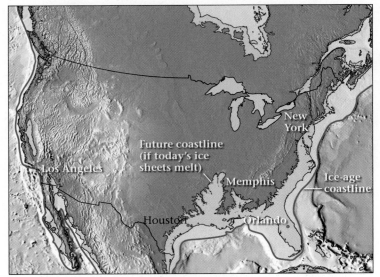

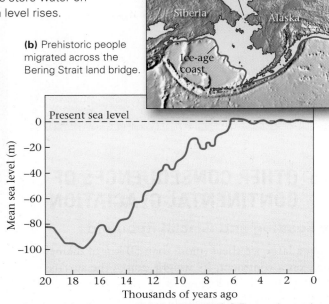

(b) Prehistoric people migrated across the Bering Strait land bridge.

(a) The red line shows the coastline during the last ice age; much of the continental shelf was dry. If present-day ice sheets melt, coastal lands will flood.

(c) Sea-level rise between 17,000 and 7,000 B.C.E. was due to the melting of ice-age glaciers.

Sea-Level Changes: The Glacial Reservoir

More of the Earth's surface and near-surface freshwater is stored in glacial ice than in any other reservoir. During the last ice age, when glaciers covered almost three times as much land area as they do today, they held significantly more water. In effect, water from the ocean reservoir transferred to the glacial reservoir and remained trapped on land. As a consequence, sea level dropped by as much as 100 m, and extensive areas of continental shelves became exposed as the coastline migrated seaward (Fig. 18.20a–c). People and animals migrated into the newly exposed coastal plains. The drop in sea level led to the appearance of land bridges across the Bering Strait between North America and northeastern Asia and between Australia and Indonesia, providing convenient migration routes for animals and people.

Ice Dams, Drainage Reversals, and Lakes

When ice freezes over a sewer opening in a street, neither meltwater nor rain can enter the drain, and the street floods. Ice sheets play a similar role in glaciated environments. The ice may block the course of a river, leading to the formation of a lake. In addition, the weight of a glacier changes the tilt of the land

FIGURE 18.21 Ice Age lakes in North America.

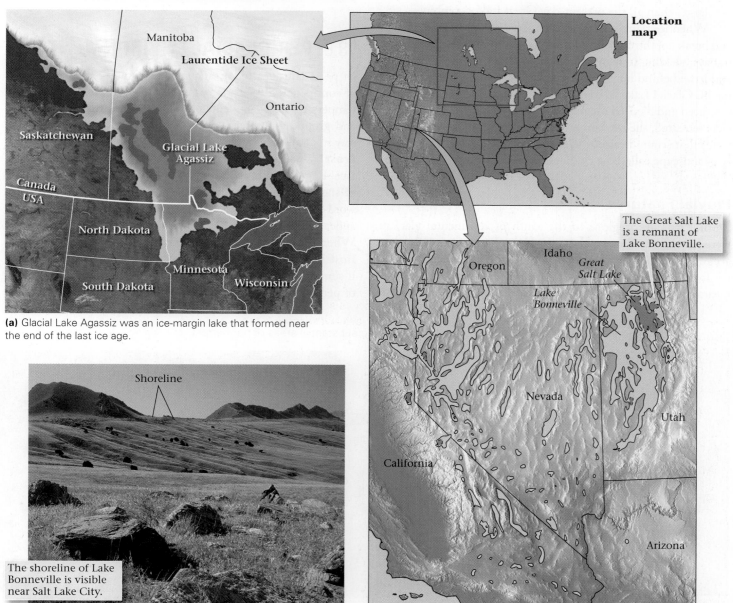

(a) Glacial Lake Agassiz was an ice-margin lake that formed near the end of the last ice age.

The shoreline of Lake Bonneville is visible near Salt Lake City.

The Great Salt Lake is a remnant of Lake Bonneville.

(b) Pluvial lakes occurred throughout the Basin and Range Province during the last ice age due to the wetter climate. The largest of these was Lake Bonneville in western Utah.

surface and therefore the gradient of streams, and glacial sediment may fill preexisting valleys. In sum, continental glaciation destroys preexisting drainage networks. While the glacier exists, streams find different routes and carve out new valleys. By the time the glacier melts away, these new streams have become so well established that old river courses may remain abandoned.

Ice-margin lakes, as we've mentioned, develop where glaciers block drainage and meltwater fills the depressed land surface at the front of the glacier. Between about 11,700 and 9,000 years ago, as the most recent phase of the last ice age came to a close and North America's vast continental glacier retreated back toward Hudson Bay, a huge ice-margin lake formed to the south of the receding glacier (Fig. 18.21a). This lake, known as Glacial Lake Agassiz, was up to 1,100 km (680 mi) wide—it was larger than all the present-day Great Lakes combined.

When ice dams that hold back large ice-margin lakes melt and break up, the contents of the lakes drain in a matter of hours to days, yielding immense floodwaters that strip the land of soil and leave behind huge ripple marks. As we discussed in Chapter 14, Glacial Lake Missoula, in Montana, filled when glaciers advanced and blocked the outlet of a large valley. When the glaciers retreated, the ice dam broke, releasing immense torrents of water that scoured eastern Washington, creating a barren, soil-free landscape called the channeled scablands.

Pluvial Features

During ice ages, regions to the south of continental glaciers were wetter than they are today. Fed by enhanced rainfall, lakes accumulated in low-lying land even at a great distance from the ice front. Such **pluvial lakes** (from the Latin word *pluvia*, meaning rain) flooded interior basins of the Basin and Range Province in Utah and Nevada (Fig. 18.21b). The largest pluvial lake, Lake Bonneville, covered almost a third of western Utah. When this lake suddenly drained after a natural dam holding it back broke, it left a bathtub ring of shorelines rimming the Wasatch Mountains near Salt Lake City. Today's Great Salt Lake itself is but a small remnant of Lake Bonneville.

TAKE-HOME MESSAGE

The weight of a continental ice sheet can cause the ground surface to sink. When the glacier melts away, the surface slowly rises. Continental ice sheets store significant amounts of water. As a consequence, growth or melting of ice sheets affects sea level.

18.6 PERIGLACIAL ENVIRONMENTS

In polar latitudes today, and in regions adjacent to the fronts of continental glaciers during the last ice age, the mean annual temperature stays low enough (below −5°C) that soil moisture and groundwater freeze and, except in the upper few meters, stay solid all year. Such permanently frozen ground, or **permafrost**, may extend to depths of 1,500 m below the ground surface. Regions with widespread permafrost but without a blanket of snow or ice are called periglacial environments (the Greek word *peri* means around, or encircling; periglacial environments appear around the edges of glacial environments) (Fig. 18.22a). When surface temperatures remain below freezing, permafrost is so hard that it can support the weight of a truck. But if summer temperatures rise above freezing, the top few meters of permafrost melt and become a gooey mire. Annual thawing

FIGURE 18.22 Periglacial regions are not ice covered but do include substantial areas of permafrost.

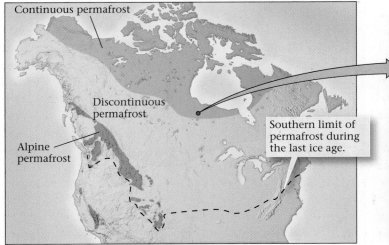

(a) The present-day distribution of periglacial environments in North America.

(b) An example of patterned ground near a pond in Manitoba, Canada. The polygons are about 15 m across.

and refreezing splits the ground into pentagonal or hexagonal shapes, creating a landscape of **patterned ground** (Fig. 18.22b). Water fills the gaps between the cracks and freezes to create wedge-shaped walls of ice.

Permafrost presents a unique challenge to people who live in polar regions or who work to extract resources from these regions. For example, heat from a building may warm and melt underlying permafrost, causing the building to sink. For this reason, buildings in permafrost regions must be placed on stilts, so that cold air can circulate beneath them to keep the ground frozen.

TAKE-HOME MESSAGE

Regions that are cold most of the year, but are not glaciated, may be underlain by permanently frozen ground (permafrost). Summer melting may turn the top few meters into mud. Freeze and thaw cycles produce patterned ground.

18.7 THE PLEISTOCENE ICE AGES

The Pleistocene Glaciers

Today, most of the land surface in New York City lies hidden beneath concrete and steel, but in the city's Central Park it's still possible to see land in a seminatural state. If you stroll through the park and study the rock outcrops, you'll find that their top surfaces are smooth and polished, and in places have been grooved and scratched (Fig. 18.10c). You are seeing evidence that an ice sheet once scraped along this ground, rasping and gouging the bedrock. Geologists estimate that the ice sheet that overrode the New York City area may have been 250 m thick, enough to bury the Empire State Building up to the 75th floor.

The fact that these glacial features decorate the surface of the Earth today means that the most recent ice age occurred fairly recently during Earth history. This ice age, responsible for the glacial landforms of North America and Eurasia, happened mostly during the Pleistocene Epoch, which began 1.8 million years ago (see Chapter 11), so it is commonly known as the Pleistocene ice age. (More recent evidence suggests that glaciations began as early as 3 million years ago.) Geologists use the name Holocene to refer to the last 11,000 years, the time since the last glaciation.

Based on their mapping of glacial striations and deposits, geologists have determined where the great Pleistocene ice sheets originated and flowed. In North America, the Laurentide ice sheet started to grow at two zones of accumulation in Canada, and eventually covered all of Canada east of the Rocky Mountains and extended southward across the border as far as southern Illinois (Fig. 18.23). At its maximum, it attained a

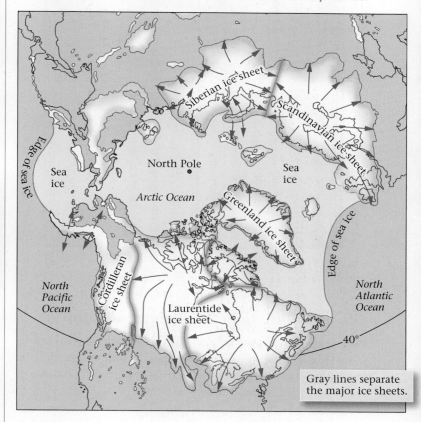

FIGURE 18.23 Pleistocene ice sheets and their flow trajectories.

Gray lines separate the major ice sheets.

thickness of 2 to 3 km. These ice sheets also eventually merged with the Greenland ice sheet to the northeast and the Cordilleran ice sheet to the west. The Cordilleran ice sheet covered the mountains of western Canada as well as the southern third of Alaska. Mountain ice caps and valley glaciers also grew in the Rocky Mountains, the Sierra Nevada, and the Cascade Mountains of the United States.

In Eurasia, a large ice sheet formed in northernmost Europe and adjacent Asia, and gradually covered all of Scandinavia and northern Russia. This ice sheet flowed southward across France until it reached the Alps and merged with Alpine mountain glaciers; it also covered all of Ireland and almost all of the United Kingdom. A smaller ice sheet grew in eastern Siberia and expanded in the mountains of central Asia. In the southern hemisphere, Antarctica remained ice covered, and mountain ice caps expanded in the Andes, but there were no continental glaciers in South America, Africa, or Australia.

In addition to continental ice sheets, sea ice in the northern hemisphere expanded to cover all of the Arctic Ocean and parts of the North Atlantic. Sea ice surrounded Iceland and approached Scotland, and also fringed most of western Canada and southeastern Alaska.

FIGURE 18.24 Climate belts during the Pleistocene.

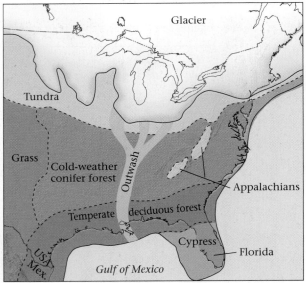

(a) Tundra covered parts of the United States, and southern states had forests like those of New England's today.

(b) Regions of Europe that support large populations today would have been barren tundra during the Pleistocene.

Life and Climate in the Pleistocene World

During the Pleistocene ice age, all climatic belts shifted southward (Fig. 18.24a, b). Geologists can document this shift by examining fossil pollen, which can survive for thousands of years, preserved in the sediment of bogs.

Fossils also tell us that numerous species of now-extinct large mammals inhabited the Pleistocene world (Fig. 18.24c). Giant mammoths and mastodons, relatives of the elephant, along with woolly rhinos, musk oxen, reindeer, giant ground sloths, bison, lions, saber-toothed cats, giant cave bears, and hyenas wandered forests and tundra in North America. Early human-like species were already foraging in the woods by the beginning of the Pleistocene Epoch, and by its end, modern *Homo sapiens* lived on every continent except Antarctica, and had discovered fire and invented tools.

Timing of the Pleistocene Ice Ages

Louis Agassiz assumed that only one ice age had affected the planet. But close examination of the stratigraphy of glacial deposits on land revealed that paleosol (ancient soil preserved in the stratigraphic record), as well as beds containing fossils of warmer-weather animals and plants, separated distinct layers of glacial sediment. This observation suggested that between episodes of glacial deposition, glaciers receded and temperate climates prevailed. During the second half of the twentieth century, when modern methods for dating geologic materials became available, the difference in ages between the different layers of glacial sediment could be confirmed. Clearly, ice-age glaciers had advanced and then retreated more than once. Times during which the glaciers grew and covered substantial areas of the continents are called glacial periods, or glaciations,

(c) Now-extinct large mammals roamed periglacial environments.

and times between glacial periods are called interglacial periods, or interglacials.

Using the on-land sedimentary record, geologists recognized five Pleistocene glaciations in Europe (named, in order of increasing age: Würm, Riss, Mindel, Gunz, and Donau) and traditionally four in the midwestern United States (Wisconsinan, Illinoian, Kansan, and Nebraskan, named after the southernmost states in which their till was deposited; Fig. 18.25). Since the mid-1980s, geologists no longer recognize Nebraskan and Kansan; they are lumped together as "pre-Illinoian." The four-stage chronology of North American glaciation, however, was turned on its head in the 1960s, when geologists began to study submarine sediment containing the fossilized shells of microscopic marine plankton. Because the assemblage of plankton species living in warm water is not the same as the assemblage living in cold water, geologists can track changes in the temperature of the ocean by studying plankton fossils. Researchers found that the succession of plankton in sediment of the last 2 million years, assuming that cold water indicates a glacial period and warm water an interglacial period, records twenty to thirty different glacial advances. The four traditionally recognized glaciations possibly represent only the largest of these—sediments deposited on land by other glaciations were eroded and redistributed during subsequent glaciations, or were eroded by streams and wind during interglacials.

Geologists refined their conclusions about the frequency of Pleistocene glaciations by examining the composition of fossil shells. Shells of many plankton species consist of calcite ($CaCO_3$). The oxygen in the shells includes two isotopes, a heavier one (^{18}O) and a lighter one (^{16}O). The ratio of these isotopes tells us about the water temperature in which the plankton grew. This is because as water gets colder plankton incorporate a higher proportion of ^{18}O into their shells. Thus, intervals in the stratigraphic record during which plankton shells have a large ratio of ^{18}O to ^{16}O define times when Earth had a colder, glacial climate. The record also indicates that over twenty of these glaciations occurred during the last 2 million years (Fig. 18.26a).

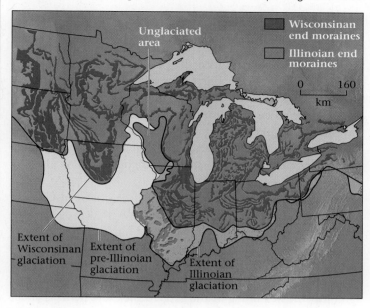

FIGURE 18.25 Pleistocene glacial deposits in the north-central United States. Curving moraines reflect the shape of glacial lobes.

FIGURE 18.26 The timing of glaciations. Ice ages have occurred at several times in the geologic past.

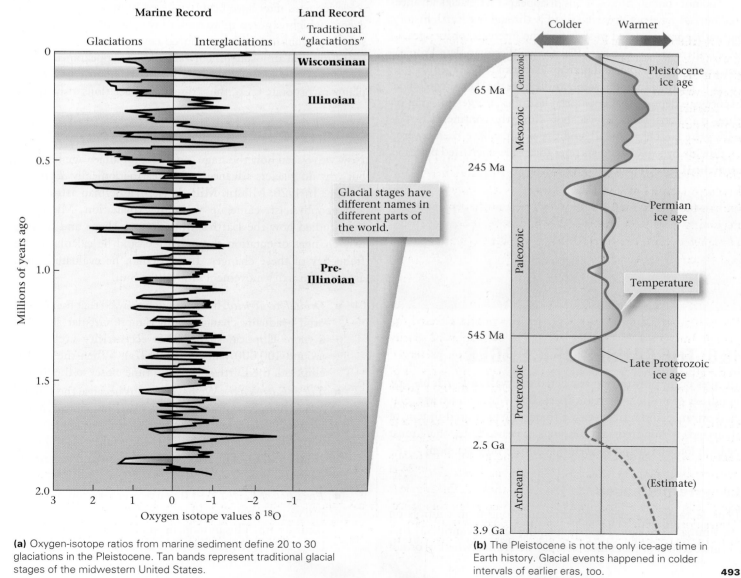

(a) Oxygen-isotope ratios from marine sediment define 20 to 30 glaciations in the Pleistocene. Tan bands represent traditional glacial stages of the midwestern United States.

(b) The Pleistocene is not the only ice-age time in Earth history. Glacial events happened in colder intervals of earlier eras, too.

Older Ice Ages During Earth History

So far, we've focused on the Pleistocene ice age, because of its importance in developing Earth's present landscape. Was this the only ice age during Earth history, or do ice ages happen frequently? To answer such questions, geologists study the stratigraphic record and search for ancient glacial deposits that have hardened into rock. These deposits, called tillites, consist of larger clasts distributed throughout a matrix of sandstone and mudstone. In many cases, tillites are deposited on glacially polished surfaces.

By using the stratigraphic principles described in Chapter 10, geologists have determined that tillites were deposited about 280 million years ago in Permian time. These are the deposits Alfred Wegener studied when he argued in favor of continental drift. Tillites were also deposited about 600 to 700 million years ago (at the end of the Proterozoic Eon), about 2.2 billion years ago (near the beginning of the Proterozoic), and perhaps about 2.7 billion years ago. Strata deposited at other times in Earth history do not contain tillites. Thus, it appears that glacial advances and retreats have not occurred steadily throughout Earth history, but rather are restricted to specific time intervals, or ice ages, of which there are four or five: Pleistocene, Permian, late Proterozoic, early Proterozoic, and perhaps Archean (Fig. 18.26b).

Of particular note, some tillites of the late Proterozoic event were deposited at equatorial latitudes, suggesting that on three occasions, for at least a short time, the continents worldwide were largely glaciated, and the sea may have been covered worldwide by ice. Geologists refer to the ice-encrusted planet as **snowball Earth** (see Chapter 11).

TAKE-HOME MESSAGE

During the last Ice Age, which began about 3 Ma, continental ice sheets advanced and retreated several times. This period includes the Pleistocene (which formally began 1.8 Ma, and ended at about 11 Ka). Other ice ages happened earlier in Earth history.

18.8 THE CAUSES OF ICE AGES

Ice ages occur only during restricted intervals of Earth history, hundreds of millions of years apart. But within an ice age, glaciers advance and retreat with a frequency measured in tens of thousands to hundreds of thousands of years. Thus, there must be both long-term and short-term controls on glaciation.

Long-Term Causes

Plate tectonics probably provides some long-term controls on the timing of ice ages. For example, the drift of continents that occurs because of plate motions may determine whether or not continents lie in appropriate latitudes and configurations to be glaciated. The arrangement of continents and volcanic arcs may also affect the global pattern of ocean currents that transport heat. This pattern, in turn, affects climate. Finally, rates of sea-floor spreading affect global sea level, and thus determine whether or not continents are above sea level and susceptible to ice formation.

The concentration of carbon dioxide in the atmosphere may also determine whether an ice age can occur. Carbon dioxide is a greenhouse gas—it traps infrared radiation rising from the Earth—so if the concentration of CO_2 increases, the atmosphere becomes warmer. Ice sheets cannot form during periods when the atmosphere has a relatively high concentration of CO_2, even if other factors favor glaciation. What might cause long-term changes in CO_2 concentration? Possibilities include changes in the population of living organisms, for many organisms extract CO_2; changes in the amount of chemical weathering on land, for weathering absorbs CO_2; and changes in the amount of volcanic activity, for volcanoes release gas. Thus, some ice ages may have been triggered by mountain-building events, which exposed more rock to weathering, and by major stages in evolution, which produced new species that extracted CO_2 from the atmosphere. For example, the appearance of coal swamps at the end of Paleozoic may have removed CO_2, for plants incorporate CO_2, thus triggering glaciations of Pangaea.

Short-Term Causes

Now we've seen how the stage can be set for an ice age to occur, but why do glaciers advance and retreat periodically *during* an ice age? In 1920, Milutin Milankovitch, a Serbian astronomer and geophysicist, came up with an explanation. Milankovitch studied how the Earth's orbit changes shape and how its axis changes orientation through time, and he calculated the frequency of these changes. In particular, he evaluated three aspects of Earth's movement around the Sun.

- *Orbital eccentricity*: Milankovitch showed that the Earth's orbit gradually changes from a more circular shape to a more elliptical shape. This eccentricity cycle takes around 100,000 years (Fig. 18.27a). When the orbit is elliptical, the Earth spends less time closer to the Sun.
- *Tilt of Earth's axis*: We have seasons because the Earth's axis is not perpendicular to the plane of its orbit. Milankovitch calculated that over time, the tilt angle varies between 22.5° and 24.5°. This cycle takes 41,000 years (Fig. 18.27b). Tilt affects the duration and severity of winter.
- *Precession of Earth's axis*: If you've ever set a top spinning, you've probably noticed that its axis gradually traces a conical path. This motion, or wobble, is called precession (Fig. 18.27c). Milankovitch determined that the Earth's

FIGURE 18.27 Milankovitch cycles influence the amount of insolation received at high latitudes.

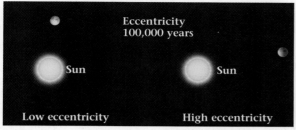

Eccentricity
100,000 years

Sun   Sun

Low eccentricity   High eccentricity

(a) Variations caused by changes in orbital shape.

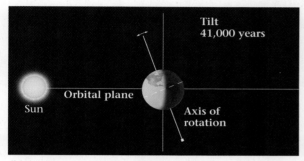

Tilt
41,000 years

Orbital plane

Sun

Axis of rotation

(b) Variations caused by changes in axis tilt.

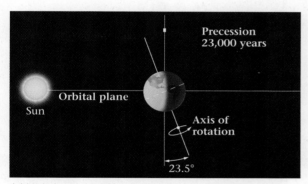

Precession
23,000 years

Orbital plane

Sun

Axis of rotation

23.5°

(c) Variations caused by the precession of Earth's axis.

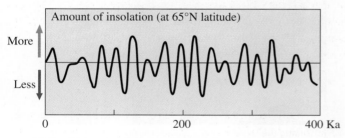

Amount of insolation (at 65°N latitude)

More

Less

0   200   400 Ka

(d) Combining the effects of eccentricity, tilt, and precession produces distinct periods of more or less insolation.

axis wobbles over the course of about 23,000 years. Precession determines the relationship between the timing of the seasons and the position of Earth along its orbit around the Sun.

Milankovitch showed that orbital eccentricity, tilt, and precession combine to create a cyclic variation in the total annual amount of insolation (exposure to the Sun's rays) and the sea-

FIGURE 18.28 Until recently, Earth's atmosphere has been gradually cooling, overall, since the Cretaceous.

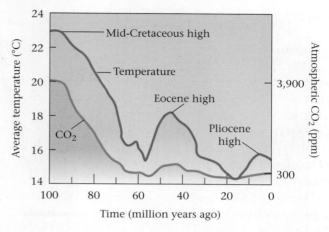

sonal distribution of insolation that the Earth receives at the high latitudes (Fig. 18.27d). For example, high-latitude regions receive more insolation when the Earth's axis is almost perpendicular to its orbital plane than when its axis is greatly tilted. According to Milankovitch, glaciers tend to advance during times of cool summers at 65°N. These insolation cycles are now called **Milankovitch cycles**. When geologists began to study the climate record, they found climate cycles with frequencies similar to the insolation cycles that Milankovitch described.

Milankovitch cycles are probably not the whole story, however. Geologists suggest that several other factors may also come into play to trigger a glacial advance.

- *Solar variability*: Changes in the radiation output of the Sun could affect the amount of energy the Earth receives.

- *A changing albedo*: When snow remains on land throughout the year, or clouds form in the sky, the albedo (reflectivity) of the Earth increases, so Earth's surface reflects incoming sunlight and thus becomes even cooler.

- *Interrupting the global heat conveyor*: As the climate cools, evaporation rates from the sea decrease, so seawater does not become so salty. And decreasing salinity might stop the system of thermohaline currents that brings warm water to high latitudes (see Chapter 15). Such changes can be very abrupt.

- *Biological processes that change CO_2 concentration*: A bloom of algae in the ocean may have amplified climate changes by decreasing the concentration of CO_2 in the atmosphere.

A Model for Pleistocene-Ice-Age History

Long-term cooling in the Cenozoic Era. Taking all of the above causes into account, we can now propose a scenario for the events that led to the Pleistocene glacial advances. Our story begins in the Eocene Epoch, about 55 million years ago (Fig. 18.28). At that time, climates were warm and balmy, not only in the tropics but even above the Arctic Circle. At the

FIGURE 18.29 The Little Ice Age and its demise. Glaciers that advanced between 1550 and 1850 have since retreated.

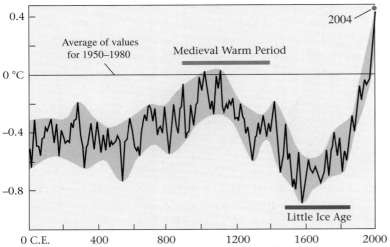

(a) A model of global temperature for the past 2000 years. Overall trends display the Medieval Warm Period followed by the Little Ice Age. Since 1850, temperatures have warmed.

(b) Skaters (c. 1600) on the frozen canals of the Netherlands during the Little Ice Age.

(c) During the Little Ice Age, a glacier filled this valley. In this 2003 photo, most of the glacier has vanished. Most of the retreat has happened in the last century.

end of the Middle Eocene (37 million years ago), the climate began to cool. These long-term climate changes may have been caused, in part, by changes in the pattern of oceanic currents that happened when India collided with Asia. The uplift of the Himalayas that resulted from this collision may also have diverted winds in a way that cooled climate. This uplift also exposed more rock to chemical weathering, perhaps leading to extraction of CO_2 from the atmosphere.

In this overall cooler global climate, the stage was set for an ice age to begin. The geologic event that finally triggered the birth of the Laurentide ice sheet, 2 to 3 million years ago, may have been the closing of the gap between North and South America by the growth of the Isthmus of Panama. When this land bridge formed, it separated the waters of the Caribbean from those of the tropical Pacific for the first time. And when this happened, warm currents that previously flowed out of the Caribbean into the Pacific were blocked and diverted northward to merge with the Gulf Stream. As the warm water moved up the Atlantic Coast, it generated warm, moisture-laden air, a source for snow. In other words, the Arctic has long been cold enough for ice caps, but until the Gulf Stream was diverted northward, significant accumulation of ice could not occur.

Short-term advances and retreats in the Pleistocene Epoch. Once the Earth's climate had cooled overall, short-term processes such as the Milankovitch cycles led to periodic advances and retreats of the glaciers. To understand how, let's look at a possible case history of a single advance and retreat of the Laurentide ice sheet. (Note that such models remain the subject of vigorous debate.)

■ *Stage 1*: During the overall cooler climates of the late Cenozoic Era, the Earth reaches a point in the Milankovitch cycle when the average mean temperature in temperate

latitudes drops. Not all of winter's snow melts away during the summer in northern Canada. The snow reflects sunlight, so the region grows still colder, and even more snow accumulates as evaporation off the Gulf Stream provides moisture. Finally, the snow at the base of the pile turns to ice, and the ice begins to spread outward under its own weight. A new continental glacier has been born.

- *Stage 2*: The ice sheet grows as more snow piles up, and as the ice sheet grows, the atmosphere cools because of the albedo effect. But now, the weight of the ice loads the continent and makes it sink, so the elevation of the glacier decreases, and its surface approaches the equilibrium line. Also, the temperature becomes cold enough that the Atlantic Ocean in high latitudes freezes over, so the amount of evaporation decreases, cutting off the source of moisture. The glacial advance chokes on its own success. The decrease in the glacier's elevation (leading to warmer summer temperatures) on the ice surface, as well as the decrease in snowfall, causes the glacier to retreat.

- *Stage 3*: As the glacier retreats, temperatures gradually increase, and the sea ice begins to melt. The supply of water to the atmosphere from evaporation increases once again, but with the warmer temperatures and lower elevations, this water precipitates as rain during the summer. The rain drastically accelerates the rate of ice melting, and the retreat progresses rapidly.

Will There Be Another Glacial Advance?

What does the future hold? Considering the periodicity of glacial advances and retreats during the Pleistocene Epoch, we may be living in an interglacial period. Pleistocene inter-glacials lasted about 10,000 years, and since the present interglacial began about 11,000 years ago, the time seems ripe for a new glaciation. The Earth actually had a brush with ice-age conditions between the 1300s and the mid-1800s, when average annual temperatures in the northern hemisphere fell sufficiently for mountain glaciers to advance significantly (Fig. 18.29a). During this period, now known as the Little Ice Age, sea ice surrounded Iceland, and canals froze in the Netherlands, leading to that country's tradition of skating (Fig. 18.29b). The Little Ice Age did not become a full-fledged ice age because during the past 150 years, temperatures have warmed. You can see evidence for this warming by visiting a mountain glacier—most mountain glaciers have retreated substantially (Fig. 18.29c). Most researchers suggest that this global-warming trend is due to the addition of carbon dioxide to the atmosphere from the burning of forests and the use of fossil fuels (see Chapter 19).

TAKE-HOME MESSAGE

Ice ages happen when the distribution of continents, ocean currents, and the concentration of atmospheric CO_2 are appropriate. Advances and retreats during an ice age are controlled by Milankovitch cycles (variations in Earth's orbit shape and in the orientation of Earth's rotation axis).

Chapter Summary

- Glaciers are streams or sheets of recrystallized ice that survive for the entire year and flow in response to gravity. Mountain glaciers exist in high regions and fill cirques and valleys. Continental glaciers (ice sheets) spread over substantial areas of the continents.

- Glaciers form when snow accumulates over a long period of time. With progressive burial, the snow first turns to firn, and then to ice.

- Glacial ice moves by basal sliding over water or wet sediment, and/or by internal flow. In general, glacial ice moves tens of meters per year.

- Whether the toe of a glacier stays fixed in position, advances farther from the glacier's origin, or retreats back toward the origin depends on the balance between the rate at which snow builds up in the zone of accumulation and the rate at which glaciers melt or sublimate in the zone of ablation.

- Glacial ice can flow over sediment or incorporate sediment. The clasts embedded in glacial ice act like a rasp that abrades the substrate.

- Mountain glaciers carve numerous landforms, including cirques, arêtes, horns, U-shaped valleys, hanging valleys, and truncated spurs. Glacially carved valleys that fill with water when sea level rises after an ice age are fjords.

- Moraines are piles or ridges of glacial till. Till is unsorted sediment, which accumulates because glaciers can transport sediment of all sizes.

- Glacial depositional landforms include moraines, knob-and-kettle topography, drumlins, eskers, meltwater lakes, and outwash plains.

- Continental crust subsides as a result of ice loading. When the glacier melts away, the crust rebounds.

- When water is stored in continental glaciers, sea level drops. When glaciers melt, sea level rises.

- During past ice ages, the climate in regions south of the continental glaciers was wetter, and pluvial lakes formed. Permafrost exists in periglacial environments.

- During the Pleistocene ice age, large continental glaciers covered much of North America, Europe, and Asia.

- The stratigraphy of glacial deposits indicates that glaciers advanced and retreated many times during the Pleistocene.

- Long-term causes of ice ages include plate tectonics and changes in the concentration of CO_2 in the atmosphere. Short-term causes include the Milankovitch cycles (caused by periodic changes in Earth's orbit and tilt).

GEOPUZZLE REVISITED

Even if people had the ability to build large cities 12,000 years ago, they couldn't have built at the localities that are now New York or Edinburgh, for the land surface at these locations lay beneath hundreds of meters of ice. 12,000 years ago marked the end of the last glacial advance when huge ice sheets covered substantial areas of North America and Eurasia.

Key Terms

arête (p. 478)

basal sliding (p. 473)

cirque (p. 473)

continental glacier (p. 473)

crevasse (p. 474)

drumlin (p. 486)

end moraine (p. 481)

equilibrium line (p. 476)

erratic (p. 470)

esker (p. 486)

fjord (p. 479)

glacial advance (p. 476)

glacial retreat (p. 476)

glacial striations (p. 477)

glacial till (p. 481)

glacier (p. 470)

hanging valley (p. 478)

horn (p. 478)

ice age (p. 470)

iceberg (p. 477)

kettle hole (p. 486)

lateral moraine (p. 481)

loess (p. 482)

medial moraine (p. 481)

Milankovitch cycles (p. 495)

moraine (p. 481)

mountain glacier (p. 473)

patterned ground (p. 491)

permafrost (p. 490)

plastic deformation (p. 473)

pluvial lake (p. 490)

polar glacier (p. 473)

roche moutonnée (p. 479)

sea ice (p. 477)

snowball Earth (p. 494)

temperate glacier (p. 493)

terminal moraine (p. 486)

tidewater glacier (p. 476)

toe (terminus) (p. 476)

U-shaped valley (p. 478)

zone of ablation (p. 476)

zone of accumulation (p. 476)

THE VIEW FROM SPACE Glaciers and glacially carved features dominate the landscape of southeastern Alaska, as seen in this infrared image. Here we see the Hubbard glacier.

Review Questions

1. What evidence did Louis Agassiz offer to support the idea of an ice age?

2. How do mountain glaciers and continental glaciers differ in terms of dimensions, thickness, and patterns of movement?

3. Describe the transformation from snow to ice.

4. Explain how arêtes, cirques, and horns form.

5. Describe the mechanisms that enable glaciers to move, and explain why they move.

6. Explain how the balance between ablation and accumulation determines whether a glacier advances or retreats.

7. How does a glacier transform a V-shaped river valley into a U-shaped valley? Discuss how hanging valleys develop.

8. Describe the various kinds of glacial deposits. Be sure to note the materials from which the deposits are made and the landforms that result from deposition.

9. How does the lithosphere respond to the weight of glacial ice?

10. How was the world different during the glacial advances of the Pleistocene ice age? Be sure to mention the relation between glaciations and sea level.

11. Were there ice ages before the Pleistocene? If so, when?

12. What are some of the long-term causes that lead to ice ages? What are the short-term causes that trigger glaciations and interglacials?

On Further Thought

1. If you fly over the barren cornfields of central Illinois during the early spring, you will see slight differences in soil color due to variations in moisture content—wetter soil is darker. These variations outline the shapes of polygons that are tens of meters across. What do these patterns represent and how might they have formed? What do they tell us about the climate of central Illinois at the end of the last ice age?

2. Recent observations suggest that glaciers of southern Greenland have started to flow much faster in the past 10 years. Researchers suggest that this change might be a manifestation of the warming of the regions climate. What mechanism could account for the acceleration of the glaciers?

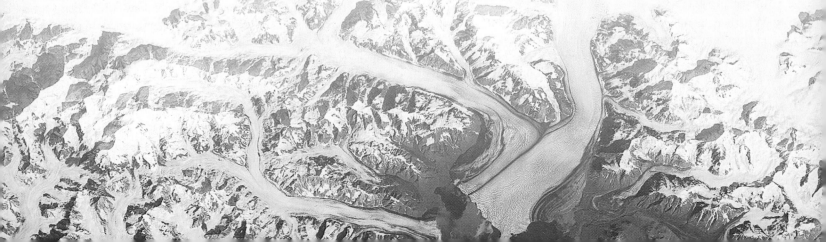

Global Change in the Earth System

This view, taken from an airplane landing at Chicago's O'Hare Airport, emphasizes the extent to which the Earth's surface has changed. Fifteen thousand years ago, the view would have been the surface of a glacier. Five hundred years ago, it would have been a tall-grass prairie. One hundred years ago, it would have been a checkerboard of farm fields.

GEOPUZZLE

Climate change has been in the news almost daily during the past few years. Why has there been increasing interest in this issue? Is climate the only aspect of the Earth System that changes through time?

All we in one long caravan
are journeying since the world began,
we know not whither, we know . . . all must go.

—Bhartrihari (Indian poet, c. 500 C.E.)

19.1 INTRODUCTION

Would the Earth's surface have looked the same in the Jurassic as it does today? Definitely not! In the Jurassic, the North Atlantic Ocean was a narrow sea and the South Atlantic Ocean didn't exist at all, so most dry land connected to form a single vast continent (Fig. 19.1). Today, both parts of the Atlantic are wide oceans, and the Earth has seven separate continents. Moreover, in the Jurassic, the call of the wild rumbled from the throats of dinosaurs; today, the largest land animals are mammals. In essence, what we see of the Earth now is just a snapshot, an instant in the life story of a constantly changing planet. This idea arguably stands as geology's greatest contribution to humanity's perception of the Universe.

Why has the Earth changed so much through time? Ultimately, it's because the Earth's asthenosphere is warm and soft enough to flow, because the Sun is close enough to heat the Earth's surface, and because gravity causes heavy objects to fall and buoyant ones to rise. A soft asthenosphere allows plate motions to happen that, in turn, drive continental drift, volcanism, and mountain building. Volcanism provides elements that make up the atmosphere and oceans, and these fluids circulate in response to solar heat and gravity. Such movement ultimately produces the streams, glaciers, waves, and wind that cause erosion and transport sediment. Gravity plays an additional role by pulling rocks and regolith down slopes, and by causing sediment to settle from fluids. And solar warmth plays an additional role by keeping water in liquid form at the Earth's surface—without liquid water, life could not have evolved. If the Earth did not have just the right mix of tectonic activity and solar heat,

FIGURE 19.1 A map of Earth's surface changes over time because of plate tectonics. Two hundred million years ago, the Americas touched Africa in Pangea. Now, the Atlantic Ocean lies in between.

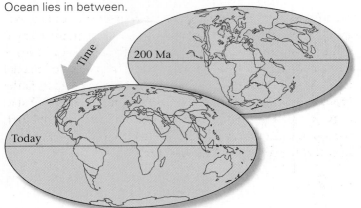

it would be a frozen dust bowl like Mars, a crater-pocked wasteland like the Moon, or a cloud-choked oven like Venus.

Changes that take place on Earth also reflect complex interactions among geologic and biological phenomena. For example, photosynthetic organisms affect the composition of the atmosphere by providing oxygen, and atmospheric composition in turn determines the nature of chemical weathering in rocks. We've referred to the global interconnecting web of physical and biological phenomena on Earth as the Earth System. We can now define **global change** as transformations or modifications of both physical and biological components in the Earth System through time.

Geologists distinguish among different types of global change, based on the rate or way in which change progresses. Gradual change takes place over long periods of geologic time (millions to billions of years), whereas catastrophic change takes place relatively rapidly in the context of geologic time (seconds to millennia). Unidirectional change involves transformations that never repeat, whereas cyclic change repeats the same steps over and over, though not necessarily with the same results. Some types of cyclic change are periodic in that the cycles happen with a definable frequency.

In this chapter, we begin by reviewing examples of global change involving phenomena discussed earlier in the book. Then we look at an example of a biogeochemical cycle, the cyclic exchange of chemicals among living and nonliving reservoirs, for some kinds of global change are due to changes in the proportions of chemicals held in different reservoirs through time. Finally, we focus on global climate change, the transformations or modifications in Earth's climate through time. We conclude this chapter, and the book, by considering hypotheses that describe the ultimate global change—the end of the Earth.

TAKE-HOME MESSAGE

By the end of this chapter, you should understand that global change encompasses many phenomena, and has been happening since the birth of this planet. You should also be familiar with the types of change, and should be aware of evidence for recent climate change.

19.2 UNIDIRECTIONAL CHANGES

The Evolution of the Solid Earth

Recall from Chapter 1 that Earth began as a fairly homogenous mass, formed by the coalescence of planetesimals. The homogeneous proto-Earth did not last long—within about 100 million years of its birth the planet began to melt, yielding liquid iron alloy that sank rapidly to the center to form the core.

FIGURE 19.2 Formation of the Moon is an example of a catastrophic unidirectional change.

A Mars-sized protoplanet collides with Earth.

Debris sprays into space.

Time

The debris coalesces to form the Moon.

This process of differentiation represents major unidirectional change: it produced a layered, onion-like planet, with an iron-alloy core surrounded by a rocky mantle.

Soon after differentiation, a Mars-sized protoplanet collided with the newborn Earth. This collision caused a catastrophic change—a significant portion of the Earth and the colliding object fragmented and vaporized, creating a ring of debris that coalesced to form the Moon (Fig. 19.2). After the collision, the Earth's mantle was probably partially molten, and its surface became a sea of magma. The Earth endured intense bombardment by asteroids and comets between 4.0 and 3.9 Ga, so any crust that had formed prior to 3.9 Ga was largely pulverized or melted. Eventually, however, bombardment ceased and our planet gradually cooled, permitting a crust to form at its surface and plate tectonics to begin operating. Over time, igneous activity produced lasting continental crust.

The Evolution of the Atmosphere and Oceans

Like its surface, the Earth's atmosphere has also changed over time. It initially formed from gases released by volcanic activity. More gases may have arrived when comets collided with our new planet. Eventually, Earth accumulated an early atmosphere composed dominantly of carbon dioxide (CO_2) and water (H_2O). Other gases, such as nitrogen (N_2), composed only a minor proportion of the early atmosphere. When the Earth's surface cooled, however, water condensed and fell as rain, collecting in low areas to form oceans. This may have first happened before 4.0 Ga, but had certainly happened by 3.8 Ga. Gradually, CO_2 dissolved in the oceans and was absorbed by chemical-weathering reactions on land, so its concentration in the atmosphere decreased. Nitrogen, which doesn't react with other chemicals, was left behind. Thus, the atmosphere's composition changed to become dominated by nitrogen. Photosynthetic organisms appeared early in the Archean. But it probably wasn't until 2.5 to 2.0 Ga that oxygen (O_2) became a significant proportion of the atmosphere.

The Evolution of Life

During most of the Hadean Eon, Earth's surface was probably lifeless, for carbon-based organisms could not survive the high temperatures of the time. The fossil record indicates that life had appeared by at least 3.8 Ga and has undergone unidirectional change (evolution) in fits and starts ever since (see Interlude E). Though simple organisms such as archaea and bacteria still exist, life evolution during the late Proterozoic and early Phanerozoic yielded multicellular plants and animals. Life now inhabits regions from a few kilometers below the surface to a few kilometers above, yielding a diverse biosphere in the Earth System (see **Geology at a Glance**, pp. 502–503).

TAKE-HOME MESSAGE

Since it first formed, the Earth has undergone major changes that are unidirectional, in that they will never repeat. Examples include the formation of the core, mantle, and moon; the evolution of the atmosphere and oceans; and the evolution of life.

19.3 PHYSICAL CYCLES

The Supercontinent Cycle

During Earth history, the map of the planet's surface has constantly changed. At times, almost all continental crust merged to form a single supercontinent, but usually the crust is distributed among several smaller continents. The process of change during which supercontinents form and later break apart is the **supercontinent cycle**. Geologists have found evidence that at least three or four times during the past 3 billion years of Earth history, supercontinents existed. The most recent one, Pangaea, formed at the end of the Paleozoic Era.

The Earth System

External energy

Sun

Thunderhead

Lightning

Mountain uplift

Rain and snow

Continental glacier

City

Ocean

Desert

Rocky coastline

Valley

Arid mountains

Mining

Lakes

Deciduous forest

Forested mountains

Beach

Tropical rain forest

Shark

Coral reef

Internal energy

The Earth's surface is the interface among the solid Earth (lithosphere); the ice and liquid water of oceans, lakes, streams, groundwater, and glaciers (the hydrosphere); and the planet's gaseous envelope (the atmosphere). Countless species of life, ranging from nearly invisible bacteria to giant whales and trees, make up the complex ecosystems of Earth's biosphere. All of these components—the lithosphere, hydrosphere, atmosphere, and biosphere—interact with each other. These components and the interactions among them constitute the Earth System.

Various materials cycle among living and nonliving components of the Earth System. In the hydrologic cycle, for example, water evaporates from the sea, rains on the land, and eventually flows back to the sea. During this process, water may be trapped temporarily in living organisms, clouds, subsurface pores, or ice sheets. Carbon dioxide can be stored in the air, dissolved in water, or trapped in plants, coal, or limestone. Some limestone forms when coral extracts ions from water. Meanwhile, the atoms that make up minerals, over the vastness of geologic time, pass through the rock cycle. New elements from the mantle may enter the cycles of the Earth System at volcanoes or black smokers. Elements at the surface may be carried back into the mantle at subduction zones. Some atoms escape from the atmosphere into space.

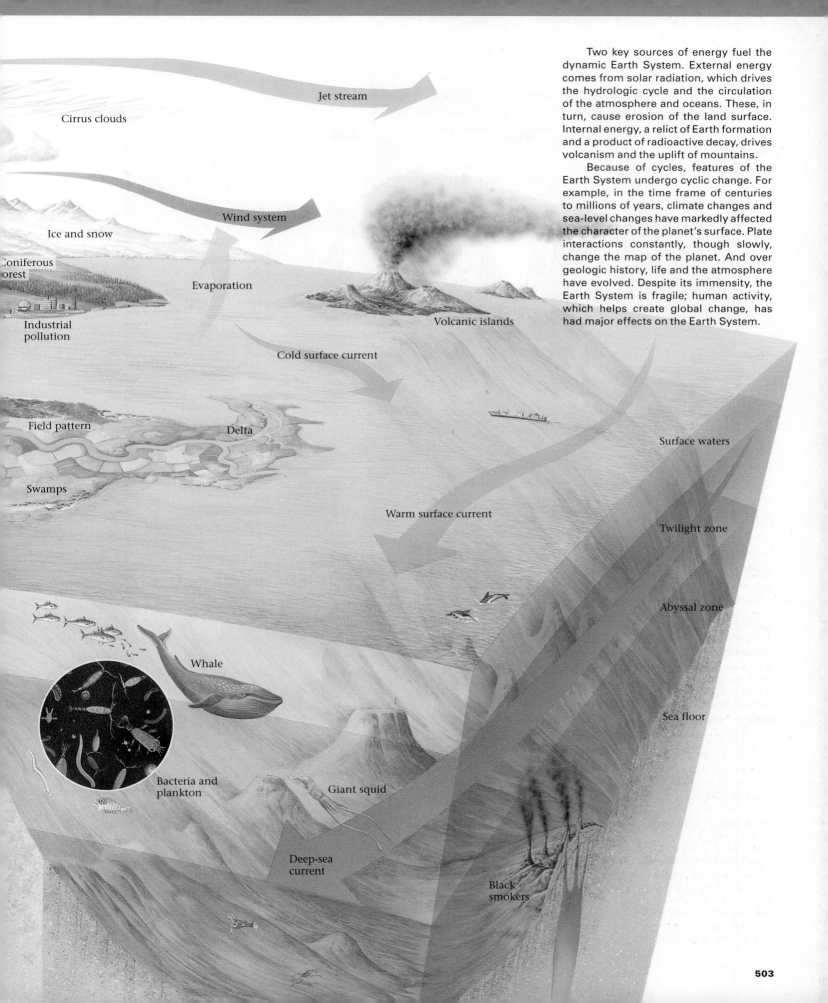

Jet stream

Cirrus clouds

Wind system

Ice and snow

Coniferous forest

Evaporation

Industrial pollution

Volcanic islands

Cold surface current

Field pattern

Delta

Surface waters

Swamps

Warm surface current

Twilight zone

Abyssal zone

Whale

Bacteria and plankton

Giant squid

Sea floor

Deep-sea current

Black smokers

Two key sources of energy fuel the dynamic Earth System. External energy comes from solar radiation, which drives the hydrologic cycle and the circulation of the atmosphere and oceans. These, in turn, cause erosion of the land surface. Internal energy, a relict of Earth formation and a product of radioactive decay, drives volcanism and the uplift of mountains.

Because of cycles, features of the Earth System undergo cyclic change. For example, in the time frame of centuries to millions of years, climate changes and sea-level changes have markedly affected the character of the planet's surface. Plate interactions constantly, though slowly, change the map of the planet. And over geologic history, life and the atmosphere have evolved. Despite its immensity, the Earth System is fragile; human activity, which helps create global change, has had major effects on the Earth System.

FIGURE 19.3 This chart provides one interpretation of sea-level change during the Phanerozoic, based on the stratigraphic record. There is not full agreement about this interpretation.

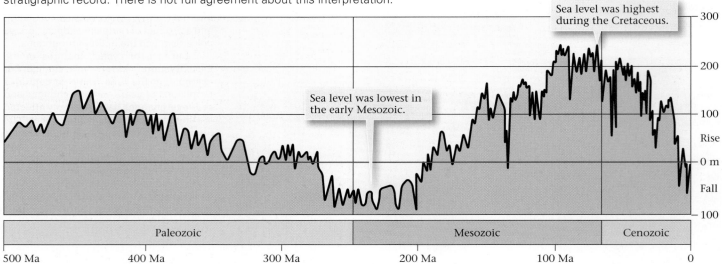

The Sea-Level Cycle

Global sea level rose and fell by as much as 300 m during the Phanerozoic, and likely did the same in the Precambrian. When sea level rises, the shoreline migrates inland, and low-lying plains in the continents become submerged. During periods of particularly high sea level, more than half of Earth's continental area can be covered by shallow seas; at such times, sediment buries continental crust. When sea level falls, the continents become dry again, and regional unconformities develop.

After studying stratigraphy around the world, geologists have pieced together a chart defining the succession of global transgressions and regressions during the Phanerozoic Eon. This global sedimentary cycle chart may largely reflect the cycles of eustatic (worldwide) sea-level change (Fig. 19.3). However, the chart probably does not give us an exact image of sea-level change, because the sedimentary record reflects other factors as well, such as changes in sediment supply. Eustatic sea-level changes may be due to a variety of factors, including advances and retreats of continental glaciers and changes in the volume of mid-ocean ridge systems.

The Rock Cycle

We learned early in this book that the crust of the Earth consists of three rock types: igneous, sedimentary, and metamorphic. Atoms making up the minerals of one rock type may later become part of another rock type. This process is the rock cycle. Each stage in the rock cycle changes the Earth by redistributing and modifying material.

TAKE-HOME MESSAGE

Some changes are cyclic in that they have stages that may be repeated. Examples include the supercontinent cycle (accumulation of continents by collision, and then dispersal by rifting), the sea-level cycle (rise and fall of the sea), and the rock cycle.

19.4 BIOGEOCHEMICAL CYCLES

A **biogeochemical cycle** involves the passage of a chemical among nonliving and living reservoirs in the Earth System, mostly on or near the surface. Nonliving reservoirs include the atmosphere, the crust, and the ocean; living reservoirs include plants, animals, and microbes. Although a great variety of chemicals participate in biogeochemical cycles, here we look at only two: water (H_2O) and carbon (C).

Some stages in a biogeochemical cycle may take only hours, some may take thousands of years, and others may take millions of years. When chemicals cycle rapidly, the transfer of a chemical from reservoir to reservoir doesn't really seem like a change in the Earth System in the way that the movement of continents or the metamorphism of rock seems like a change. In fact, for intervals of time, biogeochemical cycles attain a steady-state condition, meaning that the proportions of a chemical in different reservoirs remain fairly constant even though there is a constant flux (flow) of the chemical among reservoirs. When we speak of global change in a biogeochemical cycle, we mean a change in the relative proportions of a chemical held in different reservoirs.

The Hydrologic Cycle

As we learned in Interlude F, the hydrologic cycle involves the movement of water from reservoir to reservoir on or near the surface of the Earth. The hydrologic cycle is an example of a biogeochemical cycle, in that a chemical (H_2O) passes through both nonliving and living entities—the oceans, the atmosphere, surface water, groundwater, glaciers, soil, and living organisms. Global change in the hydrologic cycle occurs when a change in global climate alters the ratio between the amount of water held in the ocean and the amount held in continental ice sheets. For example, during an ice age, water that had been stored in oceans moves into glacial reservoirs.

The Carbon Cycle (the Movement of a Greenhouse Gas)

Most carbon in the near-surface realm of Earth originally bubbled out in the form of CO_2 gas released at volcanoes (Fig. 19.4). Once it enters the atmosphere, it can be removed in various ways. Some dissolves in seawater to form bicarbonate (HCO_3^-) ions, whereas some is absorbed by photosynthetic organisms that convert it into sugar and other organic chemicals. This carbon enters the food chain and ultimately makes up the flesh, fat, and sinew of animals. Some of the reactions that take place when rock undergoes chemical weathering incorporate atmospheric CO_2, and thus also remove carbon from the atmosphere.

Some carbon returns directly to the atmosphere, in the form of CO_2 or methane (CH_4), by the respiration of animals, by the flatulence of animals, or by the decay of dead organisms. But some can be stored for long periods of time in fossil fuels (oil and coal), in organic shale, in methane hydrates, or in limestone. This carbon either returns to the atmosphere in the form of CO_2, as a result of the burning of fossil fuels and the metamorphism of rocks containing carbonate, or returns to the sea after undergoing dissolution in river water or groundwater.

The concentrations of carbon dioxide and methane in the atmosphere play an essential role in controlling Earth's climate because these gases, along with several other trace gases (such as water), are **greenhouse gases**, meaning that they prevent heat from escaping into space from the atmosphere, much as glass traps heat in a greenhouse. An increase in their concentration warms the atmosphere, whereas a decrease cools it down.

TAKE-HOME MESSAGE

In the Earth System, chemicals—such as carbon and water—cycle through living and non-living reservoirs. For example, carbon can be stored in the atmosphere as CO_2, in sea water as bicarbonate ions, in rock as calcite, and in fossil fuels as oil, gas, and coal.

19.5 GLOBAL CLIMATE CHANGE

The **climate** is the average range of weather conditions for a given region. Global climate change, the transformations or modifications in Earth's climate through time, has happened during Earth history and is important because it affects both sea level and the distribution of climatic belts. As a result, climate change impacts the distribution of habitats, agricultural lands, and landscapes. For purposes of discussion, we distinguish between

FIGURE 19.4 In the carbon cycle, carbon transfers among various reservoirs at or near the Earth's surface. Red arrows indicate release to the air, and green arrows indicate absorption from air.

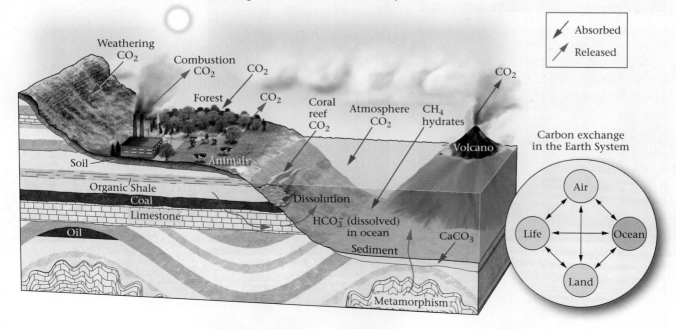

long-term climate change, which takes place over millions of years, and short-term climate change, which takes place over tens of thousands of years. If the average atmospheric and sea-surface temperature rises, we have **global warming**; if it falls, we have **global cooling**.

Methods of Study

Geologists and climatologists are working hard to define the nature of climate change that can occur, the rate at which this change takes place, and the effects it may have on the Earth System and society. There are two basic approaches to studying global climate change: (1) Researchers *measure past climate change*, or **paleoclimate**, to document the magnitude of change that is possible and the rate at which such change occurred; (2) Researchers *develop global-climate models* (**GCMs**), computer programs that calculate how factors such as atmospheric composition, topography, ocean currents, and Earth's orbit affect the climate. GCMs provide insight into when and why changes took place in the past and whether they will happen in the future.

Let's look first at how geologists study paleoclimate and document climate changes throughout Earth history. Any feature whose character depends on the climate and whose age can be determined serves as a clue to defining paleoclimate. Here are some examples:

- *The stratigraphic record*: The nature of sedimentary strata deposited at a certain location reflects the climate at that location. For example, a bed of coal represents deposits of a warm climate, whereas a bed of till represents deposits of a glacial climate.

- *Paleontological evidence*: Different assemblages of species live in different climatic belts. Thus, the succession of species in a sedimentary sequence provides clues to the changes in climate at that site. For example, a record of short-term climate change can be obtained by studying

FIGURE 19.5 Changes in the assemblage of pollen in sediment indicates a shift in climate belts.

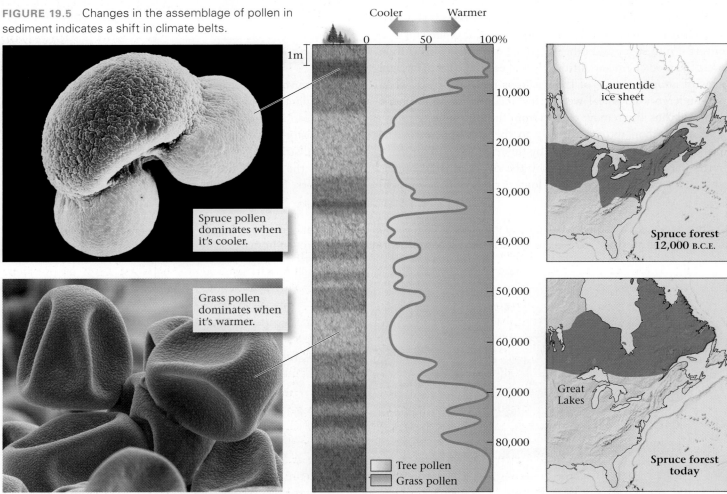

(a) The proportion of tree pollen relative to grass pollen can change in a sedimentary sequence through time. Researchers plot changes in the proportion of pollen types over time by examining samples from a column of sediment.

(b) Spruce forests (green) grew farther south 12,000 years ago than they do today.

FIGURE 19.6 The proportion of isotopes transferred between reservoirs during evaporation or precipitation depends on temperature. The $^{18}O/^{16}O$ ratio can be studied in glacial ice (H_2O) and fossil shells ($CaCO_3$).

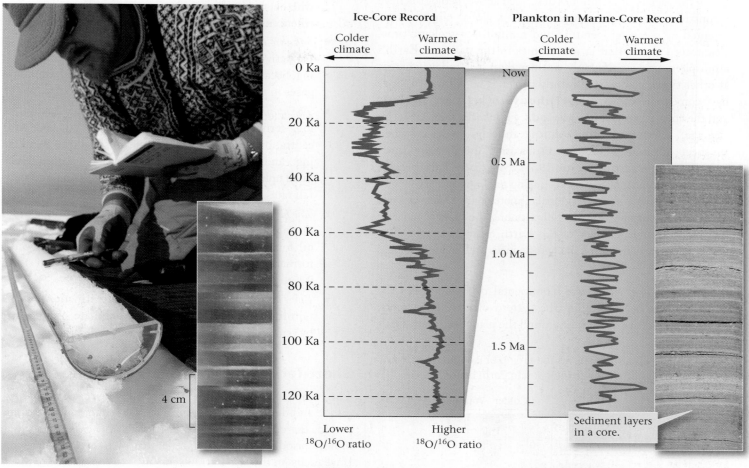

Ice-Core Record

Plankton in Marine-Core Record

(a) A researcher examines an ice core in the field. Lab photos reveal annual layers.

(b) The $^{18}O/^{16}O$ ratios in an ice core represent temperature changes.

(c) Studies of $^{18}O/^{16}O$ ratios in deep-sea cores also provide a climate record.

the succession of plankton fossils in sea-floor sediments, for cold-water species of plankton are different from warm-water species. Fossil plant pollen preserved in the mud of bogs also provides information about paleoclimate, because different plant species live in different climates (Fig. 19.5a). Pollen studies show that spruce forests, indicative of cool climates, have slowly migrated north since the ice age (Fig. 19.5b).

■ *Oxygen-isotope ratios*: Two isotopes of an element have the same atomic number but different atomic weights (see Appendix). Geologists have found that the ratio of ^{18}O to ^{16}O in glacial ice indicates the atmospheric temperature at the time when the snow that made up the ice formed: the ratio is larger in snow that forms in warmer air. Because of this relationship, the isotope ratio measured in a succession of ice layers in a glacier indicates temperature change through time. Researchers have now

obtained ice cores down to a depth of about 3 km in Antarctica and Greenland. The oldest ice in the Antarctic core may have formed over 720,000 years ago (Fig. 19.6a, b). The $^{18}O/^{16}O$ ratio in the $CaCO_3$ making up plankton shells also gives geologists an indication of past temperatures. Measurement of oxygen-isotope ratios in drill cores of marine sediment extends the record of temperature change back through millions of years (Fig. 19.6c).

■ *Growth rings*: If you've ever looked at a tree stump, you'll have noticed the concentric rings visible in the wood. Each ring represents one year of growth, and the thickness of the ring indicates the rate of growth in a given year. Trees grow faster during warmer, wetter years and more slowly during cold, dry years. Thus, the succession of ring widths provides an easily calibrated record of climate during the lifetime of the tree. Growth rings in corals and shells can provide similar information.

Long-Term Climate Change

Using the variety of techniques described above, geologists have reconstructed an approximate record of global climate, represented by mean temperature and rainfall, for geologic time. The record shows that at some times in the past, the Earth's atmosphere was significantly warmer than it is today, whereas at other times it was significantly cooler. The warmer periods have come to be known as **greenhouse** (or hothouse) **periods** and the colder as **icehouse periods**. (The more familiar term, "ice age," refers to the times during an icehouse period when ice sheets advanced and covered substantial areas of the continents.) As the chart in Figure 19.7 shows, there have been at least five major icehouse periods during geologic history.

What caused long-term global climate change? The answer may lie in the complex relationships among the various geologic and biogeochemical cycles of the Earth System, as described earlier. Let's consider some likely influences on long-term global change:

- *Positions of continents*: Continental drift influences the climate by controlling the pattern of oceanic currents, which redistribute heat around the planet's surface. Drift

also determines whether the land is at high or low latitudes (and thus how much solar radiation strikes it), and whether or not there are large continental interior regions where extremely cold winter temperatures can develop.

- *Volcanic activity*: A long-term global increase in volcanic activity may contribute to long-term global warming, because it increases the concentration of greenhouse gases in the atmosphere.

- *The uplift of land surfaces*: Tectonic events that lead to the long-term uplift of the land affect atmospheric CO_2 concentration, because such events expose land to weathering, and chemical-weathering reactions absorb CO_2. Such uplift also affects atmospheric circulation and rainfall rates.

- *The formation of coal, oil, or organic shale*: At various times during Earth history, environments suitable for coal or oil formation have been particularly widespread. Such formation removes CO_2 from the atmosphere and stores it underground.

- *Life evolution:* The appearance of, or extinction of, certain life forms may have impacted climate significantly by removing CO_2 from the atmosphere.

FIGURE 19.7 This chart shows the timing of icehouse and greenhouse (or hothouse) periods during Earth history.

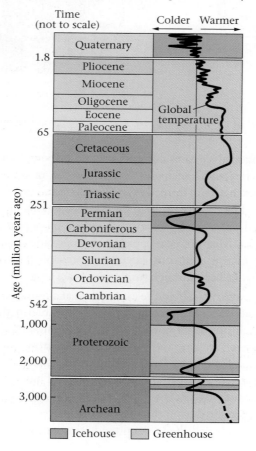

Short-Term Climate Change

The record of the past million years gives a sense of the magnitude and duration of short-term climate change. During this period, there have been about five major and twenty to thirty minor episodes of glaciation, separated by interglacial periods. If we focus on the last 15,000 years, we see that there are many short-duration ups and downs, but that overall the temperature has increased (Fig. 19.8a, b).

As a result of the warming that began 15,000 years ago, the glaciers retreated for about 4,500 years. Then there was a return to colder conditions for a few more thousand years. This interval of cooler temperature is named the Younger Dryas, after an Arctic flower that became widespread during the time. The climate then warmed, reaching a peak at 5,000 to 6,000 years ago, a period called the Holocene maximum, when average temperatures were slightly warmer than those of today. This warming peak led to increased evaporation and therefore precipitation, making the Middle East unusually wet and fertile—conditions that may partially account for the rise of civilization in this part of the world.

The temperature dipped to a low about 3,000 years ago, before returning to a high during the Middle Ages, a time called the Medieval Warm Period. During this time, when temperatures were similar to slightly above those of today, Vikings established self-supporting agricultural settlements along the coast of Greenland. The temperature dropped again from 1500 to

FIGURE 19.8 Climate during the Holocene. Measurements suggest that temperature has varied significantly.

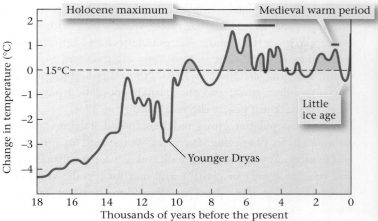

(a) A simplified model of temperature change during the past 15,000 years (the Holocene) shows warming and cooling.

(b) The Muir Glacier, Alaska, retreated by 12 km between 1941 and 2004.

about 1800, a period known as the Little Ice Age, when Alpine glaciers advanced and the canals of the Netherlands froze over in winter. Overall, the climate has warmed since the end of the Little Ice Age, and today it is as warm as it was during the Medieval Warm Period. Geologists have suggested that several natural factors explain short-term climate change; we'll discuss human impact later. These include:

- *Fluctuations in solar radiation and cosmic rays*: The amount of energy produced by the Sun varies with the sunspot cycle. This cycle involves the appearance of large numbers of sunspots (black spots thought to be magnetic storms on the Sun's surface) about every 9 to 11.5 years. There may be longer-term cycles that have not yet been identified. This variation in energy may affect the climate. Recently, some researchers have speculated that changes in the rate of influx of cosmic rays generate clouds, for cosmic rays striking the atmosphere produce clusters of ions that serve as condensation nuclei around which water molecules will congregate.

- *Changes in Earth's orbit and tilt*: As Milankovitch first recognized in 1920, the change in the tilt of Earth's axis, Earth's precession cycle, and changes in the eccentricity of Earth's orbit together cause the amount of summer heat in high latitudes to vary (see Chapter 18). These changes correlate with observed ups and downs in atmospheric and oceanic temperature.

- *Changes in volcanic emissions*: Not all of the sunlight that reaches the Earth penetrates its atmosphere and warms the ground. Some is reflected by the atmosphere. The degree of reflectivity, or albedo, of the atmosphere increases not only if cloud cover increases, as we have seen, but also if

the concentration of volcanic aerosols (microscopic liquid droplets) in the atmosphere increases.

- *Changes in ocean currents*: Recent studies suggest that the configuration of ocean currents can change quite quickly, and that this configuration affects the climate.

- *Changes in surface albedo*: Regional-scale changes in the nature of continental vegetation cover, and/or in the proportion of snow and ice on our planet's surface, affect our planet's albedo. Increasing albedo causes cooling, whereas decreasing albedo causes warming.

- *Abrupt changes in concentrations of greenhouse gases*: A sudden change in greenhouse gas concentration in the atmosphere could affect climate. One such change might happen if some of the methane hydrate that crystallized in sediment on the sea floor suddenly melted, releasing CH_4 to the atmosphere. Algal blooms and reforestation (or deforestation) conceivably could also change CO_2 concentrations significantly.

Catastrophic Climate Change and Mass-Extinction Events

The changes that happen on Earth almost instantaneously are called catastrophic changes. For example, a volcanic eruption, an earthquake, a tsunami, or a landslide can change a local landscape in seconds or minutes. But such events affect only relatively small areas. Can such catastrophes happen on a global scale? In the past decades, geoscientists have come to the conclusion that the answer is yes. The stratigraphic record shows that Earth history includes several **mass-extinction events**, when large numbers of species abruptly vanished (see Interlude E). Some of these define boundaries between geologic periods. A mass-extinction event decreases the biodiversity (the number of different species that exist at a given time) of life on Earth. The most notable examples occurred at the end of the Permian and at the end of the Cretaceous.

Geologists speculate that some mass-extinction events reflect a catastrophic change in the planet's climate, brought about by unusually voluminous volcanic eruptions or by the impact of a comet or asteroid with the Earth. Either of these events could eject enough debris into the atmosphere to block sunlight (see Chapter 11). Without the warmth of the Sun, winter-like or night-like conditions would last for weeks to years, long enough to disrupt the food chain. Either event, in addition, could eject aerosols that would turn into global acid rain, scatter hot debris that would ignite forest fires, or give off chemicals that, when dissolved in the ocean, would make the ocean either toxic or so nutritious that oxygen-consuming algae could thrive. At present, many geologists favor the hypothesis that the Permian-Triassic extinction was due to the eruption of 3 million cubic km of basalt in Siberia, and that the Cretaceous-Tertiary extinction was due to a giant meteorite impact near the Yucatán peninsula in Mexico.

TAKE-HOME MESSAGE

Earth's past climate can be studied using fossils, isotopes, pollen assemblages, and growth rings in trees. The record shows that, over geologic time, climate alternates between warm (greenhouse) conditions and cold (icehouse) conditions. Volcanic eruptions and meteorite impacts can cause catastrophic change.

19.6 HUMAN IMPACT ON THE EARTH SYSTEM

During the Stone Age, the human population worldwide was less than 10 million. By the dawn of civilization, 4000 B.C.E., it was still, at most, a few tens of millions. But by the beginning of the nineteenth century, revolutions in industrial methods, agriculture, medicine, and hygiene had substantially lowered death rates and raised living standards, so that the population began to grow at accelerating rates—it took tens of thousands of years to grow from Stone Age populations to 1 billion people worldwide in 1850, but it took only eighty years to double again, reaching 2 billion in 1930. Now, the doubling time is only forty-four years, and the human population passed the 6 billion mark just before the year 2000 (Fig. 19.9).

As the population grows and the standard of living improves, *per capita* usage of resources increases; we use land for agriculture and grazing, forests for wood, rock and dirt for construction, oil and coal for energy or plastics, and ores for metals. Without a doubt, our usage of resources has affected the Earth System profoundly, and thus humanity has become a major agent of global change (see **Geotour 19** on p. GT-40). Here, we examine some of these anthropogenic (human-induced) impacts.

The Modification of Landscapes

Every time we pick up a shovel and move a pile of dirt, we redistribute a portion of the Earth's crust. In the last century, the pace of Earth movement has accelerated, for now we have shovels in coal mines that can move 300 cubic meters of coal in a single scoop, trucks that can carry 200 tons of ore in a single load, and tankers that can transport 50 million liters of oil during a single journey. In North America, human activity now moves more sediment each year than rivers do. The extraction of rock during mining, the building of levees and dams along rivers or of sea walls along the coast, and the construction of highways and cities all involve the redistribution of Earth materials. In addition, people clear and plow fields, drain and fill wetlands, and pave over the land surface (Fig. 19.10; see also chapter-opening photo). All these activities change the landscape, the water table, and the supply of sediment.

FIGURE 19.9 Population now doubles about every forty-four years. The Black Death pandemic caused an abrupt drop that lasted for a few decades.

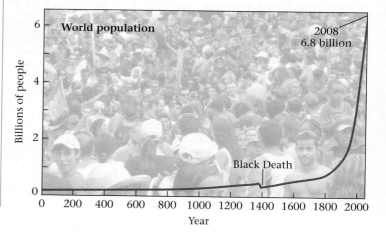

FIGURE 19.10 Excavation, agriculture, and construction modify topography, drainage, infiltration, and ecology.

Humans move vast amounts of rock.

Agriculture eliminates diverse ecosystems.

Urbanization changes the water table.

The Modification of Ecosystems

In undisturbed areas, the **ecosystem** of a region (an interconnected network of organisms and the physical environment in which they live) is the product of evolution for an extended period of time. The ecosystem's flora and fauna include species that have adapted to living together in that particular climate and on the substrate available. Human-caused deforestation, overgrazing, agriculture, and urbanization disrupt ecosystems and lead to a decrease in biodiversity.

Archaeological studies have found that the earliest example of human modification of an ecosystem occurred in the Stone Age, when hunters played a major role in causing the mass extinction of many species of large mammals (such as mammoths, giant sloths, and giant bears). Today, less than 5% of Europe retains its original habitats. The same number can be applied to the eastern United States, which lost its original forest and prairie. Tropical rain forests cover less than about half the area worldwide that they covered before the dawn of civilization, and they are disappearing at a rate of about 1.8% per year (Fig. 19.11a, b). Much of this loss comes from slash-and-burn agriculture during which farmers and ranchers destroy forest to make open land for farming and grazing (Fig. 19.11c). Unfortunately, the heavy rainfall of tropical regions removes nutrients from

FIGURE 19.11 The area of the Earth covered by forests has been shrinking.

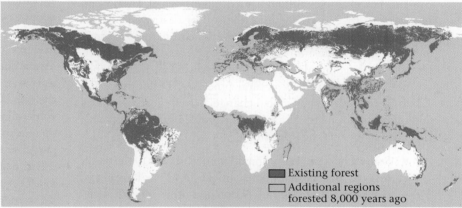

■ Existing forest
□ Additional regions forested 8,000 years ago

(a) Comparison of exisiting forest area (green) to precivilization forest area (gold) emphasizes changes in Earth's forest cover. Tropical rain forest areas are shrinking rapidly.

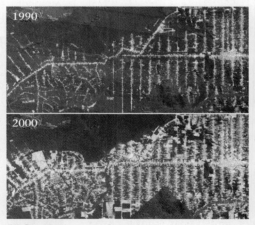

1990

2000

(b) Satellite images of a region in the Amazon show the increase in cleared areas (light).

(c) Slash-and-burn agriculture.

the soil, making the soil useless in just a few years. Overgrazing by domesticated animals can remove vegetation so completely that some grasslands have undergone desertification. And urbanization replaces the natural land surface with concrete or asphalt, a process that completely destroys an ecosystem and radically changes the amount of rain that infiltrates the land surface to become groundwater.

Pollution

The environment has always contained contaminants such as soot, dust, and the byproducts of organisms. But when human populations grew, urbanization, industrial and agricultural activity, the production of electricity, and modern modes of transportation greatly increased both the quantity and diversity of contaminants that entered the air, surface water, and groundwater. These contaminants, or **pollution**, include both natural and synthetic materials in liquid, solid, or gaseous form. They have become a problem because they are produced at a higher rate than can be naturally absorbed or modified by the Earth System. For example, although small quantities of sewage can be absorbed by clay minerals in the soil or destroyed by bacterial metabolism, large quantities overwhelm natural controls and can accumulate into destructive concentrations. Further, because many contaminants are not produced in nature, they are not easily removed by natural processes. Pollution of the Earth System is a type of global change, because it redistributes and reformulates materials. Some key problems associated with this change include the following.

- *Smog*: The term was originally coined to refer to the dank, dark air that resulted when smoke from the burning of coal mixed with fog in London and other industrial cities. Another kind of smog, called photochemical smog, is the ozone-rich brown haze that blankets cities when exhaust from cars and trucks reacts with air in the presence of sunlight.

- *Water contamination*: We dump a great variety of chemicals into surface water and groundwater. Examples include gasoline, other organic chemicals, radioactive waste, acids, fertilizers—the list could go on for pages. These chemicals affect biodiversity.

- *Acid runoff*: Dissolution of sulfide-containing minerals in ores or coal by groundwater or stream water makes the water acidic and toxic.

- *Acid rain*: When rain passes through air that contains sulfur-containing aerosols (emitted from power plants), the water dissolves the sulfur, creating sulfuric acid, or **acid rain**. Wind can carry aerosols far from a power plant, so acid rain can damage a broad region.

- *Radioactive materials*: Nuclear weapons, nuclear energy, and medical waste transfer radioactive materials from rock to Earth's surface environment.

- *Ozone depletion*: When emitted into the atmosphere, human-produced chemicals, most notably chlorofluorocarbons (CFCs), react with ozone in the stratosphere. This reaction, which happens most rapidly on the surfaces of tiny ice crystals in polar stratospheric clouds, destroys ozone molecules, thus creating an **ozone hole** over high-latitude regions, particularly during the spring (Fig. 19.12). Note that the "hole" is not really an area where no ozone is present, but rather is a region where atmospheric ozone has been reduced substantially. Ozone holes have dangerous consequences, for they affect the ability of the atmosphere to shield the Earth's surface from harmful ultraviolet radiation.

Recent Global Warming

During the past two centuries, industry, energy production, and agriculture have significantly altered the rate at which greenhouse gases, such as carbon dioxide (CO_2) and methane (CH_4), are added to the atmosphere. It appears that the rate of addition has exceeded the rate at which these gases can be absorbed by dissolution in the ocean, by incorporation in plants, or by chemical-weathering reactions. Effectively, by burning fossil fuels at the rate that we do, we transfer CO_2 from underground reservoirs (oil and coal deposits) back into the atmosphere. In 1800, the mean concentration of CO_2 in the atmosphere was 280 parts per million (ppm), in 1900 it was 295 ppm, and by 2008 it had reached about 385 ppm (Fig. 19.13a). Data on changes in CO_2 concentration during the past 10,000 years emphasizes the temporal link to industrialization (Fig. 19.13b). At the

FIGURE 19.12 Colors show the quantity of ozone in the atmosphere. The darker blue area over Antarctica is the ozone hole. A Dobson Unit represents ozone concentration.

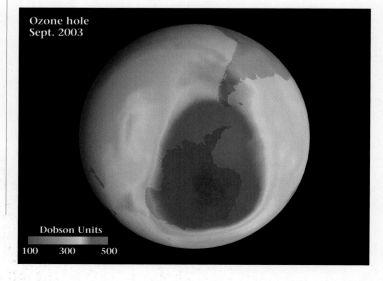

FIGURE 19.13 Indicators referred to by the Intergovernmental Panel on Climate Change as evidence of climate change.

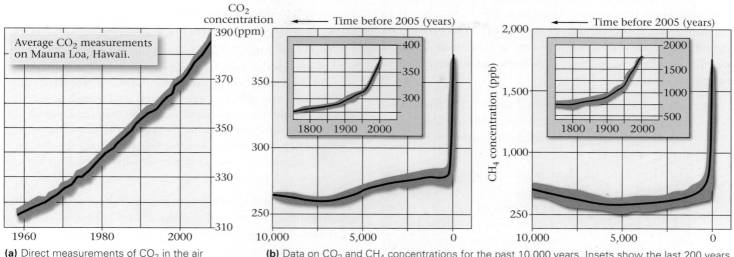

(a) Direct measurements of CO_2 in the air have been collected since 1961.

(b) Data on CO_2 and CH_4 concentrations for the past 10,000 years. Insets show the last 200 years. The black lines are averages, and the red bands are the range of measurements.

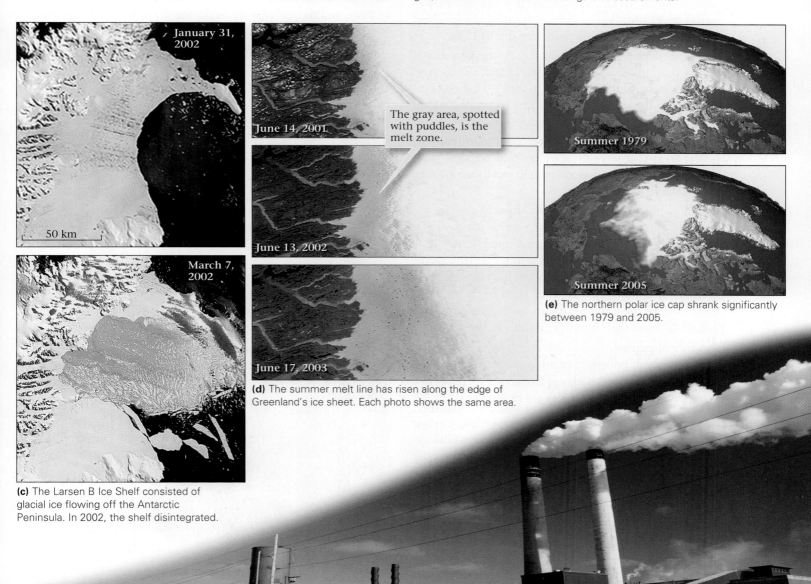

(c) The Larsen B Ice Shelf consisted of glacial ice flowing off the Antarctic Peninsula. In 2002, the shelf disintegrated.

The gray area, spotted with puddles, is the melt zone.

(d) The summer melt line has risen along the edge of Greenland's ice sheet. Each photo shows the same area.

(e) The northern polar ice cap shrank significantly between 1979 and 2005.

same time, the decay of organic material in rice paddies and the flatulence of cows have released enough methane to change the concentration of this organic chemical in the atmosphere.

In the past three decades, thousands of projects have been undertaken worldwide to collect data to determine if global warming has been happening. This work has resulted in some startling discoveries:

- Large ice shelves, such as the Larsen B Ice Shelf along the Antarctic Peninsula, and the Ayles Ice Shelf along Ellesmere Island in northernmost Canada, are breaking up (Fig. 19.13c).

- The Greenland ice sheet is melting at an accelerating pace. Studies suggest that the rate of ice loss has increased from 90 to 220 square km per year in the last 10 years and that, in places, the sheet is thinning by about 1 m per year. In addition, the annual melt zone along the margins of the ice sheet has widened dramatically because the elevation of the equilibrium line (see Chapter 18) has risen (Fig. 19.13d). Valley glaciers draining the ice sheet moved 50% faster in 2003 than they did in 1992.

- The area covered by sea ice in the Arctic Ocean has decreased substantially (Fig. 19.13e). In fact, this ice covered 14% less area in 2005 than in 2004. This observation leads to the prediction that it may be possible to sail across the Arctic Ocean within decades.

- Valley glaciers worldwide have been retreating rapidly, so that areas that were once ice covered are now bare. The change is truly dramatic in many locations (see Fig. 19.8c).

- Biological phenomena that are sensitive to climate are being disrupted. For example, the time at which sap in the maple trees of the northeastern United States starts to flow has changed, and the mosquito line (the elevation at which mosquitoes can survive) has risen substantially. Also, the average weight of polar bears has been decreasing, because the bears can no longer walk over pack ice to reach their hunting grounds in the sea.

- The area of permafrost in high latitutdes has substantially decreased, and melt ponds have replaced frozen land. Large regions that stayed frozen all year are now melting in the summer, so trees are tipping over.

To interpret these data from a broad perspective, a group of researchers founded the Intergovernmental Panel on Climate Change (IPCC) in 1988. This organization, sponsored by the World Meteorological Organization and the United Nations, reviews published research on climate change and, every 5 years, summarizes the conclusions in an assessment report. Because of the influence these reports have had, the IPCC shared the 2007 Nobel Peace Prize. The language describing the likelihood that

global warming is happening, and further that humans have contributed significantly to causing it, has become progressively less equivocal in successive versions of the report. The Fourth Assessment Report, published in 2007, states:

> Warming of the climate system is unequivocal, as is now evident from observations of increases in global average air and ocean temperatures, widespread melting of snow and ice, and rising global average sea level. . . . The understanding of anthropogenic [human-caused] warming and cooling influences on climate has improved since the Third Assessment Report, leading to *very high confidence* that the globally averaged net effect of human activities since 1750 has been one of warming.

In other words, most climate researchers have concluded that global warming is real, and that the actions of people—burning fossil fuels, cutting down forests, paving over wetlands—have played a significant role in causing it. Specifically, global mean atmospheric temperature has risen by almost 1°C during the last century, and the rate of increase during the past 50 years appears to be greater than the rate for the previous 50 years (Fig. 19.14). This 1°C temperature rise appears to be significantly more than would have taken place due to natural cycles. Also, although such a change may seem small, the magnitude of temperature change

FIGURE 19.14 Plots of temperature, relative to a reference value, against time indicates an increase in global temperature during the past two centuries. The top graph is an enlargement of the boxed area in the lower graph.

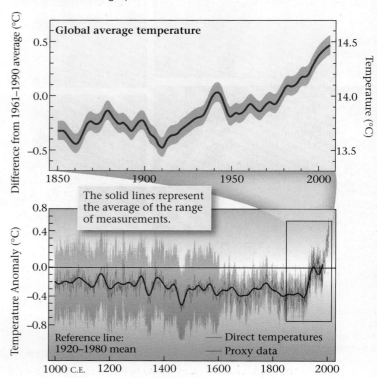

FIGURE 19.15 Examples of changes associated with global warming.

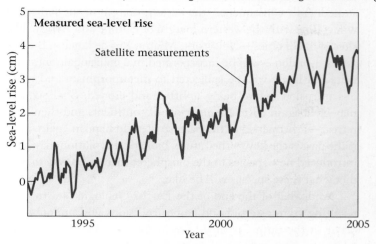

(a) Satellite measurements of global sea-level change show a gradual rise between 1995 and 2005.

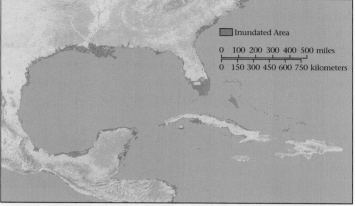

(b) The red areas indicate the areas of the Gulf of Mexico coastline that would be flooded if sea level rose by 6 m.

(c) Forest and grass fires rage in California with increasing frequency. This one occured near Los Angeles.

between the ice age and now, by comparison, was only 3°C to 5°C. A small change in average temperature may have major consequences. Of note, sea-surface temperatures also appear to be rising in many parts of the world, though not everywhere.

The effects of global warming over the coming decades to centuries remains the subject of intense debate, because predictions depend on computer models, and not all researchers agree on how to construct these models. In the worst-case scenario, global warming will continue into the future at the present rate, so that by 2050—within the lifetime of many readers of this book—the average annual temperature may have increased in some parts of the world by 1.5°C to 2.0°C. At these rates, by 2150 global temperatures may be 5°C to 11°C warmer than the present. According to many climate models, the following events might happen.

- *A shift in climate belts*: Temperate climate belts would move to higher latitudes, and vegetation belts would follow this trend. As a result, desert regions would expand, and the soil would dry out in present agricultural regions.

- *Rise in the snow line*: As temperatures increase, the snow line in mountainous areas would rise, and globally, it would move to higher latitudes. This change would impact ecosystems and even impact tourism by making it impossible for winter resorts to host snow-based activities.

- *Stronger storms*: An increase in average ocean temperatures would mean that more of the ocean could evaporate when a tropical depression passed over. This evaporation would nourish stronger hurricanes. In nondesert areas, there might be more precipitation and, therefore, flooding.

- *A rise in sea level*: The melting of ice sheets in polar regions and the expansion of water in the sea as the water becomes

warmer would make sea level rise enough to flood coastal wetlands and communities and damage deltas. Measurements suggest that there has been a rise of almost 12 cm in the past century (Fig. 19.15a). A change of several meters would inundate regions of the world where a significant percentage of the population currently lives (Fig. 19.15b).

- *An increase in wildfires*: Warmer temperatures may lead to an increase in the frequency of wildfires because the moisture content of plants is lower (Fig. 19.15c).

- *An interruption of the oceanic heat conveyor*: Oceanic currents play a major role in transferring heat across latitudes. If global warming melts polar ice, the resulting freshwater would dilute surface ocean water at high latitudes. This water could not sink, and thus thermohaline circulation would be shut off (see Chapter 18), preventing the water from conveying heat.

The potential changes described above, along with other studies that estimate the large economic cost of global warming, imply that the issue needs to be addressed seriously and soon. But what can be done? The 160 nations that signed the 1997 Kyoto Accord, at a summit meeting held in Japan, propose that the first step would be to slow the input of greenhouse gases into the atmosphere by decreasing the burning of fossil fuels and/or by forcing the CO_2 produced at power plants down wells into pore space underground. Motivating such actions, needless to say, involves controversial political decisions.

TAKE-HOME MESSAGE

Humans have become major agents of change in the Earth System. Human activities modify the landscape, disrupt ecosystems, and decrease forest cover. Evidence gathered in recent decades indicates that the release of greenhouse gases causes global warming.

19.7 THE FUTURE OF THE EARTH: A SCENARIO

Most of the discussion in this book has focused on the past, for the geologic record preserved in rocks tells us of earlier times. We bring this book to a close by facing the opposite direction and speculating what the world might look like in geologic time to come.

In the geologic near term, the future of the world depends largely on human activities. Whether the Earth System undergoes a major disruption and shifts to a new equilibrium, whether a catastrophic mass-extinction event takes place, or whether society achieves **sustainable growth** (meaning an ability to prosper within the constraints of the Earth System) will depend on our own foresight and ingenuity. Projecting thousands of years into the future, we might well wonder if the Earth will return to ice-age conditions, with glaciers growing over major cities and the continental shelf becoming dry land, or if the ice age is over for good because of global warming. No one really knows for sure.

If we project millions of years into the future, it is clear that the map of the planet will change significantly because of the continuing activity of plate tectonics. For example, during the next 50 million years or so, the Atlantic Ocean will probably become bigger, the Pacific Ocean will shrink, and the western part of California will migrate northward. Eventually, Australia will crush against the southern margin of Asia, and the islands of Indonesia will be flattened in between.

Predicting the map of the Earth beyond that is impossible, because we don't know for sure where new subduction zones will develop. Perhaps, subduction of the Pacific Ocean will lead to the collision of the Americas with Asia, to produce a supercontinent ("Amasia"). A subduction zone eventually will form

on one side of the Atlantic Ocean, and the ocean will be consumed. As a consequence, the eastern margin of the Americas will collide with the western margin of Europe and Africa. The sites of major cities—New York, Miami, Rio de Janeiro, Buenos Aires, London—will be incorporated in a collisional mountain belt, and likely will be subjected to metamorphism and igneous intrusion before being uplifted and eroded. Shallow seas may once again cover the interiors of continents and then later retreat—it happened in the past, so it could happen again. And if the past is the key to the future, biological evolution may have introduced new species to the biosphere; there is no way to predict what these species will be like.

And what of the end of the Earth? Geologic catastrophes resulting from asteroid and comet collisions will undoubtedly occur in the future as they have in the past. We can't predict when the next strike will come, but unless the object can be diverted, Earth is in for another radical readjustment of surface conditions. But it's not likely that such collisions will destroy our planet. Rather, astronomers predict that the end of the Earth will occur some 5 billion years from now, when the Sun begins to run out of nuclear fuel. When this happens, thermal pressure caused by fusion reactions will no longer be able to prevent the Sun from collapsing inward, because of the immense gravitational pull of its mass. If the Sun was a few times larger than it is, the collapse would trigger a supernova explosion that would blast matter out into space to form a new nebula. But since the Sun is not that large, the thermal energy generated when its interior collapses inward will heat the gases of its outer layers

FIGURE 19.16 In about 5 billion years, the Sun will become a red giant. When that happens, the Earth will first dry out, then vaporize; initially, it may look like a giant comet.

Today

Sun

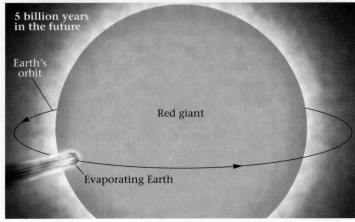

5 billion years in the future

Earth's orbit

Red giant

Evaporating Earth

sufficiently to cause them to expand. As a result, the Sun will become a red giant, a huge star whose radius will grow beyond the orbit of Earth (Fig. 19.16). Our planet will then vaporize, and its atoms will join an expanding ring of gas—the ultimate global change. If this happens, the atoms that once made Earth and all its inhabitants through geologic time may eventually be incorporated in a future solar system, where the cycle of planetary formation and evolution will begin anew.

TAKE-HOME MESSAGE

In the near term, Earth's surface will be affected by decisions of human society. Over longer time scales, the map of the Earth will change due to plate interactions and sea-level change. The end of the Earth will happen when the Sun becomes a red giant.

Chapter Summary

- We refer to the global interconnecting web of physical and biological phenomena on Earth as the Earth System. Global change involves the transformations or modifications of physical and biological components of the Earth System through time. Unidirectional change results in transformations that never repeat; cyclic change involves repetition of the same steps over and over.

- Examples of unidirectional change include the gradual evolution of the solid Earth from a homogeneous collection of planetesimals to a layered planet, the formation of the oceans, the gradual change in the composition of the atmosphere, and the evolution of life.

- Examples of physical cycles that take place on Earth include the supercontinent cycle, the sea-level cycle, and the rock cycle.

- A biogeochemical cycle involves the passage of a chemical among nonliving and living reservoirs. Examples include the hydrologic cycle and the carbon cycle. Global change occurs when factors change the relative proportions of the chemical in different reservoirs.

- Tools for documenting global climate change include the stratigraphic record, paleontology, oxygen-isotope ratios, and growth rings.

- Studies of long-term climate change show that at times in the past the Earth experienced greenhouse (warmer) periods; at other times there were icehouse (cooler) periods. Factors leading to long-term climate change include the positions of continents, volcanic activity, the uplift of land, and the formation of materials that remove CO_2, an important greenhouse gas.

- Short-term climate change can be seen in the record of the last million years. In fact, during only the past 15,000 years, we see that the climate has warmed and cooled a few times. Causes of short-term climate change include fluctuations in solar radiation and cosmic rays, changes in Earth's orbit and tilt, changes in reflectivity, and changes in ocean currents.

- Mass extinction, a catastrophic change in biodiversity, may be caused by the impact of a comet or asteroid or by intense volcanic activity.

- During the last two centuries, humans have changed landscapes, modified ecosystems, and added pollutants to the land, air, and water at rates faster than the Earth System can process.

- The addition of CO_2 and CH_4 to the atmosphere may be causing global warming, which could shift climate belts and lead to a rise in sea level.

- In the future, in addition to climate change, the Earth will witness a continued rearrangement of continents resulting from plate tectonics, and will likely suffer the impact of asteroids and comets. The end of the Earth may come when the Sun runs out of fuel in about 5 billion years and becomes a red giant.

GEOPUZZLE REVISITED

Climate change has become a "hot topic" because evidence that it is taking place and will impact human society has been growing. Phenomena such as glacial melting are occurring at rates faster than anticipated. But climate isn't the only aspect of the Earth System undergoing change. Natural geologic phenomena such as erosion, continental drift, and sea-level rise and fall have changed our planet slowly but surely over geologic time.

Key Terms

acid rain (p. 512)
biogeochemical cycle (p. 504)
climate (p. 505)
ecosystem (p. 511)
GCM (p. 506)
global change (p. 500)
global cooling (p. 506)
global warming (p. 506)
greenhouse gases (p. 505)

greenhouse period (p. 508)
icehouse period (p. 508)
mass-extinction event (p. 510)
ozone hole (p. 512)
paleoclimate (p. 506)
pollution (p. 512)
supercontinent cycle (p. 501)
sustainable growth (p. 516)

Review Questions

1. How have the Earth's crust and atmosphere changed since they first formed?
2. What processes control the rise and fall of sea level on Earth?
3. How does carbon cycle through the various Earth systems?
4. How do paleoclimatologists study ancient climate change?
5. Contrast icehouse and greenhouse conditions.
6. What are the possible causes of long-term climate change?
7. What factors explain short-term climate change?
8. Give some examples of events that cause catastrophic change.
9. Give some examples of how humans have changed the Earth.
10. What is the ozone hole, and how does it affect us?
11. Describe how carbon dioxide–induced global warming takes place, and how humans may be responsible. What effects might global warming have on the Earth System?
12. What are some of the likely scenarios for the long-term future of the Earth?

On Further Thought

1. If global warming continues, how will the distribution of grain crops change? Might this affect national economies? Why? How will the distribution of spruce forests change?
2. Currently, tropical rain forests are being cut down at a rate of 1.8% per year. At this rate, how many more years will the forests survive? In the eastern United States, the proportion of land with forest cover today has increased over the past century. In fact, most of the farmland that existed in New York State in 1850 is forestland today. Why? How might this change affect erosion rates in the region?
3. Using the library or the web, examine the change in the nature of world fisheries that has taken place in the last 50 years. Is the world's fish biomass sustainable if these patterns continue? What has happened to whale populations during the past 50 years?

THE VIEW FROM SPACE Numerous fjords make western Iceland's coast resemble the shape of a leaf. The glaciers that carved these fjords vanished over 10,000 years ago. Their growth and demise is one manifestation of climate change on Earth.

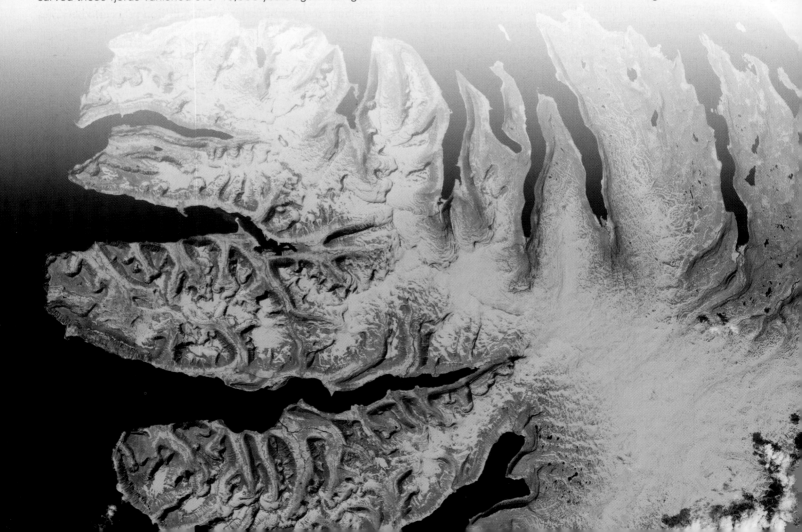

Scientific Background: Matter and Energy

a.l INTRODUCTION

In order to understand the formation and evolution of the Universe, as well as descriptions of materials that constitute the Earth and processes that shape the Earth, we must understand some basic facts about matter, energy, and heat (Fig. a.1). The following synopsis highlights key topics from physics and chemistry that serve as an essential background to this book. Our discussion utilizes units listed in Table a.1.

a.1 DISCOVERING THE NATURE OF MATTER

Matter takes up space—you can feel it and see it. We use the word **matter** to refer to any material making up the universe. The amount of matter in an object is its **mass**. An object with a greater mass contains more matter—for example, a large tree contains more matter than a blade of grass. There's a subtle but important distinction between mass and weight. **Weight**

FIGURE a.1 A lightning storm over mountains. The solid rock, the air, and the clouds consist of matter. Lightning is a manifestation of energy.

TABLE a.1 **Units Used in Scientific Discussion**

Length		

1 kilometer (km) = 0.6214 mile (mi)

1 meter (m) = 1.094 yards = 3.281 feet

1 centimeter (cm) = 0.3937 inch

1 millimeter (mm) = 0.0394 inch

1 mile (mi) = 1.609 kilometers (km)

1 yard = 0.9144 meter (m)

1 foot = 0.3048 meter (m)

1 inch = 2.54 centimeters (cm)

Mass		

1 metric ton = 2,205 pounds

1 kilogram (kg) = 2.205 pounds

1 gram (g) = 0.03527 ounce

1 pound (lb) = 0.4536 kilogram (kg)

1 ounce (oz) = 28.35 grams (g)

Area		

1 square kilometer (km^2) = 0.386 square mile (mi^2)

1 square meter (m^2) = 1.196 square yards (yd^2)

= 10.764 square feet (ft^2)

1 square centimeter (cm^2) = 0.155 square inch (in^2)

1 square mile (mi^2) = 2.59 square kilometers (km^2)

1 square yard (yd^2) = 0.836 square meter (m^2)

1 square foot (ft^2) = 0.0929 square meter (m^2)

1 square inch (in^2) = 6.4516 square centimeters (cm^2)

Pressure		

1 kilogram per square
centimeter (kg/cm^2)* = 0.96784 atmosphere (atm)

= 0.98066 bar

= 9.8067×10^4 pascals (Pa)

1 bar = 10 megapascals (Mpa)

= 1.0×10^5 pascals (Pa)

= 29.53 inches of mercury (in a barometer)

= 0.98692 atmosphere (atm)

= 1.02 kilograms per square centimeter (kg/cm^2)

1 pascal (Pa) = 1 kg/m/s^2

1 pound per square inch = 0.06895 bars

= 6.895×10^3 pascals (Pa)

= 0.0703 kilogram per
square centimeter

Volume		

1 cubic kilometer (km^3) = 0.24 cubic mile (mi^3)

1 cubic meter (m^3) = 264.2 gallons

= 35.314 cubic feet (ft^3)

1 liter (1) = 1.057 quarts

= 33.815 fluid ounces

1 cubic centimeter (cm^3) = 0.0610 cubic inch (in^3)

1 cubic mile (mi^3) = 4.168 cubic kilometers (km^3)

1 cubic yard (yd^3) = 0.7646 cubic meter (m^3)

1 cubic foot (ft^3) = 0.0283 cubic meter (m^3)

1 cubic inch (in^3) = 16.39 cubic centimeters (cm^3)

Temperature		

To change from Fahrenheit (F) to Celsius (C):

$$°C = \frac{(°F - 32°)}{1.8}$$

To change from Celsius (C) to Fahrenheit (F):

$$°F = (°C \times 1.8) + 32°$$

To change from Celsius (C) to Kelvin (K):

$$K = °C + 273.15$$

To change from Fahrenheit (F) to Kelvin (K):

$$K = \frac{(°F - 32°)}{1.8} + 273.15$$

*Note: Because kilograms are a measure of mass whereas pounds are a unit of weight, pressure units incorporating kilograms assume a given gravitational constant (g) for Earth. In reality, the gravitational constant for Earth varies slightly with location.

depends on the amount of an object's mass but also on the strength of gravity. An astronaut has the same mass on both the Earth and the Moon, but weighs a different amount in each place. Since the amount of gravitational pull exerted by an object depends on its mass, and the Moon has about one-sixth the mass of the Earth, an astronaut weighing 68 kilograms (150 pounds) on Earth would weigh only about 11 kilograms (25 pounds) on the Moon. Thus, lunar explorers have no trou-

ble jumping great distances even when burdened by a space suit and oxygen tanks.

What does matter consist of? Early philosophers deduced that the Earth and the plants and animals on it must be formed from simpler components, just as bread is made from a measured mixture of primary ingredients (such as flour, water). They initially thought that the "primary ingredients" making up matter included only earth, air, fire, and water. Then a philosopher named Democritus (ca. 460–370 B.C.E.) argued that if you were able to keep dividing matter into progressively smaller pieces, you would eventually end up with nothing, and since it doesn't seem possible to make nothing out of something, there must be a smallest piece of matter that can't be subdivided further. He proposed the name **atom** for these smallest pieces, based on the Greek word *atomos*, which means indivisible.

Our modern understanding of matter didn't become established until the seventeenth century, when chemists such as Robert Boyle (1627–1691) recognized that certain substances, like hydrogen, oxygen, carbon, and sulfur, *cannot* break down into other substances, while others, like water and salt, *can* break down into other substances (Fig. a.2). For example, water breaks down into oxygen and hydrogen. Substances that can't be broken down came to be known as **elements**, while those that can be broken down came to be known as **compounds**. An English schoolteacher, John Dalton (1766–1844), adopted the word atom for the smallest piece of an element that maintains the property of the element, and suggested that compounds consisted of combinations of different atoms. Then in 1869, a Russian chemist named Dmitri Mendeléev (1834–1907) realized that groups of elements share similar characteristics. Mendeléev organized the elements into a chart we now call the **periodic table of the elements** (Fig. a.3). In the figure, elements within each column of the table behave similarly; for example, all elements in the right-hand column are inert gases, meaning

FIGURE a.2 A compound, salt, can be subdivided to form two elements, sodium metal and chlorine gas. Neither sodium nor chlorine can be divided further.

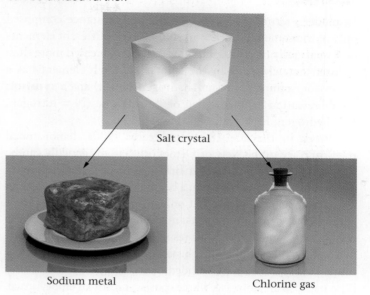

Salt crystal

Sodium metal

Chlorine gas

that they can't combine with other elements to form compounds. But Mendeléev and his contemporaries didn't know what caused the similarities and differences. An understanding of the cause would have to wait until twentieth-century physicists discovered the internal structure of the atom.

FIGURE a.3 The modern periodic table of the elements. Each column groups elements with related properties. For example, inert gases are listed in the column on the right. Metals are found in the central and left parts of the chart.

Alkali metals																	Inert gases
Symbol → He 2 / Helium / 4.002 = Atomic number / Name / Atomic weight												Nonmetals					
H 1 Hydrogen 1.007																	He 2 Helium 4.002
Li 3 Lithium 6.941	Be 4 Beryllium 9.0121											B 5 Boron 10.811	C 6 Carbon 12.011	N 7 Nitrogen 14.006	O 8 Oxygen 15.999	F 9 Fluorine 18.998	Ne 10 Neon 20.179
Na 11 Sodium 22.989	Mg 12 Magnesium 24.305	Transition elements (metals)										Al 13 Aluminum 26.981	Si 14 Silicon 28.085	P 15 Phosphorus 30.973	S 16 Sulfur 32.066	Cl 17 Chlorine 35.452	Ar 18 Argon 39.948
K 19 Potassium 39.098	Ca 20 Calcium 40.078	Sc 21 Scandium 44.955	Ti 22 Titanium 47.88	V 23 Vanadium 50.941	Cr 24 Chromium 51.996	Mn 25 Manganese 54.938	Fe 26 Iron 55.847	Co 27 Cobalt 58.933	Ni 28 Nickel 58.693	Cu 29 Copper 63.546	Zn 30 Zinc 65.39	Ga 31 Gallium 69.723	Ge 32 Germanium 72.61	As 33 Arsenic 74.921	Se 34 Selenium 78.96	Br 35 Bromine 79.904	Kr 36 Krypton 83.80
Rb 37 Rubidium 85.467	Sr 38 Strontium 87.62	Y 39 Yttrium 88.905	Zr 40 Zirconium 91.224	Nb 41 Niobium 92.906	Mo 42 Molybdenum 95.94	Tc 43 Technetium 98.907	Ru 44 Ruthenium 101.07	Rh 45 Rhodium 102.905	Pd 46 Palladium 106.42	Ag 47 Silver 107.868	Cd 48 Cadmium 112.411	In 49 Indium 114.82	Sn 50 Tin 118.710	Sb 51 Antimony 121.757	Te 52 Tellurium 127.60	I 53 Iodine 126.904	Xe 54 Xenon 131.29
Cs 55 Cesium 132.905	Ba 56 Barium 137.327	La 57 Lanthanum 138.905	Hf 72 Hafnium 178.49	Ta 73 Tantalum 180.947	W 74 Tungsten 183.85	Re 75 Rhenium 186.207	Os 76 Osmium 190.2	Ir 77 Iridium 192.22	Pt 78 Platinum 195.08	Au 79 Gold 196.966	Hg 80 Mercury 200.59	Tl 81 Thallium 204.383	Pb 82 Lead 207.2	Bi 83 Bismuth 208.980	Po 84 Polonium 208.982	At 85 Astatine 209.987	Rn 86 Radon 222.017
Fr 87 Francium 223.019	Ra 88 Radium 226.025	Ac 89 Actinium 227.027															

Ce 58 Cerium 140.115	Pr 59 Praseodymium 140.907	Nd 60 Neodymium 144.24	Pm 61 Promethium 144.912	Sm 62 Samarium 150.36	Eu 63 Europium 151.965	Gd 64 Gadolinium 157.25	Tb 65 Terbium 158.925	Dy 66 Dysprosium 162.50	Ho 67 Holmium 164.930	Er 68 Erbium 167.26	Tm 69 Thulium 168.934	Yb 70 Ytterbium 173.04	Lu 71 Lutetium 174.967
Th 90 Thorium 232.038	Pa 91 Protactinium 231.035	U 92 Uranium 238.028	Np 93 Neptunium 237.048	Pu 94 Plutonium 244.064	Am 95 Americium 243.061	Cm 96 Curium 247.070	Bk 97 Berkelium 247.070	Cf 98 Californium 251.079	Es 99 Einsteinium 252.083	Fm 100 Fermium 257.095	Md 101 Mendelevium 258.10	No 102 Nobelium 259.100	Lr 103 Lawrencium 262.11

a.2 **A MODERN VIEW OF MATTER**

Atoms

In modern terminology, an element is a substance composed only of the same kind of atoms. Ninety-two different elements occur naturally on Earth, but physicists have created more than a dozen new elements in the laboratory. Each element has a name (for example, nitrogen, hydrogen, sulfur) and a **symbol**, an abbreviation of its English or Latin name (N = nitrogen, H = hydrogen, Fe = iron, Ag = silver).

Work in the late 1800s and early 1900s demonstrated that, contrary to the view of Democritus, atoms actually can be divided. Ernest Rutherford, a British physicist, made this key discovery in 1910 when he shot a beam of atoms at a gold foil and found, to his amazement, that only a tiny fraction of the atoms bounced back; most of the mass in the beam passed through the foil as if it were invisible. This result could mean only one thing. Most of the mass in an atom clusters in a dense ball at the atom's center, and this ball is surrounded by a cloud that contains very little mass, so an atom as a whole consists mostly of empty space.

FIGURE a.4   Different ways of portraying atoms.

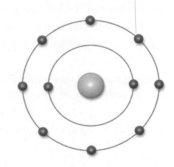

(a) A water molecule. The large ball is oxygen, and the small ones are hydrogen. The "sticks" represent chemical bonds.

(b) A diagram showing the number of electrons in the inner shells.

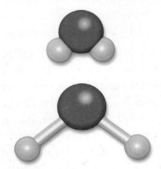

Inner electron shell   Outer electron shell

Nucleus

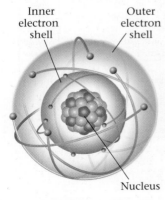

(c) An image of an atom with a nucleus orbited by electrons.

(d) A depiction of electron shells, implying that the electrons constitute a cloud.

Physicists now refer to the dense ball at the center of the atom as the **nucleus**, and the low-density cloud surrounding the nucleus as the **electron cloud**. Further study in the first half of the twentieth century led to the conclusion that the nucleus contains two types of subatomic particles: **neutrons**, which have a neutral electrical charge, and **protons**, which have a positive electrical charge (Fig. a.4a–d). The electron cloud consists of negatively charged particles, **electrons**, which are only about 1/1,836 times as massive as protons. Remember that opposite charges attract; the positive charge on the nucleus attracts the negative charge of the electrons, so the nucleus holds on to the electron cloud. (For simplicity, think of a positive charge as the "+" end of a battery and a negative charge as the "−" end.) The mass of a neutron approximately equals the sum of the mass of a proton and the mass of an electron. In the past few decades, physicists have found that protons and neutrons are made up of a myriad of even smaller particles, the smallest of which is called a **quark**.

Electron clouds have a complex internal structure. Electron are grouped in intervals called **orbitals**, energy levels, or shells. Some shells have a spherical shape, whereas others resemble dumbbells, rings, or groups of balls—for simplicity, we portray shells as circles, in cross section (Fig. a.5). Electrons in inner shells are concentrated near the nucleus, whereas those of outer shells predominate farther from the nucleus. Successively higher shells lie at progressively greater distances from the nucleus. Each shell can only contain a specific number of electrons: the lowest shell can contain only two electrons, while the next several shells each can hold 8. Electrons fill the lowest shells first, so that atoms with a small number of electrons only have the innermost shells; the outer shells do not exist unless there are electrons to fill them.

The outermost shell of electrons, in effect, defines the outer edge of an atom, so an atom with many occupied electron shells is larger than one with few occupied shells (argon, for example, with 18 electrons, contains more occupied electron shells than helium, with 2 electrons, so an argon atom is bigger than a helium atom). If we picture the nucleus of a carbon atom

FIGURE a.5   A neon atom has two complete electron shells. The inner shell has two electrons, and the outer has eight.

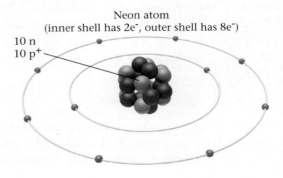

Neon atom
(inner shell has 2e⁻, outer shell has 8e⁻)

10 n
10 p⁺

as an orange, the electrons of the outermost shell would lie at a distance of over 1 km from the orange.

Atoms are so small that they can't be seen with even the strongest light microscopes. (However, by using some clever techniques, scientists have been able, in recent years, to create images of very large atoms; Fig. a.6.) Atoms are so small, in fact, that 1 gram (0.036 ounces) of helium contains 6.02×10^{23} atoms; in other words, a quantity of helium weighing little more than a postage stamp contains 602,000,000,000,000,000,000,000 atoms! The number of atoms of helium in a small balloon is approximately the same as the number of balloons it would take to replace the entirety of Earth's atmosphere.

Atomic Number, Atomic Mass, and Isotopes

We distinguish atoms of different elements from one another by their **atomic number**, the number of protons in their nucleus; hydrogen's atomic number is 1, oxygen's is 8, lead's is 82, and uranium's is 92. We write the atomic number as a subscript to the left of the element's symbol (for example, $_1$H, $_8$O, $_{82}$Pb, $_{92}$U). With the exception of the most common hydrogen nuclei, all atomic nuclei also contain neutrons. In smaller atoms, the number of neutrons equals the number of protons, but in larger atoms, the number of neutrons exceeds the number of protons. Subatomic particles are held together in a nucleus by **nuclear bonds**.

Atomic weight (or **atomic mass**) defines the amount of matter in a single atom. For a given element, the atomic weight *approximately* equals the number of protons plus the number of neutrons. (Precise atomic weights are actually slightly greater than this sum, because neutrons have slightly more mass than protons, and because the electrons have mass.) For example, helium contains 2 protons and 2 neutrons, and thus has an atomic weight of about 4, whereas oxygen contains 8 protons and 8 neutrons and thus has an atomic weight of about 16. We indicate the weight of an atom by a superscript to the left of the element's symbol (^{4}He, ^{16}O).

Some elements occur in more than one form, and these differ in atomic weight. For example, uranium 235 ($^{235}_{92}$U) contains 92 protons and 143 neutrons, while uranium 238 ($^{238}_{92}$U) contains 92 protons and 146 neutrons. Note that both forms of uranium have the *same* atomic number—they must, if they are to be considered the same element. But they have different quantities of neutrons and thus differ in atomic weight. Multiple versions of the same element, which differ from one another in atomic weight, are called **isotopes** of the element (Fig. a.7).

Ions: Atoms with a Charge

If the number of electrons (negatively charged particles) exactly equals the number of protons (positively charged particles) in an atom, then the atom is electrically neutral. But if the numbers

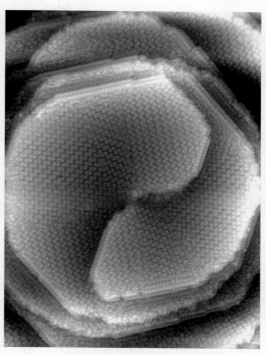

FIGURE a.6 An image of atoms taken with an electron microscope, a device that uses electrons to illuminate the surface. Each spot is an atom.

aren't equal, then the atom has a net electrical charge and is called an **ion**. For example, if an atom has two fewer electrons than protons, then we say it has a charge of 2+, and if it has two more electrons than protons, it has a charge of 2−. We write the charge as a superscript to the right of the symbol (for example, Na^{1+}). Ions with a negative charge are **anions** (pronounced ANN-eye-ons), and ions with a positive charge are **cations** (pronounced CAT-eye-ons). Ions form because atoms "prefer" to have a complete outer electron shell. Thus, an atom may give up or take on electrons in order for its outer shell to contain the proper number of electrons. As is the case with neutral atoms, the size of an ion depends on the number of shells containing electrons (Fig. a.8).

FIGURE a.7 The three isotopes of hydrogen: ordinary hydrogen, deuterium, and tritium.

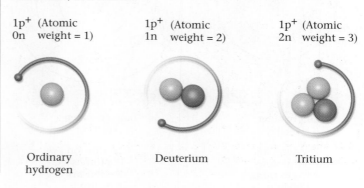

1p$^+$ (Atomic
0n   weight = 1)

1p$^+$ (Atomic
1n   weight = 2)

1p$^+$ (Atomic
2n   weight = 3)

Ordinary
hydrogen

Deuterium

Tritium

FIGURE a.8 Relative sizes of common ions making up materials in rocks at the Earth's surface.

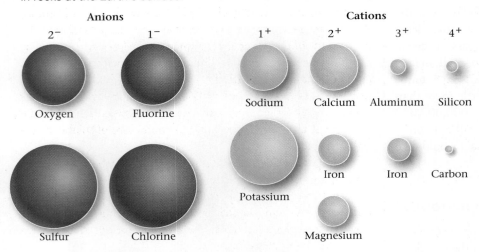

Note that oxygen ions are larger than silicon ions, even though silicon has a larger atomic number, because oxygen has gained electrons and thus uses more electron shells than does silicon, for silicon has lost electrons.

Molecules and Chemical Bonds

Most of the materials we deal with in everyday life—oxygen, water, plastic—are not composed of isolated atoms. Rather, most

FIGURE a.9 An ionic bond between sodium and chlorine. The sodium atom gives up its outer electron, and thus has one more proton than electron (and has a net positive charge), whereas the chlorine gains an electron and thus has one more electron than proton (and has a net negative charge). The two ions attract each other.

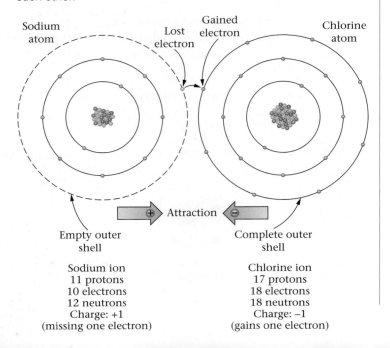

atoms tend to stick, or bond, to other atoms; two or more atoms stuck together constitute a **molecule**. Hydrogen gas, for example, consists of H_2 molecules; note that a hydrogen molecule consists of two of the same kind of atom. But many materials contain different kinds of atoms bonded together. As we noted earlier, such materials are called compounds. For example, common table salt is a compound containing sodium and chlorine atoms bonded together. Compounds generally differ markedly from the elements that make them up—salt bears no resemblance at all to pure sodium (a shiny metal) or pure chlorine (a noxious gas) (Fig. a.2). A molecule is the smallest identifiable piece of a compound, containing the correct proportion of the compound's elements. For example, a molecule of water consists of two atoms of hydrogen and one atom of oxygen. We represent water by the **chemical formula** H_2O, a concise recipe that indicates the relative proportions of different elements in the molecule (Fig. a.4a).

Chemical bonds act as the glue that holds atoms together to form molecules and holds molecules together to form larger pieces of a material. Chemical bonding results from the interaction among the electrons of nearby atoms, and can take place in four different ways. (Note that chemical bonds are not the same as the nuclear bonds that hold together protons and neutrons in a nucleus.)

- *Ionic bonds*: As an inviolate rule of nature, "like" electrical charges repel (two positive charges push each other away), while "unlike" electrical charges attract (a negative charge sticks to a positive charge). Bonds that form in this way are called **ionic bonds** (Fig. a.9). For example, in a molecule of salt, positively charged sodium ions (Na^{1+}) attract negatively charged chloride (Cl^{1-}) ions. (Chloride is the name given to ions of chlorine.)

- *Covalent bonds*: The atoms of carbon making up a diamond do not transfer electrons to one another, but rather share electrons. Bonding that involves the sharing of electrons is called **covalent bonding** (Fig. a.10). Because of the sharing, the electron shells of all the carbon atoms in a diamond are complete, and all the carbon atoms have a neutral charge. Water molecules also exist because of covalent bonding: in a water molecule, two hydrogen atoms are covalently bonded to one oxygen atom.

- *Metallic bonds*: In metals, electrons of the outer shells move easily from atom to atom and bind the atoms to each other. We call this type of bonding **metallic bonding** (Fig. a.11). Because outer-shell electrons move so freely, metals conduct electricity easily—when you connect a

FIGURE a.10 Covalent bonding. Carbon only has four electrons in its outer shell, which has a capacity of eight. Thus, the carbon atom shares electrons with four other carbon atoms.

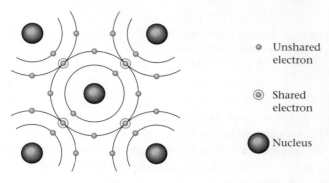

- Unshared electron
- Shared electron
- Nucleus

FIGURE a.11 In a metallically bonded material, nuclei and their inner shells of electrons float in a "sea" of free electrons. The electrons can stream through the metal.

metal wire to an electrical circuit, a current of electrons flows through the metal.

■ *Bonds resulting from the polarity of atoms or molecules*: Chemists long recognized that some materials break or split so easily that they must be held together by particularly weak chemical bonds. Eventually, they realized that these bonds are due to the permanent or temporary polarity of molecules. **Polarity** means that the molecule has a positive charge on one side and a negative charge on the other. Polarity-related bonds form because the negative side of one molecule attracts the positive side of another.

Bonds resulting from the polarity of molecules are important in many geologic materials. For example, liquid water consists of molecules in which two hydrogen atoms covalently bond to one oxygen atom. The hydrogen atoms both lie on the same

side of the oxygen atom. The oxygen nucleus, which contains 8 protons, exerts more attraction to the shared electrons than do the hydrogen nuclei, each of which contains only one proton. As a consequence, the shared electrons concentrate closer to the oxygen side of the molecule, making the oxygen side more negative than the hydrogen side. Thus, the water molecules are polar, and they attract each other. This attraction is called a **hydrogen bond** (Fig. a.12). The polarity of water molecules makes water a good solvent (meaning, it can dissolve substances), because the polar water molecules attract and surround ions of soluble materials and pull them apart.

Johannes van der Waals (1837–1923), a Dutch physicist, discovered another type of weak chemical bonding that depends on polarity. This type, now known as **van der Waals bonding**, links one covalently bonded molecule to another. The bonds exist because electrons temporarily cluster on one side of each molecule, giving it a polarity.

The Forms of Matter

Matter exists in one of four states: **solids**, which can maintain their shape for a long time without the restraint of a container;

FIGURE a.12 Water is a polar molecule, because the two hydrogen atoms lie on the same side of the oxygen atom. Thus, water molecules tend to attract one another. Polar water molecules surround chlorine and sodium ions when salt dissolves in water.

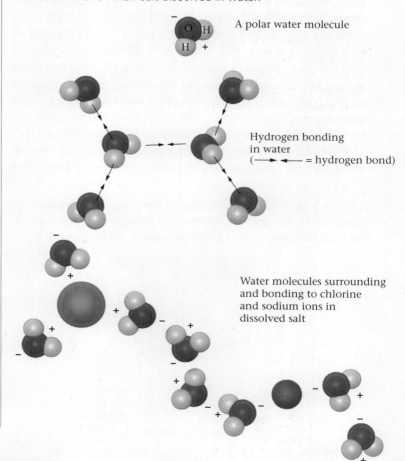

A polar water molecule

Hydrogen bonding in water (⟶ ⟵ = hydrogen bond)

Water molecules surrounding and bonding to chlorine and sodium ions in dissolved salt

FIGURE a.13 The states of matter.

Solid

(a) A solid retains its shape regardless of the size of the container.

Liquid

(b) A liquid conforms to the shape of the container, so its density does not change when the shape of the container changes.

Gas

(c) A gas expands to fill whatever volume it occupies, so its denstiy changes if the voume changes.

Plasma

(d) The Sun contains plasma, a gas-like material that is so hot that electrons have been stripped from the nuclei.

liquids, which flow fairly quickly and assume the shape of the container they fill (it's possible for a liquid to fill only part of its container); **gases**, which expand to fill the entire container they have been placed in and will disperse in all directions when not restrained; and **plasma**, an unfamiliar, gas-like mixture of positive ions and free electrons that exists only at very high temperatures (Fig. a.13a–d). Which state of matter exists at a location depends on the pressure and temperature (Fig. a.14).

To understand the differences at an atomic scale among solids, liquids, and gases—the common forms of matter on

FIGURE a.14 The state of a material depends on pressure and temperature, as depicted on this graph called a phase diagram. All three phases can exist simultaneously at the triple point.

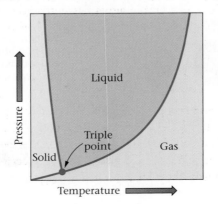

Earth—imagine a large group of students getting ready to take a 9:00 A.M. exam in a windowless auditorium. The students represent atoms. It's 8 o'clock, and some students are in their rooms, some are in the library, some are walking to the auditorium, and some are at the cafeteria. In other words, everyone is scattered all over the place; there is no order to the distribution of people, and they are too far apart to interact and bond to one another. Such a distribution is characteristic of a gas.

At ten minutes before nine, most students have arrived at the auditorium. Some are already in their seats, but many are walking in or out of the room. Such a distribution characterizes the state of a liquid, in which most of the particles are in the container, except for a slow exchange with the atmosphere (by evaporation or condensation). There is short-range order, represented by little clumps of students in their seats or chatting in the aisles, but overall there is still a lot of motion and no overall organization, or long-range order.

Suddenly the lights go off, when someone bumps against the switch, and everyone stops in place to avoid stumbling but people continue to fidget in place. Even though the migration of people has stopped, there is still no long-range order in the room. Such a situation characterizes a special kind of solid called a **glass**. Glass has a disordered arrangement of atoms, like a liquid, but cannot flow and thus behaves like a solid.

The lights go on again, and the examiner finally enters. Everyone takes a seat according to a specified plan—alternate rows and alternate seats. Even though individual students are not motionless, because they are still fidgeting in place, they are no longer moving out of an orderly arrangement. The situation now corresponds to that of a solid.

a.3 THE FORCES OF NATURE

In our everyday experience, we constantly see or feel the effect of force. Forces can squash objects, speed them up and slow them down, tear, stretch, spin, and twist them, and make them float or sink. Isaac Newton, the great British scientist who effectively founded the field of physics, was the first to describe the way forces work. He defined a **force** as simply the push or pull that causes the velocity (speed) of a mass to change in magnitude and/or direction.

We can distinguish between two categories of force. The first, which includes force applied by the movement of a mass (a hand, a hammer, the wind, waves), is called **mechanical force**, or contact force. When you push a block across the floor, you are applying a mechanical force. The second category, which includes force resulting from the action of an invisible agent, is called a **field force**, or noncontact force. If you drop a book, the invisible force of gravity pulls the book toward the floor.

Physicists further recognize four types of field forces: gravity, electromagnetic, strong nuclear, and weak nuclear. **Gravity** is a force of attraction between any two masses, and can act over large distances. The magnitude of gravitational attraction depends on the size of the masses and the distance between them. We feel a much stronger gravitational pull to the planet we walk on than we do to a baseball. **Electromagnetic force**, the force associated with electricity and magnetism, is stronger than gravity, but only operates between materials that have electrical charges or are magnetic, and only operates over short distances. Like gravity, electromagnetic force depends on the distance between objects, but unlike gravity it can be either attractive or repulsive—as mentioned previously, like charges repel and unlike charges attract. Particles within atoms are subject to two kinds of nuclear force. We'll leave the discussion of nuclear forces to a physics text.

a.4 ENERGY

To a physicist, **energy** is the ability to do work, or, in other words, the ability to apply a force that moves a mass by some distance (**work** = force × distance). According to this definition, gasoline serves as an energy source because we can burn gasoline to move heavy cars and trucks (a type of work).

Kinetic and Potential Energy

Formally, we classify energy into two basic types: **kinetic energy**, the energy of motion, and **potential energy**, the energy stored in a material. A boulder sitting at the top of a hill has potential energy, because the force of gravity will do work on it as the boulder topples down. If the boulder falls down, part of its potential energy converts into kinetic energy (Fig. a.15). Some of this kinetic energy can be transferred to a second boulder if the original boulder strikes another boulder, causing a contact force that starts the second boulder moving. Similarly, the gas in a car's tank has potential energy, which converts to kinetic energy when the gas burns and molecules start moving quickly. This kinetic energy causes the cylinders in the engine to move up and down. Now we briefly look at where energy comes from.

Energy from Chemical Reactions

A **chemical reaction** is a process whereby one or more compounds (the reactants) come together or break apart to form new

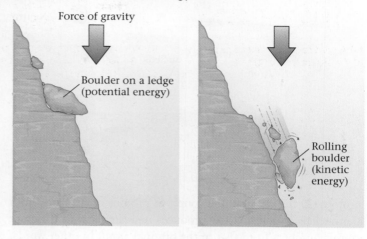

FIGURE a.15 A boulder sitting on a ledge in a gravitational field has potential energy. When the boulder starts to roll, this potential energy converts to kinetic energy.

compounds and/or ions (the products). Many chemical reactions produce energy. In the case of burning gasoline, gasoline molecules react with oxygen molecules in the air to form carbon dioxide molecules and water molecules, plus energy. We can describe this reaction in a chemical formula, which uses symbols for the molecules:

$$2C_8H_{18} + 25O_2 \rightarrow 16CO_2 + 18H_2O + \text{energy}.$$

Note that *reactants* appear on the left side of the formula (C_8H_{18} is a typical compound in gasoline), and *products* on the right side. The arrow indicates the direction in which the reaction proceeds. The energy comes from the breaking of chemical bonds—more energy is stored in the chemical bonds of gasoline and oxygen than in the chemical bonds of carbon dioxide and water.

Energy from Field Forces

When an object you place in a field force, the force acts on the object, and the object, unless it is restrained, moves. In a gravity field, as described earlier, a boulder resting on top of a hill has potential energy (it is restrained by the ground), but when the boulder starts rolling, it has kinetic energy. Similarly, an iron bar taped to a table near a magnet has potential energy, but an unrestrained bar moving toward the magnet has kinetic energy.

Fission and Fusion: Energy from Breaking or Making Atoms

During **nuclear fission**, a large atomic nucleus splits into two or more smaller nuclei and other by-products, including free neutrons (Fig. a.16). Such fission is one form of nuclear decay. Other forms include: (1) the ejection of an electron from one of the neutrons in the nucleus, transforming the neutron into a

FIGURE a.16   A uranium atom splits during nuclear fission.

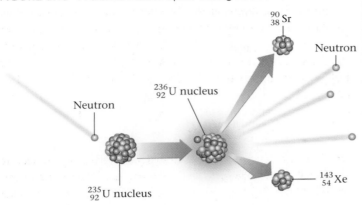

proton; (2) the ejection of a couple of protons or neutrons from the nucleus. Nuclear decay produces new atoms with different atomic numbers from that of the original atom. In other words, one element transforms into another. In effect, the ultimate goal of medieval alchemists, to turn lead into gold, can now be achieved in modern atom smashers, though at a significant cost.

When fission or other nuclear decay reactions occur, some of the matter constituting the original atom transforms into a huge amount of energy (heat and electromagnetic radiation),

FIGURE a.17   Nuclear fusion reactions.

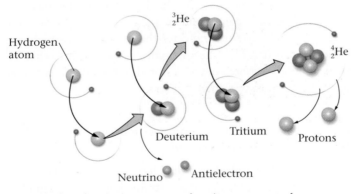

(a) In the Sun, four hydrogen atoms fuse in sequence to form a helium atom.

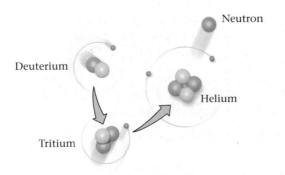

(b) In a hydrogen bomb, a deuterium and a tritium atom fuse to form a helium atom plus a free neutron.

as defined by Einstein's famous equation: $E = mc^2$, where E is energy, m is mass, and c is the speed of light. The realization that fission releases vast quantities of energy led to the rush to build atomic weapons during and after World War II. Nuclear fission provides the energy in atomic bombs, nuclear power plants, and nuclear submarines.

Nuclear fusion results when two nuclei slam together at such high velocity that they get close enough for nuclear forces to bind them together, thereby creating a new and larger atom of a different element (Fig. a.17a, b). In a manner of speaking, fusion is the opposite of fission. Fusion reactions also produce huge amounts of energy, because during fusion some of the matter converts into energy, according to Einstein's equation. Fusion reactions generate the heat in stars; in the Sun, for example, hydrogen atoms fuse together to form helium. Fusion reactions also generate the explosive energy of a hydrogen bomb—a thermonuclear device. Such reactions can only take place at extremely high temperatures (over 1,000,000°C). In fact, in order to get a hydrogen bomb to blow up, you need to use an atomic bomb as the trigger, because only an atomic explosion can generate high enough temperatures to start fusion. So far, no one has figured out how to economically sustain a controlled fusion reaction on Earth for the purpose of generating electricity.

a.5 HOT AND COLD

Thermal Energy and Temperature

The atoms and molecules that make up an object do not stay rigidly fixed in place, but rather jiggle and jostle with respect to one another. This vibration creates **thermal energy**—the faster the atoms move, the greater the thermal energy and the hotter the object. In simple terms, the thermal energy in a substance represents the sum of the kinetic energy of all the substance's atoms (this includes the back-and-forth displacements that an atom makes as it vibrates, as well as the movement of an atom from one place to another).

When we say that one object is hotter or colder than another, we are describing its temperature. It represents the average kinetic energy of atoms in the material. **Temperature** is a measure of warmth relative to some standard. In everyday life, we generally use the freezing or boiling point of water at sea level as the standard. When using the **Celsius (centigrade) scale**, we arbitrarily set the freezing point of water (at sea level) as 0°C and the boiling point as 100°C; whereas in the **Fahrenheit scale**, we set the freezing point as 32°F and the boiling point as 212°F.

The coldest a substance can be is the temperature at which its atoms or molecules stand still. We call this temperature **absolute zero**, or 0K (pronounced "zero kay"), where **K** stands

FIGURE a.18 Physical phenomena that we observe occur at a wide range of temperatures (here, as specified by the Kelvin scale).

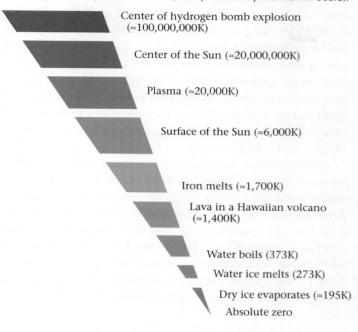

Center of hydrogen bomb explosion (≈100,000,000K)

Center of the Sun (≈20,000,000K)

Plasma (≈20,000K)

Surface of the Sun (≈6,000K)

Iron melts (≈1,700K)

Lava in a Hawaiian volcano (≈1,400K)

Water boils (373K)

Water ice melts (273K)

Dry ice evaporates (≈195K)

Absolute zero

for **Kelvin** (after Lord Kelvin [1824–1907], a British physicist), another unit of temperature. You simply can't get colder than absolute zero, meaning that you can't extract any thermal energy from a substance at 0K (–273.15°C). Degrees in the Kelvin scale are the same increment as degrees in the Celsius scale. On the Kelvin scale, ice melts at 273.15K and water boils at 373.15K (Fig. a.18).

Heat and Heat Transfer

Heat is the thermal energy transferred from one object to another. Heat can be measured in **calories**. One thousand calories can heat 1 kilogram of water by 1°C. Heat transfer takes place in four ways in the Earth System: radiation, conduction, convection, and advection.

Radiation is the process by which electromagnetic waves transmit heat into a body or out of a body (Fig. a.19a). For example, when the Sun heats the ground during the day, radiative heating takes place. Similarly, when heat rises from the ground at night, radiative heating is occurring—in the opposite direction.

FIGURE a.19 The four processes of heat transfer.

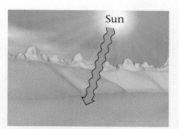

(a) Radiation occurs when sunlight strikes the ground.

40° C

Cool

Hot

90° C

(c) Convection takes place when moving fluid carries heat with it. Hot fluid rises while cool fluid sinks, setting up a convective cell.

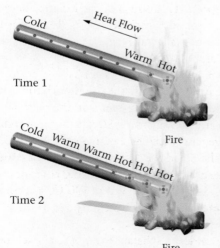

Cold

Heat Flow

Warm Hot

Time 1

Cold Warm Warm Hot Hot Hot

Time 2

Fire

Fire

(b) Conduction occurs when you heat the end of an iron bar in a flame. Heat flows from the hot region toward the cold region.

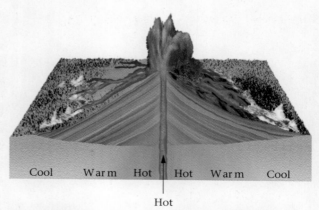

Cool   Warm   Hot   Hot   Warm   Cool

Hot

(d) During advection, a hot liquid (such as molten rock) rises into cooler material, and heat then conducts from the hot liquid into the cooler material.

Conduction takes place when you stick the end of an iron bar in a fire (Fig. a.19b). The iron atoms at the fire-licked end of the bar start to vibrate more energetically; they gradually incite atoms farther from the flame to start jiggling, and these atoms in turn set atoms even farther along in motion. In this way, heat slowly flows down the bar until you feel it with your hand. Note that even though the iron atoms vibrate more as the bar heats up, the atoms remain locked in their position within the solid. Thus, conduction does not involve actual movement of atoms from one place to another. If you place two bars of different temperatures in contact with each other, heat conducts from the hot bar into the cold one until both end up at the same temperature.

Convection takes place when you set a pot of water on a stove (Fig. a.19c). The heat from the stove warms the water at the base of the pot (it makes the molecules of water vibrate faster and move around more). As a consequence, the density of the water at the base of the pot decreases, for as you heat a liquid, the atoms move away from each other and the liquid expands. For a time, cold water remains at the top of the pot, but eventually the warm, less dense water becomes buoyant rel-ative to the cold, dense water. In a gravitational field, a buoyant material rises (like a styrofoam ball in a pool of water) if the material above is weak enough to flow out of the way. Since liquid water can easily flow, hot water rises. When this happens, cold water sinks to take its place. The new volume of cold water then heats up and then rises itself. Thus, during convection, the actual flow of the material itself carries heat. The trajectory of flow defines **convective cells**; in a convective cell, water follows a loop, with warm water rising and cold water sinking. Convection occurs in the atmosphere, the ocean, and the interior of the Earth.

Advection, a less familiar process, happens when heat moves with a fluid, flowing through cracks and pores within a solid material (Fig. a.19d). The heat brought by the fluid conductively heats up the solid that the fluid passes through. Advection takes place, for example, if you pass hot water through a sponge and the sponge itself gets hot. In the Earth, advection occurs where molten rock rises through the crust beneath a volcano and heats up the crust in the process or where hot water circulates through cracks in rocks beneath geysers and hot springs.

THE VIEW FROM SPACE   An X-ray image of the matter and energy of a supernova explosion, *Cassiopeia A*, 10,000 light years away. The light of this explosion reached the Earth 300 years ago.

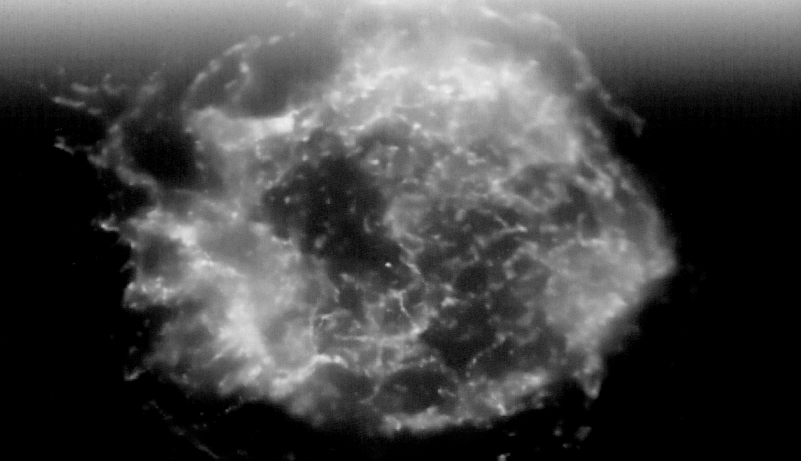

Geotours

See for yourself...

How to Use *Google Earth*™ to See Geologic Features

In earlier editions of this book, we could provide photos of landscapes, but could not convey a 3-D image of a region from a variety of perspectives. Fortunately, web-based computer tools such as *Google Earth*™, *NASA World Wind*, or *Microsoft Virtual Earth*, now permit you to tour our planet at speeds faster than a rocket. You can visit millions of locations and *see for yourself* what geologic features look like in context. These tools provide a compilation of imagery in a format that permits you to go to locations quickly and view them from any elevation, perspective, or direction. You can examine boulders on a hillside, zoom to airplane height to see the landform that the boulders rest on, and then zoom to astronaut height to see where the landform lies in a continent. You can look straight down or obliquely, and you can stay in one position, circle around a location, or fly like a bird over a landscape.

To help you use computer-visualization programs to understand geology, each chapter of this book includes a *Geotour*, with locations and descriptions of geologic features relevant to the chapter. Because only *Google Earth*™ can be used on both PC and Mac computers (as of this writing), we key the Geotours to the *Google Earth*™ format. Below, we provide a brief introduction to the use of *Google Earth*™ in the context of this book. Some of the details of procedures may change as *Google* updates the program. There's not room for a complete tutorial, but you'll find that the program is so easy to use that you will be proficient with it simply by working with it for a few minutes. The thumbnail images provided on this page are only to help identify tour sites. Go to wwnorton.com/studyspace to experience this flyover tour.

Opening *Google Earth*™

To use a Geotour, start by opening *Google Earth*™. As the program initializes, an image of the globe set in a background of stars appears in a window. The toolbar across the top provides a window icon. Click on this icon and a left sidebar appears (**Image GP.1**)—click on the icon again and the sidebar disappears so that the image becomes larger. The toolbar also provides a pushpin tool for marking locations, and a ruler tool for measuring distances. The sidebar contains information about locations and provides options for adding information to the screen image. For example, when you click on "borders," "roads," and "Populated Places" in the Layers panel, political boundaries, highways, and city names appear to provide a visual reference frame (**Image GP.2**). Any location you have marked with a pushpin appears in the sidebar within the Places panel. Double-clicking on the location in the sidebar will make the pushpin appear and will take you to the location. If you type a location in the Search panel at the top, a click on the magnifying glass icon will fly you to the location.

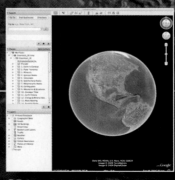

GP.1

GP.2

The Basic Tools of *Google Earth*™

In *Google Earth 4.3*™, three navigation tools appear in the upper right-hand corner of the screen (**Image GP.3**): (1) The **Look tool** (circle with an eye in the center) allows you to rotate the view. You can do this either by clicking and dragging the mouse on or within the outer ring, or by rotating the view clockwise or counterclockwise by clicking on the left or right arrows, respectively. Double-clicking the N button reorients the view with north at the top of the screen. Clicking on the top arrow of the Look tool tilts the image to show an oblique view, whereas clicking on the bottom arrow rotates the perspective back towards vertical. You can see this contrast by comparing the "Basic Tools-Vertical" and the "Basic Tools-Oblique" images of a site in the Andes Mountains of Bolivia (Lat 18°21'36.08"S, Long 66°0'13.95"W) from an altitude of 10 km (**Images GP.4** and **GP.5**). Note that, because of the nature of the imagery used by *Google Earth*™, steep cliffs are distorted. Clicking and dragging the mouse in the inner circle of the Look tool performs both actions simultaneously. (2) The **Move tool** (circle with a hand in the center) pans across the image. By clicking the arrows, you move a finite distance in the specified direction. Clicking and dragging the mouse in the inner circle acts as a joystick and translates the view in any direction. (3) The vertical **Zoom slider** zooms the view up or down. You can either move the slider or click on the end of the bar. Clicking on the **+** end takes you to a lower elevation, whereas clicking on the **−** end takes you to a higher elevation. The slider offers better control. Using the "View" menu bar, you can click to display a bar scale in the lower left portion of the screen. Note that the default setting in *Google Earth 4.3*™ is to swoop (tilt) into a perspective view as you zoom closer. This auto-tilt behavior may be turned off in the Preferences (Options) menu.

On the Earth image, you will see a hand-shaped cursor. By dragging the cursor across the screen, you can move the image. If you quickly drag the hand cursor, while holding down the mouse, and then let go, the movement will continue. For additional information, see the *Google Earth*™ User Guide.

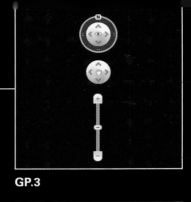

GP.3

GP.4

GP.5

Working with Location and Elevation
(Lat 43°8'18.84"N, Long 77°34'18.36"W)

On the bottom rule of the window, you will see three information items. The location of the point just beneath the hand-shaped cursor on the screen is specified on the left in terms of latitude (degrees, minutes, and seconds north or south of the equator) and longitude (degrees, minutes, and seconds east or west of the of the prime meridian). Just to the right of the Lat/Long information, a number indicates the elevation of the ground surface just below the hand-shaped cursor. The next number to the right tells you how much of the image has streamed onto your computer—an image starts out blurry, and then as streaming approaches 100%, it becomes clearer. The number on the far right ("Eye alt") indicates your viewing elevation.

To practice using these tools, enter the latitude and longitude provided above and zoom to a viewpoint 25,000 km (15,500 miles) out in space. You will see all of North America (**Image GP.6**). If you zoom down to 1800 m (5900 ft), you can see the details of Cobb's Hill Park in Rochester, New York (**Image GP.7**). Tilt your image so you just see the horizon, and rotate the image so you are looking SW (**Image GP.8**). With this perspective, you realize that the bean-shaped reservoir sits on top of an elongate ridge.

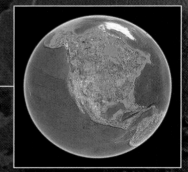

GP.6

Image Boundaries
(Lat 16°6'3.12"N, Long 75°48'25.84"E)

The globe that you see on your screen is a compilation of many separate satel-lite images merged by computer; some regions, therefore, look like a patch-work. To see this effect, fly to the coordinates provided in India and look down from an elevation of 600 km (**Image GP.9**). Image boundaries may reflect one or more of the following factors. (1) *Resolution*: The sharpness or clarity of images on *Google Earth*™ depend on the resolution of the image that is avail-able in the database. High-resolution images are clear at high magnification, whereas low-resolution images appear grainy or pixelated at high magnifica-tion. Depending on the speed of your internet connection, it may take time for the full available resolution of an image to show up. The image is fully loaded when the blue circle in the lower right-hand corner has become solid. (2) *Season*: Some images show regions in the winter, when leaves are down and the ground is visible, whereas some show regions in summer, when the canopy of trees hides the ground. (3) *Photo Tint*: Not all satellite images are taken in the same way, so not all have the same basic color tint. *Google Earth*™ constantly updates images, as new ones become available, so a region that looks like a patchwork one day might not the next.

GP.9

Finding Locations Using Latitude and Longitude
(Lat 48°52'25.26"N, Long 2°17'42.18"E)

You can find towns, parks, and landmarks on the Earth using *Google Earth*™ by entering the name in the search panel in the upper left corner. For example, enter Arc de Triomphe, hit the return button, and you fly to this well known land-mark in the center of Paris, France (**Image GP.10**). Many of the places you will visit in Geotours of this book are not, however, near a well-known landmark. Thus, to get you to a locality, we provide a latitude and longitude in the form:

Lat 48°52'25.44"N, Long 2°17'42.31"E

The first number is the latitude in degrees, minutes (60' = 1°), and seconds (60.00" = 1'). Simply enter the latitude and longitude on your screen. When typing numbers into the space provided on the program, you can abbreviate as:

48 52 25.44N, 2 17 42.31E

Note that instead of typing in latitude and longitude, you can simply copy the ".kmz" files provided in this book's website onto your computer. When you double-click on a file, it will open *Google Earth*™ and add all the locations referred to in the book to the "Temporary Places" folder. Then you simply need to click on the particular image referred to in a Geotour and the program will fly you straight there.

GP.10

Enjoy!

Please take advantage of the Geotours provided in this book. They provide you with an understanding of geology that simply can't appear on a printed page, and they are, arguably, as fun as a video game. Take control of the *Google Earth*™ tools, change your elevation and perspective, and fly around the landscape. Geology will come alive for you, and explora-tion of the Earth may become your hobby.

See for yourself...
The Variety of Earth's Surface

Google Earth™ allows you to sense what a space traveler orbiting the Earth would see if there were no clouds and almost no sea ice, and if features of the sea floor were visible. We can even see relics of meteorite impact craters. Let's orbit the Earth, and then look more closely at examples of its surface. The thumbnail images provided on this page are only to help identify tour sites. Go to wwnorton. com/studyspace to experience each flyover tour.

Distant View of the Earth

Zoom to view the planet from an elevation of about 12,000 km (about 8,000 miles). Turn on the "Lat/Long Grid" to see lines of latitude and lines of longitude (**Image G1.1**). Orient the image so that the equator is horizontal and examine the numbering scheme of these lines relative to the equator and the prime meridian. Click and drag from right to left and then let go to simulate the Earth's rotation and see the distribution of land and sea. Note that different shades of blue distinguish between deeper and shallower water, as you can see in the western Caribbean and Bahamas region, from an elevation of 3,000 km (1,864 miles) (**Image G1.2**).

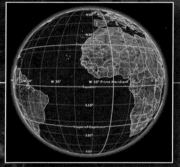

G1.1

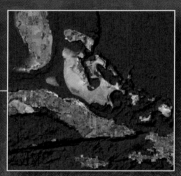

G1.2

G1.3

Eastern Greenland
(Lat 74°55'21.75"N, Long 22°5'32.62"W)

Ice covers the land surface of Greenland. Fly to the designated locality, then zoom to 3,200 km (2,000 miles) above sea level (**Image G1.3**). You can see the vast sheet of ice.

Zoom to 160 km (100 miles) above sea level. Now you can see that the ice sheet drains to the sea via slowly moving "rivers of ice" (**Image G1.4**). These valley glaciers were once longer and they carved deep valleys. Rising sea level filled the valleys to form fingers of the ocean called fjords. Note that the water in these fjords has frozen to form sea ice. Next, zoom down to 16 km (10 miles) and tilt the image so you just see the Earth's horizon (**Image G1.5**). Fly along the coast to see spectacular valleys and cliffs.

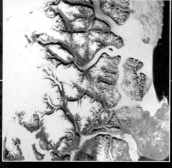

G1.4

G1.5

G1.6

G1.7

Southern Alps, New Zealand
(Lat 43°30'45.61"S, Long 170°50'41.97"E)

Zoom to an elevation of 20 km (about 12 miles) and you can see an example of rugged, mountainous topography and sediment-choked streams. Next, zoom to 16 km (10 miles), tilt so that you can see the horizon, and fly northwest up the river valley (**Image G1.6**). Turn 180° and fly down the river. Eventually, the river leaves the mountains and crosses the plains. Here, farmers have divided the land into fields (**Image G1.7**). The river eventually reaches the coast and drains into the Pacific Ocean.

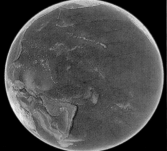

Pacific Ocean (Lat 3°43'48.84"N, Long 163°51'15.32"W)

Zoom out to an elevation of 17,000 km (10,500 miles) above sea level. The Pacific Ocean almost fills your field of view (**Image G1.8**). You can see a few islands and can note that the floor of the sea is not perfectly flat—it has seamounts, trenches, and oceanic ridges.

G1.8

Midwestern United States (Lat 39°51'02.69"N, Long 87°24'30.44"W)

Zoom to 9,000 m (30,000 feet) to see the view from a jet plane (**Image G1.9**). The image displays the Wabash River, farm fields, wooded areas along small streams, and a town (Newport, Indiana). What percentage of this landscape has been changed by human hands?

G1.9

A Sand Sea in Saudi Arabia (Lat 28°55'2.02"N, Long 39°36'59.07"E)

Zoom to 35 km (22 miles). You see golden sand, blown by the wind into ridges called dunes. Tilt the image so that the horizon just appears, and then fly slowly north (**Image G1.10**). You'll cross the occasional barren rocky hill without a tree or shrub in sight.

G1.10

Manicouagan Crater, Quebec, Canada
(Lat 51°23'58.50"N, Long 68°41'44.11"W)

Fly to the coordinates of the crater and zoom to an elevation of 800 km (500 miles). Note that the crater is really obvious even from this elevation. Now zoom to a lower elevation. At what elevation does the crater fill the field of view? Tilt the field of view to see the horizon and, using the look tool, fly around the crater to see its context (**Image G1.11**). If you descend to 4,570 m (15,000 feet), do you even realize that you are within a crater? Notice that the interior of the crater has risen, relative to the rim. This is due to the crust rebounding after the crater was excavated by impact. Manicouagan Crater formed between 206 and 214 million years ago. The preserved portion of the crater is about 70 km (43.5 miles) in diameter, but before erosion, it was probably 100 km (62 miles) in diameter. Due to the construction of a dam, the depressed outer edge of the crater filled with a lake.

G1.11

See for yourself...
Plate Boundaries

Where do you go if you want to see a plate boundary for yourself? First, it's important to realize that a plate boundary is not a simple line on the surface of the Earth,
but is a zone perhaps 40 to 200 km wide. Still, there are places where you can go to
study plate-boundary characteristics in person. Visit the following localities and
you'll see for yourself. The thumbnail images provided on this page are only to
help identify tour sites. Go to wwnorton.com/studyspace to experience each flyover tour.

Divergent Plate Boundaries

Most mid-ocean ridges that define the trace of a divergent plate boundary lie
submerged below 2 km (1.2 miles) of water. Thus, to get close to see the faults
and volcanic vents of a ridge, geologists descend in research submersibles. *Google
Earth™*, however, can help you see the shape of a mid-ocean ridge without the
submersible, by taking you to Iceland, one of the few places on this planet where a
ridge rises above sea level.

The Mid-Atlantic Ridge (Overall View)

Zoom out to about 8,000 km (5,000 miles), and orient your view to show the
north Atlantic Ocean (**Image G2.1**). See how the coastline of northwest Africa
matches that of eastern North America? Now focus on the ocean floor between
these coastlines and you'll see an image of the Mid-Atlantic Ridge—its trace
follows the curve of the coastlines. Though this image has only low resolution,
you can see the segmentation of the ridge axis and can recognize the oceanic
fracture zones that link the end of one segment to the end of the next. Fracture
zones appear to extend beyond the junctions with ridge segments. But only
the portion of a fracture zone between two ridge segments is an active transform fault.

G2.1

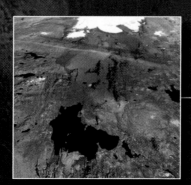

G2.2

The Mid-Atlantic Ridge in Iceland
(Lat 64°15'11.95"N, Long 20°56'12.66"W)

Trace the ridge northward to Iceland, an island that
straddles the ridge. Zoom in to an elevation of 35
km (22 miles) at the above coordinates, tilt, and look
northeast (**Image G2.2**). Along the south coast, east
of Reykjavik, you can see a set of northeast trending ridges whose faces are fault scarps along the
ridge axis, and if you look around, you'll see some
volcanoes. A glacier covers part of the ridge—after
all, it is Iceland!

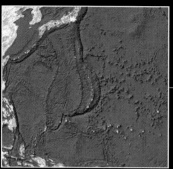

G2.3

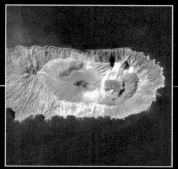

G2.4

The Mariana Trench
(Lat 16°20'49.97"N, Long 145°41'48.16"E)

Fly to the coordinates given above, and you'll find yourself over the western Pacific Ocean. Zoom to an elevation of 6,000 km (3,700 miles), and you will be looking at a curving band of dark blue, south of Japan (**Image G2.3**). This is the Mariana Trench, whose floor—the deepest point in the ocean—marks the boundary where the Pacific Plate subducts beneath the Philippine Plate. The curving chain of islands (the Marianas) to the west of the trench is the volcanic island arc on the edge of the Philippine Plate.

Zoom down to an elevation of 12 km (7.5 miles). You will be looking down on the volcano of Anatahan (**Image G2.4**). The central portion of the island has collapsed to form a depression called a caldera.

G2.5

Cascade Volcanic Arc
(Lat 46°12'25.47"N, Long 121°29'25.80"W)

Fly to the coordinates given (Mt. Adams volcano), and zoom to an elevation of 45 km (28 miles). You'll be looking down on the peak of Mt. Adams, a volcano in the Cascade volcanic arc, a continental arc. Tilt your image so you just see the northern horizon, and use the compass to reorient your image to look along the volcanic chain (**Image G2.5**). You can see the spacing between the volcanoes.

Transform Plate Boundaries

Many major earthquakes happen on the San Andreas Fault. All along its length, slip on this continental transform plate boundary has affected the landscape—valleys, elongate ponds, and narrow ridges follow the fault. Also, the fault has offset stream channels.

G2.6

San Andreas Transform Fault
(Lat 34°30'43.93"N, Long 118°01'06.19"W)

Fly to these coordinates and zoom to an elevation of 12 km (7.5 miles). You are looking down on the San Andreas Fault, southeast of Palmdale, California. This is the transform boundary between the Pacific Plate and the North American Plate. The trace of the fault is the boundary between flat land to the northeast and the hilly area to the southwest (**Image G2.6**). At this locality, you can see a gravelly river channel that has been offset at the fault. Zoom closer, tilt the image, and rotate the view so you are looking along the fault. Follow its trace and see all the buildings, canals, and roads that cross it.

See for yourself . . .
Diamond Mines

Diamonds fetch such a high price that prospectors seek them even in very remote areas. In the 1860s, attention focused on Kimberley, South Africa—at the time, a very remote area. In recent years, new deposits have been found in Arctic Canada. The thumbnail images provided on this page are only to help identify tour sites. Go to wwnorton.com/studyspace to experience this flyover tour.

G3.1

Kimberley, South Africa
(Lat 28°44'17.06"S, Long 24°46'30.77"E)

Fly to the coordinates provided and zoom out to an elevation of about 15 km (9 miles). You will see the town of Kimberley in the dry interior of South Africa. Zoom down to an elevation of 5 km (3 miles). You will be hovering over a large circular pit (**Image G3.1**). This is one of the diamond mines for which Kimberley is famous.

Mine near Yellowknife, Canada
(Lat 64°43'14.74"N, Long 110°37'32.76"W)

At these coordinates, you will see a remote region of Arctic Canada in which prospectors found diamond pipes in the early 1990s, after a 20-year search. If you zoom to an altitude of 200 km (125 miles), you can get a feel for the tundra landscape of scrubby vegetation and frigid lakes (**Image G3.2**). From an altitude of 10 km (6 miles), you can see one of the mines where a pipe is being excavated (**Image G3.3**). For the story of this discovery, please see Krajick, K., 2001, *Barren Lands*; the complete reference is at the end of the chapter.

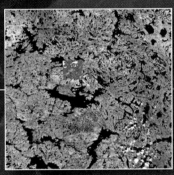

G3.2

G3.3

G3.4   G3.5

Diamond Prospect, Diamantina, Brazil
(Lat 18°15'4.35"S, Long 43°34'57.21"W)

Fly to the coordinates, and hover at 15 km (9 miles). You are looking at the town of Diamantina, nestled in the Espinhaço Range of eastern Brazil (**Image G3.4**). The town's name means "diamond," and for a good reason. Prospectors (known as *Bandeirantes*) discovered diamonds at the locale in the 1720s, and the stones have been mined ever since. These diamonds are detrital—in the Precambrian, they weathered out of kimberlites and then were transported as sediment. Eventually, they accumulated with quartz grains and pebbles to form a thick unit of quartzite and conglomerate. These rocks form the ridges of the Espinhaço. Weathering of this rock frees the diamonds once again. Zoom to 3 km (2 miles), tilt, and look north, and you see a small prospect pit (**Image G3.5**)

See for yourself...
Exposures of Igneous Rocks

What do igneous rocks look like in the field? They are exposed in many places around the world. Here, we take you on tour to see a few of the better examples. The thumbnail images provided on this page are only to help identify tour sites. Go to wwnorton.com/studyspace to experience this flyover tour.

G4.1

Yosemite National Park, California
(Lat 37°47'25.70"N, Long 119°29'16.74"W)

Fly to the coordinates provided and zoom up to an elevation of 7 km (4 miles). You can see a vast expanse of whitish-gray outcrop in Yosemite National Park in the Sierra Nevada (Image G4.1). This outcrop consists of granite and similar rocks formed by slow cooling of felsic magma many kilometers beneath the surface. The magma that became these rocks formed when Pacific Ocean floor subducted beneath North America, along a convergent plate boundary, about 80 to 100 million years ago. Subsequent uplift and erosion stripped away volcanic rocks that once lay above the granite. During the last twenty thousand years, glacial erosion polished the outcrops.

Now tilt the image so you just see the horizon, and rotate so that you are looking southwest. This points you downstream along Merced Canyon (Image G4.2). If you fly slowly down the canyon, you will pass famous mountains—Half Dome (on your left) and eventually El Capitan on the right—whose hard surfaces appeal to mountain climbers.

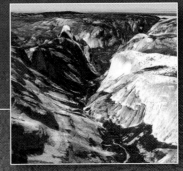
G4.2

Shiprock, New Mexico
(Lat 36°41'17.03"N, Long 108°50'09.31"W)

Fly to the coordinates provided and zoom out to an elevation of 10 km (6 miles). You are looking down on Shiprock. This is the eroded remnant of an explosive volcano that last erupted about 30 Ma (Image G4.3). Erosion stripped away layers of lava and ash that once formed the volcano, leaving behind dark intrusive rock that froze inside and just below the volcano. Shiprock consists of a rock type similar to basalt; this rock shattered into fragments as it approached the ground surface. Three dikes cut into the countryside, emanating from Shiprock like spokes from a wheel. These dikes are the remnants of wall-like intrusions.

Zoom down to 5 km (3 miles), tilt the image so you just see the horizon, and rotate so you're looking NW (Image G4.4). You can see the steep-sided mountain and two of the dikes. The mountain is 500 m (1,600 feet) in diameter and 600 m (2,000 feet) high.

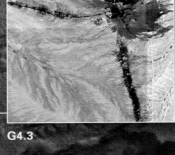
G4.3

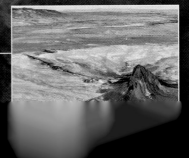

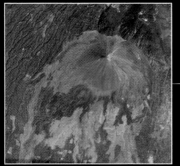

G4.5   **G4.6**

Izalco Volcano, El Salvador
(Lat 13°48'50.40"N, Long 89°37'57.74"W)

This volcano was active almost continuously from 1770 to 1958. Zoom to an elevation of 10 km (6 miles). Basalt lava flows flooded down Izalco's flanks and are still clearly visible (**Image G4.5**). Younger flows are darker colored than older flows, because the younger flows have had less time to react with the atmosphere and water to undergo weathering (see Chapter 7). The gray, evenly distributed material closer to the summit consists of tiny pellets of cooled lava. Zoom down to 7 km (4 miles), and tilt the image so that you are looking north and just see the horizon (**Image G4.6**). You can see nearby volcanoes. The summit of one, Santa Ana, has collapsed to form a circular depression called a caldera.

G4.7

Dikes, Western Australia
(Lat 22°50'13.69"S, Long 117°23'59.74"E)

Fly to this locality and zoom down to an elevation of 3.5 km (2 miles). You are hovering over the desert of northwestern Australia, an area where extensive areas of Precambrian rocks are exposed. Here, a number of dikes stand out in relief, because they are harder than the surrounding rock, which has eroded away. Tilt the image, look north, and you can see the dikes more clearly (**Image G4.7**).

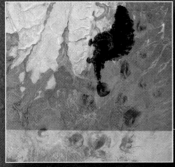

G4.8   **G4.9**

Cinder Cones, Arizona
(Lat 35°34'56.58"N, Long 111°37'55.10"W)

Fly to these coordinates and zoom to 19 km (12 miles) (**Image G4.8**). You can see many of the cinder cones that spot the landscape north of San Francisco Peak, a stratovolcano near Flagstaff, Arizona. The darkest spot is SP Crater, which erupted about 71 Ka. The black apron to the north of SP Crater is a 30 m-thick basalt flow. Descend to 2.5 km (1.5 miles), tilt and rotate so you are looking south for a better view (**Image G4.9**). At an elevation of 10 km (6 miles), fly 16.5 km (10 miles) to the SSE, to find Sunset Crater, which erupted an apron of bright orange tephra at 1 Ka. At this site (Lat 35°21'51.82"N, Long 111°30'10.87"W), zoom to 4 km, tilt the view, and look east (**Image G4.10**).

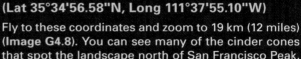

See for yourself...
Volcanic Features

There are more than 1,500 active volcanoes on Earth and thousands more volcanic landscapes. The thumbnail images provided on this page are only to help identify tour sites. Go to wwnorton.com/studyspace to experience this flyover tour.

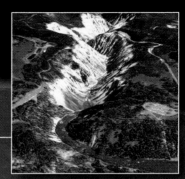

G5.1

Yellowstone Falls (Lat 44°43'4.96"N, Long 110°29'45.44"W)

At these coordinates, hover above the Grand Canyon of the Yellowstone from an elevation of 5 km (3 miles). The canyon walls expose bright yellow tuffs deposited during cataclysmic eruptions that occurred during the past 2 million years. Drop to an elevation of 3 km (2 miles), look downstream, and tilt to get a better view (Image G5.1).

Mt. Saint Helens, Washington (Lat 46°12'1.24"N, Long 122°11'20.73"W)

Box 5.1 discussed the explosion of Mt. Saint Helens. To see the damage for yourself, fly to the coordinates and zoom to 30 km (18.5 miles). The breached crater, the blowdown zone, the slumps, and the lahars are all visible (Image G5.2). To get a better sense of the devastation, descend to 13 km (8 miles), tilt your image, and fly around the mountain (Image G5.3).

G5.2

G5.3

Smoking Volcano, Ecuador (Lat 1°27'58.96"S, Long 78°26'58.24"W)

Fly to this locality and zoom to 75 km (46.5 miles). You are looking at the Andean Volcanic Arc in Ecuador, a consequence of subduction of Pacific Ocean floor beneath South America. Here, you see four volcanoes (Image G5.4). When this image was taken, Tungurahua was erupting, producing a plume of ash that winds blew to the southwest. The largest volcano in view is Chimborazo, the snow-covered peak at the western edge of the image.

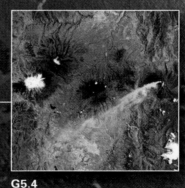

G5.4

G5.5

Mt. Etna, Sicily (Lat 37°45'5.49"N, Long 14°59'38.41"E)

At these coordinates, from an elevation of 50 km (31 miles), you can see Mt. Etna, which dominates the landscape of eastern Sicily (Image G5.5). This volcano erupts fairly frequently—in this image small clouds of volcanic smoke rise from the 3200 m-high summit. Note the numerous lava flows on its flanks. Zoom to 5 km (3 miles), rotate the image, and tilt it so you are looking east. You can see calderas on the summit (Image G5.6).

G5.6

G5.7

Mt. Vesuvius, Italy (Lat 40°49'18.24"N, Long 14°25'32.04"E)

Eruption of Mt. Vesuvius in 79 C.E. destroyed Herculaneum and Pompeii. Fly to the coordinates provided and you'll be hovering over the volcano's crater. Zoom to 50 km (31 miles), and you can see the entire bay of Naples—several volcanic calderas lie on the west side of Naples. In fact, the entire bay is a caldera. Zoom to 15 km (9 miles), tilt the view, and fly around Vesuvius (**Image G5.7**). The central peak of the volcano lies within a larger caldera (Somma) that formed 17,000 years ago.

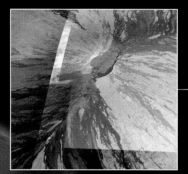

G5.8

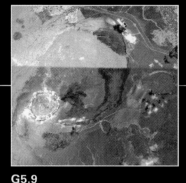

G5.9

Hawaiian Volcanoes (Lat 19°28'20.47"N, Long 155°35'32.82"W)

At these coordinates, you'll be above the caldera of Mauna Loa, one of the shield volcanoes on Hawaii. From 25 km (15.5 miles) , you can see that the caldera is elliptical, and that other, smaller calderas occur to the SW. Tilt the image, and you can see a large fissure cutting the frozen lava lake in the Caldera. Distinctive lava flows spilled out of the ends of the caldera (**Image G5.8**). Fly about 33 km (20.5 miles) ESE to find the caldera of Kilauea (**Image G5.9**). From here, fly 41 km (25.5 miles) to the NNE to cross Mauna Kea, home to an observatory (white buildings).

Mt. Shishildan, Alaska (Lat 54°45'36.59"N, Long 163°58'28.65"W)

Unimak Island, in the Aleutian chain, hosts a large stratovolcano, Mt. Shishildan. From an elevation of 35 km (22 miles), you'll note that the snowline starts halfway up the mountain, but that the peak itself is black (**Image G5.10**). That's because the volcano is active, and recent eruptions have buried and melted snow at the peak. From an elevation of 10 km (6 miles), you can see a red glow at the peak. Note that another, smaller volcano lies to the east.

G5.10

G5.11

East African Rift (Two Locations)

To understand volcanism in the East African rift, it is necessary to visit several locations. Here we provide two: (1) Mt. Kilimanjaro (Lat 3°3'53.63"S, Long 37°21'31.02"E): From a height of 10 km, you can see the caldera at the top of Africa's highest volcano (**Image G5.11**). The glaciers on the summit have been shrinking rapidly and may be gone in 20 years. Slumping has produced steep cliffs. Fly 45 km NE to find a long chain of cinder cones marking eruptions along a fault in the rift. (2) Goma region (Lat 1°39'27.40"S Long 29°14'15.27"E): At these coordinates, zoom to 80 km (50 miles), and tilt your view to look north (**Image G5.12**). The chain of lakes marks the western arm of the East African Rift. Active volcanoes lie just north of the city of Goma. Zoom to 3 km (2 miles) and tilt the image to look north—note the lava flow covering a portion of the airport runway and the city (**Image G5.13**).

G5.12

G5.13

See for yourself . . .
Sedimentary Rocks and Environments

You can see dramatic exposures of sedimentary rocks at many localities across the planet. Stratification in these exposures tends to be better where climate is drier and vegetation sparse. With a little searching, you can also find many examples of places where sediment is now accumulating. The thumbnail images provided on this page are only to help identify tour sites. Go to wwnorton.com/studyspace to experience this flyover tour.

Grand Canyon, Arizona
(Lat 36°08'00.34"N, Long 112°13'38.95"W)

Fly to these coordinates and zoom out to an elevation of 50 km (31 miles). You can see the entire width of the Grand Canyon, in northern Arizona (Image G6.1). Here, the Canyon is about 20 km (12 miles) wide. It cuts into the Colorado Plateau, a region of dominantly flat-lying Paleozoic and Mesozoic strata. Zoom down to about 6 km (4 miles), and tilt the image to see the horizon (Image G6.2). You now get a sense of the rugged topography of the canyon, and can see how the different layers of sedimentary rock stand out. More resistant rock units hold up cliffs. Fly down the Colorado River to develop a sense of the Canyon's majesty.

G6.1

G6.2

Lewis Range, Montana (Lat 47°48'02.31"N, Long 112°45'30.97"W)

Fly to these coordinates and zoom out to an elevation of 25 km (15 miles). You will be looking at a series of north–south-trending cuestas underlain by tilted layers of late Paleozoic and Cretaceous strata (Image G6.3). A cuesta is an asymmetric ridge—one side is a cliff cutting across bedding, and the other side is a slope parallel to the tilted bedding. Faulting has caused units to be repeated. Thus, some of the adjacent cuestas contain the same strata. Tilt your field of view so the horizon just appears, and you get a clear sense of how bedding can control topography. Zoom down to 6 km (4 miles) and you'll be able to pick out individual beds.

G6.3

Death Valley, California (Lat 36°20'49.15"N, Long 116°50'8.14"W)

Death Valley includes the lowest point in North America. It's a hot, dry desert area. When rain does fall, brief but intense floods carry sediment out of the bordering mountains and deposit it in alluvial fans at the foot of the mountains. Water that makes it out into the valley eventually evaporates and leaves behind salt. Fly to the coordinates given, zoom to an elevation of 10 km (6 miles), and tilt the image (Image G6.4). You can see contemporary sites of alluvial-fan and evaporite deposition.

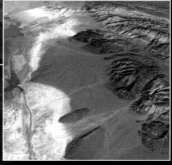
G6.4

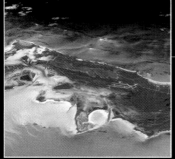

G6.5

Great Exhuma, Bahamas (Lat 23°27'03.08"N, Long 75°38'15.68"W)

Here, in the warm waters of the Bahamas, you can see numerous examples of carbonate depositional environments. If you fly to the coordinates given, zoom to an elevation of 12 km (7 miles), and then tilt the image, you'll see reefs, beaches, lagoons, and deeper-water settings (Image G6.5). The white sand consists of broken shell fragments.

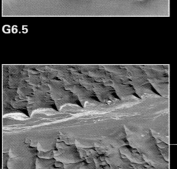

G6.6   **G6.7**

Sand Dunes, Namibia (Lat 24°44'40.42"S, Long 15°30'5.34"E)

Fly to these coordinates and zoom to an elevation of about 18 km (11 miles). Tilt your image so you gain some 3-D perspective (Image G6.6). You are looking at a sea of sand piled in huge dunes by the wind. In the field of view, you can also see the deposits of a stream that occasionally washes through the area. Were the sand dunes to be buried and transformed into rock, they would become thick layers of cross-bedded sandstone. Zoom to 5.5 km (3 miles), tilt, and look NW (Image G6.7). You can determine the wind direction.

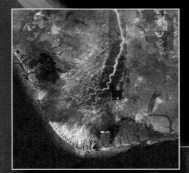

G6.8

Niger Delta, Nigeria (Lat 5°24'54.48"N, Long 6°31'26.25"E)

Head to the west coast of Africa, near the equator. The Niger River drains into the Atlantic and has built a 400 km (249 mile)-wide delta. Enter the coordinates given and zoom up to an elevation of 450 km (280 miles) to see the entire delta at once (Image G6.8). The present surface of the delta sits on top of a sedimentary accumulation that is many kilometers thick.

Now zoom down to 35 km (22 miles). You can see sandbars accumulating at bends in the river (Image G6.9). If the abundant vegetation of the dense surrounding jungle was preserved and buried, it would transform into coal.

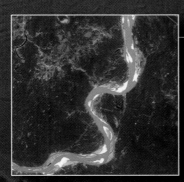

G6.9

Images provided by *Google Earth*™ mapping service/DigitalGlobe, TerraMetrics, NASA, Europa Technologies—copyright 2008.

See for yourself . . .
Precambrian Metamorphic Terranes

A metamorphic terrane is a region of crust composed of metamorphic rock. In unvegetated exposures you can see regional grain (the map trend of foliations), nonconformities between basement (metamorphic rock) and cover (overlying sedimentary strata), and contacts between different rock types. Explore these examples! The thumbnail images provided on this page are only to help identify tour sites. Go to wwnorton.com/studyspace to experience this flyover tour.

G7.1

Wind River Mountains, Wyoming
(Lat 42°51'30.49"N, Long 109°02'40.81"W)

At these coordinates, zoom to 750 km (466 miles), and you can see all of Wyoming. The NW-SE-trending Wind River Mountains lie in west-central Wyoming (**Image G7.1**). Zoom to 250 km (155 miles). A large fault bounds the SW side of the range. During mountain building, between 80 and 40 Ma, movement on the fault pushed up Precambrian metamorphic basement and warped overlying beds of Paleozoic and Mesozoic sedimentary rock. Erosion of sedimentary rock exposed the basement. Zoom to 40 km (25 miles) to see the contrast between tilted bedding of the sedimentary strata and knobby Precambrian gneiss. Tilt your field of view and look NW; the contrast becomes clearer (**Image G7.2**).

G7.2

Canadian Shield, East of Hudson Bay
(Lat 61°09'50.15"N, Long 76°42'43.88"W)

At these coordinates, zoom to 230 kilometers. You can see the east coast of Hudson Bay and a 150 km-wide swath of the Canadian shield, here covered only by sparse tundra vegetation (**Image G7.3**). The land surface displays the trend of metamorphic foliation. Note that a belt of E-W trending grain truncates a region of N-S-trending grain—the contact between these two provinces represents the contact between Archean rocks (to the south) and Proterozoic rocks (which form the E-W-trending band). This contact originally formed deep below the Earth's surface and was subsequently exhumed.

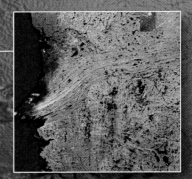

G7.3

Pilbara Craton, Western Australia
(Lat 21°14'10.00"S, Long 119°09'39.15"E)

Archean rocks underlie much of northwestern Australia, a crustal province called the Pilbara craton. Here, felsic intrusives and high-grade gneisses occur in light-colored dome-shaped bodies (circular to elliptical regions in map view). Low- to intermediate-grade "greenstone" (mafic and ultramafic metavolcanic rocks, interlayered locally with metasedimentary rocks) comprise curving belts that surround the domes. Zoom to an altitude of about 65 km (40 miles) at the coordinates given, and you can see domes surrounded by greenstone belts (**Image G7.4**). Zoom closer, and you can detect folded layering within the greenstone belts.

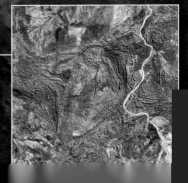

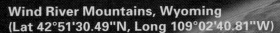

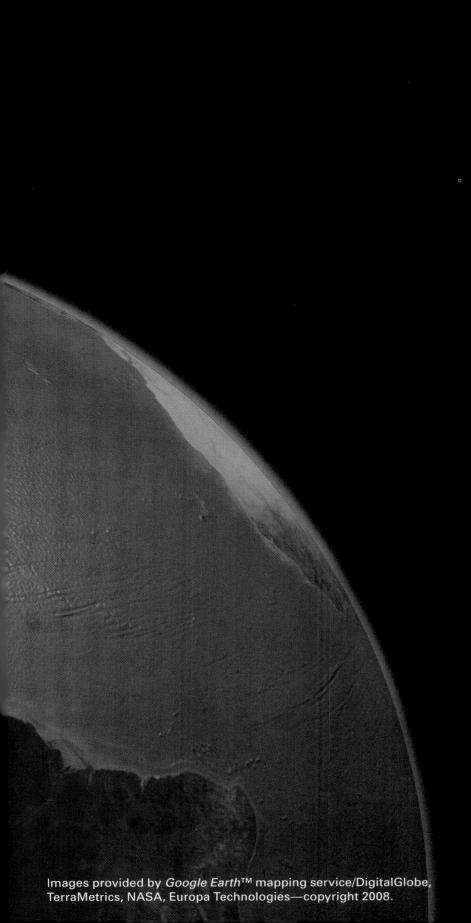

See for yourself . . .
Seismically Active Faults

In many locations, faulting during an earthquake ruptures the ground. Below, we visit examples of fault zones that have left a mark on the land surface. The journey also illustrates potential hazards associated with seismic activity. The thumbnail images provided on this page are only to help identify tour sites. Go to wwnorton. com/studyspace to experience this flyover tour.

G8.1

San Andreas Fault near Palmdale, California
(Lat 34°30'35.20"N, Long 118°1'8.12"W)

If you fly to the above coordinates and zoom up to an elevation of 350 km (220 miles), you will see the coast of California, the San Gabriel Mountains, and a triangular area of high, flatter land that comprises the western Mojave Desert (Image G8.1). The San Andreas fault is the NW-trending line at the bottom of the triangle and the Garlock fault is the NE-trending line at the top. Zoom to an elevation of 9 km (5 miles), and you'll see a stream flowing north from the San Gabriels, near Palmdale. Where the stream crosses the fault, the channel steps to the right by over 1 km (0.6 miles), due to the displacement on the fault (Image G8.2). Narrow pressure ridges and sag ponds delineate the trace of the fault. Tilt the image and reorient so you are looking northwest, along the trace of the fault. You can see canals, roads, and reservoirs that cross the fault (Image G8.3). Fly along the fault and you will get a sense of how the fault disrupts the landscape.

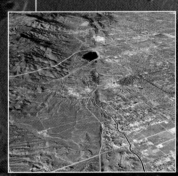

G8.2

San Andreas Fault, San Francisco
(Lat 37°34'41.53"N, Long 122°24'41.88"W)

Here, from an altitude of 20 km (13 miles), the trace of the San Andreas fault runs parallel to Highway 280, along the peninsula between San Francisco Bay on the east and the Pacific Ocean on the west. Erosion along the fault trace has produced a narrow NW-trending valley (Image G8.4). The dam straddling the fault holds back the San Andreas Reservoir. Zoom down to an elevation of 2 km (1.3 miles), tilt your view, and orient the image so that you are looking along the fault. Now, slowly fly NW, along the trace of the fault. You'll reach the end of the San Andreas Reservoir, cross housing developments, and eventually reach the ocean (Image G8.5). Fly a short distance up the coast to Lat 37°40'35.70"N, Long 122°29'35.51"W, head slightly out to sea, turn your view so you are looking east, and look back at the coast. You'll see a cliff, with waves lapping at its base (Image G8.6). This cliff formed because of tectonic uplift associated with faulting. Along the cliff, there are several spoon-shaped indentations—these are slump scarps, places where the cliff face slid down to the sea relatively recently. Places along the cliff face that are devoid of vegetation represent the most recent slumps. Slumping may be triggered by earthquakes. Note that developers have built homes up to the cliff edge, and onto the protrusions between existing slumps.

G8.3

G8.4

G8.5

G8.6

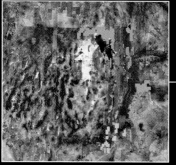

G8.7

G8.8

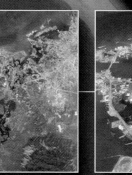

G8.9

Salt Lake City, Utah
(Lat 40°55'22.81"N, Long 111°51'52.48"W)

Fly to these coordinates just north of Salt Lake City and rise to an elevation of about 1500 km (930 miles). You can see the Basin and Range Province, an active rift that has stretched the crust between Salt Lake City and Reno (**Image G8.7**). The Colorado Plateau lies to the east, and the Sierra Nevada to the west. The Wasatch Mountains form a high ridge of uplifted land along the western edge of the Colorado Plateau. Zoom down to an elevation of 8 km (5 miles) and tilt the image (**Image G8.8**). You are looking at the Wasatch Front, an escarpment that delineates the western face of the Wasatch Mountains, just east of I-15. The Wasatch Front is a somewhat eroded fault scarp, formed when normal faulting dropped the Great Salt Lake basin down, relative to the Wasatch Mountains. Note how the scarp truncates the ridges between the streams flowing out of the mountains. Fly south along the range front to Lat 40°35'36.44"N, Long 111°47'27.01"W, and you will get another view of the Wasatch Front. Zoom to 7 km (4 miles), tilt the image to see the horizon, and look east (**Image G8.9**). This image also emphasizes how mountain building can be a long-term consequence of continued faulting.

G8.10

G8.11

Tsunami Damage, Banda Aceh
(Lat 5°30'46.75"N, Long 95°16'14.68"E)

Fly to these coordinates and zoom up to an elevation of 16 km. You will see an isthmus near the city of Banda Aceh, at the north end of Sumatra, that was attacked on both sides by the December 2004 tsunami (**Image G8.10**). The land that was submerged has turned brown, because the saltwater waves washed away most vegetation and killed the remainder. Now, zoom down to the northern part of the image (Lat 5°33'25.50"N, Long 95°17'12.52"E), drop to an elevation of 1 km (0.6 miles), and tilt the image to look north (**Image G8.11**). The seaside part of the city was totally destroyed. Rotate the image so you are looking south, and fly to about Lat 5°31'54.31"N, Long 95°17'38.25"E. Here you see the boundary between the land that was submerged and the land that was not—the unscathed land is much greener (**Image G8.12**). Note that the highway in the center of the image nearly disappears as it crosses the boundary.

See for yourself . . .
Mountains and Structures

Some of the best examples of geologic structures, and some of the most spectacular examples of mountain ranges, are in fairly remote areas. Your computer can now take you there. Note that structures control erosion, and that you can trace them out more easily in dry regions. The thumbnail images provided on this page are only to help identify tour sites. Go to wwnorton.com/studyspace to experience this flyover tour.

Erosion in the Alps
(Lat 46°12'56.50"N, Long 10°33'59.43"E)

Fly to the coordinates provided and zoom to an elevation of about 1,100 km (680 miles). You are looking down on the northern end of the Italian Peninsula. The Alps, a Cenozoic collisional mountain belt, is the snowy arc of peaks that encompasses northern Italy and parts of France and Switzerland (Image G9.1). Drop to an elevation of 47 km (29 miles), tilt the image, and look north. You'll see a typical mountain landscape—its ruggedness a consequence of erosion by glaciers and rivers (Image G9.2).

G9.1

G9.2

Mt. Everest—the Top of the World
(Lat 27°59'17.84"N, Long 86°55'18.01"E)

Ever wonder what the view looks like from the top of Mt. Everest, the highest point on Earth? Fly to the coordinates provided, zoom to 1,300 km (800 miles) and you'll be looking down on the summit of Mt. Everest, in the Himalayas, a range that formed due to the collision of India with Asia. Tilt your view so you are looking north (Image G9.3). In the foreground is the Ganges Plain, and in the distance, the dry Tibet Plateau. Zoom to 15 km (9 miles) and rotate to look southeast. You can see the surrounding peaks and glaciers and the plains in the far distance (Image G9.4). Fly around the summit to see the view in all directions.

G9.3

G9.4

Joints in Arches National Park
(Lat 38°47'52.13"N, Long 109°35'26.37"W)

Zoom to an elevation of 5 km (3 miles) at the specified coordinates in southeastern Utah and you'll see a prominent set of NW-SE-trending joints cutting horizontal white sandstone beds (Image G9.5). Weathering along the joints made them into narrow troughs. A second set of smaller NE-SW-trending joints also cut the rocks. The major joints are about 20–50 m (65–165 feet) apart. Now, fly SSE for about 5 km (3 miles) and you'll find a set of joints cutting red sandstone. Zoom to 3.3 km (2 miles), tilt the image and look NNW (Image G9.6). Arches of the park develop where holes weather through the walls between joint-controlled troughs.

G9.5

G9.6

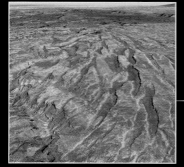

G9.7

Horsts and Grabens, Canyonlands (Lat 38°6'1.40"N, Long 109°54'40.14"W)

From an elevation of 20 km (12 miles) at these coordinates in Canyonlands Park, SE Utah, you can see a set of NE-trending grabens—troughs bordered on each side by a normal fault. Fly down to 7 km (4 miles), tilt, and look NE (Image G9.7). You'll be gazing down the axis of grabens which formed when the region southeast of the river valley stretched in a NW-SE direction. This stretching occurred because river erosion removed lateral support from a sequence of strata that lie above a weak salt layer.

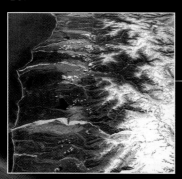

G9.8

Alpine Fault, New Zealand (Lat 43°21'15.86"S, Long 170°15'23.18"E)

At these coordinates near the west coast of South Island, New Zealand, zoom to 50 km (30 miles). You can see a high, ice-capped mountain range, called the Southern Alps, formed by motion on the Alpine Fault. Though a significant component of displacement on the fault is strike-slip, its motion also accommodates thrusting—eastern Zealand is being pushed up to the west. Tilt your image and look NE and you'll be looking along the fault (Image G9.8). The fault runs along the abrupt break in slope between the mountains to the SE, and the plains to the NW.

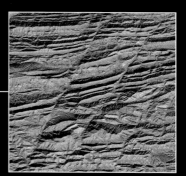

G9.9

Faults and Folds, Makran Range (Lat 25°54'0.31"N, Long 64°14'8.63"E)

Fly to the coordinates provided and zoom to 10 km (6 miles) (Image G9.9). You can see a series of ENE-trending ridges that represent beds of strata tilted by folding and faulting in the Makran Range of southern Pakistan, about 63 km (40 miles) north of the coast. Note that ridges are offset where they are cut by a series of NE-SW-trending valleys. The valleys are eroded left-lateral strike-slip faults. Zoom to 2 km (1 mile), and tilt the view so you are looking up the valley of the most prominent fault. A dry stream follows the valley—faulting broke up the rock, making it easier to erode.

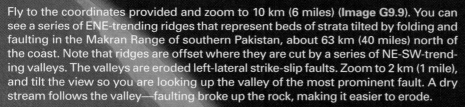

G9.10

Appalachian Fold Belt, Pennsylvania
(Lat 40°53'24.28"N, Long 77°4'28.11"W)

From an elevation of 114 km (70 miles), at these coordinates, you can see the curving traces of large folds in the Appalachian Mountains. These folds formed about 280 million years ago when compression wrinkled layers of Paleozoic sedimentary rock. The hinge zones of anticlines have been eroded; resistant ledges of sandstone define the fold limbs. Zoom to 20 km (12 miles), tilt the view, and rotate so you are looking west. You will be peering down the hinge of the folds (Image G9.10). This region is the Pennsylvania Valley and Ridge Province. Note that ridges remained forested.

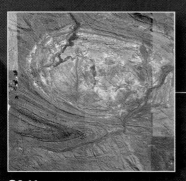

G9.11

Dome and Syncline, Western Australia
(Lat 22°52'9.85"S, Long 117°18'1.58"E)

Here we see an exposure of Precambrian crust, part of the Pilbara craton. Zoom to 50 km (30 miles) and you'll see a 30 km (18 miles)-wide circular region of light rock—a dome of granite and gneiss—surrounded by folded metasedimentary and metavolcanic rock (Image G9.11). The layering within the sedimentary and volcanic rocks stands out—resistant layers form ridges, and outlines the shape of folds. Now fly to the fold on the SW side of the dome. Reorient the view so you are looking west, and tilt the view. You are looking along the plunging hinge of a syncline.

Images provided by *Google Earth*™ mapping service/DigitalGlobe, TerraMetrics, NASA, Europa Technologies—copyright 2008.

See for yourself ...
The Strata of the Colorado Plateau

The Colorado Plateau encompasses portions of Arizona, Utah, New Mexico, and Colorado. Uplift of the region during the past 20 million years raised the land surface to elevations of 2 to 3 km. Erosion then cut down into the plateau to reveal portions of a 2 km (1.2 miles)-thick sequence of strata. These rocks, which are well exposed because of the plateau's present dryness, tell a story of changing climates and changing sea levels over the past half billion years. *Google Earth*™ allows you to fly through the canyons and along the cliffs of this magical landscape to see the record in the rocks for yourself. The thumbnail images provided on this page are only to help identify tour sites. Go to wwnorton.com/studyspace to experience this flyover tour.

Grand Canyon National Park, Arizona
(Lat 36°5'36.05"N, Long 112°7'0.88"W)

These coordinates take you to Plateau Point, on the edge of the inner gorge. You can reach the Point by trail from Park Headquarters on the south rim of the canyon. Zoom to an elevation of 4 km (2.5 miles), then tilt your view so you are looking due north (Image G10.1). You can see a stratigraphic section from Precambrian basement, through the Paleozoic to the Kaibab Limestone at the top. Try to spot the formations shown on Figure 10.11. Now, fly to the river, pivot your view, and proceed downstream (Image G10.2). You'll be able to see continuous exposure of the unconformity at the base of the Tapeats Sandstone. Note that the narrow inner gorge cuts down into hard Precambrian gneiss—the wider outer gorge cuts into sedimentary rock.

G10.1

G10.2

Vermillion Cliffs, Arizona
(Lat 36°49'4.81"N, Long 111°37'56.59"W)

Highway 89 crosses Marble Canyon at the coordinates provided. Just upstream of the bridge, rafters enter the Colorado River for a journey downstream into the Grand Canyon. Zoom to 5 km (3 miles), tilt your view, and look SSW. You can see the gash of Marble Canyon (Image G10.3) bounded by cliffs of Permian Kaibab Limestone. In the right foreground, you can see small hills surrounded by subtle rings that trace out beds of horizontal strata from the Lower Triassic Moenkopi Formation. Rotate your view to look NW and you'll see the steep escarpment of the Vermillion Cliffs, which expose a complete section of the Moenkopi, a unit composed of dark red to brown shale, siltstone, and sandstone (Image G10.4).

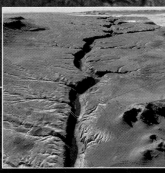

G10.3

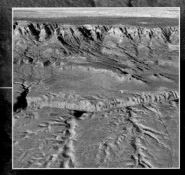

G10.4

G10.5

Monument Valley, Arizona
(Lat 36°55'25.00"N, Long 110°7'8.98"W)

The towering cliffs at these coordinates, along the northern border of Arizona, expose brilliant red strata deposited at the end of the Permian and beginning of the Triassic. Tilt your view and pivot to look due west (**Image G10.5**). Weak beds (the Organ Rock Shale) comprise a moderate slope of stairstep-like ledges at the base of the cliff. The vertical cliff consists of massive de Chelly Sandstone, and the thin-bedded unit at top is Moenkopi Shale, capped locally by beds of the Chinle Formation. Note that the Permian strata here differ from the same-age strata of Marble Canyon—this difference demonstrates that the environment of deposition can change with location at a given time.

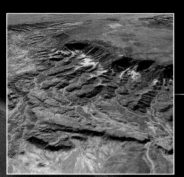

G10.6

Painted Desert, Arizona
(Lat 35°5'8.13"N, 109°47'14.78"W)

This portion of the Colorado Plateau exposes a colorful sequence of Late Triassic fluvial deposits, including a variety of varicolored sandstones and shales. Fly to the coordinates given, zoom to an elevation of 3 km (1.9 miles), tilt your view, and pivot to see the gentle escarpment (**Image G10.6**). The sequence of strata was deposited in a terrestrial setting. Numerous soil horizons formed as the strata were exposed prior to each new phase of deposition. The colors represent the paleosols—ancient soil horizons—preserved in the stratigraphic record.

G10.7

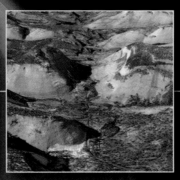

G10.8

Upper Zion Canyon Park, Utah
(Lat 37°13'34.59"N, Long 112°53'12.03"W)

Fly to these coordinates and zoom to an elevation of 10 km (6 miles). You're looking down on Route 9, traversing the upper part of Zion Park (**Image G10.7**). Note that the erosional pattern is strongly controlled by N-S-trending joints. Now zoom down to 3.5 km (2 miles), then tilt and pivot your view so you are looking west along Route 9 (**Image G10.8**). From this vantage, you see a thick sequence of cross-bedded Jurassic sandstone recording a time when much of the Colorado Plateau region was a Sahara-like desert covered by large sand dunes.

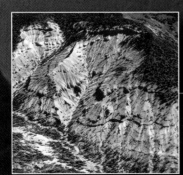

G10.9

Cedar Breaks, Utah
(Lat 37°39'12.14"N, Long 112°50'33.93"W)

At Cedar Breaks National Monument, you're at the west edge of the Colorado Plateau. A view to the west looks out over the Basin and Range Rift of western Utah. For our purpose, fly to an elevation of 2 km (1.2 miles) at the coordinates given. Tilt and pivot your view so you are gazing east, and you will see the multicolored Tertiary strata eroded along an escarpment (**Image G10.9**). Note that the section includes beds of resistant sandstone, standing out as ridges. But much of the sequence consists of weak shale, the deposits of lakes and floodplains. You would see similar rocks at Bryce Canyon, about 40 km (25 miles) to the east, but as of this writing, the imagery available for Bryce is very poor.

See for yourself…
Earth Has a History

The Earth displays the record of a long and complex geologic history, marked by the collision of continents, the rifting of crust, the building and eventual erosion of mountains, and the deposition of strata. Geologists can unravel the character and sequence of events in this history by studying accumulations of sedimentary rocks, the age and character of metamorphic and igneous rocks, and the geometry of geologic structures. Different geologic provinces display the consequences of different events. Thus, you can see the evidence of Earth's geologic history in its landscapes. The following sites provide examples from North America. The thumbnail images provided on this page are only to help identify tour sites. Go to wwnorton.com/studyspace to experience this flyover tour.

Folds of the Canadian Shield
(Lat 58°4'44.52"N, Long 69°36'31.14"W)

Let's start by looking at some of the oldest crust of this continent, which is exposed in the Canadian Shield. This region is part of the craton, a tectonically inactive region of continental crust that formed in the Precambrian. Though it had a complex geologic past, it has not endured mountain building during the last billion years—instead, it has been subjected to erosion. As a result, rocks that were formed at depths of over 10 km (6 miles) now crop out. Fly to the coordinates provided and you'll find yourself in northeastern Canada, west of Ungava Bay (Quebec). Zoom to an elevation of 70 km (45 miles) (Image G11.1). You can see a subdued landscape—none of the hills are very high, and none of the valleys are very deep. Nevertheless, the ups and downs define swirling patterns, some emphasized by dark lakes. These patterns are controlled by the folded foliation of Precambrian gneiss.

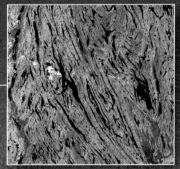

G11.1

Appalachian Mountains, Tennessee
(Lat 36°28'33.93"N, Long 83°45'28.63"W)

Fly to this locality and hover at an elevation of 100 km (62 miles). You can see a portion of the fold belt (known geographically as the Appalachian Valley and Ridge province) in northern Tennessee and southern Kentucky (Image G11.2). If you zoom down to an elevation of 20 km (12 miles) in the lower right corner of the image, tilt to see the horizon, and pivot the image so you are looking northeast, the Valley and Ridge topography is clear (Image G11.3). Resistant stratigraphic formations hold up ridges, and weaker ones erode away to form valleys. The deformation here involves Cambrian through Pennsylvania strata, so geologists deduce that the deformation is a consequence of a late Paleozoic event. This event is the Alleghanian orogeny, due to the collision of Africa with North America. This collision compressed the crust and caused it to wrinkle along its eastern margin, somewhat like a rug on a slippery floor.

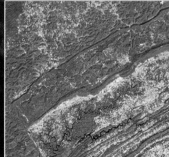

G11.2   G11.3

G11.4

Maroon Bells, Colorado
(Lat 39°4'17.65"N, Long 106°59'19.42"W)

Fly to the coordinates given and you'll find a set of peaks known as the Maroon Bells, south of Aspen, Colorado. Zoom to an elevation of 7.5 km (4.5 miles), tilt so you just see the horizon, and rotate your view so you are looking due east (Image G11.4). You can now see layering of the reddish strata that make up these mountains—fly around the mountains to get a broader perspective. Here, we have the record of two major orogenic events: (1) The late Paleozoic Ancestral Rockies event, which formed deep basins and uplifts. The reddish strata of the mountains were deposited in one of the basins. Because the event happened at roughly the same time as the Alleghanian orogeny, it might also be due to compression caused by the collision of Africa and South America; and (2) The Mesozoic/Cenozoic Laramide orogeny, during which the region was uplifted by several kilometers. Erosion subsequent to the Laramide orogeny produced the present landscape. Some valleys were carved by glaciers during the last ice age.

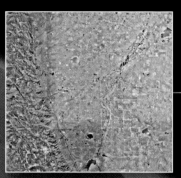

G11.5

G11.6

The Rocky Mountain Front, Colorado
(Lat 39°44'55.71"N, Long 104°59'44.74"W)

Fly to this location and zoom to an elevation of 70 km (45 miles) and you are hovering over Denver, Colorado. The field of view (Image G11.5) encompasses the Rocky Mountain front which marks the boundary between the Rocky Mountains to the west and the plains to the east. Now, zoom down to an elevation of 14 km (8.7 miles), tilt your view so you just see the horizon, and rotate so you are looking southwest. You can now see the abrupt face of the mountain front (Image G11.6). The snow-capped peak on the horizon is Mt. Evans, one of several 4.3 km (14,000 feet)-high peaks in Colorado. Fly about 12 km (7.5 miles) to the mountain front, just where highway US-285 crosses the front (Lat 39° 38'3.75"N Long 105° 10'14.28"W), drop down to an elevation of 3.5 km (2 miles), and tilt your view so you are looking north. Here you see the hogbacks at the front of the range (Image G11.7). These are asymmetric ridges composed of strata dipping toward the east at about 35°. We see two components of orogeny—uplift and deformation. These developed during the Laramide orogeny.

G11.8

G11.7

Basin and Range Rift
(Lat 38°50'30.06"N, Long 114°46'41.99"W)

Fly to these coordinates and zoom to an elevation of 150 km (93 miles). Here, you see an approximately 120 km (75 miles)-wide block of the Basin and Range Province, in Nevada, just west of the Utah border (Image G11.8). This region is a continental rift, formed during the Cenozoic, after the Laramide orogeny ceased. The tilted fault blocks now form N-S-trending narrow ranges. Because of their higher elevation, they are lightly forested. The basins between these blocks have filled with sediment. Their flat surfaces are at lower elevation and have desert climates. Streams occasionally flow along the axes of the basins. If you zoom down to 14 km (8.7 miles), tilt the image to just see the horizon, and look due east at the easternmost range in the previous view, then you can see how a range rises between two basins (Image G11.9). The floor of the basin in this view occasionally fills with water, forming a desert feature called a playa lake.

G11.9

See for yourself . . .
Sources of Energy and Mineral Resources

The geologic features associated with resources are not all visible at the ground surface, but you can see evidence of where resources are being extracted, and of how they
are being transported. The thumbnail images provided on this page are only to help
identify tour sites. Go to wwnorton.com/studyspace to experience this flyover tour.

Ackerly Oil Field near Lamesa, Texas
(Lat 32°33'18.42"N, Long 101°46'55.39"W)

Fly to this locality and zoom to 8 km (5 miles). You can see a grid of farm
fields (1.6 km, or 1 mile, wide) within which small roads lead to patches of
dirt (Image G12.1). Each patch hosts or hosted a single pump for extracting oil
from underground. In effect, an oil field viewed from the air looks like a pattern of spots connected by a web of roads. These wells are in the Ackerly Oil
field of the Permian Basin in west Texas. Drilling here began in the 1940s to
tap oil reserves in Permian sandstone beds that now lie at depths of between
1,500 and 2,500 m (5,000 and 8,200 feet). The beds tilt slightly and grade laterally into impermeable shales, so producers are exploiting stratigraphic traps.
To date, 50 million barrels of oil have been extracted from this field.

G12.1

Ghawar Oil Field, Saudi Arabia
(Lat 24°47'48.02"N, Long 49°10'42.04"E)

Fly to the coordinates provided, zoom to 4 km, and you'll find an active portion of
the field where wells feed a processing station linked to pipelines (Image G12.2).
Oil of the Ghawar field resides in an anticline trap whose reservoir rock consists
of sandstone, 2,100 m (7,000 feet) below the surface. Oil companies have drilled
3,500 wells into the anticline. So far, the field has yielded 55 *billion* barrels of oil.

G12.2

Trans-Alaska Pipeline
(Lat 64°55'38.64"N, Long 147°37'23.71"W)

The oil of the isolated North Slope oil fields near Prud-
hoe Bay moves through the Alaska pipeline across the
entire state, crossing permafrost (permanently frozen
ground) and seismic belts, to reach ports on the south
coast where it is loaded into supertankers. Fly to the
coordinates provided, zoom to 650 m (0.4 miles), tilt
your image, and look northwest. You can see the
Trans-Alaska pipeline just north of Fairbanks (Image
G12.3). At Lat 63°24'16.92"N, Long 145°44'31.72"W,
zoom to 1.5 km (0.9 miles) and tilt the image so you
are looking north. Here, you can see a section of the
pipeline near the active Denali fault (Image G12.4).
The pipeline sits on sliding pedestals that allow it to
move without breaking during an earthquake.

G12.3

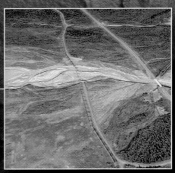

G12.4

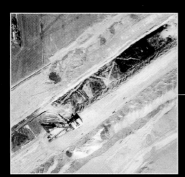

G12.5

Coal Mine, Farmersburg, Indiana
(Lat 39°14'57.59"N, Long 87°20'53.20"W)

Fly to this locality and zoom to 8 km (5 miles). You are looking at an active coal strip mine exploiting Pennsylvanian-age coal seams. Note several points about this mine. First, the active part of the mine consists of a 3.4 km (2 mile)-long trench. To produce the coal, giant bulldozers first scrape off the soil layers to expose bedrock. Then workers drill a series of holes, fill them with explosives, and blast the bedrock into fragments. The dragline scrapes out the rock and sets it aside. The dragline then piles the coal into giant trucks, which can carry as much as 150 tons in a load. Once a dragline completes a trench, bulldozers fill it back in and resurface it with topsoil. The reclaimed land can be cultivated or reforested. Zoom down to 900 m (2,950 feet) and, in the image online at the time of this writing, you can see the giant drag haul, with a tiny pickup truck nearby (**Image G12.5**).

G12.6

G12.7

Bingham Copper Mine, Utah
(Lat 40°31'14.66"N, Long 112°9'1.97"W)

Fly to the coordinates provided and zoom to 15 km (9 miles). You can see the huge open pit of the Bingham Copper mine, dug in a Cenozoic porphyritic igneous rock of the Oquirrh Mountains (**Image G12.6**). This is the largest open-pit mine in the world, and can be seen from at least 2,000 km (1,200 miles) out in space. Zoom down to 3 km (2 miles) and tilt the image to look north. You can see terraces built to stabilize the walls of the pit, and ramps that provide access for giant trucks that carry 240 tons of ore. The mine has produced 17 million tons of copper, as well as vast quantities of gold and silver. Zoom down to 3 km and tilt the image (**Image G12.7**) to get a sense of the pit's size.

G12.8

Iron Mine near Ouro Preto, Brazil
(Lat 20°9'54.45"S, Long 43°30'7.86"W)

Fly to these coordinates and zoom to 3 km (1.8 miles), tilt the image, and look north (**Image G12.8**). You are looking into an iron mine in eastern Brazil. The ore consists of banded iron formation, deposited 2.2 billion years ago. Subsequently, these rocks were deformed and metamorphosed, and now include bands of magnetite and/or hematite. The pools of iron-colored water are settling ponds. The ore from this mine is transported to factories in Brazil, China, and Japan.

See for yourself...
Examples of Landslides

Landslides cause distinctive scars on the Earth's surface, such as head scarps, hummocky ground, and/or disrupted vegetation. Examples of these features occur worldwide. The thumbnail images provided on this page are only to help identify tour sites. Go to wwnorton.com/studyspace to experience this flyover tour.

G13.1

Slumping Hawaii (Lat 19°18'21.37"N, Long 155°5'2.55"W)

These coordinates take you to Volcanoes National Park, along the south coast of the Big Island of Hawaii, where active lava flows reach the sea. From an altitude of 5 km (3 miles), you can see black lava flows that descended over the Holei Pali cliff (Image G13.1). The 250 m (820 foot)-high cliff is the head scarp of the Hilina slump. Here, we see only the upper 2.5 km (1.5 miles) of the slump on land. The rest is submerged and extends about 40 km (25 miles) offshore. Tilt your view to look east along the scarp (Image G13.2).

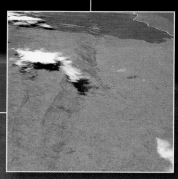

G13.2

1964 Slumps, Anchorage, Alaska (Lat 61°12'52.06"N, Long 149°54'31.39"W)

During the 1964 earthquake, several coastal slumps moved. Fly to the coordinates given and zoom to 2 km (1.2 miles)—you are hovering over downtown Anchorage (Image G13.3). Note the curving head scarps near the shore marking slumps that sank by about 7 m (21 feet) during the 1964 quake. Fly SW along the coast until you see the airport. At Lat 61°11'51.51"N, Long 149°58'25.11"W you'll find a crescent-shaped band of woods along the shore (Image G13.4). This is Earthquake Park, what's left of the Turnagain Heights neighborhood, destroyed by 1964 slumping.

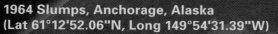

G13.3

G13.4

Roadside Landslide, Pacific Coast Highway, California (Lat 34°3'31.44"N, Long 118°58'15.85"W)

Portions of the Pacific Coast Highway lie at the base of unstable slopes—landslides down these slopes block the road. At the coordinates provided, from an elevation of 800 m (2,600 feet), you can see the subtle scar left by one of these landslides, and the remnants of debris cleared by bulldozers along the base of the cliff (Image G13.5).

G13.5

Portuguese Bend Landslide, California (Lat 33°44'46.94"N, Long 118°22'7.83"W)

The land here started to slump about 37,000 years ago. At the coordinates given, from an altitude of 5 km (3 miles), you can see the head scarp of the 3 km (1.8 mile)-wide slump (Image G13.6). In the 1950s, developers built a housing project on the hummocky land of the slump. The southeastern portion of the slump began moving again, ultimately destroying 150 homes.

G13.6

G13.7

La Conchita Mudslide, California (Lat 34°21'50.29"N, Long 119°26'46.85"W)

At La Conchita, developers built a housing project in the narrow strip of land between the cliffs and the shore. When heavy rains drench this landscape, the slope becomes unstable. In 1995 and again in 2005, mudflows buried homes. From an elevation of 2 km (1.2 miles), you can see the development (Image G13.7). Note that the cliff bounds a broad, uplifted terrace. This terrace was a wave-cut platform formed at sea level. Tectonic movements caused the terrace to rise. In addition to the mudslide, which also destroyed a road traversing the hill, you can see gullies and canyons that have been incised by successive floods.

Mt. Saint Helens Lahars, Washington (Lat 46°10'36.32"N, Long 122°10'4.44"W)

Fly to these coordinates, and you are over the southeast flank of Mt. Saint Helens. Zoom to an elevation of 12 km (7.5 miles), tilt your view, and pivot to look NW (Image G13.8). The southeast flank was not destroyed by the 1980 explosion, but rather was the site of numerous lahars. The gray lahars look like spills of gravy down the side of the mountain. In the stream valleys, the lahars traveled much farther.

G13.8

Gros Ventre Slide, Wyoming (Lat 43°37'29.78"N, Long 110°33'3.49"W)

Fly to the coordinates provided. From an elevation of 10 km (6 miles), you can see Lower Slide Lake, northeast of Jackson, Wyoming (Image G13.9). This lake formed in 1925, when the Gros Ventre slide tumbled down Sheep Mountain and blocked the Gros Ventre River. Zoom to 6 km (3.7 miles), tilt your view, and look southeast (Image G13.10). The scarp left by the landslide is obvious, as is the hummocky ground on top of the debris.

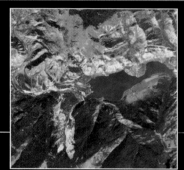

G13.9

G13.10

Debris Fall, Yungay, Peru (Lat 9°8'53.88"S, Long 77°43'20.35"W)

Remnants of the debris fall of 1970 that buried parts of Yungay, Peru, can still be seen decades later. Fly to the latitude and longitude provided, zoom to 15 km (9 miles), tilt to see the horizon, and look NE (Image G13.11). Note the towering, ice-covered peak of Nevado Huascarán in the far distance, and the valley down which the debris flow traveled in the middle distance. In the foreground, the broad apron of debris is now farmed, but it does not host many homes.

G13.11

Rockfalls, Canyonlands National Park, Utah (Lat 38°29'52.65"N, Long 110°0'50.20"W)

In southeastern Utah, at the coordinates provided, you can see evidence of rockfalls along the banks of the Green River. Zoom to an elevation of 3 km (1.8 miles) and look down (Image G13.12). Note that NW-trending joints cut the whitish, resistant Permian sandstone layer that forms the top of the mesa. When the cliff breaks away along a joint, blocks topple down the slope and break up. Zoom down to 2 km (1.2 miles), tilt to see the horizon, and pivot so you are looking east (Image G13.13). You can see a large rockfall that sent debris almost down to the bottom of the slope.

G13.12

G13.13

See for yourself . . .
Fluvial Landscapes

Streams stand out in the landscape, for they carve intricate shapes as their waters flow from high areas to low. In this Geotour, we visit a variety of landscapes whose features are a consequence of deposition and erosion in a fluvial setting. The thumbnail images provided on this page are only to help identify tour sites. Go to wwnorton.com/studyspace to experience this flyover tour.

Deep Gorge in the Himalayas (Lat 28°9'40.37"N, Long 85°25'53.21"E)

Fly to these coordinates, zoom to 10 km (6 miles), and tilt your image so you are looking up the valley to the east (Image G14.1). You are 52 km (42 miles) NNE of Katmandu, Nepal. This image illustrates how the upper reaches of a mountain stream have steep gradients and flow down deep gorges. If the rock walls of the gorge are relatively weak, the stream valley attains a V-shape.

G14.1

Headward Erosion, Canyonlands, Utah (Lat 38°17'58.16"N, Long 109°50'18.76"W)

At these coordinates, in Canyonlands National Park, fly to an altitude of 8 km (5 miles). You see a side canyon carved into horizontal strata by intermittent tributaries that flow into the Colorado River (Image G14.2). An abrupt scarp marks the upstream limit of each tributary. As the streams erode, they cut into the land by headward erosion, so the scarp migrates upstream. Zoom down to 4 km (2.5 miles), tilt the image, and look southwest to better visualize the concept of headward erosion (Image G14.3).

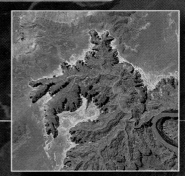

G14.2

G14.3

Meanders Along Rio Ucayali, Peru (Lat 7°27'1.80"S, Long 75°2'48.92"W)

At this locality, 560 km (347 miles) NNE of Lima, Peru, you will find a nearly horizontal landscape on the east side of the Andes. Because of the low gradient here, the Rio Ucayali has become a meandering stream. This view, from an elevation of 100 km (62 miles), shows numerous meanders within a flood plain (Image G14.4). You also see abandoned meanders, point bars, and oxbow lakes. You can find very similar meanders along the Mississippi River (USA) at Lat 31°29'29.67"N, Long 91°38'0.23"W.

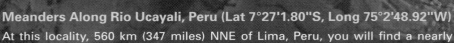
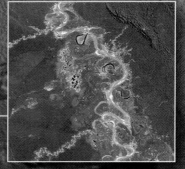

G14.4

Incised Meanders, Canyonlands, Utah (Lat 38°13'56.26"N, Long 109°49'27.64"W)

Return to Canyonlands National Park, Utah, zoom to 6.5 km (4 miles), and tilt the image to look northeast. You can see a meander of the Colorado River that has almost cut through the meander neck (Image G14.5). The meanders incised down through bedrock, and now lie at the floor of a steep-walled canyon.

G14.5

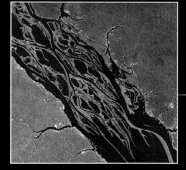
G14.6

Mid-Stream Bars, Rio Negro, Brazil
(Lat 2°43'31.82"S, Long 60°42'50.50"W)

At this locality, along the Rio Negro (a major tributary of the Amazon), 105 km (65 miles) upstream of Manaus, the river is about 15 km (9 miles) wide (**Image G14.6**). From an altitude of 60 km (37 miles), you can see numerous mid-stream bars of sediment that emerge from the river. These bars will be submerged during a flood, and may be shifted by the force of flowing water.

G14.7

G14.8

Point Bars, Trinity River, Texas
(Lat 30°10'2.73"N, Long 94°48'58.24"W)

Fly to this locality and zoom to an elevation of 14 km (8.7 miles). You are looking at a reach of the Trinity River in the vicinity of Dayton Lakes, 70 km (43 miles) to the northeast of Houston (**Image G14.7**). Along the inner curve of each meander, a bright tan point bar of sand has formed. In the green landscape surrounding the river, you will also be able to spot a number of abandoned meanders and oxbow lakes. Zoom down to 3 km (1.9 miles) and tilt so you see the horizon (**Image G14.8**). The bars and abandoned meanders stand out in the landscape.

G14.9

Trellis Drainage Pattern, Pennsylvania
(Lat 40°45'31.75"N, Long 76°50'45.41"W)

Fly to these coordinates and zoom to 150 km (93 miles). You can see the Valley and Ridge Province of Pennsylvania, produced by the erosion of folded strata. Here, we see a trellis drainage pattern—tributaries flow down valleys and intersect the Susquehanna at right angles (**Image G14.9**). The Susquehanna itself cuts across ridges.

G14.10

Radial Drainage, Mt. Shasta
(Lat 41°24'17.85"N, Long 122°11'42.87"W)

Looking straight down from an elevation of 25 km (15.5 miles), you can see the peak of Mt. Shasta, a volcano in California. The streams that flow down its slopes, away from the peak, define a radial network (**Image G14.10**). This pattern resembles the spokes of a wheel.

Dendritic Drainage, Pennsylvania (Lat 41°30'29.73"N, Long 78°14'18.04"W)

Fly to these coordinates, zoom to an elevation of 30 km (18.6 miles), and you are looking down on a region of the Appalachian Plateau near the town of Emporium, in north-central Pennsylvania. The strata here consist of flat-lying beds of shale. Streams cutting down into this fairly homogeneous and soft substrate carve a dendritic network (**Image G14.11**). This means that the pattern of streams and tributaries resembles the pattern of veins on a leaf.

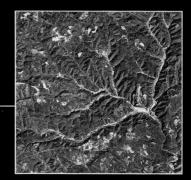
G14.11

See for yourself . . .
Landscapes of Oceans and Coasts

You can find a huge variety of different coastal and bathymetric features with *Google Earth*™ or comparable programs. In addition to visiting the stops identified in this Geotour, simply fly to a coast, tilt your view, and set the image in motion—you'll be amazed at what you can see! The thumbnail images provided on this page are only to help identify tour sites. Go to wwnorton.com/studyspace to experience this flyover tour.

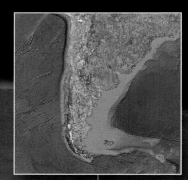

G15.2

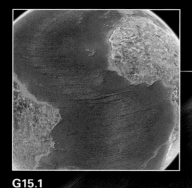

G15.1

Mid-Atlantic Ridge Bathymetry
(Lat 0°3'51.75"N, Long 24°47'35.66"W)

Fly to the coordinates provided and zoom to 8000 km (5000 miles) (Image G15.1). You can see the entire South Atlantic Ocean. Examine the bathymetry of the sea floor. Ridge segments and transforms are obvious.

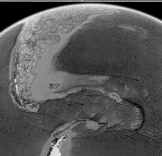

G15.3

Southern South America Bathymetry
(Lat 39°22'15.03"S, Long 65°9'1.31"W)

Zoom to 5000 km (3100 miles) at the coordinates provided. You are looking down on southern South America (Image G15.2). Compare the bathymetry of the west coast (an active convergent margin) to that of the east coast (a passive margin). Where does a broad continental shelf occur? Now, fly south to Lat 54°29'29.28"S, Long 52°1'4.39"W, zoom to 4500 km (2800 miles), and tilt to look north. You can see the Scotia Sea, between South America and Antarctica, and the Scotia volcanic arc (Image G15.3). Can you identify the pattern of plate boundaries?

G15.4

Coral Reefs, Pacific (Lat 16°47'26.92"S, Long 150°58'1.27"W)

Fly to the coordinates given and zoom to 6 km (3.7 miles). You are looking down on the coral reef and lagoon of Huahine, one of the Society Islands in the South Pacific. Zoom down to 1.5 km (1 mile), tilt, and look north (Image G15.4). Note how the reef absorbs wave energy and protects the shore. Now, fly west to Lat 16°37'25.65"S, Long 151°29'32.80"W, and zoom to 25 km (15 miles). You are looking down on the island of Tahaa, an atoll surrounded by an offshore reef (Image G15.5).

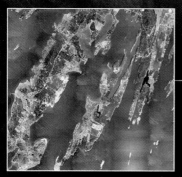

Rocky Coast, Maine
(Lat 43°46'34.77"N, Long 69°58'28.58"W)

Fly to these coordinates and zoom to 10 km (6 miles). From this viewpoint, you can see the rocky coast of Maine (Image G15.6). Islands here were carved by glaciers during the last ice age. Post-ice age sea-level rise submerged the landscape.

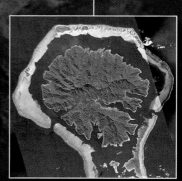

G15.5

G15.6

G15.7

Offshore Bar, near Cape Hatteras, North Carolina (Lat 35°8'24.05"N, Long 75°53'34.62"W)

At the coordinates provided, zoom to 28 km (17 miles) and tilt your view so that you are looking NW. You can see an offshore sand bar (**Image G15.7**). The bar formed due to redistribution of sand by waves and currents. Contrast this coast with the one you saw in Maine.

G15.8

Chicago Shoreline (Lat 41°54'59.76"N, Long 87°37'36.41"W)

This locality, on the west shore of Lake Michigan, provides an excellent example of how the moving sand of a beach interacts with groins. Zoom to 2.5 km (1.5 miles) and look down on Chicago's beach front (**Image G15.8**). Note that sand has accumulated in asymmetric wedges due to a south-flowing current.

Fjords of Norway (Lat 60°53'56.09"N, Long 5°12'31.84"E)

At the coordinates provided, zoom to 80 km (50 miles) and you see the intricate coast of Norway (**Image G15.9**). During the last ice age, glaciers carved deep valleys which filled with sea water when the ice melted and sea level rose. Zoom to 13 km (8 miles), tilt, and look north to see the steep cliffs bordering fjords. Now fly to Lat 61°12'34.22"N, Long 5°7'18.20"E. Here, the image resolution is better, and if you zoom to 5 km (3 miles), tilt, and look east, you can look inland, along the axis of a fjord (**Image G15.10**).

G15.9

G15.10

Organic Coast, Florida (Lat 25°7'48.94"N, Long 80°59'3.18"W)

Much of southern Florida, the Everglades, is a vast swamp, through which fresh water flows slowly south. Here, the transition from land to sea is gradual. Fly to the coordinates provided and look down from 40 km (25 miles) (**Image G15.11**). You see lagoons, swamps and bars, all colored by vegetation. Darker green areas are mangrove thickets. Zoom to 700 m (2300 feet) and tilt to look north (**Image G15.12**). You can see how the thickets stabilize the shore.

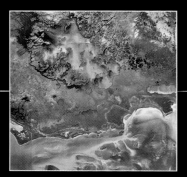

G15.11

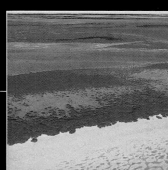

G15.12

Sandspit, Cape Cod (Lat 42°2'40.10"N, Long 70°11'31.46"W)

Cape Cod formed when a glacier deposited a 200 m (600 foot)-thick layer of sand and gravel to form a ridge called a moraine about 18,000 years ago. Eventually, sea level rose and currents began to transport sand along the shore. Fly to the coordinates provided and look down from an elevation of 20 km (12 miles) (**Image G15.13**). Note the large sand spit that protects Provincetown. On the north shore, you can see a beach and dunes.

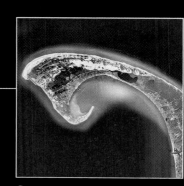

G15.13

See for yourself . . .
Evidence of Groundwater

You can see features at the surface of the Earth that result from the rise of groundwater at springs or due to pumping, or from underground erosion by groundwater. The thumbnail images provided on this page are only to help identify tour sites. Go to wwnorton.com/studyspace to experience this flyover tour.

Irrigation in the Saudi Desert (Lat 25°18'18.36"N, Long 44°28'5.72"E)

Fly to the coordinates provided and zoom to 70 km (43 miles). You can see the desert of Saudi Arabia. Note the sand dunes in the upper right corner (Image G16.1). Porous rock holds large reserves of groundwater under the surface. Each circle in the image is a field, irrigated by groundwater pumped from a well at the center. Move to Lat 25° 17'53.71"N, Long 44°28'27.17"E and zoom to an elevation of 2 km (1.2 miles) to see a field close up (Image G16.2).

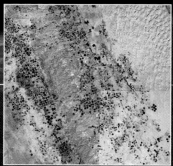

G16.1

G16.2

Water Table, Minnesota (Lat 46°34'49.28"N, Long 95°43'55.13"W)

Fly to this locality and look down from 12 km (7.5 miles) (Image G16.3). A large number of ponds fill kettles, depressions that formed at the end of the ice age when blocks of ice buried in glacial sediment melted away. The surface of the lakes represents the surface of the water table. Dry land exists where the ground surface lies slightly above the water table.

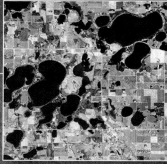

G16.3

Everglades, Florida (Lat 25°40'12.62"N, Long 80°42'1.11"W)

Zoom to 15 km (9 miles) at the coordinates provided, and you are hovering over the Everglades (Image G16.4). This swamp is actually a broad, slowly moving stream that flows from Lake Okeechobee to the tip of Florida. During the wet season, the water table effectively lies above ground, so only vegetated hammocks rise above the water table. But during the dry season, the water table lies below the surface. Why are the hammocks elongated? Some of the water has been diverted into canals, one of which occurs on the right side of this view.

G16.4

G16.5

Hot Springs, Yellowstone (Lat 44°27'44.65"N, Long 110°51'17.98"W)

Zoom to an elevation of 3 km (2 miles) at the coordinates provided and you're looking at Black Sand Basin, near Old Faithful geyser in Yellowstone National Park (Image G16.5). Magma heats the groundwater of Yellowstone. The hot water rises and either fills pools or erupts at geysers. This view shows examples of the pools. Bacteria grow in the pools and add color.

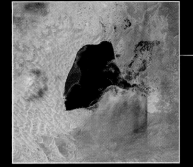

G16.6

Desert Oasis, Egypt (Lat 30°14'41.27"N, Long 28°55'58.95"E)

The Sahara Desert contains vast seas of sand. Yet here and there, ponds or vegetated areas fill depressions where springs bring groundwater to the surface. You can see such an oasis from 10 km (6 miles) at the locality specified above (**Image G16.6**).

Sinking Venice, Italy (Lat 45°26'9.01"N, Long 12°20'0.02"E)

The city of Venice covers an island in the Venice Lagoon. From 20 km (12 miles) red roofs in the city stand out (**Image G16.7**). Zoom to 2.5 km (1.5 miles), and you can see that the main thoroughfares of Venice are canals; the city has subsided in part due to groundwater removal (**Image G16.8**).

G16.7

G16.8

G16.9

Sinkholes in Florida (Lat 28°38'36.16"N, Long 81°22'18.69"W)

Fly to the coordinates provided and zoom to 6 km (4 miles). You are looking at the landscape around Interstate Highway 4. Here, the circular ponds fill sinkholes (**Image G16.9**). Superficially, they look like the kettles of Minnesota, but they form in a totally different manner—sinkholes develop when an underground cavern collapses. The water surface represents the water table.

Karst Landscape, Puerto Rico (Lat 18°24'40.75"N, Long 66°25'7.08"W)

At this locality, a view looking down from 10 km (6 miles) reveals part of an extensive karst region (**Image G16.10**). 36 km (22.3 miles) WSW of this area (Lat 18°20'39.41"N, Long 66°45'10.10"W), a sinkhole was converted into the Arecibo Observatory radio telescope (**Image G16.11**).

G16.10

G16.11

Tower Karst of China (Lat 22°32'29.97"N, Long 107°26'18.17"E)

In southeastern China, at these coordinates, a view from 20 km (12 miles) illustrates a tower karst landscape (**Image G16.12**). The thick layer of limestone now at the surface was a cave network. When the roof of the network collapsed, rooms became sinkholes, and tunnels became valleys. Joints localized solution.

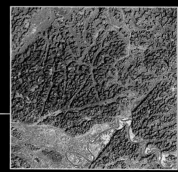

G16.12

See for yourself . . .
Desert Landscapes

In the desert, the unvegetated land surface lies exposed, so landform shapes are clear and dramatic. High-resolution images allow you to see cliffs, dunes, fans, and even individual boulders. Here, we tour several deserts to see the variety of landscapes. Zoom out into space to remind yourself of where you are on the globe. The thumbnail images provided on this page are only to help identify tour sites. Go to wwnorton.com/studyspace to experience this flyover tour.

G17.1

Sand Dunes, Namib Desert, Africa (Lat 24°43'57.60"S, Long 15°26'59.90"E)

Fly to the coordinates provided, zoom to 20 km (12 miles), and you will see a reddish sand sea in the Namib Desert on the west coast of southern Africa. The color is due to oxidized iron. A dry wash cut through the sand and washed it away. Zoom to 8 km (5 miles), and tilt to look NNE to see star dunes, formed by shifting winds (Image G17.1).

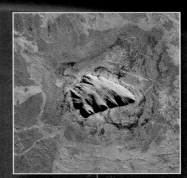

G17.2

Uluru (Ayers Rock), Australia (Lat 25°20'52.42"S, Long 131°2'24.98"E)

At these coordinates, from 10 km (6 miles), you can see the famous inselberg, Uluru (Ayers Rock) in the "red center" of Australia (Image G17.2). Zoom to an elevation of 4 km (2.5 miles), tilt your view, and look southeast. Note that the beds of sandstone comprising Uluru have a vertical dip—they are one limb of a syncline. If you fly to Lat 24°10'33.30"S, Long 132°55'19.57"E, and zoom to 20 km (12 miles), you can see another fold whose shape, without vegetation cover, stands out in the landscape (Image G17.3).

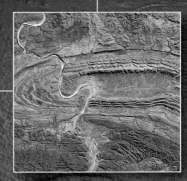

G17.3

G17.4

Atacama Desert near Chala, Peru (Lat 15°50'40.47"S, Long 74°16'8.10"W)

The Atacama Desert, on the west coast of South America, exists because a cold current flows northwards just offshore. At the coordinates provided, from an elevation of 15 km (9 miles), you can see the stark dryness of the land, right up to the shore of the blue Pacific Ocean (Image G17.4).

Tarim Basin of Western China (Lat 37°39'56.20"N, Long 82°19'7.20"E)

Fly to this location, zoom to an elevation of 60 km (37 miles), and tilt the view to look north (Image G17.5). The desert stretches as far as the eye can see. Sand dunes cover the land surface except in the green, irrigated oasis of the middle distance.

G17.5

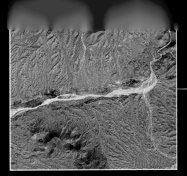

G17.6

Dry Wash in Arizona (Lat 33°39'49.26"N, Long 111°35'10.25"W)

Stream channels in deserts fill only intermittently. At other times, the water table lies below the surface of the ground so the channels are dry. At these coordinates, from 9 km (5.6 miles), you see the floor of a dry wash (**Image G17.6**). Adjacent slopes display a dendritic drainage pattern, because even though water flows only intermittently, water is still the dominant agent of erosion.

G17.7

Urbanizing a Desert, Tucson, Arizona (Lat 32°15'53.09"N, Long 110°51'19.87"W)

Much of southern Arizona lies in the Sonoran Desert. There's hardly any surface water, but the warm climate attracts a growing population, and the region's cities have ballooned. Zoom to 5 km (3 miles) and tilt your view to look north (**Image G17.7**). Suburbs of Tucson sprawl to the foot of the Catalina Mountains; intense watering keeps the golf courses green.

Playa in Death Valley, California (Lat 36°13'40.64"N, Long 116°45'55.28"W)

Here at Badwater, in Death Valley, you see the lowest point in North America (85 m or 280 feet below sea level). Zoom to an elevation of 3 km (2 miles), tilt your view, and look east (**Image G17.8**). The white surface is the salt crust of a dry playa lake. Note an alluvial fan building out onto the playa.

G17.8

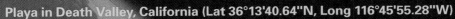

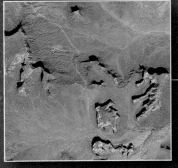

G17.9

Buttes of Monument Valley, Arizona (Lat 36°57'6.66"N, Long 110°5'19.66"W)

Fly to this locality and look down from an elevation of 10 km (6 miles) (**Image G17.9**). You are seeing the buttes and chimneys of Monument Valley. These landforms consist of massive sandstone eroding by cliff retreat when joint-bounded walls of rocks collapse.

Sand Sea in the Sahara, Egypt (Lat 26°53'42.61"N, Long 26°5'5.02"E)

The Sahara is the largest continuous desert on Earth. At this locality, from an elevation of 10 km (6 miles), you see two scales of dunes (**Image G17.10**). Very large longitudinal dunes trend from NNW to SSE. On top of these, smaller dunes developed.

See for yourself...
Glacial Landscapes

On *Google Earth*™ imagery, you can easily spot glacial erosional and depositional features. And at high latitudes and/or elevations, you can still see glacial ice. The thumbnail images provided on this page are only to help identify tour sites. Go to wwnorton.com/studyspace to experience this flyover tour.

Continental Glacier, Antarctica
(Lat 89°59'52.49"S, Long 144°50'31.19"E)

At the southern end of the Earth, zoom 8,000 km (5,000 miles) and look down to see the continental glacier that covers Antarctica (Image G18.1). Rock crops out only along the Transantarctic Mountains and along the coast. The Antarctic Peninsula protrudes toward the southern tip of South America.

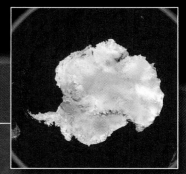

G18.1

Greenland and the Arctic Ocean
(Lat 85°4'55.25"N, Long 35°8'59.46"W)

At the northern end of the Earth, from 11,000 km (6,800 miles) out in space, you can see the Arctic Ocean. *Google Earth*™ does not generally show sea ice, so it's clear that, unlike the southern end, the northern end of the Earth is an ocean. A continental glacier covers the large island of Greenland (Image G18.2).

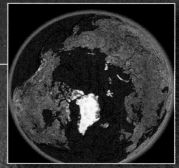

G18.2

Southern Tip of Greenland (Lat 60°18'48.05"N, Long 44°28'1.45"W)

From an elevation of 250 km (155 miles), you can see the ice cap, valley glaciers, and fjords of southernmost Greenland (Image G18.3). Zoom to 15 km (9 miles), at Lat 60°9'20.86"N, Long 44°3'34.30"W, tilt your view to look north, and you can see the effects of glacial erosion in detail.

G18.3

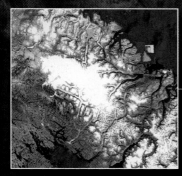

Baffin Island, Canada
(Lat 67°3'38.31"N, Long 65°33'2.88"W)

Zoom to 400 km (250 miles) and your view encompasses a portion of Baffin Island. You can see a small ice cap, and several valley glaciers ending in fjords (Image G18.4). Move to Lat 67°8'18.30"N, Long 65°0'56.99"W and zoom to 60 km (37 miles) to see valley glaciers "draining" the ice cap. Zoom down to 15 km (9 miles), tilt, and look upstream along a glacier (Image G18.5). Note how medial moraines form from the lateral moraines of valley glaciers that merge.

G18.4   G18.5

G18.6

Matterhorn, Switzerland (Lat 46°5'15.04"N, Long 7°38'2.72"E)

Ice-age glaciers sculpted the peaks that we see today. At these coordinates, zoom to 10 km (6 miles), tilt, and look south. You can see a glacially carved valley with the Matterhorn at its end. Multiple cirque formation produced the peak (**Image G18.6**).

G18.7

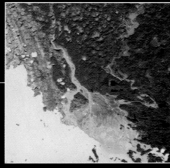

G18.8

Malaspina Glacier Area, Alaska (Lat 59°57'5.59"N, Long 140°32'54.35"W)

At the coordinates provided, zoom to 65 km (40 miles) and tilt your view to look NNE. You are gazing at the Malaspina Glacier, a piedmont glacier spreading out on a coastal plain (**Image G18.7**). A terminal moraine outlines the toe of the glacier. Note that the flow of the ice produced complex folds traced out by the pattern of debris on the glacier's surface. Zoom down to 8 km (5 miles) at Lat 59°50'35.76"N, Long 140°1'27.22"W. You can see knob-and-kettle topography and a braided outwash stream (**Image G18.8**).

Glaciated Peaks, Montana (Lat 48°54'53.37"N, Long 113°50'51.04"W)

At the coordinates provided, zoom to 7 km (4 miles), tilt your view, and look east. You can see the peak of Mt. Cleveland at the head of two cirques separated by a sharp arête (**Image G18.9**). Move a little south, to Lat 48°45'35.56"N, Long 113°37'47.02"W, and you are in Glacier National Park. Zoom to 8.5 km (5.3 miles) and tilt your view to look southwest, and you can see a U-shaped valley with a tarn at its head (**Image G18.10**).

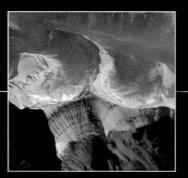

G18.9

G18.10

G18.11

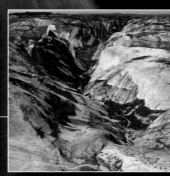

G18.12

Sierra Nevada, California (Lat 37°44'33.69"N, Long 119°16'21.06"W)

Fly to the coordinates provided and zoom to 7 km (4 miles). You can see several cirques and arêtes in the high peaks of the Sierra Nevada (**Image G18.11**). Fly to Lat 37°46'23.82"N, Long 119°30'10.68"W, zoom to 5 km (3 miles), and tilt to look SSW. You are looking down the Yosemite Valley, a U-shaped valley at the heart of Yosemite National Park. Waterfalls spill out of hanging valleys into the main valley (**Image G18.12**).

G18.13

Pluvial Lake Shore, Salt Lake City, Utah (Lat 40°31'18.45"N, Long 111°49'54.14"W)

Zoom to 3.5 km (2 miles), tilt, and look south to see the terrace of sediment that marks the beach of a pluvial lake, Lake Bonneville (**Image G18.13**). This lake covered the entire area of what is now Salt Lake City during the Pleistocene.

See for yourself . . .
Aspects of Global Change

Human activity has become a major agent of change on planet Earth. The consequences of this activity are clearly evident in the character of landscapes on the surface. The thumbnail images provided on this page are only to help identify tour sites. Go to wwnorton.com/studyspace to experience this flyover tour.

Clear Cutting the Amazon, Brazil
(Lat 4°3'51.60"S, Long 54°50'47.96"W)

At these coordinates, and an elevation of 100 km (62 miles), you can see swaths of clear cutting (Image G19.1). The pattern reflects access roads. Move east to Lat 3°51'7.92"S, Long 54°11'12.93"W, where the resolution is better, and from 5 km (3 miles), you can see clear-cut areas converted into rangeland (Image G19.2).

Long-Term Deforestation, Brazil
(Lat 16°41'18.46"S, Long 43°58'0.76"W)

At the coordinates provided, the view from 7 km (4 miles) shows hilly land (Image G19.3). Deforestation here took place decades ago. Forest remains only in valleys. Fly east to the Brazilian highlands at Lat 19°20'0.43"S, Long 41°14'25.05"W. From 25 km (15 miles), you can see that only the highest hills remain forested (Image G19.4). Yearly burning keeps the forest from regrowing.

Edge of the Everglades, Florida
(Lat 25°33'41.34"N, Long 80°35'2.75"W)

Looking down from 45 km (28 miles) at these coordinates, you see the transition between the natural Everglades on the west and intensively cultivated farmland on the east (Image G19.5). This radical change in landscape affects many aspects of the Earth System. Fly to Lat 25°53'27.04"N, Long 80°8'25.19"W, zoom to 6 km (4 miles) and look down at what was once a sandbar, but is now a forest of concrete (Image G19.6).

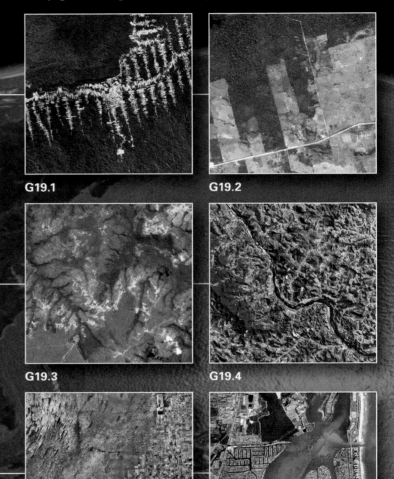

G19.1 G19.2

G19.3 G19.4

G19.5 G19.6

G19.7

Urbanizing the Desert, Arizona
(Lat 33°33'48.47"N, Long 111°48'33.37"W)

At the coordinates provided, the view from 5 km (3 miles) shows the edge of a Phoenix suburb (Image G19.7). Note that construction has nearly eliminated natural drainage. The transformation of the Sonoran Desert of Arizona into cityscapes and farmland requires huge amounts of water. Some of the water comes from pumping underground aquifers, and some from canals. A canal traverses this image.

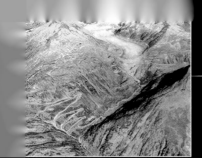

G19.8

Receding Glacier, Switzerland
(Lat 46°34'21.13"N, Long 8°22'40.61"E)

Most glaciers have been retreating at accelerating paces in recent decades. Here, near the town of Gletsch, zoom to 7 km (4 miles), tilt, and look northeast (Image G19.8). You can see the empty U-shaped valley that was occupied by ice a century ago.

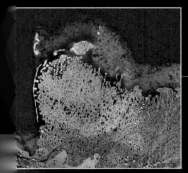
G19.9

Melting Permafrost, Siberia
(Lat 73°20'58.61"N, Long 124°43'59.76"E)

Here, on the Arctic coast of Siberia, a landscape that had been permafrost appears to be melting. Zoom to 200 km (125 miles), and you can see the pockmarks on land and beneath shallow coastal waters of the Arctic (Image G19.9). This image gives the impression that recent sea-level rise has submerged the landscape.

Intense Urbanization, Tokyo, Japan
(Lat 35°41'3.94"N, Long 139°48'35.69"E)

The view of Tokyo from an elevation of 1.5 km (1 mile) shows the results of complete transformation of a natural coastal area into a human-controlled environment (Image G19.10). Vehicles generate pollutants, the concrete absorbs and retains heat, and geologic materials have been transformed into buildings. Zoom out to 20 km (12 miles) to see how human activity has affected the coastline.

G19.10

G19.11

Village and Fields, China
(Lat 35°7'3.20"N, Long 114°27'16.86"E)

From an elevation of 5 km (3 miles), at these coordinates, you see a landscape in central China that has been transformed from forest to farmland (Image G19.11). Every square meter that does not lie beneath a house or road has been turned into a farm field.

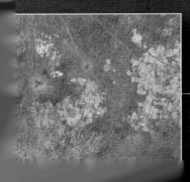

Desertification of the Sahel, Africa
(Lat 13°1'46.67"N, Long 26°3'49.89"E)

Here, the view from 12 km (7 miles) reveals a dry landscape (Image G19.12). The radial pattern of paths on the left side of the image comes from cattle walking to a tiny watering hole. Cattle compact the land, preventing water infiltration, farming removes nutrients, and grazing causes devegetation. As a result, the landscape has become more desert-like.

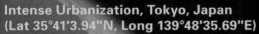

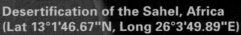

a'a' A lava flow with a rubbly surface.

abandoned meander A meander that dries out after being cut off.

ablation The removal of ice at the toe of a glacier by melting, sublimation (the evaporation of ice into water vapor), and/or calving.

abrasion The process in which one material (such as sand-laden water) grinds away at another (such as a stream channel's floor and walls).

absolute age Numerical age (the age specified in years).

absolute plate velocity The movement of a plate relative to a fixed point in the mantle.

abyssal plain A broad, relatively flat region of the ocean that lies at least 4.5 km below sea level.

Acadian orogeny A convergent mountain-building event that occurred around 400 million years ago, during which continental slivers accreted to the eastern edge of the North American continent.

accreted terrane A block of crust that collided with a continent at a convergent margin and stayed attached to the continent.

accretionary coast A coastline that receives more sediment than erodes away.

accretionary orogen An orogen formed by the attachment of numerous buoyant slivers of crust to an older, larger continental block.

accretionary prism A wedge-shaped mass of sediment and rock scraped off the top of a downgoing plate and accreted onto the overriding plate at a convergent plate margin.

acid mine runoff A dilute solution of sulfuric acid, produced when sulfur-bearing minerals in mines react with rainwater, that flows out of a mine.

acid rain Precipitation in which air pollutants react with water to make a weak acid that then falls from the sky.

active continental margin A continental margin that coincides with a plate boundary.

active fault A fault that has moved recently or is likely to move in the future.

active sand The top layer of beach sand, which moves daily because of wave action.

active volcano A volcano that has erupted within the past few centuries and will likely erupt again.

adiabatic cooling The cooling of a body of air or matter without the addition or subtraction of thermal energy (heat).

adiabatic heating The warming of a body of air or matter without the addition or subtraction of heat.

aerosols Tiny solid particles or liquid droplets that remain suspended in the atmosphere for a long time.

aftershocks The series of smaller earthquakes that follow a major earthquake.

agents of erosion Components of the Earth System (e.g., moving glaciers, rivers, wind) that cause erosion to occur.

air The mixture of gases that make up the Earth's atmosphere.

air-fall tuff Tuff formed when ash settles gently from the air.

air mass A body of air, about 1,500 km across, that has recognizable physical characteristics.

air pressure The push that air exerts on its surroundings.

albedo The reflectivity of a surface.

Alleghenian orogeny The orogenic event that occurred about 270 million years ago when Africa collided with North America.

alloy A metal containing more than one type of metal atom.

alluvial fan A gently sloping apron of sediment dropped by an ephemeral stream at the base of a mountain in arid or semi-arid regions.

alluvium Sorted sediment deposited by a stream.

alluvium-filled valley A valley whose floor fills with sediment.

amber Hardened (fossilized) ancient sap or resin.

amphibolite facies A set of metamorphic mineral assemblages formed under intermediate pressures and temperatures.

amplitude The height of a wave from crest to trough.

Ancestral Rockies The late Paleozoic uplifts of the Rocky Mountain region; they eroded away long before the present Rocky Mountains formed.

angiosperm A flowering plant.

angle of repose The angle of the steepest slope that a pile of uncemented material can attain without collapsing from the pull of gravity.

angularity The degree to which grains have sharp or rounded edges or corners.

angular unconformity An unconformity in which the strata below were tilted or folded before the unconformity developed; strata below the unconformity therefore have a different tilt than strata above.

anhedral grains Crystalline mineral grains without well-formed crystal faces.

annual probability The likelihood, expressed as a percentage, that an event (e.g., a flood of a given size) will happen in a given year.

Antarctic bottom water mass The mass of cold, dense water that sinks along the coast of Antarctica.

antecedent stream A stream that cuts across an uplifted mountain range; the stream must have existed before the range uplifted and must then have been able to downcut as fast as the land was rising.

anthracite coal Shiny black coal formed at temperatures between 200° and 300°C. A high-rank coal.

anticline A fold with an arch-like shape in which the limbs dip away from the hinge.

anticyclone The clockwise flow of air around a high-pressure mass.

Antler orogeny The Late Devonian mountain-building event in which slices of deep-marine strata were pushed eastward, up and over the shallow-water strata on the western coast of North America.

anvil cloud A large cumulonimbus cloud that spreads laterally at the tropopause to form a broad, flat top.

aphanitic A textural term for fine-grained igneous rock.

apparent polar-wander path A path on the globe along which a magnetic pole appears to have wandered over time; in fact, the continents drift, while the magnetic pole stays fairly fixed.

aquiclude Sediment or rock that transmits no water.

aquifer Sediment or rock that transmits water easily.

aquitard Sediment or rock that does not transmit water easily and therefore retards the motion of the water.

archaea A kingdom of "old bacteria," now commonly found in extreme environments like hot springs. (Also called "archaeobacteria.")

Archean Eon The middle Precambrian Eon.

Archimedes' principle The mass of the water displaced by a block of material equals the mass of the whole block of material.

arête A residual knife-edge ridge of rock that separates two adjacent cirques.

argillaceous sedimentary rock Sedimentary rock that contains abundant clay.

arkose A clastic sedimentary rock composed of sand-sized grains that include quartz and feldspar.

arroyo The channel of an ephemeral stream; dry wash; wadi.

artesian spring A place where groundwater gushes out of the ground under pressure.

artesian well A well in which water rises on its own.

ash fall Ash that falls to the ground out of an ash cloud.

ash flow An avalanche of ash that tumbles down the side of an explosively erupting volcano.

ash (volcanic) Very fine particles erupted by a volcano; they consist of glass shards formed by the rapid cooling of lava droplets erupted into the air, and/or of tiny rock particles blasted into the air by an explosion.

assimilation The process of magma contamination in which blocks of wall rock fall into a magma chamber and dissolve.

asteroid One of many millions of small, rocky, and/or metallic objects that orbit the Sun, consisting of fragments of once-larger planetesimals, or chunks of protoplanetary material; most lie in the region between Mars and Jupiter.

asthenosphere The layer of the mantle that lies between 100–150 km and 350 km deep; the asthenosphere is relatively soft and can flow when acted on by force.

atm A unit of air pressure that approximates the pressure exerted by the atmosphere at sea level.

atmosphere A layer of gases that surrounds a planet.

atoll A coral reef that develops around a circular reef surrounding a lagoon.

atomic number The number of protons in the nucleus of a given element.

atomic weight The number of protons plus the number of neutrons in the nucleus of a given element. (Also known as atomic mass.)

aurora australis The same phenomenon as the aurora borealis, but in the Southern Hemisphere.

aurora borealis A ghostly curtain of varicolored light that appears across the night sky in the Northern Hemisphere when charged particles from the Sun interact with the ions in the ionosphere.

avalanche A turbulent cloud of debris mixed with air that rushes down a steep hill slope at high velocity; the debris can be rock and/or snow.

avalanche chute A downslope hillside pathway along which avalanches repeatedly fall, consequently clearing the pathway of mature trees.

avulsion The process in which a river overflows a natural levee and begins to flow in a new direction.

axial plane The imaginary surface that encompasses the hinges of successive layers of a fold.

axial surface In the context of folds, this is the imaginary plane that contains the hinge lines of successive layers in the fold; it is the surface that divides a fold into its two separate limbs.

axial trough A narrow depression that runs along a mid-ocean ridge axis.

backscattered light Atmospheric scattered sunlight that returns back to space.

backshore zone The zone of beach that extends from a small step cut by high-tide swash to the front of the dunes or cliffs that lie farther inshore.

backswamp The low marshy region between the bluffs and the natural levees of a floodplain.

backwash The gravity-driven flow of water back down the slope of a beach.

bajada An elongate wedge of sediment formed by the overlap of several alluvial fans emerging from adjacent valleys.

Baltica A Paleozoic continent that included crust that is now part of today's Europe.

banded-iron formation (BIF) Iron-rich sedimentary layers consisting of alternating gray beds of iron oxide and red beds of iron-rich chert.

bar (1) A sheet or elongate lens or mound of alluvium; (2) a unit of air-pressure measurement approximately equal to 1 atm.

barchan dune A crescent-shaped dune whose tips point downwind.

barrier island An offshore sand bar that rises above the mean high-water level, forming an island.

barrier reef A coral reef that develops offshore, separated from the coast by a lagoon.

basal sliding The phenomenon in which meltwater accumulates at the base of a glacier, so that the mass of the glacier slides on a layer of water or on a slurry of water and sediment.

basalt A fine-grained mafic igneous rock.

base level The lowest elevation a stream channel's floor can reach at a given locality.

basement Older igneous and metamorphic rocks making up the Earth's crust beneath sedimentary cover.

basement uplift Uplift of basement rock by faults that penetrate deep into the continental crust.

base metals Metals that are mined but not considered precious. Examples include copper, lead, zinc, and tin.

basin A fold or depression shaped like a right-side-up bowl.

Basin and Range Province A broad, Cenozoic continental rift that has affected a portion of the western United States in Nevada, Utah, and Arizona; in this province, tilted fault blocks form ranges, and alluvium-filled valleys are basins.

batholith A vast composite, intrusive, igneous rock body up to several hundred kilometers long and 100 km wide, formed by the intrusion of numerous plutons in the same region.

bathymetric map A map illustrating the shape of the ocean floor.

bathymetric profile A cross section showing ocean depth plotted against location.

bathymetry Variation in depth.

bauxite A residual mineral deposit rich in aluminum.

baymouth bar A sandspit that grows across the opening of a bay.

beach A band of sand or gravel that lies along a coastline; wave action affects the profile and sediment size of a beach.

beach drift The gradual migration of sand along a beach.

beach erosion The removal of beach sand caused by wave action and long-shore currents.

beach face A steeply concave part of the foreshore zone formed where the swash of the waves actively scours the sand.

bed A distinct layer of sedimentary strata.

bedding Layering or stratification in sedimentary rocks.

bed load Large particles, such as sand, pebbles, or cobbles, that bounce or roll along a stream bed.

bedrock Rock still attached to the Earth's crust.

Bergeron process Precipitation involving the growth of ice crystals in a cloud at the expense of water droplets.

berm A horizontal or landward-sloping terrace in the backshore zone of a beach that receives sediment during a storm.

big bang A cataclysmic explosion that scientists suggest represents the formation of the Universe; before this event, all matter and all energy were packed into one volumeless point.

biochemical sedimentary rock Sedimentary rock formed from material (such as shells) produced by living organisms.

biodiversity The number of different species that exist at a given time.

biofuel Gas or liquid fuel made from plant material (biomass). Examples of biofuel include alcohol (from fermented sugar), biodiesel from vegetable oil, and wood.

biogeochemical cycle The exchange of chemicals between living and nonliving reservoirs in the Earth System.

bioremediation   The injection of oxygen and nutrients into a contaminated aquifer to foster the growth of bacteria that will ingest or break down contaminants.

biosphere   The region of the Earth and atmosphere inhabited by life; this region stretches from a few km below the Earth's surface to a few km above.

bioturbation   The mixing of sediment by burrowing animals such as clams and worms.

bituminous coal   Dull, black intermediate-rank coal formed at temperatures between 100° and 200°C.

black-lung disease   Lung disease contracted by miners from the inhalation of too much coal dust.

black smoker   The cloud of suspended minerals formed where hot water spews out of a vent along a mid-ocean ridge; the dissolved sulfide components of the hot water instantly precipitate when the water mixes with seawater and cools.

blind fault   A fault that does not intersect the ground surface.

block   In the context of igneous materials, a block is a chunk of igneous rock blasted out of a volcano.

blocking temperature   The temperature below which isotopes in a mineral are no longer free to move, so the radiometric clock starts.

blowout   A deep, bowl-like depression scoured out of desert terrain by a turbulent vortex of wind.

blue shift   The phenomenon in which a source of light moving toward you appears to have a higher frequency.

body waves   Seismic waves that pass through the interior of the Earth.

bog   A wetland dominated by moss and shrubs.

bolide   A solid extraterrestrial object such as a meteorite, comet, or asteroid that explodes in the atmosphere.

bomb   A frozen blob of rock formed when magma ejected from a volcano freezes in flight; bombs are typically streamlined.

bornhardt   An inselberg with a loaf geometry, like that of Uluru (Ayers Rock) in central Australia.

Bowen's reaction series   The sequence in which different silicate minerals crystallize during the progressive cooling of a melt.

braided stream   A sediment-choked stream consisting of entwined subchannels.

breaker   A water wave in which water at the top of the wave curves over the base of the wave.

breakwater   An offshore wall, built parallel or at an angle to the beach, that prevents the full force of waves from reaching a harbor.

breccia   Coarse sedimentary rock consisting of angular fragments; or rock broken into angular fragments by faulting.

breeder reactor   A nuclear reactor that produces its own fuel.

brine   Water that is not fresh but is less salty than seawater; brine may be found in estuaries.

brittle deformation   The cracking and fracturing of a material subjected to stress.

brittle-ductile transition (brittle-plastic transition)   The depth above which materials behave brittlely and below which materials behave ductilely (plastically); this transition typically lies between a depth of 10 and 15 km in continental crustal rock, and 60 m deep in glacial ice.

buoyancy   The upward force acting on a less dense object immersed or floating in denser material.

burial metamorphism   Metamorphism due to the increase in temperature and pressure in a rock when it has been buried to a depth of several kilometers.

butte   A medium-sized, flat-topped hill in an arid region.

caldera   A large circular depression with steep walls and a fairly flat floor, formed after an eruption as the center of the volcano collapses into the drained magma chamber below.

caliche   A solid mass created where calcite cements the soil together (also called calcrete).

calving   The breaking off of chunks of ice at the edge of a glacier.

Cambrian explosion   The remarkable diversification of life, indicated by the fossil record, that occurred at the beginning of the Cambrian Period.

Canadian Shield   A broad, low-lying region of exposed Precambrian rock in the Canadian interior.

canyon   A trough or valley with steeply sloping walls, cut into the land by a stream.

capacity   The total volume of sediment that a stream can carry.

capillary fringe   The thin subsurface layer in which water molecules seep up from the water table by capillary action to fill pores.

carbonate rocks   Rocks containing calcite and/or dolomite.

carbon-14 (^{14}C) dating   A radiometric dating process that can tell us the age of organic material containing carbon originally extracted from the atmosphere.

cast   Sediment that preserves the shape of a shell it once filled before the shell dissolved or mechanically weathered away.

catabatic winds   Strong winds that form at the margin of a glacier where the warmer air above ice-free land rises and the cold, denser air from above the glaciers rushes in to take its place.

catastrophic change   Change that takes place either instantaneously or rapidly in geologic time.

catchment   *Drainage network.*

cement   Mineral material that precipitates from water and fills the spaces between grains, holding the grains together.

cementation   The phase of lithification in which cement, consisting of minerals that precipitate from groundwater, partially or completely fills the spaces between clasts and attaches each grain to its neighbor.

Cenozoic   The most recent era of the Phanerozoic Eon, lasting from 65 Ma up until the present.

chalk   Very fine-grained limestone consisting of weakly cemented plankton shells.

change of state   The process in which a material changes from one phase (liquid, gas, or solid) to another.

channel   A trough dug into the ground surface by flowing water.

channeled scablands   A barren, soil-free landscape in eastern Washington, scoured clean by a flood unleashed when a large glacial lake drained.

chatter marks   Wedge-shaped indentations left on rock surfaces by glacial plucking.

chemical sedimentary rock   Sedimentary rock made up of minerals that precipitate directly from water solution.

chemical weathering   The process in which chemical reactions alter or destroy minerals when rock comes in contact with water solutions and/or air.

chert   A sedimentary rock composed of very fine-grained silica (cryptocrystalline quartz).

Chicxulub crater   A circular excavation buried beneath younger sediment on the Yucután peninsula; geologists suggest that a meteorite landed there 65 Ma.

chimney   (1) A conduit in a magma chamber in the shape of a long vertical pipe through which magma rises and erupts at the surface; (2) an isolated column of strata in an arid region.

cinder cone   A subaerial volcano consisting of a cone-shaped pile of tephra whose slope approaches the angle of repose for tephra.

cinders   Fragments of glassy rock ejected from a volcano.

cirque   A bowl-shaped depression carved by a glacier on the side of a mountain.

cirrus cloud   A wispy cloud that tapers into delicate, feather-like curls.

clast   A fragment of detritus (e.g., a sand grain or a pebble).

clastic (detrital) sedimentary rock   Sedimentary rock consisting of cemented-together detritus derived from the weathering of preexisting rock.

cleavage   (1) The tendency of a mineral to break along preferred planes; (2) a type of foliation in low-grade metamorphic rock.

cleavage planes   A series of surfaces on a crystal that form parallel to the weakest bonds holding the atoms of the crystal together.

cliff (or scarp) retreat   The change in the position of a cliff face caused by erosion.

climate   The average weather conditions, along with the range of conditions, of a region over a year.

cloud   A mist of tiny water droplets in the sky.

coal   An organic sedimentary rock formed from plant debris.

coalbed methane   Natural gas produced during the diagenesis of coal.

coal gasification   The transformation, by human activity, of coal into various gases.

coal rank   A measurement of the carbon content of coal; higher-rank coal forms at higher temperatures.

coal reserve   The quantities of discovered, but not yet mined, coal in sedimentary rock of the continents.

coal swamp   A swamp whose oxygen-poor water allows thick piles of woody debris to accumulate; this debris transforms into coal upon deep burial.

coast   The boundary region between land and the sea.

coastal plain   Low-relief regions of land adjacent to the coast.

coastal wetland   A vegetated area extending from the shore inland.

cold front   The boundary at which a cold air mass pushes underneath a warm air mass.

collision   The process of two buoyant pieces of lithosphere converging and squashing together.

columnar jointing   A type of fracturing that yields roughly hexagonal columns of basalt; columnar joints form when a dike, sill, or lava flow cools.

comet   A ball of ice and dust, probably remaining from the formation of the Solar System, that orbits the Sun.

compaction   The phase of lithification in which the pressure of the overburden on the buried rock squeezes out water and air that was trapped between clasts, and the clasts press tightly together.

competence   The ability of flowing water to carry sediment, as represented by the largest clast size that the stream can transport.

composite volcano   *Stratovolcano.*

compositional banding   A type of metamorphic foliation, found in gneiss, defined by alternating bands of light and dark minerals.

compressibility   The degree to which a material's volume changes in response to squashing.

compression   A push or squeezing felt by a body.

compressional waves   Waves in which particles of material move back and forth parallel to the direction in which the wave itself moves.

conchoidal fractures   Smoothly curving, clamshell-shaped surfaces along which materials with no cleavage planes tend to break.

condensation   The process of gas molecules linking together to form a liquid.

condensation nuclei   Preexisting solid or liquid particles, such as aerosols, onto which water condenses during cloud formation.

cone of depression   The downward-pointing, cone-shaped surface of the water table in a location where the water table is experiencing drawdown because of pumping at a well.

confined aquifer   An aquifer that is separated from the Earth's surface by an overlying aquitard.

conglomerate   Very coarse-grained sedimentary rock consisting of rounded clasts.

consuming boundary   *Convergent plate boundary.*

contact   The boundary surface between two rock bodies (as between two stratigraphic formations, between an igneous intrusion and adjacent rock, between two igneous rock bodies, or between rocks juxtaposed by a fault).

contact metamorphism   *Thermal metamorphism.*

contaminant plume   A cloud of contaminated groundwater that moves away from the source of the contamination.

continental crust   The crust beneath the continents.

continental divide   A highland separating drainage that flows into one ocean from drainage that flows into another.

continental drift hypothesis   The idea that continents have moved and are still moving slowly across the Earth's surface.

continental glacier   A vast sheet of ice that spreads over thousands of square kilometers of continental crust.

continental-interior desert   An inland desert that develops because by the time air masses reach the continental interior, they have lost all of their moisture.

continental lithosphere   Lithosphere topped by continental crust; this lithosphere reaches a thickness of 150 km.

continental margin   A continent's coastline.

continental rift   A linear belt along which continental lithosphere stretches and pulls apart.

continental rifting   The process by which a continent stretches and splits along a belt; if it is successful, rifting separates a larger continent into two smaller continents separated by a divergent boundary.

continental rise   The sloping sea floor that extends from the lower part of the continental slope to the abyssal plain.

continental shelf   A broad, shallowly submerged region of a continent along a passive margin.

continental slope   The slope at the edge of a continental shelf, leading down to the deep sea floor.

continental volcanic arc   A long curving chain of subaerial volcanoes on the margin of a continent adjacent to a convergent plate boundary.

contour interval   The vertical elevation difference between two adjacent contour lines on a topographic map.

contour lines   Lines on a map along which a parameter has a constant value; for example, all points along a contour line on a topographic map are at the same elevation.

control rod   Rods that absorb neutrons in a nuclear reactor and thus decrease the number of collisions between neutrons and radioactive atoms.

convection   Heat transfer that results when warmer, less dense material rises while cooler, denser material sinks.

convergence zone   A place where two surface air flows meet so that air has to rise.

convergent margin   *Convergent plate boundary.*

convergent plate boundary   A boundary at which two plates move toward each other so that one plate sinks (subducts) beneath the other; only oceanic lithosphere can subduct.

coral reef   A mound of coral and coral debris forming a region of shallow water.

core   The dense, iron-rich center of the Earth.

core-mantle boundary   An interface 2,900 km below the Earth's surface separating the mantle and core.

Coriolis effect   The deflection of objects, winds, and currents on the surface of the Earth owing to the planet's rotation.

cornice   A huge, overhanging drift of snow built up by strong winds at the crest of a mountain ridge.

correlation   The process of defining the age relations between the strata at one locality and the strata at another.

cosmic rays   Nuclei of hydrogen and other elements that bombard the Earth from deep space.

cosmology   The study of the overall structure of the Universe.

country rock (wall rock)   The preexisting rock into which magma intrudes.

crater   (1) A circular depression at the top of a volcanic mound; (2) a depression formed by the impact of a meteorite.

craton   A long-lived block of durable continental crust commonly found in the stable interior of a continent.

cratonic platform   A province in the interior of a continent in which Phanerozoic strata bury most of the underlying Precambrian rock.

creep   The gradual downslope movement of regolith.

crevasse   A large crack that develops by brittle deformation in the top 60 m of a glacier.

critical mass   A sufficiently dense and large mass of radioactive atoms in which a chain reaction happens so quickly that the mass explodes.

cross bed   A lamination inclined to the main bedding; it represents the slip face of a layer deposited in a current.

cross-cutting relation   A means of determining the relative age of rock by looking at which rock or structure cuts another; the feature that has been cut is older.

cross section   A diagram depicting the geometry of materials underground as they would appear on an imaginary vertical slice through the Earth.

crude oil   Oil extracted directly from the ground.

crust   The rock that makes up the outermost layer of the Earth.

crustal root   Low-density crustal rock that protrudes downward beneath a mountain range.

crystal   A single, continuous piece of a mineral bounded by flat surfaces that formed naturally as the mineral grew.

crystal face   The flat surface of an euhedral mineral grain.

crystal form   The geometric shape of a crystal, defined by the arrangement of crystal faces.

crystal habit   The general shape of a crystal or cluster of crystals that grew unimpeded.

crystal lattice   The orderly framework within which the atoms or ions of a mineral are fixed.

crystal structure   The internal arrangement of atoms or ions within a crystal.

crystalline   Containing a crystal lattice.

crystalline igneous rock   A rock that solidifies from a melt and consists of interlocking crystals.

cuesta   An asymmetric ridge formed by tilted layers of rock, with a steep cliff on one side cutting across the layers and a gentle slope on the other side; the gentle slope is parallel to the layering.

cumulonimbus cloud   A rain-producing puffy cloud.

cumulus cloud   A puffy, cotton-ball-shaped cloud.

current   (1) A well-defined stream of ocean water; (2) the moving flow of water in a stream.

cut bank   The outside bank of the channel wall of a meander, which is continually undergoing erosion.

cutoff   A straight reach in a stream that develops when erosion eats through a meander neck.

cyanobacteria   Blue-green algae; a type of archaea.

cycle   A series of interrelated events or steps that occur in succession and can be repeated, perhaps indefinitely.

cyclone   (1) The counterclockwise flow of air around a low-pressure mass; (2) the equivalent of a hurricane in the Indian Ocean.

cyclothem   A repeated interval within a sedimentary sequence that contains a specific succession of sedimentary beds.

Darcy's law   A mathematical equation stating that a volume of water, passing through a specified area of material at a given time, depends on the material's permeability and hydraulic gradient.

daughter isotope   The decay product of radioactive decay.

day   The time it takes for the Earth to spin once on its axis.

debris avalanche   An avalanche in which the falling debris consists of rock fragments and dust.

debris fall   A sudden collapse of disaggregated material on a steep slope.

debris flow   A downslope movement of mud mixed with larger rock fragments.

debris slide   A sudden downslope movement of material consisting only of regolith.

decompression melting   The kind of melting that occurs when hot mantle rock rises to shallower depths in the Earth so that pressure decreases while the temperature remains unchanged.

deep current   An ocean current at a depth greater than 100 m.

deep-focus earthquake   An earthquake that occurs at a depth between 300 and 670 km; below 670 km, earthquakes do not happen.

deflation   The process of lowering the land surface by wind abrasion.

deformation   A change in the shape, position, or orientation of a material, by bending, breaking, or flowing.

dehydration   Loss of water.

delta   A wedge of sediment formed at a river mouth when the running water of the stream enters standing water, the current slows, the stream loses competence, and sediment settles out.

delta plain   The low, swampy land on the surface of a delta.

delta-plain flood   A flood in which water submerges a delta plain.

dendritic network   A drainage network whose interconnecting streams resemble the pattern of branches connecting to a deciduous tree.

dendrochronologist   A scientist who analyzes tree rings to determine the geologic age of features.

density   Mass per unit volume.

denudation   The removal of rock and regolith from the Earth's surface.

deposition   The process by which sediment settles out of a transporting medium.

depositional landform   A landform resulting from the deposition of sediment where the medium carrying the sediment evaporates, slows down, or melts.

desert   A region so arid that it contains no permanent streams except for those that bring water in from elsewhere, and has very sparse vegetation cover.

desertification   The process of transforming nondesert areas into desert.

desert pavement   A mosaic-like stone surface forming the ground in a desert.

desert varnish   A dark, rusty-brown coating of iron oxide and magnesium oxide that accumulates on the surface of rock in an arid environment.

detachment fault   A nearly horizontal fault at the base of a fault system.

detritus   The chunks and smaller grains of rock broken off outcrops by physical weathering.

dewpoint temperature   The temperature at which air becomes saturated so that dew can form.

diagenesis   Changes that happen to sediment or sedimentary rock during and subsequent to lithification but at temperatures less than that of the lowest-grade metamorphism.

differential stress   A condition causing a material to experience a push or pull in one direction of a greater magnitude than the push or pull in another direction; in some cases, differential stress can result in shearing.

differential weathering   What happens when different rocks in an outcrop undergo weathering at different rates.

differentiation   In the context of planet formation, the process by which a planet separates into a metallic core and a rocky mantle very early in its history.

diffraction   The splitting of light into many tiny beams that interfere with one another.

digital elevation map (DEM)   A map depicting the topography of an area, produced by a computer that uses 3-D data; the computer assigns a location (lat/long) and elevation to each point on a map.

dike   A tabular (wall-shaped) intrusion of rock that cuts across the layering of country rock.

dimension stone   An intact block of granite or marble to be used for architectural purposes.

dip   The angle at which a layer tilts, relative to horizontal; the angle is measured in an imaginary vertical plane that trends perpendicular to the strike.

dipole   A magnetic field with a north and south pole, like that of a bar magnet.

dipole field (for Earth) The part of the Earth's magnetic field, caused by the flow of liquid iron alloy in the outer core, that can be represented by an imaginary bar magnet with a north and south pole.

dip-slip fault A fault in which sliding occurs up or down the slope (dip) of the fault.

dip slope A hill slope underlain by bedding parallel to the slope.

disappearing stream A stream that intersects a crack or sinkhole leading to an underground cavern, so that the water disappears into the subsurface and becomes an underground stream.

discharge The volume of water in a conduit or channel passing a point in one second.

discharge area A location where groundwater flows back up to the surface, and may emerge at springs.

disconformity An unconformity parallel to the two sedimentary sequences it separates.

displacement (or **offset**) The amount of movement or slip across a fault plane.

disseminated deposit A hydrothermal ore deposit in which ore minerals are dispersed throughout a body of rock.

dissolution A process during which materials dissolve in water.

dissolved load Ions dissolved in a stream's water.

distillation column A vertical pipe in which crude oil is separated into several components.

distributaries The fan of small streams formed where a river spreads out over its delta.

divergence zone A place where sinking air separates into two flows that move in opposite directions.

divergent plate boundary A boundary at which two lithosphere plates move apart from each other; they are marked by mid-ocean ridges.

diversification The development of many different species.

DNA (deoxyribonucleic acid) The complex molecule, shaped like a double helix, containing the code that guides the growth and development of an organism.

doldrums A belt with very slow winds along the equator.

dolostone A type of carbonate sedimentary rock that contains significant quantities of dolomite.

dome Folded or arched layers with the shape of an overturned bowl.

Doppler effect The phenomenon in which the frequency of wave energy appears to change when a moving source of wave energy passes an observer.

dormant volcano A volcano that has not erupted for hundreds to thousands of years but does have the potential to erupt again in the future.

downcutting The process in which water flowing through a channel cuts into the substrate and deepens the channel relative to its surroundings.

downdraft Downward-moving air.

downgoing plate (or **slab**) A lithosphere plate that has been subducted at a convergent margin.

downslope force The component of the force of gravity acting in the downslope direction.

downslope movement The tumbling or sliding of rock and sediment from higher elevations to lower ones.

downwelling zone A place where near-surface water sinks.

drag fold A fold that develops in layers of rock adjacent to a fault during or just before slip.

drainage divide A highland or ridge that separates one watershed from another.

drainage network (or **basin**) An array of interconnecting streams that together drain an area.

drawdown The phenomenon in which the water table around a well drops because the users are pumping water out of the well faster than it flows in from the surrounding aquifer.

drilling mud A slurry of water mixed with clay that oil drillers use to cool a drill bit and flush rock cuttings up and out of the hole.

dripstone Limestone (travertine in a cave) formed by the precipitation of calcium carbonate out of groundwater.

drop stone A rock that drops to the sea floor once the iceberg that was carrying the rock melts.

drumlin A streamlined, elongate hill formed when a glacier overrides glacial till.

dry-bottom (polar) glacier A glacier so cold that its base remains frozen to the substrate.

dry wash The channel of an ephemeral stream when empty of water.

dry well (1) A well that does not supply water because the well has been drilled into an aquitard or into rock that lies above the water table; (2) a well that does not yield oil, even though it has been drilled into an anticipated reservoir.

ductile (plastic) deformation The bending and flowing of a material (without cracking and breaking) subjected to stress.

dune A pile of sand generally formed by deposition from the wind.

dust storm An event in which strong winds hit unvegetated land, strip off the topsoil, and send it skyward to form rolling dark clouds that block out the Sun.

dynamic metamorphism Metamorphism that occurs as a consequence of shearing alone, with no change in temperature or pressure.

dynamo A power plant generator in which water or wind power spins an electrical conductor around a permanent magnet.

dynamothermal metamorphism Metamorphism that involves heat, pressure, and shearing.

earthquake A vibration caused by the sudden breaking or frictional sliding of rock in the Earth.

earthquake belt A relatively narrow, distinct belt of earthquakes that defines the position of a plate boundary.

earthquake engineering The design of buildings that can withstand shaking.

earthquake zoning The determination of where land is relatively stable and where it might collapse because of seismicity.

Earth System The global interconnecting web of physical and biological phenomena involving the solid Earth, the hydrosphere, and the atmosphere.

ebb tide The falling tide.

eccentricity cycle The cycle of the gradual change of the Earth's orbit from a more circular to a more elliptical shape; the cycle takes around 100,000 years.

ecliptic The plane defined by a planet's orbit.

ecosystem An environment and its inhabitants.

eddy An isolated, ring-shaped current of water.

effusive eruption An eruption that yields mostly lava, not ash.

Ekman spiral The change in flow direction of water with depth, caused by the Coriolis effect.

Ekman transport The overall movement of a mass of water, resulting from the Eckman spiral, in a direction 90° to the wind direction.

elastic-rebound theory The concept that earthquakes occur when rock elastically bends until it fractures; the fracturing generates earthquake energy and decreases the elastic energy stored in the rock.

elastic strain A change in shape of a material that disappears instantly when stress is removed.

electromagnet An electrical device that produces a magnetic field.

electron microprobe A laboratory instrument that can focus a beam of electrons on a small part of a mineral grain in order to create a signal that defines its chemical composition.

El Niño The flow of warm water eastward from the Pacific Ocean that reverses the upwelling of cold water along the western coast of South America and causes significant global changes in weather patterns.

embayment A low area of coastal land.

emergent coast A coast where the land is rising relative to sea level or sea level is falling relative to the land.

end moraine (terminal moraine) A low, sinuous ridge of till that develops when the terminus (toe) of a glacier stalls in one position for a while.

energy The capacity to do work.

energy resource Something that can be used to produce work; in a geologic context, a material (such as oil, coal, wind, flowing water) that can be used to produce energy.

eon The largest subdivision of geologic time.

epeirogenic movement The gradual uplift or subsidence of a broad region of the Earth's surface.

epeirogeny An event of epeirogenic movement; the term is usually used in reference to the formation of broad mid-continent domes and basins.

ephemeral (intermittent) stream A stream whose bed lies above the water table, so that the stream flows only when the rate at which water enters the stream from rainfall or meltwater exceeds the rate at which water infiltrates the ground below.

epicenter The point on the surface of the Earth directly above the focus of an earthquake.

epicontinental sea A shallow sea overlying a continent.

epoch An interval of geologic time representing the largest subdivision of a period.

equant A term for a grain that has the same dimensions in all directions.

equatorial low The area of low pressure that develops over the equator because of the intertropical convergence zone.

equilibrium line The boundary between the zone of accumulation and the zone of ablation on a glacier.

equinox One of two days out of the year (September 22 and March 21) in which the Sun is directly overhead at noon at the equator.

era An interval of geologic time representing the largest subdivision of the Phanerozoic Eon.

erg Sand seas formed by the accumulation of dunes in a desert.

erosion The grinding away and removal of Earth's surface materials by moving water, air, or ice.

erosional coast A coastline where sediment is not accumulating and wave action grinds away at the shore.

erosional landform A landform that results from the breakdown and removal of rock or sediment.

erratic A boulder or cobble that was picked up by a glacier and deposited hundreds of kilometers away from the outcrop from which it detached.

esker A ridge of sorted sand and gravel that snakes across a ground moraine; the sediment of an esker was deposited in subglacial meltwater tunnels.

estuary An inlet in which seawater and river water mix, created when a coastal valley is flooded because of either rising sea level or land subsidence.

Eubacteria The kingdom of "true bacteria."

euhedral crystal A crystal whose faces are well formed and whose shape reflects crystal form.

eukaryotic cell A cell with a complex internal structure, capable of building multicellular organisms.

eustatic sea-level change A global rising or falling of the ocean surface.

evaporate The process of transforming a liquid into a gas.

evapotranspiration The sum of evaporation from bodies of water and the ground surface and transpiration from plants and animals.

evolution (biological) The change over time of populations of organisms, due to survival of the fittest.

exfoliation The process by which an outcrop of rock splits apart into onion-like sheets along joints that lie parallel to the ground surface.

exhumation The process (involving uplift and erosion) that returns deeply buried rocks to the surface.

exotic terrane A block of land that collided with a continent along a convergent margin and attached to the continent; the term "exotic" implies that the land was not originally part of the continent to which it is now attached.

expanding Universe theory The theory that the whole Universe must be expanding because galaxies in every direction seem to be moving away from us.

explosive eruption Violent volcanic eruption that produces clouds and avalanches of pyroclastic debris.

external energy (geology) Energy that comes to Earth from the Sun.

external process A geomorphologic process—such as downslope movement, erosion, or deposition—that is the consequence of gravity or of the interaction between the solid Earth and its fluid envelope (air and water). Energy for these processes comes from gravity and sunlight.

extinction The death of the last members of a species so that there are no parents to pass on their genetic traits to offspring.

extinct volcano A volcano that was active in the past but has now shut off entirely and will not erupt in the future.

extraordinary fossil A rare fossilized relict, or trace, of the soft part of an organism.

extrusive igneous rock Rock that forms by the freezing of lava above ground, after it flows or explodes out (extrudes) onto the surface and comes into contact with the atmosphere or ocean.

eye The relative calm in the center of a hurricane.

eye wall A rotating vertical cylinder of clouds surrounding the eye of a hurricane.

facet The flat surface of a cut gemstone; facets are produced by grinding.

facies (1) Sedimentary: a group of rocks and primary structures indicative of a given depositional environment; (2) Metamorphic: a set of metamorphic mineral assemblages formed under a given range of pressures and temperatures.

failure surface The detachment or sliding horizon on which downslope movement of rock or debris occurs.

fault A fracture on which one body of rock slides past another.

fault-block mountains An outdated term for a narrow, elongate range of mountains that develops in a continental-rift setting as normal faulting drops down blocks of crust, or tilts blocks.

fault breccia Fragmented rock in which angular fragments were formed by brittle fault movement; fault breccia occurs along a fault.

fault creep Gradual movement along a fault that occurs in the absence of an earthquake.

fault gouge Pulverized rock consisting of fine powder that lies along fault surfaces; gouge forms by crushing and grinding.

faulting Slip events along a fault.

fault scarp A small step on the ground surface where one side of a fault has moved vertically with respect to the other.

fault system A grouping of numerous related faults.

fault trace (or line) The intersection between a fault and the ground surface.

felsic An adjective used in reference to igneous rocks that are rich in elements forming feldspar and quartz.

Ferrel cells The name given to the middle-latitude convection cells in the atmosphere.

fetch The distance across a body of water along which a wind blows to build waves.

fine-grained A textural term for rock consisting of many fine grains or clasts.

firn Compacted granular ice (derived from snow) that forms where snow is deeply buried; if buried more deeply, firn turns into glacial ice.

fission track A line of damage formed in the crystal lattice of a mineral by the impact of an atomic particle ejected during the decay of a radioactive isotope.

fissure A conduit in a magma chamber in the shape of a long crack through which magma rises and erupts at the surface.

fjord A deep, glacially carved, U-shaped valley flooded by rising sea level.

flank eruption An eruption that occurs when a secondary chimney, or fissure, breaks through the flank of a volcano.

flash flood A flood that occurs during unusually intense rainfall or as the result of a dam collapse, during which the floodwaters rise very fast.

flexing The process of folding in which a succession of rock layers bends and slip occurs between the layers.

flocculation The clumping together of clay suspended in river water into bunches that are large enough to settle out.

flood An event during which the volume of water in a stream becomes so great that it covers areas outside the stream's normal channel.

flood basalt Vast sheets of basalt that spread from a volcanic vent over an extensive surface of land; they may form where a rift develops above a continental hot spot, and where lava is particularly hot and has low viscosity.

floodplain The flat land on either side of a stream that becomes covered with water during a flood.

floodplain flood A flood during which a floodplain is submerged.

flood stage The stage when water reaches the top of a stream channel.

flood tide The rising tide.

floodway A mapped region likely to be flooded, in which people avoid constructing buildings.

flow fold A fold that forms when the rock is so soft that it behaves like weak plastic.

flowstone A sheet of limestone that forms along the wall of a cave when groundwater flows along the surface of the wall.

fluvial deposit Sediment deposited in a stream channel, along a stream bank, or on a floodplain.

flux Flow.

focus The location where a fault slips during an earthquake (hypocenter).

fog A cloud that forms at ground level.

fold A bend or wrinkle of rock layers or foliation; folds form as a consequence of ductile deformation.

fold axis An imaginary line that, when moved parallel to itself, can trace out the shape of a folded surface.

fold-thrust belt An assemblage of folds and related thrust faults that develop above a detachment fault.

foliation Layering formed as a consequence of the alignment of mineral grains, or of compositional banding in a metamorphic rock.

foraminifera Microscopic plankton with calcitic shells, components of some limestones.

foreland sedimentary basin A basin located under the plains adjacent to a mountain front, which develops as the weight of the mountains pushes the crust down, creating a depression that traps sediment.

foreshocks The series of smaller earthquakes that precede a major earthquake.

foreshore zone The zone of beach regularly covered and uncovered by rising and falling tides.

formation *Stratigraphic formation.*

fossil The remnant, or trace, of an ancient living organism that has been preserved in rock or sediment.

fossil assemblage A group of fossil species found in a specific sequence of sedimentary rock.

fossil correlation A determination of the stratigraphic relation between two sedimentary rock units, reached by studying fossils.

fossil fuel An energy resource such as oil or coal that comes from organisms that lived long ago, and thus stores solar energy that reached the Earth then.

fossiliferous limestone Limestone consisting of abundant fossil shells and shell fragments.

fossilization The process of forming a fossil.

fossil succession The principle that the assemblage of fossil species in a given sequence of sedimentary strata differs from that found in older sequences or in younger sequences; a given species appears at a certain level and then disappears (goes extinct) at a higher level.

fractional crystallization The process by which a magma becomes progressively more silicic as it cools, because early-formed crystals settle out.

fracture zone A narrow band of vertical fractures in the ocean floor; fracture zones lie roughly at right angles to a mid-ocean ridge, and the actively slipping part of a fracture zone is a transform fault.

fragmental igneous rock Fragments of igneous material that have been stuck together to form a coherent mass.

fresh rock Rock whose mineral grains have their original composition and shape.

friction Resistance to sliding on a surface.

fringing reef A coral reef that forms directly along the coast.

front The boundary between two air masses.

frost wedging The process in which water trapped in a joint freezes, forces the joint open, and may cause the joint to grow.

fuel rod A metal tube that holds the nuclear fuel in a nuclear reactor.

Fujita scale A scale that distinguishes among tornadoes on the basis of wind speed, path dimensions, and possible damage.

Ga Billions of years ago (abbreviation).

gabbro A coarse-grained, intrusive mafic igneous rock.

Gaia The term used for the Earth System, with the implication that it resembles a complex living entity.

galaxy An immense system of hundreds of billions of stars.

gas-giant planet The outer planets (Jupiter, Saturn, Uranus, Neptune) that are very large and consist mostly of volatile elements.

gas hydrate An ice-like solid consisting of water and methane.

gem A mineral or form of a mineral that is particularly beautiful and/or rare, and thus has value.

gene An individual component of the DNA code that guides the growth and development of an organism.

genetics The study of genes and how they transmit information.

geocentric Universe concept An ancient Greek idea suggesting that the Earth sat motionless in the center of the Universe while stars and other planets and the Sun orbited around it.

geochronology The science of dating geologic events in years.

geode A cavity in which euhedral crystals precipitate out of water solutions passing through a rock.

geographical pole The locations (north and south) where the Earth's rotational axis intersects the planet's surface.

geoid A shape representing the pull of gravity as a function of location on the Earth; it is the shape that the surface of the Earth would have if the surface were entirely covered by the ocean, so that elevation represented the strength of the pull of gravity.

geologic column A composite stratigraphic chart that represents the entirety of the Earth's history.

geologic history The sequence of geologic events that has taken place in a region.

geologic map A map showing the distribution of rock units and structures across a region.

geologic structures Bends, breaks, and fabrics in rocks that form as a result of deformation.

geologic time The span of time since the formation of the Earth.

geologic time scale A scale that describes the intervals of geologic time.

geologists Scientists who study the Earth.

geology   The study of the Earth, including our planet's composition, behavior, and history.

geotherm   The change in temperature with depth in the Earth.

geothermal energy   Heat and electricity produced by using the internal heat of the Earth.

geothermal gradient   The rate of change in temperature with depth.

geothermal region   A region of current or recent volcanism in which magma or very hot rock heats up groundwater, which may discharge at the surface in the form of hot springs and/or geysers.

geyser   A fountain of steam and hot water that erupts periodically from a vent in the ground in a geothermal region.

glacial abrasion   The process by which clasts embedded in the base of a glacier grind away at the substrate as the glacier flows.

glacial advance   The forward movement of a glacier's toe when the supply of snow exceeds the rate of ablation.

glacial drift   Sediment deposited in glacial environments.

glacial incorporation   The process by which flowing ice surrounds and incorporates debris.

glacial marine   Sediment consisting of ice-rafted clasts mixed with marine sediment.

glacial outwash   Coarse sediment deposited on a glacial outwash plain by meltwater streams.

glacially polished surface   A polished rock surface created by the glacial abrasion of the underlying substrate.

glacial plucking (or **quarrying**)   The process by which a glacier breaks off and carries away fragments of bedrock.

glacial rebound   The process by which the surface of a continent rises back up after an overlying continental ice sheet melts away and the weight of the ice is removed.

glacial retreat   The movement of a glacier's toe back toward the glacier's origin; glacial retreat occurs if the rate of ablation exceeds the rate of supply.

glacial striations   Scratches or troughs carved into rock by the sediment embedded in ice at the base of a flowing glacier.

glacial subsidence   The sinking of the surface of a continent caused by the weight of an overlying glacial ice sheet.

glacial till   Sediment transported by flowing ice and deposited beneath a glacier or at its toe.

glaciation   A period of time during which glaciers grew and covered substantial areas of the continents.

glacier   A river or sheet of ice that slowly flows across the land surface and lasts all year long.

glass   A solid in which atoms are not arranged in an orderly pattern.

glassy igneous rock   Igneous rock consisting entirely of glass, or of tiny crystals surrounded by a glass matrix.

glide horizon   The surface along which a slump slips.

global change   The transformations or modifications of both physical and biological components of the Earth System through time.

global circulation   The movement of volumes of air in paths that ultimately take it around the planet.

global climate change   Transformations or modifications in Earth's climate over time.

global cooling   A fall in the average atmospheric temperature.

global positioning system (GPS)   A satellite system people can use to measure rates of movement of the Earth's crust relative to one another, or simply to locate their position on the Earth's surface.

global warming   A rise in the average atmospheric temperature.

gneiss   A compositionally banded metamorphic rock typically composed of alternating dark- and light-colored layers.

Gondwana   A supercontinent that consisted of today's South America, Africa, Antarctica, India, and Australia. Also called Gondwanaland.

graben   A down-dropped crustal block bounded on either side by a normal fault dipping toward the basin.

graded bed   A layer of clastic sediment or sedimentary rock in which clast size progressively decreases from the base to the top of the bed; graded beds form by deposition from a turbidity current.

gradualism   The theory that evolution happens at a constant, slow rate.

grain   A fragment of a mineral crystal or of a rock.

grain rotation   The process by which rigid, inequant mineral grains distributed through a soft matrix may rotate into parallelism as the rock changes shape owing to differential stress.

granite   A coarse-grained intrusive silicic igneous rock.

granulite facies   A set of metamorphic mineral assemblages formed at very high pressures and temperatures.

gravitational energy   The force of attraction that one mass has on another.

gravitational spreading   A process of lateral spreading that occurs in a material because of the weakness of the material; gravitational spreading causes continental glaciers to grow and mountain belts to undergo orogenic collapse.

graywacke   An informal term used for sedimentary rock consisting of sand-size up to small-pebble-size grains of quartz and rock fragments all mixed together in a muddy matrix; typically, graywacke occurs at the base of a graded bed.

greenhouse conditions (greenhouse period)   Relatively warm global climate leading to the rising of sea level for an interval of geologic time.

greenhouse effect   The trapping of heat in the Earth's atmosphere by carbon dioxide and other greenhouse gases, which absorb infrared radiation; somewhat analogous to the effect of glass in a greenhouse.

greenhouse gases   Atmospheric gases, such as carbon dioxide and methane, that regulate the Earth's atmospheric temperature by absorbing infrared radiation.

greenschist facies   A set of metamorphic mineral assemblages formed under relatively low pressures and temperatures.

greenstone   A low-grade metamorphic rock formed from basalt; if foliated, the rock is called greenschist.

Greenwich mean time (GMT)   The time at the astronomical observatory in Greenwich, England; time in all other time zones is set in relation to GMT.

Grenville orogeny   The orogeny that occurred about 1 billion years ago and yielded the belt of deformed and metamorphosed rocks that underlie the eastern fifth of the North American continent.

groin   A concrete or stone wall built perpendicular to a shoreline in order to prevent beach drift from removing sand.

ground moraine   A thin, hummocky layer of till left behind on the land surface during a rapid glacial recession.

groundwater   Water that resides under the surface of the Earth, mostly in pores or cracks of rock or sediment.

groundwater contamination   The introduction of harmful and/or non-natural chemicals to groundwater.

group   A succession of stratigraphic formations that have been lumped together, making a single, thicker stratigraphic entity.

growth ring   A rhythmic layering that develops in trees, travertine deposits, and shelly organisms as a consequence of seasonal changes.

gusher   A fountain of oil formed when underground pressure causes the oil to rise on its own out of a drilled hole.

guyot   A seamount that had a coral reef growing on top of it, so that it is now flat-crested.

gymnosperm   A plant whose seeds are "naked," not surrounded by a fruit.

gyre   A large circular flow pattern of ocean surface currents.

Hadean Eon   The oldest of the Precambrian eons; the time between Earth's origin and the formation of the first rocks that have been preserved.

Hadley cells   The name given to the low-latitude convection cells in the atmosphere.

hail   Falling ice balls from the sky, formed when ice crystallizes in turbulent storm clouds.

hail streak   An approximately 2-by-10-km stretch of ground, elongate in the direction of a storm, onto which hail has fallen.

half-graben   A wedge-shaped basin in cross section that develops as the hanging-wall block above a normal fault slides down and rotates; the basin develops between the fault surface and the top surface of the rotated block.

half-life   The time it takes for half of a group of a radioactive element's isotopes to decay.

halocline   The boundary in the ocean between surface-water and deep-water salinities.

hamada   Barren rocky highlands in a desert.

hand specimen (geology)   A chunk of rock, about the size of a fist, typically collected to serve as a sample for further study.

hanging valley   A glacially carved tributary valley whose floor lies at a higher elevation than the floor of the trunk valley.

hanging wall   The rock or sediment above an inclined fault plane.

hardness   In mineralogy, hardness refers to the resistance of a mineral to scratching; a harder mineral can scratch a softer mineral.

hard water   Groundwater that contains dissolved calcium and magnesium, usually after passing through limestone or dolomite.

head   (1) The elevation of the water table above a reference horizon; (2) the edge of ice at the origin of a glacier.

headland   A place where a hill or cliff protrudes into the sea.

head scarp   The distinct step along the upslope edge of a slump where the regolith detached.

headward erosion   The process by which a stream channel lengthens up its slope as the flow of water increases.

headwaters   The beginning point of a stream.

heat   Thermal energy resulting from the movement of molecules.

heat capacity   A measure of the amount of heat that must be added to a material to change its temperature.

heat flow   The rate at which heat rises from the Earth's interior up to the surface.

heat-transfer melting   Melting that results from the transfer of heat from a hotter magma to a cooler rock.

heliocentric Universe concept   An idea proposed by Greek philosophers around 250 B.C.E. suggesting that all heavenly objects including the Earth orbited the Sun.

Hercynian orogen   The late Paleozoic orogen that affected parts of Europe; a continuation of the Alleghenian orogen.

heterosphere   A term for the upper portion of the atmosphere in which gases separate into distinct layers on the basis of composition.

hiatus   The interval of time between deposition of the youngest rock below an unconformity and deposition of the oldest rock above the unconformity.

high-altitude westerlies   Westerly winds at the top of the troposphere.

high-grade metamorphic rocks   Rocks that metamorphose under relatively high temperatures.

high-level waste   Nuclear waste containing greater than 1 million times the safe level of radioactivity.

hinge   The portion of a fold where curvature is greatest.

hogback   A steep-sided ridge of steeply dipping strata.

Holocene   The period of geologic time since the last glaciation.

Holocene climatic maximum   The period from 5,000 to 6,000 years ago, when Holocene temperatures reached a peak.

homosphere   The lower part of the atmosphere, in which the gases have stirred into a homogenous mixture.

hoodoo   The local name for the brightly colored shale and sandstone chimneys found in Bryce Canyon National Park in Utah.

horn   A pointed mountain peak surrounded by at least three cirques.

hornfels   Rock that undergoes metamorphism simply because of a change in temperature, without being subjected to differential stress.

horse latitudes   The region of the subtropical high in which winds are weak.

horst   The high block between two grabens.

hot spot   A location at the base of the lithosphere, at the top of a mantle plume, where temperatures can cause melting.

hot-spot track   A chain of now-dead volcanoes transported off the hot spot by the movement of a lithosphere plate.

hot-spot volcano   An isolated volcano not caused by movement at a plate boundary, but rather by the melting of a mantle plume.

hot spring   A spring that emits water ranging in temperature from about 30°C to 104°C.

hummocky surface   An irregular and lumpy ground surface.

hurricane   A huge rotating storm, resembling a giant spiral in map view, in which sustained winds blow over 119 km per hour.

hurricane track   The path a hurricane follows.

hyaloclastite   A rubbly extrusive rock consisting of glassy debris formed in a submarine or sub-ice eruption.

hydration   The absorption of water into the crystal structure of minerals; a type of chemical weathering.

hydraulic conductivity   The coefficient K in Darcy's law; hydraulic conductivity takes into account the permeability of the sediment or rock as well as the fluid's viscosity.

hydraulic gradient   The slope of the water table.

hydraulic head   The elevation to which groundwater rises in a pipe drilled into the ground; in non-artesian systems, the head is the water table.

hydrocarbon   A chain-like or ring-like molecule made of hydrogen and carbon atoms; petroleum and natural gas are hydrocarbons.

hydrocarbon reserve   An accumulation of accessible oil and gas.

hydrocarbon system   The association of source rock, migration pathway, reservoir rock, seal, and trap geometry that leads to the occurrence of a hydrocarbon reserve.

hydrologic cycle   The continual passage of water from reservoir to reservoir in the Earth System.

hydrolysis   The process in which water chemically reacts with minerals and breaks them down.

hydrosphere   The Earth's water, including surface water (lakes, rivers, and oceans), groundwater, and liquid water in the atmosphere.

hydrothermal deposit   An accumulation of ore minerals precipitated from hot-water solutions circulating through a magma or through the rocks surrounding an igneous intrusion.

hydrothermal metamorphism   The change that occurs in a rock due to interaction with high-temperature water solutions.

hypocenter (or **focus**)   The point below the Earth's surface where the energy is produced during an earthquake.

hypothesis   A reasonable idea that has the possibility of being correct, but has not yet been proven.

hypsometric curve   A graph that plots surface elevation on the vertical axis and the percentage of the Earth's surface on the horizontal axis.

ice age   An interval of time in which the climate was colder than it is today, glaciers occasionally advanced to cover large areas of the continents, and mountain glaciers grew; an ice age can include many glacials and interglacials.

iceberg   A large block of ice that calves off the front of a glacier and drops into the sea.

icehouse conditions (icehouse period)   A period of time when the Earth's temperature was cooler than it is today and ice ages could occur.

ice-margin lake   A meltwater lake formed along the edge of a glacier.

ice-rafted sediment   Sediment carried out to sea by icebergs.

ice sheet   A vast glacier that covers the landscape.

ice shelf   A broad, flat region of ice along the edge of a continent formed where a continental glacier flowed into the sea.

ice stream   A portion of a glacier that travels much more quickly than adjacent portions of the glacier.

ice tongue   The portion of a valley glacier that has flowed out into the sea.

igneous rock   Rock that forms when hot molten rock (magma or lava) cools and freezes solid.

ignimbrite   Rock formed when deposits of pyroclastic flows solidify.

inactive fault   A fault that last moved in the distant past and probably won't move again in the near future, yet is still recognizable because of displacement across the fault plane.

inactive sand   The sand along a coast that is buried beneath a layer of active sand and moves only during severe storms or not at all.

incised meander   A meander that lies at the bottom of a steep-walled canyon.

index minerals   Minerals that serve as good indicators of metamorphic grade.

induced seismicity   Seismic events caused by the actions of people (e.g., filling a reservoir, that lies over a fault, with water).

industrial minerals   Minerals that serve as the raw materials for manufacturing chemicals, concrete, and wallboard, among other products.

inequant   A term for a mineral grain whose length and width are not the same.

inertia   The tendency of an object at rest to remain at rest.

infiltrate   Seep down into.

injection well   A well in which a liquid is pumped down into the ground under pressure so that it passes from the well back into the pore space of the rock or regolith.

inner core   The inner section of the core 5,155 km deep to the Earth's center at 6,371 km, and consisting of solid iron alloy.

inselberg   An isolated mountain or hill in a desert landscape created by progressive cliff retreat, so that the hill is surrounded by a pediment or an alluvial fan.

insolation   Exposure to the Sun's rays.

interglacial   A period of time between two glaciations.

interior basin   A basin with no outlet to the sea.

interlocking texture   The texture of crystalline rocks in which mineral grains fit together like pieces of a jigsaw puzzle.

intermediate magma   A silicate melt that contains a moderate amount (≈60%) of silica.

internal energy (geology)   Energy provided by heat sources within the Earth.

internal process   A process in the Earth System, such as plate motion, mountain building, or volcanism, ultimately caused by Earth's internal heat.

intertidal zone   The area of coastal land across which the tide rises and falls.

intertropical convergence zone   The equatorial convergence zone in the atmosphere.

intraplate earthquake   Earthquake that occurs away from plate boundaries.

intrusive contact   The boundary between country rock and an intrusive igneous rock.

intrusive igneous rock   Rock formed by the freezing of magma underground.

ionosphere   The interval of Earth's atmosphere, at an elevation between 50 and 400 km, containing abundant positive ions.

iron catastrophe   The proposed event very early in Earth history when the Earth partly melted and molten iron sank to the center to form the core.

isobar   A line on a map along which the air has a specified pressure.

isograd   (1) A line on a pressure-temperature graph along which all points are taken to be at the same metamorphic grade; (2) A line on a map making the first appearance of a metamorphic index mineral.

isostasy (or **isostatic equilibrium**)   The condition that exists when the buoyancy force pushing lithosphere up equals the gravitational force pulling lithosphere down.

isostatic compensation   The process in which the surface of the crust slowly rises or falls to reestablish isostatic equilibrium after a geologic event changes the density or thickness of the lithosphere.

isotherm   Lines on a map or cross section along which the temperature is constant.

isotopes   Different versions of a given element that have the same atomic number but different atomic weights.

isotopic dating   The science of dating geologic events in years by measuring the ratio of parent radioactive atoms to daughter product atoms.

jet stream   A fast-moving current of air that flows at high elevations.

jetty   A manmade wall that protects the entrance to a harbor.

joints   Naturally formed cracks in rocks.

joint set   A group of systematic joints.

Jovian   A term used to describe the outer gassy, Jupiter-like planets (gas-giant planets).

kame   A stratified sequence of lateral-moraine sediment that's sorted by water flowing along the edge of a glacier.

karst landscape   A region underlain by caves in limestone bedrock; the collapse of the caves creates a landscape of sinkholes separated by higher topography, or of limestone spires separated by low areas.

kerogen   The waxy molecules into which the organic material in shale transforms on reaching about 100°C. At higher temperatures, kerogen transforms into oil.

kettle hole   A circular depression in the ground made when a block of ice calves off the toe of a glacier, becomes buried by till, and later melts.

knob-and-kettle topography   A land surface with many kettle holes separated by round hills of glacial till.

K-T boundary event   The mass extinction that happened at the end of the Cretaceous Period, 65 million years ago, possibly because of the collision of an asteroid with the Earth.

laccolith   A shallow igneous intrusion that has a blister-like shape.

lag deposit   A surface accumulation of coarser sediment that forms where wind has blown away finer sediment.

lagoon   A body of shallow seawater separated from the open ocean by a barrier island.

lahar   A thick slurry formed when volcanic ash and debris mix with water, either in rivers or from rain or melting snow and ice on the flank of a volcano.

landform   A distinct feature of a landscape, as defined by its shape and character.

landscape   A region of the land surface, with definable characteristics.

landslide   A sudden movement of rock and debris down a nonvertical slope.

landslide-potential map   A map on which regions are ranked according to the likelihood that a mass movement will occur.

land subsidence   Sinking elevation of the ground surface; the process may occur over an aquifer that is slowly draining and decreasing in volume because of pore collapse.

La Niña   Years in which the El Niño event is not strong.

lapilli   Marble-to-plum-sized fragments of pyroclastic debris.

Laramide orogeny   The mountain-building event that lasted from about 80 Ma to 40 Ma, in western North America; in the United States, it formed the Rocky Mountains as a result of basement uplift and the warping of the younger overlying strata into large monoclines.

latent heat of condensation   The heat released during condensation, which comes only from a change in state.

lateral moraine   A strip of debris along the side margins of a glacier.

laterite soil   Soil formed over iron-rich rock in a tropical environment, consisting primarily of a dark-red mass of insoluble iron and/or aluminum oxide.

Laurentia   A continent in the early Paleozoic Era composed of today's North America and Greenland.

Laurentide ice sheet   An ice sheet that spread over northeastern Canada during the Pleistocene ice age(s).

lava Molten rock that has flowed out onto the Earth's surface.

lava dome A dome-like mass of rhyolitic lava that accumulates above the eruption vent.

lava flows Sheets or mounds of lava that flow onto the ground surface or sea floor in molten form and then solidify.

lava lake A large pool of lava produced around a vent when lava fountains spew forth large amounts of lava in a short period of time.

lava tube The empty space left when a lava tunnel drains; this happens when the surface of a lava flow solidifies while the inner part of the flow continues to stream downslope.

leach To dissolve and carry away.

leader A conductive path stretching from a cloud toward the ground, along which electrons leak from the base of the cloud, and which provides the start for a lightning flash to the ground.

lightning flash A giant spark or pulse of current that jumps across a gap of charge separation.

light year The distance that light travels in one Earth year (about 6 trillion miles or 9.5 trillion km).

lignite Low-rank coal that consists of 50% carbon.

limb The side of a fold, showing less curvature than at the hinge.

limestone Sedimentary rock composed of calcite.

liquefaction The transformation of seemingly solid sediment into a liquid-like slurry, in response to ground shaking.

lithification The transformation of loose sediment into solid rock through compaction and cementation.

lithologic correlation A correlation based on similarities in rock type.

lithosphere The relatively rigid, nonflowable, outer 100- to 150-km-thick layer of the Earth; constituting the crust and the top part of the mantle.

lithosphere plate A portion of the outer, relatively rigid layer of the Earth; most seismic activity happens at the boundaries of plates, while the interior of a plate is relatively stable.

little ice age A period of cooler temperatures, between 1500 and 1800 C.E., during which many glaciers advanced.

local base level A base level upstream from a drainage network's mouth.

lodgment till A flat layer of till smeared out over the ground when a glacier overrides an end moraine as it advances.

loess Layers of fine-grained sediments deposited from the wind; large deposits of loess formed from fine-grained glacial sediment blown off outwash plains.

longitudinal (seif) dune A dune formed when there is abundant sand and a strong, steady wind, and whose axis lies parallel to the wind direction.

longitudinal profile A cross-sectional image showing the variation in elevation along the length of a river.

longshore current A current that flows parallel to a beach.

lower mantle The deepest section of the mantle, stretching from 670 km down to the core-mantle boundary.

low-grade metamorphic rocks Rocks that underwent metamorphism at relatively low temperatures.

low-velocity zone The asthenosphere underlying oceanic lithosphere in which seismic waves travel more slowly, probably because rock has partially melted.

luster The way a mineral surface scatters light.

L-waves (love waves) Surface seismic waves that cause the ground to ripple back and forth, creating a snake-like movement.

Ma Millions of years ago (abbreviation).

macrofossil A fossil large enough to be seen with the naked eye.

mafic A term used in reference to magmas or igneous rocks that are relatively poor in silica and rich in iron and magnesium.

magma Molten rock beneath the Earth's surface.

magma chamber A space below ground filled with magma.

magma contamination The process in which flowing magma incorporates components of the country rock through which it passes.

magmatic deposit An ore deposit formed when sulfide ore minerals accumulate at the bottom of a magma chamber.

magnetic anomaly The difference between the expected strength of the Earth's magnetic field at a certain location and the actual measured strength of the field at that location.

magnetic declination The angle between the direction a compass needle points at a given location and the direction of true north.

magnetic dipole A magnetic entity that has a north and south end.

magnetic field The region affected by the force emanating from a magnet.

magnetic field lines The trajectories along which magnetic particles would align, or charged particles would flow, if placed in a magnetic field.

magnetic force The push or pull exerted by a magnet.

magnetic inclination The angle between a magnetic needle free to pivot on a horizontal axis and a horizontal plane parallel to the Earth's surface.

magnetic pole The north or south ends of a magnet; field lines point straight down at the pole.

magnetic reversal The change of the Earth's magnetic polarity; when a reversal occurs, the field flips from normal to reversed polarity, or vice versa.

magnetic-reversal chronology The history of magnetic reversals through geologic time.

magnetization The degree to which a material can exert a magnetic force.

magnetometer An instrument that measures the strength of the Earth's magnetic field.

magnetosphere The region protected from the electrically charged particles of the solar winds by Earth's magnetic field.

magnetostratigraphy The comparison of the pattern of magnetic reversals in a sequence of strata, with a reference column showing the succession of reversals through time.

magnitude Any numerical representation of the size of an earthquake as determined by measuring the amplitude of ground motion.

manganese nodules Lumpy accumulations of manganese-oxide minerals precipitated onto the sea floor.

mantle The thick layer of rock below the Earth's crust and above the core.

mantle plume A column of very hot rock rising up through the mantle.

marble A metamorphic rock composed of calcite and transformed from a protolith of limestone.

mare The broad darker areas on the Moon's surface, which consist of flood basalts that erupted over 3 billion years ago and spread out across the Moon's lowlands.

marginal sea A small ocean basin created when sea-floor spreading occurs behind an island arc.

marine magnetic anomaly An unusually strong or unusually weak magnetic field, as measured over the sea floor; in map view, they look like stripes that are parallel to the mid-ocean ridge.

maritime tropical air mass A mass of air that originates over tropical or subtropical oceanic regions.

marsh A wetland dominated by grasses.

mass extinction event A time when vast numbers of species abruptly vanish.

mass movement (or **mass wasting**) The gravitationally caused downslope transport of rock, regolith, snow, or ice.

matrix Finer-grained material surrounding larger grains in a rock.

meander A snake-like curve along a stream's course.

meandering stream A reach of stream containing many meanders (snake-like curves).

meander neck A narrow isthmus of land separating two adjacent meanders.

principle of inclusions If a rock contains fragments of another rock, the fragments must be older than the rock containing them.

principle of original continuity Sedimentary layers, before erosion, formed fairly continuous sheets over a region.

principle of original horizontality Layers of sediment, when originally deposited, are fairly horizontal.

principle of superposition In a sequence of sedimentary rock layers, each layer must be younger than the one below, for a layer of sediment cannot accumulate unless there is already a substrate on which it can collect.

principle of uniformitarianism The physical processes we observe today also operated in the past in the same way, and at comparable rates.

prograde metamorphism Metamorphism that occurs as temperatures and pressures are increasing.

Proterozoic Eon The most recent of the Precambrian eons.

protocontinent A block of crust composed of volcanic arcs and hot-spot volcanoes sutured together.

protolith The original rock from which a metamorphic rock formed.

protoplanet A body that grows by the accumulation of planetesimals but has not yet become big enough to be called a planet.

protoplanetary disk The flattened cloud of dust, gas, and ice that orbits a nascent star prior to the formation of planets.

protoplanetary nebula A ring of gas and dust that surrounded the newborn Sun, from which the planets were formed.

protostar A dense body of gas that is collapsing inward because of gravitational forces and that may eventually become a star.

pumice A glassy igneous rock that forms from felsic frothy lava and contains abundant (over 50%) pore space.

punctuated equilibrium The hypothesis that evolution takes place in fits and starts; evolution occurs very slowly for quite a while and then, during a relatively short period, takes place very rapidly.

P-waves Compressional seismic waves that move through the body of the Earth.

P-wave shadow zone A band between 103° and 143° from an earthquake epicenter, as measured along the circumference of the Earth, inside which P-waves do not arrive at seismograph stations.

pycnocline The boundary between layers of water of different densities.

pyroclastic debris Fragmented material that sprayed out of a volcano and landed on the ground or sea floor in solid form.

pyroclastic flow A fast-moving avalanche that occurs when hot volcanic ash and debris mix with air and flow down the side of a volcano.

pyroclastic rock Rock made from fragments blown out of a volcano during an explosion that were then packed or welded together.

quarry A site at which stone is extracted from the ground.

quartzite A metamorphic rock composed of quartz and transformed from a protolith of quartz sandstone.

quenching A sudden cooling of molten material to form a solid.

quick clay Clay that behaves like a solid when still (because of surface tension holding the water-coated clay flakes together), but that flows like a liquid when shaken.

radial network A drainage network in which the streams flow outward from a cone-shaped mountain, and define a pattern resembling spokes on a wheel.

radioactive decay The process by which a radioactive atom undergoes fission or releases particles, thereby transforming into a new element.

radioactive isotope An unstable isotope of a given element.

rain band A spiraling arm of a hurricane radiating outward from the eye.

rain shadow The inland side of a mountain range, which is arid because the mountains block rain clouds from reaching the area.

range (for fossils) The interval of a sequence of strata in which a specific fossil species appears.

rapids A reach of a stream in which water becomes particularly turbulent; as a consequence, waves develop on the surface of the stream.

reach A specified segment of a stream's path.

recessional moraine The end moraine that forms when a glacier stalls for a while as it recedes.

recharge area A location where water enters the ground and infiltrates down to the water table.

recrystallization The process in which ions or atoms in minerals rearrange to form new minerals.

rectangular network A drainage network in which the streams join each other at right angles because of a rectangular grid of fractures that breaks up the ground and localizes channels.

recurrence interval The average time between successive geologic events.

red giant A huge red star that forms when Sun-sized stars start to die and expand.

red shift The phenomenon in which a source of light moving away from you very rapidly shifts to a lower frequency; that is, toward the red end of the spectrum.

reef bleaching The death and loss of color of a coral reef.

reflected ray A ray that bounces off a boundary between two different materials.

reflection When energy (e.g., light) bounces off a surface.

refracted ray A ray that bends as it passes through a boundary between two different materials.

refraction The bending of a ray as it passes through a boundary between two different materials.

reg A vast stony plain in a desert.

regional metamorphism *Dynamothermal metamorphism*; metamorphism of a broad region, usually the result of deep burial during an orogeny.

regolith Any kind of unconsolidated debris that covers bedrock.

regression The seaward migration of a shoreline caused by a lowering of sea level.

relative age The age of one geologic feature with respect to another.

relative humidity The ratio between the measured water content of air and the maximum possible amount of water the air can hold at a given condition.

relative plate velocity The movement of one lithosphere plate with respect to another.

relief The difference in elevation between adjacent high and low regions on the land surface.

renewable resource A resource that can be replaced by nature within a short time span relative to a human life span.

reservoir rock Rock with high porosity and permeability, so it can contain an abundant amount of easily accessible oil.

residence time The average length of time that a substance stays in a particular reservoir.

residual mineral deposit Soils in which the residuum left behind after leaching by rainwater is so concentrated in metals that the soil itself becomes an ore deposit.

resource A supply of usable material.

resurgent dome The new mound, or cone, of igneous rock that grows within a caldera as an eruption begins anew.

retrograde metamorphism Metamorphism that occurs as pressures and temperatures are decreasing; for retrograde metamorphism to occur, water must be added.

return stroke An upward-flowing electric current from the ground that carries positive charges up to a cloud during a lightning flash.

reversed polarity   Polarity in which the paleomagnetic dipole points north.

reverse fault   A steeply dipping fault on which the hanging-wall block slides up.

Richter magnitude scale   A scale that defines earthquakes on the basis of the amplitude of the largest ground motion recorded on a seismogram.

ridge axis   The crest of a mid-ocean ridge; the ridge axis defines the position of a divergent plate boundary.

ridge-push force   The force that drives plates away from a mid-ocean ridge; it is caused by the fact that the ridge is elevated relative to the regions of oceanic plate away from the ridge.

rifting   The process by which continental lithosphere stretches and breaks apart; rifting produces normal faults and, commonly, volcanism.

right-lateral strike-slip fault   A strike-slip fault in which the block on the opposite fault plane from a fixed spot moves to the right of that spot.

rip current   A strong, localized seaward flow of water perpendicular to a beach.

ripple marks   Wave-like ridges and troughs on the surface of a sedimentary layer formed during deposition in a current.

riprap   Loose boulders or concrete piled together along a beach to absorb wave energy before it strikes a cliff face.

roche moutonnée   A glacially eroded hill that becomes elongate in the direction of flow and asymmetric; glacial rasping smoothes the upstream part of the hill into a gentle slope, while glacial plucking erodes the downstream edge into a steep slope.

rock   A coherent, naturally occurring solid, consisting of an aggregate of minerals or a mass of glass.

rock burst   A sudden explosion of rock off the ceiling or wall of an underground mine.

rock cycle   The succession of events that results in the transformation of Earth materials from one rock type to another, then another, and so on.

rockfall   A sudden collapse of bedrock down a steep slope.

rock flour   Fine-grained sediment produced by glacial abrasion of the substrate over which a glacier flows.

rock glacier   A slow-moving mixture of rock fragments and ice.

rock slide   A sudden downslope movement of rock.

rocky coast   An area of coast where bedrock rises directly from the sea, so beaches are absent.

Rodinia   A proposed Precambrian supercontinent that existed around 1 billion years ago.

rotational axis   The imaginary line through the center of the Earth around which the Earth spins.

R-waves (*Rayleigh waves*)   Surface seismic waves that cause the ground to ripple up and down, like water waves in a pond.

sabkah   A region of formerly flooded coastal desert in which stranded seawater has left a salt crust over a mire of mud that is rich in organic material.

salinity   The degree of concentration of salt in water.

saltation   The movement of a sediment in which grains bounce along their substrate, knocking other grains into the water column (or air) in the process.

salt dome   A rising bulbous dome of salt that bends up the adjacent layers of sedimentary rock.

salt wedging   The process in arid climates by which dissolved salt in groundwater crystallizes and grows in open pore spaces in rocks and pushes apart the surrounding grains.

sand spit   An area where the beach stretches out into open water across the mouth of a bay or estuary.

sandstone   Coarse-grained sedimentary rock consisting almost entirely of quartz.

sand volcano (or **sand blow**)   A small mound of sand produced when sand layers below the ground surface liquify as a result of seismic shaking, causing the sand to erupt onto the Earth's surface through cracks or holes in overlying clay layers.

saprolite   A layer of rotten rock created by chemical weathering in warm, wet climates.

Sargasso Sea   The center of North Atlantic Gyre, named for the tropical seaweed sargassum, which accumulates in its relatively noncirculating waters.

saturated solution   Water that carries as many dissolved ions as possible under given environmental conditions.

saturated zone   The region below the water table where pore space is filled with water.

scattering   The dispersal of energy that occurs when light interacts with particles in the atmosphere.

schist   A medium-to-coarse-grained metamorphic rock that possesses schistosity.

schistosity   Foliation caused by the preferred orientation of large mica flakes.

scientific laws   Principals of science that must be true; if they were violated, the physical world would not behave as we see it.

scientific method   A sequence of steps for systematically analyzing scientific problems in a way that leads to verifiable results.

scientists   People who study the character of the natural world, and try to come up with ideas to explain how it operates, in the context of physical laws.

scoria   A glassy mafic igneous rock containing abundant air-filled holes.

scouring   A process by which running water removes loose fragments of sediment from a stream bed.

sea arch   An arch of land protruding into the sea and connected to the mainland by a narrow bridge.

sea-floor spreading   The gradual widening of an ocean basin as new oceanic crust forms at a mid-ocean ridge axis and then moves away from the axis.

sea ice   Ice formed by the freezing of the surface of the sea.

seal   A relatively impermeable rock, such as shale, salt, or unfractured limestone, that lies above a reservoir rock and stops the oil from rising further.

seam   A sedimentary bed of coal interlayered with other sedimentary rocks.

seamount   An isolated submarine mountain.

seasonal floods   Floods that appear almost every year during seasons when rainfall is heavy or when winter snows start to melt.

seasonal well   A well that provides water only during the rainy season when the water table rises below the base of the well.

sea stack   An isolated tower of land just offshore, disconnected from the mainland by the collapse of a sea arch.

seawall   A wall of riprap built on the landward side of a backshore zone in order to protect shore cliffs from erosion.

second   The basic unit of time measurement, now defined as the time it takes for the magnetic field of a cesium atom to flip polarity 9,192,631,770 times, as measured by an atomic clock.

secondary enrichment   The process by which a new ore deposit forms from metals that were dissolved and carried away from preexisting ore minerals.

secondary porosity   New pore space in rocks, created some time after a rock first forms.

secondary recovery technique   A process used to extract the quantities of oil that will not come out of a reservoir rock with just simple pumping.

sediment   An accumulation of loose mineral grains, such as boulders, pebbles, sand, silt, or mud, that are not cemented together.

sedimentary basin   A depression, created as a consequence of subsidence, that fills with sediment.

sedimentary rock   Rock that forms either by the cementing together of fragments broken off preexisting rock or by the precipitation of mineral crystals out of water solutions at or near the Earth's surface.

sedimentary sequence   A grouping of sedimentary units bounded on top and bottom by regional unconformities.

sedimentary structure A characteristic of sedimentary deposits that pertains to the character of bedding and/or the surface features of a bed.

sediment budget The proportion of sand supplied to sand removed from a depositional setting.

sediment load The total volume of sediment carried by a stream.

sediment maturity The degree to which a sediment has evolved from a crushed-up version of the original rock into a sediment that has lost its easily weathered minerals and become well sorted and rounded.

sediment sorting The segregation of sediment by size.

seep A place where oil-filled reservoir rock intersects the ground surface, or where fractures connect a reservoir to the ground surface, so that oil flows out onto the ground on its own.

seiche Rhythmic movement in a body of water caused by ground motion.

seismic belts (or **zones**) The relatively narrow strips of crust on Earth under which most earthquakes occur.

seismicity Earthquake activity.

seismic-moment magnitude scale A scale that defines earthquake size on the basis of calculations involving the amount of slip, length of rupture, depth of rupture, and rock strength.

seismic ray The changing position of an imaginary point on a wave front as the front moves through rock.

seismic-reflection profile A cross-sectional view of the crust made by measuring the reflection of artificial seismic waves off boundaries between different layers of rock in the crust.

seismic tomography Analysis by sophisticated computers of global seismic data in order to create a three-dimensional image of variations in seismic-wave velocities within the Earth.

seismic velocity The speed at which seismic waves travel.

seismic-velocity discontinuity A boundary in the Earth at which seismic-wave velocity changes abruptly.

seismic waves Waves of energy emitted at the focus of an earthquake.

seismogram The record of an earthquake produced by a seismograph.

seismograph (seismometer) An instrument that can record the ground motion from an earthquake.

semipermanent pressure cell A somewhat elliptical zone of high or low atmospheric pressure that lasts much of the year; it forms because high-pressure zones tend to be narrower over land than over sea.

Sevier orogeny A mountain-building event that affected western North America between about 150 Ma and 80 Ma, a result of convergent margin tectonism; a fold-thrust belt formed during this event.

shale Very fine-grained sedimentary rock that breaks into thin sheets.

shatter cones Small, cone-shaped fractures formed by the shock of a meteorite impact.

shear strain A change in shape of an object that involves the movement of one part of a rock body sideways past another part so that angular relationships within the body change.

shear stress A stress that moves one part of a material sideways past another part.

shear waves Seismic waves in which particles of material move back and forth perpendicular to the direction in which the wave itself moves.

shear zone A fault in which movement has occurred ductilely.

sheetwash A film of water less than a few millimeters thick that covers the ground surface during heavy rains.

shield An older, interior region of a continent.

shield volcano A subaerial volcano with a broad, gentle dome, formed either from low-viscosity basaltic lava or from large pyroclastic sheets.

shocked quartz Grains of quartz that have been subjected to intense pressure such as occurs during a meteorite impact.

shock metamorphism Solid-state changes in rock that result from the extreme pressure accompanying a meteorite impact.

shoreline The boundary between the water and land.

shortening The process during which a body of rock or a region of crust becomes shorter.

short-term climate change Climate change that takes place over hundreds to thousands of years.

Sierran arc A large continental volcanic arc along western North America that initiated at the end of the Jurassic Period and lasted until about 80 million years ago.

silica SiO_2.

silicate minerals Minerals composed of silicon-oxygen tetrahedra linked in various arrangements; most contain other elements too.

silicate rock Rock composed of silicate minerals.

siliceous sedimentary rock Sedimentary rock that contains abundant quartz.

silicic Rich in silica with relatively little iron and magnesium.

silicon-oxygen tetrahedron The basic building block of silicate minerals; it consists of one silicon atom surrounded by four oxygen atoms.

sill A nearly horizontal table-top-shaped tabular intrusion that occurs between the layers of country rock.

siltstone Fine-grained sedimentary rock generally composed of very small quartz grains.

sinkhole A circular depression in the land that forms when an underground cavern collapses.

slab-pull force The force that downgoing plates (or slabs) apply to oceanic lithosphere at a convergent margin.

slate Fine-grained, low-grade metamorphic rock, formed by the metamorphism of shale.

slaty cleavage The foliation typical of slate, and reflective of the preferred orientation of slate's clay minerals, that allows slate to be split into thin sheets.

slickensides The polished surface of a fault caused by slip on the fault; lineated slickensides also have groves that indicate the direction of fault movement.

slip face The lee side of a dune, which sand slides down.

slip lineations Linear marks on a fault surface created during movement on the fault; some slip lineations are defined by grooves, some by aligned mineral fibers.

slope failure The downslope movement of material on an unstable slope.

slump Downslope movement in which a mass of regolith detaches from its substrate along a spoon-shaped sliding surface and slips downward semicoherently.

smelting The heating of a metal-containing rock to high temperatures in a fire so that the rock will decompose to yield metal plus a nonmetallic residue (slag).

snotite A long gob of bacteria that slowly drips from the ceiling of a cave.

snowball Earth Our planet during periods in the Precambrian when its entire surface was ice covered.

snow line The boundary above which snow remains all year.

soda straw A hollow stalactite in which calcite precipitates around the outside of a drip.

soil Sediment that has undergone changes at the surface of the Earth, including reaction with rainwater and the addition of organic material.

soil erosion The removal of soil by wind and runoff.

soil horizon Distinct zones within a soil, distinguished from each other by factors such as chemical composition and organic content.

soil moisture Underground water that wets the surface of the mineral grains and organic material making up soil, but lies above the water table.

soil profile A vertical sequence of distinct zones of soil.

Solar System The Sun and all the objects that orbit it (including planets, moons, comets, and asteroids).

solar wind   A stream of particles with enough energy to escape from the Sun's gravity and flow outward into space.

solid-state diffusion   The slow movement of atoms or ions through a solid.

solifluction   The type of creep characteristic of tundra regions; during the summer, the uppermost layer of permafrost melts, and the soggy, weak layer of ground then flows slowly downslope in overlapping sheets.

solstice   A day on which the polar ends of the terminator (the boundary between the day hemisphere and the night hemisphere) lie 23.5° away from the associated geographic poles.

Sonoma orogeny   A convergent-margin mountain-building event that took place on the western coast of North America in the Late Permian and Early Triassic periods.

sorting   (1) The range of clast sizes in a collection of sediment; (2) the degree to which sediment has been separated by flowing currents into different-size fractions.

source rock   A rock (organic-rich shale) containing the raw materials from which hydrocarbons eventually form.

southeast tradewinds   Tradewinds in the Southern Hemisphere, which start flowing northward, deflect to the west, and end up flowing from southeast to northwest.

southern oscillation   The oscillating of atmospheric pressure cells back and forth across the Pacific Ocean, in association with El Niño.

specific gravity   A number representing the density of a mineral, as specified by the ratio between the weight of a volume of the mineral and the weight of an equal volume of water.

speleothem   A formation that grows in a limestone cave by the accumulation of travertine precipitated from water solutions dripping in a cave or flowing down the wall of a cave.

sphericity   The measure of the degree to which a clast approaches the shape of a sphere.

spreading boundary   *Divergent plate boundary.*

spreading rate   The rate at which sea floor moves away from a mid-ocean ridge axis, as measured with respect to the sea floor on the opposite side of the axis.

spring   A natural outlet from which groundwater flows up onto the ground surface.

spring tide   An especially high tide that occurs when the Sun is on the same side of the Earth as the Moon.

stable air   Air that does not have a tendency to rise rapidly.

stable slope   A slope on which downward sliding is unlikely.

stalactite   An icicle-like cone that grows from the ceiling of a cave as dripping water precipitates limestone.

stalagmite   An upward-pointing cone of limestone that grows when drips of water hit the floor of a cave.

standing wave   A wave whose crest and trough remain in place as water moves through the wave.

star   A large sphere, composed dominantly of hydrogen and helium, in which fusion reactions are producing energy.

star dune   A constantly changing dune formed by frequent shifts in wind direction; it consists of overlapping crescent dunes pointing in many different directions.

stellar wind   Particles that have been ejected from a star and are shooting through space.

stick-slip behavior   Stop-start movement along a fault plane caused by friction, which prevents movement until stress builds up sufficiently.

stone rings   Ridges of cobbles between adjacent bulges of permafrost ground.

stoping   A process by which magma intrudes; blocks of wall rock break off and then sink into the magma.

storm   An episode of severe weather in which winds, precipitation, and in some cases lightning become strong enough to be bothersome and even dangerous.

storm-center velocity   A storm's (hurricane's) velocity along its track.

storm surge   Excess seawater driven landward by wind during a storm; the low atmospheric pressure beneath the storm allows sea level to rise locally, increasing the surge.

strain   The change in shape of an object in response to deformation (i.e., as a result of the application of a stress).

strata   A succession of sedimentary beds.

stratified drift   Glacial sediment that has been redistributed and stratified by flowing water.

stratigraphic column   A cross-section diagram of a sequence of strata summarizing information about the sequence.

stratigraphic formation   A recognizable layer of a specific sedimentary rock type or set of rock types, deposited during a certain time interval, that can be traced over a broad region.

stratigraphic sequence   An interval of strata deposited during periods of relatively high sea level, and bounded above and below by regional unconformities.

stratopause   The temperature pause that marks the top of the stratosphere.

stratosphere   The stable, stratified layer of atmosphere directly above the troposphere.

stratovolcano   A large, cone-shaped subaerial volcano consisting of alternating layers of lava and tephra.

stratus cloud   A thin, sheet-like, stable cloud.

streak   The color of the powder produced by pulverizing a mineral on an unglazed ceramic plate.

stream   A ribbon of water that flows in a channel.

stream bed   The floor of a stream.

stream capacity   The total quantity of sediment a stream carries.

stream competence   The maximum particle size that a stream can carry.

stream gradient   The slope of a stream's channel in the downstream direction.

stream piracy (or **capture**)   The situation in which headward erosion causes one stream to intersect the course of another, previously independent stream, so that the intersected stream starts to flow down the channel of the first stream.

stream rejuvenation   The renewed downcutting of a stream into a floodplain or peneplain, caused by a relative drop of the base level.

stream terrace   A flat surface, underlain by alluvium, that borders a stream; terraces form when the stream cuts down into the alluvium that it had deposited previously.

stress   The push, pull, or shear that a material feels when subjected to a force; formally, the force applied per unit area over which the force acts.

stretching   The process during which a layer of rock or a region of crust becomes longer.

strike   The compass trend of an imaginary horizontal line on a plane.

strike-slip fault   A fault in which one block slides horizontally past another (and therefore parallel to the strike line), so there is no relative vertical motion.

strip mining   The scraping off of all soil and sedimentary rock above a coal seam in order to gain access to the seam.

stromatolite   Layered mounds of sediment formed by cyanobacteria; cyanobacteria secrete a mucuous-like substance to which sediment sticks, and as each layer of cyanobacteria gets buried by sediment, it colonizes the surface of the new sediment, building a mound upward.

structural control   The condition in which geologic structures, such as faults, affect the distribution and drainage of water or the shape of the land surface.

subaerial   Pertaining to land regions above sea level (i.e., under air).

subduction   The process by which one oceanic plate bends and sinks down into the asthenosphere beneath another plate.

subduction zone   The region along a convergent boundary where one plate sinks beneath another.

sublimation   The evaporation of ice directly into vapor without first forming a liquid.

submarine canyon   A narrow, steep canyon that dissects a continental shelf and slope.

submarine fan   A wedge-shaped accumulation of sediment at the base of a submarine slope; fans usually accumulate at the mouth of a submarine canyon.

submarine slump   The underwater downslope movement of a semicoherent block of sediment along a weak mud detachment.

submergent coast   A coast at which the land is sinking relative to sea level.

subpolar low   The rise of air where the surface flow of a polar cell converges with the surface flow of a Ferrel cell, creating a low-pressure zone in the atmosphere.

subsidence   The vertical sinking of the Earth's surface in a region, relative to a reference plane.

substrate   A general term for material just below the ground surface.

subtropical high (subtropical divergence zone)   A belt of high pressure in the atmosphere at 30° latitude formed where the Hadley cell converges with the Ferrel cell, causing cool, dense air to sink.

subtropics   Desert climate regions that lie on either side of the equatorial tropics between the lines of 20° and 30° north or south of the equator.

summit eruption   An eruption that occurs in the summit crater of a volcano.

sunspot cycle   The cyclic appearance of large numbers of sunspots (black spots thought to be magnetic storms on the Sun's surface) every 9 to 11.5 years.

supercontinent cycle   The process of change during which supercontinents develop and later break apart, forming pieces that may merge once again in geologic time to make yet another supercontinent.

supernova   A short-lived, very bright object in space that results from the cataclysmic explosion marking the death of a very large star; the explosion ejects large quantities of matter into space to form new nebulae.

superplume   A huge mantle plume.

superposed stream   A stream whose geometry has been laid down on a rock structure and is not controlled by the structure.

superposition   The principle that younger layers of sediment are deposited on older layers of sediment; thus, in a sequence of strata, the oldest layer is at the base.

surface current   An ocean current in the top 100 m of water.

surface load   Sediment that rolls or bounces in wind along the ground surface.

surface waves   Seismic waves that travel along the Earth's surface.

surface westerlies   The prevailing surface winds in North America and Europe, which come out of the west or southwest.

surf zone   A region of the shore in which breakers crash onto the shore.

surge (glacial)   A pulse of rapid flow in a glacier.

suspended load   Tiny solid grains carried along by a stream without settling to the floor of the channel.

sustainable growth   Enlargement of society and the economy in ways that can be supported indefinitely by renewable natural resources.

swamp   A wetland dominated by trees.

swash   The upward surge of water that flows up a beach slope when breakers crash onto the shore.

S-waves   Seismic shear waves that pass through the body of the Earth.

S-wave shadow zone   A band between 103° and 180° from the epicenter of an earthquake inside of which S-waves do not arrive at seismograph stations.

swelling clay   Clay possessing a mineral structure that allows it to absorb water between its layers and thus swell to several times its original size.

symmetry   The condition in which the shape of one part of an object is a mirror image of the other part.

syncline   A trough-shaped fold whose limbs dip toward the hinge.

systematic joints   Long planar cracks that occur fairly regularly throughout a rock body.

tabular intrusions   Sheet intrusions that are planar and of roughly uniform thickness.

Taconic orogeny   A convergent mountain-building event that took place around 400 million years ago, in which a volcanic island arc collided with eastern North America.

tailings pile   A pile of waste rock from a mine.

talus apron   A wedge-shaped pile of rock fragments that accumulates at the base of a cliff.

tar   Hydrocarbons that exist in solid form at room temperature.

tarn   A lake that forms at the base of a cirque on a glacially eroded mountain.

tar sand   Sandstone reservoir rock in which less viscous oil and gas molecules have either escaped or been eaten by microbes, so that only tar remains.

taxonomy   The study and classification of the relationships among different forms of life.

temperate glacier   A glacier that exists in regions where it is warm enough for liquid water to occur in films between the grains of ice.

tension   A stress that pulls on a material and could lead to stretching.

tephra   Unconsolidated accumulations of pyroclastic grains.

terminal moraine   The end moraine at the farthest limit of glaciation.

terminator   The boundary between the half of the Earth that has daylight and the half experiencing night.

terrace   The elevated surface of an older floodplain into which a younger floodplain had cut down.

terrestrial   A term used to describe the inner, Earth-like planets.

thalweg   The deepest part of a stream's channel.

theory   A scientific idea supported by an abundance of evidence that has passed many tests and failed none.

theory of evolution   An idea that explains how the assemblage of species on Earth and the character of species on Earth have changed over time, as a consequence of the survival of the fittest; the idea is a "theory" because it has successfully explained many observations, can make testable predictions, and has not failed any test.

theory of plate tectonics   The theory that the outer layer of the Earth (the lithosphere) consists of separate plates that move with respect to one another.

thermal metamorphism   Metamorphism caused by heat conducted into country rock from an igneous intrusion.

thermocline   A boundary between layers of water with differing temperatures.

thermohaline circulation   The rising and sinking of water driven by contrasts in water density, which is due in turn to differences in temperature and salinity; this circulation involves both surface and deep-water currents in the ocean.

thermosphere   The outermost layer of the atmosphere containing very little gas.

thin section   A 3/100-mm-thick slice of rock that can be examined with a petrographic microscope.

thin-skinned deformation   A distinctive style of deformation characterized by displacement on faults that terminate at depth along a subhorizontal detachment fault.

thrust fault   A gently dipping reverse fault; the hanging-wall block moves up the slope of the fault.

tidal bore   A visible wall of water that moves toward shore with the rising tide in quiet waters.

tidal flat   A broad, nearly horizontal plain of mud and silt, exposed or nearly exposed at low tide but totally submerged at high tide.

tidal reach   The difference in sea level between high tide and low tide at a given point.

tide   The daily rising or falling of sea level at a given point on the Earth.

tide-generating force   The force, caused in part by the gravitational attraction of the Sun and Moon, and in part by the centrifugal force created by the Earth's spin, that generates tides.

tidewater glacier   A glacier that has flowed into the sea.

till   A mixture of unsorted mud, sand, pebbles, and larger rocks deposited by glaciers.

tillite   A rock formed from hardened ancient glacial deposits and consisting of larger clasts distributed through a matrix of sandstone and mudstone.

toe (terminus)   The leading edge or margin of a glacier.

tombolo   A narrow ridge of sand that links a sea stack to the mainland.

topographic map   A map that uses contour lines to represent variations in elevation.

topography   Variations in elevation.

topsoil   The top soil horizons, which are typically dark and nutrient-rich.

tornado   A near-vertical, funnel-shaped cloud in which air rotates extremely rapidly around the axis of the funnel.

tornado swarm   Dozens of tornadoes produced by the same storm.

tower karst   A karst landscape in which steep-sided residual bedrock towers remain between sinkholes.

transform fault   A fault marking a transform plate boundary; along mid-ocean ridges, transform faults are the actively slipping segment of a fracture zone between two ridge segments.

transform plate boundary   A boundary at which one lithosphere plate slips laterally past another.

transgression   The inland migration of shoreline resulting from a rise in sea level.

transition zone   The middle portion of the mantle, from 400 to 670 km deep, in which there are several jumps in seismic velocity.

transpiration   The release of moisture as a metabolic by-product.

transverse dune   A simple, wave-like dune that appears when enough sand accumulates for the ground surface to be completely buried, but only moderate winds blow.

trap   In the context of hydrocarbons, a trap is a geologic configuration that accumulates and holds oil underground.

travel-time curve   A graph that plots the time since an earthquake began on the vertical axis, and the distance to the epicenter on the horizontal axis.

travertine   A carbonate rock formed by precipitation of carbonate minerals from water at springs or on the surface of caves.

trellis network   A drainage system that develops across a landscape of parallel valleys and ridges so that major tributaries flow down the valleys and join a trunk stream that cuts through the ridge; the resulting map pattern resembles a garden trellis.

trench   A deep elongate trough bordering a volcanic arc; a trench defines the trace of a convergent plate boundary.

triangulation   The method for determining the map location of a point from knowing the distance between that point and three other points; this method is used to locate earthquake epicenters.

tributary   A smaller stream that flows into a larger stream.

triple junction   A point where three lithosphere plate boundaries intersect.

tropical depression   A tropical storm with winds reaching up to 61 km per hour; such storms develop from tropical disturbances, and may grow to become hurricanes.

tropical disturbance   Cyclonic winds that develop in the tropics.

tropopause   The temperature pause marking the top of the troposphere.

troposphere   The lowest layer of the atmosphere, where air undergoes convection and where most wind and clouds develop.

truncated spur   A spur (elongate ridge between two valleys) whose end was eroded off by a glacier.

trunk stream   The single larger stream into which an array of tributaries flow.

tsunami   A large wave along the sea surface triggered by an earthquake or large submarine slump.

tuff   A pyroclastic igneous rock composed of volcanic ash and fragmented pumice, formed when accumulations of the debris cement together.

tundra   A cold, treeless region of land at high latitudes, supporting only species of shrubs, moss, and lichen capable of living on permafrost.

turbidite   A graded bed of sediment built up at the base of a submarine slope and deposited by turbidity currents.

turbidity current   A submarine avalanche of sediment and water that speeds down a submarine slope.

turbulence   The chaotic twisting, swirling motion in flowing fluid.

typhoon   The equivalent of a hurricane in the western Pacific Ocean.

ultimate base level   Sea level; the level below which a trunk stream cannot cut.

ultramafic   A term used to describe igneous rocks or magmas that are rich in iron and magnesium and very poor in silica.

unconfined aquifer   An aquifer that intersects the surface of the Earth.

unconformity   A boundary between two different rock sequences representing an interval of time during which new strata were not deposited and/or were eroded.

unconsolidated   Consisting of unattached grains.

undercutting   Excavation at the base of a slope that results in the formation of an overhang.

undersaturated   A term used to describe a solution capable of holding more dissolved ions.

uniformitarianism   The principle that the same physical processes observed today are responsible for the formation of ancient geologic features; put concisely, "the present is the key to the past."

Universe   The sum of all matter and energy making up the hundreds of billions of known galaxies.

unsaturated zone   The region of the subsurface above the water table.

unstable air   Air that is significantly warmer than air above and has a tendency to rise quickly.

Credits

Book Cover Art

Book Cover: Stephen Marshak; **Inside Front Cover:** USGS.

Unnumbered Photos and Art

ii: Stephen Marshak; **1:** Stephen Marshak; **9:** R. Williams (STScI), the Hubble Deep Field Team and NASA; **18–19:** original artwork by Gary Hincks; **32:** NASA; **33:** Courtesy of Thomas N. Taylor; **62–63:** original artwork by Gary Hincks; **67:** Christopher Hormann; **68:** Javier Trueba/MSF/Photo Researchers, Inc.; **84:** © 1998 Jeff Scovil; **85:** Getty Images/Altrendo; **92:** Stephen Marshak; **105:** original artwork by Gary Hincks; **110:** Jacques Descloitres, MODIS Team, NASA Visible Earth; **111:** National Geographic/Getty Images; **122–123:** original artwork by Gary Hincks; **138:** Stephen Marshak; **144–145:** original artwork by Gary Hincks; **151:** United States Department of Agriculture/Natural Resources Conservation Services; **152:** William Manning/Corbis; **168–169:** original artwork by Gary Hincks; **173:** Serge Andrefouet, University of South Florida/NASA; **175:** Stephen Marshak; **188–189:** original artwork by Gary Hincks; **194 (all):** Stephen Marshak; **198–199:** original artwork by Gary Hincks; **200:** Anjum Naveed/AP Images; **204–205:** original artwork by Gary Hincks; **231:** Jacques Descloitres, MODIS Rapid Response Team, NASA/GSFC/Visible Earth; **240:** U.S. Geological Survey; **256–257:** original artwork by Gary Hincks; **265:** NationalAtlas.gov/USGS; **266:** Stephen Marshak; **271:** Stephen Marshak; **276:** Stephen Marshak; **290–291:** original artwork by Gary Hincks; **297:** USGS National CVenter of EROS and NASA, Landsat Project Science office; **298:** Stephen Marshak; **316–317:** original artwork by Gary Hincks; **320:** Jacques Descloitres, MODIS Team, NASA Visible Earth; **321:** Mario Tama/Getty Images; **350:** Stephen Marshak; **351:** JPL/NASA; **358–359:** original artwork by Gary Hincks; **360:** Image courtesy of USGS National Center for EROS and NASA Landsat Project Science Office; **361:** Rob Varela/Ventura Country Star/Corbis; **370–371:** original artwork by Gary Hincks; **377:** NASA/Calvin J. Hamiton; **378:** Stephen Marshak; **400–401:** original artwork by Gary Hincks; **403:** NASA; **404:** Stephen Marshak; **420–421:** original artwork by Gary Hincks; **429:** NASA/GSFC/LaRC/JPL, MISR Team; **430:** Yann Arthus Bertrand/Corbis; **446–447:** original artwork by Gary Hincks; **446 (top):** Lois Kent; **446 (bottom):** ML Sinibaldi/CORBIS; **447 (right):** Ashley Cooper/CORBIS; **447 (left):** Stephen Marshak; **450:** GeoEye/ Photo Researchers; **451:** Stephen Marshak; **462–463:** original artwork by Gary Hincks; **468:** U.S. Geological Survey; **469:** Stephen Marshak; **484–485:** original artwork by Gary Hincks; **498:** U.S. Geological Survey; **499:** Stephen Marshak; **502–503:** original artwork by Gary Hincks; **A–12:** NASA/CXC/SAO.

Numbered Photos and Art

Prelude

P.1b: Stephen Marshak; **P.1c:** Stephen Marshak; **P.2:** Stephen Marshak; **P.3:** AP Images; **Box P.1 fig. 1:** Stephen Marshak; **Box P.1 fig. 1 (inset):** Courtesy Peter Fiske.

Chapter 1

1.1a: Rare Book Division, The New York Public Library, Astor, Lenox and Tilden Foundation; **1.1b:** Mary Evans Picture Library/Alamy; **1.3:** NASA; **1.6:** J. Hester and P. Scowen/NASA; **1.7:** European Space Agency/Photo Researchers; **1.8:** J. Hester and P. Scowen/NASA; **1.9:** Photograph by Pelisson, SaharaMet; **Box 1.1 fig. 1:** Fred Espenak, Photo Researchers, Inc.; **Box 1.1 fig. 2:** Images provided by *Google Earth*™ mapping service/DigitalGlobe, TerraMetrics, NASA, Europa Technologies; **1.10:** Jet Propulsion Laboratory/NASA; **1.11d:** NASA/Photo Researchers, Inc.; **1.12a:** Johnson Space Center/NASA; **1.15:** Tate Gallery, London/Art Resource, NY.

Chapter 2

2.1a: Alfred Wegener Institute for Polar and Marine Research; **2.17b:** University of Texas Institute of Geophysics; **2.18:** Photo by Dudley Foster, Woods Hole Oceanographic Institute; **2.22d:** Kevin Schafer/Alamy; **2.26c:** Johnson Space Center/NASA.

Chapter 3

3.1: Crown copyright. Historic Royal Palaces; **3.2a:** Ken Lucas/Visuals Unlimited; **3.2b:** Stephen Marshak; **3.3a:** Jay Schomer; **3.3b:** Wally McNamee/Corbis; **3.5c:** Charles O'Rear/Corbis; **3.5d:** Courtesy of John A. Jaszczak, Michigan Technological University; **3.7d:** Copyright 1996 Jeff Scovil; **3.8a:** Richard P. Jacobs/JLM Visuals; **3.8b:** Richard P. Jacobs/JLM Visuals; **3.8c:** Breck P. Kent/JLM Visuals; **3.8d:** Richard P. Jacobs/JLM Visuals; **3.8e (left):** Andrew Silver/USGS; **3.8e (right):** Andrew Silver/USGS; **3.8f:** Charles D. Winters/Photo Researchers, Inc.; **3.8g:** Scientifica/Visuals Unlimited/Alamy; **3.9a:** Richard P. Jacobs/JLM Visuals; **3.9b:** © 1992 Jeff Scovil; **3.9c:** Marli Miller/Visuals Unlimited; **3.9d:** Richard P. Jacobs/JLM Visuals; **3.9e:** Richard P. Jacobs/JLM Visuals;

3.9f: Ann Bryant/Geology.com; **3.9f:** Ann Bryant/Geology. com; **3.11:** 1996 Smithsonian Institution; **Box 3.2 fig 1:** 1985 Darrel Plowes; **Box 3.2 fig 2:** Ken Lucas/Visuals Unlimited; **3.12a:** A.J. Copley/Visual Unlimited; **3.12a (inset):** Stephen J. Krasemann/Photo Researchers, Inc.

Interlude A

A.1a (left): Richard P. Jacobs/JLM Visuals; **A.1a (left):** Courtesy David W. Houseknecht, U.S. Geological Survey; **A.1b (left):** sciencephotos/Alamy; **A.1b (right):** Courtesy of Kent Ratajeski, Department of Geology and Geophysics, University of Wisconsin, Madison; **A.2a (left):** Stephen Marshak; **A.2b (left):** Stephen Marshak; **A.2c (left):** Stephen Marshak; **A.3a (right):** Stephen Marshak; **A.3b (right):** Stephen Marshak; **A.3c (right):** Stephen Marshak; **A.5a (left):** Tom Bean; **A.5b (left):** H. Johnson; **A.6c:** Stephen Marshak; **A.6d:** Scenics & Science/Alamy; **A.7:** Courtesy of Bradley Hacker, Department of Geology University of California, Santa Barbara.

Chapter 4

4.1a: Jim Sugar Photography/Corbis; **4.1b:** Photo by J.D. Griggs/ U.S. Geological Survey; **4.1c:** Stephen Marshak; **4.1d:** Stephen Marshak; **4.2b:** Uri ten Brink (U.S. Geological Survey); **4.8a:** Stephen Marshak; **4.8b:** Stephen Marshak; **4.8c:** U.S. Geological Survey; **4.8d:** Stephen Marshak; **4.9a:** Stephen Marshak; **4.9b:** Stephen Marshak; **4.9c:** 1998 Tom Bean; **4.10a:** Stephen Marshak; **4.10c:** Stephen Marshak; **4.11d:** Stephen Marshak; **4.12a (left):** Courtesy of Dr. Kent Ratajeski Department of Geology and Geophysics, University of Wisconsin, Madison; **4.12a(middle):** Public Domain; **4.12a (right):** Courtesy of Dr. Kent Ratajeski, Department of Geology and Geophysics, University of Wisconsin, Madison; **4.12b:** Doug Sokell/Visuals Unlimited; **4.12c (left top):** Dirk Wiersma/Photo Researchers, Inc.; **4.12c (left middle):** Stephen Marshak; **4.12c(left bottom):** John S. Shelton; **4.12c(right top):** Stephen Marshak; **4.12c(right middle):** Dirk Wiersma/Photo Researchers, Inc.; **4.12c(right bottom):** Stephen Marshak; **4.14a:** Stephen Marshak; **4.14b:** Stephen Marshak.

Chapter 5

5.1a: Mount Vesuvius in Eruption, 1817 (w/c on paper), Joseph Mallord William Turner (1775–1851)/Yale Center for British Art, Paul Mellon Collection, USA,/The Bridgeman Art Library; **5.1b:** Stephen Marshak; **5.1c:** Stephen Marshak; **5.3a:** Roger Ressmeyer/CORBIS; **5.3b:** Stephen Marshak; **5.3c:** Stephen Marshak; **5.3d:** Stephen Marshak; **5.3e:** Thomas Hallstein/ Alamy; **5.3f:** Stephen Marshak; **5.4:** Marli Miller/Visuals Unlimited; **5.5a:** AP Images; **5.5b:** Stephen Weaver; **5.5b (inset):** Stephen Marshak; **5.6a:** Stephen and Donna O'Meara, Volcano Watch International; **5.6b:** A.M. Sarna-Wojcicki/U.S. Geological Survey; **5.6c:** Ken Wagner/Visuals Unlimited; **5.6d:** Associated Press Antigua Sun; **5.6e:** Roger Ressmeyer/CORBIS; **5.6f:** Stephen Marshak; **5.7a:** Stephen Marshak; **5.7b:** Stephen Marshak; **5.8b:** N. Banks/U.S. Geographical Survey; **5.9:** Marli Miller/Visuals Unlimited; **5.10a:** Marli Miller/Visuals Unlimited;

5.10b: 1985 Tom Bean; **5.10c:** Roger Ressmeyer/Corbis; **Box 5.1 fig 3 inset:** Gary Braasch/Corbis; **5.12(inset):** Barry W. Eakins/USGS; **5.13d:** Stephen Marshak; **5.14:** Stephen Marshak; **5.15b:** David Sandwell/Scripps Institute of Oceanography; **5.16a:** Antonio Parrinello/Reuters/Corbis; **5.16b:** Antonio Parrinello/ Reuters/Corbis; **5.16c:** AP Images; **5.16d:** Stephen Marshak; **5.16e:** AP Images; **5.16g:** Philippe Bourseiller; **5.16f:** U.S. Geographical Survey; **5.17 (left):** J. Marso/U.S. Geographical Survey; **5.17:** Peter Turnley/Corbis; **5.18b:** Stephen Marshak; **5.20:** photo by Sigurgeir Jonasson/Frank Lane Picture Agency/ Corbis; **5.21a:** Julian Baum/Photo Researchers, Inc.; **5.21b:** NASA/JPL; **5.21c:** NASA/JPL; **5.21d:** NASA/JPL; **5.21e:** NASA/JPL; **5.21e (inset):** NASA/JPL.

Interlude B

B.12: Jim Richardson/Corbis; **B.2:** Stephen Marshak; **B.3:** Stephen Marshak; **B.4a:** Marli Miller/Visuals Unlimited; **B.4b:** Stephen Marshak; **B.4b (inset):** Stephen Marshak; **B.5b:** 1998 Tom Bean; **B.5c:** Stephen Marshak; **B.6b:** Stephen Marshak; **B.8b:** Stephen Marshak; **B.8c:** Stephen Marshak; **B.8d:** Stephen Marshak; **B.9c:** Stephen Marshak.

Chapter 6

6.1a: Stephen Marshak; **6.4a (left):** Stephen Marshak; **6.4a (right):** Stephen Marshak; **6.4b (left):** Stephen Marshak; **6.4b (right):** Stephen Marshak; **6.4c (left):** Stephen Marshak; **6.4c (right):** Scottsdale Community College; **6.4d (left):** Stephen Marshak; **6.4d (right):** E.R. Degginger/Color-Pic, Inc.; **6.4e (left):** Stephen Marshak; **6.4e (right):** Stephen Marshak; **6.5a:** Stephen Marshak; **6.5b:** Stephen Marshak; **6.5c:** Stephen Marshak; **6.6a:** Stephen Marshak; **6.6b:** Stephen McCutcheon/ Visuals Unlimited; **6.7a:** Gerald and Buff Corsi/Visuals Unlimited; **6.8a:** Marli Miller/Visuals Unlimited; **6.8b:** 1998 M.W. Schmidt; **6.9a:** Stephen Marshak; **6.9b:** Stephen Marshak; **6.10:** Stephen Marshak; **6.11a:** Stephen Marshak; **6.11b:** © 1980 by the Grand Canyon Natural History Association; **6.12a:** Stephen Marshak; **6.12b:** Stephen Marshak; **6.13b:** Stephen Marshak; **6.14 Back Ground:** Stephen Marshak; **6.14c:** Marli Miller/ Visuals Unlimited; **6.15a:** Stephen Marshak; **6.15b:** Stephen Marshak; **6.16a:** Emma Marshak; **6.16b:** Stephen Marshak; **6.16c:** Marli Miller/Visuals Unlimited; **6.16d:** Stephen Marshak; **6.16e:** Stephen Marshak; **6.16f:** John S. Shelton; **6.18b:** Stephen Marshak; **6.19b:** Yann Arthus-Bertrand/Corbis; **6.20a:** Belgian Federal Science Policy Office; **6.20b:** G.R. Roberts NSIL.

Chapter 7

7.1a: Stephen Marshak; **7.1a:** L.S. Stephanowicz/Visuals Unlimited; **7.1b:** Stephen Marshak; **7.1b:** Courtesy Kurt Freihauf, Kutztown University of Pennsylvannia; **7.1b (inset):** Stephen Marshak; **7.4a:** Ludovic Maisant/Corbis; **7.4a (inset):** Emma Marshak; **7.4b:** Stephen Marshak; **7.5a:** Stephen Marshak; **7.5b:** Stephen Marshak; **7.5c:** Stephen Marshak; **7.6:** Stephen Marshak; **7.7a:** Stephen Marshak; **7.7b:** Stephen Marshak; **7.7c:** Stephen Marshak; **Box 7.2:** Stephen Marshak; **7.10a:** David Muench/Corbis; **7.10b:** Stephen Marshak.

Chapter 8

8.4a: National Geophysical Data Center/NOAA; **8.4b:** Photo Courtesy of Paul "Kip" Otis-Diehl, USMC, 29 Palms CA; **8.15:** Patrick Robert/Corbis Sygma; **8.16b:** Corbis; **8.16c:** George Hall/Corbis; **8.18:** Courtesy State Historical Society of Missouri, Columbia; **8.20a:** AP Images; **8.20b:** Pacific Press Service/Alamy; **8.20c:** M. Celebi, U.S. Geographical Survey; **8.20d:** Reuters; **8.21a:** Barry Lewis/Alamy; **8.21b:** Bettmann/Corbis; **8.21c:** NGDC; **8.21d:** Karl V. Steinbrugge Collection, Earthquake Engineering Research Center; **8.22:** James Mori, Research Center for EarthquakePrediction, Disaster Prevention Institute, Kyoto University; **8.23:** U.S. Geological Survey; **8.25a:** Vasily V. Titov, Associate Director, Tsunami Inundation Mapping Efforts (TIME), NOAA/PMEL-UW/JISAO, USA; **8.25b:** AFP/Getty Images; **8.25c:** David Rydevik; **8.25d:** Space Imaging; **8.29c:** NOAA/NOA Center for Tsunami Research.

Interlude D

Box D.1: ESA; **D.10a:** John Q. Thompson, courtesy of Dawson Geophysical Company; **D.10b:** John Q. Thompson, courtesy of Dawson Geophysical Company.

Chapter 9

9.1: National Geophysical Data Center/NOAA; **9.3a:** Stephen Marshak; **9.3b:** Stephen Marshak; **9.3c:** Stephen Marshak; **9.5b:** Stephen Marshak; **9.5d:** Stephen Marshak; **Box 9.1:** Stephen Marshak; **9.7a:** Galen Rowell/Corbis; **9.7b:** Stephen Marshak; **9.7c:** Stephen Marshak; **9.9a:** John S. Shelton; **9.9b:** Stephen Marshak; **9.9c:** U.S. Geological Survey; **9.10a–c:** Stephen Marshak; **9.12a:** Stephen Marshak; **9.12b:** Stephen Marshak; **9.12c:** Stephen Marshak; **9.12d:** Stephen Marshak; **9.12e:** John S. Shelton; **9.13a:** Stephen Marshak; **9.13b:** Stephen Marshak; **9.15:** Stephen Marshak; **9.16c:** Stephen Marshak; **9.17a:** Stephen Marshak; **9.17b:** Stephen Marshak; **9.18b:** Stephen Marshak; **9.19b:** Based on a photo by J. Beckett of a model by J. Malavielle © American Museum of Natural History; **9.20b:** Stephen Marshak; **9.21:** U.S. Geological Survey; **9.23:** Courtesy E. Norabuena, et al. 1998. *Science* 279: 359.

Interlude E

E.1a: Stephen Marshak; **E.1b:** Richard T. Nowitz/Corbis; **E.1c:** Photo by William L. Jones from the Stones & Bones Collection, http://www.stones-bones.com; **E.3a:** Zoological Institute of Russian Academy of Sciences and Zoological Museum; **E.3b:** Doug Lundeberg (www.ambericawest.com); **E.3c:** Stephen Marshak; **E.3d:** Stephen Marshak; **E.3e:** Stephen Marshak; **E.3f:** Depatment of Earth and Space Sciences, UCLA; **E.3g:** Kevin Schafer/Corbis; **E.3h:** Stephen Marshak; **E.3i:** Stephen Marshak; **E.4a:** Courtesy of Senckenberg, Messel Research Department; **E.4b:** Humboldt-Universität zu Berlin, Museum für Naturkunde. Photo by W. Harre.

Chapter 10

10.1a left: Stephen Marshak; **10.1a right:** Stephen Marshak; **10.1b:** Stephen Marshak; **10.1c:** Stephen Marshak; **10.4:** Stephen Marshak; **10.6:** Stephen Marshak; **10.11:** Stephen Marshak; **10.11:** Charles Preitner/Visuals Unlimited; **10.11:** Stephen Marshak; **10.11:** Stephen Marshak; **10.13:** Stephen Marshak; **10.14:** Lonnie G. Thompson, Byrd Polar Research Center, the Ohio State University, Columbus.

Chapter 11

11.1: Bonestell Space Art; **11.3a:** Courtesy of Dr. J. William Schopf/UCLA; **11.3b:** Stephen Marshak; **11.7b:** Lisa-Ann Gershwin/U.C. Museum of Paleontology; **11.7c:** Courtesy of Dr. Paul Hoffman, Harvard University; **11.8b:** Ronald C. Blakey; Colorado Plateau Geosystems, Inc.; **11.8c:** Ronald C. Blakey; Colorado Plateau Geosystems, Inc.; **11.9:** Jonathan J. Havens, Irving Materials, Inc.; **11.10b:** Ronald C. Blakey; Colorado Plateau Geosystems, Inc.; **11.10c:** AFP/Getty Images; **11.12a:** Images provided by *Google Earth*™ mapping service/DigitalGlobe, TerraMetrics, NASA, Europa Technologies—copyright 2008; **11.13:** E.R. Degginger/Color-Pic, Inc.; **11.13 (inset):** Dan Osipov; **11.16:** Ronald C. Blakey; Colorado Plateau Geosystems, Inc.; **11.16:** Ronald C. Blakey; Colorado Plateau Geosystems, Inc.; **11.15:** Richard Bizley FIAAA; **11.17b, c:** Ronald C. Blakey; Colorado Plateau Geosystems, Inc.; **11.18:** Don Davis/NASA.

Chapter 12

12.5b: Stephen Marshak; **12.5e:** Robert Holden/Photophile; **12.5 inset:** Malcolm Fife/Getty Images; **12.5d:** Stephen Marshak; **12.6a:** Ron Testa/The Field Museum, Chicago; **12.6d:** Courtesy University of Melbourne; **12.8a:** Stephen Marshak; **12.8b:** Roger Ressmeyer/CORBIS; **12.9:** James Blank/Photophile; **12.10:** G.R. NSIL; **12.11a:** Stephen Marshak; **12.11b:** Stephen Marshak; **12.11c:** Stephen Marshak; **12.13:** G.R. Roberts NSIL; **12.14:** Stephen Marshak; **12.14 (inset):** Layne Kennedy/Corbis; **12.15a:** Richard P. Jacobs/JLM Visuals; **12.15b:** Richard P. Jacobs/JLM Visuals; **12.17:** Stephen Marshak; **12.19:** Stephen Marshak; **12.20a:** Richard P. Jacobs/JLM Visuals; **12.20b:** Connie Toops; **12.21a:** Doug Sokell; **12.21b:** A.J. Copley/Visual Unlimited.

Interlude F

F.1a: Stephen Marshak; **F.1b:** Stephen Marshak; **F.1c:** Stephen Marshak; **F.1d:** Stephen Marshak; **F.2a:** G.R. Roberts NSIL; **F.2b:** Photo by Davie Pierson/St. Petersburg Times; **F.3a:** Stephen Marshak; **F.3b:** Stephen Marshak; **F.3c:** Stephen Marshak; **F.4a:** JSC/NASA (AS17-137-21011); **F.4b:** Dr. David Smith, NASA Goddard Space Flight Center/MOLA Science Team; **F.4c:** NASA/JPL; **Box F.2:** NASA.

Chapter 13

13.1a: Lloyd Cluff/Corbis; **13.1b:** Lloyd Cluff/Corbis; **13.2c (inset):** Colorado Geological Society/U.S. Geological Survey; **13.2d:** Marli Miller/Visuals Unlimited; **13.3a:** Tom Myers/Photo Researchers, Inc.; **13.4a:** Cascades Volcano Observatory/USGS; **13.4b:** Cascades Volcano Observatory/USGS; **13.5:** AFP/Getty images; **13.6a:** AFP/Getty images; **13.6b:** Joan M. Jacobs/JLM Visuals; **13.7a:** Stephen Marshak; **13.7b:**

Stephen Marshak; **13.9:** Stephen Marshak; **13.13:** Breck P. Kent/JLM Visuals.

Chapter 14

14.1: Courtesy Johnstown Area Heritage Association; **14.3d:** Stephen Marshak; **14.8a:** Stephen Marshak; **14.8b:** Rex Clark; **14.9a:** Stephen Marshak; **14.9b:** Stephen Marshak; **14.13a:** Stephen Marshak; **14.13b:** Stephen Marshak; **14.14b:** James Blank/Photophile; **14.15a:** Marli Miller/Visuals Unlimited; **14.15b:** Stephen Marshak; **14.16:** 1998 Tom Bean; **14.17c:** Images provided by *Google Earth*™ mapping service/DigitalGlobe, TerraMetrics, NASA, Europa Technologies—© 2008; **14.20:** 1997 Tom Bean; **14.24a (top):** NASA images created by Jesse Allen, Earth Observatory, using data provided courtesy of the Landsat Project Science Office—© 2008; **14.24a (bottom):** NASA images created by Jesse Allen, Earth Observatory, using data provided courtesy of the Landsat Project Science Office—© 2008; **14.24b:** Stephen Marshak; **14.24c:** Binsar Bakkara/AP Images; **14.25:** U.S. Geological Survey; **14.26:** Stephen Marshak; **14.28:** Photo courtesy of the Bureau of Reclamation.

Chapter 15

15.3: University of New Brunswick Ocean Mapping Group; **15.4:** NASA; **15.7c:** Michael St. Maur Sheil/Corbis; **15.7d:** Michael St. Maur Sheil/Corbis; **15.8:** Stephen Marshak; **15.9a:** G.R. Roberts NSIL; **15.9b:** Stephen Marshak; **15.10a:** Stephen Marshak; **15.10b:** Stephen Marshak; **15.10e:** NASA; **15.10f:** Stephen Marshak; **15.11a:** G.R. Roberts NSIL; **15.11b:** Stephen Marshak; **15.11d (left):** Stephen Marshak; **15.11d (right):** Stephen Marshak; **15.12a:** Stephen Marshak; **15.12b:** Stephen Marshak; **15.13a:** Stephen Marshak; **15.13a(inset):** Stephen Marshak; **15.15:** Cody Duncan/Alamy; **15.19a:** Annie Griffiths Belt/Corbis; **15.19b:** Stephen Marshak; **15.20:** Stephen Marshak; **15.21a:** SVS/NASA; **15.21d:** NHC/NOAA; **15.22a:** NOAA; **15.22b:** AP Images; **15.22c (inset):** New York Times Graphics; **15.22d:** Vincent Laforet/Pool/Reuters/Corbis; **15.22e:** Stephen Marshak.

Chapter 16

16.1a: GeoPhoto Publishing Company; **16.1c:** GeoPhoto Publishing Company; **16.2b:** Society for Sedimentary Geology/Earth Science World Image Bank; **16.2c:** Stephen Marshak; **16.7:** Photo courtesy of USDA Natural Resources Conservation Service; **16.8:** Stephen Marshak; **16.9b:** Stephen Marshak; **16.9c:** Stephen Marshak; **16.9d:** Stephen Marshak; **16.14a:** Paul F. Hudson, University of Texas; **16.14b:** Photo by David Parker, courtesy of the National Astronomy and Ionosphere Center-Arecibo Observatory, a facility of the NSF; **16.16a:** Stephen Marshak; **16.18b:** Stephen Marshak.

Chapter 17

17.1: Stephen Marshak; **17.2:** Stephen Marshak; **17.4:** Andre Danderfer; **17.5a:** Stephen Marshak; **17.5b:** Gorden Whitten/Corbis; **17.6a:** Stephen Marshak; **17.6b:** Stephen Marshak; **17.8:** Stephen Marshak; **17.10:** Stephen Marshak; **17.11a:** Stephen Marshak; **17.11b:** Marli Miller/Visuals Unlimited; **17.12a:** Stephen Marshak; **17.12b:** Stephen Marshak; **17.12b (inset):** Stephen Marshak; **17.13b:** Stephen Marshak; **17.13c:** Stephen Marshak; **17.14b:** Stephen Marshak; **17.14c:** Stephen Marshak; **17.15:** Sinclair Stammers/SPL/Photo Researchers, Inc.; **17.16b:** Charles and Josette Lenars/Corbis; **17.16c:** Corbis; **17.17:** Image courtesy Jacques Descloitres MODIS Rapid Response Team.

Chapter 18

18.1a: Tom Bean/Corbis; **18.2a:** Photos courtesy Kenneth G. Libbrecht, www.snowcrystals.com; **18.2b (left):** Stephen Marshak; **18.2b (right):** Stephen Marshak; **18.2c (top):** Emma Marshak; **18.2c (bottom):** National Geographic Image Collection; **18.3b:** Stephen Marshak; **18.3c:** Stephen Marshak; **18.3d:** 1996 Galen Rowell; **Box 18.1 (left):** NASA Jet Propulsion Laboratory (NASA-JPL); **Box 18.1 (right):** NASA; **18.5:** Harry M. Walker; **18.9b:** Ralph A. Clevenger/Corbis; **18.9c:** Stephen Marshak; **18.9d:** ESA; **18.10a:** 1986 Jack Olson; **18.10b:** Stephen Marshak; **18.10c:** Stephen Marshak; **18.10d:** Stephen Marshak; **18.11a:** Stephen Marshak; **18.11b:** Stephen Marshak; **18.11c:** Stephen Marshak; **18.11d:** 1986 Keith S. Walklet/Quietworks; **18.12b:** Marli Miller; **18.13:** Wolfgang Meier/zefa/Corbis; **18.14:** Stephen Marshak; **18.15 (background):** Stephen Marshak; **18.15a:** Stephen Marshak; **18.15b:** Stephen Marshak; **18.15c (left):** Stephen Marshak; **18.15c (right):** Stephen Marshak; **18.15d:** Stephen Marshak; **18.15e(left):** Stephen Marshak; **18.15e(right):** Kevin Schafer/Alamy; **18.15e (left):** Stephen Marshak; **18.15e (right):** Kevin Schafer/Alamy; **18.17b:** Stephen Marshak; **18.17e:** Glenn Oliver/Visuals Unlimited; **18.18:** Tom Bean/Corbis; **18.21:** Scott T. Smith/Corbis; **18.22:** Lynda Dredge/Geological Survey of Canada; **18.24:** Detail of mural by Charles R. Knight, © American Museum of Natural History.

Chapter 19

19.5 (top): ISM/PhotoTakeUSA.com; **19.5 (bottom):** Bob Sacha/Corbis; **19.6a:** Nick Cobbing/Alamy; **19.6a (inset):** NOAA; **19.6c (inset):** Columbia.edu; **19.8:** NASA/Goddard Space Flight Center Scientific Visualization Studio; **19.9:** Stephen Marshak; **19.10 (left):** Courtesy P&H Mining Equipment; **19.10 (middle):** Richard Hamilton Smith/Corbis; **19.10 (right):** Stephen Marshak; **19.11b:** M-SAT LTD/SPL/Photo Researchers, Inc.; **19.11c:** Kennan Ward/Corbis; **19.12:** NASA/Goddard Space Flight Center Scientific Visualization Studio; **19.13c:** NASA/Goddard Space Flight Center Scientific Visualization Studio; **19.13d:** Jacques Descloitres, MODIS Rapid Response Team, NASA/GSFC; **19.13e:** Scientific Visualizations Studio/NASA/GSFC.

Appendix

a.1: Robert Glusic/PhotoDisc; **a.6:** Image courtesy of Arthur Smith and Randall Feenstra, Carnegie Mellon University; **a.13:** Visuals Unlimited.

Geotours

All thumbnail images for the Geotours were provided by *Google Earth*™ mapping service/DigitalZGlobe, TerraMetrics, NASA, Europa Technologies—© 2009.

Index